POLLEN RECORDS OF LATE-QUATERNARY NORTH AMERICAN SEDIMENTS

edited by

Vaughn M. Bryant, Jr.
and
Richard G. Holloway

Published by the American Association of
Stratigraphic Palynologists Foundation

Additional copies of this book may be obtained by sending $35.00 to: Robert T. Clarke, Treasurer, AASP Foundation, c/o Mobil Research & Development Corp., P.O. Box 819047, Dallas, Texas 75381-9047. Make check payable to AASP Foundation.

Phototypeset by Hart Information Systems
Printed in the United States by Hart Graphics, Inc.
8000 Shoal Creek Blvd., Austin Texas 78758

This book is dedicated to the Founding Members of the American Association of Stratigraphic Palynologists who in 1967 had the foresight to establish a new professional organization open to palynologists in all subfields of the discipline.

O. Ben Bourn
Robert T. Clarke
Fritz H. Cramer
William C. Elsik
George R. Fournier
John F. Grayson
George F. Hart
Richard W. Hedlund
William S. Hopkins
Arthur E. LeBlanc
Dennis R. Logan
D. Colin McGregor
David R. Mishell
William C. Myers
Paul W. Nygreen
Eart T. Peterson

Kenneth M. Piel
Delbert E. Potter
William A. S. Sarjeant
Mart P. Schemel
Bernard L. Shaffer
Howard M. Simpson
Lewis E. Stover
Herbert J. Sullivan
Robert L. Tabbert
Mel W. Thompson
Alfred Traverse
Charles F. Upshaw
James B. Urban
Logan L. Urban
William F. von Almen
Graham L. Williams

TABLE OF CONTENTS

FOREWORD

(with apologies to Perrault, H. C. Anderson, and the Grimm Brothers)

Once upon the time there was a powerful king who ruled over a kingdom created by himself. A benevolent ruler, he took care that all his subjects knew him and each other and that everybody knew how everbody else lived and worked. In this, he had a good helper in his prime minister who, albeit out of grace, worked hard for the benefit of the kingdom.

As the kingdom grew in extent, fiefdoms sprang up at the periphery, some of them rather independent, but the general position did not change very much during his lifetime. Some of the subjects also were opposed to his reign and wanted to practice some new laws of their own. Although he said that he himself would abide by the old rules and laws, he had no objection against the new ones if anybody liked them better. He was a very wise ruler.

Shortly after, he died, and the kingdom split into a number of independent dukedoms and earldoms, which lived on peaceful terms with each other in spite of adhering to different legislative systems. They still so do, but they have become so many that nobody knows any more how the inhabitants of the other states live and work. Moreover, most of the dukes and earls have died, too, and their heirs, who never knew the old king, have taken over.

When Erdtman wrote his list of "The literature on pollen-statistics published before 1927" he enumerated *ca.* 30 authors from *ca.* 15 countries. Between them, they had published some 130 papers. Madeleine van Campo's latest annual bibliography comprises *ca.* 1,200 papers from 60 countries or non-political areas. The number of authors must be around 500.

Obviously, it is impossible to keep abreast of to-day's wave of palynological literature, even within a narrow field, like a thin slice of the chronological cake: the Quaternary or even the late-Quaternary, or a narrow geographical wedge like Middle and North Europe. Palynology has spread geographically from its Scandinavian birth-place which is easy to understand: a powerful technique like that should be useful elsewhere. Equally simple to grasp is the temporal spread: a technique that yields so much in the Quaternary should be useful also in older epochs. The third field of extension is equally obvious: once you start looking down the microscope for one type of microfossils you cannot help noticing others that behave more or less in the same way: palynology develops into micropaleontology. As a matter of fact these other microfossils had been noticed since the remote birth of palynology in the first half of the 19th century. Only the quantum leap represented by pollen analysis changed the picture for some time.

So simple and obvious were these transitions that they have occurred without anybody realizing that they involve conceptual revolutions. Pollen analysis was conceived as a stratigraphic tool in a region dominated by anemophilous plants—most of the few dominating entomophilous ones also spread their pollen lavishly into the air. The concept of *the general pollen rain* is the philosophical excuse for the percentage calculations which actually presume that all relevant plants are represented in the pollen rain. However, a major part of the Earth's surface is covered by vegetation mainly or exclusively formed by entomophilous taxa, often with intricate pollination syndromes that keep the pollen flow in a closed circuit and gives very little, if anything, out into the ambient air. In such vegetation, there is nothing like a diagnostic pollen rain in the sense of "von Postian" pollen analysis theory. There are ways to get around the problem, but it still awaits its theoretical solution. The art of interpreting pollen spectra from regions covered by entomophilous plants has not yet developed to any degree of perfection.

The second point: for purely stratigraphic research microfossils are markers only—multicolored plastic beads would in many ways have been better than the sometimes cumbersome pollen grains. However, potentially pollen grains are much more than stratigraphic markers: in addition they are more or less well-defined ecologic markers, indicating the conditions under which the vegetation lived. Pollen analysis is therefore also an ecologic tool—as long as we know the ecology of those taxa represented by the pollen types. In the Quaternary we may safely say that we do know it, not always in very precise terms, though. Even in the Tertiary we are on less safe ground, and in the older formations our primary ecological background verges towards nil. Instead of providing prima facie ecologic information, the microfossil taxa of Mesozoic and Palaeozoic vegetation are defined ecologically by the external circumstances of the find: exactly the opposite way round in comparison with Quaternary pollen analysis. As compared with analyses from older formations modern Quaternary diagrams possess an ecological dimension, which was absent from the stratigraphic diagrams of classical pollen analysis. (However, von Post was too much of a botanist to forget about this all the time!) On the other hand, the secondary ecological information that can be established for pre-Tertiary palynological assemblages may be used to transfer ecologic data from one stratigraphic series to another. Here, again, is a field which I feel could be and certainly will be further developed practically and theoretically in the future. Just now, we are too busy looking for stratigraphic markers for the oil industry to have sufficient time to investigate their ecologic value.

A field in which the pollen rain concept is usually meaningless is that of strongly anthropogenic deposits from open-air sites to grain-fields, to caves and even to latrines. All such deposits can yield very valuable information not available by other means. However, such information cannot be utilized under the classical theory even if it is often presented in a graphic form that is deceptively similar. The fundamentally different genesis of such pollen assemblages, the selection processes going on before and at deposition, must always be kept in mind in their interpretation, which is subject to the primary paleontological rule: never to conclude from negative evidence. The latter is obviously done by the percentage calculations.

The earliest pollen analyses from North America were published in 1927 (not many remember them today!), from South America in 1930 and from Mexico in 1944. Since then, palynology in all its aspects has proliferated, especially in the North, like everywhere else. Many problems that were fairly simple in the impoverished floras of North and Middle Europe are formidable in the much richer floras—even of the same major plant communities—in America, and have necessitated much work on methods and interpretation. An almost obsessive preoccupation with methodology has been characteristic of much American palynology. Sometimes it is tempting to remember von Post's old dictum: The results justify the methods.

After a somewhat reluctant start, North American palynology soon gained momentum with scientists like H. P. Hansen, Potzger, Sears, L. R. Wilson, and others whose period of activity spanned some 50 years. After them, during the same period, came a host of younger workers of note, and today palynologic literature dealing even with the Quaternary of North and Middle America is too vast for any single scientist to master fully: a comprehensive view is needed. It is definitely needed outside North America, and I strongly suspect that also American colleagues will benefit highly from a balanced account. It is needed for a synoptic view of where similarities occur and where there are differences, both between North American sites, between American and European/Asiatic sites and cross-Equatorially in America. There is still a long way to go before we understand the physical forces behind the changes our diagrams present; this volume should have the potential to help us in that direction. Taking stock of the past is the fundament of the future.

Knut Faegri
Botanisk Museum
P.O. Box 12
N-5014, Bergen, Norway

PREFACE

The idea for this book was conceived during the last decade while teaching classes in Quaternary palynology. One of the common complaints in those courses, from both faculty and students, was the lack of a synthetic treatment of late-Quaternary palynological data which covered all areas of the North American continent in a single reference source. There have been many excellent summaries of late-Quaternary pollen data published during the last ten years but many of them were not published at the same time nor in the same reference sources. In addition, some geographical regions have generally been omitted while others were repeatedly studied. And finally, as was often true, only selected pollen studies were included as one aspect in books dealing with a whole range of Quaternary investigations.

In selecting the authors for the chapters in this volume, we attempted to solicit the help of individuals who are recognized palynological authorities in their respective geographical regions. There were only two changes in authors from the list originally selected. One of those changes pertained to the chapter on the Great Lakes. The author orginally selected to write this chapter withdrew late in 1984 due to unforeseen circumstances. Since the book was nearing completion by that date, we undertook to write this chapter to insure that all geographical regions of North America would be covered. We are not attempting to imply expertise in this region, but as editors we felt we had no other choice at that "eleventh hour" but to write the chapter ourselves.

The choice of geographical regions may appear to the reader as somewhat arbitrary. We recognize that physiographic regions do not overlap the geopolitical boundaries of the states and provinces in North America as one might assume by reading the chapter headings. For convenience, we divided the regions along state and provincial boundaries. At the same time we encouraged the authors to overlap their discussions to include the border regions of adjacent areas. Thus, for example, sites in western Pennsylvania and New York were covered in the chapters by both Gaudreau and Webb, and Holloway and Bryant; some of the sites in Oklahoma were covered by Baker and Waln in their chapter and also were discussed by Bryant and Holloway in their chapter on Texas. There are other similar instances of overlap which will be obvious to the reader throughout the book.

The organizational framework for each of the chapters is slightly different. This aspect was planned since we preferred to allow the authors as much latitude as possible when writing the summary reviews for each region. Also, the reader will note that some authors were able to discuss their regions using a longer chronological sequence than was possible for some of the other regions. The length of the time chronology covered was left to the discretion of each author, and in most cases, depended on the amount of data available and the expertise of the authors to venture back into earlier portions of the late-Quaternary Period.

At this point, we wish to thank Robert T. Clarke of the American Association of Stratigraphic Palynologists Foundation for his support and efforts in helping this project reach completion. We especially appreciate his careful and thorough review of the completed book prior to printing and his constructive comments. Also we wish to thank Mrs. Celinda A. Stevens for the many long hours she spent putting much of the original data onto computer disks. And finally, we wish to thank each of the authors for their efforts and patience.

Vaughn M. Bryant, Jr. and
Richard G. Holloway
College Station, Texas
May 14, 1985

INTRODUCTION

The late-Quaternary is one of the most often discussed and studied periods of geologic history. The late-Quaternary represents that critical time when man first entered North America, when large scale extinctions of the large Pleistocene mammals occurred just prior to the onset of the Holocene, and when boreal vegetation of tundra and spruce woodlands covered large areas of the North American landscape south of the continental ice sheets. Late-Quaternary sediments provide a plethora of information for a number of disciplines. Remains of late-Quaternary animals are abundant and thus provide paleontologists with relatively large populations with which to work. Glacial geologists can readily obtain data on the effects of the large continental ice masses as evidenced in the distribution of lakes, moraines, and U-shaped valleys which appeared as the glaciers retreated northward. Most scientists agree that man did not migrate into the North American continent prior to the late-Quaternary Period and thus these sediments contain much information on the adaptations and culture of these people. Finally, late-Quaternary age sediments contain data which are useful to botanists who are attempting to identify plant migrational routes and changes in the composition of late-Quaternary plant communities. Among botanists it is the studies of fossil pollen records that have provided some of the key information which we currently use to reconstruct suspected late-Quaternary vegetation types on the North American continent.

Not surprisingly, we have few fossil pollen records dating from some of the earlier glacial advances which covered North America during the pre- and Early Wisconsin time periods. Some of the more important studies of these older deposits include those of Adam (this volume) who has recently obtained a fossil pollen record from California extending back over 130,000 yrs. B.P. In addition, a few isolated fossil pollen records of pre- and Early Wisconsin age sites are available from the Great Lakes Region (Holloway and Bryant, this volume), the Central United States (Baker and Waln, this volume), and the upper sediments of a 1.6 m.y. record from the San Augustin Plains, New Mexico (Markgraf et al., 1983).

The most recent glacial advance in North America during the late-Quaternary (Wisconsin), is the best documented and has provided data obtained from a large number of sites. The Late Wisconsin full-glacial period (ca. 25,000-16,000 yrs. B.P.) is characterized by the presence of the large continental Laurentide and Cordilleran Ice Sheets which covered large areas of North America. Most of Canada and the northern tier of states in the United States were buried under this ice mass. The ice also extended beyond the present shoreline along the Northeast Coast of North America with a concomitant lowering of the mean sea level. Vegetationally, the late-Quaternary full-glacial biomes were diverse. Plant communities existed which have no modern analog as far as we know, and were characterized by unique compositions of plant taxa moving in response to the changing climatic regimes of that period. Existing data suggest that in many areas of North America the Life zones were depressed, but by exactly how much is still the subject of an ongoing debate.

By 18,000 yrs. B.P. the climatic and vegetational effects of the Late Wisconsin full-glacial period were beginning to ameliorate. For example, evidence is beginning to accumulate that ice recession was in progress by this time along the contact edges between the Laurentide and Cordilleran Ice Sheets in southwestern Alberta (Ritchie, this volume). This is reflected in the vegetational changes which are evident during this period, especially from those areas of North America furthest removed from the glacial front, such as the southeastern United States (Delcourt and Delcourt, this volume).

In other regions of North America the evidence of vegetational shifts may have begun slowly during the later stages of the Late Wisconsin full-glacial but are not fully evident until the late-glacial period (16,000-10,000 yrs. B.P.) when large scale reductions in the continental ice sheets were in progress. By late-glacial times the fossil pollen records suggest the presence of significant vegetational changes occurring on newly exposed areas along the ice margins (Gaudreau and Webb, this volume; Holloway and Bryant, this volume) as well as in other regions throughout the southern United States (Delcourt and Delcourt, this volume; Bryant and Holloway, this volume; Hall, this volume). However, these and other data from the late-glacial and Holocene transition period such as the rapid retreat of the Laurentide Ice Sheet and the subsequent vegetational response have served to provide a conflicting data base for the dating of this episode. Because of the importance of these two time periods and the events which occurred during each period, we feel a brief discussion of the importance of dating the transition period is needed.

The establishment of a precise time period for delimiting the late-glacial and Holocene periods has been an active area of research for a number of years. This research activity has resulted in several scientific meetings and has led to the establishment of an International Congress to attempt to find solutions. This problem is complicated by the time-transgressive character of the late-glacial/Holocene transition. As Mörner (1976) has observed, the transition was effectively completed by 13,000 yrs. B.P. in North Africa, while the Laurentide Ice Sheet in Northern Canada did not completely disintegrate until 6,000 yrs. B.P. Recently, a number of proposals have been offered for the placement of the transition (Broecker *et al.,* 1960; Mercer, 1972; Nilsson, 1965; Terasmae, 1972; Fairbridge, 1983; among others). The date assigned by these contributors center around 14,000, 12,000, 10,000 and 7,500 yrs. B.P. As Mörner (1976) observed, most of these estimated dates were based upon interior continental sections, thus amplifying the effects of temporal differences. Accordingly, a report issued by the IX INQUA Congress in New Zealand (1973) recommended the establishment of the Pleistocene/Holocene boundary at 10,000 B.P. measured in radiocarbon years (Mörner, 1976). The stratotype proposed for this boundary is located in the Botanical Garden at Gothenberg, Sweden. At that location, evidence from a continuous core of sediment suggests that at 10,000 yrs. B.P. there occurred a lithologic change, a palynological change, and a distinct paleomagnetic intensity maximum (Mörner, 1976). Thus by agreement, we have available a useable date for this boundary marker.

In this volume we have elected to allow each author the discretion of assigning his/her own dates to the transition zone between the late-glacial and Holocene periods. At least by implication, all the authors have subscribed to the 10,000 yrs. B.P. date for the initiation of the Holocene. As expected, regional differences in the timing of successional vegetational changes do occur. For example, in Texas sediments Bryant and Holloway (this volume) feel that an important vegetational transition period existed between 14,000 and 10,000 yrs. B.P. Thus, in Texas there does not appear to be a sudden vegetational shift marking the traditional end of the late-glacial and the beginning of the Holocene. Instead, what seems to be evident is that by 10,000 yrs. B.P. the Holocene character of the vegetation in that region of North America is firmly established. In another region of North America, the Great Lakes region, (Holloway and Bryant, this volume) the close proximity of that geographical area to the retreating ice sheets created a much quicker climate shift which is expressed vegetationally as a rapid change over the relatively short period of time of approximately 1,000 years. Yet based on the currently available evidence, even in that region of North America the major late-glacial to Holocene vegetational transition is completed by a date of 10,000 yrs. B.P. The remainder of the other regions of North America covered by chapters in this book likewise agree, at least tentatively, to a completion date of around 10,000 yrs. B.P. for the transition.

As seen in the discussion of each chapter in this book, the early-Holocene period represented a time when vegetational communities began to respond to the opening up of new areas to be colonized. As the ice sheets retreated northward, freshly exposed surfaces in those northern latitudes were colonized initially by transitional vegetational communities which were quickly replaced by coniferous forests. One of the major remaining problems is the distinct recognition of many of these late-glacial/Holocene transitional communities. Many reports from the Great Lakes Region interpret forest communities in close proximity to the ice sheets. Recently, with improved chronologic control and the use of pollen influx values we can begin to see that these same communities were principally open, composed primarily of low pollen producing plants, and the significant arboreal pollen present was deposited as a result of long distance transport. Thus, the improved techniques of recognition based upon fossil pollen data help us to understand the vegetation dynamics of this early Holocene Period.

In the Upper Midwest, the early-Holocene period is marked by the replacement of *Picea* dominated communities by prairie. This replacement occurred fairly early, in some areas by 11,000 yrs. B.P. and is admirably documented through the mapping of changes in the pollen assemblages through time (Webb *et al.* 1983; Gaudreau and Webb, this volume). Further evidence of this prairie replacement is indicated from sites in the southern plains of Oklahoma where by 11,000 yrs. B.P. a dominant prairie was established (Bryant and Holloway, this volume). The sequence of pollen data reveals that these plant communities were dynamic with rapidly changing eastern and western boundaries. Similar vegetational trends caused by the general warming of the postglacial climate are also seen in the fossil pollen records from other southern regions of North America such as in the record of the American Southwest (Hall, this volume).

Most geographical regions surveyed in this book report a distinct vegetational shift during the mid-Holocene period. This is reflected most dramatically in the migration of the prairie in the Upper Midwestern region of the United States. This period of inferred warmth is variously defined as the mid-Holocene, the climatic optimum, or the Hypsithermal. Evidence for the magnitude and duration of such a period is still equivocal at best. Again, the most conclusive evidence is seen in the Upper Midwest where pollen records reveal a major eastward extension of the prairie (the Prairie Peninsula) during this time period. However, radiocarbon dating of this interval suggests that it was time transgressive. Authors also disagree on the direction of the proposed changes. For example, a current controversy exists as to the wet vs. dry interpretation of the Hypsithermal from the Southwestern United States. Additionally, based on deep sea cores Thomas (1966) has recently questioned the existence of this interval. Only in the Upper Midwest is there unqualified data suggesting the presence of this period. In other areas such as the Southeastern United States and the Pacific Northwest, the evidence is meager, if present at all. This is one of the ongoing areas of active research which with additional environmental data, will hopefully soon be resolved.

As revealed in the chapters of this book, the fossil pollen records of the mid-to late-Holocene period are more abundant than for the earlier periods since in many cases those sediments have been easier to sample and more often contain fossil pollen which has not yet become destroyed by the agents of time. In addition, more pollen records exist for this final portion of the geologic record since this represents the time when prehistoric man in the New World was rapidly expanding his range and population. Thus, it is the time period which has drawn the concentrated attention not only of paleoecologists but also of archaeologists, many of whom have collected fossil pollen data to help them resolve questions concerning the paleoenvironments which might have affected the subsistence and economies of prehistoric man.

The late-Quaternary period remains a mystery. Even after the publication of the recent two volume summary of late-Quaternary environments (Porter, 1983; Wright, 1983) and the publication of this volume, many questions still remain as to the precise events that occurred during the late-Quaternary. However, as each new body of research is completed and published, we gain a larger body of data upon which to base our hypotheses and speculations. Thus, like other recent books on this subject what you read herein represents a compilation of the most up-to-date information available on fossil pollen records for the late-Quaternary period in North America. However, like other books, this volume will soon be replaced by new texts which will build on this information and present new hypotheses and speculations focused upon larger and more nearly complete data bases which will become available in the future.

References Cited

BROECKER, W.S. *et al.*
    1960   Evidence for an abrupt change in climate close to 11,000 years ago. *American Journal of Science* 258:420-448.

FAIRBRIDGE, R.W.
    1983   The Pleistocene-Holocene boundary. *Quaternary Science Reviews* 1:215-244.

MARKGRAF, V., BRADBURY, J.P., FORESTER, R.M., McCOY, W., SINGH, G., and STERNBERG, R.
    1983   Paleoenvironmental reassessment of the 1.6 million-year-old record from San Augustin Basin, New Mexico. *New Mexico Geological Society Guidebook* II:291-297.

MERCER, J.H.
    1972   The lower boundary of the Holocene. *Quaternary Research* 2:15-24.

MÖRNER, N.-A.
    1976   The Pleistocene/Holocene boundary: a proposed boundary stratotype in Gothenburg, Sweden. *Boreas* 5:193-274.

NILSSON, T.
    1965   The Pleistocene-Holocene boundary and the subdivision of the late Quaternary in southern Sweden. *INQUA Report of the VIth Congress, Warsaw, 1961* 1:479-494.

PORTER, S.C.
    1983   *Late Quaternary environments of the United States, Volume 1, the late Pleistocene.* University of Minnesota Press, Minneapolis, MN.

TERASMAE, J.
    1972   The Pleistocene-Holocene boundary in Canadian context. *24th International Geological Congress Series* 12:120-125.

THOMAS, C.W.
    1966   The Post-Pleistocene Hypsithermal interval: is it fact of fiction? IN: Vierck, E.G. (ed.) *Proceedings of the 16th Alaskan Science Conference. AAAS Alaskan Division.* pp. 138-149.

WRIGHT, H.E.
    1983   *Late Quaternary Environments of the United States, Volume 2, the Holocene.* University of Minnesota Press, Minneapolis, MN.

RICHARD G. HOLLOWAY
Palynology Laboratory
Department of Anthropology
Texas A&M University
College Station, TX 77843

VAUGHN M. BRYANT JR.
Department of Anthropology and
Department of Biology
Texas A&M University
College Station, TX 77843

QUATERNARY PALYNOLOGY AND VEGETATIONAL HISTORY OF THE SOUTHEASTERN UNITED STATES

HAZEL R. DELCOURT
Program for Quaternary Studies of the Southeastern United States
Department of Botany and Graduate Program in Ecology
University of Tennessee
Knoxville, Tennessee 37996

PAUL A. DELCOURT
Program for Quaternary Studies of the Southeastern United States
Department of Geological Sciences and Graduate Program in Ecology
University of Tennessee
Knoxville, Tennessee 37996

Abstract

The region of eastern North America east of 100° W longitude and south of 40° N latitude has provided a challenge to Quaternary palynologists because of the diversity of its physiographic regions, the complexity of vegetation types, and the richness of species within the flora. Over the past 55 years, palynologists have developed successful strategies for location of late-Quaternary stratigraphic sites in the southeastern United States. Since A.D. 1950, progressive refinements in the understanding of vegetational dynamics in this region have resulted from the application of isotopic dating techniques, the development of taxonomic reference collections and keys for identification of plant fossils, and the quantification of modern pollen-vegetation relationships on a variety of spatial scales.

We present a concise historical perspective concerning the contributions of Quaternary palynologists in the Southeast. We document the major radiocarbon-dated palynological sequences and literature, describe the modern pollen-vegetation relationships for important tree taxa, characterize the vegetational history for the past 20,000 years, and indicate key problems for future research in Quaternary paleoecology of the southeastern United States.

INTRODUCTION

The southeastern United States is an important region that offers many exciting opportunities for research in Quaternary palynology. The territory south of the maximum extent of continental glaciation by the Laurentide Ice Sheet provided a principal full-glacial refuge for plant and animal taxa that have subsequently recolonized deglaciated landscapes during interglacial times. The nature of full-glacial biotic communities, the timing, magnitude, and rate of changes in climatic and vegetational gradients during deglaciation, the rates, directions, and processes of post-glacial plant migrations, and the extent of Indian and Euro-American impacts upon the natural vegetation are all significant questions to be answered with palynological data from appropriate sites in the southeastern United States.

Over the past 55 years, Quaternary palynologists have actively sought to understand the history and development of the flora and vegetation of the Southeast. From the beginnings of Quaternary pollen analysis in North America, the Southeast has posed a series of challenges to paleoecologists because of its diversity of physiographic regions and plant communities, its richness in species of both woody and herbaceous vascular plants, and its large number of endemic plant species. However, for many years, lake and bog environments suitable for preservation of late-Quaternary plant fossils were considered to be extremely scarce (if not nonexistent) south of the glacial margin, with the notable exceptions of the "Carolina Bay" lakes along the Atlantic Coastal Plain (Buell, 1939, 1945a, 1945b, 1946) and of the karst ponds in the lake district of central peninsular Florida. Outside those two regions, most Quaternary plant-fossil studies were restricted initially to coastal peat deposits (J. Davis, 1946), river-terrace deposits with organic lenses associated with Pleistocene megafaunal assemblages (Brown, 1938), or preliminary analysis of isolated peat bogs (Sears and Couch, 1932; Sears, 1935; Potzger and Tharp, 1943, 1947).

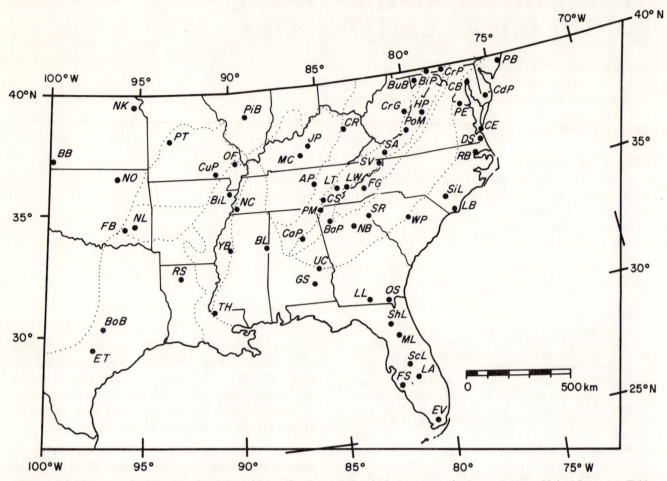

Figure 1. Location map of radiocarbon-dated sites with late-Quaternary palynological sequences in the southeastern United States (see Table 1 for letter code to sites on map). The dotted lines correspond with boundaries of major physiographic regions.

While instrumental in opening a new field of biostratigraphic inquiry in the Southeast, these pioneering studies were limited by the lack of absolute time control, minimal reference collections for identification of modern and fossil pollen types, sparse literature on pollen taxonomy, and difficulties in accurately determining stratigraphic, ecologic, and hydrologic contexts for the plant-fossil deposits.

With the advent of radiocarbon dating in A.D. 1950 (Arnold and Libby, 1950), attention focused upon documenting the extent of climatic cooling during the last glacial maximum through determining the maximum southward displacement of boreal species and locating the full-glacial refuge areas for deciduous forest species. These biogeographic questions became central issues for debates that continued over a thirty-year time span (Deevey, 1949; Braun, 1950, 1951, 1955; Martin and Harrell, 1957; Martin, 1958a; Whitehead, 1965a, 1973, 1981; Watts, 1970, 1975a, 1975b, 1979, 1980a, 1980b; Watts and Stuiver, 1980; Bryant, 1977, 1978; Delcourt and Delcourt, 1975, 1977, 1979, 1981; H.

Delcourt, 1979; P. Delcourt, 1980; P. Delcourt et al., 1980; M. Davis, 1976, 1981). As the density of well-dated late-Pleistocene and Holocene sites has increased (Fig. 1), the broad patterns of late-Quaternary vegetational and climatic change have been established in the Southeast (Whitehead, 1973; M. Davis, 1976, 1981, 1983; Watts, 1980a, 1983; Stone and Brown, 1983; Delcourt and Delcourt, 1979, 1981, 1983, 1984a, 1984b, 1985). Certain major biogeographic questions still remain unresolved; for example, what were the full-glacial distributional limits of tundra species in the southern Appalachian Mountains and along the southern margin of the Laurentide Ice Sheet in the central Midwest? In addition to the continuing quest for full-glacial sites in previously unstudied areas, research questions currently being addressed in Quaternary palynology are shifting in emphasis. Quantitative paleoecologic approaches now permit more detailed examination of the dynamic responses of late-Pleistocene and Holocene vegetation to a variety of environmental changes.

Contemporary research in Quaternary palynology in the southeastern United States includes exploration of these new paleoecological lines of investigation:

1) development of comprehensive calibrations of pollen percentages with modern vegetational composition, both locally around fossil-pollen sites and across broad regions, in order to reconstruct and map changes in vegetational composition through time;

2) application of dissimilarity coefficients for objective location and evaluation of potential modern analogues for full-glacial and late-glacial pollen assemblages;

3) study of changes in geomorphic processes and frequency of disturbances as agents influencing changes in vegetational composition through the late-Pleistocene and Holocene intervals;

4) examination of the nature of floristic gradients and changes in composition of plant communities across significant late-Pleistocene and Holocene vegetational ecotones;

5) evaluation of the extent to which late-Quaternary climatic changes have influenced patterns of diversity and stability of the vegetation; and

6) analysis of pollen and charcoal particles in sediments from basins located near archaeological sites to resolve the extent of impact on native vegetation by American Indians prior to Euro-American settlement.

In the following sections of this paper, we review the history of development of the field of Quaternary palynology in the southeastern United States, including a comprehensive bibliography and summary of information concerning sites published or described in graduate theses through A.D. 1984. We include late-Quaternary sites that are both located in eastern North America south of 40° N latitude and south of the Laurentide glacial margin at 18,000 B.P., as well as situated east of 100° W longitude (the approximate boundary between eastern forests and prairie) (Fig. 1). For discussion of the time intervals represented by the fossil-pollen sequences in these late-Quaternary sites, we use the following terms: (1) the Sangamonian interglacial interval, from 125,000 to 80,000 B.P.; (2) the Early Wisconsin interval, from 80,000 to 28,000 B.P.; (3) the Middle Wisconsin interval, from 28,000 to 23,000 B.P.; (4) the full-glacial portion of the Late Wisconsin interval, from 23,000 to 16,500 B.P.; (5) the late-glacial portion of the Late Wisconsin interval, from 16,500 to 12,500 B.P.; (6) the early-Holocene interval, from 12,500 to 8,500 B.P.; (7) the middle-Holocene interval, from 8,500 to 4,000 B.P.; and (8) the late-Holocene interval, from 4,000 B.P. to

the present day. We summarize the regional quantitative relationships of pollen and vegetation assemblages based upon comprehensive data sets of the modern pollen rain and continuous forest inventories. Within the major physiographic regions of the Southeast, we discuss successful strategies for locating lacustrine, alluvial, and bog environments that are suitable for testing of specific hypotheses in Quaternary paleoecology and archaeology. We then review current interpretations of vegetational change from the last full-glacial period to the present, based upon radiocarbon-dated pollen sequences from southeastern North America. This discussion includes specific topical areas of current research that constitute breakthroughs in the application of Quaternary palynology in the southeastern United States.

HISTORICAL PERSPECTIVE

The first palynological study of late-Quaternary deposits south of the glacial margin in the eastern United States (Fig. 2) was that of a sediment core taken from Dismal Swamp (Fig. 1), Virginia (Lewis and Cocke, 1929). This study was published only two years after the first paper dealing with Quaternary palynology within North America (Auer, 1927). By the early 1930's (Fig. 2), Paul Sears and his students (Sears and Couch, 1932; Sears, 1935) had published pollen spectra from a number of bogs and peat lenses scattered from western Arkansas to East Tennessee. Although lacking techniques for obtaining absolute chronologies for these sites, Sears attempted to summarize post-glacial forest history along east-west and north-south transects (Sears, 1935), and he was the first to compile maps depicting the relative sequence of northward post-glacial migrations of important tree genera across eastern North America (Sears, 1942).

The problems of differential pollen production, dispersal, and preservation were appreciated early through the experience of European Quaternary palynologists (literature summarized in Cain, 1939). The first study in North America comparing representation of pollen from surface moss-polster samples with quantitative forest composition was conducted in the spruce-fir forest on Mt. Collins, in the Great Smoky Mountains of East Tennessee, by Gladys Carroll (1943), a student of Stanley Cain (Fig. 2) at the University of Tennessee.

The most significant discoveries in the first two decades of Quaternary plant-fossil studies in the Southeast (Fig. 2) were reports of boreal conifers such

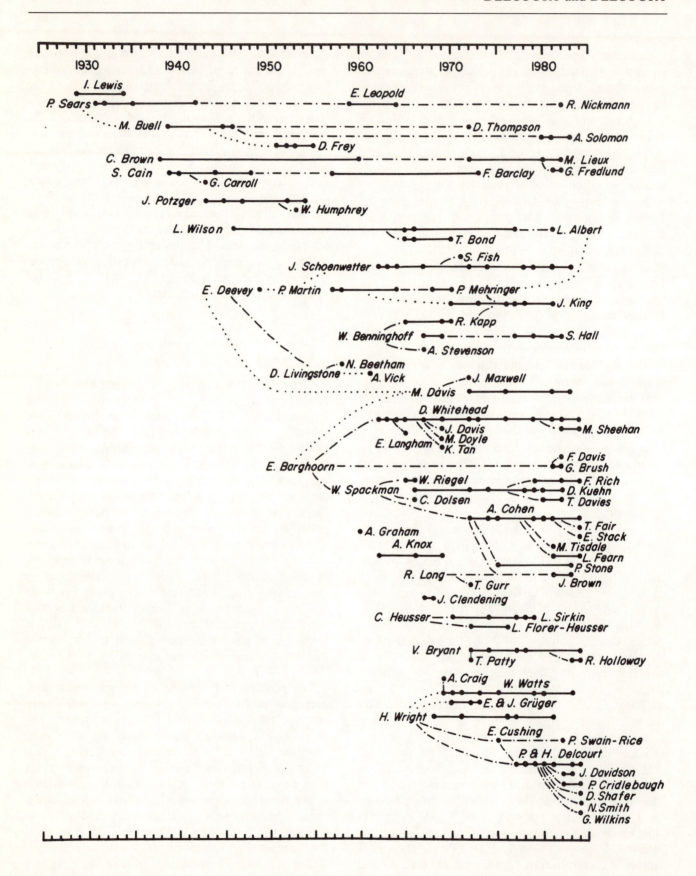

as spruce *(Picea)*, larch *(Larix)*, and fir *(Abies)* from Pleistocene-age deposits on the coastal plains of North Carolina (Buell, 1939, 1945a, 1945b, 1946), Florida (L. R. Wilson *in* J. Davis, 1946), Louisiana (Brown, 1938), and east-central Texas (Potzger and Tharp, 1943, 1947, 1954). These findings spurred a major controversy (Fig. 2) concerning the magnitude of climatic change and displacement of the biota south of the glacial margin during full-glacial episodes (literature summarized in Deevey, 1949; Martin, 1958a; Martin and Harrell, 1957; and Delcourt and Delcourt, 1979).

With the development of the technique of radiocarbon dating, the previous era of biostratigraphic correlation of Quaternary deposits was transformed. One of the first palynological sites in North America to be radiocarbon-dated was Cranberry Glades (Fig. 1), West Virginia (Darlington, 1943; Arnold and Libby, 1950). Through the application of radiometric dating, an absolute temporal framework could be established for the palynological biostratigraphy from each late-Quaternary site, allowing for synthesis of biogeographic patterns in the development of vegetation across the region (use of radiocarbon dating anticipated by Deevey, 1949; Martin, 1958a).

Progress in biogeography proceeded in tandem with development of paleoecological studies. D. G. Frey (Fig. 2) studied the development of both aquatic and terrestrial ecosystems based on microfossil analysis of cores from Singletary Lake (Fig. 1), North Carolina (1951, 1953, 1955). These paleolimnologic and paleoecologic analyses were designed to identify the mode of origin of Carolina Bays (shallow, oriented, elliptical basins distributed across the Atlantic Coastal Plain and Piedmont). D. R. Whitehead (Fig. 2) expanded the paleoecological investigations of the Carolina Bays of North Carolina (Fig. 1) with refinements in pollen taxonomy, studies of modern pollen-vegetation relationships, and use of indicator plant species to generate sophisticated reconstructions of late-Quaternary coastal-plain environments (Whitehead, 1963, 1964, 1965a, 1965b, 1967, 1973, 1981; Whitehead and Tan, 1969).

Holocene peat deposits of the Okefenokee Swamp (Fig. 1), southeastern Georgia, and the Everglades

(Fig. 1) of southern Florida have been studied intensively by W. Spackman and his students (Fig. 2) as modern analogues for Carboniferous coal-forming environments (Cohen, 1975; Cohen and Spackman, 1972; Cohen *et al.*, 1984; Kuehn, 1980; Rich, 1979; Rich and Spackman, 1979; Rich *et al.*, 1982; Riegel, 1965; Spackman *et al.*, 1966, 1974).

The effective field team of W. A. Watts and H. E. Wright, Jr., opened new geographic regions to paleoecologic research within the Southeast, beginning in the 1960's (Fig. 2). Among the innovations Professor Watts has contributed to Quaternary paleoecology in the southeastern United States was the combined use of systematic pollen and plant-macrofossil analysis of lacustrine sediments. Watts and his student, Alan Craig, examined late-Quaternary sites (Fig. 1) within the Ridge and Valley Physiographic Province (Craig, 1969; Watts, 1970, 1973), the central Appalachian Mountains (Watts, 1979), and the Atlantic Coastal Plain (Watts, 1979, 1980a, 1980b). In addition, Watts sought full-glacial refugia for temperate deciduous trees in the karst terrain of southern Georgia (Watts, 1971) and Florida (Watts, 1969, 1971, 1975a; Watts and Stuiver, 1980).

The study of Buckle's Bog (Fig. 1) on the Allegheny Plateau of Maryland (Maxwell and Davis, 1972), provided the first pollen-influx data for a late-Quaternary site in the unglaciated Southeast (Fig. 2). Pollen assemblages from Buckle's Bog documented the existence of tundra in the central Appalachian Mountains during the full-glacial and late-glacial intervals, supporting the earlier interpretations of Martin (1958b), and foreshadowing the evidence of Watts (1979) for extensive alpine tundra within the central Appalachians.

The classic biogeographic controversy between Deevey (1949) and Braun (1950, 1951, 1955) was resolved with the analysis of a late-Quaternary site, Anderson Pond (Fig. 1), located within the heartland of the modern Mixed Mesophytic Forest Region in Middle Tennessee (H. Delcourt, 1979). Anderson Pond, the first site in the Southeast with diagrams for both pollen influx and plant-macrofossil influx (Fig. 2), provided conclusive evidence for full-glacial boreal forest and significant climatic cooling south of

Figure 2. Genealogical chart of Quaternary palynologists in the southeastern United States, 1929 through 1984. Palynologists represented on this chart are individuals who have published two or more scientific papers concerning southeastern Quaternary palynology, served as palynological mentors for subsequent researchers in the region, or are students within an established palynological center or school with a primary research focus in the Southeast. For each palynologist, solid dots correspond to specific years of his or her publications (as listed in Appendix 1) and the solid horizontal line indicates the length of time devoted to Quaternary palynological research in the Southeast. A palynological mentor (*e.g.*, teacher/student relationship) is noted by a line constructed of alternating dots and dashes; a line series of dots indicates a strong professional interaction between two palynologists.

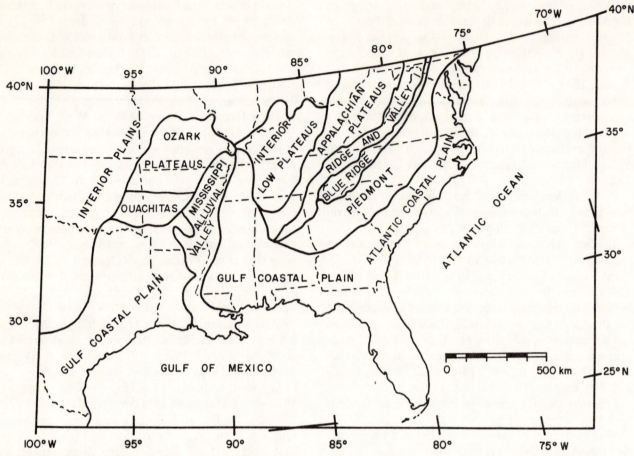

Figure 3. Physiographic regions of the southeastern United States.

the glacial limit to 36° N latitude and west of the Appalachian Mountains (H. Delcourt, 1979). Anderson Pond spans continuously the time interval of the past 19,000 years, and has been used for the first application in the Southeast of a series of quantitative techniques for the reconstruction of vegetational history, including forest-stand simulation modeling (Solomon *et al.*, 1980, 1981), taxon calibrations, and dissimilarity coefficients for determining modern analogues for fossil pollen spectra (Delcourt and Delcourt, 1985).

The first full-glacial refugium identified for temperate deciduous tree populations was documented by the pollen and plant-macrofossil evidence from Nonconnah Creek (Figs. 1 and 2), southwestern Tennessee (P. Delcourt *et al.*, 1980). The Nonconnah Creek site verified that the Blufflands, *i.e.*, the loess-mantled uplands east of the Lower Mississippi Alluvial Valley, served as a major refugium for temperate deciduous forest. In addition to Nonconnah Creek, other sites recently discovered with evidence for persistence of temperate deciduous

forest throughout the full-glacial interval include Goshen Springs (Fig. 1), south-central Alabama (P. Delcourt, 1980), and Sheelar Lake (Fig. 1), northern Florida (Watts and Stuiver, 1980).

Recent refinements in pollen taxonomy offer opportunities for additional breakthroughs in understanding Quaternary biogeographic patterns (Birks and Peglar, 1980; Lieux, 1980a, 1980b; Lieux and Godfrey, 1982; Solomon 1983a, 1983b). Holloway and Bryant (1984) have applied scanning electron microscopy recently (Fig. 2) to identification of fossil pollen grains of *Picea* from Boriack Bog (Fig. 1), in order to identify the species of spruce present in east-central Texas during the late-glacial interval.

MODERN POLLEN-VEGETATION RELATIONSHIPS

The diversity of topography and physiography of the southeastern United States (Fig. 3) is reflected in the high species richness of its flora and the great number of plant communities distributed across the

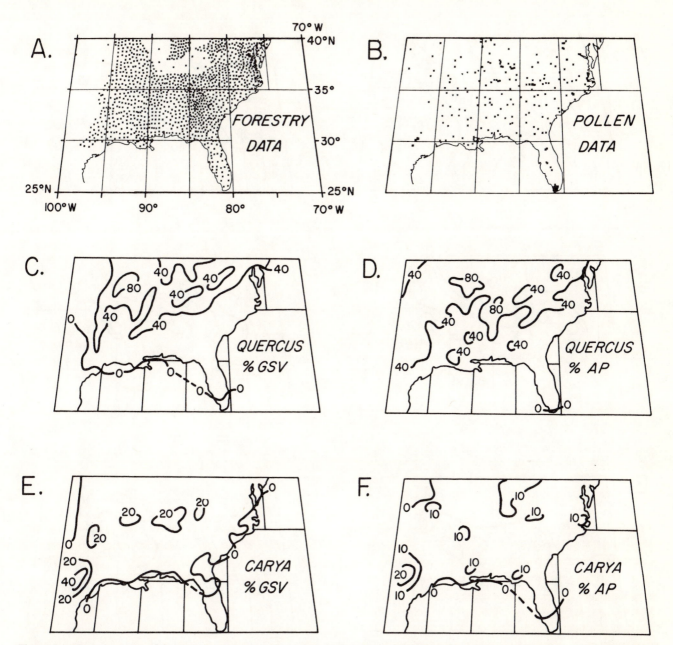

Figure 4. a) Location map of forestry data (percent of growing stock volume, %GSV) used to prepare contoured maps of percent tree dominance (isophyte maps). b) Location map of modern pollen samples used to prepare contoured maps of percent arboreal pollen (%AP, isopoll maps). c) Isophyte map for *Quercus*. d) Isopoll map for *Quercus*. e) Isophyte map for *Carya*. f) Isopoll map for *Carya*. (Maps modified from P. Delcourt *et al.*, 1983).

region. As many as fifty arboreal pollen types are represented in a single Holocene lacustrine deposit (H. Delcourt *et al.*, 1983), and pollen grains of many additional tree taxa are found in Pleistocene-age sediments in the Southeast. Late-Quaternary vegetational reconstructions are greatly facilitated by an appreciation of the quantitative relationships between modern pollen assemblages derived from surface samples and vegetation composition of extant forests. P. Delcourt *et al.* (1983) presented both contoured isophyte maps representing the dominance

distributions of 19 tree taxa, based upon county summaries of growing-stock volume (%GSV) from the Continuous Forest Inventories of the USDA-Forest Service (H. Delcourt *et al.*, 1981), and corresponding isopoll maps of arboreal pollen percentages (%AP), based upon 250 surface pollen samples distributed across the Southeast (Fig. 4a, 4b). Mapping was extended to cover the majority of the ranges of dominant, subdominant, and common tree taxa across eastern North America (P. Delcourt *et al.*, 1984), and the subcontinental-scale data sets

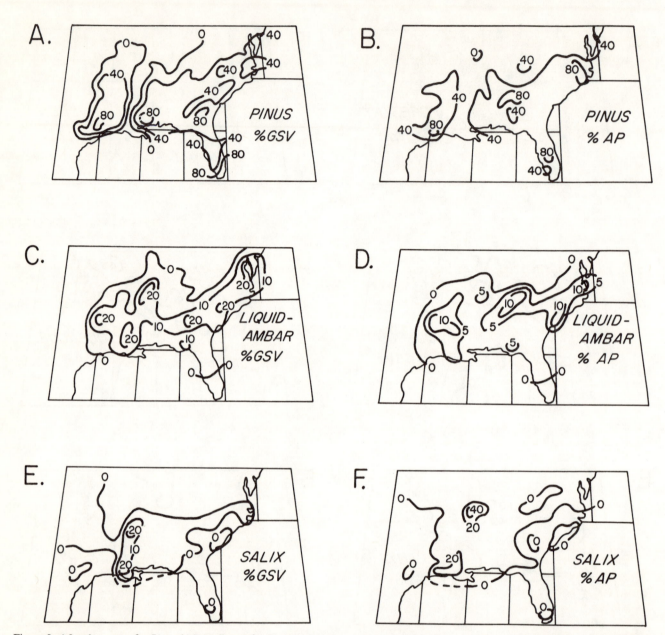

Figure 5. a) Isophyte map for *Pinus*. b) Isopoll map for *Pinus*. c) Isophyte map for *Liquidambar styraciflua*. d) Isopoll map for *Liquidambar styraciflua*. e) Isophyte map for *Salix*. f) Isopoll map for *Salix*. (Maps modified from P. Delcourt *et al.*, 1983).

provided taxon calibrations appropriate for reconstructing changes in tree dominances through time across the region.

In the Southeast today, several distinctive patterns of dominance are evident for different groups of tree species. Some genera are widespread as forest dominants and subdominants. For example, oak (*Quercus*, Fig. 4c, 4d) and hickory (*Carya*, Fig. 4e, 4f) are genera with numerous species that occupy a broad range along gradients of moisture and topography, thus having broad apparent ecological amplitudes. Population centers for *Quercus* are located today on

the Gulf Coastal Plain of eastern Texas, in the Ouachita and Ozark Mountains of Arkansas and Missouri, the Cumberland Plateau of Tennessee and Alabama, and the Ridge and Valley and Blue Ridge provinces of the Appalachian Mountains. Both *Quercus* trees and pollen are abundant in the Piedmont from the Carolinas to Alabama but decrease to the south across the Gulf Coastal Plain into Florida. On the isophyte map (%GSV) for *Carya*, population centers are evident on the Gulf Coastal Plain of eastern Texas, the Ozark Mountains of Missouri, and the Interior Low Plateaus of Middle Tennessee and

Kentucky. The isopoll map (%AP) for *Carya* illustrates that population centers for hickory are generally mirrored in the modern pollen rain, with a strong correspondence also for the southern limit in the range of the genus depicted on both maps. *Carya* is found throughout the Piedmont and Atlantic Coastal Plain but is typically a dominant in the forest canopy there primarily on dry sites. Water oak *(Carya aquatica)* and pecan *(Carya illinoensis)* occur in alluvial backswamps within the Mississippi Embayment (Fowells, 1965).

Eleven native species of pine (*Pinus*, Fig. 5a, 5b) grow in the southeastern United States today (Little, 1971). Many of the species require dry upland sites; however, several coastal plain species are tolerant of seasonally wet conditions (Fowells, 1965). Both the isophyte and isopoll maps (Fig. 5a, 5b) reflect the predominance of *Pinus* in the sandy uplands of the Gulf and southern Atlantic coastal plains, and the virtual absence of pines in the annually inundated backswamp environments of the Mississippi Alluvial Valley. Percentages of both *Pinus* trees and pollen decrease northward away from the coastal plains. A strong correspondence in mapped patterns of isophyte and isopoll contours for *Pinus* is evident in the Southeast despite the tendency for *Pinus* to be overrepresented because of its high pollen productivity. In part, this correspondence occurs because the landscape is generally forested, decreasing the distance over which pine pollen is transported away from source trees, and in part because of the high pollen productivity of other regionally dominant tree taxa such as *Quercus* (P. Delcourt *et al.*, 1983).

Sweetgum (*Liquidambar styraciflua*, Fig. 5c, 5d) is found today as an understory tree throughout the Southeast, with the exception of higher elevations of the Blue Ridge Province and the southern Florida peninsula. Sweetgum is especially abundant on loess bluffs east of the Mississippi River Valley and on the Gulf Coastal Plain of Louisiana; the isopoll map reflects the ubiquitous occurrence of *Liquidambar* pollen grains in surface samples of modern pollen rain from throughout the Piedmont and the Atlantic and Gulf coastal plains.

Several genera of southeastern trees are primarily characteristic of poorly-drained alluvial habitats that occur extensively along major river valleys in the coastal plains from Texas to Florida and northward to Maryland, and within large swamp complexes such as the Everglades. For example, willow (*Salix*, Fig. 5e, 5f) is a commercially important tree within the Lower Mississippi Alluvial Valley, and its pollen percentages reflect its importance within bottomland

habitats across the southeastern region. The genus *Nyssa* (Fig. 6a, 6b) consists of three species existing primarily in poorly-drained alluvial flats, sloughs, and seepage ponds, although one species, black gum (*Nyssa sylvatica*), is characteristic of more well-drained soils of mid- to upper slopes throughout the region. Occurrence of *Nyssa* pollen generally corresponds with the predominance of tupelo gum (*Nyssa aquatica*) and Ogeechee tupelo (*Nyssa ogeche*) in major coastal swamps, and only as trace occurrences across the uplands. Species of magnolias (*Magnolia*, Fig. 6c, 6d) occur both on moist, well-drained soils along streams and in swamps of the coastal plains and on mesic slopes of the southern Appalachian Mountains. Magnolias are generally minor constituents of the forests, and their insect-pollinated flowers shed little pollen; it is only occasionally recorded as more than trace occurrences in pollen surface samples.

Two plant families, Cupressaceae and Taxodiaceae, are generally represented by only one pollen type in either surface samples or fossil preparations. These families include three species of commercially important trees within the southeastern United States, red cedar (*Juniperus virginiana*), Atlantic white cedar (*Chamaecyparis thyoides*), and bald cypress (*Taxodium distichum*) (Little, 1971). The primary contributor to the pollen type is *Taxodium*, a dominant within alluvial backswamp environments of major river systems and swamps such as the Everglades. *Juniperus* is locally abundant on dry, shallow soils of the cedar glades of central Kentucky and Tennessee (Fig. 6e, 6f).

A number of tree species characteristic of northern conifer-hardwoods forest (hemlock, *Tsuga canadensis*; sugar maple, *Acer saccharum*; beech, *Fagus grandifolia*; and yellow birch, *Betula alleghaniensis*) are important constituents of forest communities of the Southeast only at middle and high elevations of the central and southern Appalachian Mountains. Boreal conifers such as spruce (*Picea*) and fir (*Abies*) occur infrequently in the central Appalachian Mountains and only above an elevation of 1,500 m on ten mountain peaks in the southern Blue Ridge Province (Delcourt and Delcourt, 1984a). Broad-scale calibrations for these taxa are best derived from pollen and vegetation data from regions to the north where major gradients occur in their abundances on modern landscapes (maps and calibrations for these taxa across eastern North America are included in P. Delcourt *et al.*, 1984).

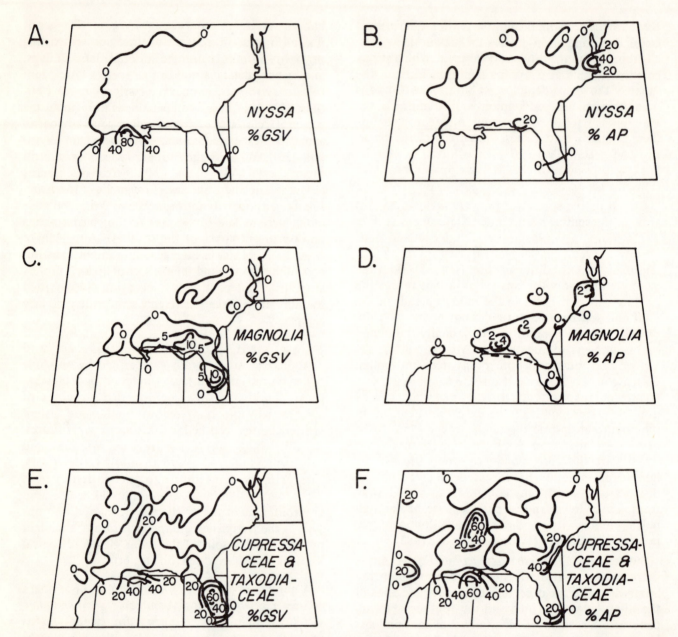

Figure 6. a) Isophyte map for *Nyssa*. b) Isopoll map for *Nyssa*. c) Isophyte map for *Magnolia*. d) Isopoll map for *Magnolia*. e) Isophyte map for Cupressaceae & Taxodiaceae. f) Isopoll map for Cupressaceae & Taxodiaceae. (Maps modified from P. Delcourt *et al.*, 1983).

STRATEGIES FOR LOCATING LATE-QUATERNARY PALYNOLOGICAL SITES

The unglaciated Southeast is a region in which many lacustrine, alluvial, and peat-bog sites may be found that span the past 20,000 years or more, because these sites are not constrained by the timing for glacial retreat that dictated the post-glacial age for kettle lakes on formerly glaciated terrain. The polygenetic origins of late-Quaternary pollen sites in the

Southeast, however, pose special problems in location, recovery, and interpretation of fossil pollen assemblages.

Lacustrine deposits typically are used for interpretation of vegetation history because the processes of deposition and preservation of plant microfossils are best understood for temperate lake basins with continuous sediment deposition (Delcourt and Delcourt, 1980; Hall, 1981; Jacobson and Bradshaw, 1981). Lacustrine sites with Holocene and late-Pleistocene sediment accumulations occur in three principal regions south of the glacial border in eastern

North America: (1) the "Carolina Bay" lakes, which number one-half million and extend along the Atlantic seaboard from southeast New Jersey to northeast Florida (Prouty, 1952); (2) the karst lake region of the panhandle and peninsular Florida (White, 1970); and (3) karst sinkholes of the Interior Low Plateaus of Tennessee and Kentucky (Kemmerly, 1982), and the Ozark Plateaus of Missouri and Arkansas (Vineyard and Feder, 1982).

"Carolina Bay" lakes are shallow, elliptical depressions oriented along a northeast-southwest axis that have originated predominantly during the Middle Wisconsin and during the late-glacial/early-Holocene transition (Thom, 1970; Kaczorowski, 1977; Whitehead, 1981). North of about 39° N latitude, the bays may have originated as thermokarst depressions during times of extreme periglacial conditions (Pewe, 1983); southward, between 33° and 39° N, many bay depressions represent deflation hollows that expanded in size during full-glacial times with cold and windy climatic conditions (Whitehead, 1973, 1981; Denny *et al.*, 1979; Watts, 1980b). South of 33° N, development of humate-cemented sands served to perch the groundwater table near the land surface and facilitated the generation of late-Quaternary Carolina Bays within temperate climatic conditions in the southern Atlantic seaboard (Thom, 1967; Kaczorowski, 1977).

Karst solution ponds of coastal Florida contain lacustrine sediments dating from interstadial and interglacial times. During the Wisconsin full-glacial and late-glacial periods, lowered sea levels caused the groundwater table to drop by as much as 20 m along the Gulf Coastal region of southern Georgia and the panhandle of northern Florida, resulting in the drying of many lake basins and hiatuses in the accumulation of organic-rich lacustrine sediments (Watts, 1969, 1980b). In peninsular Florida, only sinkholes that today contain more than 20 m of water have had continuous sedimentation through the Wisconsin full-glacial period (Watts, 1975a, 1980b; Watts and Stuiver, 1980). Along upland interfluves unconnected with the regional water table, such as on early-Pleistocene terraces of the Conecuh River in southcentral Alabama, karst collapse at depth beneath the Quaternary deposits may create sag ponds with perched water tables that contain lacustrine sediments dating beyond the limit of radiocarbon dating (P. Delcourt, 1980). Sites such as Goshen Springs, Alabama, are rare, and sediment accumulation is relatively slow in small, isolated, closed basins located on upland interfluves (P. Delcourt, 1980).

Thousands of karst depressions have developed by solution of underlying limestones and dolomites of Cambrian age through Mississippian age in the Interior Low Plateaus regions, extending from the Appalachian Mountains to the Ozark Plateaus. The sinkholes range in age from those actively forming today to those containing lacustrine sediments exceeding the limits of radiocarbon dating. Development of a karst lake basin with permanently-standing water depends upon the development of a clay or shale seal that plugs outflow from springs that feed the basin (H. Delcourt, 1979; H. Delcourt *et al.*, 1983; Smith, 1984; Wilkins, 1985). Karst lake basins with lacustrine sediments deposited over the past 20,000 to 40,000 years and located between approximately 33° and 40° N latitude typically have had rapid accumulation of sediments during late-Pleistocene time and much slower, although uniform, rates of deposition during the Holocene. During the full-glacial and late-glacial periods, these karst depressions contained relatively deep, open ponds. Under Holocene climatic conditions, these lacustrine sites became generally shallow, with water levels that fluctuate as much as several meters throughout the year because of high rates of evaporation during the summer months (H. Delcourt, 1979; Smith, 1984; Wilkins, 1985; P. Delcourt, 1985).

In physiographic regions such as the Lower Mississippi Alluvial Valley where the only large, permanent lakes available are oxbow lakes representing cutoffs of river meanders, alluvial sections provide the primary source of material for paleoecological analysis. Holocene sediments may be obtained from the several meander trains of the Mississippi River (Holloway and Valastro, 1983); sediments of Pleistocene age are found in stream terraces exposed by Holocene downcutting of the Mississippi River and its tributaries (Delcourt and Delcourt, 1977; P. Delcourt *et al.*, 1980). Stream terrace deposits can be found along every major river system in the Southeast (Markewich and Christopher, 1982). Comparatively little is known about the processes of transport, sorting, deposition, and preservation of palynomorphs along river systems and their influence upon the palynological record preserved within oxbow lakes, such as B. L. Bigbee Oxbow in east-central Mississippi (Whitehead and Sheehan, 1984). Interpretation of such alluvial sites requires development of special data sets for calibration of pollen and modern vegetation to take these factors into account (Engelhardt, 1963).

About 500 m elevation in the central and southern Appalachian Mountains, late-Quaternary palynological sites are primarily restricted to karst "sag ponds" (Craig, 1969; Watts, 1979) or peat deposits of poorly-drained fens or "glades" such as Buckle's Bog, Maryland (Maxwell and Davis, 1972), Cranberry Glades, West Virginia (Darlington, 1943; Watts, 1979), Flat Laurel Gap, North Carolina (Shafer, 1984), and the bogs along Boone Fork and in Long Hope Valley, North Carolina (H. Delcourt, 1985). Additional sources of palynological data that can provide information concerning the Holocene development of local forest communities within the Southern Appalachians include humus soils of mesic forests, organic soils of heath balds and grassy balds (Beetham, 1958), and woodland-hollow pools such as Lake in the Woods in Cades Cove, Great Smoky Mountains (Davidson, 1983).

Additional late-Quaternary pollen records may be obtained from bogs and buried soils in the Piedmont of the Carolinas and Georgia (Cain, 1939; Whitehead and Barghoorn, 1962; Humphrey, 1953; Carbone *et al.*, 1982), from solution ponds located on the surface of salt domes in Louisiana (Kolb and Fredlund, 1981), spring deposits located at the contact of geologic formations of contrasting lithology, such as the Carizzo Aquifer in east-central Texas (Potzger and Tharp, 1943, 1947, 1954; Larson *et al.*, 1972; Bryant, 1977), and former back-barrier lagoons along coastal barrier-ridge systems (Frey, 1952; Whitehead and Davis, 1969; Whitehead and Doyle, 1969; Thompson, 1972; Cronin *et al.*, 1981; Tidsdale, 1981; Nickmann and Demarest, 1982). The key to locating suitable late-Quaternary sites within the southeastern United States for palynological investigations lies in having a diverse set of strategies when searching within any given physiographic region. The potential of the region for advances in Quaternary palynological research has just begun to be tapped.

**LATE-QUATERNARY HISTORY OF
VEGETATIONAL CHANGE**

Of the published palynological sequences available from southeastern North America and the adjacent oceans, only one marine core, from the Northern Gulf of Mexico, spans the Tertiary-Quaternary boundary and documents changes in fossil pollen assemblages throughout the entire Quaternary (Elsik, 1969). Unfortunately, no terrestrial site south of the glacial limit in eastern North America has been discovered that spans a comparable period of time.

To date, no late-Quaternary pollen sequence is known from the unglaciated Southeast that records changes in terrestrial vegetation continuously over the past 125,000 years, since the Sangamonian interglacial. Pittsburg Basin, a former kettle lake developed on Illinoian till in south-central Illinois, provides a unique palynological record of biotic and environmental change following the retreat of the Illinoian continental glacier (E. Gruger, 1972a, 1972b). Many palynological sequences reflect only limited portions of the Quaternary, and their correlation in absolute time is not possible without independent dating techniques such as radiocarbon analysis, thermoluminescence dating, or other isotopic methods (Coleman and Pierce, 1977). In Table 1, the locations, time spans, depositional environments, pollen analysts, and bibliographic citations are listed for the principal radiocarbon-dated sites (Fig. 1) that provide primary palynological evidence making possible the interpretation of changes in broad-scale floristic and vegetational patterns during the late-Pleistocene and Holocene intervals.

The Late Wisconsin Full-glacial Interval (23,000 to 16,500 B.P.)

Ten radiocarbon-dated sites in unglaciated southeastern North America continuously span at least the last 17,000 years. These include eight published sites, listed alphabetically: Anderson Pond, Tennessee (H. Delcourt, 1979); Buckle's Bog, Maryland (Maxwell and Davis, 1972); Goshen Springs, Alabama (P. Delcourt, 1980); Pigeon Marsh, Georgia (Watts, 1975b); Quicksand Pond, Georgia (Watts, 1970); Rockyhock Bay, North Carolina (Whitehead, 1981); Singletary Lake (Frey, 1953, 1955); and White Pond, South Carolina (Watts, 1980a). The remaining two sites are Jackson Pond, Kentucky (Wilkins, 1985) and Cupola Pond, Missouri (Smith, 1984). Many additional radiocarbon-dated sites provide palynological information concerning the nature of full-glacial vegetation during the time interval of 23,000 to 16,500 B.P. (Table 1). Palynological data from many of these sites were used in preparation of a full-glacial vegetation map (Fig. 7a), with map units depicting major plant formations and vegetational types (map generalized from Delcourt and Delcourt, 1979, 1981). Concerning the full-glacial interval, three important biogeographic questions remain central issues today: (1) what was the distribution and vegetational composition of full-glacial refugia for mesic, temperate deciduous forest species; (2) what was the southern limit of the continuous distribution of boreal forest species; and (3) what was the distribution of alpine tundra in the Appalachian Mountains?

Evidence for refugial locations of mixed communities of mesic, temperate, hardwood tree species is at present restricted to only three full-glacial localities in southeastern North America. These mesic arboreal taxa today characteristic of the Mixed Mesophytic Forest Region (*sensu* Braun, 1950) include beech (*Fagus grandifolia*), sugar maple (*Acer saccharum*), basswood (*Tilia*), walnut (*Juglans*), buckeye (*Aesculus*), tulip poplar (*Liriodendron tulipifera*), chestnut (*Castanea dentata*), and certain mesic species of ash (*Fraxinus*), hickory (*Carya*), and oak (*Quercus*) (Braun, 1950). Not all of these taxa are consistently represented in modern or presettlement pollen samples, and not all of them are distinguishable to species based upon pollen morphology. Detection of their refugial areas depends upon finding lacustrine or stream-terrace deposits located in close proximity to suitable mesic habitats such as ravines in karst regions, fire-protected spring-fed basins situated along interfluves, dissected valley slopes along major river systems, or mesic loessial soils associated with upland ravines. The three full-glacial sites with definitive plant-fossil evidence for a large assemblage of these mesic trees include Nonconnah Creek, Tennessee (P. Delcourt *et al.*, 1980), Goshen Springs, Alabama (P. Delcourt, 1980), and Sheelar Lake, Florida (Watts and Stuiver, 1980; Watts, 1983). Apparently, full-glacial refugia for mesic, temperate hardwood trees was widely dispersed across the Gulf Coastal Plain in a diversity of habitats, including loess-capped uplands, ravines or "hammocks" dissected within sandy interfluves, and irregular topography of karst terrain. The majority of fire-prone sandy uplands probably was occupied continuously by warm-temperate species of pine, oak, and hickory, within a persistent Southeastern Evergreen forest region (Fig. 7a) (Delcourt and Delcourt, 1983).

Table 1.
Primary Radiocarbon-dated Fossil-Pollen Sites In The Southeastern United States

Site Name, State	Code	Latitude North	Longitude West	Time Range	Depositional Environment	Pollen Analyst	Publications
Anderson Pond, TN	AP	36°02'	85°30'	Middle Wisconsin to late-Holocene	Karst sink	H. Delcourt	H. Delcourt (1979)
Bartow County Ponds, GA (Bob Black, Quicksand, and Green ponds)	BaP	34°12'	84°52'	Early Wisconsin to late-Holocene	Karst sinks	W. Watts	Watts (1970, 1973)
Big Basin, KS	BB	37°15'	100°00'	Late-Holocene	Karst sink	C. Shumard	Shumard (1974)
Big Pond, PA	BiP	39°46'	78°33'	Early- to late-Holocene	Pond with perched water table	W. Watts	Watts (1979)
B. L. Bigbee Oxbow, MS	BL	33°34'	88°28'	Early- to late-Holocene	Oxbow lake	M. Sheehan	Whitehead and Sheehan (1984)
Big Lake, AR	BiL	35°54'	90°07'	Late-Holocene	Sunk Land (earthquake-generated basin)	J. King	King (1978)
Boriack Bog, TX	BoB	30°21'	97°08'	Late-glacial to late-Holocene	Spring	V. Bryant	Bryant (1977)
Buckle's Bog, MD	BuB	39°34'	79°16'	Full-glacial to late-Holocene	Montane bog or glade	J. Maxwell	Maxwell and Davis (1972)
Cahaba Pond, AL	CaP	33°34'	86°32'	Early- to late-Holocene	Karst sink	H. Delcourt	H. Delcourt *et al.*(1983)
Central Delmarva Peninsula, MD	CdP	38°20'	75°20'	Early Wisconsin to early-Holocene	Carolina Bays	L. Sirkin	Sirkin *et al.* (1977); Denny *et al.* (1979)
Chattanooga sites, TN (Friar Branch and Boyd Buchanan School)	CS	35°10'	85°13'	Early-to middle-Holocene	Alluvial	H. Delcourt	DeSelm and Brown (1978); H. Delcourt (1979)
Chesapeake Bay, VA	CB	39°33'	76°04'	Late-Holocene	Estuary	F. Davis	F. Davis (1982)
Chesapeake Bay Entrance, VA	CE	36°58'	76°07'	Late-glacial to late-Holocene	Estuary	J. Terasmae	Harrison *et al.* (1965)
Cloudsplitter Rockshelter, KY	CR	37°50'	83°37'	Early- to late-Holocene	Cave (archaeological) site	C. Cowan	Cowan *et al.* (1981)
Cranberry Glades, WV	CrG	38°12'	80°17'	Late-glacial to late-Holocene	Montane bog or glade	P. Sears; P. Martin; W. Watts	Darlington (1943); Arnold and Libby (1950); Guilday *et al.*(1964); Watts (1979)

Table 1. Continued
Primary Radiocarbon-dated Fossil-Pollen Sites In The Southeastern United States

Site Name, State	Code	Latitude North	Longitude West	Time Range	Depositional Environment	Pollen Analyst	Publications
Crider's Pond, PA	CrP	39°58′	77°33′	Late-glacial to late-Holocene	Pond with perched water table	W. Watts	Watts (1979)
Cupola Pond, MO	CuP	36°48′	91°06′	Full-glacial to late-Holocene	Karst sink	N. Smith	Smith (1984)
Dismal Swamp, VA	DS	36°35′	76°26′	Early- to late-Holocene	Coastal swamp	I. Lewis; A. Vick; D. Whitehead	Lewis and Cocke (1929); Cocke et al. (1934); Vick (1961); Whitehead (1972); Whitehead and Oaks (1979)
East-central Texas bogs, TX (Hershop and Soefje bogs)	ET	29°35′	97°37′	Early- to late-Holocene	Springs	A. Graham; T. Patty; V. Bryant	Graham and Heimsch (1960); Larson et al. (1972); Bryant (1977)
Everglades, FL	EV	25°15′	80°52′	Late-Holocene	Coastal swamp	W. Riegel	Riegel (1965); Spackman et al. (1966)
Ferndale Bog, OK	FB	34°25′	95°49′	Middle- to late-Holocene	Spring	L. Albert	Albert and Wycoff (1981)
Flat Laurel Gap, NC	FG	35°24′	82°45′	Late-Holocene	Montane bog or glade	D. Shafer	Shafer (1984)
Florida Springs, FL (Little Salt Spring and Warm Mineral Spring)	FS	27°03′	82°16′	Early- to late-Holocene	Slough and karst sink	J. Brown; J. King	Clausen et al. (1979); Brown (1981); Sheldon (1977)
Goshen Springs, AL	GS	31°43′	86°08′	Early Wisconsin to late-Holocene	Spring	P. Delcourt	P. Delcourt (1980)
Hack Pond, VA	HP	37°59′	79°00′	Late-glacial to late-Holocene	Karst sink	A. Craig	Craig (1969)
Jackson Pond, KY	JP	37°27′	85°43′	Full-glacial to late-Holocene	Karst sink	G. Wilkins	Wilkins (1985)
Lake Annie, FL	LA	27°12′	81°21′	Early Wisconsin, late-glacial to late-Holocene	Karst sink	W. Watts	Watts (1975, 1980, 1983)
Lake in the Woods, TN	LW	35°35′	83°50′	Middle- to late-Holocene	Woodland hollow	J. Davidson	Davidson (1982, 1983, 1984)
Lake Louise, GA	LL	30°44′	83°16′	Early Wisconsin, middle- to late-Holocene	Karst sink	W. Watts	Watts (1971)
Little Tennessee River Ponds, TN (Black and Tuskegee Ponds)	LT	35°35′	84°13′	Late-Holocene	Karst sink and terrace scour pool	P. Cridlebaugh	Cridlebaugh (1982, 1984)
Long Beach, NC	LB	33°55′	78°10′	Early Wisconsin	Eolian-deflation basin	D. Whitehead; M. Doyle	Whitehead and Doyle (1969)
Mammoth Cave System, KY (Mammoth and Salts caves)	MC	37°11′	86°04′	Late-Holocene	Cave (archaeological) sites	W. Benninghoff; V. Bryant; J. Schoenwetter	Bryant (1974); Schoenwetter (1974); Yarnell (1969)
Mud Lake, FL	ML	29°18′	81°52′	Early Wisconsin, middle- to late-Holocene	Karst sink	W. Watts	Watts (1969)
Natural Lake, OK	NL	34°39′	95°24′	Late-Holocene	Oxbow lake	L. Albert	Albert and Wycoff (1981)
Nodoroc Bog (Winder Bog), GA	NB	34°00′	83°43′	Full-glacial to early-Holocene, late-Holocene	Bog with perched water table	W. Humphrey; M. Sheehan	Humphrey (1953); Carbone et al. (1982)
Nonconnah Creek, TN	NC	35°05′	89°55′	Early Wisconsin to late-glacial	Alluvial	P. Delcourt	P. Delcourt et al. (1980)
Northeast Kansas marshes, KS (Arrington and Muscotah marshes)	NK	39°32′	95°31′	Middle Wisconsin to late-Holocene	Springs	W. Horr; J. Gruger	Horr (1955); J. Gruger (1973)
Northeast Oklahoma sites, OK (Big Hawk Shelter, Cut Finger Cave, Little Caney River Valley, and Painted Shelter)	NO	36°25′	96°20′	Late-Holocene	Alluvial, cave, and archaeological sites	S. Hall	Hall (1977, 1981, 1982); Henry (1978)

Table 1. Continued
Primary Radiocarbon-dated Fossil-Pollen Sites In The Southeastern United States

Site Name, State	Code	Latitude North	Longitude West	Time Range	Depositional Environment	Pollen Analyst	Publications
Okefenokee Swamp, GA	OS	30°40′	82°12′	Middle- to late-Holocene	Coastal swamp	A. Cohen; F. Rich; L. Fearn	Cohen (1974, 1975); Rich and Spackman (1979); Fearn (1981); Cohen *et al.* (1984)
Old Field, MO	OF	37°07′	89°50′	Early- to late-Holocene	Swamp on braided-stream terrace	J. King	King and Allen (1977)
Pine Barrens Bog, NJ	PB	39°43′	74°30′	Early- to late-Holocene	Alluvial swamp	L. Florer	Florer (1972)
Pigeon Marsh, GA	PM	34°50′	85°24′	Full-glacial to late-Holocene	Marsh with perched water table	W. Watts	Watts (1975)
Pittsburg Basin, IL	PiB	38°53′	89°12′	Late Illinoian, Sangamonian, Early Wisconsin to late-Holocene	Kettle block basin in area of Illinoian-age glaciation	E. Gruger	E. Gruger (1970, 1972, 1973)
Pomme de Terre Springs, MO (Boney, Jones, Kirby, and Trolinger springs)	PT	38°07′	93°22′	Middle Wisconsin to late-glacial	Springs	P. Mehringer; J. King	Mehringer *et al.* (1968); King (1973); King and Lindsey (1976)
Potomac Estuary, VA-MD	PE	38°15′	77°00′	Late-Holocene	Estuary	G. Brush	Brush *et al.* (1982)
Potts Mountain Pond, VA	PoM	37°36′	80°08′	Early- to late-Holocene	Pond with perched water table	W. Watts	Watts (1979)
Rayburn Salt Dome, LA	RS	32°28′	93°10′	Full- to late-glacial, middle- to late-Holocene	Solution pool on salt dome	G. Fredlund	Kolb and Fredlund (1981)
Rockyhock Bay, NC	RB	36°10′	76°41′	Early Wisconsin to late-Holocene	Carolina Bay	D. Whitehead	Whitehead (1973, 1981)
Saltville, VA	SA	36°52′	81°46′	Late-glacial to late-Holocene	Montane bog or glade	W. Benninghoff; H. Delcourt	Ray *et al.* (1967); McDonald *et al.* (1983)
Savannah River Valley, SC-GA	SR	34°08′	82°42′	Early- to middle-Holocene	Alluvial	D. Whitehead; M. Sheehan	Carbone *et al.* (1982)
Scott Lake, FL	ScL	27°58′	81°57′	Middle- to late-Holocene	Karst sink	W. Watts	Watts (1971)
Shady Valley Bog, TN	SV	36°32′	81°56′	Early- to late-Holocene	Montane bog or glade	P. Sears; F. Barclay	Sears (1935); Barclay (1957)
Sheelar Lake, FL	ShL	29°31′	82°00′	Middle Wisconsin to late-Holocene	Karst sink	W. Watts	Watts and Stuiver (1980)
Singletary Lake, NC	SiL	34°36′	78°28′	Early Wisconsin to late-Holocene	Carolina Bay	D. Frey; D. Whitehead	Frey (1951, 1953, 1955); Whitehead (1963, 1964, 1967)
Tunica Hills, LA (Percy Bluff and Tunica Bayou)	TH	30°56′	91°30′	Late-glacial to late-Holocene	Alluvial	H. Delcourt	Delcourt and Delcourt (1977)
Uphapee Creek Terrace, AL	UC	32°26′	85°40′	Middle Wisconsin, middle- to late-Holocene	Alluvial	R. Christopher	Markewich and Christopher (1982)
White Pond, SC	WP	34°10′	80°47′	Full-glacial to late-Holocene	Eolian-deflation basin (?)	W. Watts	Watts (1980)
Yazoo Basin, MS	YB	33°10′	90°38′	Late-glacial to late-Holocene	Alluvial	R. Holloway	Holloway and Valastro (1983)

During the full-glacial interval, the southern limit of the boreal forest region (Fig. 7a) was more than 1,200 km south of its modern southern border in Canada. Jack pine (*Pinus banksiana*), today a dominant tree within the southern boreal forest of Manitoba and east-central Ontario, dominated the pollen records of a number of sites between 34° N and 37° N latitude during the full-glacial interval (H. Delcourt, 1979; Watts, 1970, 1980a, 1980b, 1983; Whitehead, 1981; Wright, 1981; Smith, 1984). Comparison of full-glacial fossil pollen assemblages from North Carolina (Whitehead, 1981) and Tennessee (H. Delcourt, 1979; Delcourt and Delcourt, 1985) with a large array of modern pollen samples from the

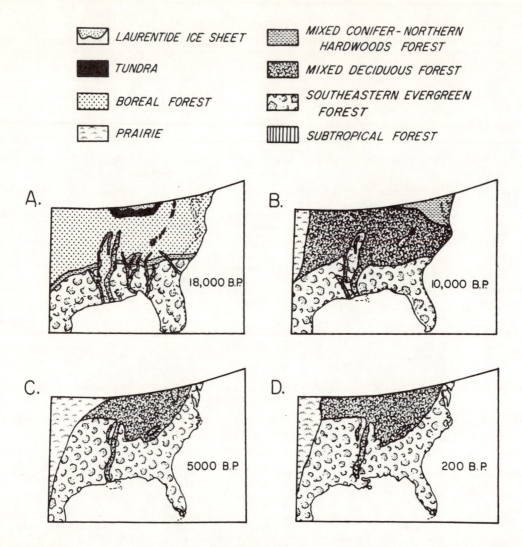

Figure 7. a) Paleovegetation map for 18,000 B.P. b) Paleovegetation map for 10,000 B.P. c) Paleovegetation map for 5,000 B.P. d) Paleovegetation map for 200 B.P. (Maps modified from Delcourt and Delcourt, 1981).

boreal forest region of eastern Canada (Webb and McAndrews, 1976; Davis and Webb, 1975; P. Delcourt *et al.,* 1984), produced very low coefficients of dissimilarity as calculated by Euclidean Distance. In addition, at Anderson Pond, Chord Distance and Standardized Euclidean Distance measures yielded low coefficients of dissimilarity for the full-glacial time interval (Fig. 8). These empirical indices of analog (Prentice, 1982) confirm the existence of close modern analogs for full-glacial vegetation of mid-latitudes of eastern North America for the interval from 20,000 to 16,500 B.P. At Anderson Pond, all three dissimilarity measures provided compatible results (Fig. 8) that were consistent with paleovegetation reconstructed using taxon calibrations of modern pollen-vegetation relationships (Delcourt and Delcourt, 1985). From the results of the application

of dissimilarity coefficients, we can infer that community composition of full-glacial boreal forests at 36° N latitude was within the range of variation in species occurrence and abundance within modern boreal forests of southern Manitoba and east-central Ontario.

Taxon calibrations can be applied to fossil pollen assemblages from full-glacial sites across the Southeast to map contoured values of percent dominance of major tree taxa across their former ranges. Paleo-isophyte maps for 18,000 B.P. (Fig. 9) illustrate the gradients in dominance of tree taxa. The main range of spruce (*Picea*) occurred north of 34° N latitude in eastern North America (Fig. 9a). South of 34° N latitude, white spruce (*Picea glauca*) occurred locally in limited populations in braided-stream habitats of

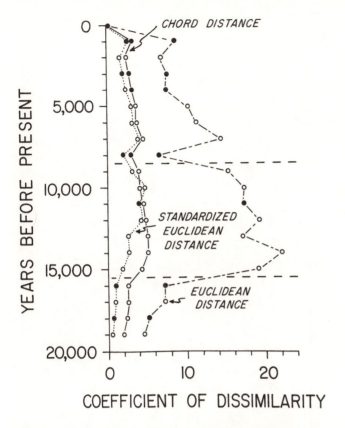

Figure 8. Dissimilarity coefficients calculated at 1,000-year intervals from 19,000 B.P. to the present between fossil-pollen spectra from Anderson Pond (White County, Tennessee) and a modern array of 1,684 modern-pollen samples (from Delcourt and Delcourt, 1985). Black dots indicate instances in which two or more of the dissimilarity coefficients identified the same modern-pollen sample as closest modern analogue for a fossil-pollen spectrum (diagram from Delcourt and Delcourt, 1985)

the lower Mississippi Alluvial Valley and its tributaries (Delcourt and Delcourt, 1977). Primary population centers for spruce were located in the uplands west of the Mississippi Alluvial Valley and in the central and southern Appalachian Mountains. As in the modern boreal forest, spruce increased in percent dominance toward the northern limits of its distribution during the full-glacial period (Wilkins, 1985). The paleo-isophyte map for fir (*Abies*) illustrates a prominent, although narrow, latitudinal band situated along the southern limit of the boreal forest 18,000 B.P. (Fig. 9b). Today, balsam fir (*Abies balsamea*) is most abundant in the eastern maritime provinces of Canada, where annual precipitation exceeds 75 cm liquid equivalent (Halliday and Brown, 1943). We interpret the full-glacial pattern of fir distribution to reflect boreal climatic conditions with abundant moisture throughout the growing season (Delcourt and Delcourt, 1984a). During the full-glacial interval, population centers for jack pine and red pine, *Pinus resinosa* (the two species of

northern Diploxylon *Pinus*), were located on the Interior Low Plateaus west of the Appalachian Mountains and along the central Atlantic seaboard.

Today, the southern limit of boreal forest corresponds with the mean winter position of the climatic Polar Frontal Zone, which separates prevailing Arctic and Pacific airmasses from the predominance of the Maritime Tropical Airmass (Bryson, 1966; Bryson and Wendland, 1967). The southern limit of the full-glacial boreal forest at about 34° N latitude represented a steep vegetational ecotone between boreal and temperate ecosystems (Fig. 7a). We infer this boundary to coincide with a relatively stable climatic boundary at which the Polar Frontal Zone separated the prevailing westerly flow of the Pacific Airmass, situated to the north, from the Maritime Tropical Airmass to the south (Delcourt and Delcourt, 1983, 1984b). The full-glacial Polar Frontal Zone would have anchored the position of storms along a narrow latitudinal belt, providing increased available moisture that would have favored populations of fir.

Alpine tundra environments persist today in the northern Appalachian Mountains, *e.g.*, Mt. Washington in the Presidential Ranges of New Hampshire (Davis *et al.*, 1980). Buckle's Bog, Maryland, is the only radiocarbon-dated site in the Appalachian Mountains with both full-glacial and late-glacial plant-fossil evidence for tundra (Maxwell and Davis, 1972). Late-glacial pollen and plant-macrofossil evidence for tundra communities has been documented near the periphery of the full-glacial limit of the Laurentide Ice Sheet in sites such as Longswamp and Tannersville, Pennsylvania (Watts, 1979), and at mid-elevations in the central Appalachian Mountains as much as 300 km south of the glacial border at Cranberry Glades, West Virginia (Watts, 1979). Geomorphological evidence of severe periglacial environments characterized by frequent freeze-thaw churning of the regolith is preserved in the form of relict, sorted patterned-ground features, including stone polygons, ellipses, and stripes, in the central and southern Appalachian Mountains (Clark, 1968; Pewe, 1983). Patterned-ground features with polygon diameters of 2 to 10 m reflect climatic conditions with mean annual temperatures of −4° C or colder (Clark, 1968). Comparison of plant-fossil evidence for tundra with the occurrence of relict Pleistocene patterned-ground features along the Appalachian Mountains (Fig. 10) indicates that the lower elevational limit of the Pleistocene periglacial deposits represents a conservative estimate of the lower limit of the occurrence of tundra species in the late-Pleistocene interval. From this relationship, we

A.   **B.**

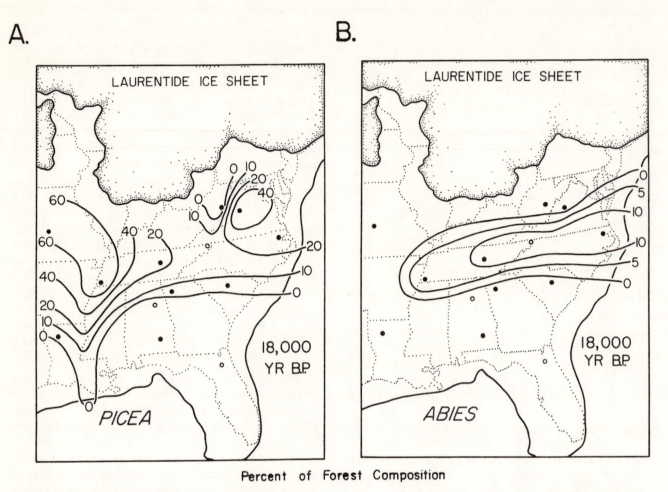

Figure 9. a) Paleo-isophyte map for *Picea* 18,000 B.P. b) Paleo-isophyte map for *Abies* 18,000 B.P. (Maps from Delcourt and Delcourt, 1984a).

extrapolate the full-glacial alpine tundra zone to extend into the southernmost Appalachian Mountains at elevations generally above 1,450 m (Fig. 7a). Mid-elevation sites in the central Appalachians with evidence for tundra vegetation are characterized by very low rates of pollen influx and relatively high influx of mineral sediments (Maxwell and Davis, 1972). Extensive mass-wastage of surficial sediments during the late-glacial interval resulted in the rapid filling of topographic depressions with inorganic mineral sediments. At high elevations within the southern Appalachians, montane basins such as Flat Laurel Gap, North Carolina (Fig. 10) were inundated by late-glacial solifluction lobes and boulder streams; initiation of peat development began only after stabilization of the colluvial slopes and establishment of the modern alluvial glade environment in the Holocene (Shafer, 1984).

The Late Wisconsin Late-glacial Interval (16,500 to 12,500 B.P.)

At sites located near or just to the north of the southern full-glacial limit of the boreal forest, vegetational response to late-glacial climatic amelioration occurred as early as 16,500 B.P. (Delcourt and Delcourt, 1984b). During the late-glacial interval, at sites such as White Pond, Rockyhock Bay, and Anderson Pond (Fig. 11a, 11b, 11c) a major decline in population size (*i.e.*, percentages of dominance in the vegetation) of northern Diploxylon *Pinus* was accompanied by increasing populations of mesic boreal and cool-temperate deciduous taxa. The values of percent dominance for spruce and fir (Fig. 11) reflect an expansion in their populations during the late-glacial interval. We interpret this expansion in spruce and fir to indicate the persistence of cool climatic conditions and the increased availability of precipitation during the summer growing season (Delcourt and Delcourt, 1984b). During the late-glacial interval, oaks and hickories expanded their populations in adjustment to increasing length of growing season and to increasing mean-annual temperatures. Along the Blufflands, the loess-mantled uplands adjacent to the Lower Mississippi Alluvial Valley, fossil evidence from Nonconnah Creek documents the increase in dominance of cool-temperate deciduous trees during the late-glacial interval as well as the northward migration of warm-temperate taxa along this Pleistocene pathway (P. Delcourt *et al.*, 1980). Thus, in the Southeast between 34° and 35° N latitude, the first biotic response to climatic amelioration at the full-glacial/late-glacial transition, at 16,500 B.P., occurred at the population level with the expansion in populations of deciduous tree species from refugial areas and the beginning of their northward migrations. These trees began to colonize habitats vacated by less-tolerant boreal conifers such as jack pine that could no longer compete effectively within the rapidly changing climatic conditions. A second level of biotic response to climatic change occurred at the transition from Pleistocene to Holocene conditions, approximately 12,500 B.P., with a changeover in structural dominants from boreal to temperate plant communities. This ecosystem-level response occurred as a result of changing environmental conditions that exceeded the thresholds of tolerance of the boreal forest species.

The Early-Holocene Interval (12,500 to 8,500 B.P.)

During the early-Holocene interval, the rapid northward migration of cool-temperate, mesic tree species was followed by their expansion in dominance throughout the mid-latitudes of the southeastern United States (Fig. 7b). From sites with continuous records of sedimentation spanning the time interval from about 12,500 to 8,500 B.P., such as Anderson Pond, White Pond, and Cahaba Pond, Alabama (H. Delcourt *et al.*, 1983), it is evident that the forest communities of the early-Holocene period differed in species composition and dominance from those that developed in the middle- and late-Holocene intervals. At Anderson Pond, hornbeam (*Ostrya/Carpinus* type) reached up to 30% of the arboreal pollen spectra during this time interval (H. Delcourt, 1979). Forests on the watershed surrounding Cahaba Pond were dominated by beech from 12,000 to 10,200 B.P., with substantial representation of hornbeam, oak, hickory, elm, and ash. At 10,000

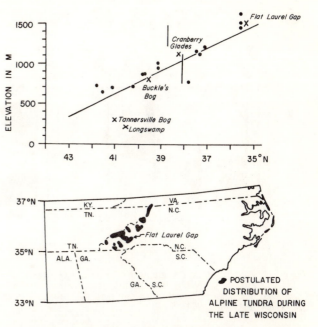

Figure 10. Distribution of relict, periglacial sorted, patterned-ground features (designated by solid dots) and selected montane bog sites (designated by X's), plotted along an elevational and latitudinal transect in the central and southern Appalachian Mountains. In the upper graph, the diagonal line denotes the full-glacial altitudinal boundary between alpine tundra and boreal forest along the axis of the Appalachian Mountains.

B.P., several species that are not found together today co-occurred in a mixed coniferous/broadleaf deciduous forest at Cahaba Pond (Fig. 12). In the early-Holocene interval, bald cypress (*Taxodium distichum,* identified on the basis of papillate pollen grains), today a coastal species, extended inland, and eastern white pine (*Pinus strobus*) and hemlock (*Tsuga*) extended southward of their present ranges in the Appalachian Mountains to the southern Ridge and Valley Province of central Alabama. Other taxa exhibiting these distributional patterns occurred at Cahaba Pond between 12,000 and 10,000 B.P., including striped maple (*Acer pensylvanicum*), and mountain maple (*Acer spicatum*). The modern floristic regions thus became defined at 34° N latitude only in the middle- to late-Holocene intervals as climatic conditions there shifted from cool-temperate to warm-temperate (H. Delcourt *et al.*, 1983).

The Middle-Holocene Interval (8,500 to 4,000 B.P.)

In the Midwestern United States, the middle-Holocene or Hypsithermal Interval was marked by the eastward expansion of prairie at the expense of forest (Fig. 7c). The effects of Hypsithermal warming and drying extended into the mid-latitudes of the

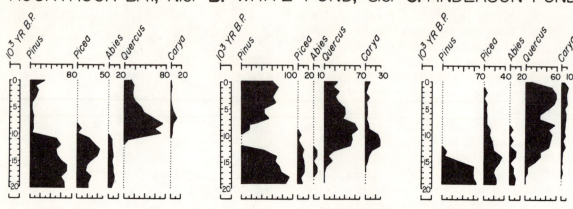

Figure 11. Paleovegetation diagrams for the past 19,000 years for three sites situated in mid-latitudes of the Southeast: (a) Rockyhock Bay, North Carolina; (b) White Pond, South Carolina; and (c) Anderson Pond, Tennessee (from Delcourt and Delcourt, 1984b). Curves for selected tree types represent changes in percent dominance in the forests, based upon taxon calibrations developed for 20 major tree taxa (P. Delcourt et al., 1984).

Southeast west of the Appalachian Mountains; in Middle Tennessee, forest communities became species-poor and xeric between 8,500 and 4,000 B.P. (H. Delcourt, 1979).

Farther south and east, in the southern Appalachian Mountains and the northern Gulf Coastal Plain, middle-Holocene vegetation reflected warm but wet regional climate. By 5,000 B.P., coastal-plain species characteristic of wetland environments had dispersed successfully into sag ponds in the Ridge and Valley of central Alabama (H. Delcourt et al., 1983) and northwestern Georgia (Watts, 1970). In Cades Cove, Great Smoky Mountains National Park, East Tennessee, a woodland-hollow pond formed between 7,000 and 6,500 B.P. because of a water table perched over colluvium (Davidson, 1983). This restricted and poorly-drained environment at 500 m elevation in the mountains today harbors numerous species characteristic of southern coastal plains. The pollen sequence from Lake in the Woods indicates that coastal-plain taxa had migrated into Cades Cove by 6,500 B.P. during a period of warm, wet climate and have persisted there to the present time. Thus, the high species richness of the Great Smoky Mountains reflects both the intermingling of floristic elements characteristic of alpine tundra, boreal forest, temperate deciduous forest, and southeastern evergreen forest during late-Quaternary times of changing climates, and the persistence of these species in unique relict habitats (Davidson, 1983).

During the middle-Holocene interval, a shift in dominance occurred among tree taxa of the Southeastern Evergreen Forest (Fig. 7c). Previously dominated by xeric species of oak and hickory, coastal-plain forests became dominated by species of southern pine by 5,000 B.P. (P. Delcourt, 1980; Watts, 1969, 1975a; Watts and Stuiver, 1980). The shift to dominance of southern pines may have reflected an increase in fire frequency accompanying a strengthening of the influence of the Maritime Tropical Airmass and intensification of hurricane frequency (P. Delcourt, 1980; Delcourt and Delcourt, 1984b). Alternatively, Late-Archaic Indians may have set fires in the sandy uplands, which would have both maintained large tracts of open pine forest and increased the carrying capacity of faunal resources such as populations of white-tailed deer (P. Delcourt, 1980). Available pollen evidence indicates that, on a plant-formational level, the Southeastern Evergreen Forest has remained intact on upland interfluves of the Gulf Coastal Plain over the last glacial/interglacial cycle; within this plant formation, the aerial dominance of the constituent community types has changed because of subtle changes in effective precipitation and fire frequency (Delcourt and Delcourt, 1983; Delcourt et al., 1983).

The Late-Holocene Interval (4,000 B.P. to the Present)

During the past 4,000 years, vegetational adjustments have occurred in ecotonal and other climatically-sensitive areas of the Southeast. In response to minor climatic cooling, spruce and fir have locally re-

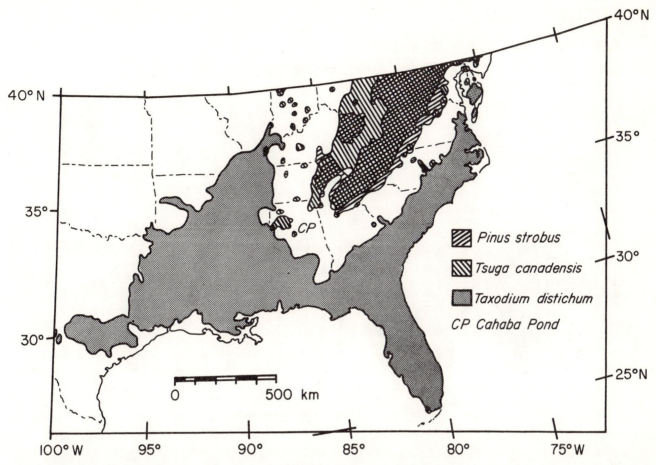

Figure 12. Modern distributional limits of *Pinus strobus, Tsuga canadensis,* and *Taxodium distichum* (after Little, 1971) in relation to their fossil occurrence during the early Holocene interval at Cahaba Pond, St. Clair County, Alabama.

expanded their populations at mid- and high elevations in the central and southern Appalachian Mountains (Barclay, 1957; Watts, 1979; Shafer, 1984; Delcourt and Delcourt, 1984a; H. Delcourt, 1985). Northward migration and expansion in abundance of the American chestnut (*Castanea dentata*) in the late-Holocene interval resulted in the development of areally extensive oak-chestnut forest throughout the central and southern Appalachian Mountains (Delcourt and Delcourt, 1981). In the Ozarks of Missouri and eastern Oklahoma, shortleaf pine (*Pinus echinata*) expanded its range northward (Albert and Wycoff, 1981; Smith, 1984). Along the Atlantic Coastal Plain, Carolina Bays filled in with peat to form extensive "pocosin" wetlands (Whitehead, 1965, 1973, 1981). Coastal swamps expanded under the influence of rising sea levels through the Holocene (Spackman *et al.*, 1966; Whitehead and Oaks, 1979; Cohen *et al.*, 1984).

The effects of American Indians on native vegetation of the Southeast have been largely undetected by Quaternary paleoecologists. Occasional grains of maize (*Zea mays*) pollen have been recorded in

sediments dating as much as 2,000 years old in sites such as B. L. Bigbee Oxbow (Whitehead and Sheehan, 1984) and the Dismal Swamp (Whitehead, 1965b; Whitehead and Oaks, 1979). Archaeological evidence of human occupation along river systems throughout the Southeast extends throughout the Holocene and includes ethnobotanical evidence of cultigens including gourd (*Lagenaria siceraria*), squash (*Cucurbita pepo*), beans (*Phaseolus vulgaris*), and maize introduced during the late-Holocene interval (Chapman *et al.*, 1982; Cridlebaugh, 1984).

In the lower Little Tennessee River Valley of East Tennessee, the pollen diagram from Tuskegee Pond (Fig. 13) illustrates the magnitude of local anthropogenic impact on bottomland, stream-terrace, and upland plant communities in the vicinity of major Indian occupation sites (Cridlebaugh, 1982, 1984). Maize was cultivated locally at Tuskegee Pond from at least 1,500 B.P. to the present. High percentages of ragweed (*Ambrosia* type) pollen, along with pollen of herbs indicative of disturbed, open ground (*Chenopodium* type, *Iva*, Portulacaceae, *Plantago* spp. and *Rumex*), are evidence for open landscapes on low and

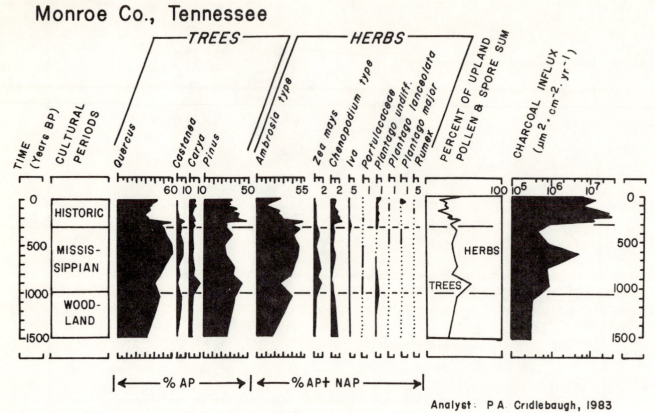

Figure 13. Indian impact on native ecosystems, reflected by selected curves for pollen percentages and charcoal influx at Tuskegee Pond, Monroe County, Tennessee, for the past 1,500 years (from Cridlebaugh, 1984).

mid-level terraces adjacent the Little Tennessee River (Fig. 13). In contrast, a pollen diagram from Black Pond, located 4 km away from the river, is dominated by arboreal pollen through the late-Holocene interval, with only a few percent of *Ambrosia* type and other upland herbs represented (Cridlebaugh, 1982, 1984). Influx of charcoal particles to Tuskegee Pond sediments increased by an order of magnitude from Woodland to Mississippian cultural periods (Fig. 13), indicating an intensification of the use of fire by Indians for cooking of food and clearing of land. With the arrival of Euro-Americans in the A.D. 1700's, both the influx of mineral sediments and charcoal particles to the ponds increased by another order of magnitude, reflecting the conflicts between British soldiers and the Cherokee, the final deforestation of the local landscape to provide palisades for nearby British Colonial Fort Loudon in A.D. 1756, and introduction of an improved technology for cultivation, the steel plow (Cridlebaugh, 1982, 1984). We speculate that anthropogenic impacts have been widespread on arable lands along all major river systems in the southeastern United States during the last several thousand years, although Indian influences on native ecosystems have been most intense within a several-km radius of the principal Indian agricultural settlements and major ceremonial centers.

KEY PROBLEMS FOR FUTURE RESEARCH

Increasingly, as the regional framework is established as a broad context and as geologic strategies are developed for location of suitable sites, Quaternary paleoecologists will be able to use palynology as a tool in order to test and refine specific ecologic, archaeologic, geologic, and climatologic hypotheses. Future problems must be defined within a context of appropriate temporal and spatial scales to be addressed (H. Delcourt *et al.*, 1982).

The sequence of long-term late-Quaternary vegetational change, on the order of the past 50,000 to 500,000 years, is virtually unknown in the Southeast (Cronin *et al.*, 1981). Questions of importance on this time scale include the degree of stability of coastal-plain vegetation and the degree of analogy between the vegetational changes during the Holocene and those that occurred during previous interglacial intervals. The fundamental biogeographic patterns have been identified for the post-glacial routes and rates of migration of plant species as they expanded northward across deglaciated terrain (Davis, 1981), but little information yet exists with which to map the sequence of southward migrations and areal restrictions of species during the onset of Late Wisconsin glaciation. Clearly, a continuing objective involves both the location of new sites with sedimentary sequences covering long time spans and documentation of their temporal framework provided by adequate chronologic control.

For the time interval from Late Wisconsin full-glacial to present, application of pollen-vegetation calibrations to quantitative reconstruction and mapping of changes in plant population centers and range limits will soon be possible. Application of ecological techniques such as ordination and classification of vegetation to the paleovegetational data will make possible refinement of previous paleovegetation maps.

For the full-glacial interval, the nature of vegetational zonation from jack pine-dominated boreal forest at 36° N to spruce-dominated boreal forest at 38° N to taiga and possibly tundra near the margin of the Laurentide Ice Sheet has yet to be defined, but can be approached through pollen-influx studies of well-dated sinkhole-pond sites such as Cupola Pond, Missouri (Smith, 1984) and Jackson Pond, Kentucky (Wilkins, 1985).

Another full-glacial dilemma is the location of refuge areas for many of the tree taxa today characteristic of the mixed conifer-northern hardwoods communities of the Great Lakes region, including northern white cedar (*Thuja occidentalis*), Canadian hemlock (*Tsuga canadensis*), eastern white pine (*Pinus strobus*), and red pine (*Pinus resinosa*). Until additional sites are studied along the full-glacial ecotone between boreal and temperate forests (Fig. 7a) or at low elevations in the Southern Appalachian region, these refuge areas will remain elusive.

Problems in reconstructing Holocene paleoenvironments in the Gulf Coastal Plain will be in part resolved through studies of influx of charcoal particles to lake sediments. The interplay of fire, man, and development of coastal-plain ecosystems such as pocosins will require palynological investigations (including analysis of charcoal particles) that are tied to specific archaeological sites with detailed, radiocarbon-dated chronologies.

Glade, bog, and sinkhole sites in the southeastern United States often harbor rare and endangered plant species and typically also contain significant late-Quaternary palynological sequences. Many of these sites are included on the registry of National Natural Landmarks, are located within National Parks, or are of interest for purchase by organizations such as the Audubon Society or the Nature Conservancy (*e.g.,* Cupola Pond, Missouri, Anderson Pond, Tennessee, Lake in the Woods, Tennessee, Flat Laurel Gap Bog, North Carolina, and Long Hope Valley bogs, North Carolina). Paleoecologists have the opportunity to contribute significantly to conservation efforts by providing documentation of the late-Quaternary development of these rare-plant habitats. Understanding the influences of changing climates, shifting disturbance regimes, and the role of man in modifying the physical and biological landscape is essential for preservation of native ecosystems of the southeastern United States.

ACKNOWLEDGMENTS

This research was sponsored by the Ecology Program of the National Science Foundation through grants DEB-80-04168, BSR-83-06915, and BSR-83-00345 to the University of Tennessee. We wish to thank the following individuals for their constructive suggestions for improvement of this manuscript: F. H. Barclay, J. G. Brown, G. S. Brush, V. M. Bryant, Jr., H. F. Buell, A. D. Cohen, E. J. Cushing, S. A. Hall, R. G. Holloway, A. Graham, J. E. King, D. A. Livingstone, R. E. McLaughlin, P. S. Martin, J. Schoenwetter, P. B. Sears, A. J. Sharp, L. A. Sirkin, W. Spackman, P. A. Stone, and L. R. Wilson. Contribution No. 30, Program for Quaternary Studies of the Southeastern United States, University of Tennessee, Knoxville, Tennessee. United States contribution to IGCP Project 158 B.

References Cited

ALBERT, L. E., and WYCOFF, D. G.

1981   Ferndale Bog and Natural Lake: five thousand years of environmental change in southeastern Oklahoma. *Oklahoma Archaeological Survey, Studies in Oklahoma's Past,* 7: 1-125.

ARNOLD, J. R., and LIBBY, W. F.

1950   Radiocarbon Dates. University of Chicago, Institute for Nuclear Studies, 15 p.

AUER, V.

1927   Stratigraphical and morphological investigations of peat bogs of southeastern Canada. *Comm. ex. Inst. Quaest. Forest. Finlandiae editae,* 12: 1-62.

BARCLAY, F. H.

1957   The natural vegetation of Johnson County, Tennessee, past and present. Ph.D. Dissertation, University of Tennessee, Knoxville, 147 p.

BEETHAM, N. M.

1958   Pollen studies of forest and bald soils from the mountains of North Carolina. M.A. Thesis, Duke University, Durham, North Carolina, 153 p.

BIRKS, H. J. B., and PEGLAR, S. M.

1980   Identification of *Picea* pollen of late Quaternary age in eastern North America: a numerical approach. *Canadian Journal of Botany,* 58: 2043-2058.

BRAUN, E. L.

1950   *Deciduous Forests of Eastern North America.* Hafner Press, MacMillan Publishing Company, 596 p. (reprinted in 1974).

1951   Plant distribution in relation to the glacial boundary. *Ohio Journal of Science,* 51: 139-146.

1955   The phytogeography of unglaciated eastern United States and its interpretation. *Botanical Review,* 21: 297-375.

BROWN, C. A.

1938   The flora of Pleistocene deposits in the western Florida parishes, West Feliciana Parish, and East Baton Rouge Parish, Louisiana. *In: Contributions to the Pleistocene history of the Florida parishes of Louisiana.* Louisiana Department of Conservation, Geological Bulletin Number 12, p. 59-96.

BRYANT, V. M., Jr.

1977   A 16,000 year pollen record of vegetational change in central Texas. *Palynology,* 1: 143-156.

1978   Palynology: a useful method for determining paleoenvironment. *Texas Journal of Science,* 30: 25-42.

BRYSON, R. A.

1966   Air masses, streamlines, and the boreal forest. *Geographical Bulletin,* 8: 228-269.

BRYSON, R. A., and WENDLAND, W. M.

1967   Tentative climatic patterns for some late glacial and post-glacial episodes in central North America. *In: Mayer-Oakes, W. J. (ed.), Life, Land, and Water, Proceedings of the 1966 Conference of Environmental Studies of the Glacial Lake Agassiz Region,* University of Manitoba Press, Winnipeg, p. 271-298.

BUELL, M. F.

1939   Peat formation in the Carolina Bays. *Bulletin of the Torrey Botanical Club,* 66: 483-487.

1945a   Late Pleistocene forests of southeastern North Carolina. *Torreya,* 45: 117-118.

1945b   The age of Jerome Bog, "A Carolina Bay". *Science,* 103: 14-15.

1946   Jerome Bog, a peat-filled "Carolina Bay". *Bulletin of the Torrey Botanical Club,* 73: 24-33.

CAIN, S. A.

1939   Pollen analysis as a paleo-ecological research method. *Botanical Review,* 5: 627-654.

CARBONE, V. A., SEGOVIA, A., FOSS, J., SHEEHAN, M., WHITEHEAD, D. R., and JACKSON, S.

1982   Paleoenvironmental investigations along the Savannah River Valley, Richard B. Russell Dam and Lake Project. Paper presented at the 39th Annual Southeastern Archaeological Conference, Memphis, Tennessee, October 28-30, 1982.

CARROLL, G.

1943   The use of Bryophytic polsters and mats in the study of recent pollen deposition. *American Journal of Botany,* 30: 361-366.

CHAPMAN, J., DELCOURT, P. A., CRIDLEBAUGH, P. A., SHEA, A. B., and DELCOURT, H. R.

1982   Man-land interaction: 10,000 years of American Indian impact on native ecosystems in the Lower Little Tennessee River Valley, Eastern Tennessee. *Southeastern Archaeology,* 1: 115-121.

CLARK, G. M.

1968   Sorted patterned ground: new Appalachian localities south of the glacial border. *Science,* 161: 355-356.

COHEN, A. D.

1975   Peats from the Okefenokee Swamp-Marsh complex. *Geoscience and Man,* 11: 123-131.

COHEN, A. D., and SPACKMAN, W.

1972   Methods in peat petrology and their application to reconstruction of paleoenvironments. *Geological Society of America Bulletin,* 83: 129-142.

COHEN, A. D., CASAGRANDE, D. J., ANDREJKO, M. J., and BEST, G. R.

1984   *The Okefenokee Swamp: Its Natural History, Geology, and Geochemistry.* Wetland Surveys, Los Alamos, New Mexico.

COLEMAN, S. M., and PIERCE, K. L.

1977   Summary table of Quaternary dating methods. *United States Geological Survey Miscellaneous Field Studies Map,* MF-904.

CRAIG, A. J.

1969   Vegetational history of the Shenandoah Valley, Virginia. *Geological Society of America Special Paper,* 123: 283-296.

CRIDLEBAUGH, P. A.

1982　Macrofossil and pollen evidence for paleoenvironmental change in the lower Little Tennessee River Valley, East Tennessee. *Association of Southeastern Biologists Bulletin*, 29: 56.

1984　American Indian and Euro-American impact on Holocene vegetation in the Lower Little Tennessee River Valley, East Tennessee. Ph.D. Dissertation, University of Tennessee, Knoxville, 225 p.

CRONIN, T. M., SZABO, B. J., AGER, T. A., HAZEL, J. E., and OWENS, J. P.

1981　Quaternary climates and sea levels of the U. S. Atlantic Coastal Plain. *Science*, 211: 233-240.

DARLINGTON, H. C.

1943　Vegetation and substrate of Cranberry Glades, West Virginia. *Botanical Gazette*, 104: 371-393.

DAVIDSON, J. L.

1983　Paleoecological analysis of Holocene vegetation, Lake in the Woods, Cades Cove, Great Smoky Mountains National Park. M.S. Thesis, University of Tennessee, Knoxville, 100 p.

DAVIS, J. H., Jr.

1946　The peat deposits of Florida, their occurrence, development, and uses. *Florida Department of Conservation, Geological Survey, Geological Bulletin*, 30: 1-247.

DAVIS, M. B.

1976　Pleistocene biogeography of temperate deciduous forests. *Geoscience and Man*, 13: 13-26.

1981　Quaternary history and the stability of forest communities. *In:* West, D. C., Shugart, H. H., and Botkin, D. B. (eds.), *Forest Succession, Concepts and Application*, Springer-Verlag, New York, p. 132-153.

1983　Holocene vegetational history of the eastern United States. *In:* Wright, H. E., Jr. (ed.), *Late-Quaternary Environments of the United States, Vol. 2, The Holocene*, University of Minnesota Press, Minneapolis, p. 166-181.

DAVIS, M. B., SPEAR, R. W., and SHANE, L. C. K.

1980　Holocene climate of New England. *Quaternary Research*, 14: 240-250.

DAVIS, R. B., and WEBB III, T.

1975　The contemporary distribution of pollen in eastern North America: a comparison with the vegetation. *Quaternary Research*, 5: 395-434.

DEEVEY, E. S., Jr.

1949　Biogeography of the Pleistocene, Part I: Europe and North America. *Bulletin of the Geological Society of America*, 60: 1315-1416.

DELCOURT, H. R.

1979　Late-Quaternary vegetation history of the eastern Highland Rim and adjacent Cumberland Plateau of Tennessee. *Ecological Monographs*, 49: 255-280.

1985　Holocene vegetational changes in the Southern Appalachian Mountains, U.S.A. *Ecologia Mediterranea*, in press.

DELCOURT, P. A.

1980　Goshen Springs: late-Quaternary vegetation record for southern Alabama. *Ecology*, 61: 371-386.

1985　The influence of late-Quaternary climatic and vegetational change on paleohydrology in unglaciated Eastern North America. *Ecologia Mediterranea*, in press.

DELCOURT, H. R., and DELCOURT, P. A.

1975　The Blufflands: Pleistocene pathway into the Tunica Hills. *American Midland Naturalist*, 94: 385-400.

1984a　Late-Quaternary history of the Spruce-Fir Ecosystem in the Southern Appalachian Mountain Region. *In:* White, P. S. (ed.), *The Southern Appalachian spruce-fir ecosystem: its biology and threats*. U.S. National Park Service Research Resource Management Report Ser-71: 22-35.

1985　Comparison of taxon calibrations, modern analog techniques, and forest-stand simulation models for the quantitative reconstruction of past vegetation. *Earth Surface Processes and Landforms*, 10:293-304.

DELCOURT, P. A., and DELCOURT, H. R.

1977　The Tunica Hills, Louisiana-Mississippi: late glacial locality for spruce and deciduous forest species. *Quaternary Research*, 7: 218-237.

1979　Late Pleistocene and Holocene distributional history of the deciduous forest in the southeastern United States. *Veroffentlichungen des Geobotanischen Institutes der ETH, Stiftung Rubel (Zurich)*, 68: 79-107.

1980　Pollen preservation and Quaternary environmental history in the southeastern United States. *Palynology*, 4: 215-231.

1981　Vegetation maps for eastern North America: 40,000 yr BP to the present. *In:* Romans, R. (ed.), *Geobotany II*. Plenum Press, New York, p. 123-166.

1983　Late-Quaternary vegetational dynamics and community stability reconsidered. *Quaternary Research*, 19: 265-271.

1984b　Late-Quaternary paleoclimates and biotic responses in Eastern North America and the Western North Atlantic Ocean. *Palaeogeography, Palaeoclimatology, Palaeoecology*, 48: 263-284.

DELCOURT, H. R., WEST, D. C., and DELCOURT, P. A.

1981　Forests of the southeastern United States: quantitative maps for aboveground woody biomass, carbon, and dominance of major tree taxa. *Ecology*, 62: 879-887.

DELCOURT, H. R., DELCOURT, P. A., and SPIKER, E. C.

1983　A 12,000-year record of forest history from Cahaba Pond, St. Clair County, Alabama. *Ecology*, 64: 874-887.

DELCOURT, H. R., DELCOURT, P. A., and WEBB, T. III

1982　Dynamic plant ecology: the spectrum of vegetational change in space and time. *Quaternary Science Reviews*, 1: 153-175.

DELCOURT, P. A., DELCOURT, H. R., BRISTER, R. C., and LACKEY, L. E.

1980　Quaternary vegetation history of the Mississippi Embayment. *Quaternary Research*, 13: 111-132.

DELCOURT, P. A., DELCOURT, H. R., and DAVIDSON, J. L.

1983　Mapping and calibration of modern pollen-vegetation relationships in the southeastern United States. *Review of Palaeobotany and Palynology*, 39: 1-45.

DELCOURT, P. A., DELCOURT, H. R., and WEBB, T. III

1984　Atlas of mapped distributions of dominance and modern pollen percentages for important tree taxa of Eastern North America. *American Association of Stratigraphic Palynologists, Contributions Series No. 14*: 1-131.

DENNY, C. S., OWENS, J. P., SIRKIN, L. A., and RUBIN, M.

1979　The Parsonburg Sand in the Central Delmarva Peninsula, Maryland and Delaware. *United States Geological Survey Professional Paper*, 1067B: B1-B16.

ELSIK, W. C.
1969   Late Neogene palynomorph diagrams, northern Gulf of Mexico. *Transactions of the Gulf Coast Association of Geological Societies,* 19: 509-528.

ENGELHARDT, D. W.
1963   Palynology of recent sediments in southeastern United States and the adjacent area of the Gulf of Mexico. *Pan American Petroleum Corporation Research Department Report No,* M63-G-51: 1-22.

FOWELLS, H. A.
1965   Silvics of forest trees of the United States. *United States Department of Agriculture Handbook,* 271: 1-762.

FREY, D. G.
1951   Pollen succession in the sediments of Singletary Lake, North Carolina. *Ecology,* 32: 518-533.
1952   Pollen analysis of the Horry Clay and a seaside peat deposit near Myrtle Beach, South Carolina. *American Journal of Science,* 250: 212-225.
1953   Regional aspects of the late-glacial and post-glacial pollen succession of southeastern North Carolina. *Ecological Monographs,* 23: 289-313.
1955   A time revision of the Pleistocene pollen chronology of southeastern North Carolina. *Ecology,* 36: 762-763.

GRUGER, E.
1972a   Pollen and seed studies of Wisconsinan vegetation in Illinois, U.S.A. *Geological Society of America Bulletin,* 83: 2715-2734.
1972b   Late Quaternary vegetation development in south-central Illinois. *Quaternary Research,* 2: 217-231.

HALL, S. A.
1981   Deteriorated pollen grains and the interpretation of Quaternary pollen diagrams. *Review of Palaeobotany and Palynology,* 32: 193-206.

HALLIDAY, W. E. D., and BROWN, A. W. A.
1943   The distribution of some important forest trees in Canada. *Ecology,* 24: 353-373.

HOLLOWAY, R. G., and BRYANT, V. M., Jr.
1984   *Picea glauca* pollen from late glacial deposits in Central Texas. *Palynology,* 8: 21-32.

HOLLOWAY, R. G., and VALASTRO, S.
1983   Palynological investigations along the Yazoo River. *In:* Thorne, R. M., and Curry, H. K. (eds.), *Cultural Resources Survey of Items 3 and 4, Upper Yazoo River Projects, Mississippi, with a Paleoenvironmental Model of the Lower Yazoo Basin.* Archaeological Papers of the Center for Archaeological Research, Number 3, Center for Archaeological Research, University of Mississippi, University, p. 159-257.

HUMPHREY, W. F.
1953   Pollen analysis of three Georgia peat bogs. M.S. Thesis, University of Georgia, Athens, 39 p.

JACOBSON, G. L., Jr., and BRADSHAW, R. H. W.
1981   The selection of sites for paleovegetational studies. *Quaternary Research,* 16: 80-96.

KACZOROWSKI, R. T.
1977   The Carolina Bays: a comparison with modern oriented lakes. *University of South Carolina, Columbia, South Carolina, Department of Geology, Coastal Research Division, Technical Report,* 13-CRD: 1-124.

KEMMERLY, P. R.
1982   Spatial analysis of a karst depression population: clues to genesis. *Geological Society of America Bulletin,* 93: 1078-1086.

KOLB, C. R., and FREDLUND, G. G.
1981   Palynological studies, Vacherie and Rayburn's Domes, North Louisiana Salt Dome Basin, topical report. *Louisiana State University, Baton Rouge, Institute for Environmental Studies, Report,* E530-02200-T-2: 1-50.

KUEHN, D.
1980   Offshore transgressive peat deposits of southwest Florida: evidence for late Holocene rise of sea level. M.S. Thesis, Pennsylvania State University, University Park.

LARSON, D. A., BRYANT, V. M., Jr., and PATTY, T. S.
1972   Pollen analysis of a central Texas bog. *American Midland Naturalist,* 88: 358-367.

LEWIS, I. F., and COCKE, E. C.
1929   Pollen analysis of Dismal Swamp peat. *Journal of the Elisha Mitchell Scientific Society,* 45: 37-58.

LIEUX, M. H.
1980a   An atlas of pollen of trees, shrubs, and woody vines of Louisiana and other southeastern states, Part I. Ginkgoaceae to Lauraceae. *Pollen et Spores,* 22: 2-57.
1980b   An atlas of pollen of trees, shrubs, and woody vines of Louisiana and other southeastern states, Part II. Platanaceae to Betulaceae. *Pollen et Spores,* 22: 191-243.

LIEUX, M. H., and GODFREY, W. M.
1982   An atlas of pollen of trees, shrubs, and woody vines of Louisiana and other southeastern states, Part III. Polygonaceae to Ericaceae. *Pollen et Spores,* 24: 21-64.

LITTLE, E. L., Jr.
1971   Conifers and important hardwoods. Volume 1. Atlas of United States trees. *United States Department of Agriculture, Forest Service Miscellaneous Publication,* 1146: 8 p., 200 maps.

MARKEWICH, H. W., and CHRISTOPHER, R. A.
1982   Pleistocene (?) and Holocene fluvial history of Uphapee Creek, Macon County, Alabama. *United States Geological Survey Bulletin,* 1522: 1-16.

MARTIN, P. S.
1958a   Taiga-tundra and the full-glacial period in Chester County, Pennsylvania. *American Journal of Science,* 256: 470-502.
1958b   Pleistocene ecology and biogeography of North America. *In:* Hubbs, C. S. (ed.), *Zoogeography.* American Association for the Advancement of Science Publication 51, p. 375-420.

MARTIN, P. S., and HARRELL, B. E.
1957   The Pleistocene history of temperate biotas in Mexico and eastern United States. *Ecology,* 38: 468-480.

MAXWELL, J. A., and DAVIS, M. B.
1972   Pollen evidence of Pleistocene and Holocene vegetation on the Allegheny Plateau, Maryland. *Quaternary Research,* 2: 506-530.

NICKMANN, R. J., and DEMAREST, J. M., II.
1982   Pollen analysis of some Mid-Pleistocene Interglacial lagoonal sediments from southern Delaware. *Quaternary Research,* 17: 93-104.

PEWE, T. L.
1983   The periglacial environment in North America during Wisconsin time. *In:* Porter, S. C. (ed.), *Late-Quaternary Environments of the United States, Vol. 1, The Late Pleistocene.* University of Minnesota Press, Minneapolis, p. 157-189.

POTZGER, J. E., and THARP, B. C.
 1943 Pollen record of Canadian spruce and fir from Texas bog. *Science,* 98: 584.
 1947 Pollen profile from a Texas bog. *Ecology,* 28: 274-280.
 1954 Pollen study of two bogs in Texas. *Ecology,* 35: 462-466.

PRENTICE, I. C.
 1982 Calibration of pollen spectra in terms of species abundance. *In:* Berglund, B. (ed.), *Palaeohydrological Changes in the Temperate Zone in the Last 15,000 Years, I.G.C.P. 158, Subproject B, Lakes and Mire Environments, Project Guide, Volume III.* Department of Quaternary Geology, Lund University, Lund, Sweden, p. 25-51.

PROUTY, W. F.
 1952 Carolina bays and their origin. *Geological Society of America Bulletin,* 63: 167-224.

RICH, F. J.
 1979 The origin and development of tree islands in the Okefenokee Swamp, as determined by peat petrography and pollen stratigraphy. Ph.D. Dissertation, Pennsylvania State University, University Park, 301 p.

RICH, F. J., and SPACKMAN, W.
 1979 Modern and ancient pollen sedimentation around tree islands in the Okefenokee Swamp. *Palynology,* 3: 219-226.

RICH, F. J., KUEHN, D., and DAVIES, T. D.
 1982 The paleoecological significance of *Ovoidites. Palynology,* 6: 19-28.

RIEGEL, W. L.
 1965 Palynology of environments of peat formation in southwestern Florida. Ph.D. Dissertation, Pennsylvania State University, University Park, 189 p.

SEARS, P. B.
 1935 Types of North American pollen profiles. *Ecology,* 16: 488-499.
 1942 Postglacial migration of five forest genera. *American Journal of Botany,* 29: 684-691.

SEARS, P. B., and COUCH, G. C.
 1932 Microfossils in an Arkansas peat and their significance. *Ohio Journal of Science,* 32: 63-68.

SHAFER, D. S.
 1984 Late-Quaternary paleoecologic, geomorphic, and paleoclimatic history of Flat Laurel Gap, Blue Ridge Mountains, North Carolina. M.S. Thesis, University of Tennessee, Knoxville, 148 p.

SMITH, E. N., Jr.
 1984 Late-Quaternary vegetational history at Cupola Pond, Ozark National Scenic Riverways, Southeastern Missouri. M.S. Thesis, University of Tennessee, Knoxville, 115 p.

SOLOMON, A. M.
 1983a Pollen morphology and plant taxonomy of white oaks in eastern North America. *American Journal of Botany,* 70: 481-494.
 1983b Pollen morphology and plant taxonomy of red oaks in eastern North America. *American Journal of Botany,* 70: 495-507.

SOLOMON, A. M., WEST, D. C., and SOLOMON, J. A.
 1981 Simulating the role of climate change and species immigration in forest succession. *In:* West, D. C., Shugart, H. H., and Botkin, D. B. (eds.), *Forest Succession, Concepts and Application.* Springer-Verlag, New York, p. 154-177.

SOLOMON, A. M., DELCOURT, H. R., WEST, D. C., and BLASING, T. J.
 1980 Testing a simulation model for reconstruction of prehistoric forest-stand dynamics. *Quaternary Research,* 14: 275-293.

SPACKMAN, W., DOLSEN, C. P., and RIEGEL, W.
 1966 Phytogenic organic sediments and sedimentary environments in the Everglades-Mangrove complex. Part I: Evidence of a transgressing sea and its effects on environments of the Shark River area of southwestern Florida. *Sonder-abdruck aus Palaeontographica, Beitrage Zur Naturgeschichte Der Vorzeit,* 117(B): 135-152.

SPACKMAN, W., COHEN, A. D., GIVEN, P. H., and CASAGRANDE, D. J.
 1974 A field guidebook to aid in the comparative study of the Okefenokee Swamp and the Everglades-mangrove swamp-marsh complex of southern Florida. *Geological Society of America Preconvention Field Trip,* 6: 1-265.

STONE, P. A., and BROWN, J. G.
 1983 The pollen record of Pleistocene and Holocene paleoenvironmental conditions in southeastern United States. *In:* Colquhoun, D. J. (ed.), *Variation in Sea Level on the South Carolina Coastal Plain,* Department of Geology, University of South Carolina, Columbia, p. 169-206.

THOM, B. G.
 1967 Humate and coastal geomorphology. *Louisiana State University, Coastal Studies Bulletin,* 1: 15-17.
 1970 Carolina Bays in Horry and Marion Counties, South Carolina. *Geological Society of America Bulletin,* 81: 783-814.

THOMPSON, D. E.
 1972 Paleoecology of the Pamlico Formation, Saint Mary's County, Maryland. Ph.D. Dissertation, Rutgers University, New Jersey, 194 p.

TISDALE, M. G.
 1981 Sedimentology and paleoecology of a buried, Pleistocene, carbonaceous sand from Myrtle Beach State Park, Horry County, South Carolina. M.S. Thesis, University of South Carolina, Columbia, 62 p.

VINEYARD, J. D., and FEDER, G. L.
 1982 Springs of Missouri. *Missouri Department of Natural Resources, Division of Geology and Land Survey Water Resources Report,* 29: 1-26.

WATTS, W. A.
 1969 A pollen diagram from Mud Lake, Marion County, north-central Florida. *Geological Society of America Bulletin,* 80: 631-642.
 1970 The full-glacial vegetation of northwestern Georgia. *Ecology,* 51: 17-33.
 1971 Postglacial and interglacial vegetation history of southern Georgia and central Florida. *Ecology,* 52: 676-690.
 1973 The vegetation record of a Mid-Wisconsin Interstadial in northwest Georgia. *Quaternary Research,* 3: 257-268.
 1975a A late-Quaternary record of vegetation from Lake Annie, south-central Florida. *Geology,* 3: 344-346.
 1975b Vegetation record for the last 20,000 years from a small marsh on Lookout Mountain, northwestern Georgia. *Geological Society of America Bulletin,* 86: 287-291.
 1979 Late Quaternary vegetation of central Appalachia and the New Jersey Coastal Plain. *Ecological Monographs,* 49: 427-469.

1980a   The late Quaternary vegetation history of the southeastern United States. *Annual Reviews of Ecology and Systematics,* 11: 387-409.

1980b   Late-Quaternary vegetation history at White Pond on the Inner Coastal Plain of South Carolina. *Quaternary Research,* 13: 187-199.

1983   Vegetational history of the eastern United States 25,000 to 10,000 years ago. *In:* Porter, S. C. (ed.), *Late-Quaternary Environments of the United States, Vol. 1, The Late Pleistocene.* University of Minnesota Press, Minneapolis, p. 294-310.

WATTS, W. A., and STUIVER, M.

1980   Late Wisconsin climate of northern Florida and the origin of species-rich deciduous forest. *Science,* 210: 325-327.

WEBB III, T., and McANDREWS, J. H.

1976   Corresponding patterns of contemporary pollen and vegetation in central North America. *Geological Society of America Memoir,* 145: 267-299.

WHITE, W. A.

1970   The geomorphology of the Florida Peninsula. *Florida Department of Natural Resources, Division of Interior Resources, Bureau of Geology, Geological Bulletin,* 51: 1-164.

WHITEHEAD, D. R.

1963   "Northern" elements in the Pleistocene flora of the Southeast. *Ecology,* 44: 403-406.

1964   Fossil pine pollen and full-glacial vegetation in southeastern North Carolina. *Ecology,* 45: 767-776.

1965a   Palynology and Pleistocene phytogeography of unglaciated eastern North America. *In:* Wright, H. E., Jr., and Frey, D. G. (eds.), *The Quaternary of the United States.* Princeton University Press, Princeton, p. 417-432.

1965b   Prehistoric maize in southeastern Virginia. *Science,* 150: 881-883.

1967   Studies of full-glacial vegetation and climate in southeastern United States. *In:* Cushing, E. J., and Wright, H. E., Jr. (eds.), *Quaternary Paleoecology.* Yale University Press, New Haven, p. 237-248.

1973   Late-Wisconsin vegetational changes in unglaciated eastern North America. *Quaternary Research,* 3: 621-631.

1981   Late-Pleistocene vegetational changes in northeastern North Carolina. *Ecological Monographs,* 51: 451-471.

WHITEHEAD, D. R., and BARGHOORN, E. S.

1962   Pollen analytical investigations of Pleistocene deposits from western North Carolina and South Carolina. *Ecological Monographs,* 32: 347-369.

WHITEHEAD, D. R., and DAVIS, J. T.

1969   Pollen analysis of an organic clay from the interglacial Flanner Beach Formation, Craven County, North Carolina. *Southeastern Geology,* 10: 149-164.

WHITEHEAD, D. R., and DOYLE, M. V.

1969   Late-Pleistocene peats from Long Beach, North Carolina. *Southeastern Geology,* 10: 1-16.

WHITEHEAD, D. R., and OAKS, R. Q., Jr.

1979   Development history of the Dismal Swamp. *In:* Kirk, P. W., Jr. (ed.), *The Great Dismal Swamp.* University Press of Virginia, Charlottesville, p. 25-43.

WHITEHEAD, D. R., and SHEEHAN, M. C.

1984   Holocene vegetational changes in the Tombigbee River Valley, eastern Mississippi. *American Midland Naturalist,* in press.

WHITEHEAD, D. R., and TAN, K. W.

1969   Modern vegetation and pollen rain in Bladen County, North Carolina. *Ecology,* 50: 235-248.

WILKINS, G. R.

1985   Late-Quaternary vegetational history at Jackson Pond, Larue County, Kentucky. M.S. Thesis, University of Tennessee, Knoxville, 172 p.

WRIGHT, H. E., Jr.

1981   Vegetation east of the Rocky Mountains 18,000 years ago. *Quaternary Research,* 15: 113-125.

APPENDIX: BIBLIOGRAPHY OF QUATERNARY PALYNOLOGY IN THE SOUTHEASTERN UNITED STATES

ALBERT, L. E., and WYCOFF, D. G.

1981 Ferndale Bog and Natural Lake: five thousand years of environmental change in southeastern Oklahoma. *Oklahoma Archaeological Survey, Studies in Oklahoma's Past,* 7: 1-125.

BALSAM, W. L., and HEUSSER, L. E.

1976 Direct correlation of sea surface paleotemperatures, deep circulation, and terrestrial paleoclimates: foraminiferal and palynological evidence from two cores off Chesapeake Bay. *Marine Geology,* 21: 121-147.

BARCLAY, F. H.

1957 The natural vegetation of Johnson County, Tennessee, past and present. Ph.D. Dissertation, University of Tennessee, Knoxville, 147 p.

1973 Analysis of arboreal pollen of Shady Valley Bog. *Proceedings of the 1973 International Geobotany Conference,* University of Tennessee, Knoxville.

BEETHAM, N. M.

1958 Pollen studies of forest and bald soils from the mountains of North Carolina. M.A. Thesis, Duke University, Durham, North Carolina, 153 p.

BENNINGHOFF, W. S., and STEVENSON, A. L.

1967 Pollen analysis of cave breccia from Ladds Locality, Bartow County, Georgia. *Bulletin of the Georgia Academy of Science,* 25: 188-191.

BOND, T. A.

1965 Ephedran pollen grains in Pleistocene sediments of central and southeastern Oklahoma. *Oklahoma Geological Survey, Oklahoma Geology Notes,* 25: 302-307.

1966 Palynology of Quaternary terraces and floodplains of the Washita and Red Rivers, central and southeastern Oklahoma. Ph.D. Dissertation, University of Oklahoma, Norman, 96 p.

1970 Preliminary pollen analysis and radiocarbon dates from Okefenokee Swamp, Carlton County, Georgia. *Bulletin of the Georgia Academy of Science,* 28: 18.

BROOK, G. A., and DAVIS, J. D.

1979 Paleoenvironmental implications of deposits in Red Spider Cave, northwest Georgia. *Association of American Geographers, Southeastern Division Annual Conference, Nashville, Tennessee, Program Abstracts,* 1979: 12.

BROOK, G. A., SWAIN, P. C., and WENNER, D. B.

1982 A paleoenvironmental history of northwest Georgia for the last 40,000 years from oxygen isotope and pollen analysis of speleothems in Red Spider Cave. *Geological Society of America, Abstracts with Programs,* 14: 452.

BROWN, C. A.

1938 The flora of Pleistocene deposits in the western Florida parishes, West Feliciana Parish, and East Baton Rouge Parish, Louisiana. *In: Contributions to the Pleistocene history of the Florida parishes of Louisiana.* Louisiana Department of Conservation, Geological Bulletin Number 12, p. 59-96.

1960 *Palynological techniques.* Louisiana State University Press, Baton Rouge, 188 p.

BROWN, C. W., BURK, C. J., and CURRAN, H. A.

1984 Palynological analyses of a dated peat core from the North Carolina Outer Banks. *Bulletin of the Association of Southeastern Biologists,* 31: 51-52.

BROWN, J. G.

1981 Palynologic and petrographic analyses of Bayhead Hammock and marsh peats at Little Salt Spring Archaeological Site (8So18), Florida. M.S. Thesis, University of South Carolina, Columbia, 52 p.

BROWN, J. G., and COHEN, A. D.

1985 Palynologic and petrographic analyses of peat deposits, Little Salt Spring. *National Geographic Research,* 1: 21-31.

BRUSH, G. S., and DeFRIES, R. S.

1981 Spatial distributions of pollen in surface sediments of the Potomac estuary. *Limnology and Oceanography,* 26: 295-309.

BRUSH, G. S., MARTIN, E. A., DeFRIES, R. S., and RICE, C. A.

1982 Comparisons of 210-Pb and pollen methods for determining rates of estuarine sediment accumulation. *Quaternary Research,* 18: 196-217.

BRYANT, V. M., Jr.

1974 Pollen analysis of prehistoric human feces from Mammoth Cave. *In:* Watson, P. J. (ed.), *Archaeology of the Mammoth Cave Area.* Academic Press, p. 203-209.

1977 A 16,000 year pollen record of vegetational change in central Texas. *Palynology,* 1: 143-156.

1977 Pollen analysis of the Shiver Site, Davies County, Missouri. *In: The Archaeology of the Shiver Site.* Anthropology Department, University of Missouri, Publication, p. 85-101.

1978 Palynology: a useful method for determining paleoenvironment. *Texas Journal of Science,* 30: 25-42.

BUELL, M. F.

1939 Peat formation in the Carolina Bays. *Bulletin of the Torrey Botanical Club,* 66: 483-487.

1945 Late Pleistocene forests of southeastern North Carolina. *Torreya,* 45: 117-118.

1945 The age of Jerome Bog, "A Carolina Bay". *Science,* 103: 14-15.

1946 A size-frequency study of *Pinus banksiana* pollen. *Journal of the Elisha Mitchell Scientific Society,* 62: 211-228.

1946 Size frequency study of fossil pine pollen with herbarium-preserved pollen. *American Journal of Botany,* 33: 510-515.

1946 Jerome Bog, a peat-filled "Carolina Bay". *Bulletin of the Torrey Botanical Club,* 73: 24-33.

CAIN, S. A.

1939   Pollen analysis as a paleo-ecological research method. *Botanical Review,* 5: 627-654.

1940   The identification of species in fossil pollen of *Pinus* by size-frequency determination. *American Journal of Botany,* 27: 301-308.

1944   Pollen analysis of some buried soils, Spartanburg County, South Carolina. *Bulletin of the Torrey Botanical Club,* 71: 11-22.

1944   Size-frequency characteristics of *Abies fraseri* pollen as influenced by different methods of preparation. *American Midland Naturalist,* 31: 232-236.

CAIN, S. A., and CAIN, L. G.

1944   Size-frequency studies of *Pinus palustris* pollen. *Ecology,* 25: 229-232.

1948   Size-frequency characteristics of *Pinus echinata* pollen. *Botanical Gazette,* 110: 325-330.

CARBONE, V. A., SEGOVIA, A., FOSS, J., SHEEHAN, M., WHITEHEAD, D. R., and JACKSON, S.

1982   Paleoenvironmental investigations along the Savannah River Valley, Richard B. Russell Dam and Lake Project. Paper presented at the 39th Annual Southeastern Archaeological Conference, Memphis, Tennessee, October 28-30, 1982.

CARROLL, G.

1943   The use of Bryophytic polsters and mats in the study of recent pollen deposition. *American Journal of Botany,* 30: 361-366.

CLAUSEN, C. J., COHEN, A. D., EMILIANI, C., HOLMAN, J. A., and STIPP, J. J.

1979   Little Salt Spring, Florida: a unique underwater site. *Science,* 203: 609-614.

CLENDENING, J. A., RENION, J. J., and PARSONS, B. M.

1967   Preliminary palynological and mineralogical analyses of a Lake Monongahela (Pleistocene) Terrace deposit at Morgantown, West Virginia. *West Virginia Geological and Economic Survey, Circular Series* 4: 1-18.

COCKE, E. C., LEWIS, I. F., and PATRICK, R.

1934   A further study of Dismal Swamp peat. *American Journal of Botany,* 21: 374-395.

COHEN, A. D.

1974   Possible influences of subpeat topography and sediment type upon the development of the Okefenokee swamp-marsh complex of Georgia. *Southeastern Geology,* 15: 141-151.

1975   Peats from the Okefenokee Swamp-Marsh complex. *Geoscience and Man,* 11: 123-131.

COHEN, A. D., and CORVINUS, D.

1980   Palynology and petrography of peats from the Snuggedy Swamp of South Carolina. *Palynology,* 4: 237.

COHEN, A. D., and SPACKMAN, W.

1972   Methods in peat petrology and their application to reconstruction of paleoenvironments. *Geological Society of America Bulletin,* 83: 129-142.

COHEN, A. D., CASAGRANDE, D. J., ANDREJKO, M. J., and BEST, G. R.

1984   *The Okefenokee Swamp: Its Natural History, Geology, and Geochemistry.* Wetland Surveys, Los Alamos, New Mexico.

COLEMAN, J. M.

1966   Recent coastal sedimentation: central Louisiana coast. *Louisiana State University Press, Coastal Studies Series,* 17: 73 p.

COMANOR, P. L.

1964   Arboreal hammock vegetation correlated with surface soil pollen spectra in northcentral Florida. M.S. Thesis, University of Florida, Gainesville.

CONN, D.

1982   Pollen analysis. *In:* Waselkov, G. A., Wood, B. M., and Herbert, J. M. (eds.), *Colonization and Conquest: the 1980 Archaeological Excavations at Fort Toulouse and Fort Jackson, Alabama.* Auburn University Archaeological Monograph 4, Auburn University at Montgomery, p. 349-358.

COWAN, C. W., JACKSON, H. E., MOORE, K., NICKELHOFF, A., and SMART, T. L.

1981   The Cloudsplitter Rockshelter, Menifee County, Kentucky: a preliminary report. *Southeastern Archaeological Conference Bulletin,* 24: 60-76.

COX, D. D.

1968   A late-glacial pollen record from the West Virginia-Maryland border. *Castanea,* 33: 137-149.

CRAIG, A. J.

1969   Vegetational history of the Shenandoah Valley, Virginia. *Geological Society of America Special Paper,* 123: 283-296.

CRIDLEBAUGH, P. A.

1982   Macrofossil and pollen evidence for paleoenvironmental change in the lower Little Tennessee River Valley, East Tennessee. *Association of Southeastern Biologists Bulletin,* 29: 56.

1984   American Indian and Euro-American impact on Holocene vegetation in the Lower Little Tennessee River Valley, East Tennessee. Ph.D. Dissertation, University of Tennessee, Knoxville, 225 p.

CRONIN, T. M., SZABO, B. J., AGER, T. A., HAZEL, J. E., and OWENS, J. P.

1981   Quaternary climates and sea levels of the U. S. Atlantic Coastal Plain. *Science,* 211: 233-240.

CUSHING, E. J.

1975   Late-Quaternary pollen stratigraphy at a site in the central Ozarks. *Bulletin of the Ecological Society of America,* 56: 63.

DARLINGTON, H. C.

1943   Vegetation and substrate of Cranberry Glades, West Virginia. *Botanical Gazette,* 104: 371-393.

DARRELL, J. H. II

1973   Statistical evaluation of palynomorph distribution in the sedimentary environments of the modern Mississippi River Delta. Ph.D. Dissertation, Louisiana State University, Baton Rouge, 491 p.

DAVIDSON, J. L.

1982   Paleoecological analysis of Holocene vegetation, Lake in the Woods, Cades Cove, Great Smoky Mountains National Park. *Association of Southeastern Biologists Bulletin,* 29: 57.

1983   Holocene vegetational and climatic changes in Cades Cove, Great Smoky Mountains National Park, East Tennessee. *Ecological Society of America Bulletin,* 64: 132.

1983   Paleoecological analysis of Holocene vegetation, Lake in the Woods, Cades Cove, Great Smoky Mountains National Park. M.S. Thesis, University of Tennessee, Knoxville, 100 p.

DAVIES, T. D.
1980   Peat formation in Florida Bay and its significance in interpreting the Recent vegetational and geological history of the bay area. Ph.D. Dissertation, Pennsylvania State University, University Park, 316 p.

DAVIES, T. D., and SPACKMAN, W.
1980   The palynology of the peats of Florida Bay. *Palynology,* 4: 238.

DAVIS, F. W.
1982   The history of submerged aquatic vegetation at the head of Chesapeake Bay: a stratigraphic study. Ph.D. Dissertation, John Hopkins University, Baltimore, Maryland.

DAVIS, J. H., Jr.
1946   The peat deposits of Florida, their occurrence, development, and uses. *Florida Department of Conservation, Geological Survey, Geological Bulletin,* 30: 1-247.

DAVIS, M. B.
1976   Pleistocene biogeography of temperate deciduous forests. *Geoscience and Man,* 13: 13-26.
1981   Quaternary history and the stability of forest communities. *In:* West, D. C., Shugart, H. H., and Botkin, D. B. (eds.), *Forest Succession, Concepts and Application,* Springer-Verlag, New York, p. 132-153.
1983   Holocene vegetational history of the eastern United States. *In:* Wright, H. E., Jr. (ed.), *Late-Quaternary Environments of the United States, Vol. 2, The Holocene,* University of Minnesota Press, Minneapolis, p. 166-181.

DAVIS, R. B., and WEBB III, T.
1975   The contemporary distribution of pollen in eastern North America: a comparison with the vegetation. *Quaternary Research,* 5:395-434.

DEEVEY, E. S., Jr.
1949   Biogeography of the Pleistocene, Part I: Europe and North America. *Bulletin of the Geological Society of America,* 60: 1315-1416.

DeSELM, H. R., and BROWN, J. L.
1977   Final report of the fossil flora of the Friar Branch and Boyd Buchanan School sites. Hensley-Schmidt, Inc., Chattanooga, Tennessee, 22 p.
1978   Fossil flora of the Friar Branch and Boyd Buchanan School sites. *Association of Southeastern Biologists Bulletin,* 25: 83.

DELCOURT, H. R.
1978   Late-Quaternary vegetation history of the eastern Highland Rim and adjacent Cumberland Plateau of Tennessee. Ph.D. Dissertation, University of Minnesota, Minneapolis, 210 p.
1979   Late-Quaternary vegetation history of the eastern Highland Rim and adjacent Cumberland Plateau of Tennessee. *Ecological Monographs,* 49: 255-280.
1985   Holocene vegetational changes in the Southern Appalachian Mountains, U.S.A. *Ecologia Mediterranea,* in press.

DELCOURT, P. A.
1978   Quaternary vegetation history of the Gulf Coastal Plain. Ph.D. Dissertation, University of Minnesota, Minneapolis, 244 p.
1980   Goshen Springs: late-Quaternary vegetation record for southern Alabama. *Ecology,* 61: 371-386.
1985   The influence of late-Quaternary climatic and vegetational change on paleohydrology in unglaciated Eastern North America. *Ecologia Mediterranea,* in press.

DELCOURT, H. R., and DELCOURT, P. A.
1984   Late-Quaternary history of the Spruce-Fir Ecosystem in the Southern Appalachian Mountain Region. *In:* White, P. S. (ed.), *The Southern Appalachian spruce-fir ecosystem: its biology and threats.* U.S. National Park Service Research/Resource Management Report Ser-71: 22-35.
1985   Comparison of taxon calibrations, modern analog techniques, and forest-stand simulation models for the quantitative reconstruction of past vegetation. *Earth Surface Processes and Landforms,* 10:293-304.
1985   Late Quaternary vegetational history of the central Atlantic States. *In:* McDonald, J., and Bird, S. O. (eds.), *The Quaternary of Virginia.* Virginia Division of Mineral Resources Special Publication, in press.

DELCOURT, P. A., and DELCOURT, H. R.
1977   The Tunica Hills, Louisiana-Mississippi: late glacial locality for spruce and deciduous forest species. *Quaternary Research,* 7: 218-237.
1979   Late Pleistocene and Holocene distributional history of the deciduous forest in the southeastern United States. *Veroffentlichungen des Geobotanischen Institutes der ETH, Stiftung Rubel (Zurich),* 68: 79-107.
1980   Pollen preservation and Quaternary environmental history in the southeastern United States. *Palynology,* 4: 215-231.
1981   Vegetation maps for eastern North America: 40,000 yr BP to the present. *In:* Romans, R. (ed.), *Geobotany II.* Plenum Press, New York, p. 123-166.
1983   Late-Quaternary vegetational dynamics and community stability reconsidered. *Quaternary Research,* 19: 265-271.
1984   Late-Quaternary paleoclimates and biotic responses in Eastern North America and the Western North Atlantic Ocean. *Palaeogeography, Palaeoclimatology, Palaeoecology,* 48: 263-284.

DELCOURT, H. R., DELCOURT, P. A., and SPIKER, E. C.
1983   A 12,000-year record of forest history from Cahaba Pond, St. Clair County, Alabama. *Ecology,* 64: 874-887.

DELCOURT, H. R., DELCOURT, P. A., and WEBB III, T.
1982   Dynamic plant ecology: the spectrum of vegetational change in space and time. *Quaternary Science Reviews,* 1: 153-175.

DELCOURT, P. A., DELCOURT, H. R., BRISTER, R. C., and LACKEY, L. E.
1980   Quaternary vegetation history of the Mississippi Embayment. *Quaternary Research,* 13: 111-132.

DELCOURT, P. A., DELCOURT, H. R., and DAVIDSON, J. L.
1983   Mapping and calibration of modern pollen-vegetation relationships in the southeastern United States. *Review of Palaeobotany and Palynology,* 39: 1-45.

DELCOURT, P. A., DELCOURT, II. R., and WEBB III, T.
1984   Atlas of mapped distributions of dominance and modern pollen percentages for important tree taxa of Eastern North America. *American Association of Stratigraphic Palynologists, Contributions Series No. 14:* 1-131.

DENNY, C. S., OWENS, J. P., SIRKIN, L. A., and RUBIN, M.
1979   The Parsonburg Sand in the Central Delmarva Peninsula, Maryland and Delaware. *United States Geological Survey Professional Paper,* 1067B: B1-B16.

ELSIK, W. C.
1969   Late Neogene palynomorph diagrams, northern Gulf of Mexico. *Transactions of the Gulf Coast Association of Geological Societies,* 19: 509-528.

ENGELHARDT, D. W.

1963   Palynology of recent sediments in southeastern United States and the adjacent area of the Gulf of Mexico. *Pan American Petroleum Corporation Research Department Report No.,* M63-G-51: 1-22.

FAIR, T. P.

1984   Palynology of a Minnie's Lake transect in the Okefenokee Swamp, Georgia. M.S. Thesis, University of South Carolina, Columbia.

FEARN, L. B.

1981   A paleoecological reconstruction of Okefenokee Swamp based on pollen stratigraphy and peat petrography. M.S. Thesis, University of South Carolina, Columbia, 123 p.

FEARN, L. B., and COHEN, A. D.

1984   Palynologic investigations of six sites in the Okefenokee Swamp. *In:* Cohen, A. D., Casagrande, D. J., Andrejko, M. J., and Best, G. R. (eds.), *The Okefenokee Swamp: Its Natural History, Geology, and Geochemistry,* Wetland Surveys, Los Alamos, New Mexico, p. 423-443.

FIELD, M. E., MEISBURGER, E. P., STANLEY, E. A., and WILLIAMS, S. J.

1979   Upper Quaternary peat deposits on the Atlantic inner shelf of the United States. *Geological Society of America Bulletin, Part 1,* 90: 618-628.

FISH, S. K.

1971   Archaeological pollen analysis of the Powers phase, Southeastern Missouri. Manuscript on file at the Museum of Anthropology, University of Michigan, Ann Arbor.

FLORER, L. E.

1972   Palynology of a postglacial bog in the New Jersey Pine Barrens. *Bulletin of the Torrey Botanical Club,* 99: 135-138.

FREDLUND, G. G., and JOHNSON, W. C.

1982   Palynology of the Bug Hill archaeological site, Clayton Lake, Oklahoma. *In:* Altschul, J., *et al.* (eds.), *The Archaeological Reinvestigation of the Bug Hill Site (34PU116), Pushmataha County, Oklahoma.* New World Research, Inc.

FREY, D. G.

1951   Pollen succession in the sediments of Singletary Lake, North Carolina. Ecology, 32: 518-533.

1952   Pollen analysis of the Horry Clay and a seaside peat deposit near Myrtle Beach, South Carolina. *American Journal of Science,* 250: 212-225.

1953   Regional aspects of the late-glacial and post-glacial pollen succession of southeastern North Carolina. *Ecological Monographs,* 23: 289-313.

1955   A time revision of the Pleistocene pollen chronology of southeastern North Carolina. *Ecology,* 36: 762-763.

GILLESPIE, W. H., and CLENDENING, J. A.

1968   A flora from proglacial Lake Monongahela. *Castanea,* 33: 267-300.

GISH, J. W.

1980   Pollen results from submerged archaeological sites off the West Coast of Florida. Unpublished report to Department of Anthropology, Arizona State University, Tempe, 37 p.

GLEASON, P. J., STONE, P. A., GOODRICK, R., GUERIN, G., and HARRIS, L.

1975   The significance of paleofloral studies and ecological aspects of floating peat islands to water management in the Everglades Conservation Areas. Unpublished report, South Florida Water Management District, West Palm Beach, Florida, 176 p.

GRAHAM, A., and HEIMSCH, C.

1960   Pollen studies of some Texas peat deposits. *Ecology,* 41: 751-763.

GRUGER, E.

1970   The development of the vegetation of southern Illinois since late Illinoian time (preliminary report). *Revue Geographie Physique et Geologie Dynamique,* 12: 143-148.

1972   Pollen and seed studies of Wisconsinan vegetation in Illinois, U.S.A. *Geological Society of America Bulletin,* 83: 2715-2734.

1972   Late Quaternary vegetation development in south-central Illinois. *Quaternary Research,* 2: 217-231.

1973   Vegetation and climate of the last glacial period in central United States. *In:* Grichuk, V. P. (ed.), *Palynology of Pleistocene and Pliocene.* Proceedings of the 3rd International Palynological Conference (Novosibirsk 1971), p. 171-174.

GRUGER, J.

1973   Studies on the late Quaternary vegetation history of northeastern Kansas. *Geological Society of America Bulletin,* 84: 239-250.

GUILDAY, J. E., MARTIN, P. S., and McCRADY, A. D.

1964   New Paris No. 4: a late Pleistocene cave deposit in Bedford County, Pennsylvania. *National Speleological Society Bulletin,* 26: 121-194.

GURR, T. M.

1972   The geology of a central Florida peat bog, Sect. 26, Town. 30S, Range 25E, Polk County, Florida. M.S. Thesis, University of South Florida, 86 p.

HALE, H. S.

1983   An interpretation of a pollen core from Hontoon Island: a submerged shell midden on the St. John's River in Florida. *Paper Presented at the 40th Annual Southeastern Archaeological Conference, University of South Carolina, Columbia, November 4, 1983,* 14 p.

HALL, S. A.

1977   Geology and palynology of archaeological sites and associated sediments. *In:* Henry, D. O. (ed.), *The Prehistory of the Little Caney River, 1976 Field Season.* Contributions to Archaeology, 1, Laboratory of Archaeology, University of Tulsa, Oklahoma, p. 13-41.

1977   Holocene geology and paleoenvironmental history of the Hominy Creek Valley. *In:* Henry, D. O. (ed.), *The Prehistory and Paleoenvironment of Hominy Creek Valley.* Contributions to Archaeology, 2, Laboratory of Archaeology, University of Tulsa, Oklahoma, p. 12-42.

1977   Geological and paleoenvironmental studies. *In:* Henry, D. O. (ed.), *The Prehistory of Birch Creek Valley.* Contributions to Archaeology, 3, Laboratory of Archaeology, University of Tulsa, Oklahoma, p. 11-31.

1981   Deteriorated pollen grains and the interpretation of Quaternary pollen diagrams. *Review of Palaeobotany and Palynology,* 32: 193-206.

1982   Late Holocene paleoecology of the Southern Plains. *Quaternary Research,* 17: 391-407.

HARRISON, W., MALLOY, R. J., RUSNAK, G. A., and TERAS-MAE, J.
1965   Possible late Pleistocene uplift: Chesapeake Bay entrance. *Journal of Geology,* 73: 201-229.

HENRY, D. O.
1978   Big Hawk Shelter in northeastern Oklahoma: environmental, economic and cultural changes. *Journal of Field Archaeology,* 5: 269-287 (pollen analysis by S. A. Hall).

HENRY, D. O., BUTLER, B. H., and HALL, S. A.
1979   The late prehistoric human ecology of Birch Creek Valley, northeastern Oklahoma. *Plains Anthropologist,* 24: 207-238.

HOLLOWAY, R. G., and BRYANT, V. M., Jr.
1984   *Picea glauca* pollen from late glacial deposits in Central Texas. *Palynology,* 8: 21-32.

HOLLOWAY, R. G., and VALASTRO, S.
1983   Palynological investigations along the Yazoo River. *In:* Thorne, R. M., and Curry, H. K. (eds.), *Cultural Resources Survey of Items 3 and 4, Upper Yazoo River Projects, Mississippi, with a Paleoenvironmental Model of the Lower Yazoo Basin.* Archaeological Papers of the Center for Archaeological Research, Number 3, Center for Archaeological Research, University of Mississippi, p. 159-257.

HORR, W. H.
1955   A pollen profile study of the Muscotah Marsh. *University of Kansas Science Bulletin,* 37: 143-149.

HUMPHREY, W. F.
1953   Pollen analysis of three Georgia peat bogs. M.S. Thesis, University of Georgia, Athens, 39 p.

KAPP, R. O.
1965   Illinoian and Sangamon vegetation in southwestern Kansas and adjacent Oklahoma. *Contributions from the Museum of Paleontology, The University of Michigan,* 19: 167-255.
1969   *How to Know Pollen and Spores.* W. C. Brown Co., Dubuque, Iowa, 249 p.
1970   Pollen analysis of Pre-Wisconsin sediments from the Great Plains. *In:* Dort, W., Jr., and Jones, J. K., Jr. (eds.), *Pleistocene and Recent Environments of the Central Great Plains.* University of Kansas, Department of Geology Special Publication 3, p. 143-155.

KING, J. E.
1973   Late Pleistocene palynology and biogeography of the western Missouri Ozarks. *Ecological Monographs,* 43: 539-565.
1978   New evidence on the history of the Saint Francis Sunklands, northeastern Arkansas. *Geological Society of America Bulletin,* 89: 1719-1722.
1981   Late Quaternary vegetational history of Illinois. *Ecological Monographs,* 51: 43-62.

KING, J. E., and ALLEN, W. H., Jr.
1977   A Holocene vegetation record from the Mississippi River Valley, southeastern Missouri. *Quaternary Research,* 8: 307-323.

KING, J. E., and LINDSAY, E. H.
1976   Late Quaternary biotic records from spring deposits in western Missouri. *In:* Wood, W. R., and McMillan, R. B. (eds.), *Prehistoric Man and His Environments, a Case Study in the Ozark Highland.* Academic Press, New York, p. 63-78.

KNOX, A. S.
1962   Pollen from the Pleistocene terrace deposits of Washington, D.C. *Pollen et Spores,* 4: 356-358.
1966   The Walker Interglacial Swamp, Washington, D. C. *Journal of the Washington Academy of Sciences,* 56: 1-8.
1969   Glacial age marsh, Lafayette Park, Washington, D. C. *Science,* 165: 795-797.

KOLB, C. R., and FREDLUND, G. G.
1981   Palynological studies, Vacherie and Rayburn's Domes, North Louisiana Salt Dome Basin, topical report. *Louisiana State University, Baton Rouge, Institute for Environmental Studies Report* E530-02200T2: 50 p.

KUEHN, D.
1980   Offshore transgressive peat deposits of southwest Florida: evidence for late Holocene rise of sea level. M.S. Thesis, Pennsylvania State University, University Park, 104 p.

KURMANN, M. H.
1981   An opal phytolith and palynomorph study of extant and fossil soils in Kansas. M.S. Thesis, Kansas State University, 81 p.

LARSON, D. A., BRYANT, V. M., and PATTY, T. S.
1972   Pollen analysis of a central Texas bog. *The American Midland Naturalist,* 88: 358-367.

LEOPOLD, E. B.
1959   Pollen, spores, and marine microfossils. *In:* Malde, H. E. (ed.), *Geology of the Charleston Phosphate Area, South Carolina. United States Geological Survey Bulletin 1079,* p. 49-53.

LEWIS, I. F., and COCKE, E. C.
1929   Pollen analysis of Dismal Swamp peat. *Journal of the Elisha Mitchell Scientific Society,* 45: 37-58.

LIEUX, M. H.
1972   A melissopalynological study of 54 Louisiana (U.S.A.) honeys. *Review of Palaeobotany and Palynology,* 13: 95-124.
1980   An atlas of pollen of trees, shrubs, and woody vines of Louisiana and other southeastern states, Part I. Ginkgoaceae to Lauraceae. *Pollen et Spores,* 22: 2-57.
1980   An atlas of pollen of trees, shrubs, and woody vines of Louisiana and other southeastern states, Part II. Platanaceae to Betulaceae. *Pollen et Spores,* 22: 191-243.

LIEUX, M. H., and GODFREY, W. M.
1982   An atlas of pollen of trees, shrubs, and woody vines of Louisiana and other southeastern states, Part III. Polygonaceae to Ericaceae. *Pollen et Spores,* 24: 21-64.

McDONALD, J., DELCOURT, H. R., and DELCOURT, P. A.
1983   A report of the Saltville, Virginia, site. Paper presented at the 43rd Annual Meeting, Society of Vertebrate Paleontology, University of Wyoming, October 27-29, 1983.

MARKEWICH, H. W., and CHRISTOPHER, R. A.
1982   Pleistocene (?) and Holocene fluvial history of Uphapee Creek, Macon County, Alabama. *United States Geological Survey Bulletin,* 1522: 1-16.

MARTIN, P. S.
1958   Pleistocene ecology and biogeography of North America. *In:* Hubbs, C. S. (ed.), *Zoogeography.* American Association for the Advancement of Science Publication 51, p. 375-420.
1958   Taiga-tundra and the full-glacial period in Chester County, Pennsylvania. *American Journal of Science,* 256: 470-502.

MARTIN, P. S., and HARRELL, B. E.
1957   The Pleistocene history of temperate biotas in Mexico and eastern United States. *Ecology,* 38: 468-480.

MAXWELL, J. A., and DAVIS, M. B.
1972   Pollen evidence of Pleistocene and Holocene vegetation on the Allegheny Plateau, Maryland. *Quaternary Research,* 2: 506-530.

MEHRINGER, P. J., Jr., KING, J. E., and LINDSAY, E. H.
1970   A record of Wisconsin-age vegetation and fauna from the Ozarks of western Missouri. *In:* Dort, W., Jr., and Jones, J. K., Jr. (eds.), *Pleistocene and Recent Environments of the Central Great Plains.*University of Kansas, Department of Geology Special Publication 3, p. 173-183.

MEHRINGER, P. J., Jr., SCHWEGER, C. E., WOOD, W. R., and McMILLAN, B. R.
1968   Late-Pleistocene boreal forest in the western Ozark Highlands. *Ecology,* 49: 567-568.

NICKMANN, R. J., and DEMAREST, J. M., II.
1982   Pollen analysis of some Mid-Pleistocene Interglacial lagoonal sediments from southern Delaware. *Quaternary Research,* 17: 93-104.

OWENS, J. P., STEFANSSON, K., and SIRKIN, L. A.
1974   Chemical, mineralogic, and palynologic character of the upper Wisconsinan-lower Holocene fill in parts of Hudson, Delaware, and Chesapeake estuaries. *Journal of Sedimentary Petrology,* 44: 390-408.

PETERSON, G. M.
1978   Pollen spectra from surface sediments of lakes and ponds in Kentucky, Illinois, and Missouri. *The American Midland Naturalist,* 100: 333-340.

POTZGER, J. E.
1945   The pine barrens of New Jersey: a refugium during Pleistocene times. *Butler University Botanical Studies,* 7: 182-196.
1952   What can be inferred from pollen profiles of bogs in the New Jersey pine barrens. *Bartonia,* 26: 20-27.

POTZGER, J. E., and THARP, B. C.
1943   Pollen record of Canadian spruce and fir from Texas bog. *Science,* 98: 584.
1947   Pollen profile from a Texas bog. *Ecology,* 28: 274-280.
1954   Pollen study of two bogs in Texas. *Ecology,* 35: 462-466.

RAY, C. E., COOPER, B. N., and BENNINGHOFF, W. S.
1967   Fossil mammals and pollen in a late Pleistocene deposit at Saltville, Virginia. *Journal of Paleontology,* 41: 608-622.

REED, J. C., Jr., BRYANT, B., LEOPOLD, E. B., and WEILER, L.
1964   A Pleistocene section at Leonards cut, Burke County, North Carolina. *United States Geological Survey Professional Paper,* 475-D: D38-D42.

RICH, F. J.
1979   The origin and development of tree islands in the Okefenokee Swamp, as determined by peat petrography and pollen stratigraphy. Ph.D. Dissertation, Pennsylvania State University, University Park, 301 p.
1984   Ancient flora of the eastern Okefenokee Swamp as determined by palynology. *In:* Cohen, A. D., Casagrande, D. J., Andrejko, M. J., and Best, G. R. (eds.), *The Okefenokee Swamp: Its Natural History, Geology, and Geochemistry,* Wetlands Surveys, Los Alamos, New Mexico, p. 410-422.

1984   Development of three tree islands in the Okefenokee Swamp, as determined by palynostratigraphy and peat petrography. *In:* Cohen, A. D., Casagrande, D. J., Andrejko, M. J., and Best, G. R. (eds.), *The Okefenokee Swamp: Its Natural History, Geology, and Geochemistry,* Wetlands Surveys, Los Alamos, New Mexico, p. 444-455.

RICH, F. J., and SPACKMAN, W.
1979   Modern and ancient pollen sedimentation around tree islands in the Okefenokee Swamp. *Palynology,* 3: 219-226.

RICH, F. J., KUEHN, D., and DAVIES, T. D.
1982   The paleoecological significance of *Ovoidites. Palynology,* 6: 19-28.

RIEGEL, W. L.
1965   Palynology of environments of peat formation in southwestern Florida. Ph.D. Dissertation, Pennsylvania State University, University Park, 189 p.

SCHAETZL, R. L., and JOHNSON, W. C.
1983   Pollen and spore stratigraphy of a mollic hapludalf (degraded chernozem) in northeastern Kansas. *The Professional Geographer,* 35: 183-191.

SCHOENWETTER, J.
1962   A late postglacial pollen chronology from the Central Mississippi Valley. *In:* Fowler, M. L. (ed.), *American Bottoms Archaeology, First Annual Report,* Illinois Archaeological Survey, University of Illinois.
1963   Survey of Palynological results. *In:* Fowler, M. L. (ed.), *American Bottoms Archaeology, Second Annual Report,* Illinois Archaeological Survey, University of Illinois.
1964   Pollen report. *In:* Fowler, M. L. (ed.), *American Bottoms Archaeology, Third Annual Report,* Illinois Archaeological Survey, University of Illinois.
1964   Preliminary palynological report on the Carlyle Reservoir, Appendix. *Southern Illinois University Museum Archaeological Salvage Report No.* 17.
1967   Pollen studies in southern Illinois. Report submitted to M. L. Fowler, Southern Illinois University.
1971   Pollen analysis of the Koster Site: first report. Report submitted to S. Struever, Northwestern University.
1972   Pollen studies at the Macoupin Site, Appendix. *In:* Rackerby, F., *The Macoupin Site: A Middle Woodland Community in the Lower Illinois River Valley,* Ph.D. Dissertation, Northwestern University, Chicago.
1974   Pollen analysis of sediments from Salts Cave Vestibule. *In:* Watson, P. J. (ed.), *Archaeology of the Mammoth Cave Area,* Academic Press, New York, p. 97-105.
1974   Pollen analysis of human paleofeces from Upper Salts Cave. *In:* Watson, P. J. (ed.), *Archaeology of the Mammoth Cave Area,* Academic Press, New York, p. 49-59.
1974   Principal results of palynological studies at Koster: summary statement. Report submitted to S. Struever, Northwestern University.
1976   Appendix C, Palynological investigations of Cache River Project sediment samples. *In:* Schiffer, M. B., and House, J. H. (eds.), *The Cache River Archaeological Project, Arkansas Archaeological Survey, Publication in Archaeology Research Series No.* 8.
1978   Surface and archaeological sediment pollen studies in the Mammoth Cave National Park study area: a methodological and interpretive report. Report submitted to P. J. Watson, Washington University, St. Louis, Missouri, 55 p.

1979   Appendix I: Archaeological pollen analysis of Copan Reservoir sediment samples. *In:* Vehik, S. C., and Pailes, R. A. (eds.), *Excavations in the Copan Reservoir of Northeastern Oklahoma and Southeastern Kansas, University of Oklahoma, Archaeological Research and Management Center, Research Series No.* 4.

1981   Contributions of pollen analysis in investigations of New World agriculture. *International Palynological Conference, Lucknow, India (1976-1977),* 3: 269-278.

1983   Appendix VIII, Palynological report. *In:* Neitzel, S. (ed.), *The Grand Village of the Natchez Revisited,* Mississippi Department of Archives and History, *Archaeological Report No.* 12: 178.

SEARS, E. O'R.

1982   Pollen analysis. *In:* Sears, W. H. (ed.), *Fort Center, an Archaeological Site in the Lake Okeechobee Basin.* University Presses of Florida, Gainesville, p. 118-129.

SEARS, E. O'R., and SEARS, W. H.

1976   Preliminary report on prehistoric corn pollen from Fort Center, Florida. *Bulletin of the Southeastern Archaeological Conference,* 19: 53-56.

SEARS, P. B.

1931   Recent climate and vegetation, a factor in the mound-building cultures? *Science,* 73: 640-641.

1932   The archaeology of environment in eastern North America. *American Anthropologist,* 34: 610-622.

1935   Types of North American pollen profiles. *Ecology,* 16: 488-499.

1942   Postglacial migration of five forest genera. *American Journal of Botany,* 29: 684-691.

SEARS, P. B., and COUCH, G. C.

1932   Microfossils in an Arkansas peat and their significance. *Ohio Journal of Science,* 32: 63-68.

1935   Humus stratigraphy as a clue to past vegetation in Oklahoma. *Proceedings of the Oklahoma Academy of Science,* 15.

SHAFER, D. S.

1984   Late-Quaternary paleoecologic, geomorphic, and paleoclimatic history of Flat Laurel Gap, Blue Ridge Mountains, North Carolina. M.S. Thesis, University of Tennessee, Knoxville, 148 p.

SHELDON, E. S.

1977   Reconstruction of prehistoric environment and its useful plants: Warm Mineral Springs (8So-19), Florida. Paper presented at the Society of Economic Botany, June 11 to 15, 1977, Coral Gables, Florida (pollen data by J. E. King).

SIRKIN, L. A., DENNY, C. S., and RUBIN, M.

1977   Late Pleistocene environment of the central Delmarva Peninsula. *Geological Society of America Bulletin,* 88: 139-142.

SIRKIN, L. A., OWENS, J. P., MINARD, J. P., and RUBIN, M.

1970   Palynology of some upper Quaternary peat samples from the New Jersey coastal plain. *United States Geological Survey Professional Paper,* 700-D: D77-D87.

SMITH, E. N., Jr.

1984   Late-Quaternary vegetational history at Cupola Pond, Ozark National Scenic Riverways, Southeastern Missouri. M.S. Thesis, University of Tennessee, Knoxville, 115 p.

SOLOMON, A. M.

1983   Pollen morphology and plant taxonomy of white oaks in eastern North America. *American Journal of Botany,* 70: 481-494.

1983   Pollen morphology and plant taxonomy of red oaks in eastern North America. *American Journal of Botany,* 70: 495-507.

SOLOMON, A. M., WEST, D. C., and SOLOMON, J. A.

1981   Simulating the role of climate change and species immigration in forest succession. *In:* West, D. C., Shugart, H. H., and Botkin, D. B. (eds.), *Forest Succession, Concepts and Application.* Springer-Verlag, New York, p. 154-177.

SOLOMON, A. M., DELCOURT, H. R., WEST, D. C., and BLASING, T. J.

1980   Testing a simulation model for reconstruction of prehistoric forest-stand dynamics. *Quaternary Research,* 14: 275-293.

SPACKMAN, W., DOLSEN, C. P., and RIEGEL, W.

1966   Phytogenic organic sediments and sedimentary environments in the Everglades-Mangrove complex. Part I: Evidence of a transgressing sea and its effects on environments of the Shark River area of southwestern Florida. *Sonder-abdruck aus Palaeontographica, Beitrage Zur Naturgeschichte Der Vorzeit,* 117(B): 135-152.

SPACKMAN, W., COHEN, A. D., GIVEN, P. H., and CASAGRANDE, D. J.

1974   A field guidebook to aid in the comparative study of the Okefenokee Swamp and the Everglades-mangrove swamp-marsh complex of southern Florida. *Geological Society of America Preconvention Field Trip,* 6: 1-265.

STACK, E.

1984   Palynology of a Chesser Prairie transect in the Okefenokee Swamp, Georgia. M.S. Thesis, University of South Carolina, Columbia.

STANLEY, E. A.

1965   Abundance of pollen and spores in marine sediments off the eastern coast of the United States. *Southeastern Geology,* 7: 25-33.

STINGELIN, R. W.

1965   Late-glacial and post-glacial vegetational history in the north central Appalachian region. Ph.D. Dissertation, Pennsylvania State University, University Park, 192 p.

STONE, P. A., and BROWN, J. G.

1983   The pollen record of Pleistocene and Holocene paleoenvironmental conditions in southeastern United States. *In:* Colquhoun, D. J. (ed.), *Variation in Sea Level on the South Carolina Coastal Plain,* Department of Geology, University of South Carolina, Columbia, p. 169-206.

STONE, P. A., and GLEASON, P. J.

1983   Environmental and paleoenvironmental significance of organic sediments (peats) in southeastern United States. *In:* Colquhoun, D. J. (ed.), *Variation in Sea Level on the South Carolina Coastal Plain.* Department of Geology, University of South Carolina, Columbia, p. 121-141.

THOMPSON, D. E.

1972   Paleoecology of the Pamlico Formation, Saint Mary's County, Maryland. Ph.D. Dissertation, Rutgers University, New Jersey, 194 p.

TISDALE, M. G.
1981   Sedimentology and paleoecology of a buried, Pleistocene, carbonaceous sand from Myrtle Beach State Park, Horry County, South Carolina. M.S. Thesis, University of South Carolina, Columbia, 62 p.

VICK, A. R.
1961   Some pollen profiles from the Coastal Plain of North Carolina. Ph.D. Dissertation, Syracuse University, New York, 73 p.

WATTS, W. A.
1969   A pollen diagram from Mud Lake, Marion County, north-central Florida. *Geological Society of America Bulletin,* 80: 631-642.

1970   The full-glacial vegetation of northwestern Georgia. *Ecology,* 51: 17-33.

1971   Postglacial and interglacial vegetation history of southern Georgia and central Florida. *Ecology,* 52: 676-690.

1973   The vegetation record of a Mid-Wisconsin Interstadial in northwest Georgia. *Quaternary Research,* 3: 257-268.

1975   A late-Quaternary record of vegetation from Lake Annie, south-central Florida. *Geology,* 3: 344-346.

1975   Vegetation record for the last 20,000 years from a small marsh on Lookout Mountain, northwestern Georgia. *Geological Society of America Bulletin,* 86: 287-291.

1979   Late Quaternary vegetation of central Appalachia and the New Jersey Coastal Plain. *Ecological Monographs,* 49: 427-469.

1980   Late-Quaternary vegetation history at White Pond on the Inner Coastal Plain of South Carolina. *Quaternary Research,* 13: 187-199.

1980   The late Quaternary vegetation history of the southeastern United States. *Annual Reviews of Ecology and Systematics,* 11: 387-409.

1983   Vegetational history of the eastern United States 25,000 to 10,000 years ago. *In:* Porter, S. C. (ed.), *Late-Quaternary Environments of the United States, Vol. 1, The Late Pleistocene.* University of Minnesota Press, Minneapolis, p. 294-310.

WATTS, W. A., and STUIVER, M.
1980   Late Wisconsin climate of northern Florida and the origin of species-rich deciduous forest. *Science,* 210: 325-327.

WEBB III, T., and McANDREWS, J. H.
1976   Corresponding patterns of contemporary pollen and vegetation in central North America. *Geological Society of America Memoir,* 145: 267-299.

WHITEHEAD, D. R.
1963   "Northern" elements in the Pleistocene flora of the Southeast. *Ecology,* 44: 403-406.

1964   Fossil pine pollen and full-glacial vegetation in southeastern North Carolina. *Ecology,* 45: 767-776.

1965   Palynology and Pleistocene phytogeography of unglaciated eastern North America. *In:* Wright, H. E., Jr., and Frey, D. G. (eds.), *The Quaternary of the United States.* Princeton University Press, Princeton, p. 417-432.

1965   Prehistoric maize in southeastern Virginia. *Science,* 150: 881-883.

1967   Studies of full-glacial vegetation and climate in southeastern United States. *In:* Cushing, E. J., and Wright, H. E., Jr. (eds.), *Quaternary Paleoecology.* Yale University Press, New Haven, p. 237-248.

1972   Development and environmental history of the Dismal Swamp. *Ecological Monographs,* 42: 301-315.

1973   Late-Wisconsin vegetational changes in unglaciated eastern North America. *Quaternary Research,* 3: 621-631.

1981   Late-Pleistocene vegetational changes in northeastern North Carolina. *Ecological Monographs,* 51: 451-471.

WHITEHEAD, D. R., and BARGHOORN, E. S.
1962   Pollen analytical investigations of Pleistocene deposits from western North Carolina and South Carolina. *Ecological Monographs,* 32: 347-369.

WHITEHEAD, D. R., and CAMPBELL, S. K.
1976   Palynological studies of the Bull Creek peat, Horry County, South Carolina: geomorphological implications. *Southeastern Geology,* 17: 161-174.

WHITEHEAD, D. R., and DAVIS, J. T.
1969   Pollen analysis of an organic clay from the interglacial Flanner Beach Formation, Craven County, North Carolina. *Southeastern Geology,* 10: 149-164.

WHITEHEAD, D. R., and DOYLE, M. V.
1969   Late-Pleistocene peats from Long Beach, North Carolina. *Southeastern Geology,* 10: 1-16.

WHITEHEAD, D. R., and LANGHAM, E. J.
1965   Measurement as a means of identifying fossil maize pollen. *Bulletin of the Torrey Botanical Club,* 92: 7-20.

WHITEHEAD, D. R., and OAKS, R. Q., Jr.
1979   Development history of the Dismal Swamp. *In:* Kirk, P. W., Jr. (ed.), *The Great Dismal Swamp.* University Press of Virginia, Charlottesville, p. 25-43.

WHITEHEAD, D. R., and SHEEHAN, M. C.
1984   Holocene vegetational changes in the Tombigbee River Valley, eastern Mississippi. *American Midland Naturalist,* in press.

WHITEHEAD, D. R., and TAN, K. W.
1969   Modern vegetation and pollen rain in Bladen County, North Carolina. *Ecology,* 50: 235-248.

WILKINS, G. R.
1985   Late-Quaternary vegetational history at Jackson Pond, LaRue County, Kentucky. M.S. Thesis, University of Tennessee, Knoxville, 172 p.

WILSON, L. R.
1965   *Rhizophagites,* a fossil fungus from the Pleistocene of Oklahoma. *Oklahoma Geological Survey, Oklahoma Geology Notes,* 25: 257-260.

1966   Palynology of the Domebo Site. Great Plains Historical Association, Lawton, Oklahoma, *Contributions of the Museum of the Great Plains No. 1:* 44-50.

1977   Palynological study of Feature 4, Mi-63, from an Early Caddoan village in eastern Oklahoma. Oklahoma Highway Archaeological Survey, Oklahoma City, Oklahoma, *Papers in Highway Archaeology, No. 3:* 243-257.

WOLFE, F. A. and NEASE, F. R.
1970   Nonpetrified fossils from the site of a phosphate mine in eastern North Carolina. *The Journal of the Elisha Mitchell Scientific Society,* 86: 44-48.

WRIGHT, H. E., Jr.
1968   The roles of pine and spruce in the forest history of Minnesota and adjacent areas. *Ecology,* 49: 937-955.

1971   Late Quaternary vegetational history of North America. *In:* Turekian, K. K. (ed.), *The Late Cenozoic Glacial Ages.* Yale University Press, p. 425-464.

1976   The dynamic nature of Holocene vegetation, a problem in paleoclimatology, biogeography, and stratigraphic nomenclature. *Quaternary Research,* 6: 581-596.

1977   Quaternary vegetation history—some comparisons between Europe and America. *Annual Reviews of Earth and Planetary Science,* 5: 123-158.

1981   Vegetation east of the Rocky Mountains 18,000 years ago. *Quaternary Research,* 15: 113-125.

YARNELL, R. A.
1969   Contents of human paleofeces. *In:* Watson, P. J. (ed.), *The Prehistory of Salts Cave, Kentucky, Illinois State Museum, Springfield, Illinois, Report of Investigations,* 16: 41-64 (pollen analysis by W. A. Benninghoff, p. 51).

A LATE-QUATERNARY PALEOENVIRONMENTAL RECORD OF TEXAS: AN OVERVIEW OF THE POLLEN EVIDENCE

VAUGHN M. BRYANT, JR.
Department of Anthropology and
Department of Biology
Texas A&M University
College Station, Texas 77843

RICHARD G. HOLLOWAY
Palynology Laboratory
Department of Anthropology
Texas A&M University
College Station, Texas 77843

Abstract

Paleoenvironmental records spanning more than 30,000 years are used to reconstruct an overview of vegetational changes occurring in Texas during the late-Quaternary. Primary emphasis is devoted to reconstructions based on published and unpublished palynological data. Those records are supplemented, where appropriate, by supporting information gained from geological, archaeological, paleontological, and non-palynological botanical sources. Briefly, the combined record indicates that elements of the northern and southeastern United States flora, including *Picea glauca*, invaded regions of east and central Texas during the late-Quaternary period and remained part of the regional vegetation until late-glacial times. Although vast regions of west Texas were probably covered by conifer forests during the full-glacial period, those forests were not as widespread as some earlier reports have indicated. The area of south Texas revealed that a mosaic vegetational pattern probably was well established throughout the post-glacial period and perhaps even longer. And, it appears that regions of north Texas contained grassland savannas during the post-glacial and part of the late-glacial period.

INTRODUCTION

The state of Texas occupies a large geographical area and contains a wide variety of ecological habitats which makes its analysis, from a paleoenvironmental point of view, somewhat difficult. On the other hand, the critical importance of Texas' geographical location is essential to the understanding of continental vegetational interpretations during the late-Pleistocene and Holocene. For example, Texas is presently located between the deciduous forests of the southeastern United States, the semi arid and arid regions in the southwestern United States, the prairies and hills of the Ozarks, and the arid grasslands of south Texas and northern Mexico.

Deevey (1949) was one of the earliest investigators to recognize the critical location of Texas in reference to possible ancient plant migrations when he suggested that during the Quaternary period major plant migrations between areas of Mexico and the southeastern United States may have occurred across areas of south Texas during cycles of glacial advance and retreat. Lucy Braun (1955) offered an alternative model consisting of little or no suggested vegetational or climatic changes in Texas during the entire Quaternary period. Dillon (1956), and later Martin and Harrell (1957), proposed still other phytogeographic reconstructions suggesting that only moderate vegetational and climatic changes took place in Texas during portions of the Quaternary period.

Zoologists such as Blair (1958) and paleontologists such as Graham (1976), Lundelius (1967, 1974), Lundelius *et al.* (1983), and Slaughter (1963) also have offered their interpretations of Quaternary vegetational changes in Texas. Based upon the fossil evidence of extinct and extant fauna these authors have suggested that more than 10,000 years ago certain regions of Texas probably were covered by large wooded areas while in other regions there were savannas. Other clues revealed by the fossil animal record suggest that there was probably a north-south temperature gradient change during portions of the Quaternary period which undoubtedly affected the range and distribution of both plants and animals.

The recovery and analysis of wood rat *(Neotoma)* nests found in montane areas of west Texas (Wells

1966; Van Devender *et al.* 1977, 1978) also have provided data useful for the interpretation of late-Quaternary vegetational changes. Wood rats make their nests from objects they find in their environment. Often these items include fruits, nuts, and leaves which later can be identified and dated. Using information recovered from wood rat nests in the Chisos Mountains of west Texas, Wells (1966) found that during the full-glacial period (*ca.* 20,000 yrs. B.P.) yellow pine woodlands probably covered large areas of the Chisos Mountains and may have expanded outward into regions of lower elevation as well. In other wood rat nests from the Guadalupe Mountain area Van Devender *et al.* (1977) found traces of spruce and Douglas fir needles, suggesting that these plants formed part of the full-glacial age woodlands in that region of west Texas.

An alternative approach to the study of late-Quaternary displacement, migration, and reconstruction of paleovegetations in Texas has been through the study of sequential changes in the fossil pollen record. Although those types of data are difficult to obtain, they often are more reliable as a paleoenvironmental indicator than are the data collected from archaeological or paleontological sources. Fossil pollen grains are considered more useful as indicators since they tend to be more evenly distributed and thus can be directly linked to the plant taxa which produced them. Aboriginal groups and animals are mobile and thus their remains in archaeology or paleontological sites sometimes can be found in ecological areas outside their optimum habitats.

The earliest attempts to obtain fossil pollen data and to apply them to problems related to the paleoenvironment of Texas were made during the 1940's and early 1950's by Potzger and Tharp (1943, 1947, 1954) when they analyzed a series of peat deposits in east and central Texas. Their analysis of the Patschke peat bog in Lee County revealed the presence of fossil fir, spruce, pine, maple, basswood, birch, chestnut, walnut, and alder pollen in the lowermost deposits. Variations in the pollen composition of subsequent deposits from that same bog were interpreted by Potzger and Tharp (1943, 1947) to represent a late-Quaternary vegetational sequence that began with an early cool-moist period, progressed through a warm-dry period, changed to a warm-moist period, and finally ended in modern times with a warm-dry interval characterized by the establishment of the present Post Oak Savanna vegetational zone (Gould, 1975) in east-central Texas. Potzger and Tharp (1954) later analyzed other

Texas peat deposits in nearby Milam and Robertson Counties and determined that those pollen records were similar to the ones they had recovered at the Patschke Bog. Although useful, their records were undated. They did suggest, however, that their earliest pollen records represented part of the Wisconsin full-glacial period and that the deposits probably were about 20,000 years old.

No additional fossil pollen studies in Texas were conducted until the 1960's when Graham and Heimsch (1960) discovered and sampled the Soefje Peat Bog in Gonzales County located in south-central Texas. Graham and Heimsch also reexamined the original peat samples which were collected in the late 1940's by Potzger and Tharp from the Gause Bog in Milam County. These new studies led Graham and Heimsch (1960) to propose their own paleoenvironmental interpretation for the central Texas late-Quaternary period. Their study provided a more detailed analytical survey than did the previous studies conducted by Potzger and Tharp and led Graham and Heimsch to reject the original four-stage climatic sequence proposed by Potzger and Tharp (1947, 1954). Instead, Graham and Heimsch proposed that central Texas environments were cool and moist prior to 12,500 years ago and that after that time there was a gradual warming trend that went through several minor fluctuations until it ended in the warm-dry conditions found in central Texas today.

The earliest pollen studies of late-Quaternary age sediments conducted in west Texas began in the late 1950's and resulted in Hafsten (1961) publishing a pollen analytical record of sediments obtained from a series of playa lakes located on, or near the Llano Estacado. His study was significant since it provided a partially dated chronology covering a time span estimated to be older than 35,000 years and resulted in a proposed four-stage vegetational sequence for areas of west Texas and the Llano Estacado. His sequence consisted of: (1) in interval more than 30,000 years ago when grasslands were common in west Texas and on the Llano Estacado, (2) a period between 22,500 - 14,000 years ago when many of the grassland regions of west Texas were replaced by conifer woodlands, (3) a transition period between 14,000 - 10,000 years ago when many of the west Texas conifer woodlands decreased in area and were being replaced by scrub grasslands, and (4) the establishment of the present grasslands in the lower elevations of west Texas and on the Llano Estacado around 10,000 years ago.

Later, studies by Oldfield and Schoenwetter (1964, 1975) of archaeological sediments and other playa lake deposits provided additional information concerning suspected vegetational changes in west Texas during the past 30,000 years. Although important, the studies by Oldfield and Schoenwetter did not significantly alter any of the major paleoenvironmental reconstructions originally proposed by Hafsten (1961).

In the years since those initial studies were conducted, other investigators have expanded our knowledge of the Texas late-Quaternary paleoenvironmental record through their studies of plant macrofossil remains, faunal deposits, archaeological records, geomorphology, and additional pollen records. Instead of trying to incorporate all of those reports into our discussion of this paper, we have selected to focus our attention on the fossil pollen records and use other supporting data primarily when fossil pollen records for a region are meager or absent. Fossil pollen records from lake sediments, archaeological deposits, peat bogs, and fossil wood rat middens are the primary sources of data we have used to propose a tentative vegetational sequence for areas of Texas covering portions of the late-Quaternary period. The majority of those data have been collected from fossil pollen records in areas of central, west, and southwest Texas. Unfortunately, only a few fossil pollen records are yet available for regions of northeast, south-central or south Texas. This lack of pollen information from those regions is due to several factors. First, there has been only a limited number of attempts to recover fossil pollen from deposits in those regions and second, what few attempts have been made have generally resulted in a failure to find preserved, fossilized pollen.

In spite of these limitations, we shall devote the remainder of this paper to a discussion of what is presently known about the paleoenvironmental record of Texas. We shall discuss these late-Quaternary environmental changes using the time periods originally proposed by Hafsten (1961). In our paper when we refer to a "woodland" we are interpreting it to mean an open canopy forest containing an understory of herbaceous shrubs or grasses. Likewise, we view a "parkland" as consisting of a grassland which is periodically interrupted by isolated trees and/or large clumps of trees. Our term "savanna" refers to a grassland containing a few scattered trees and shrubs but still indicates an area where the continuity of the grassland is basically uninterrupted. Finally, our term "scrub grasslands" is used to refer to a prairie-like grassland vegetation containing large numbers of scrubby plants and an occasional tree or small stands of trees in the more favorable locales.

MIDDLE WISCONSIN PERIOD—33,500-22,500 YEARS B.P.

None of the peat bog or archaeological site deposits thus far examined in Texas contain a fossil record older than full-glacial age. Thus, the only Texas fossil pollen evidence that is currently available for the Middle Wisconsin period comes from a few playa lake deposits located in northwest Texas (Fig. 1, Table 1). In the late 1950's as part of a two-year study of the late-Quaternary paleoecology on the Llano Estacado Hafsten conducted fossil pollen studies of a series of playa lake deposits. In that original study Hafsten (1961) examined sediments from Arch Lake and Rich Lake (Fig. 2) both of which contained radiocarbon dated sediments from what he termed as the Wisconsin Interpluvial period (Middle Wisconsin period). Based upon the fossil pollen record containing high percentages of grass, cheno-am, and composite pollen, Hafsten proposed that much of the Llano Estacado region in Texas was covered by a fairly stable scrub grassland during most of the Middle Wisconsin period prior to about 22,500 yrs. B.P. In addition, his low arboreal pollen counts of spruce and pine led Hafsten (1961) to suggest that trees were either totally absent or were restricted to a few isolated and protected locales on the High Plains of Texas during this early period. Based upon inferences derived from the vegetational reconstruction, Hafsten (1961) proposed that much of west Texas probably enjoyed a cool and moist climate during most of the Middle Wisconsin period.

Using Hafsten's (1961) work as an initial guide, Oldfield and Schoenwetter (1964, 1975) restudied a number of locations in northwest Texas on the Llano Estacado. Their fossil pollen analysis of sediments from Rich Lake, White Lake, and Illusion Lake (Fig. 1) enabled them to construct a longer vegetational sequence for the Middle Wisconsin period which they divided into a series of pollen analytical episodes beginning with the Brownfield Oscillation and continuing through the Arch Lake Interpluvial, the Terry Subpluvial, and terminating with the Rich Lake Interpluvial (Fig. 2). In their study (Oldfield and Schoenwetter, 1975) the oldest deposits examined came from cores extracted from Rich Lake and were assigned to their Brownfield Oscillation pollen analytical episode. These sediments were not dated, yet stratigraphically were known to predate the cores analyzed by Hafsten (1961) which came from a

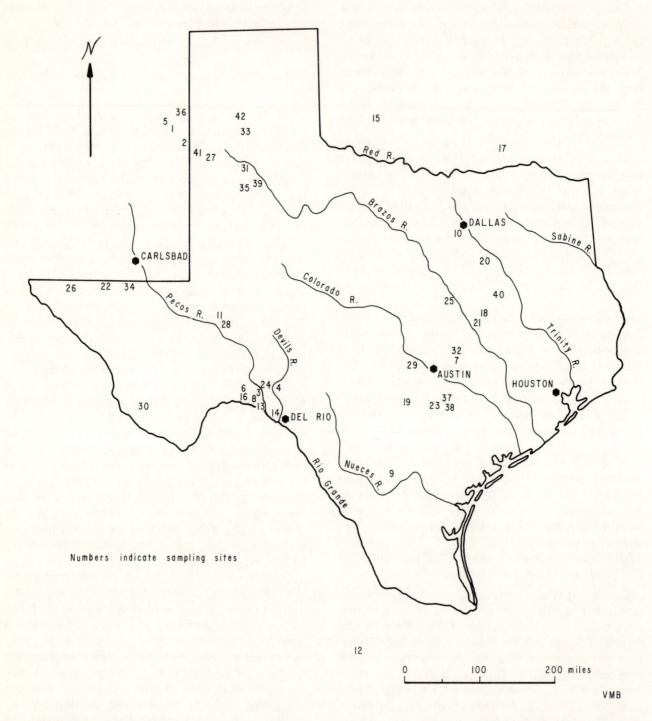

Figure 1. Map of Texas noting the location of sites discussed in the text and mentioned in Table 1. Numbers on the map correspond to site numbers listed in Table 1.

period between 22,500 and 33,500 yrs. B.P. According to Oldfield and Schoenwetter, the Brownfield Oscillation is characterized in the fossil pollen record by sequences of alternating pine and grass pollen dominance suggesting a period when the Llano Estacado was covered by a mixed prairie and pine parkland vegetation that oscillated back and forth as each of those plant communities responded to changes in the available moisture.

TABLE 1: Late-Quaternary Sites from Texas and Adjacent Areas Mentioned in the Text and Identified in Figure 1.

Site	Site Number	Reference to Original Research
Anderson Basin, New Mexico	1	Oldfield & Schoenwetter, 1975
Arch Lake, New Mexico	2	Hafsten, 1961
Arenosa Shelter	3	Bryant, 1967
Baker Cave	4	Hester, 1982
Blackwater Draw, New Mexico	5	Oldfield & Schoenwetter, 1975
Bonfire Shelter	6	Hevly, 1966; Bryant, 1969; Bryant and Shafer, 1977; Bryant and Holloway, ms.
Boriack Bog	7	Bryant, 1969, 1977b; Holloway and Bryant, 1984
Centipede Cave	8	Johnson, 1963
Choke Canyon Reservoir	9	Holloway, 1985; Steele, 1985
Coob-Poole Site	10	Raab and Woosley, 1982
Crane Lake	11	Hafsten, 1961; Foreman and Clisby, 1961
Cueva de la Zona de Derrumbes, Mexico	12	Bryant and Riskind, 1980
Damp Cave	13	Johnson, 1963
Devils Mouth Site	14	Johnson, 1964; Bryant and Larson, 1968
Domebo Site, Oklahoma	15	Wilson, 1966
Eagle Cave	16	McAndrews and Larson, 1966
Ferndale Bog, Oklahoma	17	Albert, 1981; Holloway and Ferring, 1985
Franklin Bog	18	Potzger & Tharp, 1954
Freisenhan Cave	19	Evans, 1961; Graham, 1976; Hall, unpublished
Frossard Site	20	Raab et al., 1982
Gause Bog	21	Potzger and Tharp, 1954; Graham and Heimsch, 1960; Bryant, 1969, 1977b
Guadalupe Mountains	22	Van Devender et al., 1979
Hershop Bog	23	Larson et al., 1972
Hinds Cave	24	Shafer and Bryant, 1977; Bryant, 1977a; Dering, 1977, 1979; Williams-Dean, 1978; Dean, 1983; Lord, 1983
Horn Shelter	25	Bryant, unpublished report
Hueco Mountains	26	Van Devender and Riskind, 1979
Illusion Lake	27	Oldfield and Schoenwetter, 1975
Juan Cordova Lake	28	Hafsten, 1961
Levi Site	29	Bryant, 1969
Livingstone Hills	30	Van Devender et al., 1978
Lubbock Lake	31	Hafsten, 1961; Oldfield and Schoenwetter, 1975; Hall, unpublished report; Holliday et al., 1983, Holliday et al., 1985
Patschke Bog	32	Potzger and Tharp, 1943, 1947, 1954; Patrick, 1946
Plainview	33	Oldfield and Schoenwetter, 1975
Pratt Cave,	34	Bryant, 1969, 1983
Rich Lake	35	Hafsten, 1961; Oldfield and Schoenwetter, 1975
San Jon Site, New Mexico	36	Hafsten, 1961; Oldfield and Schoenwetter, 1975
Soefje Bog	37	Graham and Heimsch., 1960
South Soefje Bog	38	Bryant, 1969, 1977b
Tahoka Lake	39	Oldfield and Schoenwetter, 1975
Weakly Bog	40	Holloway et al., 1985
White Lake	41	Oldfield and Schoenwetter, 1975
Vigo Park Lake	42	Oldfield and Schoenwetter, 1975

Following the Brownfield Oscillation came the Arch Lake Interpluvial which Oldfield and Schoenwetter (1975) characterized as reflecting a shift in the regional vegetation from prairies mixed with pine parklands to prairies mixed with pine savannas. In addition, they noted that throughout the Arch Lake Interpluvial the fossil record was dominated by low percentages of pine pollen combined with high *Artemisia,* grass and composite pollen counts suggesting a cool-dry climate.

During the next phase, the Terry Subpluvial, which also is undated but is believed to be slightly older than 33,000 yrs. B.P., the pollen record indicates that the vegetation probably consisted of a mosaic containing elements of savanna, prairie, and parklands with both pine and spruce trees present (Oldfield and Schoenwetter, 1975). The last pollen analytical episode in the Oldfield and Schoenwetter chronology is the Rich Lake Interpluvial which corresponds to the Wisconsin Interpluvial episode mentioned by Hafsten (1961). In their analysis Oldfield and Schoenwetter (1975) found high percentages of grass and *Artemisia* pollen combined with low amounts of pine pollen. Essentially, they agreed with Hafsten's (1961) original reconstruction and stated that the vegetation during the Rich Lake Interpluvial was dominated by widespread areas of grassland and *Artemisia* mixed with some *Juniperus* and *Pinus* elements located in favorable habitats.

Since fossil pollen records are lacking for other areas of Texas during this same time period it is difficult to reconstruct the precise vegetational composition for the rest of the state. On the other hand, we would like to propose a few speculations. A fossil pollen study of deposits in south-central Missouri by Mehringer, *et al.,* (1970) revealed that that region was dominated by an open vegetation consisting of grasses and herbaceous plants with only a few scattered conifers around 30,000 years ago. By projecting a similar pattern for areas south of the Ozarks we find that north and central Texas also might have been covered by a similar scrub and grassland vegetation during the same Middle Wisconsin period. Whether or not this is an accurate reconstruction of the central and north Texas vegetation during this early period is unknown and must await confirmation based on future fossil pollen studies. However, for the present it seems to represent a reasonable possibility.

The vegetation of south and east Texas during the Middle Wisconsin period is an even greater mystery.

Even though no fossil pollen records exist for those regions of Texas it is worth examining the related work in nearby regions of the southeastern United States for possible correlations. Fossil pollen reconstructions from Goshen Springs, Alabama (Delcourt, 1980) and from Nonconna Creek, Tennessee (Delcourt *et al.,* 1980; Delcourt and Delcourt, 1981) suggest that much of the Gulf Coastal plain from Georgia and the Carolinas in the east to regions of southeast Texas in the west may have been covered by an oak-hickory-pine forest. Their reconstruction of the Middle Wisconsin period vegetation suggests moister conditions for regions of east and southeast Texas than those indicated for north and central Texas based on the results of Mehringer's *et al.* (1970) study. If the Delcourts' (Delcourt and Delcourt, 1981) projection for Texas are correct, then it suggests that certain major vegetational zones may have been distributed latitudinally and that a broad grassland zone may have existed in regions of north and central Texas and that a forested zone may have covered the Gulf Coastal regions of east and southeast Texas around 30,000 years ago.

An alternate hypothesis would be that east Texas may have represented an ecotonal region between widespread grasslands and scrubby vegetation to the west and deciduous forests in the southeastern United States. Ecologically, the latter reconstruction is the easiest to hypothesize since it would fit the view that south and east Texas formed transitional vegetational zones between the southeastern flora of the United States and west Texas areas.

WISCONSIN FULL-GLACIAL PERIOD— 22,500-14,000 YEARS B.P.

The fossil pollen records of full-glacial age deposit in Texas reveal evidence of a mesic vegetation, and by inference, a cooler and perhaps wetter climate. In west Texas the grassland steppe conditions of the previous period were replaced in some areas by conifer forests. In southwest Texas there was an expansion of pinyon and juniper parklands while in nearby central Texas a mixed deciduous forest with some conifer elements dominated. In south Texas rich grasslands and oak scrublands formed a mosaic and in north Texas we suspect that a mixed deciduous forest containing spruce formed the region's main vegetation.

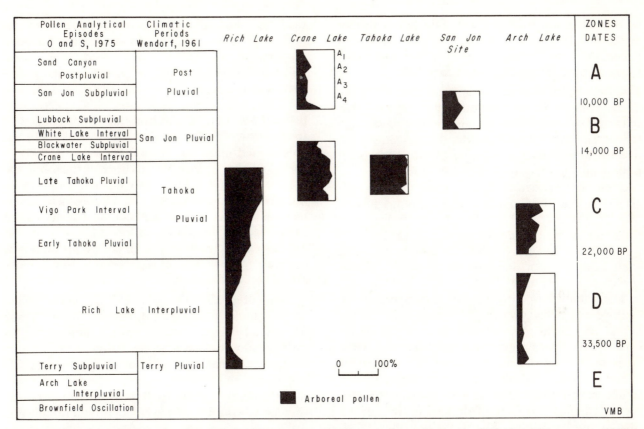

Figure 2. Composite diagram showing the relationship of the pollen analytical episodes as discussed by Oldfield and Schoenwetter (1975) to the climatic intervals defined by Wendorf (1961). Pollen diagrams indicate the AP/NAP ratio determined by Hafsten (1961) in his initial study of deposits on the Llano Estacado. Dates represent determinations made by Hafsten (1961) based on radiocarbon dates and comparisons of his fossil pollen data with fossil records from other nearby sites. Zone designations (A-E) are those originally used by Hafsten (1961).

West Texas

In deposits from Rich and Arch lakes (Fig. 1) located on the Llano Estacado, Hafsten (1961) recovered a fossil pollen record showing a steady and rapid rise in the percentages of both spruce and pine pollen. In conjunction with this rise in conifer pollen was a decrease in grass and herbaceous pollen. Similar pollen records from nearby Crane Lake (Fig. 2) also showed that by 15,000 years ago conifers completely dominated the fossil pollen rain in a large area on, and around, the southern edge of the Llano Estacado. In some sediments, such as those from Tahoka Lake, pine accounted for over 90% of all fossil pollen recovered in certain strata from this time period (Fig. 2). Hafsten (1961) interpreted this information to imply that there was a woodland vegetation of spruce and pine covering much of the Llano Estacado region north of the Pecos River. Hafsten also stated that he believed most of the pines were primarily ponderosa

pine and that although spruce trees were undoubtedly present, they were probably fairly scarce. In addition, Hafsten did not feel that fir *(Abies)* trees were part of the Wisconsin full-glacial period vegetation on any part of the Llano Estacado in Texas.

Oldfield and Schoenwetter (1975) collected new sediment cores and reexamined some of the same full-glacial age playa lake deposits which Hafsten (1961) had sampled and analyzed in the early 1960's. Oldfield and Schoenwetter also sampled new west Texas locations such as White Lake, Illusion Lake, and Vigo Park Lake (Fig. 1). Once they had concluded their pollen analyses, Oldfield and Schoenwetter (1975) divided the west Texas full-glacial period into three distinct pollen analytical episodes (Fig. 2). They termed their earliest full-glacial episode the Early Tahoka Pluvial and noted that its vegetational reconstruction was based primarily upon the fossil pollen records from Vigo Park Lake, Illusion Lake, Rich Lake, White Lake, and Tahoka Lake. Their second episode was called the Vigo Park Interval and its fossil pollen record came mainly from sediment cores taken from Vigo Park

Lake and Illusion Lake. The last of the three episodes is termed the Late Tahoka Pluvial and it was documented mainly from deposits in Vigo Park Lake, Illusion Lake, White Lake, and Crane Lake.

According to Oldfield and Schoenwetter (1975) the Early Tahoka Pluvial vegetation in west Texas consisted primarily of continuous boreal forests over large areas of the Llano Estacado. Based mainly upon fossil pollen records containing over 95% arboreal pollen and percentages of 8-15% spruce pollen, Oldfield and Schoenwetter proposed an early full-glacial vegetation consisting of continuous forests over large areas of the Llano Estacado extending from higher elevations in the north down to approximately 3,000 feet above sea level and as far south as at least Rich Lake (Fig. 1). Based on percentages of fossil spruce pollen around 10% and on available data noting that spruce pollen is highly underrepresented, Oldfield and Schoenwetter (1975) proposed that the boreal forests in the northern part of the Llano Estacado contained a ratio of 40% spruce trees while areas south of the Lubbock Lake Site contained conifer forests composed of approximately 25% spruce trees. Finally, Oldfield and Schoenwetter also predicted that the Early Tahoka Pluvial spruce species on the Llano Estacado was either *Picea engelmanii* or *P. pungens*.

The Vigo Park Interval was designated as a pollen analytical episode by Oldfield and Schoenwetter (1975) because they believed that it reflected a brief warming and drying phase during the west Texas full-glacial period. Reduced levels of arboreal pollen from percentages of over 95% during the Early Tahoka Pluvial, to less than 50% during the Vigo Park Interval led Oldfield and Schoenwetter to suggest that the previous boreal forests were briefly thinned to a conifer parkland containing grasses and a variety of herbaceous taxa. They also saw added support for their parkland hypothesis in the fossil pollen record of Illusion Lake which contained high percentages of Chenopodiaceae pollen during the Vigo Park Interval. Oldfield and Schoenwetter (1975) believed that the Chenopodiaceae pollen rise was closely related to lowered lake levels at Vigo Park which encouraged a rich growth of plants such as *Atriplex* and *Suaeda* along the newly exposed lake margins.

After the brief Vigo Park Interval the vegetational characteristics of the Late Tahoka Pluvial were very similar to those of the Early Tahoka Pluvial with some minor exceptions (Oldfield and Schoenwetter, 1975). High percentages of conifer pollen from deposits at White Lake, Vigo Park Lake, and Illusion Lake led Oldfield and Schoenwetter to suggest that during the Late Tahoka Pluvial episode the northern part of the Llano Estacado, down to at least the Lubbock Lake region, was covered by a closed cover, boreal forest. However, they believed that from that region south towards Crane Lake the boreal forest thinned to a light, less continuous pine parkland at elevations below 2,500 feet. In addition, lower percentages of fossil spruce pollen in the Late Tahoka Pluvial episode led Oldfield and Schoenwetter (1975) to speculate that spruce populations throughout the southern part of the Llano Estacado (south of Lubbock Lake) were thinning and losing their importance as a major vegetational component during the final stages of the full-glacial period. Like Hafsten, they also agreed that the true fir *(Abies)* probably was not present in any area of west Texas during the full-glacial period.

It was originally believed that high conifer pollen counts, similar to the ones discovered in west Texas full-glacial deposits, were reliable indicators of a widespread woodland type vegetation. However, this interpretation has been questioned by Martin (1964) and Martin and Mehringer (1965) who suggested that west Texas full-glacial pollen records may reflect widespread conifer parklands rather than woodlands.

In an extensive study of the modern pollen rain in grassland and conifer woodland areas of eastern Washington and western Idaho, Mack and Bryant (1974) noted that percentages of pine pollen as high as 50% could be recovered in grassland areas approximately 30 miles from the nearest pine sources. In general, however, the average percentage of pine pollen for most of the vast, treeless grassland areas in eastern Washington was 30-40%. In other studies of the modern pollen rain distribution in Washington State, Mack et al. (1978) found that percentages of pine pollen could reach as high as 80% in surface samples collected in conifer parklands composed primarily of pines and grasses. They also noted that percentages of pine pollen could reach as high as 70% in surface samples collected in scrub grasslands where only a few isolated pine trees were present. It should be pointed out, however, that Mack and Bryant (1974) were examining the modern pollen rain distribution in and around the Columbia Basin region of Washington state which is surrounded by extensive conifer forests composed of pines, spruce, Douglas fir and true fir. Thus, long distance transport of conifer pollen was solely responsible for almost all of the elevated percentages of conifer pollen in the nearby grasslands region of the Columbia Basin.

Using the above information as a possible corollary for the west Texas area we would like to suggest a modification of the full-glacial vegetational interpretations proposed by Hafsten (1961) and Oldfield and Schoewetter (1975). The high percentages of fossil pine pollen may, as they suggested, represent a continuous conifer forest covering much of west Texas including the Llano Estacado. On the other hand, we believe that their same pollen data could be interpreted to represent a full-glacial west Texas vegetation characterized by a mixture of conifer forests confined mainly to elevated regions, conifer parklands in the foothill regions, and perhaps areas of conifer savannas and scrub grasslands in the lower elevations of the Llano Estacado.

Southwest Texas

Fossil pollen records from Bonfire Shelter (Fig. 3) provide one of the only potential sources of full-glacial vegetational records in southwest Texas (Bryant, 1969). Although no radiocarbon dates are yet available from the lowermost pollen bearing strata in Bonfire Shelter, we agree with Dr. David Dibble (personal communication) that the rock spalls in those deposits were produced primarily by severe ice wedging that loosened pieces of limestone from the roof and walls of the shelter during the full-glacial period. High percentages of mostly haploxylon pine pollen recovered from those spall zone deposits suggest that during the late full-glacial period southwest Texas may have been covered by a mosaic vegetation containing woodlands, pinyon parklands, and scrub grasslands with some junipers present. We believe that junipers were a component of the southwest Texas full-glacial period vegetation even though its pollen was not recovered at Bonfire Shelter. Juniper pollen, like the pollen of certain other conifers such as Douglas fir, is fragile and does not preserve well in alkaline sediments similar to the ones found in Bonfire Shelter (Holloway, 1981). Therefore, the absence of juniper pollen in the fossil record of southwest Texas does not necessarily mean that juniper trees were not present.

We suspect that the proposed full-glacial conifer forests in the higher elevations of west Texas (Hafsten, 1961; Oldfield & Schoenwetter, 1975) did not invade the southwest Texas region. In addition, the fossil pollen record from Bonfire Shelter also suggests that spruce trees were not present in the southwest Texas region during any portion of the full-glacial period.

A plant macrofossil record of full-glacial deposits has been reported from areas of extreme southwest Texas. This record was derived from fossil wood rat middens found in the Livingstone Hills region of Presidio County (Fig. 1) by Van Devender *et al.* (1978). These deposits were radiocarbon dated as being approximately 16,000 years old and reflect evidence of a pinyon-juniper woodland habitat. That vegetational reconstruction is consistent with full-glacial fossil pollen data from the Llano Estacado to the north (Hafsten, 1961; Oldfield and Schoenwetter, 1975) and information derived from deposits in Bonfire Shelter to the east (Bryant, 1969). According to van Devender *et al.* (1978) the wood rat midden macrofossils were dominated by *Pinus edulis* and were associated with 26 additional species including, *Quercus pungens, Q. hinckleyi, Sophora* nr. *gypsophila, Yucca* cf. *rostrata, Rhus virens,* and *Opuntia imbricata.*

Geological and hydrological studies of sediments on the Llano Estacado by Reeves (1965a, 1965b, 1966) also can be used to infer cool and humid conditions for the full-glacial period in west Texas. Reeves used his resulting data to estimate that during the full-glacial period midsummer temperature averages were at least 5°C. cooler than at present and were accompanied by an annual precipitation rate of approximately 838-863 mm with a maximum annual evaporation rate no higher than 1,117 mm per year (this compares with the present rates of 304-508 mm annual rainfall and an annual evaporation rate of 1,524 mm). He indicated that it would require at least those estimated rates to maintain the lake levels of west Texas playa basins at their known full-glacial depths.

Central Texas

Today, central Texas is not represented by either a uniform soil or vegetation composition (Gould, 1975) and we suspect that some of these same differences probably existed as far back as the late full-glacial period. In the western part of central Texas fossil pollen data are lacking for the Edwards Plateau region (Shaw *et al.,* 1980) which today represents a hilly area of grasslands mixed with junipers, oaks, and mesquite. The eastern part of central Texas is today represented by blackland prairies and post oak savannas yet available fossil pollen data suggest that the region was heavily forested during the full-glacial period.

Full-glacial age fossil pollen records from deposits in central Texas are few in number. The sediments from Boriack Bog, (Figs. 4A and 4B) revealed a

BONFIRE SHELTER 41 VV 218

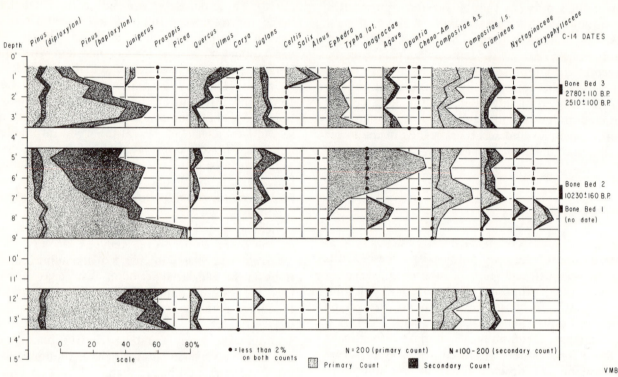

Figure 3. Relative fossil pollen counts from deposits in Bonfire Shelter. Shaded areas represent: 1) an initial primary pollen count of all taxa, and 2) a secondary pollen count in which Compositae and *Ephedra* pollen were excluded. Sampling was conducted at six-inch intervals. Blank areas in the diagram indicate zones which did not contain sufficient fossil pollen for analytical purposes. Primary spall zone is located between the 7.5 - 15 foot depth.

record of *Picea, Populus, Betula, Corylus, Fraxinus, Acer, Alnus, Cornus, Carya, Tilia, Quercus,* and *Pinus* in deposits radiocarbon dated as older than 15,000 yrs. B.P. (Bryant, 1977b). That evidence, in addition to an earlier undated study from nearby Patschke Bog (Potzger and Tharp, 1943, 1947) suggests that the invasion of full-glacial forest elements into the state was not restricted to regions of west Texas.

Boriack Bog full-glacial age peat deposits are dominated by high percentages of alder pollen (Bryant, 1977b). Alder plants can either be shrub species or tree species thus it is possible that the alder pollen found in Boriack Bog could have come from either alder trees or from the shrub species of this plant. However, since alder shrubs are commonly found around the margins of northern peat bogs, we suspect that the alder pollen in Boriack Bog is from shrubby species of this plant. Furthermore, since alder tends to produce great quantities of airborne pollen, its pollen is often overrepresented in any fossil record (Janssen, 1966). The secondary pollen

count (Fig. 4B) shows pollen frequencies when alder pollen is excluded.

Other tree pollen types such as basswood, maple, poplar, and spruce pollen were also present but are not as well represented in the fossil record of full-glacial age deposits in Boriack Bog. This may be the result of under-representation of their pollen in the fossil record. Unlike alder, those plants produce either low quantities of pollen or their pollen grains are fragile and become easily destroyed by microorganisms and other mechanisms of destruction before they can become preserved. We suspect that each of these plant genera were an important component in the regional vegetation of the eastern portion of central Texas during the full-glacial period even though they are weakly represented in the fossil pollen record.

The pine and oak pollen percentages in these early deposits from Boriack Bog range from a low of 4% to a maximum of 15% yet in actuality these plants may have been weakly represented in the vegetation of that time period. Like alder, both taxa produce great

BORIACK BOG
Primary Pollen Count

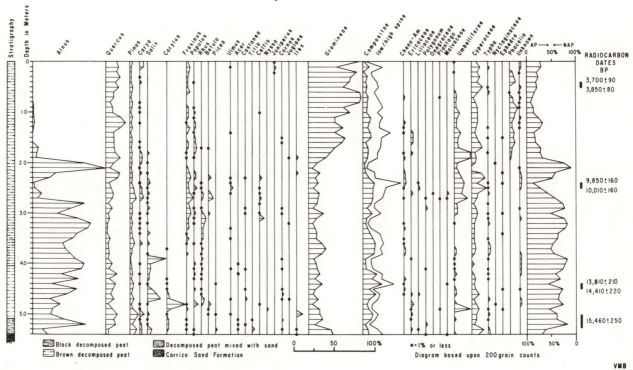

Figure 4A. Primary fossil pollen counts of deposits in Boriack Bog.

quantities of airborne pollen which are easily dispersed and are capable of traveling great distances. In fact, modern pollen studies (Bryant, 1977b) demonstrated that most, if not all, of the fossil pine pollen in Boriack Bog probably came from large stands of loblolly pines located 50 km southwest of the bog near the town of Bastrop. We have found no evidence to suggest that large groups of pines were growing any nearer than 50 km away from Boriack Bog during the full-glacial period.

Fossil *Picea* pollen was found in percentages ranging from 1-3% in the full-glacial age deposits of Boriack Bog (Bryant, 1977b). Potzger and Tharp (1943, 1947) also reported the occurrence of spruce pollen from theoretically equivalent sediments in Patschke Bog. Potzger and Tharp (1943, 1947) assumed on the basis of overall size that those pollen grains represented either *Picea mariana,* or *Picea glauca* which have a present distribution restricted to areas of Canada north of the 21°C mean July isotherm.

Recently, Holloway and Bryant (1984) succeeded in identifying the full-glacial spruce pollen from

Boriack Bog. Based on morphological comparisons made with a scanning electron microscope, they identified the spruce pollen as belonging to *Picea glauca*. Because of the low percentages of spruce pollen in Boriack Bog, however, we feel that spruce trees probably were restricted to a few isolated and protected areas. If we are correct, then their presence infers a mean July temperature decrease of approximately 5.5°C. for central Texas during the full-glacial period.

Implications concerning the paleoenvironmental record of the Texas full-glacial period also can be based upon the presence, or absence, of certain animals which are known to prefer specific types of plant habitats. Slaughter (1963), for example, has found the faunal remains of the giant beaver *(Castoroids)* in northeast Texas full-glacial deposits suggesting that that region could have been forested and may have had a cool or cold full-glacial climate. Further south at Friesenhahn Cave (Fig. 1) Evans (1961) and later Graham (1976) recovered the remains of the long-nosed peccary *(Mylohyus)* which is a forest dweller, the mastodon *(Mammut)* which

BORIACK BOG
Secondary Pollen Count

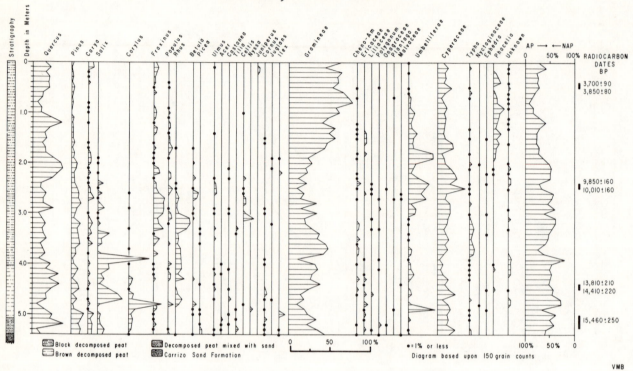

Figure 4B. Secondary fossil pollen counts of Boriack Bog.

preferred forested areas when they were available, and the tapir *(Tapirus)* which is usually associated with humid forest environments. All of those fauna were found in south Texas full-glacial deposits. An analysis of the deposits from Friesenhahn Cave revealed a fossil pollen record (Hall, personal communications) containing high percentages of grass and oak pollen, suggesting a mosaic vegetation of prairie grasslands, woodlands, and oak dominated parklands. Lundelius (1967, 1974) also noted that a number of animal species such as the masked shrew *(Sorex cinereus)* and the bog lemming *(Synaptomys cooperi)* lived in areas of north and central Texas during the full-glacial period and up through the early part of the post-glacial period before they disappeared. Both of these animals currently live in more northern areas of North America and are today found in cool, wet habitats. Their earlier presence, and their current absence, in Texas infers earlier climatic conditions similar to those reflected by the fossil pollen record of Boriack Bog.

In summary, it appears that most of east, central, north, and west Texas was considerably cooler and more humid than today and that those areas were

covered by grasslands, woodlands, and parklands during most of the full-glacial period. Few data are available for south Texas during the full-glacial period but the pollen records from Friesenhahn Cave do contain higher percentages of grass pollen than pollen records of similar age deposits in central, southwest and west Texas. Based upon that information we would conclude that from Bexar County south the full-glacial period vegetation probably was composed of large areas of grass and scrublands and few areas of parklands or woodlands.

LATE GLACIAL PERIOD—14,000-10,000 YEARS B.P.

The late-glacial period in Texas represents a transitional period characterized by a slow climatic deterioration which is noted in the fossil pollen record by the gradual loss of woodland and parkland areas in many regions of the state. In west Texas and in regions of the Llano Estacado in northwest Texas the late-glacial is characterized as a period when existing areas of conifers at the lower elevations were replaced by open grasslands while conifer forests at higher elevations remained more or less stable. In southwest Texas the existing vegetation developed a broad mosaic pattern during the late-glacial period with scrub grasslands beginning to cover larger areas of the landscape at the expense of the remaining pinyon-juniper woodland and parkland regions. In central Texas the deciduous woodland regions began to disappear and were replaced by grasslands and oak savannas. And, in east Texas the existing deciduous woodlands probably lost certain key taxa such as *Picea* and *Corylus* yet the region remained forested with a wide variety of deciduous tree taxa.

West Texas

In west Texas Hafsten's (1961) work at Crane Lake and the San Jon site (Fig. 2) revealed a pollen record characterized by declines in pine and spruce pollen accompanied by rises in grass, composite and Cheno-podiaceae pollen. This led Hafsten to conclude that the late-glacial period in west Texas could be characterized vegetationally as a time when large areas of grasslands developed and climatically as being drier with warmer summers than the preceding full-glacial period. Oldfield and Schoenwetter's (1975) work in the same region resulted in a similar conclusion. Their analysis of late-glacial sediments from Crane Lake, the San Jon site, Tahoka Lake, Illusion Lake, White Lake, Blackwater Draw, Lubbock Reservoir, and the Anderson Basin (Fig. 1) led them to conclude that the late-glacial represented a period during which the vegetation and climate in west Texas began to deteriorate. They noted that although there were minor regional fluctuations, the overall fossil pollen record showed a trend towards a reduction in conifer tree cover with pines and spruce being primarily restricted to small stands in protected locales and to higher elevations. In addition, high frequencies of Chenopodiaceae and composite pollen at some locales and persistent percentages of *Ephedra* (Morman tea) and grass pollen at other locals led them to conclude that there was a general expansion of grasslands at the expense of pine parklands in west Texas during most of the late-glacial period.

Like pollen, plant macrofossils recovered from wood rat middens provide valuable evidence useful for interpreting the late-glacial vegetational sequence in west Texas. Van Devender *et al.* (1977) found macrofossils of spruce and Douglas fir *(Pseudotsuga)* in late-glacial age wood rat nests located in the Guadalupe Mountains region of Texas. The presence of Douglas fir macrofossil remains in those wood rat nests confirms that *Pseudotsuga* was a vegetational

component of the late-glacial period in west Texas even though no traces of its pollen were found in any of the west Texas deposits. Absence of *Pseudotsuga* in the fossil pollen record is not unexpected since the genus produces relatively low quantities of pollen and is highly underrepresented in the pollen rain of the Chisos Mountains (Meyer, 1977). It is also underrepresented when it is a dominant component in a mixed pine, spruce, and Douglas fir forest (Mack *et al.*, 1978). Based upon the above evidence we feel confident that Douglas fir was an important component of the late-glacial vegetation above 2,000 m in regions of west Texas and probably was present in restricted locales at higher elevations on the Llano Estacado. On the other hand, without additional evidence we find it difficult to estimate precisely how widespread Douglas fir may have been in west Texas during the late-glacial period.

Other wood rat middens provided macrofossil evidence that confirms the west Texas late-glacial vegetational sequence reconstructed from the fossil pollen records (Hafsten, 1961; Oldfield and Schoenwetter, 1975). Van Devender *et al.*, (1977) found macrofossils suggesting a forest habitat containing *Pinus, Juniperus, Pseudotsuga, Picea, Quercus, Rubus* (dewberry), and *Arctostaphylos* (manzanita) from wood rat middens at elevations of 2,000 m in the Guadalupe Mountains. In other wood rat middens of the same age at lower elevations (1,420 m) in the nearby Hueco Mountains Van Devender and Riskind (1979) found macrofossils of pinyon pine and juniper but not Douglas fir suggesting somewhat more xeric conditions at lower elevations.

Additional wood rat middens dating between 11,000-12,000 yrs. B.P. at the 2,000 m elevation in the Guadalupe Mountains (Van Devender *et al.*, 1977) contained macrofossils of *Pseudotsuga, Pinus edulis,* and *Ostrya knowltonii* (hophornbeam) but were missing macrofossils of some of the previous more mesic types (*Picea, Rubus,* and *Arctostaphylos*) found in that region several thousand years earlier. Likewise, during the 11,000-12,000 yrs. B.P. interval lower elevation (1,500 m) wood rat middens (Van Devender and Riskind, 1979) contained macrofossils of *Pinus edulis, Juniperus, Quercus, Celtis* (hackberry), *Robinia* (locust), and *Prunus* (plum) suggesting even more xeric conditions at the lower elevations near the end of the late-glacial period.

In summary, the wood rat midden data suggest that around 14,000 years ago higher elevations (2,000 m) in the Guadalupe Mountains were covered by woodlands composed primarily of conifers while elevations approximately 500 m lower contained open woodlands of pinyons and juniper. Several thousand years later around 11,000-12,000 yrs. B.P. the higher elevation (2,000 m) woodlands could no longer support the growth of spruce and contained a variety of less mesic taxa while the lower elevation (1,500 m) vegetation was being replaced by open scrub vegetation suggesting a drier and/or warmer condition than existed at the higher elevations.

Southwest Texas

Fossil pollen records recovered from Bonfire Shelter (Fig.3) in southwest Texas are not well dated below the post-glacial period. Nevertheless, suspected late-glacial deposits at that site reflect regional trends similar to those found in other areas in west Texas (Bryant, 1969). Decreasing percentages of pine pollen accompanied by rises in grass, composite and *Ephedra* pollen just prior to the onset of the post-glacial period at Bonfire Shelter suggest nearby areas of pinyon and juniper woodlands were being reduced in size while the amount of grasslands and scrublands was steadily increasing. This proposed shift in the vegetational composition of the lower Pecos River area during the late-glacial period may have resulted from a variety of factors including a suspected reduction in river discharge and reduced availability of groundwater moisture caused by higher evaporation rates resulting from warmer summer temperatures.

We suspect that much of the vegetational change probably resulted from a reduced amount of available moisture. As the continental glaciers of the previous Wisconsin Period melted and began their retreat northward, wind patterns and the jet stream that flows across the mid-continental United States was probably altered. Such an alteration could have had sweeping effects. For example, Reeves (1973) has noted that many of the pluvial lakes in west Texas, and as far south as Crane Lake in the Lower Pecos region, remained permanent lakes until midway through the late-glacial period. By 12,500 years ago, however, most of them were reduced to seasonal lakes containing water only during parts of each year. Haynes (1975) believes that a reduction in ground water and a lowering of the water table in west Texas and the Lower Pecos region played a key role in the disappearance of the pluvial lakes and the accompanying desiccation of the vegetation in some regions. Reeves (1976) disagrees with Haynes' interpretation and feels that a sharp reduction in precipitation caused by the retreat of the continental ice mass was the primary causal factor in the desiccation of the many west Texas and Lower Pecos pluvial lakes. Regardless of cause, the results are clear; the Lower Pecos region underwent a gradual vegetational change during the late-glacial period which culminated in a reduction of forested regions, which had been prominent during the previous Wisconsin full-glacial period, and an overall thinning of local pinyon and juniper populations.

Central Texas

As in west Texas, the central Texas regional vegetation reflects an apparent steady warming and/or drying trend during the late-glacial period. Fossil pollen records recovered from peat bog deposits at a number of central Texas localities (Fig. 1) indicate that the fragile balance between the open deciduous forests and grassland regions was changing throughout the late-glacial in favor of the grasslands. The pollen records suggest that at first these changes were subtle but that by the end of the late-glacial period there were increased losses of major arboreal taxa as well as losses of major forested areas in the central Texas region.

Potzger and Tharp (1947) were the first to propose a late-glacial age vegetational reconstruction for central Texas based upon fossil pollen records. In their initial study of the undated sediments in Patschke Bog they assumed that the fossil record extended from the full-glacial period to the present. After examining the Patschke Bog fossil pollen record Potzger and Tharp (1947) devoted most of their attention to an interpretation of why boreal conifers *(Picea* and *Abies)* were present in central Texas during the full-glacial period. In a brief published discussion on the fossil pollen sequence from Patschke Bog the authors (Potzger and Tharp, 1947) noted that following the initial cool-wet period, characterized by boreal conifer pollen, was a warm-dry cycle indicated by high percentages of oak and grass fossil pollen. Although we are not certain, we suspect that their warm-dry cycle equates to the late-glacial period.

In the 1950s Potzger and Tharp (1954) reported on two additional fossil pollen records from central Texas bog deposits. In undated deposits from the Gause Bog (Fig. 1) Potzger and Tharp found fossil spruce and fir pollen in the lowest levels followed by rises in oak and grass pollen in subsequent levels. Those data from the Gause Bog convinced them that their initial reconstruction (Potzger and Tharp, 1947) of the central Texas full-glacial vegetation was correct and that suspected boreal elements (spruce and fir) were widespread and not localized. Potzger and Tharp also hoped to gain an additional full-glacial vegetational record from deposits in the Franklin Bog (Potzger and Tharp, 1954) located near the Gause Bog (Fig. 1). On the other hand, after their fossil pollen analysis was completed they found that the Franklin Bog lacked deposits from either the initial cool-moist climatic period or the subsequent warm-dry period.

One of the important late-glacial fossil pollen records comes from the Boriack Bog location in Lee County, Texas (Bryant, 1977b). Although the fossil record from Boriack Bog does not show a sudden or dramatic reduction in the ratio of aboreal to non-arboreal pollen during the late-glacial period, the number of represented arboreal taxa does show a steady reduction. *Picea, Corylus* (hazelnut), *Myrica* (waxmyrtle), *Tilia* (basswood), and *Acer* (maple) all disappear from the fossil pollen record of Boriack Bog during the late-glacial period suggesting that those taxa were no longer components of the regional deciduous forests. In addition, pollen from other arboreal taxa such as *Salix* (willow), *Fraxinus* (ash), and *Betula* (birch) show reductions during the late-glacial period but their pollen remains weakly represented up into the post-glacial period.

The apparent late-glacial loss of arboreal taxa is also seen in the fossil pollen record of Hershop Bog located in Gonzales County, Texas (Larson *et al.,* 1972). Peat deposits radiocarbon dated as being near the end of the late-glacial period in Hershop Bog note an overall reduction in arboreal pollen and a total loss of *Betula* pollen. Detailed transmission electron microscope studies of the fossil birch pollen in Hershop Bog revealed that they belong to *Betula nigra* (Larson *et al.,* 1972). Currently, *Betula nigra* grows in regions of east Texas where the annual rainfall is in excess of 1,016-1,270 mm. In addition, the closest known present stands of *Betula nigra* are 415 km northeast of Hershop Bog. Those data suggest that during the late-glacial period, south-central Texas climates probably were wetter and perhaps cooler than the climate in that region today.

Although undated by radiocarbon methods, the Gause Bog (Fig. 5) contains a fossil pollen record which probably begins during the late-glacial period (Bryant, 1977b). In the lower levels of the Gause Bog the percentage of arboreal pollen exceeds 50% and a variety of arboreal taxa are represented in the fossil pollen record. Somewhat later, at a level estimated to represent the end of the late-glacial period, the Gause Bog pollen record notes the loss of selected mesic and arboreal taxa such as *Picea, Corylus, Tilia,* and *Ostrya/Carpinus* (hornbeam). In addition, the percentages of grass and oak pollen began to rise after the end of the late-glacial period suggesting a gradual replacement of the established deciduous forest with an oak savanna after 10,000 yrs. B.P.

One of the important aspects of the late-glacial period in central Texas is the final disappearance of *Picea* pollen from the fossil record. As mentioned earlier, scanning electron microscopy has determined that the fossil spruce pollen most probably is *Picea glauca* (Holloway and Bryant, 1984). Other reports of spruce in the fossil records of the late-glacial come from nearby Louisiana where Delcourt and Delcourt (1977) and Kolb and Fredlund (1981) have noted spruce pollen, wood, and needles of *Picea glauca* in deposits recovered from the Tunica Hills region. Although not yet found in Texas deposits, both larch *(Larix)* pollen and larch wood were recovered from late-glacial age deposits in the Tunica Hills region of Louisiana (Delcourt and Delcourt, 1977).

North Texas

No late-glacial age fossil pollen records exist for north Texas but some inferences can be drawn from studies conducted in nearby Oklahoma. In a recent reexamination of the Ferndale Bog (Fig. 6) Holloway and Ferring (in preparation) collected sediment cores extending from the late-glacial period to the present. Fossil pollen records from those cores, which date from approximately 12,000 yrs. B.P., contain weak traces of spruce pollen (<1%) and high percentages of grass pollen suggesting that in southeastern Oklahoma, as in central Texas, grasslands were replacing the previous deciduous conifer woodlands. More importantly, these new data by Holloway and Ferring suggest an earlier establishment of late-glacial grasslands in the Ouachita Mountains of Oklahoma than was originally proposed for that region by Delcourt and Delcourt (1977, 1981). Likewise, we suspect that the higher elevations of the eastern Ouachita Mountains may have served as a migration route for boreal conifers as the ameliorating climate allowed an expansion of grasslands into southeastern Oklahoma. We also suspect that by the time of the earliest deposits in the Ferndale Bog (*ca.* 12,000 yrs. B.P.) spruce trees probably were growing no closer than 160 km (Hall, personal communication).

The palynological data from Ferndale Bog are very similar to equivalent age pollen records recovered

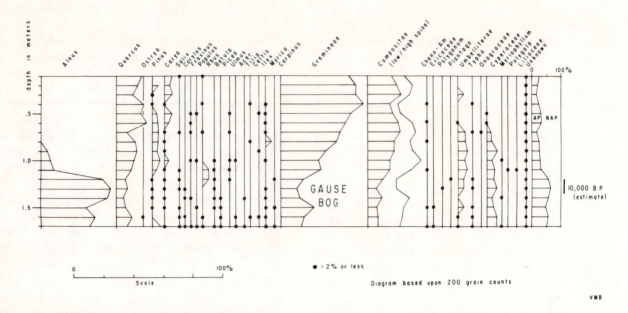

Figure 5. Relative fossil pollen counts from deposits in Gause Bog. Estimated date is based on similarities between the *Alnus* pollen decline as recorded in Boriack Bog (Fig. 4A) and Gause Bog.

from the Domebo Site (Wilson, 1966) located in Caddo County, Oklahoma. At both sites late-glacial pollen assemblages are dominated by low percentages of arboreal pollen and high counts of grass and composite pollen. Trace elements of *Picea* pollen were found in late-glacial deposits at both sites but the meager amounts suggest that spruce trees probably were restricted to a few isolated habitats and soon disappeared from the vegetational communities of both regions. This interpretation differs slightly from Wilson's (1966) original paleoenvironmental reconstruction for the Domebo Site. In his reconstruction of the terminal late-glacial period Wilson (1966) postulated that all of the *Pinus* and *Picea* pollen found in the Domebo Site deposits were products of long distance transport. While we would agree with Wilson's conclusions that the low percentages of *Pinus* pollen (*ca*. 10%) probably came from long distance transport sources, we feel that the *Picea* pollen at the Domebo Site came from local sources. Hall's (this volume) recent study of the modern pollen rain in the southern Rocky Mountains and the southern High Plains indicates that *Picea* pollen drops to less than 3% of the modern pollen rain at a distance of only 16 km from its source. Hall also noted that beyond 16 km from its source spruce pollen consistently contributes less than 2% to the regional pollen rain. Beyond 75 km from its source Hall found that spruce pollen is rarely found in a regional pollen rain. Based on Hall's data and our knowledge that fossil spruce pollen from the Domebo

Site was present in amounts reaching 3%, we suspect that at least a few spruce trees may have grown in locales no further than 16 km from the site during the late-glacial period.

East Texas

Thus far, east Texas sediments have not yielded a single lengthy pollen record. Over the past 20 years I (Bryant) have examined and sampled a significant number of east Texas archaeological sites and swamp locales in an effort to obtain a fossil pollen record from that region. To date none of the examined samples contained sufficient fossil pollen to conduct statistically valid analyses of 200 or more pollen grains per sample.

The apparent absence of fossil pollen in most east Texas locations may result from a variety of factors. First, the soils in that region consist mainly of oxisols and alfisols which are characterized by high rates of oxidation and low percentages of organic matter; conditions under which pollen rarely preserves (Bryant and Holloway, 1983). Second, the high regional rainfall average repeatedly wets and dries the soils of east Texas. As Holloway (1981) has demonstrated under experimental laboratory conditions, cycles of wetting and drying cause structural weakening of the pollen exine and encourages mechanical breakdown of the pollen wall. Third, high rates of

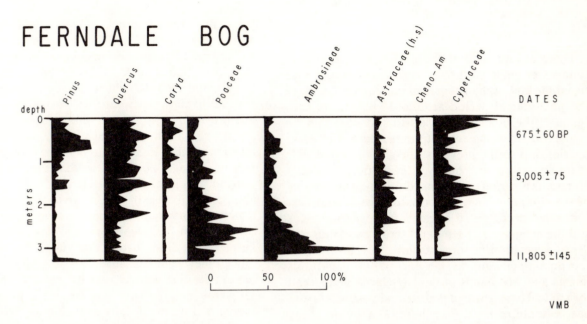

Figure 6. Relative fossil pollen counts from Ferndale Bog illustrating percentages of the dominant pollen types. Percentages are calculated from counts of over 200 fossil pollen grains per stratum.

microbal activity in the leaf litter layer probably causes the loss of certain pollen types and damages others (Golstein, 1960).

Pollen records from the nearby Tunica Hills region of Louisiana, contain fossil pollen of *Picea, Pinus, Quercus,* and other deciduous taxa suggesting that western Louisiana, and perhaps east Texas, still was covered by a deciduous woodland during the late-glacial period. We suspect that the only major late-glacial vegetational changes were in the actual percentages of certain taxa and absence of other genera. For example, we suspect that genera such as *Picea, Corylus,* and *Alnus* either were already absent from those deciduous woodlands or were severely reduced in percentages. Although difficult to prove in the absence of fossil pollen evidence, we suspect that the proportion of certain taxa such as *Acer, Betula, Fagus* (beech), and *Carpinus* may have been reduced while other taxa such as *Quercus, Liquidambar* (sweetgum), and *Pinus* increased.

Other Evidence

The fossil fauna records from late-glacial age deposits in Texas reveal the loss of many animals suggesting that their vegetational habitats and the accompanying climates were changing. Graham (1976) noted that the most significant changes in the prehistoric record of central Texas occurred between his records for Groups I and II fossil fauna. His "Group I" refers to the faunal record up through the late-glacial period and his "Group II" refers to the post-glacial period. The significant changes that he refers to at the end of Group I are the sequential disappearances of animals found today only in areas of North America where climates are considerably cooler and more moist than the present climates of Texas. Graham's list of animals in Group I that disappeared during the late-glacial period in central Texas includes the masked shrew *(Sorex cinereus),* the ermine *(Mustela erminea),* the meadow vole *(Microtus pennsylanicus),* the long-tailed shrew *(Sorex vagrans),* and the hare *(Lepus townsendii).* Similar losses of these and other Pleistocene fauna also were noted for the late-glacial period from regions of north and west Texas (Lundelius 1967, 1974). In deposits at the Lubbock Lake Site (Fig. 1) Johnson (1974) found faunal evidence to suggest that by the end of the late-glacial period that region probably contained large open grasslands. Missing from those terminal late-glacial deposits at Lubbock Lake were certain woodland vertebrates whose remains had been found associated with earlier Lubbock Lake late-glacial deposits but which disappeared just prior to 10,000 years ago. The invertebrate faunal remains recovered from the Lubbock Lake Site (Johnson, 1974) also imply a similar west Texas vegetational and climatic trend towards drier and warmer conditions throughout the late-glacial period.

POST-GLACIAL PERIOD—10,000 YEARS B.P. TO PRESENT

There are no sudden or significant vegetational changes or geologic disconformities in Texas that signal the end of the late-glacial and the beginning of the post-glacial period around 10,000 years ago. Instead, the changes are subtle and only can be visualized when the entire time period is viewed in relationship to the previous late-glacial period. For example, in west and southwest Texas the post-glacial represents a period of continued warming and drying of the climate which is expressed vegetationally by continued decreases in forested and parkland areas and the emergence of semi-arid scrub grasslands and desert vegetation in some regions. In central Texas the post-glacial marks the end of the deciduous forests and woodlands of the late-glacial period and the gradual emergence of prairies and post-oak savannas in some regions. Elsewhere, as they do today (Shaw *et al.*, 1980), juniper and oaks probably dominated in areas west of the Balcones Escarpment where the higher elevations of the Edward's Plateau form a hilly topography. In south Texas the region continued to be dominated by grasslands mixed with oaks in the more favorable habitats during the post-glacial period while in east Texas the pine dominated forested areas of the post-glacial period remained a mixture of pines and junipers mixed with the remnants of the previous late-glacial deciduous woodlands.

Southwest Texas

In southwest Texas the inferred late-glacial mosaic vegetation of woodlands, parklands, and scrub grasslands gradually was being replaced by larger areas of scrub grasslands between 10,000-7,000 years ago. This interpretation is based upon the fossil pollen records at the Devil's Mouth Site (Bryant and Larson, 1968), Eagle Cave (McAndrews and Larson, 1966), Bonfire Shelter (Hevly, 1966; Bryant, 1969), and Hinds Cave (Bryant, 1977a) which show gradual reductions in the percentages of both diploxylon and haploxylon fossil pine pollen coupled with rises in pollen from herbaceous plants and grasses. However, sufficient fossil pinyon pine pollen was recovered from southwest Texas deposits during this 3,000 year interval to suggest that there were still some of those trees growing in protected canyons and in limited upland locales drained by the Pecos River.

Preliminary analyses of plant remains recovered from Hinds Cave and Baker Cave in the Lower Pecos region (Dering, 1977, 1979) demonstrated that by 8,500 yrs. B.P. local aboriginal groups already were exploiting plants such as *Agave, Yucca, Dasylirion* (sotol), and *Opuntia* which are generally associated with fairly xeric environments. Furthermore, Dering did not recover any plant macrofossil remains of pinyon nuts in those deposits which suggests that by 8,500 yrs. B.P. pinyon pines perhaps no longer grew locally. Instead, he suggested that they had already retreated to higher elevations beyond the limits of the aboriginal's food gathering range in the Lower Pecos region. An alternative hypothesis would be that for some reason the prehistoric peoples of the Lower Pecos region did not collect or use pinyon nuts even though they were available. This alternate hypothesis is possible, yet we feel it is improbable that those early groups would have ignored such a valuable food source had it been locally available for exploitation (Shafer and Bryant, 1977).

There are only limited fossil pollen records available from areas of southwest Texas during the next 3,000 year interval from 7,000-4,000 yrs. B.P. The fossil pollen records from Centipede and Damp Cave (Johnson, 1963) are incomplete and based largely upon inadequate pollen counts of less than 200 grains per sample (Barkley, 1934). In spite of those shortcomings, the data from Centipede and Damp Caves must be considered since they represent one of the few fossil pollen sequences available for any portion of that 3,000 year time interval. During his analysis of those deposits Johnson noted that there did not appear to be any dramatic changes in either the vegetational composition or climate of the Lower Pecos region. On the other hand, inspection of the pollen data suggest a progressive deterioration of the previous mesic vegetation, and by inference, an elevation of moisture evaporation rates and/or a reduction in rainfall. Johnson also noted an increase in *Agave* pollen around the end of this 3,000 year period. However, since the *Agave* plant is entomophilous and Johnson's samples came from cultural levels of archaeological sites, a culture bias rather than a climatic shift may be the cause. The rise in *Agave* pollen does support Dering's (1979) suggestion that *Agave* plants were becoming a common member of the regional vegetation after 8,000 yrs. B.P. and thus were available for exploitation by aboriginal groups.

The apparent increasing emphasis upon the use of plants and animals from xeric habitats by people living in the Lower Pecos region during the early post-glacial also has been noted by Williams-Dean

(1978), Lord (1983), and Dean (1984). In Williams-Dean's (1978) study of human coprolites recovered from horizons in Hind's Cave radiocarbon dated as being 6,000 years old, she found a heavy emphasis upon the use of cactus as a primary food source. In a similar study of the faunal bones recovered from Hind's Cave, Lord (1983) noted that between 6,800-6,000 yrs. B.P. there was an apparent shift away from eating deer and rabbits towards the catching and eating of small rodents such as the wood rat and mice. When we combine the plant macrofossil record (Dering, 1979), the pollen record (Johnson, 1963; Bryant, 1977a), the coprolite record (Williams-Dean, 1978) and the faunal record (Lord, 1983), what emerges is a convincing argument that around 6,000 years ago local aboriginal groups were forced to adjust to vegetational and climatic conditions that were becoming increasingly more xeric and drier.

Studies of Hinds Cave deposits include pollen analyses by Dering (1979) and an additional study of eight sediment levels (Bryant, 1977a) that span a record from approximately 8,700-6,000 yrs. B.P. (Fig. 7). The fossil pollen data from Hinds Cave reveal high levels of grass accompanied by low percentages of pine and other arboreal pollen types between 8,700-7,000 yrs. B.P. Also present in these early deposits are significant percentages of economic pollen from semi-arid xerophytic plants such as *Agave, Dasylirion, Opuntia,* and *Prosopis.* These data suggest: 1) the possibility that the local environment already was progressing towards arid conditions and that the regional vegetation could be characterized as containing some areas of grasslands mixed with other areas where thinner topsoil and less moisture favored the growth of desert xerophytes, or 2) that the early deposits at Hinds Cave are unreliable environmental indicators since their fossil pollen records were skewed by the cultural influences of the cave's human inhabitants.

Decreases in the percentages of grass pollen, rises in the percentage of pine pollen, and an overall increase in arboreal pollen between 7,000-6,000 yrs. B.P. may reflect the beginning of widespread soil erosion in the Lower Pecos area caused by reductions in the percentages of plant cover. Likewise, the increase in arboreal pollen during the same 1,000 year interval may be an artifact reflecting higher rates of long distance transported pollen rather than indicating climatic changes favoring the increased growth of mesic arboreal plants. An alternative hypothesis is that changes in the cultural use of grass (*i.e.,* bedding, matting, linings for cooking pits, etc.) within Hinds Cave may have caused a reduction in

the percentage of introduced grass pollen thereby allowing higher percentages of the natural pollen rain to be represented in the deposits of this later time period.

Until more paleoenvironmental data can be obtained from additional fossil pollen records for the time span between 7,000-4,000 yrs. B.P., we must rely upon other types of evidence that already exist. This other evidence consists primarily of data which suggest that areas of southwest Texas along the Rio Grande and Pecos Rivers were subjected to intervals of severe flooding during much of this 3,000 year period (Patton, 1977). These periods of erosion and flooding are clearly marked in the alluvial terraces and sediments of archaeological sites in the Lower Pecos region such as the Devil's Mouth Site (Johnson, 1964) located on the banks of the Rio Grande and Arenosa Shelter (Dibble, 1967) located on the alluvial terraces of the Pecos River. Of the 22 identified floods in the deposits at Arenosa Shelter dating from 4,500 yrs. B.P. to the present, ten of them occurred between 3,200-4,500 yrs. B.P. (Patton, 1977).

The causes of erosion and flooding during this 1,300 year period are not fully understood. One possible explanation is that perhaps minor rises in summer temperatures or short periods of drought may have led to partial removal of the upland vegetation thereby causing increased rainfall runoff resulting in higher levels of river discharge. Another possible explanation could be an increase in precipitation during the later portion of this period, as suggested by Haynes (1968). If true, the possible increase in annual precipitation may have been caused by a series of active frontal systems moving further inland than usual. Although no direct paleoclimatological nor climatological evidence exists that would suggest this type of frontal system phenomenon occurred in the past, these storm systems do move through the Lower Pecos area today and can release great amounts of moisture in a short period of time. During the summer of 1975, for example, a frontal system released over 13 cm of rainfall in the Lower Pecos region in less than eight hours. The resulting runoff filled many streams that were normally dry and caused significant erosional activity along the alluvial banks of many ephemeral streams and the Rio Grande River.

Still another possibility for the widespread alluvial erosion between 3,200-4,500 yrs. B.P. could have come from short periods of intense rainfall associated with the aftermath of a hurricane that may have moved unusually far inland from either the Pacific of

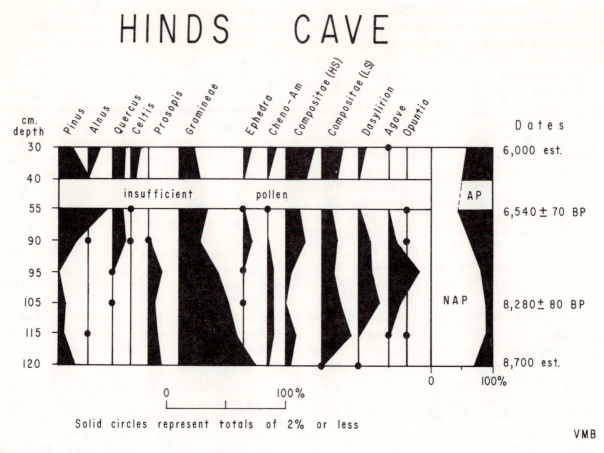

Figure 7. Relative fossil pollen counts recovered from deposits of Area D of Hinds Cave. Percentages are based on fossil pollen counts of over 200 grains per stratum. Dates represent ^{14}C dates of those deposits.

Gulf Coast area. An example of that phenomenon occurred in 1954 when Hurricane Alice moved inland over the Rio Grande Valley. The resulting rainfall almost completely filled the near empty Falcon Reservoir (located between Laredo and McAllen, Texas) in less than one week. Based upon available records, Patton (1977) calculated that at the height of that 1954 flood Areonsa Shelter, located on the Pecos River, was more than nine meters under water and that the normally shallow Pecos River reached a peak depth of over 24 m.

The last 4,000 years of the post-glacial period in southwest Texas is represented by fossil pollen records indicating a gradual and continual trend towards increased aridity. Only once around 2,500 years ago was this apparent trend interrupted. At Bonfire Shelter and at the Devil's Mouth Site fossil pollen records dating from around 2,500 years ago show marked increases in the percentages of both pine and grass pollen suggesting a brief return of somewhat cooler and more mesic conditions (Bryant and Larson, 1968). This apparent mesic interval,

however, was short-lived and soon the trend toward increased aridity was resumed and has continued in southwest Texas until the present.

West Texas

The post-glacial period pollen studies from west Texas reveal a record of ever increasing xeric conditions similar to those recorded for the post-glacial of southwest Texas. Fossil pollen records containing higher percentages of *Artemisia* (sagebrush), other composites, grass, *Ephedra,* and Chenopodiaceae pollen coupled with reduced percentages of *Pinus* pollen recovered from deposits in Crane Lake, the San Jon Site, and Juan Cordova Lake led Hafsten (1961) to suggest that the west Texas post-glacial climatic sequence could be characterized as drier and warmer than the previous late-glacial period. Specifically, Hafsten noted that in an area extending from the watershed of the upper Pecos River north to areas on the Llano Estacado the post-glacial period vegetation probably responded to a continual trend towards drier and warmer conditions by developing into a dry steppe environment dominated by grasses mixed with chenopods, composites, *Artemisia,* and *Ephedra.* He also suspected that most of the *Quercus* and *Juniperus* trees continued to become more restricted to a few west Texas locales and were not common in the grassland areas. Hafsten (1961) indicated that by the late post-glacial period most of the arboreal vegetation in west Texas probably was confined strictly to higher elevations, protected canyons, and along major river drainages.

In the west Texas fossil pollen records examined by Oldfield and Schoenwetter (1975) they began their post-glacial sequence with the Yellow House pollen analytical episode and ended it with the Sand Canyon Postpluvial episode. During the post-glacial period, which Oldfield and Schoenwetter reconstructed from fossil pollen samples collected at the San Jon Site, White Lake, Lubbock Reservoir, and the Blackwater Draw sites, they noted gradual declines in the frequencies of pine pollen and increases in the percentages of grass, composite, chenopods, and *Ephedra* pollen. Those data led them to suggest that the post-glacial period on the Llano Estacado could be characterized by gradual, but continuous, declines in the extent of existing pine parklands coupled with the spread of stable grasslands mixed with some herbaceous plants and scrub vegetation.

Another important aspect regarding the understanding of the west Texas post-glacial period is the interpretation of vegetational conditions relating to the Lubbock Subpluvial pollen analytical episode. According to Wendorf (1970), the Lubbock Subpluvial episode is radiocarbon dated to a time period between 10,300-10,600 yrs. B.P. which would place it at the very end of the late-glacial. In deposits of Lubbock Subpluvial age at the Lubbock Lake Site Oldfield and Schoenwetter (1975) found very high percentages of pine in association with some spruce pollen. However, in similar deposits assigned to the Lubbock Subpluvial episode at White Lake, Crane Lake, Rich Lake, and the San Jon Site Oldfield and Schoenwetter's (1975) analyses revealed no sudden rises in the percentage of pine pollen which instead was only moderately represented in those sediments. In their subsequent interpretation of the Lubbock Subpluvial episode they state that the unusually high percentages of pine pollen in some of the deposits assigned to the period probably represented and artificial overrepresentation caused by selective destruction of pollen types other than pine thereby

leaving pine pollen as one of the only types which remained preserved (Oldfield and Schoenwetter, 1975).

Two recent studies may help to clarify the reasons why high percentages of pine pollen were found in Lubbock Subpluvial sediments at the Lubbock Lake Site. Current chronological and stratigraphic studies (Holliday *et al.,* 1983) indicate that the Lubbock Lake Site samples originally analyzed by Oldfield and Schoenwetter (1975) were collected from a buried soil A horizon that formed on top of substratum 2B. According to Holliday *et al.* (1985), the buried soil A horizon represents a stable land surface that was exposed between 6,000-8,000 yrs. B.P. and should not be confused with, or included with, the immediate underlying sediments associated with the Lubbock Subpluvial episode. In a related study of Lubbock Lake sediments Steve Hall (personal communications) found that soil samples collected from the buried soil A horizon in substratum 2B contained primarily pine pollen which was badly degraded suggesting that other pollen types may have been missing and that only the most durable types remained preserved. Based on his analysis from those samples Hall, like Oldfield and Schoenwetter (1975), believed that the high conifer pollen percentages resulted primarily from over-representation created by differential pollen preservation and should not be used for vegetation or climatic reconstructions.

If the assumption is correct that the original pollen samples collected from the Lubbock Lake Site came from a buried soil A horizon located on top of substratum 2B and not from a discrete sedimentary unit of stratum 2B, as noted by Wendorf (1970), then the original argument made for the presence of a spruce and pine boreal forest on the Llano Estacado during the Lubbock Subpluvial is invalid. Instead, we feel that it is more appropriate to base an interpretation of the vegetational conditions during the Lubbock Subpluvial episode on the fossil pollen records collected and analyzed from Crane Lake, White Lake, the San Jon Site, and Rich Lake (Hafsten, 1961; Oldfield and Schoenwetter, 1975) which show an absence of forested conditions on the Llano Estacado at the very end of the late-glacial period.

Hafsten's (1961) fossil pollen data from Crane Lake record what appears to be a brief mesic interval characterized by increases in the percentages of pine and composite pollen during the deposition of his post-glacial sub-zone A$_2$ (Fig. 2). In explaining the pine and composite pollen rise Hafsten stated that he felt those increases resulted strictly from changes in the percentage of pollen from long-distance transport sources rather than from any type of local or regional climatic change. Later, when Oldfield and Schoenwetter's (1975) study of fossil pollen from the Blackwater Draw region was completed they also noted a similar rise in pine and composite pollen in the jointed sand levels which they chronologically assigned to a time period younger than 4,000 yrs. B.P. Unlike Hafsten, Oldfield and Schoenwetter suggested that the brief pine and composite pollen maximum reflected an environmental fluctuation which only temporarily interrupted the overall regional climatic trend towards increased aridity.

Although difficult to prove, we feel that the late post-glacial pine pollen maximum recorded in the fossil records of the Devil's Mouth Site, Bonfire Shelter, Crane Lake, and Blackwater Draw all reflect a brief, but fairly widespread, mesic interval around 2,500 yrs. B.P. One final point worth noting is that it was during that suspected mesic interval that a large number of *Bison bison* were killed at Bonfire Shelter in at least three separate bison jumps (Dibble and Lorraine, 1968). It is possible that the mesic interval around 2,500 yrs. B.P. may have caused heavier than usual snowfall in the grasslands of the Llano Estacado which in turn forced bison from that region further south into the Lower Pecos watershed in search of winter feeding grounds. Once in the Lower Pecos area the bison would have been available for exploitation by local aboriginal groups or by nomadic hunting bands which may have followed the bison southward into the region.

Although located in the Guadalupe Mountains near Carlsbad, New Mexico, the fossil pollen study from Pratt Cave (Bryant, 1983) contains a limited chonology of post-glacial vegetation in that region near the Texas border (Fig. 1). The Pratt Cave deposits were not radiocarbon dated yet based upon the recovered data Shroeder (1983) has suggested that the cave deposits are probably no older than 2,500 yrs. B.P. If Shroeder's age estimate is correct, then the fossil pollen sequence from those deposits can be utilized to demonstrate that protected areas at higher elevations in west Texas remained wooded throughout the last 2,500 years of the post-glacial period. Bryant's (1983) fossil pollen assemblage from Pratt Cave reflects evidence of a fairly stable vegetation dominated by *Quercus* and *Arbutus* (madrone), and to a lesser degree by *Celtis, Juniperus,* and *Fraxinus* in protected canyons where groundwater moisture was present throughout most of the year. The fossil pollen data also suggest that grasses, *Artemisia,* other composites, and a variety of herbaceous plants

dominated the drier, upland regions above the canyons. As a single record consisting of only 10 fossil pollen spectra the Pratt Cave record cannot be used effectively to make sweeping interpretations about the late post-glacial vegetation in the Guadalupe Mountain region of the Llano Estacado. On the other hand, the Pratt Cave fossil pollen record does reinforce our conclusion that there were no major post-glacial climatic or vegetational changes indicated in west Texas after the end of the brief mesic period around 2,500 yrs. B.P.

South Texas

No post-glacial fossil pollen records exist for the region of south Texas. Although parts of south Texas are currently incorporated within the Gulf Prairie and Marshes vegetational zone as defined by Gould (1975), the majority of this region presently consists of broad grasslands mixed with scrub vegetation dominated by acacias and mesquite. The only exception to this vegetational pattern is seen along the banks of major streams and rivers and in some shallow canyon areas where riparian vegetation consisting of *Carya, Salix, Celtis, Quercus,* and *Platanus* (sycamore) flourish.

The south Texas region lacks peat bogs, suitable lakes containing lengthy deposits, and dry caves where palynologists might be able to obtain sufficient fossil pollen to reconstruct vegetational chronologies. In addition, the high soil pH, low soil organic content, and poorly drained soils of the south Texas region have yielded only meager traces of badly degraded fossil pollen in quantities too small for analytical purposes from a large number of test samples collected at archaeological sites and processed by us.

Several non-palynological studies are available for the region of south Texas and they can be used to formulate a few generalized statements about the post-glacial vegetation. Holloway (1985) has examined a number of charcoal samples recovered from archaeological deposits spanning the last 6,000 years in the Choke Canyon region (Fig. 1) south of San Antonio, Texas. In that study Holloway found that the primary fuel sources used by local aboriginal groups consisted of *Acacia* and *Prosopis* (mesquite). To a lesser degree these same aboriginal groups also used firewood from riparian sources which included *Salix* (willow), *Carya* (pecan), and perhaps *Diospyros* (persimmon). Based upon those findings Holloway (1985) suggested that during the last 6,000 years of the post-glacial period the Choke Canyon region of south Texas contained a vegetation very similar to that area's current vegetation and that no apparent major vegetational changes occurred during that period.

Steele's (1985) analysis of faunal remains recovered from some of the same Choke Canyon archaeological sites studied by Holloway (1985) revealed a mixture of animal usage by aboriginal groups during the past 6,000 years. Recovered faunal remains from those sites included taxa such as the raccoon *(Proyon locto)*, opposum *(Didelphis virgineanus)*, and muskrat *(Ondatra zibethicus)* which represent types generally associated with wooded areas similar to some of the current south Texas riparian habitats. In addition, other faunal remains from those same archaeological sites came from species such as the pronghorn *(Anilocapra americana)*, bison *(Bison bison)*, peccary *(Dicotyles taja)*, and jackrabbit *(Lepus californicus)* which reflect a grassland and scrub habitat similar to what we currently find in the drier regions of south Texas away from riverine locales.

In summary, we find that we must base our post-glacial interpretations of the south Texas vegetational chronology upon a very limited amount of information. We suspect that although the climate may have been quite stable during the post-glacial period in the southern portion of Texas, slight changes in temperature, combined with different weather and rainfall patterns, could have created significant temporary alterations in local and regional post-glacial vegetational communities. Fossil pollen studies of the Northern Chihuahuan Desert (Meyer, 1972, 1973; Brown, this volume) and the Cueva de la Zona de Derrumbes Site (Fig. 1) located several hundred miles south of the Rio Grande in the state of Nuevo Leon, Mexico, show a record of vegetational stability in that region during the last 5,000 years of the post-glacial period (Bryant and Riskind, 1980). In addition, the Bryant and Riskind study noted that present south Texas and northeastern Mexico vegetations are sensitive to change and are controlled to a great extent by factors of temperature, exposure, elevation, moisture availability, and edaphic conditions. Thus, Steele's (1985) data, like Holloway's (1985) data, reflect evidence for a mosaic post-glacial vegetational pattern in south Texas consisting of some limited wooded areas dispersed within a larger grassland and scrub vegetational zone.

Gunn *et al.* (1982) have recently proposed an alternative model for climate change in the region of south Texas. Utilizing current climatic records as their data base and projecting these trends into the past, Gunn *et at.* (1982) have proposed a series of alternating wet and dry periods which may have been operating to control Holocene climatic variations, and thus vegetation. Much of their empirical evidence is based on phytolith data (Robinson, 1979, 1982), which relies heavily on the identification of a definable Hypsithermal interval in south Texas. However, neither the radiocarbon dated pollen sequences from central Texas (Graham and Heimsch, 1960; Bryant, 1977b) nor the plant macrofossil records (Holloway, 1985) nor the faunal (Steele, 1985) evidence from south Texas support the identification of such a discreet Hypsithermal unit within that geographical area. Instead, the central Texas pollen and plant macrofossil record, and faunal data from south Texas suggest a gradual trend toward aridity for at least the last 6,000 years. Based on the combined evidence of charcoal, fauna, and pollen data we suggest that a re-evaluation of the Gunn *et al.* (1982) model is warranted.

Central Texas

Like other regions in Texas, the post-glacial period in central Texas reflects a gradual warming and drying trend throughout the past 10,000 years. Post-glacial sediments from peat deposits (Fig. 1) in the Patschke Bog (Potzger and Tharp, 1947), Franklin Bog (Potzger and Tharp, 1954), Gause Bog (Potzger and Tharp, 1954; Graham and Heimsch, 1960; Bryant, 1977b), Soefje Bog (Graham and Heimsch, 1960), South Soefje Bog (Bryant, 1969), Hershop Bog (Larson *et al.,* 1972), and Boriack Bog (Bryant, 1977b) all show overall declines in arboreal pollen types (except *Quercus*) coupled with increases in herbaceous and grass pollen.

The fossil pollen evidence consisting of stable percentages of *Quercus,* grass and composite pollen from the southern portion of central Texas suggests that a remarkably stable vegetation, similar to the present vegetation, was in place throughout most of the post-glacial period. Graham and Heimsch (1960) were the first to conduct fossil pollen studies in that region. Their analysis of the Soefje Bog consisted of a 4.7 m core which was radiocarbon dated as being slightly younger than 8,000 yrs. B.P. Although their fossil pollen record contained minor fluctuations in most pollen types, the overall percentages of each pollen taxon remained fairly constant throughout the entire span of their 8,000 year record. Those data led them to conclude that the present oak-hickory savanna vegetation in that region was probably established fairly quickly after the end of the late-glacial period and that it has remained essentially unchanged ever since. Bryant (1969) revisited the Soefje Bogs during the mid 1960's and extracted an undated 4.0 m core from the South Soefje Bog (Fig. 8) located 800 m south of the original Soefje Bog examined by Graham and Heimsch. Bryant's (1969) results corroborated the earlier study by Graham and Heimsch but revealed no new information that might alter Graham and Heimsch's (1960) original interpretations.

Larson *et al.* (1972) examined a 5.4 m peat core from the Hershop Bog located approximately 3 km southwest of the Soefje Bogs. Their core was radiocarbon dated as being slightly older than 10,000 yrs. B.P. and contained a fossil pollen record similar to the ones obtained by Graham and Heimsch (1960) and Bryant (1969). Like the previous studies, Larson *et al,* (1972) concluded that during the last 10,000 years the south-central Texas vegetation was a mosaic of fairly stable grassland savannas mixed with pockets of oaks and hickory in the uplands and areas of riparian woodlands along the banks of major rivers. They also noted that their pollen record revealed no indication of any major plant migrations either into, or out of, the south-central Texas region during the post-glacial period.

The paleoenvironmental record in central Texas has received repeated attention during the past 40 years. In the earliest study, Potzger and Tharp (1943) published a brief article noting the presence of *Picea* and *Abies* pollen in undated peat deposits which they believed were from the full-glacial period. Later, Potzger and Tharp (1947) completed their study of a fossil pollen record recovered from the Patschke Bog. In that study they presented a four-stage climatic sequence for central Texas which ended with a warm-dry post-glacial interval characterized by high percentages of grass pollen coupled with increases in both *Quercus* and *Carya* pollen. After locating and coring several additional central Texas bog deposits, Potzger and Tharp (1954) published the last of their three articles on the paleoenvironment of that region. Fossil pollen recovered from the Gause Bog and the Franklin Bog convinced Potzger and Tharp (1954) that their initial vegetational sequence was correct and that the uppermost portion of all three central Texas bog deposits represented the post-glacial period which they characterized as being warm-dry with a predominant savanna vegetation of grasses mixed with some *Quercus* and *Carya* trees.

Bryant's (1977b) palynological re-analysis of the Gause Bog (Fig. 5) and analysis of the Boriack Bog (Fig. 4) presented new information concerning the post-glacial vegetational sequence in central Texas. Based upon the recovered fossil pollen in both bogs Bryant (1977b) concluded that the central Texas post-glacial period was characterized by a continual progression towards less mesic conditions. He derived that interpretation based on fossil pollen data showing overall decreases in most arboreal pollen types (except *Quercus*) and corresponding rises in non-arboreal pollen such as composites and grass. In addition, Bryant (1977b) concluded that the establishment of the present oak-savanna vegetation in central Texas (Gould, 1975) probably occurred no later than 3,000 years ago and that the vegetation in that region has remained fairly stable ever since. However, as discussed later in this article, recent pollen analyses of deposits in Weakly Bog (Holloway *et al.,* 1985) suggest that the establishment of the present oak-savanna vegetation in central Texas may not have occurred until around 1,500 years ago. Finally, Bryant (1977b) also noted that, unlike the southwest Texas and west Texas pollen records, the central Texas bog deposits did not reveal any evidence of a brief mesic period occurring approximately 2,500 years ago.

Unfortunately, many of the central and south-central Texas peat bog deposits do not contain a complete fossil pollen record for the last portion of the post-glacial period. Larson *et al.* (1972) noted that the upper samples in Hershop Bog representing the last 2,000 years of the post-glacial period did not contain preserved pollen. In Bryant's (1977b) study of Boriack Bog his uppermost radiocarbon date of 3,400 yrs. B.P. comes from the 40 cm level suggesting that a record of the last several thousand years of deposition is missing. Gause Bog deposits that were reexamined by Bryant (1977b) also lacked samples from the late post-glacial period. The inability to obtain a late post-glacial fossil pollen record from any of those bog deposits is attributable to different reasons. Attempts to drain the Hershop Bog during the 1950's areated the upper portion and led to a complete loss of those fossil pollen assemblages. Draining of the upper part of the Boriack Bog during the 1940's led to an uncontrolled surface fire that burned for days before finally being extinguished by locally heavy rains (C. Boriack, personal communications). And, commercial peat mining operations in central Texas during the 1940's focused on a number of bog locations including the Gause Bog (originally called Atkinson Bog) where an estimated 3-4 meters

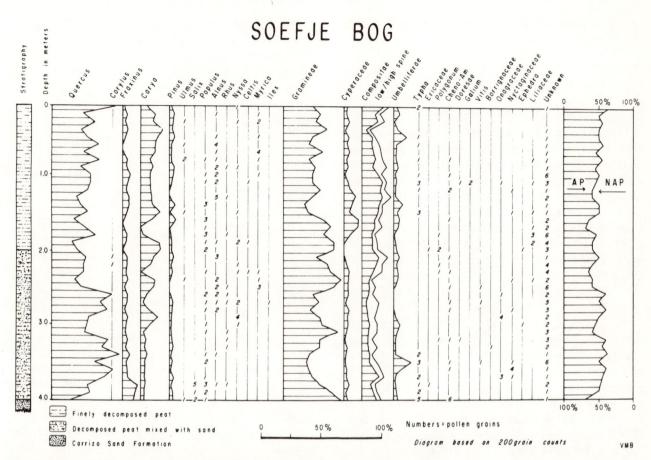

Figure 8. Relative fossil pollen counts from deposits in South Soefje Bog. Percentages are based on counts of 200 grains per stratum.

of peat were removed from the upper portion of the bog and sold to plant nurseries as bedding material (Plummer, 1945).

Fortunately, the recent fossil pollen analysis of Weakly Bog (Fig. 1), located in Leon County, Texas, by Holloway *et al.* (1985) can now be used as a key to central Texas vegetational conditions during the late post-glacial period. A radiocarbon date from the basal portion of the 1.4 m Weakly bog core revealed that the deposits span the last 2,400 years of the post-glacial period. The relative fossil pollen counts from those deposits indicate the same types of fossil pollen spectra as are seen in the uppermost levels of the nearby Boriack Bog. The relative pollen frequencies would have been interpreted as representing a stable oak-savanna vegetational habitat during the last 2,400 years were it not for the pollen influx analyses of those same deposits. Based upon pollen influx studies, Holloway *et al.* (1985) noted, in the peat segment from Weakly Bog dated from 2,400-1,500 yrs. B.P., that *Quercus* pollen was represented by very high influx values. After 1,500 yrs. B.P. the influx values for *Quercus* decreased rapidly and were replaced by high influx values of grass pollen. These

influx data suggest that from 2,400-1,500 yrs. B.P. the central Texas vegetation probably consisted of an open woodland dominated by oaks with some open grassland areas. However, around 1,500 yrs. B.P. the present oak-savanna vegetation of central Texas established itself and replaced the previous oak woodlands in that region. That proposed vegetational shift is interpreted to reflect a prolonged period of drier, and perhaps warmer, climatic conditions in central Texas after 1,500 yrs. B.P.

Related data by Patrick (1946) suggest the vegetational shift, and inferred climatic change, noted at Weakly Bog (Holloway *et al.,* 1985) was regional, rather than local. In the study of the diatom frustules recovered from sediments in Patschke Bog, Patrick noted a shift from mesic to more xeric types in a stratigraphic zone four feet below the bog surface. Based upon an examination of the fossil pollen data from that same zone by Potzger and Tharp (1947), we suspect that the four-foot level represents a time period between 1,500-2,000 years ago. If so, then the diatoms in Patschke Bog, like the plants represented by the pollen deposits in Weakly Bog, responded to the same suspected environmental changes in central Texas around 1,500 yrs. B.P.

North Texas

North of the Texas border in Atoka County, Oklahoma, Albert (1981) examined deposits recovered from Ferndale Bog which spanned the last one-half of the post-glacial period. In conducting her analysis, Albert (1981) extracted and analyzed several peat cores which she correlated incorrectly on the basis of radiocarbon chronology and site stratigraphy (Hall, 1983). In spite of those discrepancies, her overall data reveal that from approximately 5,000 yrs. B.P. until 1,700 yrs. B.P. the fossil pollen record is dominated by high percentages of grass and herbaceous pollen with some *Quercus* and *Carya* pollen present. Although minor fluctuations in pollen types occurred during that 3,300 year record, the overall trend reflects vegetational stability with no indication of major plant taxa migration. By 1,700 yrs. B.P. the fossil pollen record shows a shift towards higher percentages of pine pollen suggesting a migration of *Pinus* into the regional vegetation in the vicinity of Ferndale Bog. From that time period until the present the fossil pollen chronology from Ferndale Bog records a series of fluctuations between the percentages of *Pinus, Quercus,* and *Carya* pollen. Although Albert (1981) interpreted those fluctuations to represent repeated climatic changes as evidenced first by the increase of *Pinus,* then an expansion of *Quercus,* and finally a dominance of *Carya,* a careful examination of Albert's original data reveals that those fluctuations are associated with layers of charcoal in the bog sediments. We suspect that the fluctuations in arboreal taxa, which Albert attributed to climatic variations, are instead a series of secondary vegetational secessional patterns resulting from local forest fires.

Recently, Holloway and Ferring began a reinvestigation of the sediments in Ferndale Bog. Their testing of the bog deposits resulted in the collection of several new sediment cores over 3 m in depth. Preliminary dating of those sediments revealed a radiocarbon chronology extending from the present back into the middle late-glacial period. Fossil pollen extracted from the Ferndale Bog sediment cores indicate that for several thousand years after the end of the late-glacial period the upper elevations of the western Ouachita Mountains in Oklahoma probably were dominated by a grassland vegetation. Those new palynological data confirm the original late-glacial

vegetational reconstruction proposed for central Oklahoma by Wilson (1966) in his report of the Domebo Site and support some of the conclusions mentioned in the Ferndale Bog report by Albert (1981).

Fossil pollen data from the new studies of Ferndale Bog by Holloway and Ferring reveal that by approximately 9,000 yrs. B.P. (Fig. 6) *Quercus* began invading the grasslands in the Ferndale Bog region and created first an oak savanna, later an oak woodland (which apparently lasted until pines began migrating into the region around 2,100 yrs. B.P.), and finally an oak-hickory-pine forest after about 1,200 yrs. B.P. One of the reasons we believe that the oak-hickory-pine forest did not become established until 1,200 yrs. B.P. is that the percentages of pine pollen are moderate and fairly stable between 2,100-1,200 yrs. B.P. and it does not increase dramatically until after 1,200 yrs. B.P.

In spite of some radiocarbon dating discrepancies, both Albert's (1981) fossil pollen study and the renewed study by Holloway and Ferring show the same general series of vegetational events spanning the last portion of the post-glacial period at Ferndale Bog. Both suggest that an initial post-glacial grassland vegetation was replaced first by an oak savanna, then an oak woodland by mid post-glacial times, and finally by a migration of pines into the region around 1,200 yrs. B.P. However, because the percentage of fossil pine pollen in the deposits of Ferndale Bog never excedes 30%, and the percentage of oak pollen remains high throughout the late post-glacial period, we suspect that the region remained primarily an oak-hickory-pine forest where pines were present but never became a major vegetational element (Davis, 1969).

During the past several decades there have been a number of attempts to recover fossil pollen records from archaeological sites in north and central Texas (Bryant, 1969; Edelman, 1976; Lynott and Peter, 1977; Lynott, 1975; Gallager and Beardon, 1976; Hall, 1980, 1982a, 1982b). For the most part, these attempts have failed to recover sufficient fossil pollen for statistically valid analyses. Two notable exceptions have been Woosley's pollen analytical work from both the Cobb-Poole site (Raab and Woosley, 1982) and from the Frossard site (Raab, *et al.,* 1982). In the pollen analysis of the Cobb-Poole site surface sediments and deposits at the 78 cm level, Raab and Woosley (1982) reported the only known occurrence of *Larix* pollen in Texas Quaternary sediments. In other samples their recovery of *Scirpus,* Juncaceae, and Nymphaceae pollen from Cobb-Poole sediments dating from approximately 1,500 yrs. B.P. led them to suggest the presence of widespread swamp and marsh habitats in that region of north Texas during the late post-glacial period. And, finally, they concluded from their percentages of fossil oak and grass pollen that the late post-glacial vegetation in north Texas consisted primarily of an oak savanna.

At the second of these sites, the Frossard Site, Woosley's fossil pollen, counts included an unusually large variety and high amounts of entomophilous pollen, many of which represented plants generally restricted to very mesic habitats (Raab, *et al.*, 1982). Subsequent pollen studies at the Frossard Site by Holloway (n.d.) in 1982 focused on the analysis of a second sediment column collected from the same exposed profile that was originally sampled and analyzed by Woosley (Raab *et al.*, 1982). Holloway's (n.d.) analysis of the second sediment column revealed the presence of only a few badly degraded pollen grains in contrast to statements in the Frossard Site report by Raab *et al.* (1982:99) which noted, "Good grain preservation and high grain counts were obtained from all samples drawn from the project." In addition, Holloway's (n.d.) study failed to find the large variety of pollen taxa mentioned in the Frossard Site report (Raab *et al.*, 1982). In an effort to resolve the apparent discrepancy between pollen data presented in the Frossard Site report (Raab *et al.*, 1982) and the pollen data presented by Holloway (n.d.), Bryant was asked to examine four of the original pollen residues processed and counted by Woosley as part of her analysis of the Frossard Site. Bryant's reexamination of those four samples failed to confirm Woosley's original data and supported results similar to those noted by Holloway (n.d.) in his studies of the second sediment column from the Frossard Site.

Since no other fossil pollen evidence dating from the late post-glacial period in north Texas has been able to confirm the presence of *Larix* pollen, the inferred presence of marsh or swamp habitats, or the recovery and identification of certain fragile entomophilous pollen types, both the Cobb-Poole and Frossard Site analyses should be used with caution in reconstructing the north Texas fossil pollen record.

SUMMARY

Trying to reconstruct and understand the paleoenvironmental conditions that existed in Texas during the late-Quaternary is of importance not only to those of us who live and/or conduct research in the State, but to others as well. The geographical location of Texas between the lush, mixed deciduous and conifer forests of the southeastern United States; the arid and semi-arid flora of the American Southwest; the grasslands and scrub woodlands of the south-central United States' prairies and hills; and the tropical flora of Mexico places Texas at the crossroads of any major vegetational migrations which may have occurred between any of those areas. In addition, any late-Quaternary southward vegetational movement in the south-central portion of North America would have been encouraged or halted by then existing edaphic, climate, or vegetational patterns in Texas.

The information we have presented in this article has attempted to summarize what is currently known about the late-Quaternary environments of Texas as revealed primarily through evidence from the fossil pollen record. As mentioned earlier, some regions of Texas have yielded longer and better palynological records than other areas of the State and therein lies the problem. There is currently an imbalance of late-Quaternary paleoenvironmental data that has been created partly by accident and partly by design. Recovering fossil pollen records from Texas sediments is an expensive procedure in terms of both time and money. Thus, to obtain an adequate fossil pollen record from a given Texas locale one needs: 1) access to sampling areas and sediments which contain preserved fossil pollen; 2) trained personnel to collect samples, conduct laboratory extractions, and complete the resulting analysis; 3) an adequate level of funding; 4) establishment of a chronological time sequence for the collected samples; and 5) publication of the results so that others will have access to the information.

Unlike some states, most of the land in Texas is privately owned and fenced. This aspect creates a significant problem for individuals seeking places to collect fossil pollen records and is one of the reasons why some areas of the State have not been adequately sampled. In some areas of North America when a fossil pollen record is needed from a region, convenient access to public lands makes it easy to search along stream and river valleys for potential fossil pollen sampling sites. This is not true for Texas where trying to gain access to sampling locales and trying to obtain permission to search for new sampling locales have been prime reasons why so few sites have been analyzed in certain regions of the State. Local residents' suspicion of outsiders and absentee land ownership often make pollen transect and fossil site sampling difficult and time consuming tasks.

Gaining permission to sample a Texas locale does not always insure eventual success. As we discussed in the main text of this article there are many regions of Texas which contain soil conditions that are not ideal for the preservation of fossil pollen. Why some soils in Texas contain well preserved fossil pollen while other soils do not is a problem that has yet to be fully understood. On the other hand, experimental research conducted by Holloway (1981) suggests that fossil pollen destruction is caused by a variety of factors which react alone or in combination with one another. Over the years our examination of hundreds of Texas sediment samples collected from a variety of locales within each of the major geographical regions of the State (Gould, 1975) has revealed a few general trends pertaining to the potential of recovering fossil pollen. These include:

1. Alluvial sediments from east, central, north, and south Texas generally do not contain preserved pollen in quantities sufficient for the reconstruction of valid paleoenvironmental studies. Likewise, alluvial sediments from sites located in southwest and west Texas are not ideal sources of preserved fossil pollen even though a few locales such as the Lubbock Lake Site (Oldfield and Schoenwetter, 1975), the Devil's Mouth Site (Bryant and Larson, 1968), and Arenosa Shelter (Bryant, 1967) have yielded marginal amounts of poorly preserved fossil pollen.

2. Sediments from rockshelter sites in southwest Texas such as Bonfire Shelter (Hevly, 1966; Bryant, 1969), Eagle Cave (McAndrews and Larson, 1966), and Hinds Cave (Bryant, 1977a) have contained sufficient fossil pollen from most levels for valid analyses. On the other hand, similar attempts to recover fossil pollen from central Texas rockshelters such as the Levi Site (Bryant, 1969) and Horn Shelter (Bryant, n.d.) have resulted in failure. Unfortunately, we have not made sufficient tests of sediments from rockshelter sites elsewhere in Texas to offer any speculations as to their potential for fossil pollen preservation.

3. Most of the west Texas playa lakes sampled by Hafsten (1961) and Oldfield and Schoenwetter (1975) contained sufficient pollen in most levels for valid analytical counts. To our knowledge no other lake deposits in Texas have been sampled.

4. Peat bog deposits in south-central and central Texas have contained the best preserved and highest concentrations of fossil pollen yet found in the State. Hopefully, additional peat bog deposits will be located and sampled in other areas of Texas. They also should contain excellently preserved fossil pollen.

5. Wood rat middens have yielded fossil pollen records from areas of west and southwest Texas yet it seems doubtful that similar sites will be found in other wetter regions of Texas.

6. Fossil pollen counts have been recovered from human coprolite samples found in rockshelter sediments of southwest Texas yet those pollen records have provided little information useful for interpreting the paleoenvironmental record.

As we have attempted to explain, the diversity of the physiographic provinces of Texas provides a complex set of problems for any paleoenvironmentalist to try to understand. Some regions of the State could provide critically needed information yet those areas have not been examined or have not contained suitable fossil records. Other areas of Texas have been poorly studied and the result has been a series of meager records which need further verification. Hopefully, future research of Texas sediments will attempt to clarify and expand the existing late-Quaternary paleoenvironmental reconstruction which we have proposed in this article.

ACKNOWLEDGEMENTS

We would like to thank Dr. Stephen A. Hall for his valuable comments and suggestions regarding earlier drafts of this paper. In addition, Mrs. Celinda Stevens is graciously thanked for her patience and assistance in typing this manuscript.

References Cited

ALBERT, L. E.
1981   Ferndale Bog and Natural Lake: five thousand years of environmental change in southeastern Oklahoma. *Oklahoma Archaeological Survey,* 7:1-127.

BARKLEY, F. A.
1934   The statistical theory of pollen analysis. *Ecology,* 47:439-447.

BLAIR, W. F.
1958   Distributional patterns of vertebrates in the southern United States in relation to past and present environments. *In:* Hubbs, C. (ed.), *Zoogeography.* American Association for the Advancement of Science, 51:433-468.

BRAUN, E. L.
1955   The phytogeography of unglaciated eastern United States and its interpretation. *Botanical Review,* 21:297-375.

BRYANT, V. M., JR.
1967   Pollen analysis of sediments in Arenosa Shelter. *In:* Dibble, D., *Excavations at Arenosa Shelter, 1965-1966.* Progress report submitted to the National Park Service by the Texas Archaeological Salvage Project, Austin, Texas: 77-85.

1969   Late full-glacial and post-glacial pollen analysis of Texas sediments. Ph.D. dissertation, The University of Texas, Austin, Texas, 168 p.

1977a Preliminary pollen analysis of Hinds Cave. *In:* Shafer, H. J. and Bryant, V. M. (eds.), Archaeological and botanical studies at Hinds Cave, Val Verde County, Texas. *Texas A&M University Anthropology Laboratory Special Series,* 1:70-80.

1977b A 16,000 year pollen record of vegetational change in central Texas. *Palynology,* 1:143-156.

1983   Pollen analysis of Pratt Cave. *In:* Schroeder, A. (assmb.), The Pratt Cave studies. El Paso Archaeological Society, *The Artifact,* 21:161-166.

n.d.   Unpublished report on the pollen studies conducted at Horn Shelter located near Waco, Texas (1975).

BRYANT, V. M., JR. and HOLLOWAY, R. G.
1983   The role of palynology in archaeology. *In:* Schiffer, M. (ed.), *Advances in Archaeological Method and Theory,* 6. Academic Press, New York, New York: 191-224.

ms.   Pollen analysis of Bonfire Shelter, Val Verde County, Texas. Manuscript on file at the Texas A&M University Palynology Laboratory. 26 p.

BRYANT, V. M., JR., and LARSON, D. L.
1968   Pollen analysis of the Devil's Mouth Site, Val Verde County, Texas. *In:* Sorrow, W., The Devil's Mouth Site: The Third Season. *Papers of the Texas Archaeological Salvage Project,* 14:57-70.

BRYANT, V. M., JR., and RISKIND, D. H.
1980   The paleoenvironmental record for northeastern Mexico: A review of the pollen evidence. *In:* Epstein J. F., Hester T. R., and Graves C. (eds.), Papers on the Prehistory of Northeastern Mexico and Adjacent Texas. *Center for Archaeological Research Special Report,* 9 (San Antonio, Texas): 7-31.

BRYANT, V. M., JR., and SHAFER, H. J.
1977   The late-Quaternary paleoenvironments of Texas: A model for the archaeologist. *Bulletin of the Texas Archeological Society,* 48: 1-25.

DAVIS, M. B.
1969   Palynology and environmental history during the Quaternary period. *American Scientist,* 57: 317-332.

DEAN, G.
1984   Putting dinner on the table in the Texas Archaic Chihuahuan Desert. *Discovery,* 15:10-13.

DEEVEY, E. S.
1949   Biogeography of the Pleistocene. *Geological Society of America Bulletin,* 60:1315-1416.

DELCOURT, P. A.
1980   Goshen Springs: Late-Quaternary vegetation record for southern Alabama. *Ecology,* 61:371-386.

DELCOURT, P. A. and DELCOURT, H. R.
1977   The Tuncia Hills, Louisiana-Mississippi: Late-glacial locality for spruce and deciduous forest species. *Quaternary Research,* 7:218-237.

1981   Vegetation maps for Eastern North America: 40,000 yr. BP to the present. *In:* Romans, R. C. (ed.), *Geobotany II.* Plenum Press, New York, New York: 123-166.

DELCOURT, P. A., DELCOURT, H. R., BRISTER, R. C., and LACKEY, L. E.
1980   Quaternary vegetation history of the Mississippi embayment. *Quaternary Research,* 13:111-132.

DERING, J. P.
1977   Plant Macrofossil Study: a progress report. *In:* Shafer, H. J., and Bryant, V. M. (eds.), Archaeological and botanical studies at Hinds Cave, Val Verde County, Texas. *Texas A&M University Anthropology Laboratory Special Series,* 1:84-103.

1979   Pollen and plant macrofossil vegetation record recovered from Hinds Cave, Val Verde County, Texas. M.S. Thesis, Texas A&M University, College Station, Texas, 79 p.

DIBBLE, D. S.
1967   Excavations at Arenosa Shelter, 1965-1966. Progress report submitted to the National Park Service by the Texas Archaeological Salvage Project, Austin, Texas.

DIBBLE, D. S., and LORRAIN, D.
1968   Bonfire Shelter. A stratified bison kill site, Val Verde County, Texas. *Texas Memorial Museum Miscellaneous Papers,* 1:1-78.

DILLON, L. S.
1956   Wisconsin climate and life zones in North America. *Science,* 123:167-176.

EDELMAN, D.
1976   Palynological data: the Arnold site. *In:* Doehner, C., and Larsen, R. (eds.), Archaeological research at the proposed Cooper Lake, northeast Texas. *Southern Methodist University Archaeology Research Program Research Report* 75:158.

EVANS, G. L.
1961   The Friesenhahn Cave. (Part I). *Bulletin of the Texas Memorial Museum,* 2:1-22.

FOREMAN, F., and CLISBY, K. H.
 1961   Crane Lake pollen and sediment correlation. *In:* Wendorf, F. (ed.), Paleoecology of the Llano Estacado. *Fort Burgwin Research Center,* 1. Museum of New Mexico Press, Santa Fe, New Mexico : 92-93.

GALLAGHER, J. G., and BEARDEN, S. E.
 1976   The Hopewell school site: A late archaic campsite in the central Brazos River valley. *Southern Methodist University Contributions in Anthropology,* 19, 129 p.

GOLDSTEIN, S.
 1960   Destruction of pollen by Phycoycetes. *Ecology,* 41:543-545.

GOULD, F. W.
 1975   Texas plants: A checklist and ecological summary. *Texas Agricultural Experiment Station Publication* MP-585, Texas A&M University, College Station, Texas, 121 p.

GRAHAM, A. and HEIMSCH, C.
 1960   Pollen studies of some Texas peat deposits. *Ecology,* 41:785-790.

GRAHAM, R. W.
 1976   Pleistocene and Holocene mammals, taphonomy, and paleoecology of the Friesenhahn Cave local fauna, Bexar County, Texas. Ph.D. dissertation, The University of Texas. Austin, 121 p.

GUNN, J., HESTER, T. R., JONES, R., ROBINSON, R. L., and MAHULA, R. A.
 1982   Climate change in southern Texas. *In:* Hall, G., Black, S., and Graves C. (eds.), Archaeological investigations at Choke Canyon Reservoir, South Texas: The Phase I findings. Center for Archaeological Research, San Antonio, Texas, *Choke Canyon Series,* 5:578-597.

HALL, S. A.
 1980   Pollen studies of the archaeological sites. *In:* Henry, D., Kirby, F., Justen, A., and Hays, T. (eds.), *The Prehistory of Hog Creek. An Archaeological Investigation of Bosque and Coryell Counties, Texas.* University of Tulsa Laboratory of Archaeology, Tulsa, Oklahoma: 126-130.
 1982a Pollen studies at Granger Reservoir. *In:* Hays, T. (ed.), Archaeological investigations at the San Gabriel Reservoir District, Central Texas, 2: 5-18 Archaeology Program, Institute of Applied Sciences, North Texas State University, Denton, Texas.
 1982b Late Holocene paleoecology of the Southern Plains. *Quaternary Research,* 17:391-407.
 1983   Book review of: *Ferndale Bog and Natural Lake: Five Thousand Years of Environmental Change in Southeastern Oklahoma* (L.E. Albert). *Plains Anthropologist,* 28:100.

HAYNES, C. V.
 1968   Geochronology of late Quaternary alluvium. *In:* Morrison, R. B., and Wright, H. E. (eds.), *Means of Correlation of Quaternary Successions,* University of Utah Press, Salt Lake City, Utah :591-631.
 1975   Pleistocene and recent stratigraphy. *In:* Wendorf, F. and Hester, J. (eds.), Late Pleistocene Environments of the Southern High Plains, *Fort Burgwin Research Center Publication,* 9:576-96.

HAFSTEN, U.
 1961   Pleistocene development of vegetation and climate in the southern high plains as evidence by pollen analysis. *In:* Wendorf, F. (ed.), Paleoecology of the Llano Estacado. *Fort Burgwin Research Center,* 1. Musuem of New Mexico Press, Santa Fe, New Mexico : 59-91.

HESTER, T. R.
 1982   Late Paleo-indian occupations at Baker Cave, Southwestern Texas. *Bulletin of the Texas Archeological Society,* 53:101-119.

HEVLY, R. H.
 1966   A preliminary pollen analysis of Bonfire Shelter. *In:* Story, D. A., and Bryant, V. M., Jr., (eds.) *A Preliminary Study of the Paleoecology of the Amistad Reservoir Area.* Report submitted to the National Science Foundation, Department of Anthropology, University of Texas at Austin : 165-178.

HOLLIDAY, V. T., JOHNSON, E., HASS, H., and STUCKENRATH, R.
 1983   Radiocarbon ages from the Lubbock Lake Site, 1950-1980: Framework for culture and ecological change on the southern high plains. *Plains Anthropologist,* 28:165-182.

HOLLIDAY, V. T., JOHNSON, E., HALL, S. A., AND BRYANT, V. M.
 1985   Re-evaluation of the Lubbock Subpluvial. *Current Research in the Pheistocene,* 2:119-121.

HOLLOWAY, R. G.
 1981   Preservation and experimental diagenesis of the pollen exine. Ph.D. dissertation, Texas A&M University, College Station, Texas, 317 p.
 1985   Macrobotanical analyses of charcoal materials from the Choke Canyon Reservior area, Texas. *In:* Hall, G. D., Hester, T. R., and Black, S. L. (eds.), The Prehistoric Sites at Choke Canyon Reservoir, Southern Texas: Results of the Phase II Archaeological Investigations. Center for Archaeological Research, San Antonio, Texas, *Choke Canyon Series* No. 10. In press.
 n.d.   Late-Holocene paleoenvironmental studies in north-central Texas. Manuscript on file, Archaeological Research Program, Anthropology Department, Southern Methodist University, Dallas, Texas. 51 p.

HOLLOWAY, R. G., and BRYANT, V. M.
 1984   *Picea glauca* pollen from late-glacial deposits in central Texas. *Palynology,* 8:21-32.

HOLLOWAY, R. G., and FERRING, C. R.
 1985   Ferndale Bog: A 12,000 year pollen record from the southern plains. In prep.

HOLLOWAY, R. G., RAAB, L. M., and STUCKENRATH, R.
 1985   Pollen analysis of Late Holocene sediments from a central Texas bog. *Texas Journal of Science.* In press.

JANSSEN, C. R.
 1966   Recent pollen spectra from the deciduous and coniferous-deciduous forests of northeastern Minnesota: A study in pollen dispersal. *Ecology,* 47:804-825.

JOHNSON, E.
 1974   Zooarchaeology and the Lubbock Lake Site. *In:* Black C. (ed), *History and Prehistory of the Lubbock Lake Site.* Texas Tech University, Lubbock, Texas :107-122.

JOHNSON, L.
 1963   Pollen analysis of two archaeological sites at Amistad Reservoir, Texas. *The Texas Journal of Science,* 15:225-230.
 1964   The Devil's Mouth Site: A stratified campsite at Amistad Reservoir, Val Verde County, Texas. Department of Anthropology, University of Texas, Austin, *Archaeology Series,* 6:1-115.

KOLB, C. R., and FREDLUND, G. G.
1981   Palynological studies Vacherie and Rayburn's domes North Louisiana salt dome basin. Louisiana State University, Baton Rouge, Louisiana, *Institute for Environmental Studies Report E530-02200-T-2.* 50 p.

LARSON, D. A., BRYANT, V. M., and PATTY, T. S.
1972   Pollen analysis of a central Texas bog. *American Midland Naturalist,* 88:358-367.

LORD, K. J.
1983   The zooarchaeology of Hinds Cave (41 VV 456). Ph.D. dissertation, The University of Texas, Austin, 296 p.

LUNDELIUS, E. L.
1967   Late Pleistocene and Holocene faunal history of central Texas. *In:* Martin, P. (ed.), *Pleistocene Extinctions: The Search for a Cause.* Yale University Press, New Haven, Connecticut: 289-320.

1974   The last 15,000 years of faunal change in North America. *The Museum Journal,* 15:141-160.

LUNDELIUS, E. L., GRAHAM, R. W., ANDERSON E., GULIDAY, J., HOLMAN, J. A., STEADMAN, D. W., and WEBB, S. D.
1983   Terrestrial vertebrate fauna. *In:* Porter, S. C. (ed.), *Late-Quaternary Environments of the United States, Vol. 1: The Late Pleistocene.* University of Minnesota Press, Minneapolis, Minnesota :311-354.

LYNOTT, M. J.
1975   Archaeological excavations at Lake Lavon 1974. *Southern Methodist University Contributions in Anthropology,* 16. 136p.

LYNOTT, M. J., and PETER, D. E.
1977   1975 archaeological investigations at Aquilla Lake, Texas. Southern Methodist University Archaeology Research Program, Dallas, Texas, *Research Report,* 100. 287 p.

MACK, R. N., and Bryant, V. M.
1974   Modern pollen spectra from the Columbia Basin, Washington. *Northwest Science,* 48:183-194.

MACK, R. N., BRYANT, V. M., and PELL, W.
1978   Modern Forest spectra from Washington and northern Idaho. *Botanical Gazette,* 139 (2) :249-255.

MARTIN, P. S.
1964   Pollen analysis and the full-glacial landscape. *In:* Hester, J. and Schoenwetter, J. (eds.), The reconstruction of past environments. *Fort Burgwin Research Center,* 3. Museum of New Mexico Press, Santa Fe, New Mexico :66-74.

MARTIN, P. S., AND HARRELL, B. F.
1957   Pleistocene history of temperate biotas in Mexico and the eastern United States. *Ecology,* 38:468-480.

MARTIN, P. S., and MEHRINGER, P. J.
1965   Pleistocene pollen analysis and biogeography of the southwest. *In:* Wright, H. E. and Frey, D. G. (eds.), *The Quaternary of the United States.* Princeton University Press, Princeton, New Jersey :433-451.

MCANDREWS, J. H., and LARSON, D. A.
1966   Pollen analysis of Eagle Cave. *In:* Story, D. A., and Bryant, V. M. (eds.) *A Preliminary Study of the Paleoecology of the Amistad Reservoir Area.* Report submitted to the National Science Foundation, Department of Anthropology, University of Texas at Austin :178-184.

MEHRINGER, P. J., KING, J. E., and LINDSAY, E. H.
1970   A record of Wisconsin-age vegetation and fauna from the Ozarks of western Missouri. *In:* Dort, W. and Jones, K. (eds.) *Pleistocene and Recent Environments of the Central Great Plains.* University of Kansas Press, Lawrence, Kansas :173-184.

MEYER, E. R.
1972   Late Quaternary paleoecology of the Cuatro Cienegas Basin, Coahuila, Mexico. Ph.D. dissertation, Arizona State University, Tempe, Arizona, 268 p.

1973   Late Quaternary paleoecology of the Cuatro Cienegas Basin, Mexico. *Ecology,* 54:982-995.

1977   A reconnaissance survey of pollen rain in Big Bend National Park, Texas: Paleoenvironmental study. *In:* Wauer, R. and Riskind, D. (eds.), Transactions of the symposium on the biological resources of the Chihuahuan Desert Region, United States and Mexico. *National Park Service Transactions and Proceedings Series,* 3 :115-123.

OLDFIELD, F., and SCHOENWETTER, J.
1964   Late Quaternary environments and early man on the southern high plains. *Antiquity,* 38:226-229.

1975   Discussion of the pollen analytical evidence. *In:* Wendorf, F. and Hester, J. (eds.), Late Pleistocene environments in the southern high plains. *Fort Burgwin Research Center Publication,* 9, Ranchos de Taos, New Mexico: 149-178.

PATTON, P. C.
1977   Geomorphic criteria for estimating the magnitude and frequency of flooding in central Texas. Ph.D. dissertation, The University of Texas, Austin, 218 p.

PATRICK, R.
1946   Diatoms from the Patschke Bog, Texas. *Notulae Naturae,* 170, Academy Natural Sciences, Philadelphia: 1-7.

PLUMMER, F. B.
1945   Progress report on peat deposits in Texas. *The University of Texas, Bureau of Economic Geology Mineral Resource Circular,* 36:1-8.

POTZGER, J. E. AND THARP, B. C.
1943   Pollen Record of Canadian Spruce and Fir from a Texas Bog. *Science,* 98:584.

1947   Pollen profile from a Texas bog. *Ecology,* 28:274-280.

1954   Pollen Study of two Bogs in Texas. *Ecology,* 35:462-466.

RAAB, L. M., and WOOSLEY, A. I.
1982   A terrace habitat and late prehistoric settlement in north-central Texas: Pollen and geological evidence. *Plains Anthropologist,* 27-97:185-193.

RAAB, L. M., MCGREGOR, D., and BRUSETH, J. E.
1982   New Pollen and geological data on the paleoenvironment of the southern plains margin: First season results Richland Creek Archaeological Project. *In:* Raab, M. (ed.), Settlement of the prairie margin: Archaeology of the Richland Creek Reservoir, Navarro and Freestone Counties, Texas 1980-1981. Southern Methodist University Archaeology Research Program, *Archaeological Monographs,* 1: 99-110.

REEVES, C. C.
1965a   Pleistocene climate of Llano Estacado. *Journal of Geology,* 73:181-189.

1965b   Chronology of west Texas pluvial lake dunes. *Journal of Geology,* 73:503-508.

1966   Pleistocene climate of the Llano Estacado II. *Journal of Geology,* 74:642-647.

1973   The full-glacial climate of the southern High Plains, west Texas. *Journal of Geology,* 81:693-704.

1976   Quaternary stratigraphy and geologic history of the southern high plains, Texas and New Mexico. *In:* Mahaney, W. C. (ed.), *Quaternary Stratigraphy of North America.* Dowden, Hutchinson & Ross, Stroudsburg, Pennsylvania: 213-234.

ROBINSON, R. L.

1979   Biosilica and climatic change at 41GD21 and 41GD21A. *In:* Fox, D. E. (ed.), Archaeological investigations of two prehistoric sites on the Coleto Creek Drainage, Goliad County, Texas. Center for Archaeological Research, San Antonio, Texas, *Archaeological Survey Report,* 69:126-138.

1982   Biosilica analysis of three prehistoric archaeological sites in the Choke Canyon Reservoir, Live Oak County, Texas: Preliminary summary of climatic implications. *In:* Hall, G., Black, S., and Graves, C. (eds), Archaeological Investigations at Choke Canyon Reservoir, South Texas: the Phase I findings. Center for Archaeological Research, San Antonio, Texas, *Choke Canyon Series,* No. 5:597-610.

SCHROEDER, A.

1983   The History of the Project. *In:* Schroeder, A. (assmb.), The Pratt Cave Studies: Guadalupe Mountains National Park, Texas. Archaeological Society of El Paso, Texas, *The Artifact,* 21:1-9.

SHAFER, H. J., and BRYANT, V. M.

1977   Archaeological and botanical studies at Hinds Cave Val Verde County, Texas. *Texas A&M University Anthropology Laboratory Special Series,*1, College Station, Texas, 137 p.

SHAW, R. B., VOLMAN, K. C., and SMEINS, F.

1980   Modern pollen rain and vegetation on the Edwards Plateau, Texas. *Palynology,* 4 :205-213.

SLAUGHTER, B. H.

1963   Some observations concerning the genus *Smilodon* with special references to *Smilodon fatalis. The Texas Journal of Science,* 10:68-81.

STEELE, D. G.

1985   Analysis of vertebrate faunal remains from 41MK201. *In:*Hall, G. D., Hester, T. R., and Black, S. L. (eds.), 1985 The Prehistoric Sites at Choke Canyon Reservoir, Southern Texas: Results of the Phase II Archaeological Investigations. Center for Archaeological Research, San Antonio, Texas, *Choke Canyon Series,* No. 10. In press.

VAN DEVENDER, T. R., and RISKIND, D. H.

1979   Late Pleistocene and early Holocene plant remains from Hueco Tanks State Historical Park: the development of a refugium. *Southwestern Naturalist,* 24:127-140.

VAB DEVENDER, T. R., FREEMAN, C. E., and WORTHINGTON, R. D.

1978   Full glacial and recent vegetation of Livingston Hills, Presidio County, Texas. *Southwestern Naturalist,* 23:289-302.

VAN DEVENDER, T. R., MARTIN, P. S., PHILLIPS III, A. M., and SPAULDING, W. G.

1977   Late Pleistocene biotic communities from the Guadalupe Mountains, Culberson County, Texas. *In:* Wauer, R. and Riskind D. (eds.), Transactions of the symposium on the biological resources of the Chihuahuan Desert Region, United States and Mexico. *National Park Service Transactions and Proceedings Series,* 3:107-113.

WELLS, P. V.

1966   Late Pleistocene vegetation and degree of pluvial climatic change in the Chihuahuan Desert. *Science,* 153:970-975.

WENDORF, F.

1961   An interpretation of late Pleistocene environments of the Llano Estacado. *In:* Wendorf, F. (ed.), Paleoecology of the Llano Estacado. *Fort Burgwin Research Center,* 1. Museum of New Mexico Press, Santa Fe: 115-133.

1970   The Lubbock Subpluvial. *In:* Dort, W. and Jones, K. (eds.), *Pleistocene and Recent Environments of the Central Great Plains.* University of Kansas Press, Lawrence, Kansas: 23-35.

WILLIAMS-DEAN, G.

1978   Ethnobotany and cultural ecology of prehistoric man in southwest Texas. Ph.D. dissertation, Texas A&M University, College Station, Texas, 286 p.

WILSON, L. R.

1966   Palynology of the Domebo site. *In:* Leonhardy, F. C. (ed.), Domebo: A paleo-indian mammoth kill in the prairie plains. *Contributions of the Museum of the Great Plains,* 1, Lawton, Oklahoma: 44-51.

A SUMMARY OF LATE-QUATERNARY POLLEN RECORDS FROM MEXICO WEST OF THE ISTHMUS OF TEHUANTEPEC.

R. BEN BROWN
Tumamoc Hill
Department of Geosciences
University of Arizona
Tucson, Arizona 85721

Abstract

This article summarizes the late-Quaternary marine, archaeological, and lacustrine pollen records that have been recovered from Mexico west of the Isthmus of Tehuantepec. While the quality and the presentation of the data vary greatly, the following conclusions may be drawn.

Prior to 9,500 yrs. B.P. the marine pollen record indicates an arid and cool environment that about 6,500/5,500 yrs. B.P. gave way to a period of quite varied moisture conditions and a relative cool-warm gradient from south to north. From 6,500/5,500 to 3,000 yrs. B.P. the marine pollen record suggests wetter, and possibly, warmer conditions. After 3,000 yrs. B.P. the pollen record from the Gulf of California is interpreted to indicate a drying trend, or a decreasing trend in summer rains while from 3,000 to 2,000 yrs. B.P., the pollen record on the Guerrero coast indicates a moist period and the pollen record off the coast of Oaxaca indicates a dry period. After 2,000 yrs B.P. the marine pollen record is more coherent, suggesting an arid period between *ca.* 2,000 and 1,800 years ago followed by a moist period that lasted until *ca.* 700–1,000 yrs. B.P. The last 700–1,000 years are interpreted as a period of relative aridity.

The archaeological pollen record indicates a pattern of high levels of cheno-am, composite and grass pollen that are associated with the high levels of human disturbance associated with the late Preclassic expansion and the late Classic-early Postclassic interface.

The lacustrine pollen record is the most extensive and indicates a change from cold and dry to warmer and wetter conditions at about 12,000 yrs. B.P. in the Texcoco core, and 9,500 yrs. B.P. in other cores, that continues until *ca.* 6,500 yrs. B.P. From 6,500 yrs. B.P. there is an increase in moisture that continues until 5,000/4,500 yrs. B.P. at the majority of sites and 3,500/3,000 yrs. B.P. in the easternmost sites. Most sites reflect aridity from 5,000/4,500 yrs. B.P. until 3,500/3,000 yrs. B.P. After 3,500/3,000 yrs. B.P. maize pollen is ubiquitous and heavy human disturbance associated with the rise of Mesoamerican civilization is indicated by increases in cheno-am, composite and grass pollen. In the cultural core zones the heavy human disturbance continues into modern times but changes in the periphery about 1,000 years ago.

Together these pollen records indicate a cool and dry environment prior to 9,500 yrs. B.P. that becomes wetter and warmer until about 6,000 yrs. B.P. From *ca.* 6,000 to 3,000 yrs. B.P. moisture and temperature conditions are quite varied. After 3,000 yrs. B.P. human disturbance dominates the archaeological and lacustrine records although the marine record indicates periods of aridity from *ca.* 2,000 to 1,800 yrs. B.P., and *ca.* 1,000 yrs. B.P. to present, and an intervening moist period.

INTRODUCTION

Environmental change and its relationship with cultural activity has been of considerable concern since Ellsworth Huntington's early speculative articles (Huntington, 1913; 1914; 1917) first stimulated interest in the relationships between climate, paleoecology and human activity in Mexico. Partially in response, various archaeological (Vaillant, 1944) and geographical (Sauer and Brand, 1932) studies were undertaken in the thirties, but it was not until the 1940's, as appropriate paleoecological approaches developed and diffused, that any paleoecological data were gathered (Deevey, 1943; 1944; 1957; Schulman, 1944; Sears, 1947; 1948; 1951a; 1951b; 1952a; 1952b; 1953a; 1953b; 1955). The initial wave of research was followed by a hiatus of about 20 years.

In the 1960's and 1970's paleoecological research blossomed not only as an adjunct to archaeological research but as a worthy undertaking in and of itself. Dieter Ohngemach (1973; 1977), under the aegis of the Fundación Alemana para la Investigación Científica, recovered various pollen cores from around the volcano La Malinche, in the Puebla-Tlaxcala Basin; Lauro González-Quintero (1980; González-Quintero and Fuentes Mata, 1980) studied various coastal and highland pollen profiles; William A. Watts and J. Platt Bradbury (1982) analyzed the pollen and diatoms in cores recovered from the Mexico City and Pátzcuaro Basins; Hellen Perlstein Pollard (1979) studied the ecology of the Pátzcuaro Basin and the rise of the Tarascan Empire; Thomas R. Van Devender (1979) analyzed the biological fossils from ancient packrat middens found in the Chihuahuan Desert (Bermejillo and near Cuatro Cienegas); Edward R. Meyer (1973; 1975) analyzed the pollen recovered from Cuatro Cienegas, Coahuila; Stuart

Scott lead a multi-disciplinary project on the west Mexican Coastal Plain (Scott, 1967; 1968; 1969; 1970; 1971; 1972; 1973; 1974) that included pollen work by Les Sirkin (1974) and Bonnie Fine Jacobs (1982); Socorro Lorzano (1979a; 1979b) analyzed the modern pollen rain of San Luis Potosi and Palacios Chavez and Arreguín (1980) undertook pollen studies in the San Juan del Rio area.

In the 1980's Brown and Jacobs (Brown, 1980; 1984; Brown and Jacobs, in prep.) analyzed various pollen cores from Nayarit, Jalisco, Guanajuato and San Luis Potosi. Sarah Metcalfe (Metcalfe and Harrison, 1983) and Alayne F. Street-Perrot (Street-Perrot *et al.,* 1982) are studying the diatoms, pollen and geomorphology of the upper Rio Lerma drainage, in and around the Pátzcuaro and Zacapu Basins.

This article will summarize and discuss the results of the pollen studies from Mexico west of the Isthmus of Tehuantepec, in other words, that part of Mexico that is geographically part of North America. Such a summary is necessary since many of the individual records are scattered in disparate and obscure publications. While completeness has been a goal of this study, there are studies without any published results just as there are bound to be studies that have been missed altogether. The material presented in this article covers the late-Pleistocene as well as the Holocene and clearly indicates the tremendous impact of the Mesoamerica civilizations on their environment.

The late-Quaternary pollen record from Mexico west of the Isthmus of Tehuantepec is derived from three types of pollen data (marine, archaeological and lacustrine) and is supplemented by geomorphological and radiometric data.

MARINE STUDIES

The marine pollen record around Mexico (Fig. 1) is derived from the study of modern pollen on the sea bed of the Gulf of California and pollen from short cores recovered from the Pacific coast. Interpretation and comparison of this material has been complicated by the ways that those data are presented and by the presentation of diagrams and data.

Gulf of California

The study of the sea bottom pollen recovered from the Gulf of California (Cross, Thompson and Zaitzeff, 1966; Cross, 1972, 1973) has identified few clear relationships. The most noteworthy relationship is based on the observation that since river transport provides the major pollen input, pollen influx values correspond to the relative amounts of upland run-off.

As part of a broad ranging oceanographic study sponsored by the National Science Foundation, various deep sea cores have been recovered from the Gulf of California. Heusser (1982) and Byrne (1982) have studied the quaternary pollen record from hole 480. While climatic interpretation of the relevant sections of this core has been hindered by dating problems (Byrne, 1982) and, "the incomplete understanding of the present vegetation interrelationships between pollen, vegetation and climatic parameters in the Sonoran Desert." (Heusser, p. 1222, 1982), Heusser and Byrne agree that the higher values of cheno-ams towards the top of the core probably indicate a mid-Holocene *ca.* 6,000-3,000 yrs. B.P. period of higher winter rainfall. Because high pine and grass pollen values are replaced by high values of composite pollen, Heusser (1982) postulates the reduction of pine forest and grasslands in the Late Wisconsin (*ca.* 20,000-10,000 years ago), and the expansion of chaparral communities in the early-Holocene (*ca.* 10,000-6,000 years ago). Since she associates higher values of composites with greater summer rainfall the declining trend of the composites throughout the Holocene is interpreted as an overall increase in aridity. This contrasts with Byrnes' (1982) interpretation. On the basis of changes in the TCT morphological group, which is composed of members of Taxodiaceae, Cupressaceae, and Taxaceae (Adams, 1964), and around the Gulf of California is probably limited to *Juniperus,* as well as changes in spruce and sagebrush pollen values, which are barely present in Heussers' (1982) diagrams, Byrne (1982) postulates a reduction in temperature. These two divergent analyses of the same core will only be resolved when the core is subjected to better dating control and there is fuller agreement as to the nature of the pollen profile.

Middle American Trench

Two marine cores (V18-338 [7.5 m long and radiometrically dated]; V18-339 [10 m long and stratigraphically cross-dated]) were recovered from about 100 km due south of Salina Cruz, Oaxaca in the Middle American Trench (MAT) (Habib, *et al.,* 1970). On the basis of geomorphological, palynological (percentage and influx values) and radiometric data, they have been divided into five stratigraphic

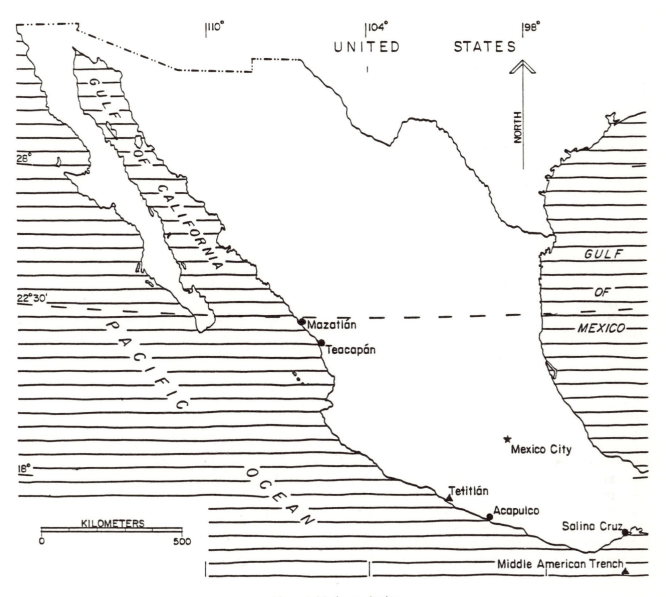

Figure 1. Marine study sites.

zones. The superficial zone indicates a drying trend over the last 1,000 years preceded by a moist period of 750 years. The driest period is from 1,750 to 3,000 yrs. B.P., while the period from 3,000 to 5~6,000 yrs. B.P. is the warmest and wettest period. The basal zone, which extends back to about 8~9,000 yrs. B.P. is depicted as a mixture of cool-dry and cool moist conditions but is overall almost as dry as the dry period between 1,750 and 3,000 yrs. B.P.

Tetitlán

Two 1.9 m pollen cores were recovered for the Tetitlán coastal lagoon which lies about 100 km to northwest of Acapulco, Guerrero (González-Quintero, 1980). The second core has two radiocarbon ages at 0.9 m (I-9020; 2,070±150 yrs. B.P.) and 1.8 m (I-9021; 3,170±280 yrs. B.P.). From the pollen curve of core #2, González-Quintero (1980) derived a temperature curve, precipitation curve, climate curve and a limnic curve. The temperature curve is based on the relative proportions of three ecological groups; low elevation or tropical elements (excluding mangrove), mid elevation elements such as oaks and high elevation elements such as pines; temperate elements such as oak, and pines. The temperature curve is divided into a warming trend over the last 1,000 years which is preceded by a 2,000 year cooling trend. The final 500 years of the cooling trend are identified as temperate, or colder than today. Prior to this, the temperature, while declining, is envisaged as warmer than today. The moisture curve, based on the relative percentages of mangrove, cat-tail, sedge and cheno-ams, indicates from 3,000 to 2,000 yrs. B.P. a period of constant and then declining moisture. This is followed by a 300 year period of slight aridity which is represented by a period of increased precipitation which reaches its maximum about 700 yrs. B.P. and achieves its minimum about 450 yrs. B.P. Combining these two curves, González-Quintero (1980) creates a climate curve which identifies a hot-wet period prior to 2,500 yrs. B.P., a hot dry period from 2,500 to 1,800 yrs. B.P., a temperate period from about 1,800 to 1,000 yrs. B.P. and the subsequent modern period with increasing trends for both precipitation and temperature.

Teacapán Estuary

The pollen record (Sirkin and Gilbert, 1980; Sirkin, 1984) is presented as a partial presence/absence list rather than a percentage or pollen influx diagram. It is divided into five zones that reflect tectonic activity and changes in sea level as much as climatic considerations. The lower four zones correspond to the development of the Cañas soil (*ca.* 8,700 yrs. B.P.) on the Initial Beach (*ca.* 14,500 yrs. B.P.), Early Beach (*ca.* 6,300 yrs. B.P.), the subsequent breaching of the associated lagoon, and the growth of the Teacapán soil (*ca.* 5,400 yrs. B.P.) (Connally, 1984). The pollen data suggest this part of the sequence began with a low but rising sea level followed by an increase in riverine input and then local non-arborescent pollen. In a similar manner the subsequent changes may correspond to the development of the later beaches but the partial nature of the published data preclude any solid conclusions.

Summary

Three sets of marine cores and a collection of sea bottom surface samples present a pollen record that is dominated by pine pollen with values that are a function of upland precipitation and run-off. Although each core responds distinctly there are seven major change over points; 10,000/9,500, 8,700/8,500, 6,500/5,500, *ca.* 3,000, *ca.* 2,000, *ca.* 1,800 and *ca.* 1,000 yrs. B.P.

Prior to 10,000/9,500 yrs. B.P., the Gulf of California material indicates a cooler temperature (Bryne,

1982), on the basis of the presence of spruce, juniper and sagebrush, and an increasing trend in aridity (Heusser, 1982), on the basis of a decreasing trend in pines and grasses. The relatively high values of composites suggest significant summer rain (Heusser, 1982). The development of the Initial Beach at Teacapán (14,500 yrs. B.P.) (Connally, 1984) supports Heussers' assertion of a drying trend.

In the early-Holocene (10,000/9,500 to *ca.* 6,500 yrs. B.P.) the Gulf of California core is interpreted as indicating a drier (Heusser, 1982) and warmer (Bryne, 1982) period. MAT indicates a change from cool/moist to cool/dry for the latter half of this period while the Teacapán material indicates a cycle of soil and beach development that may indicate drier intervals about 8,700 and 6,500 yrs. B.P.

From 6,500/5,500 to 3,000 yrs. B.P. the high cheno-am values in the Gulf of California core suggest high winter rain (Heusser, 1982) while MAT provides a warmer and wetter record (Habib *et al.,* 1970).

Over the last 3,000 years, the Gulf of California core indicates a decreasing trend in composites which is interpreted as a decreasing trend in summer rains (Heusser, 1982), while the other cores present contradictory evidence. For the period between 3,000 and 2,000 yrs. B.P. MAT indicates a dry period and the Tetitlán cores indicate a moist period. For a brief period about 2,000 to 1,700 yrs. B.P., both the MAT and Tetitlán cores suggest a period of aridity which is followed by a period of increased moisture. The last 700–1,000 years is a period of relative aridity.

ARCHAEOLOGICAL STUDIES

The first archaeological pollen studies were undertaken in 1941 as adjuncts to limnological studies that focused on Lake Pátzcuaro (Deevey, 1943; 1944). Deevey (1943; 1944) discussed four major problems; the questions of climatically *versus* culturally derived vegetation change, pollen preservation and site selection, establishment of regional sequences, and the need for greater knowledge of regional plant ecology.

The archaeological pollen record has been developed from samples collected within archaeological stratigraphic contexts or ponds within archaeological sites. By their nature, archaeological sites demonstrate high levels of human disturbance which often register as high levels of cheno-ams and composites. This disturbance can mask any climatic signal.

The sites will be considered in sequence from the northwest to southeast (Fig. 2).

Casas Grandes

The pollen profile (Kelso, 1976) derived from the sediments recovered from Reservoir #2 within the site of Casas Grandes, Chihuahua (Di Peso et al., 1976), is dominated by high values of cheno-ams (40-60%) and composites (10-20%). Cheno-ams had their maximum values during the period of human occupation (A.D. 1260-1340). Since the reservoir still contained over a meter of water when the site was abandoned, these values are interpreted as the result of the human activity and abandonment of the site rather than any climatic signal. The few grains of recovered *Zea* pollen from this feature correspond to the earliest occupation levels. Clusters of cheno-am and grass pollen together with relatively high values of cucurbit pollen in archaeological features interpreted as macaw pens are used to substantiate this identification and identify the major feeds as cheno-ams, grasses and squashes (Kelso, 1976).

San Juan del Río

The pollen profiles derived from six archaeological sites in the Valle de San Juan del Río, Querétaro (Palacios Chavez and Arréguin, 1980) demonstrate high frequency signals that provided no clear patterns although the maize, ambrosia, composite and cheno-am pollen peaks imply changes in human land use. Data from the San Juan del Río core are interpreted as evidence of widespread clearing and agriculture after 500 B.C. Data from the Los Cerritos, La Trinidad, and Santa Rosa Xajaji cores are interpreted to indicate a change in land use patterns between 800 and 1,000 yrs. B.P. (Palacios Chavez and Arréguin, 1980).

Tula

An undated pollen core that was extracted from the bank of the Río Tula, Hidalgo, provides a history of the ebbs and flows of the river and the local human activity (González-Quintero and Montufar, 1980). The presence of maize from 1.6 to 0.9 m coupled with the dramatic increase in low spine composite values and the overall increase in grass and cheno-am pollen values above 1.4 m imply an expansion of agriculture followed by a decreasing pattern of disturbance. González-Quintero and Montufar (1980)

present temperature and precipitation curves based on the relative percentages of pine *versus* oak, and cheno-ams *versus* Cyperaceae plus *Typha*.

Teotihuacán

In association with archaeological studies of Teotihuacan and its purported irrigation system (Sanders et al., 1979), Anton Kovar (1970) cored a small spring called El Tular, about 3 km southwest of San Juan Teotihuacan, State of Mexico. At a depth of 3 m this profile demonstrates a dramatic drop in sedge pollen values and a concomitant increase in pine pollen values that may indicate a change from a more moist to a drier local environment. However, there are no changes in the pollen values of composites nor cheno-ams as would be expected with either increased disturbance around the spring or dessication. The absence of *Typha* and *Taxodium* pollen and the meager presence of maize pollen above this change are considered noteworthy (Kovar, 1970).

Kovar (1970) recovered an alluvial section from the excavation site at Cuanalan that vaguely resembles the lower portion of the El Tular core. On the basis of this tenuous relationship and the stratigraphy of the Cuanalan section, Sanders, Parsons, and Santley (1979) propose that the sedge/pine changes in the El Tular core represent the incorporation of the El Tular spring into the Teotihuacan irrigation network by 150 B.C., if not 600 B.C.

Mexico City Basin

In 1941 Deevey (1943; 1944) collected suites of samples from archaeological sites such as Zacatenco, Copilco, Cuicuilco, and Ticoman and concluded that they contained insufficient pollen for interpretation. In 1948 Sears (1952a; 1952b) collected samples from archaeological sites near Xico, Culhuacán, Tlatelolco, Zacatenco, Copilco, and El Tepalcate. In conjunction with alluvial and lacustrine sequences collected in 1949 and 1950, Sears (1952b) concluded that the early Preclassic was wet (pine values > 90%) but was followed by a drying trend (oak values >10%) which was well established by 500/400 B.C. This dry period continued through the Classic Period until A.D. 900/800 when a moist period began and continued until the Spanish invasion.

Oaxaca

The easternmost pollen studies to be mentioned in this paper are those of James Schoenwetter in the Valley of Oaxaca (Schoenwetter, 1974; Flannery and Schoenwetter, 1970). On the basis of material recovered from the Guila Naquitz cave, just above the town of Mitla in the eastern end of the eastern arm of the Oaxaca valley, Schoenwetter reconstructs a cooler and more xeric climate from *ca.*10,000 to 8,000 yrs. B.P. Not only is this conclusion limited by three factors, the pollen counts are limited to 100 grains, the basal radiocarbons measurements are out of sequence and there is a long and unexplained depositional hiatus between 8,000 and 1,000 yrs. B.P., but since the samples were gathered from living floors, the pollen changes could be the result of human activity within the cave rather than the climate outside the cave. Schoenwetter (1974) tries to overcome this limitation by analyzing a suite of modern surface samples that allows him to identify four vegetation communities but do not closely resemble the pollen from Guila Naquitz. Schoenwetter (1974) identifies maize pollen in the level dated to 8,000 yrs. B.P. although *Zea* macrofossils are not known in other parts of the valley until about 3,000 yrs. B.P. (Flannery and Schoenwetter, 1970).

Other suites of samples collected, from four archaeological sites (Huitzo, San Jose Mogote, Monte Alban, and Zaachila) in the western arms of the valley, are interpreted (Flannery and Schoenwetter, 1970) on the basis of the relative values of cheno-ams and composites, as indicating a dry period from 3,200 to 2,900 yrs. B.P., a moister period quite like the modern climate from 2,900 to 2,400 yrs. B.P., and from 2,400 to 2,000 yrs. B.P. a period wetter than today.

Summary

The pattern of high values for cheno-am, composite and grass pollen, plus the presence of maize pollen, is indicative of the human disturbance related to Mesoamerican cultural activities. The archaeological studies provide only two common break points. The first (*ca.* 2,500 yrs. B.P.) coincides with the expansion of late Preclassic cultures and the second (*ca.* 1,000 or 800 yrs. B.P.) coincides with changes in the patterns of land use that relate to the interface between the late Classic cultures and the establishment of the early Postclassic cultures.

LACUSTRINE STUDIES

Since pollen generally preserves best under either extremely dry or anaerobic conditions, lake beds can provide ideal situations for the sampling of ancient pollen. Although it must be recognized that there are problems with differential production, transport and preservation, pollen usually represents the regional vegetation and provides a proxy for the regional climate. Many of the diagrams in this section are simplified versions of those presented by the original authors. For further details the reader is urged to consult the original documents.

Again, sites will be considered from north to south (Fig. 3).

Cuatro Cienegas

Two sediment cores were recovered from the floor of the Cuatro Cienegas Basin, Coahuila (27°N; 104°W) which lies at 740 m above sea level (Meyer, 1973; 1975). The resultant pollen profiles are quite complacent and do not suggest any significant vegetation changes in the area. However, Meyer interprets his data to indicate a lowering of the border between the pine forests and the upland woodlands during the Middle Wisconsin. Or in his words "regional climate was cooler and perhaps more moist than it is now" but "vegetation ecologically equivalent, if not identical, to that now present, occupied the basin floor" (Meyer, p. 982, 1973). This interpretation is supported by Van Devenders' (1979; Brown, 1980; 1984) analysis of Late Wisconsin packrat middens from Durango and Coahuila.

Ojo de San Juan

A short pollen core was extracted from a spring, the Ojo de San Juan near Villa Juarez, San Luis Potosi, which is located at N22°17′30′′;W100°16′ and is 1,100 meters above sea level. The Ojo de San Juan, along with other springs, previously fed a substantial lake, Laguna de Buenavista, that has been drained and the lake bed farmed. The regional vegetation has been described by Puig (1976; 1979) and Rzedowski (1966; 1981), and the regional pollen rain has been described by Lorzano (1979a; 1979b). The pollen profile, limited by small counts and wide sampling intervals, reveals a complacent 6,100 year record (Brown, 1984).

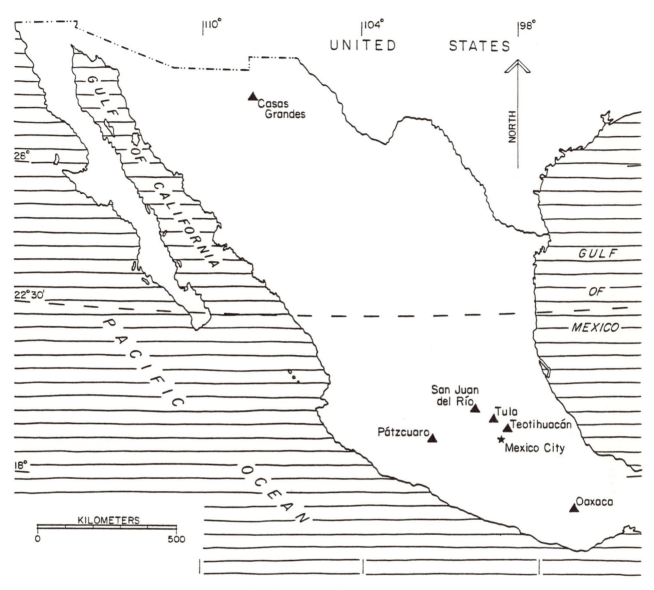

Figure 2. Archaeological study sites.

The pollen profile from the Ojo de San Juan is dominated by the TCT group which in this case is thought to be mainly composed of degraded *Taxodium* (Brown, 1984). The source of *Taxodium* pollen is probably *Taxodium muchronatum,* a riparian component at the Ojo de San Juan. The continued presence of *Taxodium* pollen and macrofossils indicate that the Ojo and its associated lake did not completely dry out for any extended period.

Assuming that the variation in the other taxa is masked by statistical constraint of the TCT, their values were recalculated excluding TCT from the sum. Consequently, it is deduced that although the Ojo de San Juan has been continually moist, the surrounding vegetation has undergone at least one period (post A.D. 1,000) of increased aridity as indicated by the changes in pollen values of high spine composites and grasses. The inverse relationship between oak and cheno-ams suggests the amount of oaks in the region declined concomitant with an expansion in agricultural disturbance or a period of aridity. The presence of *Zea* at an extrapolated age of 4,500 yrs. B.P. suggests an early diffusion of maize agriculture from the Sierra de Tamaulipas (Mangelsdorf *et al.,* 1964).

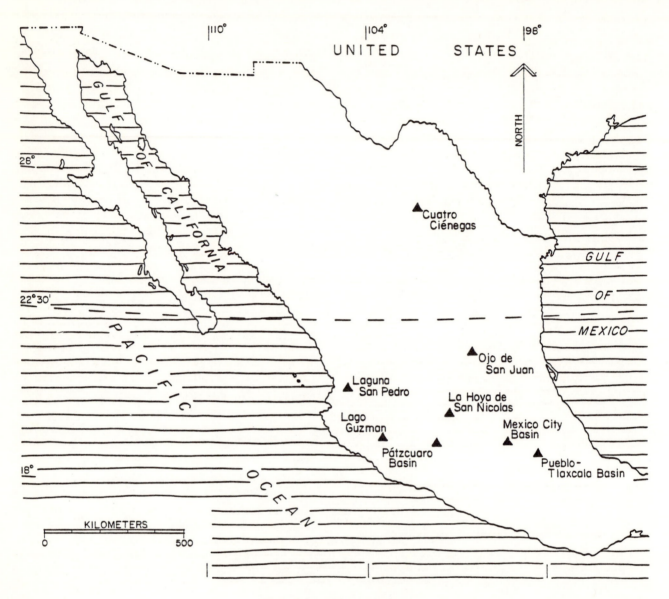

Figure 3. Lacustrine study sites.

Laguna San Pedro

Laguna San Pedro, Nayarit, is located at N21°12′30″; W104°45′ in a wide alluvial basin that drains into the Río Ameca. The lake is just below the 1,300 m contour and is surrounded by mountains that rise to over 1,900 m. The regional vegetation is classified as tropical deciduous forest (Rzedowski, 1981; Flores-Mata, *et al.*, 1971). The change from mixed pine-oak parkland to an oak parkland about 5,000 years ago is indicative of an increase in aridity. The appearance of maize *ca.* 3,000 yrs. B.P. indicative of maize agriculture and the beginning of a

period of agricultural disturbance that ends *ca.* 800 yrs. B.P. and gives way to a re-expansion of the mixed pine-oak grassland or an expansion of the coniferous forest (Brown, 1984; Brown and Jacobs, in prep.).

La Hoya de San Nicolas de Parangueo

The Hoya de San Nicolas de Parangueo is a crater lake that lies close to the confluence of the Ríos Lerma and Laja near the town of Valle de Santiago. The Hoya de San Nicolas de Parangueo is located at N20°23′;W101°17′. The top of the crater wall is about 1,780 m and the lake bed is about 1,700 m. The lake has a drainage area of about 1 km². Analysis of the pollen core was based on its division into four zones (Fig. 4) (Brown, 1984).

Zone IV (11,000 to 3,000 yrs. B.P.) is dominated by high pine pollen values (55 to 97%) but includes grass and cheno-am pollen peaks. Zone IV is interpreted as representing a period strongly dominated by the presence of pine forests on the surrounding hills and beginning with a period of disturbance followed by the establishment of a grassland in the valley bottoms which were subsequently invaded by pine forest.

Between *ca.* 10,000 and 9,000 yrs. B.P. the cheno-am pollen peak is interpreted as part of the succession resulting from the Pleistocene-Holocene interface or a response to the posited early-Holocene increase in insolation (Kutzbach, 1981). This is followed (*ca.* 9,000 to 7,000 yrs. B.P.) by the establishment of grasslands at lower elevations and the continued presence of pine forests at higher elevations. Between 7,000 and 3,000 yrs. B.P., the pine forests moved down slope, replacing the grassland.

In Zone III, from 3,000 to 1,700 yrs. B.P., pine pollen values drop considerably as oak, cheno-am and grass pollen increase suggesting an oak parkland replacing the pine forest or selective land clearing in conjunction with the introduction of agriculture associated with the Chupícuaro culture.

Zone II, dating from 1,700 to 1,000 yrs. B.P., exhibits a strong pine recovery (pine < 85%) followed by a reversal that seems to be in concert with the regions heaviest archaeological occupation.

Zone I, covering the last 1,000-800 years, presents an anomalous series of inverted radiocarbon dates that are most parsimoniously explained by abrupt and massive soil erosion caused by the introduction of European agricultural practices.

The Pátzcuaro Basin

Since the establishment of the Limnological Research Station in the 1930's (Batalla, 1940; De Buen, 1941; 1944) the Pátzcuaro Basin (N19°35'; W101°35') (2,044 m above sea level) has been a center of ecological and paleoecological studies. In 1941 Edward S. Deevey (1943; 1944; 1957) took a number of pollen cores from Central Mexico, of which the two from Lake Pátzcuaro were considered the most productive. One (P-1) was about three meters long and was recovered from about 4.5 m of water while the other (P-3) was just over six meters long and was recovered from about 3 m of water.

Deevey (1944) presented his results, based on counts of at least 200 grains per sample. Pine pollen

values vary from 50 to 90%, and 50 to 70% in the shorter and longer cores, respectively. Both cores have a small percentage of fir pollen in the lower two thirds portion. In both cores oak pollen has a slight decreasing trend, roughly decreasing from 20 to 10%, although there is more variation in P-1 than in P-3. Alder pollen shows a stronger decreasing trend. Grass pollen has fairly constant values of about 10% although they increase slightly and then decrease again towards the top of both cores. Cheno-am pollen values are complacent in the longer P-3 core but show a dramatic peak in the upper one-third of the shorter P-1 core with values going from less than 10 to over 40%. Maize (*Zea*) pollen is rarely present towards the top of the cores. While oak and alder pollen generally trend together, the only dramatic change is the cheno-am peak towards the top of P-1.

Interpretation of these pollen profiles is difficult because they are not dated and only the change in the cheno-am pollen values of P-1 clearly lie outside the range of variation expected in an unchanging vegetation. Deevey divides the profiles into three zones, and interprets the lower zone as representing pine-oak forest with alder and fir, the middle zone as representing a reduction in the lake level due to increased aridity and the upper zone as pine-oak forest disturbed by agricultural practices.

Later, the chemistry and fossil algal flora of these cores were analyzed (Hutchinson *et al.,* 1956) and the pollen was re-analyzed in conjunction with the resultant data. Hutchinson *et al.,* (1956) used the Sears Humidity Index (Sears, 1952a; Sears and Clisby, 1952) to interpret these data but concluded that the samples were so dominated by pine pollen, that small variations in the values of pine pollen created relatively large changes in the percentage values of the other pollen taxa. However, changes in the values of the calcium carbonate in the short core are seen as support for the assertion that the cheno-am pollen peak represents a dry period.

Deevey (1956; Hutchinson *et al.,* 1956) compared this sequence with those which Sears (1951a; 1951b; 1952a; 1952b; 1953a; 1953b) developed for the Mexico City Basin even though neither is dated. Based more on the Mexico City data than the Pátzcuaro data, he concluded that prior to 1,500 B.C., the late-Holocene environment was dry; from 1,500-500 B.C. it was moist; from 500 B.C. to A.D. 900 it was quite varied but with one very dry spell; and, from A.D. 900 to 1,521 it was moist again.

In the early seventies another core was recovered from Lake Pátzcuaro (Watts and Bradbury, 1982) (Fig. 5). It has a basal radiometric age of 44,000 yrs.

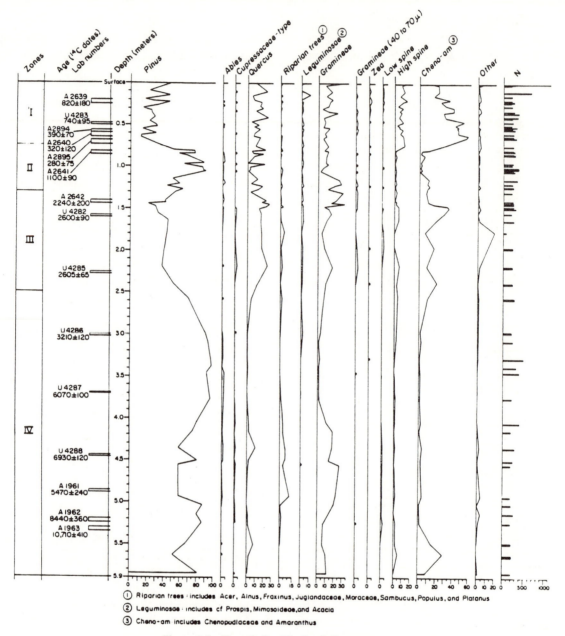

Figure 4. La Hoya de San Nicolas de Parangueo.

B.P. The lower portion, between 44,000 and 9,500 yrs. B.P., is dominated by about 50% (from 25 to 65%) pine pollen and lesser percentages of alder and oak pollen. Fairly constant but low percentages of grass pollen (5 to 10%), juniper (5 to 10%), sagebrush (<5%), fir (1%), and ragweed (1%) pollen complete the diagram. While the percentage values of the last three taxa are not high, their disappearance at 9,500 yrs. B.P. gives them greater significance. The removal of these taxa is interpreted as representing the interface between the Pleistocene and the Holocene; a change from a dry, cold environment to one of greater moisture and warmth.

Between about 9,500 and 5,000 yrs. B.P., pine continued to dominate with grass, alder and oak pollen varying slightly. After 5,000 yrs. B.P. alder values decrease and cheno-am pollen values increase. This change is interpreted as indicating the onset of agriculture with the replacement of riparian alder with agricultural plots and a concomittant increase in weedy cheno-ams or the deliberate manipulation of *Chenopodium* as a grain crop. The appearance of *Zea* pollen grains after 3,600 yrs. B.P. indicates agriculture in place. The accompanying increase in sedges suggests an increase in habitats favorable to sedges either as erosion filled in the shallows of Lake

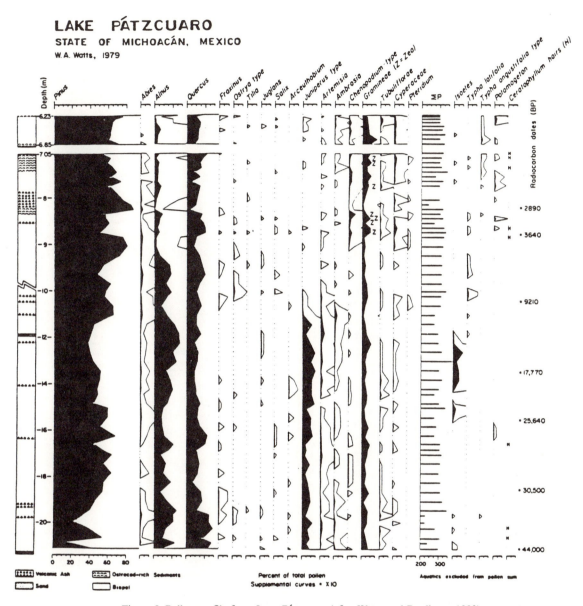

Figure 5. Pollen profile from Lago Pátzcuaro (after Watts and Bradbury, 1982).

Pátzcuaro or the changing land use created swamps and ponds in the surrounding hillsides. The absence of maize pollen from the top 60 cm may be the result of colonial, or more modern, changes in population levels and agricultural practices such as the abandonment of *chinampas,* while the disappearance of fir pollen is quite possibly due to lumbering (Watts and Bradbury, 1982).

The pollen data from Lake Pátzcuaro indicate a complacent vegetation in the late-Pleistocene that was dominated by pine with substantial components of oak, fir, alder, juniper, sagebrush and ragweed. The removal of juniper, ragweed and sagebrush in the early-Holocene suggests an increase in moisture and possibly temperature. Changes in the pollen profile subsequent to about 5,000 yrs. B.P. seem to be the result of human rather than climatic impact. The period of erosion that has been radiometrically dated to 2,300±60 yrs. B.P. (SRR 1862) (Street-Perrott *et al.,* 1982), would seem to relate to late Preclassic developments. Subsequently there is enough detail in these data to discuss the interaction between the human occupants of the Pátzcuaro Basin and their environment (Pollard, 1979; 1980; 1982).

The Mexico City Basin

Stimulated by Deevey's (1943; 1944; 1947) and Schulman's (1944) work, Paul B. Sears took a number of cores from various archaeological sites (Sears, 1951a; 1951b; 1952a; 1952b; 1953a; and 1953b) before he gained access to two long cores recovered from the center of Mexico City (Zeevaert, 1952).

These lacustrine cores were selected to increase time depth and to minimize the impact of human activity on the pollen record. Assuming that arboreal pollen (AP) reflects regional pollen production, non-arboreal pollen (NAP) reflects the local pollen production, and that human disturbance is a local phenomenon, it was reasoned that the NAP would reflect the natural or background environment (Sears 1952b). However, such cultural activities as farming and deforestation have been so extensive throughout Mesoamerica that they have effected pollen production at both the local and regional levels.

This attempt to elucidate the Holocene and Pleistocene climates of Central Mexico is weakened by the absence of Zone I, small sample size below Zone II, and the lack of dating control (Sears, 1955; Foreman, 1955; Clisby and Sears, 1955; Sears and Clisby, 1952; 1955). Zone I was considered recent cultural fill and discarded without any study. Zone II is dominated by an unusual combination of pine pollen, and Maydeae, pollen which disappears by the bottom of the zone. At the top of Zone II pollen counts reach 200 grains but at the bottom of Zone II the total count generally hovers just over 100 pollen grains per sample. While Barkely (1934) demonstrated that counts of such a size were sufficient to reliably (at a 0.85 level of significance) compare subsamples within a given sample, he did not consider at what size, nor at what statistical level, a sample reliably reflects the universe from which it is drawn. Martin (1963) helped codify the 200 grain count as the modern minimum standard when he graphically demonstrated that a 200 grain count had twice the taxonomic diversity of a 100 grain count and half the diversity of a 2,000 grain count. Subsequently others have presented information that suggests considerably larger counts are warranted (Maher, 1972; Faegri and Iversen, 1975; Duffield and King, 1979).

Bradbury (1971) redrew (Fig. 6) Clisby and Sears' (1955) Madero diagram and the resultant figure clearly demonstrates the virtual absence of pollen from sediments other than weathered ash and clay that effectively excludes Zones IV (34 to 37m), VI

and VII (48 to 70 m). Zone V (37 to 48 m) demonstrates high ($\pm$ 70%) values of pine pollen in association with low ($\pm$ 15%) values of oak plus the presence of alder, grass and composite pollen. The pollen profile for Zone III (8 to 34 m) is quite similar to Zone V but with a clearer and stronger inverse relationship between pine and oak pollen. However, in both of these zones the pollen sums barely pass 100 grain counts. In the lower portion of Zone II, there is a break in the pollen sequence brought about by a change in sediments. In the meter or so of sand and silt, overlain by an ostracod marl between 7 and 6 m, there is virtually no pollen. Above the 6 m level pollen counts reach 200 grains for the first time. Pine pollen values are lower than before, ranging from 20 to 60%, but grass, cheno-am, and composite pollen values are much higher than before. The most dramatic change is the appearance of maize pollen which goes from nothing to over 20%.

The presence of maize pollen in conjunction with the relatively high values of grass and cheno-am pollen is a clear indicator of maize agriculture. Since maize pollen falls close to the plant (Raynor et al., 1972), the high percentage of cereal pollen implies that the core was taken through a midden that included a lot of household refuse, a *chinampa* that was used to grow maize, or a field that was partially built up with maize stalks. Due to the location of the coring site, either of the latter speculations are the most likely (Calnek, 1971; González Aparicio, 1973; Adams, 1977).

Even though Sears and his collegues have published extensively about the paleo-climate of the Mexico City Basin, a hard inspection of their material reveals that very little of their data can be used today. Sears, Clisby, and Foreman are to be commended for publishing their data in sufficient detail that it can be reanalyzed and discussed, and for their pioneering role in Mexican pollen studies.

The most recent major pollen profile from the Mexico City Basin (González-Quintero and Fuentes Mata, 1980) was recovered from Lake Texcoco and can be conveniently divided into three zones (Fig. 7). Figure 7 is a summary diagram that has simplified the published data and presented two radiocarbon dates not included in the original publication. As such, it has been redrawn from the various diagrams presented by González-Quintero and Fuentes Mata (1980). The reader requiring more detail should consult González-Quintero and Fuentes Mata (1980).

Zone I in Figure 7 corresponds to the González-Quintero and Fuentes Mata's (1980) zones A2 through A5; Zone II corresponds to their zones A1,

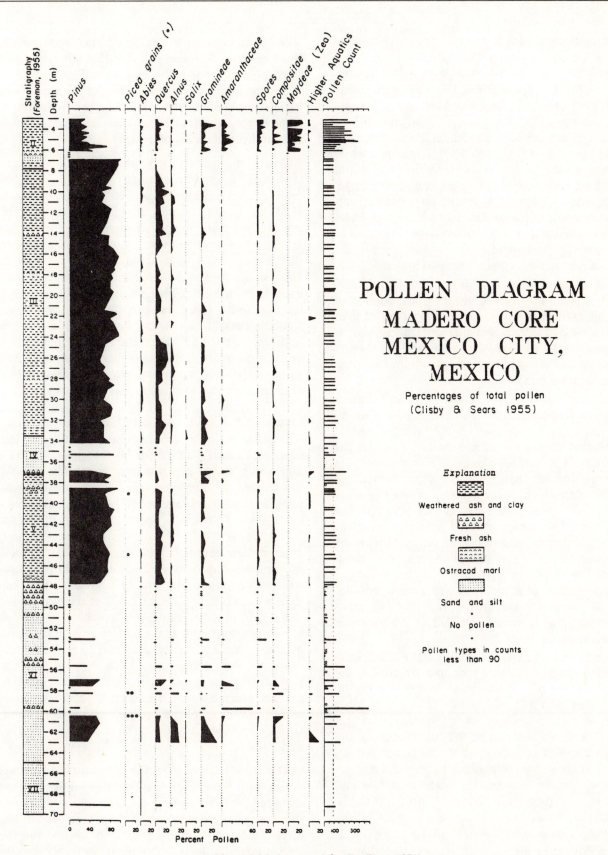

POLLEN DIAGRAM
MADERO CORE
MEXICO CITY,
MEXICO

Percentages of total pollen
(Clisby & Sears 1955)

Explanation

Weathered ash and clay

Fresh ash

Ostracod marl

Sand and silt

No pollen

Pollen types in counts
less than 90

Figure 6. Madero core (after Bradbury, 1971).

and Zone III corresponds to their B1 through B6. The finer divisions used by González-Quintero and Fuentes Mata are partially based on pollen separations within the genus *Pinus* that are problematic (Ting, 1965; 1966; Martin, 1963; Mack, 1971; Hansen and Cushing, 1973). For example, Martin (1963, p. 20, fig. 9) presents body and bladder length ranges for eleven species of pine which indicate that while ponderosa pine pollen tends to be the largest and piñon pine pollen tends to be the smallest, the ranges overlap considerably. If the total pine pollen percentage is presented in the column labelled *suma polinica* of the unlabelled diagram between pages 116 and 117 (González-Quintero and Fuentes Mata, 1980), the value of pine pollen remains constant and high.

Zone III, the oldest zone, goes from about 30,000 to 12,000 yrs. B.P. It is dominated by pine pollen and has higher values of oak, spruce and grass pollen than Zone II. Fir and alder pollen are constantly present at low percentage values. Cheno-am and composite pollen are erratically present in low values and grass pollen values decline.

While the authors' identification of pines led them to believe that they could make detailed ecological analyses, a more conservative interpretation would suggest that during the time period of Zone III the environment around the Mexico City Basin was quite stable and cooler than today. In other words, the volcanic slopes were covered with a mixture of pine-oak and pine forest with a greater proportion of pine than today. This view is derived from the consistent values of alder, fir, and oak pollen coupled with the small erratic values of composite and cheno-am pollen. The relatively high values of grasses suggest either an oak or pine parkland was a major component of the lower elevations and that the grass component was slowly supplanted as the parkland developed into a forest.

Zone II, from 12,000 to 9,500 yrs. B.P., is characterized by high values of pine and fir pollen, the virtual absence of oak and spruce pollen, the erratic presence of alder and grass pollen as well as reduced values of composites and other disturbance indicators. This would seem to be the culmination of the warmer and wet period. Towards the upper portion of Zone II spruce and oak pollen make erratic appearances which may represent climatic reversals and argue against prior extirpation of spruce from the Mexico City Basin.

In Zone I, from 9,500 yrs. B.P. to the present, pine values dominate although they are lower than in Zones II or III. Zone I is characterized by the return of oak and alder pollen along with a substantial increase in composite, grass and cheno-am pollen. Between 0.6-0.3 m pine pollen values fall below 50% as composite, grass and cheno-am pollen values rise correspondingly. Since *Zea* pollen values are at their highest levels between 0.6-0.3 m this increase in disturbance indicators suggests an increase in farming activity from about 3,000 to 1,500 yrs. B.P.

Zone I may be further divided into four subzones. If Zone I is given an estimated basal age of 9,500 years, each 0.1 m represents about 480 years (Brown, 1984). Subzone Id extends from 2 to 1.75 m, or 9,500 to 8,300 yrs. B.P. Pine pollen values drop from about 100% to 60% and recover to 80% while oak pollen values rise from zero to about 25% and fall back to 10%. As the pine values drop the alder pollen values increase from 5% to 10%. Composite pollen is irregularly present while grass and cheno-am pollen are regularly present, but at very low values. In subzone Ic, from 8,300 to 6,000 yrs. B.P., pine pollen values decrease from about 80% to 60% and recover to 90% while oak pollen is constant between 10% and 15% until the very end of this subzone when they drop to 5%. Composite, grass and cheno-am pollen are present in low values but demonstrate no particular trend. *Zea* pollen makes its first appearance in this subzone. In subzone Ib, from 6,000 to 3,400 yrs. B.P., pine pollen values decrease from about 90% to 40%, oak values rise to between 5 and 10%, while composite, grass and cheno-am pollen hold low but consistent values. Subzone Ia begins at a depth of 0.75 m (*ca.* 3,400 yrs. B.P.) as composite, cheno-am and grass pollen begin to increase and achieve their maximum values between 3,000 and 1,500 years ago, while pine values continue to decrease and reach their minimum, of 25%, at 2,500 years ago. Above this depth, pine values recover to achieve estimated values of about 80% at the surface as all other taxa decline commensurately. Subzone Ia is thought to represent the period from 3,400 years ago to the present and the major disturbance would seem to cover the period from 2,500 to 1,500 years ago, which within the limits of the accuracy of age determination, corresponds to the rise and apogee of Teotihuacan.

The late-Pleistocene in the Mexico City Basin is defined by the presence of coniferous forest that included pine, fir and spruce, and an environment that was generally cooler and drier than today. The early-Holocene, with the absence of spruce and the almost total dominance of pine was probably warmer and wetter than the preceding or subsequent periods. For the mid- and late-Holocene, climate exhibits a

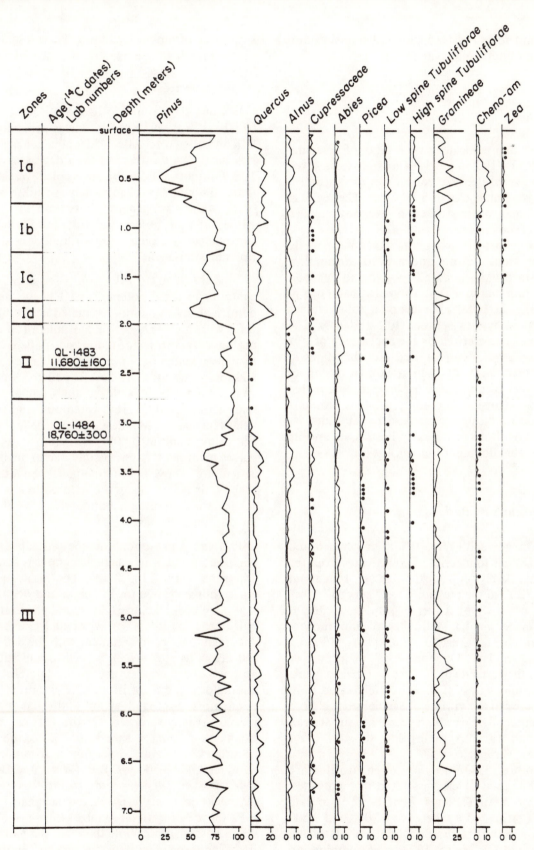

Figure 7. Texcoco core (after González-Quintero and Fuentes Mata, 1980).

drying trend followed by cycles of expansion and contraction of human activity.

Lago Guzman

Lago Guzman is located to the south of Guadalajara (N19°45′;W103°30′) at 1,500 m above sea level. While the size of the lake varies with the season, it has an approximate area of 3 by 7 km. The regional vegetation has been studied by Rzedowski and McVaugh (1966).

The pollen profile from Lago Guzman, while limited by small counts and a need for smaller sampling intervals, is dated by two radiocarbon measurements that give it an estimated basal age of 1,300 years. The basal zone has pine pollen values >50% and oak values between 10% and 20%. The middle zone characterized by the appearance of *Zea* and a cheno-am pollen peak that probably dates to A.D. 1,200 (*ca.* A.D. 1,000 to 1,500). In the upper zone, recovery of up to 20% haploxylon pine pollen suggests an expansion of piñon pine. This profile is interpreted to indicate the increase, or introduction, of maize agriculture in the early Postclassic and its subsequent abandonment (Brown, 1984).

The Puebla-Tlaxcala Basin

Dieter Ohngemach (1973, 1977; Ohngemach and Straka, 1978) has four series of pollen cores from the Puebla-Tlaxcala region that extend in age from about 35,000 yrs. B.P. to the present. The "Tlaloqua series" (Fig. 8) of cores were recovered from the Tlaloc or Tlaloqua crater on La Malinche volcano and are reported to have a ^{14}C age of *ca.* 8,000 yrs. B.P. at a depth of 1.7 m. The "Oriental series" of cores were taken from the east side of the Pico de Orizaba, while the "Jalasquillo series" (Fig. 9) were extracted from the maar between the villages of San Salvador el Seco and Zacatepec, all of which are located in the state of Puebla, and the "Acuitlapilco series" from Acuitlapilco, Tlaxcala.

At Tlaloqua (3,100 m) the pollen profiles are dominated by pine, alder, grasses and/or composites. The lowest zone of Tlaloqua, Zone IV (2.55 to 2.85 m) with an estimated age between 12,000 and 9,500 years ago represents the end of the Pleistocene (Ohngemach, 1973; Heine, 1973). The combination of high values of grass pollen (up to 50%), composite pollen (up to 20%), plus the presence of spruce, is suggestive of alpine meadows. The presence of the taxon Maydeae seems out of place in Zone IV. Moving from Zone IV to Zone III grass and composite pollen decrease.

In Zone III, from 2.55 to 1.5 m, or *ca.* 9,500 to 6,000 yrs. B.P., pine pollen values generally exceed 50% and often reach 60%. Alder pollen increases from less than 10% to 40% and then decreases to about 10%. For most of Zone III, grass pollen values are less than 10% and Maydeae values are nil. Composite pollen values are generally less than 1%. The high values of pine pollen suggest that Tlaloqua was surrounded by a pine forest with alder as the major riparian component.

Zone II (1.5 to 0.9 m) is composed of pumice and contains no fossil pollen. Zone 1 begins above the pumice at 0.9 m and an estimated age of 6,000 yrs. B.P., with high values of grass pollen (up to 70%), negligible quantities of composite pollen, low (less than 30%) values of pine pollen, negligible oak pollen and low values of alder and fir pollen. Grass pollen values remain high as alder values remain constant until about 4,500 yrs. B.P. when pine and fir pollen values increase as Maydeae and oak pollen reappear. Fir values peak at 3,000 yrs. B.P. as grass pollen values plummet to 10% possibly due to the impact of agriculture. Above 0.4 m pine pollen values decrease slightly, fir pollen values are reduced to less than 5% while grass, plantago, composite and cheno-am pollen values increase.

The early presence, subsequent absence and reappearance of Maydeae pollen grains is anomalous. Ohngemach (1977, fig. 2) describes this pollen type as grass grains over 60µm in length and demonstrates that it includes all maize (*Zea mays*) grains and about half of the teosinte (*Zea mexicana*) grains present. Although there is no argument with the definition of this taxon, its presence in the Pleistocene, subsequent absence and reappearance in the late-Holocene is anomalous in and of itself. The situation becomes even more confused when Tlaloquas' elevation (3,100 m) is considered. Of the various potential explanations, none is satisfactory. Since most pollen diagrams show the appearance of maize pollen between 4,000-3,500 yrs. B.P. (Watts and Bradbury, 1982; Brown, 1984) and the macro-fossil evidence (Mangelsdorf *et al.*, 1964) is younger than 7,000 yrs. B.P., the identification of maize pollen prior to 7,000 yrs. B.P. is suspect. While there are claims for earlier finds, they are not without controversy (Bartlett *et al.*, 1969; Beadle, 1981; Mangelsdorf, 1974; Wilson, 1974).

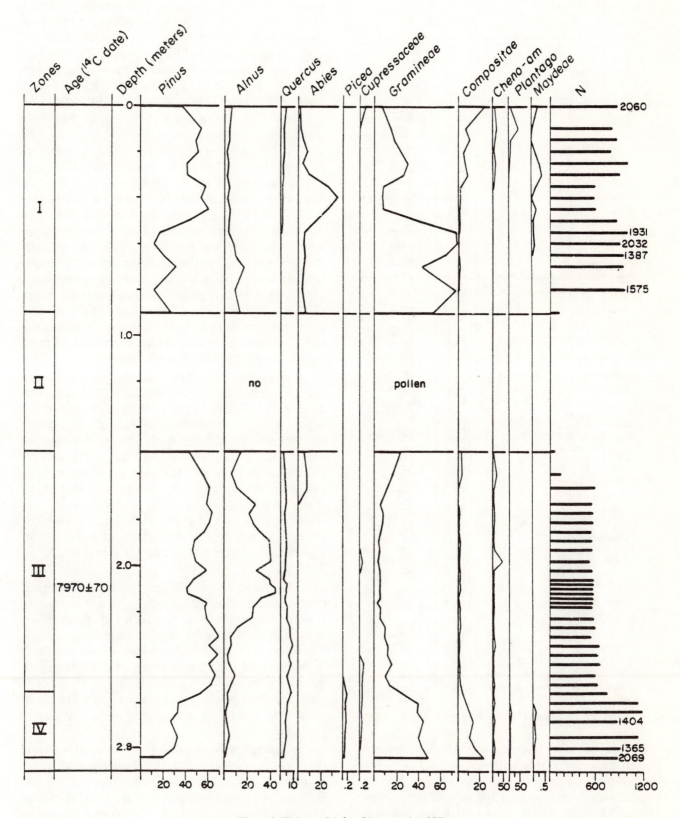

Figure 8. Tlaloqua I (after Ohngemach, 1977).

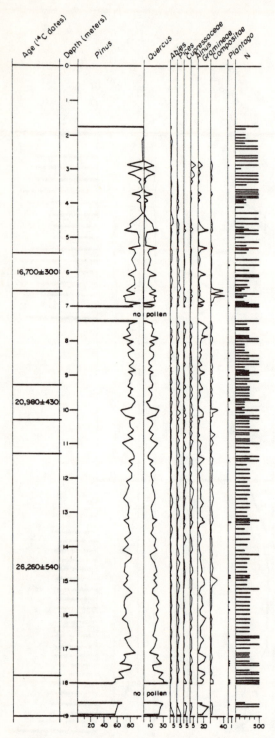

Figure 9. Jalaspaquillo (after Ohngemach, 1977).

(Raynor *et al.,* 1972). Since ethnographic example (Lumholtz, 1904) creates a precedent for the ritual use of maize and teosinte pollen in a lake to propitiate the forces of fecundity, it is possible that the late-Holocene Maydeae pollen represents ritual offerings. However, this does not explain the presence of these grains in the late-Pleistocene or early-Holocene.

Ohngemach (Ohngemach and Straka, 1978) is of the opinion that the change from Zone IV to Zone III in the Tlaloqua series represents the interface between the late-Pleistocene and the early-Holocene, if not the end of the M III glacial period of La Malinche (Heine, 1973). Ohngemach posits that during the M III glaciation the Tlaloqua crater was surrounded by alpine grasslands and that the upper treeline was depressed by 600 to 800 m relative to its modern position. As the glacier retreated the upper treeline rose to surround and encompass the crater. Ohngemach further posits the leader of this migrations to be *Pinus hartwegii* because it is the modern high elevation pine in central Mexico. Ohngemach (1973) supports this view with the presence of fossil pollen from a mistletoe specific to this pine.

Consequently, the Tlaloqua cores suggest that in the late-Pleistocene the top of La Malinche was covered in alpine grasslands that were rapidly replaced by the upward movement of a pine forest about 9,500 yrs. B.P. The early-Holocene vegetation was a stable pine forest that was disturbed by volcanic activity at an estimated age of 6,000 years ago. After this volcanic activity, the area around the crater was re-established as a grassland which was replaced by a mixed pine-oak forest. Subsequent changes in the profile suggest the introduction of farming in the Puebla-Tlaxcala basin between 4,500-3,000 yrs. B.P.

The undated profile for Oriental is quite complacent. It can be divided into two zones. The bottom zone is dominated by pine pollen but includes low percentages of oak, alder, spruce, fir and grass pollen. The upper zone is 100% pine pollen. Consequently, it is interpreted that the bottom zone represents a mixed coniferous forest that dates to the Pleistocene and that the upper zone represents pure stands of pine that could be associated with the early-Holocene.

The Jalapasquillo site (2,400 m) revealed a profile (Fig. 9) that covers the late-Pleistocene and early-Holocene, with a basal age greater than 26,000 yrs. B.P. and an estimated terminal age of 5,000 years. This profile suggests a stable Pleistocene vegetation dominated by pine with some oak, spruce, and alder; *i.e.,* a climate that was colder and possibly wetter

The subsequent reappearance of the Maydeae pollen type correlates with the spread of maize agriculture between 4,000 and 3,000 yrs. B.P. If maize and teosinte were grown on the bottom lands of the Puebla-Tlaxcala valley their pollen could not be expected to be transported the estimated minimum distance of 5,000 m laterally and 500 m vertically than today's climate. In the upper portion, where pine is virtually the sole component of the pollen spectrum, it is quite possible that any change is the result of statistical constraint rather than an ecological event. When the spruce disappears at an estimated age of 9,500 yrs. B.P., alder pollen quickly disappears in conjunction with cheno-am, grass and sagebrush pollen. The top subzone of this profile is almost pure pine pollen with little fir at an estimated age of 7,000 or 6,000 yrs. B.P. It is concluded that for the time period covered by the core the Jalapasquillo site was surrounded by pine forest.

The material from Acuitlapilco (2,200 m) covers the last 600 years. The least complacent core, Acuitlapilco III, is still quite complacent with pine pollen rarely less than 70%. Oak and alder pollen complete the rest of the pollen total except at the top of the column when Maydeae, large grass, composite and cheno-am pollen become prominent. This is probably due to the implementation of a new farming technology.

In summary, pollen evidence from the Puebla-Tlaxcala basin indicates that during the late-Pleistocene the Tlaloqua crater was in an alpine meadow. This implies that the upper treeline was 600 to 800 m lower than today, and, if there was no zonal compaction, that much of the valley bottoms were covered with pine and mixed pine-oak forests. This is clearly corroborated by the lower elevations sites such as Oriental and Jalaspaquillo.

The interface between the Pleistocene and the Holocene is assigned an age of about 9,500 yrs. B.P. The upper elevation profiles suggest that the early-Holocene upper tree line rose to surround the Tlaloqua pond with a pine forest and an alder riparian community. At the lower elevations sites the pine forest continues.

The mid-Holocene high elevation record, after a volcanic erruption about 6,000 yrs. B.P., varies from dry to cold and wet to cold and wetter still. At low elevations a warmer and wetter period follows a brief wet and cold period. After about 2,000 yrs. B.P., cultural disturbance is seen at Tlaloqua despite its elevation.

Summary

The lacustrine series indicates seven turning points beginning with a change from cold and dry to warmer and wetter at 12,000 yrs. B.P. as indicated by the Texcoco core. On the other cores of sufficient age, this change is not noted until 9,500 yrs. B.P. In the Pátzcuaro Basin Watts and Bradbury (1982) identify this change with a reduction of juniper, ragweed and sagebrush pollen. In the Mexico City Basin, (González-Quintero and Fuentes Mata, 1980) this change can be identified with the absence of spruce, and an increase in oak, composite, grass and cheno-am pollen as well as bull rush and reed pollen. In the Puebla-Tlaxcala Basin Ohngemach (1973, 1977) identifies this change with a 600 m or more rise in the upper treeline as his sites on La Malinche go from alpine meadows to mixed pine forests. While these changes are not without contradictions, when taken together, they are interpreted to indicate an increase in temperature and probably moisture.

From 9,500 to *ca.*6,500 yrs. B.P., the overall consensus is for a warmer and wetter climate but the Texcoco core indicates fluctuations between wet and dry while the Hoya de San Nicolas core indicates drier conditions. This period includes the first indications of maize in the Mexico City Basin.

From *ca.* 6,500 to 5,000/4,500 yrs. B.P. the trend is from moist to dry and ends with the appearance of maize at two more sites. From 5,000/4,500 to 3,500/3,000 yrs. B.P. the overall trend is one of continued dryness although the two easternmost series are moist. Corn is ubiquitous and heavy disturbance is indicated by the increase in values of cheno-am, composite, grass pollen as well as other ruderal taxa. Niederberger (1976, 1979) had good archaeological evidence for an agricultural village in the Mexico City Basin by 4,000 yrs. B.P. The pattern of agricultural disturbance continues to modern times in such metropolitan areas as the Mexico City and Puebla-Tlaxcala basins but changes in the periphery about 1,000 yrs. B.P.

SUMMARY

While many of the problems first elucidated by Deevey (1943, 1944) still exist, three types of palynological data have been linked to develop an overall regional sequence. As is the case of any sequence that purports to be standard for such a large area, there are some gross simplifications. The synthetic sequence has eight basic change over points.

The Pleistocene-Holocene interface is the first turning point. This interface dates to 12,000 yrs. B.P. in the Texcoco core and 9,500 yrs. B.P. in other cores. The change is from cold and dry, to warmer and wetter, conditions as identified by the dissappearance of spruce, and sometimes juniper and ambrosia pollen, followed by a succession that involves both fir and alder pollen in the development of a mixed pine forest at higher elevations. At lower elevations this event is not clearly defined.

The early-Holocene (9,500 to *ca.* 6,500 yrs. B.P.) is considered warmer than the preceding Pleistocene. On the basis of moisture it can be subdivided into and earlier (9,500 to *ca.* 8,500 yrs. B.P.) and later (*ca.* 8,500 to *ca.* 6,500 yrs. B.P.) periods. Most sequences indicate the earlier period was relatively dry and the latter wet although MAT indicates cool and moist followed by cool and dry conditions and Teacapán goes thru two cycles of aridity at *ca.* 8,500 and *ca.* 6,500 yrs. B.P.

During the mid-Holocene (*ca.* 6,500 to *ca.* 3,500 yrs. B.P.) the marine profiles indicate warmer and wetter conditions as the Pátzcuaro and San Nicolas cores indicate continued moist conditions. Changes in grass and pine values in the Tlaloqua material suggest dry conditions from *ca.* 6,500 to *ca.* 4,500 yrs. B.P. and wetter conditions from *ca.* 4,500 to *ca.* 3,500 yrs. B.P. In the Texcoco core the decrease in pine and

the corresponding increase in oak values suggest increasing aridity although the continued presence of alder belies this conclusion. Corn pollen appears in the Mexico City Basin at the beginning of this period and is widespread by the periods' end.

During the late-Holocene (*ca.* 3,500 yrs. B.P. to present) the climatic message in the terrestrial pollen record is considerably distorted by the impact of human activity. Agriculture, first seen in the Mexico City Basin about 6,000 yrs. B.P., is widespread by 3,500 yrs. B.P. and subsequently distorts the terrestrial record. However from the marine record it is possible to identify a moist period from *ca.* 3,500 to *ca.* 2,000 yrs. B.P. which was followed by a brief (*ca.* 300 year) arid period and another moist period until about 1,000 yrs. B.P. After 1,000 yrs. B.P. all three types of studies indicate environmental changes that are probably due to changes in land use patterns but only the marine record gives clear indications of a general drying trend.

References Cited

ADAMS, D. P.
  1964 Exploratory Palynology in the Sierra Nevada, California. *Interim Research Report.* #4, Geochronological Laboratories, University of Arizona.
ADAMS, R. E. W.
  1977 *Prehistory in Mesoamerica.* Little, Brown, Boston.
BARKLEY, F. A.
  1934 The Statistical Theory of Pollen Analysis. *Ecology,* 15:283-289.
BARTLETT, A. S., BARGHOORN, E. S., and BERGER, R.
  1969 Fossil Maize from Panama. *Science,* 152:642-643.
BATALLA, M. A.
  1940 Estudios morfológicos de los granos de polen de las plantas vulgares del Valle de México. Published M.A. thesis, Departamento de Biología, Faculdad de Ciencias, Universidad Nacional Autónoma de México.
BEADLE, G. W.
  1981 Origin of Corn: Pollen Evidence. *Science,* 213:890-892.
BRADBURY, J. P.
  1971 Paleolimnology of Lake Texcoco, Mexico: Evidence from Diatoms. *Limnology and Oceanography,* 16:180-200.
BROWN, R. B.
  1980 A Preparatory Statement to a Paleoecological Study on the Northern Frontier of Mesoamerica. Manuscript on file with the Arizona State Museum Library.
  1984 The Paleoecology of the Northern Frontier of Mesoamerica. Unpublished Ph.D. thesis, Department of Anthropology, University of Arizona, Tucson.

BROWN, R. B., and JACOBS, B. F.
  In prep. Analysis and Interpretation of the Pollen from Two West Mexican Lakes.
BYRNE, R.
  1982 Preliminary Pollen Analysis of Deep Sea Drilling Project Leg 64, Hole 480, Cores 1-11. *Initial Report of the Deep Sea Drilling Project.,* 64(2):1225-1235.
CALNECK, E. E.
  1972 Settlement Pattern and Chinampa Agriculture at Tenochtitlan. *American Antiquity,* 37(1):104-115.
CONNALLY, G. G.
  1984 Soil Stratigraphy and Inferred Tectonic History of the West Mexican Coastal Plain. *In: Neotectonics and sea level variations in the Gulf of California area, a Symposium.* (V. Malpica-Cruz, S. Celis-Gutiérrez, J. Guerrero-Garcia and L. Ortlieb, eds.), Instituto de Geologia, Universidad Nacional Autónoma de México, México.
CROSS, A. T.
  1972 Recycled Fossil Palynomorphs in Marine Sediments. *Abstracts with Programs 1972.,* 4(7):480. Geological Society of America.
  1973 Source and Distribution of Palynomorphs of the Gulf of California. *Geosciences and Man,* XI:156.
CROSS, A. T., THOMPSON, G. G., and ZAITEFF, J. B.
  1966 Source and Distribution of Palynomorphs in Bottom Sediments, Southern Part Gulf of California. *Marine Geology,* 4:467-524.

CLISBY, K. H., and SEARS, P. B.
1955 Palynology of Southern North America. III: Microfossil profiles under Mexico City correlated with the sedimentary profiles. *Bulletin Geological Society of America*, 66:511-520.

DE BUEN, F.
1941 El Lago de Pátzcuaro. *Instituto Panameño de Geografía e Historia, Revista Geográfica*, 1:20-44.
1944 Limnobiología de Pátzcuaro. *Anales del Instituto de Biología*, 15:261-312.

DEEVEY, E. S.
1943 Intento Para Datar Las Culturas Medias Del Valle de México Mediante Análisis de Polen. *Ciencia*, 4(4 & 5):97-105.
1944 Pollen Analysis and Mexican Archaeology; an attempt to apply the method. *American Antiquity*, 10:135-149.
1957 Limnological Studies in Middle America. *Transactions of the Connecticut Academy of Arts and Sciences*, 39:213-328.

DiPESO, C. C., RINALDO, J. B., and FENNER, G. J.
1976 *Casas Grandes*, Northland Press, Flagstaff, Arizona.

DUFFIELD, R., and KING, J. E.
1979 Sample Size and Palynology: A Midwestern Test. *Transactions of the Illinois Academy of Science*, 72(2):1-7.

FAEGRI, K., and IVERSEN, J.
1975 *Textbook of Palynology*. Hafner, New York.

FLANNERY, V. K., and SCHOENWETTER, J.
1970 Climate and Man in Formative Oaxaca. *Archaeology*, 23(2):144-152.

FLORES MATA, G., J. JIMÉNEZ LÓPEZ, X. MADRIGAL SÁNCHEZ, F. MONCAYO RUÍZ and F. TAKAKI TAKAKI
1971 *Tipos de vegetación de la República Mexicana*. Secretaría de Recursos Hidráulicos. Mexico City.

FOREMAN, F.
1955 Palynology in Southern North America. II: Study of Two Cores from Lake Sediments of the Mexico City Basin. *Bulletin Geological Society of America*, 66:475-510.

GONZÁLEZ APARICIO, L.
1973 *Plano Reconstructivo de la Región de Tenochtitlán*. Instituto Nacional de Antropología e Historia. Mexico City.

GONZÁLEZ-QUINTERO, L.
1980 Paleoecologia de un Sector Costero de Guerrero, México (3,000 años). *In: Memorias III Coloquio sobre Paleobotánica y Palinología*. (Coordinación, Fernando Sánchez). *Colección Científica Prehistoria*, 86:133-158. Instituto Nacional de Antropología e Historia, Mexico D. F.

GONZÁLEZ-QUINTERO, L., and FUENTES MATA, M.
1980 El Holoceno de la porcón central de la Cuenca del Valle de México. *In: Memorias, III Coloquio sobre Paleobotánica y Palinología*, (Coordinación, Fernando Sánchez). *Colección Científica Prehistoria*, 86:113-132. Instituto Nacional de Antropología e Historia, Mexico D. F.

HABIB, D., THURBER, D., ROSS, D., and DONAHUE, J.
1970 Holocene Palynology of the Middle American Trench near Tehuantepec, Mexico. *Memoirs of the Geological Society of America*, 126:233-261.

HANSEN, B. S., and CUSHING, E. J.
1973 Identification of Pine Pollen of Late Quaternary Age from the Chuska Mountains, New Mexico. *Bulletin Geological Society of America*, 84:1181-1200.

HEUSSER, L. E.
1982 Pollen Analysis of Laminated and Homogeneous Sediment from the Guaymas Basin, Gulf of California. *Initial Report of the Deep Sea Drilling Project*, 64(2):1217-1223.

HEINE, K.
1973 Variaciones mas importantes del clima durante los últimos 40,000 años en México. *Comunicaciones*, 7/1973, 51-58. Puebla.

HUNTINGTON, E.
1913 Shifting Climatic Zones As Illustrated in Mexico. *Bulletin of the American Geographical Society*, 45(1):1-12 and 45(2):107-116.
1914 The Climactic Factor as Illustrated in Arid America. *Publication No. 192*. Carnegie Institute of Washington.
1917 Maya Civilization and Climactic Change. *XIXth International Congress of Americanists*. Washington, D. C.

HUTCHINSON, G. E., PATRICK, R., and DEEVEY, E. S.
1956 Sediments of Lake Patzcuaro, Michoacan, Mexico. *Bulletin Geological Society of America*, 67: 1491-1504.

JACOBS, B. F.
1982 Modern Pollen Spectra from Surface Soil Samples, Northern Nayarit, Southern Sinaloa, Mexico. *Journal of the Arizona-Nevada Academy of Sciences*, 17:1-14.

KELSO, G.
1976 Pollen Analysis of Reservoir 2 and Macaw Nesting Boxes. *In: Casas Grandes*. (Charles C. DiPeso, John B. Rinaldo and Gloria Fenner, eds.), 4:34-36. Northland Press, Flagstaff, Arizona.

KOVAR, A.
1970 The Physical and Biological Environment of the Basin of Mexico. *In: The Natural Environment, Contemporary Occupation and 16th. Century of the Valley, the Teotihuacan Valley Project Final Report*, Vol. 1. *Occasional Papers in Anthropology*, 3:13-67. Department of Anthropology, Pennsylvania State University, University Park, Pennsylvania.

KUTZBACK, J. E.
1981 Monsoon Climate of the Early Holocene: Climate Experiment with the Earth's Orbital Parameters for 9,000 Years Ago. *Science*, 214:59-61.

LORZANO, M. DEL S.
1979a *Première Approche de l'Analyse Pollinque Dans la Region de San Luis Potosi (Mexique)*. Docteur de Troisieme Cycle. Université D'Aix-Marseille.
1979b Atlas de Polen de San Luis Potosí, México. *Pollen et Spores*, 21(3):287-336.

LUMHOLTZ, C.
1904 *El México Desconocido*, tomo II. Charles Scribner's Sons, New York.

MACK, R. N.
1971 Pollen Size Variation in Some Western North American Pines as Related to Fossil Pollen Identification. *Northwest Science*, 45(4):257-269.

MAHLER, L. J.
1972 Nomograms for Computing 0.95 Confidence Limits of Pollen Data. *Review of Paleobotany and Palynology*, 13:85-93.

MANGELSDORF, P. C.
1974 *Corn*. Harvard University Press. Cambridge.

MANGELSDORF, P. C., MACNEISH, R. S., and WILLEY, G. R.

1964   Origins of Agriculture in Middle America. *In: Natural Environments and Early Cultures* (edited by Robert C. West), pp. 413-445. *Handbook of Middle American Indians, vol. 1,* Robert Wauchope, general editor. University of Texas, Austin.

MARTIN, P. S.

1963   *The Last 10,000 Years.* University of Arizona, Tucson, Arizona.

METCALFE, S. and HARRISON, H. E.

1983   Preliminary reconstructions of the Late Quaternary environmental changes as recorded by lake margin deposits in The Basin of Zacapu, Michoacán. Ms. on file, The Tropical Paleoenvironments Research Group, School of Geography. University of Oxford, Oxford.

MEYER, E. R.

1973   Late Quaternary Paleoecology of the Cuatro Cienegas Basin, Coahuila, Mexico. *Ecology,* 54(5):982-995.

1975   Vegetation and Pollen Rain in the Cuatro Ciengas Basin, Coahuila, Mexico. *The Southwestern Naturalist,* 20(2):215-224.

NIEDERBERGER, C.

1976   *Zohapilco.* Colección Científica Arqueología, No. 30. Instituto Nacional de Antropología e Historia, Mexico City.

NIEDERBERGER, C.

1979   Early Sedentary Economy in the Basin of Mexico. *Science,* 203:131-140.

OHNGEMACH, D.

1973   Análisis polínico de los sedimentos del pleistoceno reciente y del holoceneo en la región de Puebla-Tlaxcala. *Comunicaciones,* 7/1973:40-45.

1977   Pollen sequence of the Tlaloqua crater (La Malinche Volcano, Tlaxcala, Mexico). *Boletín de la Sociedad Botánica de México,* 36:33-40.

OHNGEMACH, D., and STRAKA, H.

1978   La Historia de la Vegetación de la región de Puebla-Tlaxcala durante el cuaternario tardío. *Comunicaciones,* 15/1978:189-204.

PALACIOS CHÁVEZ, R., and ARREGUÍN, M. DE LA L.

1980   Analisís polínico de algunos sitios de interés arqueológico en el valle de San Juan del Rio, Querétaro. *In: Memorias, III Coloquio sobre Paleobotanica y Palinología.* (Coordinación Fernando Sánchez). *Colección Científica Prehistoria,* 86:179-184. Instituto Nacional de Antropología e Historia, México, D. F.

POLLARD, H. P.

1979   Paleoecology of the Lake Patzcuaro Basin: Implications for the Development of the Tarascan State. *Paper presented at the 43rd International Congress of Americanists, Vancouver.*

1980   Central places and cities: a consideration of the protohistoric Tarascan state. *American Antiquity,* 45:677-696.

1982   Water and Politics: Paleoecology and the Centralization of the Tarascan State. *Paper presented at the 44th International Congress of Americanists, Manchester.*

PUIG, H.

1976   *Vegetation de la Huasteca, Mexique.* Mission Archéologique et Ethnologique Francaise au Mexique. *Collection de Etudes Mesoamericaines,* No. 5. Mexico City.

1979   Notice de la feuille Guadalajara-Tampico. Carte Internationale du Tapis Végétal. *Extrait des Travaux de la Section Scientifique et Technique de l'Institut Français de Pondichéry.* Hors série No. 16-1979. Institut Français Pondichéry, India.

RAYNOR, G. S., OGDEN, E. C., and HAYES, J. V.

1972   Dispersion and Deposition of Corn Pollen from Experimental Sources. *Agronomy Journal,* 64:420-427.

RZEDOWSKI, J.

1966   Vegetación del Estado de San Luis Potosí. *Acta Cientifica Potosina,* 5(1 & 2):1-291.

1981   *La Vegetación de México.* Limusa. Mexico City.

RZEDOWSKI, J., and MCVAUGH, R.

1966   La Vegetación de Nueva Galicia. *Contributions of the University of Michigan Herberium,* 9:1-123. Southern Illinois University, Carbondale, Illinois.

SANDERS, W. T., PARSONS, J. R., and SANTLEY, R. S.

1979   *The Basin of Mexico.* Academic Press, New York.

SCHOENWETTER, J.

1974   Pollen Records of Guila Naquitz Cave. *American Antiquity,* 39(2):292-303.

SCHULMAN, E.

1944   The Possibilities of Dendrochronology in Mexico. *El Norte de México y el Sur de Estados Unidos,* pp. 305-307. Mexico City.

SCOTT, S. D. (editor)

1967   *West Mexican Prehistory, part 1.* Department of Anthropology, State University of New York at Buffalo.

1968   *West Mexican Prehistory, part 2.* Department of Anthropology, State University of New York at Buffalo.

1969   *West Mexican Prehistory, part 3.* Department of Anthropology, State University of New York at Buffalo.

1970   *West Mexican Prehistory, part 4.* Department of Anthropology, State University of New York at Buffalo.

1971   *West Mexican Prehistory, part 5.* Department of Anthropology, State University of New York at Buffalo.

1972   *West Mexican Prehistory, part 6.* Department of Anthropology, State University of New York at Buffalo.

1973   *West Mexican Prehistory, part 7.* Department of Anthropology, State University of New York at Buffalo.

1974   *West Mexican Prehistory, part 8.* Department of Anthropology, State University of New York at Buffalo.

SAUER, C. O., and BRAND, D. D.

1932   Azatlan. *Ibero-Americana,* #1.

SEARS, P. B.

1947 Notes on correlated pollen profiles and glacial substages. *Revista Mexicana de Estudios Antropológicos,* 9(1, 2 & 3):165-168.

1948 Forest sequence and climactic change in northeastern North America since early Wisconsin time. *Ecology,* 29:326-333.

1951a Palynology in North America. *Svensk Botanisk Tidskrift,* 45(1):241-246.

1951b Pollen profiles and culture horizons in the Basin of Mexico. *29th International Congress of Americanists,* 1:57-61. University of Chicago.

1952a El Análisis de Polen en la Investigación Arqueológica. *Tlatoani,* 1(3 & 4):29-30.

1952b Palynology in Southern North America. I: Archaeological horizons in the basins of Mexico. *Bulletin of the Geological Society of America,* 63:241-254.

1953a An Ecological View of Land Use in Middle America. *CEIBA,* 3:157-165.

1953b The Interdependence of Archaeology and Ecology with Examples from Middle America. *Annals of the New York Academy of Sciences, Series II,* 15:113-117.

1955 Palynology in Southern Northern America. Introduction and Acknowledgements. *Bulletin of the Geological Society of America,* 66:471-474.

SEARS, P. B., and CLISBY, K. H.

1952 Two long climactic records. *Science,* 166:176-178.

1955 Palynology of Southern North America. IV: Pleistocene Climate in Mexico. *Bulletin of the Geological Society of America,* 66:21-530.

SIRKIN, L.

1974 A Palynologic Model for Reconstructing Vegetation and Environments in the Marismas Nacionales, Sinaloa, Mexico. *In: West Mexican Prehistory,* 8:22-32. Stuart Scott, editor.

1984 Late Pleistocene Stratigraphy and Environments of the West Mexican Coastal Plain. *Neotectonics and Sea Level Variations in the Gulf of California Area, a Symposium* (V. Malpica-Cruz, S. Celis-Gutiérrez, J. Guerrero-Garcia and L. Ortlieb, editors). Instituto de Geología, Universidad Nacional Autónoma de México.

SIRKIN, L., and GILBERT, D.

1980 Holocene Palynology of the West Mexican Coastal Plain 1. The Teacapan Estuary Region. *Palynology,* 4:252.

STREET-PERROT, F. A., PERROTT, R. A.,

and HARKNESS, D. D.

1982 Holocene Environments and Man in Central Mexico—Some Preliminary Results. *Paper presented at the 44th International Congress of Americanists, Manchester.*

TING, W. S.

1965 The saccate pollen grains of Pinaceae mainly from California. *Grana Palynologica,* 6:270-289.

1966 Determination of *Pinus* species by pollen statistics. *University of California Publications in Geological Science,* No. 58.

VAILLANT, G. C.

1944 *The Aztecs of Mexico.* Doubleday, Doran and Co., New York.

VAN DEVENDER, T. R.

1979 *Reconstruction of late Pleistocene and Holocene vegetation and climate in the Chihuahuan Desert.* Research proposal submitted to the Climate Dynamics Office of the National Science Foundation, Washington, D. C.

WATTS, W. A., and BRADBURY, J. P.

1982 Paleoecological Studies at Lake Patzcuaro on the West-Central Mexican Plateau and at Chalco in the Basin of Mexico. *Quaternary Research,* 17(1):56-70.

WILSON, D.

1974 *The New Archaeology.* New American Library. New York.

ZEEVAERT, L.

1952 Estratigrafía y problemas de ingeniería en los depósitos de arcilla lacustres de la cuidad de México. *Revista Inginieria,* 25(1):12-28.

QUATERNARY POLLEN ANALYSIS AND VEGETATIONAL HISTORY OF THE SOUTHWEST

STEPHEN A. HALL
Department of Geography
University of Texas at Austin
Austin, Texas 78712

Abstract

In the Southwest, the level of maturity of pollen analytical-based vegetational history is two decades behind that of other regions in North America. Much of the pollen analytical research has been conducted at archeological sites where pollen assemblages may be biased by human disturbance and may not accurately depict the local or regional vegetation. Nevertheless, a number of studies provide a picture, albeit a weak one, of Southwest desert and alpine vegetational history for the past 40,000 years. During early- and full-glacial time, vegetation zones were lowered 900 to 1,400 meters. The transition from glacial to post-glacial vegetation and climate occurred between 14,000 and 12,000 years ago. Early-Holocene vegetation was characterized by decreasing woodlands, culminating in a marked shrinkage of woodland vegetation during the middle-Holocene during an episode of extremely warm and dry climate dated about 7,000 to 5,000 years ago. The late-Holocene vegetation was characterized in general by increased abundance of woodlands. Incompleteness and uncertainty in the late-glacial and mid-Holocene pollen records stem from changes in sedimentation rates, lowering of lake levels, drying of bogs, erosion, and soil formation. As a result, pollen sequences typically begin or end during the late-glacial or middle Holocene. Some of the problems and questions concerning Southwest vegetational history, such as pollen deterioration, fluvial palynology, pollen and climatic seasonality, middle-Holocene climate, and the pollen record and wood rat midden macrofossil record, are discussed. A bibliography of 549 references on Southwestern Quaternary palynology is included in a separate chapter of this volume.

INTRODUCTION

The primary focus of this review is the vegetational history of Arizona, Colorado, New Mexico, and Utah, as indicated by pollen analysis. The paper is a critical appraisal and survey of selected pollen sites which, ideally, come close to meeting the following criteria: the stratigraphic section from which the close-interval pollen samples were taken represents at least 2,000 years and is well described and well dated, pollen counts are moderately large, a usable pollen diagram is presented or, better still, pollen counts are presented in a table, and, finally, the study is published. Unfortunately, circumstances are such that many studies of important sites do not meet the above criteria, including some published papers as well as theses, dissertations, and abstracts of papers presented at meetings. The vegetational history outlined in this review includes consideration of pollen but not plant macrofossil data from wood rat middens. Finally, although vegetation is tied closely to climate, especially apparent in the habitat extremes of the Southwest, discussions of paleoclimate are held at a minimum.

HISTORICAL SKETCH

The field of Quaternary palynology in the American Southwest was initiated by Paul B. Sears when he examined alluvial samples that had been collected by Ernst Antevs from Tsegi Canyon, northeastern Arizona, in 1935 and 1936. Sears reported the results in 1937 (and 1961), demonstrating that a pollen sequence could be developed in that arid region. By 1959, however, only 14 papers in Southwestern pollen analysis had been published.

During the late 1950's and early 1960's, Paul S. Martin and his students and associates at the University of Arizona, Tucson, began a series of intensive investigations at a number of alluvial, lacustrine, and archeological sites throughout the arid and semiarid Southwest. At the same time, during the late 1950's, H. E. Wright, Jr., and his students at the University of Minnesota (Minneapolis) cored lake deposits in the San Juan Mountains of southwestern Colorado and the Chuska Mountains of northwestern New Mexico, producing the first pollen analyses of montane sites in the Southwest.

The Arizona-Colorado-New Mexico-Utah bibliography of Quaternary palynology (this volume) shows that 1960 was the turning point of published pollen analytical research in the Southwest (Fig. 1). Citations in the bibliography are subdivided into (a) cultural resource management or "contract archeology" reports, (b) ethnobotany, (c) paleovegetation and paleoecology, and (d) techniques, modern pollen, and aerobiology, all of which show steady increases in numbers during the 1960's, 1970's, and 1980's.

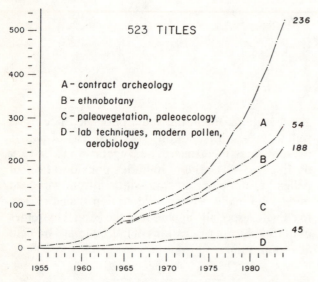

Figure 1. Cumulative number of Quaternary and modern pollen references from Arizona, Colorado, New Mexico, and Utah for the period 1937 to 1984 (see A Bibliography of Quaternary Palynology in Arizona, Colorado, New Mexico, and Utah; this volume).

Beginning in the 1960's, principally through the efforts of James Schoenwetter, a student of Paul Martin, pollen analytical studies became recognized as an important area of research in archeology. Pollen service work for archeology took on an expanded role during the increase in salvage archeology activities of the 1970's. In 1975, the number of new pollen analytical reports from contract archeology surged, and by 1984 these pollen studies accounted for nearly one-half of the entire Southwest palynological literature, largely an "invisible literature" of unpublished or limited distribution reports to institutions, government agencies, and private contracting firms. Eighty-five percent of this large literature has been produced by the laboratories of Linda J. Scott (Palynological Analysts, 386 Gladiola Street, Golden, Colorado 80401), Jannifer W. Gish (Quaternary Palynology Research Facility, 6000 West Canyon Drive, Littleton, Colorado 80123), Anne C. Cully and Karen H. Clarey (Castetter Laboratory for Ethnobotanical Studies, Department of Biology, University of New Mexico, Albuquerque, New Mexico 87131), and James Schoenwetter (Department of Anthropology, Arizona State University, Tempe, Arizona 85281). Many of the pollen studies deal with ethnobotanical

aspects of archeological sites; only a few have been published. The pollen data in these reports are important resources of ethnobotanical as well as paleoenvironmental information that has yet to be brought together in a comprehensive review.

MODERN POLLEN-MODERN VEGETATION

The Southwest flora is rich and its plant communities are varied (Kearney and Peebles, 1960; Martin and Hutchins, 1980; Weber, 1976; Brown, 1982; Benson and Darrow, 1981). However, pollen assemblages, both modern and fossil, are low in diversity compared with spectra from other regions. For example, studies of modern pollen surface samples in arid and semiarid areas generally record fewer than 40 types of which 5 or 6, such as *Pinus, Juniperus,* Gramineae, Chenopodiineae (Chenopodiaceae-*Amaranthus),* and short-spine and long-spine Compositae, may account for 90% of the pollen counts (Hevly, 1968; Hevly *et al.,* 1965; King and Sigleo, 1973; Martin, 1963a, 1963b; Mehringer *et al.,* 1971; Potter, 1967; Potter and Rowley, 1960; Schoenwetter, 1967; Schoenwetter and Doerschlag, 1971). In alpine areas, modern pollen surface samples are generally dominated by *Pinus, Picea, Abies, Juniperus, Quercus,* and *Artemisia* (Bent and Wright, 1963; Dixon, 1962; King, 1967; Maher, 1963, 1972a; Martin, 1963a).

The modern pollen surface samples collected in the above-cited studies are generally composed of 10 to 20 pinches of surface duff or mineral sediment from a small area, generally less than 50 square meters. Multiple pinches are mixed so as to avoid over-representation of pollen from individual plants in the collection area. Separate analysis of individual pinches has shown that a mix of at least 5 pinches of surficial pollen are necessary to characterize the local vegetation (Adam and Mehringer, 1975).

In an attempt to quantify pollen grain dispersal and influx, a 320-km transect of Tauber pollen traps was set up in the Southern Rocky Mountains and southwest Plains from 3,650 to 920 m elevation (Fig. 2). Pollen influx data from 1981 and 1982 show very different results (Fig. 3) that are related to long distance pollen grain dispersal and to annual variability in tree pollen production (Hall, 1984a). In a continuing Tauber trap study of pollen influx in Owens Valley, California, Solomon and Harrington (1979) have found that dispersal of diploxylon and haploxylon *Pinus* pollen is dissimilar. It is anticipated that these pollen dispersal and influx studies will provide useful background information that will aid in the interpretation of Southwest pollen diagrams.

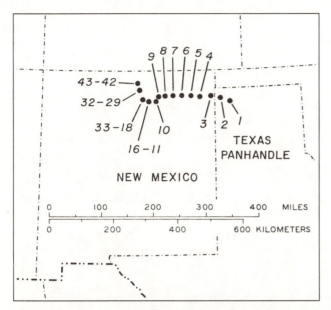

Figure 2. Location of pollen influx Tauber trap stations in New Mexico and the Texas Panhandle; pollen influx data shown in Fig. 3.

Pollen analytical studies have been greatly aided by a series of scanning electron microscope photomicrographs of various Southwestern pollen types (Martin, 1969; Martin and Drew, 1969, 1970; Solomon *et al.,* 1973). Other sources of helpful illustrations of Southwest pollen types are found in Kapp (1969), Mehringer (1967a), and Lewis *et al.* (1983).

Because of the varied ecologic and geographic characteristics of the pines and because *Pinus* pollen is found in virtually every modern and fossil sample in the Southwest, differentiation of *Pinus* pollen grains to species can be important in paleovegetation reconstruction. Investigations of size difference of *Pinus* pollen grains (Martin, 1963a, p. 20) show significant overlap in grain sizes of many pine species, indicating that size alone as a criterion for species differentiation may be misleading. Hansen and Cushing (1973) illustrate and describe qualitative and quantitative morphologic characters for pollen grains of *Pinus edulis, P. artistata* (California population), *P. flexilis, P. ponderosa,* and *P. contorta.* The differentiation of fossil pine pollen to species has resulted in important details in reconstructing past vegetation that would not otherwise have been possible (Wright *et al.,* 1973; Hall, 1977; Jacobs, 1983).

LABORATORY TECHNIQUES

Traditional laboratory methods used in processing peat (Faegri and Iversen, 1975) do not work well for arid land sediments that are dominated by quartz sand and silt. Early experimentation with various

techniques (Arms, 1960; Kurtz and Turner, 1957; Hafsten, 1959) resulted in a sequence of washes in hydrochloric, hydrofluoric, and nitric acids (Mehringer, 1967a) that has been used by many workers. Another laboratory technique that demands fewer chemicals has been developed. After hydrochloric and hydrofluoric acid washes, the organic pollen-bearing fraction of the sample is isolated by heavy liquid separation, using zinc bromide or zinc chloride with a specific gravity of 2.0 (Hall, 1977; Scott, 1983). Both techniques facilitate the processing of large volumes of sediment that contain low concentrations of fossil pollen.

POLLEN DETERIORATION

Reliable vegetation reconstruction may be dependent upon pollen grain preservation. In the arid Southwest, pollen is often not well preserved. Indeed, pollen is absent from many Quaternary age deposits, including archeological sites. Pollen deterioration may occur rapidly, and various experiments and observations indicate that pollen grains of different plant species are not destroyed at the same rate (Sangster and Dale, 1964; Holloway, 1981). Some pollen types are more susceptible to destruction than others, resulting in over-representation of some hardier pollen grains due to differential preservation. *Pinus* and *Picea* pollen, for example, are more resistant to deterioration than are pollen grains from nonconifers. The near-100% *Pinus* pollen frequencies in the upper 3.5 m at Willcox Playa, southeastern Arizona (Martin, 1963b), and the high percentages of *Pinus* and *Picea* at Estancia Basin (Fig. 4), central New Mexico (Bachhuber, 1971), can be explained, not by the presence of a nearby pine-spruce forest, but rather by differential destruction of non-*Pinus* and non-*Picea* pollen grains, resulting in artificially high percentages of *Pinus* and *Picea.*

Reliability of pollen assemblages can be partially evaluated by (a) tabulating the type of deterioration of each individual grain, using the corroded and degraded categories of Cushing (1967), as a monitor of the degree of deterioration and the percentage of grains deteriorated, (b) tabulating the number of indeterminable pollen grains that cannot be identified because of their advanced stage of deterioration, and (c) calculating the pollen concentration of each successive sample; a low concentration may result from poor preservation (Delcourt and Delcourt, 1980; Hall, 1981a). Bachhuber (1971) found that, at Estancia Basin, high percentages of *Pinus* and *Picea* occur in a zone characterized by high frequencies of

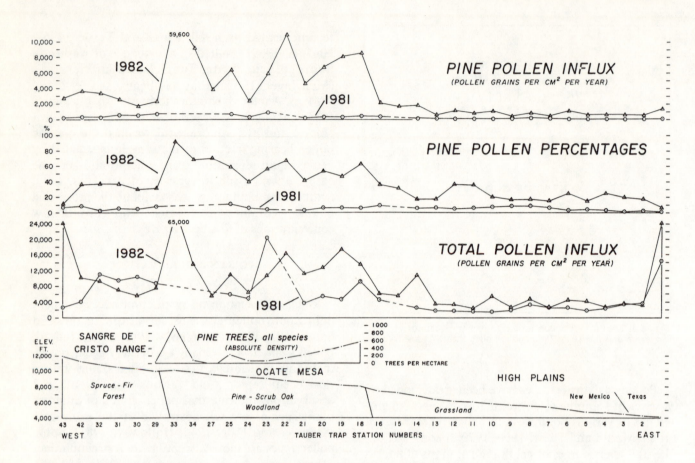

Figure 3. Pine pollen influx, pine percentages, and total pollen influx along a 320-km transect of Tauber traps, northeastern New Mexico and the Texas Panhandle; location of Tauber trap stations shown in Fig. 2 (from Hall, 1984a).

corroded grains and low pollen concentrations (Fig. 4).

ISSUES IN SOUTHWEST POLLEN ANALYSIS

A number of questions and issues dominate Southwestern pollen analysis and cloud reconstructions of late-Quaternary vegetation and environment. These issues include (1) fluvial palynology, (2) the pollen record and climatic seasonality, (3) middle-Holocene vegetation and environments, and (4) the pollen record and the wood rat midden macrofossil record. The following is a summary of the author's views on these topics.

Fluvial Palynology

The pollen record from alluvium depends ultimately upon (a) pollen origins, (b) pollen content of surface materials, (c) processes of sedimentation, and (d) post-depositional changes in alluvial pollen content. While some of the principles of pollen incorporation into alluvium developed in other regions (*i.e.,* Crowder and Cuddy, 1973; Peck, 1973) may

apply to the arid Southwest, significant differences may also be expected because of special arid land conditions such as sparse plant cover, torrential storms, and ephemeral streams. For this reason only Southwestern studies are discussed.

Pollen origins. The origins of pollen in alluvium are dependent upon a region's vegetation, climate, and geomorphology. Working in the Sonoran Desert of southeastern Arizona, Martin (1963a, p. 14-15) concluded that most alluvial pollen originates from plants on the floodplain, observing that pollen surface samples correspond closely to plant communities at the sampling point and that only small percentages of *Pinus, Quercus,* and *Ephedra* pollen from montane and upper bajada sources are found in alluvium. From studies within the City of Tucson in the Sonoran Desert, Solomon *et al.* (1982) also concluded that alluvial pollen originates principally from floodplain plants, citing as evidence the similar high frequencies of Chenopodiineae, *Ambrosia* type, *Aster* type, and Gramineae pollen in both freshly deposited alluvium and adjacent floodplain surface sediments. Floodplain plant communities do produce

Estancia Valley, New Mexico

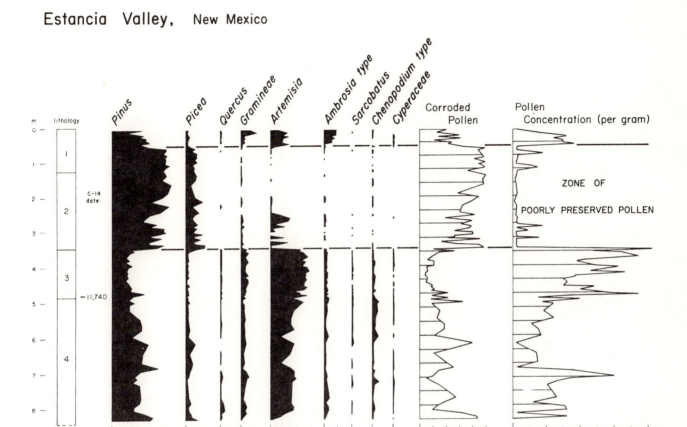

Figure 4. Summary relative frequency pollen diagram from Estancia Basin, central New Mexico, documenting regional dominance of sagebrush and pine vegetation during the Wisconsinan full- and late-glacial. The upper portion of the pollen diagram shows effects of differential preservation from post-depositional deterioration; the higher percentages of *Pinus* and *Picea* pollen are due to their greater resistance to destruction (modified from Bachhuber, 1971; Hall, 1981a).

large amounts of chenopod-composite-grass pollen, but non-floodplain lower and upper bajada communities in the Sonoran Desert also produce pollen assemblages similar to the floodplain pollen spectra discussed by Martin and Solomon, with equally high frequencies of chenopods-composites-grasses and low amounts of *Pinus, Quercus,* and *Ephedra* grains (Hevly *et al.,* 1965; Schoenwetter and Doerschlag, 1971). In fresh alluvium it may not be possible to distinguish, for example, Chenopodiineae pollen grains of adjacent floodplain plants from those transported by winds from upper bajada plant communities or by runoff from surficial sediments of upland slopes. Regardless, similar pollen assemblages characterize both floodplain and non-floodplain desert grassland communities of the Sonoran Desert. Because of the wide similarity of pollen assemblages, the region may not be ideal for investigating the processes of fluvial palynology.

In other areas of the Southwest, outside the Sonoran desert, non-floodplain plant communities may dominate alluvial pollen assemblages. A comparative study in northeastern Arizona of modern pollen surface samples from along the rim of Canyon del Muerto and freshly dried mud curls from the canyon floor has shown that most of the canyon floor pollen probably originates from the vegetation of the rim and plateau 200 to 500 m above the canyon floor (Fall, 1981). In another study, at Chaco Canyon, northwestern New Mexico, pollen analysis of historic-age sediment deposited since 1935 as inner fill in Chaco Wash arroyo shows that *Pinus* is about 50% of the pollen content at a locality over 5 km from the nearest pine tree. *Pinus, Juniperus, Quercus,* and *Artemisia* together comprise about 60% of the pollen spectra from the historic alluvium while none of these plants are found in the arroyo or on the canyon floor; rather, they occupy the escarpment slopes far upstream from the area where the pollen samples were collected. Abundant willow *(Salix),* salt cedar *(Tamarix,* historic introduction), greasewood *(Sarcobatus),* and some cottonwood *(Populus)* grow in Chaco Wash arroyo and on the canyon floor today, yet are collectively less than 5% of the historic age pollen assemblages. These riparian plants are likewise poorly represented by pollen in prehistoric alluvium at Chaco Canyon (Hall, 1977).

The Canyon del Muerto and Chaco Canyon studies indicate that the greater portion of pollen found in alluvium may originate from non-floodplain plant communities. Quantitative field measurements and experiments that would show how much of the non-floodplain-originating pollen is washed versus blown into floodplain areas have yet to be initiated in the Southwest. Solomon *et al.* (1982) specifically reject the possible role of pollen transported by overland flow, stating that pollen grains would be destroyed by abrasion during the process. They cite as evidence a 60% frequency of crumpled-degraded-corroded pollen in 5 alluvial samples from the Tucson area study in a ratio of 3-1-1, representing a total of 24% degraded and/or corroded grains in the samples. While this may appear to non-Southwestern pollen analysts as an exceptionally high degree of pollen deterioration, it is not atypical. At Chaco Canyon, for example, 12% of *Pinus* pollen grains in modern surface samples are corroded (Hall, 1977) and, at the Garnsey Bison Kill site in southeastern New Mexico, 30% of all pollen grains in the surface samples are degraded and/or corroded (Hall, 1984b). Clearly, however, if pollen surficial materials are a major source of the pollen that is incorporated into alluvium, then preservation of alluvial pollen will be no better than that of surface materials.

Surface samples and time. Time may also be an important characteristic of modern pollen assemblages, therefore having an important effect on alluvial pollen content. It is generally thought, although unverified, that modern surface samples represent many years, perhaps several decades, of pollen influx and mixing. Two separate studies of pollen surface samples collected from the same stations month-by-month throughout the year show that, while different plant species flower and release pollen, the surface pollen spectra remain similar (Hevly *et al.,* 1965; Solomon *et al.,* 1982), providing indirect evidence that surface pollen may represent many years of influx. Additional indirect evidence supporting multiple year accumulation comes from the high concentrations of pollen in surface sediment. At Chaco Canyon, surface pollen concentrations range from 19,000 to 38,200 grains per gram (Hall, 1977) and at the Garnsey Bison Kill site concentrations range from 26,300 to 69,300 grains per gram of dry surface sediment (Hall, 1984b). Pollen concentration analyses have not been widely undertaken in the Southwest and it is not known if the above values are

representative. Nevertheless, the large amounts of pollen in surface sediments at Chaco and Garnsey suggest a number of years of accumulation and mixing.

If indeed a major portion of alluvial pollen originates by sheet erosion of surficial materials containing pollen that has accumulated over many years, then the resolving power of vegetation change through analysis of alluvial pollen may be limited by the number of years of pollen influx and mixing in the surface material.

Processes of pollen sedimentation. The processes of alluvial pollen deposition in the arid Southwest are poorly known and can only be guessed, based on general sedimentologic principles. Pollen grains may be viewed as sedimentary particles, falling within the size range of medium silt to very fine sand. The low densities and flattened shapes of pollen may give them the settling properties of silt and clay. In a stream table study in which the substrate was sand and water velocity was 33 cm per sec (a very low velocity more characteristic of lake or ocean currents than a stream), pollen grain distribution in the water column was found to be constant for each of 14 pollen types introduced into the flume (Brush and Brush, 1972). Floating pollen grains, including *Pinus,* were below the concentration values of dispersed pollen in the water column. Thus it appears that, even in the most sluggish stream, pollen grains will be transported in suspension and dispersed evenly throughout the water column.

Fluvial deposition of pollen grains is another matter. It is evident from the Chaco Canyon study (Hall, 1977) that higher concentrations of pollen occur in finer grained prehistoric aluvium than are found in coarse deposits, although there is a great deal of variation. The principal question in alluvial pollen analysis is whether or not pollen grains have been sorted by size, hence by pollen type. If they have, the question then concerns the degree of differential sorting and its influence on pollen counts and interpretations of vegetation. Inspection of the historic alluvial pollen diagram from Chaco Canyon (Hall, 1977) shows that *Pinus* and Chenopodiineae frequencies correspond to high total pollen concentrations. It is not clear whether or not pollen sorting has occurred in this instance. In a study of prehistoric alluvium at Canyon del Muerto, without pollen concentration analysis, comparatively high frequencies of *Pinus* pollen correspond to sedimentary layers containing high percentages of silt and clay (Fall, 1981), indicating that some degree of sorting may have occurred. However, the historic alluvium at

Chaco Canyon and the prehistoric alluvium at Canyon del Muerto are comparatively coarse grained deposits with crossbedding and ripples, representing levees, point bars, and channel fills. These contrast with fine grained alluvium deposited by overbank sedimentation over a broad floodplain surface. Both types of alluvial valley fill are found in the Southwest. If pollen sorting has occurred, it will most likely be in coarse grained alluvium.

Post-depositional changes. Post-depositional alteration of alluvial pollen assemblages generally involves pollen destruction and loss through oxidation, which can occur by weathering and soil formation, by repeated wetting during fluctuations in a water table, or by water percolation through pervious sediment (Holloway, 1981). Even though water table levels may have been far below some Southwest alluvial valley floors throughout the Holocene, coarse-grained alluvium may have a long history of water percolation. If pollen grains are already poorly preserved prior to incorporation into alluvium, they may be more susceptible to deterioration after deposition. Also, while clays may be displaced by water fronts moving through coarse sediment, it is unlikely that pollen grains, equivalent in size from medium silt to very fine sand, would be moved through any but the coarsest, well-sorted sand or gravel.

In summary, as far as can be determined, Southwest fluvial pollen originates by overland flow and sheet erosion of pollen-bearing surficial materials; these surface pollen assemblages may represent many years of atmospheric influx and accumulation. Pollen incorporated in a flowing stream is transported in a uniform suspension and deposited downstream in overbank or channel alluvium. The ideal alluvial setting for pollen analytical studies would be a tributary valley with a small drainage basin encompassing a single vegetation type. The alluvium itself would be unweathered fine grained silts and clays.

The Pollen Record and Climatic Seasonality

In Southwestern tradition, seasonality generally refers to the time of the year of the greatest rainfall. Today's climate in Arizona, New Mexico, southern Utah, and most of Colorado is characterized by a heavy summer-dominant (July, August, September) rainfall while a spring-dominant (April, May, June) precipitation occurs in the northern half of Utah in the Great Basin (Visher, 1950).

Interpretations of shifts between light-summer and heavy-summer rainfall patterns have been based on relationships between plant ecology, stream geomorphology, and modern pollen surface data, especially from Empire Cienega in southeastern Arizona. Empire Cienega is an undissected valley which, when plant populations were surveyed and modern pollen surface samples collected in 1962, contained colonies of perennial ragweed *(Ambrosia psilostachys)* and giant ragweed *(A. trifida)* on the south side of the valley (Martin, 1963a, p. 21, 23). Short-spine Compositae pollen percentages in surface samples beneath the *Ambrosia* colonies were found to be as high as 80%. On the north side of the transect, *Atriplex* was present, resulting in over 80% Chenopodiineae pollen in surface samples. Martin (1963a, p. 59) and Schoenwetter (1962, p. 197) concluded from the Empire Cienega study that short-spine Compositae pollen (from *Ambrosia)* characterizes undissected wet meadow or cienega environments with a high water table and that, conversely (and ignoring the *Atriplex* from Empire Cienega), high frequencies of Chenopodiineae pollen are characteristic of dissected or eroding floodplains with more alkaline soils and a low water table. They further propose that Compositae-dominated cienegas are maintained during periods of light-summer rainfall while Chenopodiineae-dominated eroding floodplains are formed during periods of heavy-summer rainfall. Martin and Schoenwetter also cite the pollen diagram from late-Holocene alluvium at Cienega Creek (Martin, 1963a, p. 32; 1983, p. 43) as supporting evidence for high Compositae percentages corresponding to cienega environments. Two cienega zones occur in the alluvial pollen section. However, Compositae pollen frequencies are highly variable; both high and low frequencies were found in the cienega clays as well as the non-cienega sands.

The light- versus heavy-summer rainfall seasonality model has not been addressed by new investigations since its proposal in the early 1960's. However, a number of criticisms can be suggested. First, the model requires that most of the pollen incorporated into alluvium originate from floodplain plant communities. This issue was discussed in the section on Fluvial Palynology where it was concluded that most pollen in Southwest alluvium originates from non-floodplain sources. Specifically concerning the Sonoran Desert area studies, it was pointed out that non-floodplain sources could be the main supplier of pollen to alluvium but, owing to the strong similarity of floodplain and bajada pollen surface assemblages, pollen origins there may not be easily determined.

The pollen basis for the light/heavy-summer rainfall model is the high percentage of short-spine

Compositae pollen associated with ragweed plants in the undissected Empire Cienega. But, as Martin's pollen transect diagram shows, high frequencies of Chenopodiineae pollen are also present at Empire Cienega. Furthermore, the evidence from the Cienega Creek pollen diagram, discussed above, does not support the proposed Compositae pollen-cienega association.

The geomorphic aspects of the model call for channel trenching initiated by heavy summer rainfall. Winter rainfall, Schoenwetter (1962, p. 197) explains, is from low energy storms and is less likely to initiate channel trenching than are intense summer cloudbursts. While the cause of channel trenching has been of interest to geologists for a long time (Bryan, 1925), the search for a single explanation has not met with success. Even though periodic increases of rainfall intensity (with and without accompanying increases in total annual rainfall) have been shown to correspond to brief episodes of historic channel trenching in the Southwest (Leopold, 1951; Leopold *et al.,* 1966; Bull, 1964), it remains to be determined whether or not this explanation of historic trenching holds for other regions or can be applied to Holocene alluvial sequences.

Schoenwetter (1964, 1967) expanded the seasonality model to include winter-dominant rainfall in his studies at Navajo Reservoir and in the western San Juan Basin, northwestern New Mexico. He argued (1964, p. 101-102; 1967, p. 101-102) that high energy summer-dominant rainfall runs off quickly, eroding topsoil and resulting in floods that erode and trench valley fill and lower the water table. The ecologic consequence of this is reduced effective moisture availability with a corresponding reduction in plant growth. Low energy winter-dominant rainfall results in higher effective moisture availability, increased plant growth, and alluvial deposition, explains Schoenwetter. In both cases, Schoenwetter equates effective moisture with plant growth which in turn is equated to arboreal pollen percentages. In other words, greater effective moisture promotes plant growth which results in increased percentages of arboreal pollen percentages (mostly *Pinus* and *Juniperus).* Schoenwetter (1964, p. 102) further explains that high arboreal pollen frequencies (high effective moisture) occurring during alluvial deposition are caused by winter-dominant rainfall; high arboreal pollen occurring during valley fill erosion is caused by summer-dominant rainfall; low arboreal pollen during alluviation is caused by a decline in total annual precipitation. Valley fill erosion is still regarded as a result of summer rainfall (1967, p. 102).

Interpretation of the character and changes of climatic seasonality is an important goal, and while the intensity or dominance of rainfall could have a marked effect on vegetation and stream valley processes, the writer feels that the model relies too heavily on a rigid and unverified relationsip between rainfall pattern, alluviation, and erosion. Indeed, in application of the model, one would not be able to determine the paleoenvironmental-geomorphic relationship because that relationship is predetermined by the model. The model places similar constraints on the interpretation of past vegetation. Climatic, vegetation, and fluvial conditions have without doubt changed during the Holocene, but not necessarily in the manner proposed by Martin or Schoenwetter.

Middle-Holocene Vegetation and Climate

Mid-post-glacial time in virtually every part of the globe has long been regarded as a period of drier and warmer climate. In Europe, where the dry-warm interval is recognized by pollen analysis and other criteria, it is called the Climatic Optimum (Brooks, 1949). In eastern North America it is known as the Hypsithermal (Deevey and Flint, 1957). The Southwestern equivalent of the arid period was named the Altithermal by Ernst Antevs (1948) who also referred to it as the "Long Drought." From various lines of evidence in the early days of radiocarbon dating, the period of dry climate in the western United States was thought to extend from 7,500 to 4,000 yrs. B.P. (Antevs, 1955).

In the early 1960's, in contradiction to long-held views of a dry altithermal, Paul S. Martin proposed that mid-post-glacial time in the Southwest was characterized by increased summer rainfall, based on alluvial pollen evidence from southeastern Arizona (Martin, 1963a, 1963c). The stratigraphy and chronology of the alluvial sections along Whitewater Draw, where Martin obtained the wet altithermal pollen data, have been re-evaluated and indicate that the general period 5,000 to 7,000 yrs. B.P. is missing from the record due to the presence of an erosional unconformity (Antevs, 1962; Sayles, 1965; Haynes, 1968, p. 606-607; Mehringer, 1967b, p. 99-100; Waters, 1984). Accordingly, Martin's interpretation of a wet altithermal is mistaken because sediments from that time interval are absent. Although the circumstances which led to the mistaken proposal of a wet altithermal have been adequately described by the above-cited authors in the 1960's, a middle-Holocene wet interval continues to be cited as an established fact (Wells, 1970, p. 195; Van Devender and Spaulding, 1979, p. 709).

The Double Adobe I site (Fig. 5) is one of the principal alluvial pollen localities on which Martin (1963a, 1963c) based and defended the concept of a wet altithermal. It also is the site that Sayles (1965) and Haynes (1968) criticised for misinterpretations of stratigraphic and chronologic correlations. At Double Adobe I, Martin's "white silt, clay" is a buried calcic soil developed in alluvium dated about 7,500 to 9,500 yrs. B.P. (Haynes, 1968, p. 606). The top of the "white silt, clay" unit is an erosional unconformity representing the period about 7,500 to 4,000 or 5,000 yrs. B.P.; Haynes correlates the undated overlying

"indurated silt" with his Deposition D unit (4,000 to 2,000 yrs. B.P.). Thus, the middle of the Double Adobe I pollen section is marked by an erosional unconformity and a buried soil that together represent the entire middle-Holocene altithermal. Furthermore, as Haynes points out, when the 3,000-year unconformity is taken into account, the pollen succession of decreasing *Pinus* and increasing Chenopodiineae frequencies indicates onset of greater aridity. Regardless, pollen evidence has now accumulated from across the Southwest, and the record indicates that without qualification the middle-Holocene was a period of exceptionally warm and dry climate.

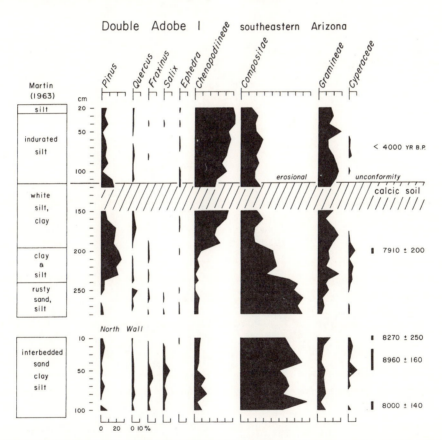

Figure 5. Summary pollen relative frequency diagram from Double Adobe I, southeastern Arizona, documenting early-Holocene vegetation. The Double Adobe I alluvial pollen site is the principal evidence Martin (1963a,c) cited for a wet mid-Holocene (or wet altithermal) climate in the Southwest. It was later determined that the middle-Holocene is missing due to the presence of an erosional unconformity and soil development; the period 7,000 to 4,000 yrs. B.P. may be missing from the alluvial record at Double Adobe I (Haynes, 1968). Pollen diagram modified from Martin (1963a, 1963c).

Pollen and Wood Rat Middens

Ancient wood rat *(Neotoma)* nests and middens are preserved in protected rock overhangs and fissures along escarpments throughout low elevation areas of the arid and semiarid Southwest. Analysis and radiocarbon dating of well-preserved plant macrofossils from middens have led to reconstructions of late-Quaternary vegetation and phytogeography of the western United States, summarized by Spaulding *et al.* (1983) and Baker (1983).

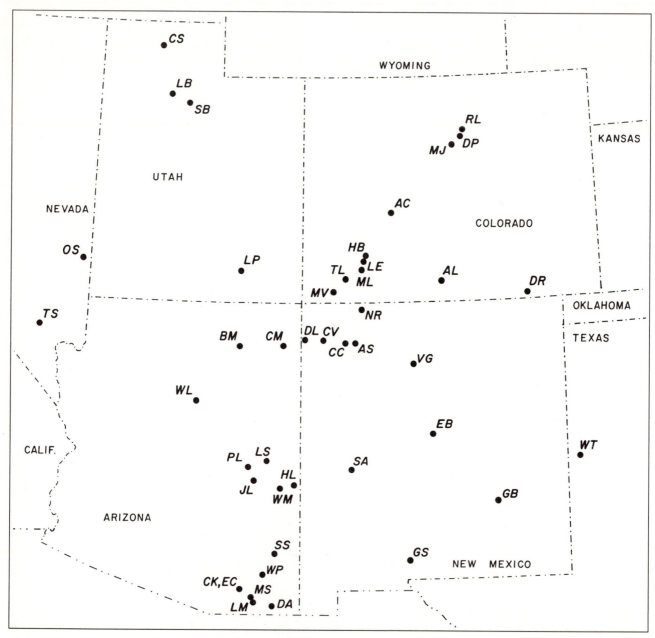

Figure 6. Map of Quaternary pollen sites in Arizona, Colorado, New Mexico, and Utah mentioned in text (see Table 1 for complete list with references).

Pollen analysis of fossil wood rat middens is in a beginning stage although a number of small studies have been reported (Van Devender and King, 1971; King and Van Devender, 1977; Hall, 1981b, 1982; Cole, 1982, 1983; Van Devender and Toolin, 1983). Pollen grains occur abundantly in wood rat middens, from 3,000 to 46,000 grains per cc in one study (Van Devender and Toolin, 1983) and 17,000 to 281,000 grains per gram dry weight in another (Hall, 1981b); grain preservation is generally excellent.

Pollen recovered from wood rat middens originates (a) by atmospheric transport directly to the midden,

(b) by grain impact on plant stems brought to middens by wood rats, and (c) from occasional flowers and cones also brought to middens by wood rat activity. Incorporation of flowers or cones should result in over-representation of pollen grains from those plants in comparison with expected pollen frequencies from modern surface samples. Analyses completed so far indicate that juniper has a strong potential for being over-represented by pollen in middens, probably because cones are terminal on stems (King and Van Devender, 1977; Hall, 1981b). Fecal pellets are another source of potential over-representation of some pollen types in wood rat middens. Pollen in fecal material may reflect *Neotoma* diet rather than the local plant community (Van Devender and King, 1971). If fecal pellets are broken up and incorporated in midden matrix, pollen analytical results may be biased. Generally, however, pollen over-representation can be recognized owing to extraordinary high percentages of one pollen type. The processes of pollen incorporation in wood rat middens and the possible sources of pollen over-representation or under-representation with respect to the local and regional plant communities are areas of continuing research.

The principal issue of wood rat midden palynology concerns the source of incongruence between pollen and plant macrofossil data in vegetation reconstruction. At present, many of the pollen studies of midden material result in vegetation interpretations that differ significantly from interpretations based on plant macrofossils. King and Van Devender (1977) were the first to point out dissimilarities in pollen and macrofossil data for the same wood rat middens, concluding that, while macrofossils provide detailed information on vegetation in the vicinity of wood rat nests, pollen analysis reflects more of the regional vegetation. Other pollen and macrofossil investigations of late-Pleistocene and Holocene middens show a pattern of divergence in interpretations: macrofossil remains are generally viewed as woodland or forest vegetation while pollen spectra from the same middens are interpreted as indicating fewer trees and a greater component of shrubs in the vegetation (Cole, 1982, 1983; Van Devender and Toolin, 1983; Hall, 1981b, 1982).

At Chaco Canyon, New Mexico, plant macrofossils from a series of Holocene wood rat middens were interpreted as pinyon and juniper woodland (Betancourt and Van Devender, 1981). Pollen analysis of the same middens showed that *Pinus* pollen frequencies (about 20%) were generally no greater than those found in modern pollen surface samples at Chaco Canyon where pine trees are today absent; the pollen record is interpreted as a shrub grassland vegetation (Hall, 1981b, 1982). Two independent pollen analytical studies at Chaco Canyon, one of the canyon fill (Hall, 1977, 1983) and the other of cave sediments (Fredlund, 1984; Fredlund and Johnson, 1984), both indicate middle-Holocene reduction of regional pine populations and increased abundance of shrubs in agreement with the pollen results from the wood rat middens. It is concluded from the Chaco Canyon data that the pinyon pine microfossils, even though common in most middens studied,

represent small stands or isolated individuals along the canyon escarpment that were insufficiently abundant to significantly affect the pollen rain.

Vegetation interpretations from wood rat midden macrofossils can be criticised as potentially misleading because of (a) behavioral differences among various species of *Neotoma* in selection of food and nest materials, such as documented in the classic study of Colorado wood rats by Finley (1958), (b) preservation of fossil wood rat middens only along escarpments (adult wood rats forage no more than 50 meters from their nests) where atypical microhabitats frequently result in escarpment plant communities that differ from non-escarpment local and regional vegetation, (c) nearly complete absence of calibration studies of plant remains from modern middens with adjacent plant communities, and (d) the possibility that some plant remains in middens are relict species from a former time when another plant community type dominated the landscape under different climatic conditions. Finally, until calibration studies of modern midden plant materials and modern vegetation in a variety of situations can be completed, plant macrofossil information can only be regarded as presence-absence data. Even though fossil plant fragments provide positive evidence for the presence of many species, interpretations of vegetation from the macrofossils alone are at present without firm foundation and may be in error. Analysis of both pollen and plant macrofossils together promises to be the best approach, macrofossils providing a minimum list of species present at the midden site and pollen indicating the abundance and importance of those and other plants in the local and regional vegetation.

QUATERNARY VEGETATIONAL HISTORY

Only a few summaries of Southwestern pollen analysis and vegetational history have appeared. The first was a brief summary by Paul S. Martin (1964) of eight full-glacial records located from Texas to California, followed by a more comprehensive synthesis by Martin and Mehringer (1965) that has long served as a regional standard. A brief but important critical survey of the important pollen studies of the early 1960's was written by Peter J. Mehringer, Jr. (1967b). Some Southwest records have been discussed in the context of broader regional summaries (Wright, 1971; McAndrews and King, 1976; Heusser, 1977; Baker, 1983). A selection of Colorado pollen studies, mainly lake and bog localities, has been recently summarized by Nichols (1982). A compilation of Southwestern glacial vegetation and

ABBREV.	SITE, STATE*	TIME RANGE	REFERENCES
AC	Alkali Creek, CO	late-glacial to late-Holocene	Markgraf and Scott, 1981
AL	Alamosa local fauna, CO	middle-Pleistocene	Price, 1971; Rogers, 1984
AS	Ashislepah Shelter, NM	late-Holocene	Fredlund, 1984; Fredlund and Johnson, 1984
BM	Black Mesa, AZ	late-Holocene	Euler et al., 1979
CC	Chaco Canyon, NM	late-Holocene	Hall, 1977, 1983
CK	Cienega Creek, AZ	late-Holocene	Martin, 1963a, 1983
CM	Canyon del Muerto, AZ	late-Holocene	Fall, 1981
CS	Crescent Spring, UT	early- to full-glacial	Mehringer, 1977
CV	Chuska Valley, NM	late-Holocene	Schoenwetter, 1967
DA	Double Adobe I, AZ	early-Holocene	Martin, 1963a, 1963c
DM	Dead Man Lake, NM	early- to full-glacial; Holocene (incomplete)	Bent and Wright, 1963; Wright et al., 1973
DP	Devlins Park, CO	full-glacial	Legg and Baker, 1980
DR	Donnelly Ranch, CO	Sangamonian	Hager, 1975
EB	Estancia Basin, NM	full-glacial to early-Holocene	Bachhuber, 1971
EC	Empire Cienega, AZ	modern	Martin, 1963a
GB	Garnsey Bison Kill site, NM	late-Holocene	Hall, 1984b
GS	Gardner Spring, NM	late-Holocene	Freeman, 1972
HB	Hurricane Basin, CO	early- and late-Holocene	Andrews et al., 1975
HL	Hay Lake, AZ	early-glacial to Holocene	Jacobs, 1983
JL	Jacob Lake, AZ	full-glacial to Holocene	Jacobs, 1983
LB	Lake Bonneville, UT	early-Pleistocene to late-Holocene	Madsen and Kay, 1982
LE	Lake Emma, CO	full-glacial to late-Holocene	Carrara et al., 1984
LM	Lehner Mammoth Kill site, AZ	late-glacial to late-Holocene	Mehringer and Haynes, 1965; Mehringer et al., 1971
LP	Lake Pagahrit, UT	late-Holocene	Lipe et al., 1975
LS	Laguna Salada, AZ	late-glacial to late-Holocene	Hevly, 1962, 1964
MJ	Mary Jane site, CO	early-glacial to late-Holocene (early-Holocene missing)	Millington, 1976 (in Nichols, 1982); Nelson et al., 1979
ML	Molas Lake and Molas Pass Bog, CO	full-glacial to late-Holocene	Maher, 1961, 1972b
MS	Murray Springs, AZ	late-Holocene	Mehringer et al., 1967
MV	Mesa Verde, CO	late-Holocene	Martin and Byers, 1965; Wyckoff, 1977
NR	Navajo Reservoir, NM	late-Holocene	Schoenwetter, 1964
OS	O'Malley Shelter, NV	late-Holocene	Madsen, 1973
PL	Potato Lake, AZ	full-glacial to late-Holocene	Whiteside, 1965
RL	Redrock Lake, CO	early-Holocene to late-Holocene	Maher, 1972a, 1973
SA	San Agustin Plains, NM	early-Pleistocene to late-Holocene	Clisby and Sears, 1956; Clisby et al., 1957; Markgraf et al., 1983, 1984
SB	Snowbird Bog, UT	late-glacial to late-Holocene	Madsen and Currey, 1979
SS	Safford-San Simon valleys, AZ	early-Pleistocene (?)	Gray, 1961
TL	Twin Lakes, CO	Holocene	Petersen and Mehringer, 1976
TS	Tule Springs, NV	early-glacial to Holocene (full-glacial missing)	Mehringer, 1965, 1967a
VG	Valle Grande, NM	pre-Wisconsinan (?)	Sears and Clisby, 1952
WL	Walker Lake, AZ	early-glacial to late-Holocene	Berry et al., 1982; Hevly, 1984
WT	White Lake, TX	early-glacial to Holocene (late-glacial missing)	Oldfield, 1975
WM	White Mountains, AZ (3 sites)	early-glacial to Holocene	Batchelder and Merrill, 1976
WP	Willcox Playa, AZ	Illinoian (?) to early-glacial	Hevly and Martin, 1961; Martin, 1963b; Martin and Mosimann, 1965

*Pollen site locations shown on map, Fig. 6.

plant biogeography, based primarily on plant macrofossils from wood rat middens, has been presented by Spaulding, Leopold, and Van Devender (1983).

The Southwest Quaternary pollen record is discussed within the categories of Wisconsinan Early-Glacial (>24,000 B.P.), Wisconsinan Full-Glacial (24,000-14,000 B.P.), Wisconsinan Late-Glacial (14,000-10,000 B.P.), Early-Holocene (10,000-7,000 B.P.), and Late-Holocene (7,000-0 B.P.). The pollen study sites mentioned in the text are located on the map (Fig. 6) and are listed in Table 1.

Early- and Middle-Pleistocene

Only a handful of pre-Wisconsinan Pleistocene localities, including core sites, have yielded pollen records. Most Pleistocene sediments in the arid Southwest are too coarse textured for pollen grains to have been deposited, or the pollen has been destroyed by post-depositional oxidation. The circumstances resulting in the small number of pollen-bearing Pleistocene sequences may be little different from those that characterize other regions of the world (Kremp, 1978), such as East Africa where only a small proportion of the late Neogene sites investigated contain pollen assemblages suitable for vegetation reconstruction (Bonnefille, 1984). Nevertheless, the Southwest is rich in Pleistocene stratigraphy containing volcanic ashes that can provide a valuable chronology (Izette and Wilcox, 1982). It is anticipated that, with additional reconnaissance, an early- and middle-Pleistocene vegetational history of the Southwest can be developed.

Early-Pleistocene pollen records in the Southwest are rare. Among the presently identified sites with a good potential for early-Pleistocene vegetation reconstruction are the San Agustin Plains, New Mexico (Clisby and Sears, 1956; Markgraf *et al.,* 1983, 1984), and Lake Bonneville, Utah (Eardley and Gvosdetsky, 1960; Madsen and Kay, 1982), where cores extending to 1.6 and 3.0 m.y., respectively, have already been obtained.

Middle-Pleistocene pollen is documented at the Safford-San Simon valleys in southeastern Arizona (Gray, 1961) and the Alamosa local fauna of southern Colorado (Price, 1971; Rogers, 1984). The Alamosa site, in contrast to other pre-Wisconsinan records obtained exclusively from cores, is a 16-meter thick outcrop at Hansen Bluff in the San Luis Valley. It contains an extensive vertebrate and invertebrate fauna and spans the Matuyama-Brunhes polarity-reversal horizon. Pollen evidence for elevational

shifts of forest and sagebrush-grassland vegetation correlates with the marine paleotemperature curve derived from oxygen isotope analysis (K. L. Rogers, C. A. Repenning, S. A. Hall, and others, in preparation). Both the Arizona and Colorado records indicate substantial shifts in vegetation zones, approaching the magnitude of that documented for the Wisconsinan.

Other possible pre-Wisconsinan pollen sites include the core records from Valle Grande, north-central New Mexico (Sears and Clisby, 1952), and Willcox Playa, southeastern Arizona (Martin, 1963b). A Sangamonian pollen record from southeastern Colorado is reported from the Donnelly Ranch vertebrate fauna (Hager, 1975). Additional Sangamonian and Illinoian pollen records have been recovered from southwestern Kansas and adjacent Oklahoma (Kapp, 1965).

Wisconsinan, Early-Glacial (>24,000 yrs. B.P.)

Only five pollen sequences of certain early-glacial age (>24,000 yrs. B.P.) are available for comparison: Tule Springs, Nevada (Mehringer, 1965, 1967a), Willcox Playa, Arizona (Martin, 1963b; Hevly and Martin, 1961; Martin and Mosimann, 1965), Dead Man Lake in the Chuska Mountains, New Mexico (Bent and Wright, 1963; Wright *et al.,* 1973), the San Agustin Plains, New Mexico (Clisby and Sears, 1956; Clisby *et al.,* 1957), and the Mary Jane site, central Colorado (Millington, 1976; Nelson *et al.,* 1979). The records indicate an expansion of sagebrush communities into low elevational areas where sagebrush today is absent and regional increases in pine, spruce, and fir as a result of the lowering of vegetation zones, variously estimated to have been 900 to 1,400 meters below present-day ranges. At Tule Springs, Mehringer (1967a) concludes that the vegetation was 1200 meters lower than today with sagebrush occupying the broad valley floor, a yellow pine parkland on both high ground and the lower bajada, and stands of fir on the upper bajada. In the Chuska Mountains, vegetation was lowered a minimum of 900 meters and spruce-fir and ponderosa pine communities were compressed; a sagebrush and pinyon-juniper vegetation occupied the floor of the San Juan Basin. Sagebrush also dominated the early-glacial vegetation of the western High Plains as recorded by the pollen record at White Lake, Texas (Oldfield, 1975). Unpublished early-glacial records in Arizona have been obtained from the White Mountains (Batchelder and Merrill, 1976), Walker Lake (Berry *et al.,* 1982; Hevly, 1984), and Hay Lake (Jacobs, 1983).

Wisconsinan, Full-Glacial (24,000 to 14,000 yrs. B.P.)

Full-glacial pollen records are more numerous than early-glacial, most from lake sediments deposited during the cool-moist glacial climate. The majority of the records are dominated by *Artemisia* and *Pinus* with minor amounts of *Picea* and *Abies* pollen and are generally interpreted as representing a cool sagebrush steppe or sagebrush-pine woodland vegetation, a vegetation type that is not widespread today in the Southwest. The sagebrush vegetation occupied lower elevation areas, which during much of the Holocene and today are dry shrub grasslands that produce assemblages dominated by Gramineae, Chenopodiineae, and Compositae pollen. Increased growth of sagebrush in low elevation areas resulted in a regional increase in *Artemisia* pollen that was wind-drifted to low and high elevation sites alike.

Pollen sites with full-glacial records dominated by *Artemisia* and *Pinus* or *Picea* include Potato Lake (Whiteside, 1965), Jacob Lake (Jacobs, 1983), Walker Lake (Berry *et al.*, 1982; Hevly, 1984), Devlins Park (Legg and Baker, 1980), Molas Lake (Maher, 1961, 1972b), Lake Emma (Carrara *et al.*, 1984), Estancia Basin (Bachhuber, 1971), Dead Man Lake (Wright *et al.*, 1973), San Agustin Plains (Markgraf *et al.*, 1983, 1984), White Lake (Oldfield, 1975), and Crescent Spring (Mehringer, 1977). Elevational differences among the above pollen sites may account for some of the observed variation in the pollen sequences. At high elevational sites, treeless herbaceous vegetation expanded in area as the treeline was depressed. The treeless alpine vegetation was probably a low pollen producer, and pollen grains from spruce-pine forests and sagebrush-pine woodlands at lower elevations were wind-drifted upslope, inundating the pollen influx at alpine sites (Markgraf, 1980; Hall, 1984a).

In general, the vegetation reconstructions from early-glacial and full-glacial appear to be similar. Continuous early- and full-glacial pollen sites include Hay Lake and Jacob Lake (Jacobs, 1983), White Mountain sites (Batchelder and Merrill, 1976), Walker Lake (Berry *et al.*, 1982; Hevly, 1984), San Agustin Plains (Clisby *et al.*, 1957), Dead Man Lake (Wright *et al.*, 1973), White Lake (Oldfield, 1975), and Crescent Spring (Mehringer, 1977). However, only the Dead Man Lake site meets the criteria outlined in the Introduction. The Dead Man Lake record (Fig. 7), in the early-to full-glacial transition, shows a decrease in *Picea* with a moderate increase in *Pinus* and with slight increases in *Abies, Quercus, Juniperus,* non-*Ambrosia* Compositae, and non-

Sarcobatus Chenopodiineae. Gramineae, *Artemisia,* and *Ambrosia*-type percentages remain about the same. Differentiation of *Pinus* pollen grains to species indicates that the pine supplying half of the early-glacial *Pinus* pollen influx was *P. ponderosa.* At the transition to full-glacial, the increase in *Pinus* pollen was principally *P. edulis* and *P. flexilis,* with a marked decrease in *P. ponderosa.* The upper forest limit was lowered 900 m during early-glacial time and, during full-glacial time was perhaps 1,000 m below its present position (a pre-34,000 yrs. B.P. zone of vegetation similar to that of full-glacial time is interpreted by Wright *et al.,* 1973, from the Dead Man Lake record). The 1,000 m lowering of upper forest limits during the full-glacial was accompanied by a compression of the ponderosa pine zone, resulting in the San Juan Basin being occupied by a sagebrush and pinyon-juniper woodland vegetation and accounting for the large amount of *Artemisia* pollen and large proportion of differentiated *Pinus edulis* grains wind-drifted upslope to the Chuska Mountain lake sites. A decrease in *Abies, Quercus, Juniperus,* Compositae, and Chenopodiineae frequencies and an increase in *Pinus ponderosa* percentages during the last part of the full-glacial, about 18,500 to 13,500 yrs. B.P., is interpreted as representing a 100 m rise in the upper treeline. Late-glacial time is largely absent from the Chuska Mountain sites or is characterized by very slow sedimentation rates and is unrecognizable in the pollen sequence.

The earlier interpretation of widespread ponderosa pine parkland at low elevational areas in the Southwest, including the Texas Plains, was based largely on high percentages of *Pinus* pollen at Rich Lake, Crane Lake, and Willcox Playa (Martin, 1964; Martin and Mehringer, 1965). As discussed elsewhere in this review, high frequencies of *Pinus,* such as observed at the above sites, are likely due to differential preservation.

Wisconsinan Late-Glacial (14,000 to 10,000 yrs. B.P.)

The late-glacial period represents the transition from glacial to post-glacial vegetation. Many Southwest records, especially those from lakes and bogs, either begin or end with the late-glacial. Other records that transcend the glacial/post-glacial boundary are marked by changes in sedimentation rates. The explanation for this lies in part with the changing hydrology of lakes as the climate became warmer/drier. Diminished water depth, decreased lake surface area,

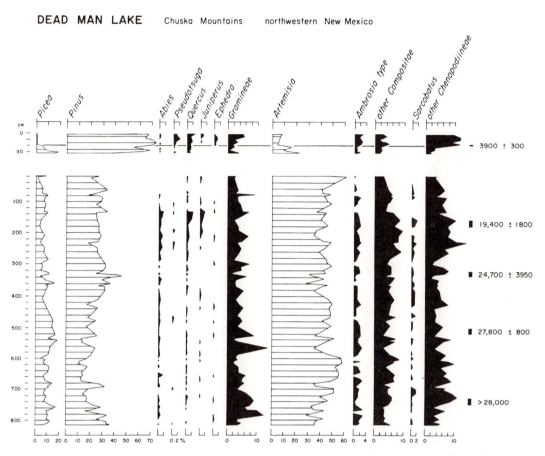

Figure 7. Summary pollen relative frequency diagram from Dead Man Lake, northwestern New Mexico, documenting regional dominance during the Wisconsinan of sagebrush grassland vegetation with spruce and pine (modified from Wright *et al.,* 1973).

and slowing of sediment deposition rates may affect pollen sedimentation in lake basins. Lowering of lake levels also exposes new areas to colonization by species in the Chenopodiineae and Compositae, resulting in vast increases in pollen production originating from shoreline weeds that may mask climate-related changes in nearby forest vegetation. Pollen influx analysis, based on multiple close-interval radiocarbon dates, will help resolve regional versus local vegetation patterns as well as the particular nature of vegetational changes during the glacial/post-glacial transition.

All of the useable pollen records indicate that the principal change from glacial to post-glacial vegetation in the Southwest took place between 14,000 and 12,000 yrs. B.P. At Tule Springs, southern Nevada, a major shift from a juniper-sagebrush to a sagebrush-shadescale vegetation occurred about 12,000 yrs. B.P. (Mehringer, 1967a). At Potato Lake, eastern Arizona, a major change from an undifferentiated spruce/fir/Douglas fir and sagebrush vegetation to a pine-oak and grass vegetation occurred stratigraphically just above a zone radiocarbon dated 14,400 ± 300 yrs.

B.P. (Whiteside, 1965), similar to a vegetation shift at nearby Jacob Lake where a pine-spruce-sagebrush was replaced by an open pine-grass vegetation about 14,700 ± 1600 yrs. B.P. (Jacobs, 1983). In the White Mountains of eastern Arizona, Batchelder and Merrill (1976) report a major change in vegetation between 14,000 and 12,000 yrs. B.P. In the Estancia Basin (Fig. 4), central New Mexico, an increase in *Artemisia* and a decrease in *Pinus* and *Picea* occurs stratigraphically below a horizon dated 11,740 ± 900 yrs. B.P. (Bachhuber, 1971). In southeastern Arizona at the Lehner mammoth kill site dated 11,200 yrs. B.P., the late-glacial desert grassland is similar to the post-glacial desert grassland vegetation, characterized by only a few inches of precipitation (present seasonality) and slightly lower temperatures, equivalent to no more than 200 to 400 m higher elevation (Mehringer and Haynes, 1965; Mehringer *et al.,* 1971).

In the high mountains of Colorado, Lake Emma (Carrara *et al.,* 1984) is evidently insensitive to the late-glacial shift in regional vegetation. At lower elevation, the Alkali Creek basin site shows a somewhat homogeneous pollen sequence with decreasing *Picea* from about 13,000 to 10,000 years ago and does not provide a clear late-glacial record; high *Picea* frequencies at the base of the pollen section (Markgraf and Scott, 1981) may represent the last of a full-glacial spruce forest vegetation. The Snowbird Bog site in Utah (Madsen and Currey, 1979) is similarly unclear on the nature and chronology of late-glacial alpine vegetation changes.

Early-Holocene (10,000 to 7,000 yrs. B.P.)

By the end of late-glacial time and the beginning of the Holocene, the major shift from glacial to postglacial vegetation is evident in most Southwestern pollen records. The only continuous, well-dated records for the Holocene come from montane bogs. Most of these are interpreted by comparison of *Picea/Pinus* pollen ratios from the bog deposit with pollen ratios from surface samples collected along vegetational and elevational gradients, a method developed by L. J. Maher, Jr., (1963) in the San Juan Mountains of southwestern Colorado. *Picea/Pinus* pollen ratios and spruce macrofossils from sites near present-day timberline provide a record of relative elevational position of the upper limit of spruce krummholtz and tree growth. Fluctuations in treeline elevation also indicate the elevational position of vegetation zones on lower mountain slopes, assuming that modern alpine tree and shrub species are displaced in concert. In addition, since the upper treeline is regarded as generally related to temperature, fluctuations in timberline provide paleoclimatic information.

Several lake and bog sites from the San Juan and La Plata mountains of southwestern Colorado have produced continuous Holocene records: Molas Lake and Molas Lake Bog (Maher, 1961, 1972b); Hurricane Basin (Andrews *et al.,* 1975); Twin Lakes (Petersen and Mehringer, 1976); and Lake Emma (Carrara *et al.,* 1984). The Lake Emma site appears to be sensitive to elevational shifts of vegetation zones during the Holocene. High *Picea/Pinus* pollen ratios and the presence of spruce macrofossils indicate that the krummholtz limit, and presumably timberline, was at least 70 m higher than present during the period 9,500 to 3,000 yrs. B.P., evidence also that the climate was warmer during that period. Hurricane Basin shows a similar pattern with low *Picea/Pinus* pollen ratios, indicating a regionally lower timberline until about 7,000 to 3,000 yrs. B.P. when the timberline was above its present-day position. The widely fluctuating *Picea/Pinus* pollen ratios from Twin Lakes, located today within the spruce-fir forest, may

be less sensitive to elevational changes in vegetation zones, especially at timberline.

The Redrock Lake pollen record (Maher, 1972a) from the Colorado Front Range is the first site in the Southwest for which pollen influx has been calculated. It has been regarded as a standard for the region. The pollen sequence is compared with modern *Picea/Pinus* pollen ratios from surface samples collected along an elevational gradient. The early-Holocene record is similar to that from other sites, indicating that vegetational zones were lower and that the climate was cooler/moister than present until about 7,600 to 6,700 yrs. B.P. when vegetational zones and climate were like the present. The pollen sequence from Alkalai Creek basin (Markgraf and Scott, 1981) shows a late-glacial decrease in *Picea* and the persistence of pine forest vegetation through the early-Holocene until about 4,000 yrs. B.P. when sagebrush expanded; however, the site is inadequately dated and carbonate concretions at the 7,800-year horizon may indicate the presence of an unconformity.

In Arizona, early-Holocene pollen sites include the White Mountains (Batchelder and Merrill, 1976), Laguna Salada (Hevly, 1964), Hay and Jacob lakes (Jacobs, 1983), the Lehner mammoth kill site (Mehringer and Haynes, 1965; Mehringer *et al.,* 1971), and Double Adobe I (Martin, 1963a, 1963c). The sequences from the White Mountains, Laguna Salada, the Lehner mammoth kill site, and Double Adobe I indicate a greater abundance of plant communities characteristic of cooler/moister climate. The Holocene records at Hay and Jacob lakes are incompletely preserved.

In an important study, the Holocene pollen record at the Lehner Mammoth Kill site has been interpreted by comparison of results from canonical analysis of a series of surface pollen samples collected from the Chihuahuan Desert, desert grassland, and oak grassland of southern Arizona (Adam, 1970; Mehringer *et al.,* 1971). It was found that the highest percentages of *Pinus* pollen occur at the greatest distance from pine forests below 1,675 m elevation in Chihuahuan Desert surface samples. The desert plant communities are dominated by insect-pollinated species that produce only small amounts of pollen. *Pinus* pollen, carried by long-distance atmospheric dispersal to the Chihuahuan Desert, result in *Pinus* percentages as high as 11%. Over-representation of *Pinus* pollen in the arid Southwest had been noted earlier by Martin and Gray (1962) who reported that sediment from the bottom of stock tanks in the Sonoran Desert and desert grassland of southern

Arizona contained 0.5 to 12% *Pinus*. The application of the canonical analysis of southern Arizona surface samples indicates that the vegetation at the Lehner site during the late-glacial and early-Holocene was a desert grassland with a climatic regime equivalent to about 200 to 400 m above that of today.

The Double Adobe I pollen sequence shows the persistence of sedges, indicating a high water table in the alluvial valley, and the comparatively high percentages of *Pinus* pollen were at one time regarded as evidence for a wet altithermal (Martin, 1963a, 1963c). The high percentages of Compositae are also thought to be related to an aggrading cienega, characterized by a high water table. Results from the canonical analysis of southern Arizona surface samples support the general correspondence of high Compositae/low Chenopodiineae with slightly cooler/moister desert grassland vegetation and the correspondence of low Compositae/high Chenopodiineae pollen percentages with slightly warmer/drier desert grassland vegetation (Mehringer *et al.,* 1971). Accordingly, the shift to high Chenopodiineae pollen about 8,000 yrs. B.P. at the Double Adobe I site (Fig. 5) could be in response to a change to warmer/drier climate following an early Holocene period of cooler/moister climate than today. On the other hand, even if the *Pinus* pollen is attributed to over-representation, a warmer/drier regional climate seems inconsistent with the presence of sedges.

Late-Holocene (7,000 to 0 yrs. B.P.)

Most Southwestern pollen studies have dealt with late-Holocene material. Many of these records are from archeological sites and, because of prehistoric cultural activities involving plants, paleovegetation and paleoenvironmental interpretations are difficult to evaluate. Only a few late-Holocene localities, principally non-archeological sites, are discussed.

The middle-Holocene, dated from about 7,000 to 5,000 yrs. B.P. and broadly corresponding to Antevs' altithermal, is missing from many Southwest pollen records. Where a record is present or can be inferred, however, it is clear that the middle-Holocene was a period of warmer/drier climate relative to both the Holocene average and to the historic average climate. Where the mid-Holocene is absent, it is characterized by lowered water tables, fluvial erosion, and soil formation. Direct pollen evidence for a warm/dry climate during the middle-Holocene is recorded (see Table 1) at Ashislepah Shelter, Chaco Canyon, Hurricane Basin, Laguna Salada, Lake Bonneville, Lake Emma, O'Malley Shelter, Snowbird Bog, Walker

Lake, and the White Mountains; the pollen sequences at Gardner Spring, Redrock Lake, and Twin Lakes also suggest a middle-Holocene warm/dry climate. Initiation of a warm/dry episode prior to 7,000 yrs. B.P. is indicated at Double Adobe I, Hurricane Basin, Lake Emma, and Snowbird Bog. Continuation of the warm/dry climate after 5,000 yrs. B.P. is suggested by the pollen records at Ashislepah Shelter, Chaco Canyon, Gardner Spring, Laguna Salada, and O'Malley Shelter. The present accumulation of evidence from many widely distributed sites, from alpine bogs to desert alluvial deposits, shows conclusively that a warm/dry climate dominated the Southwest during middle-Holocene time.

One of the well known and important contributions from the 1960's is the alluvial pollen study at Murray Springs (Fig. 8), dated from about 4,500 yrs. B.P. to present (Mehringer *et al.,* 1967). The record of sedges and cattail pollen led to the interpretation that the period 5,000 to 4,000 yrs. B.P. was moister than today. The comparatively high abundance of *Pinus* pollen was also cited as indicating expanded pine forests in the distant mountains in response to cooler/moister climate. The implication of the Murray Springs record was that the mid-Holocene was also wet (Martin, 1963a). Mehringer was careful to point out, however, that a middle-Holocene pollen record 7,000 to 5,000 yrs. B.P. was absent from the alluvial sequences of southern Arizona, including Murray Springs (Mehringer, 1967b; Mehringer *et al.,* 1967). The Murray Springs pollen diagram shows a dramatic change in relative abundances of Chenopodiineae and short-spine Compositae. High percentages of Chenopodiineae occur in the same zone with the sedge and cattail pollen that was interpreted as indicating moist conditions. If the same line of reasoning developed with the Double Adobe I pollen record, discussed above, is applied to Murray Springs, the high frequencies of Chenopodiineae should instead indicate lower water tables, erosion, and dry climate. The significance of alternating high and low frequencies of Chenopodiineae and Compositae in southeastern Arizona pollen records is not yet resolved.

The upper part of the Murray Springs diagram, dated less than 1,000 yrs. B.P., shows an increase in *Pinus* and *Ephedra* pollen percentages (Fig. 8). The higher *Pinus* pollen in the lower part of the section pre-4,000 yrs. B.P. is thought to indicate cooler/moister conditions, but the concurrent abundance of both *Pinus* and *Ephedra* together are taken to represent drier desert vegetation. Mehringer *et al.,* (1967) point out that, in southeastern Arizona at low

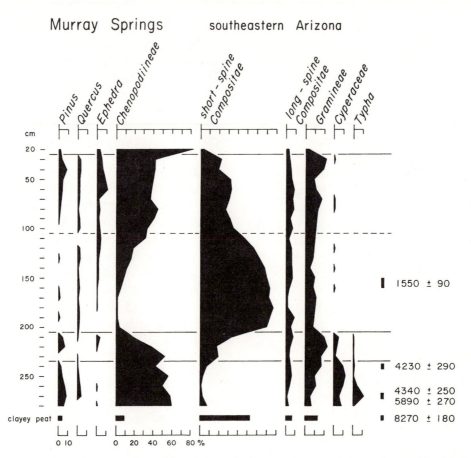

Figure 8. Summary pollen relative frequency diagram from Murray Springs, southeastern Arizona, documenting late-Holocene vegetation (modified from Mehringer *et al.,* 1967). The 5,890 ± 270 yrs. B.P. radiocarbon date is on humates and may be too old; the 4,340 ± 250 yrs. B.P. date is on the organic residue from the same sample and is a more reliable age.

elevation sites below 1675 m, highest *Pinus* and *Ephedra* frequencies occur in the Chihuahuan Desert, *Pinus* pollen grains over-represented by slightly higher percentages in the desert vegetation. Thus, the last 1,000 years at Murray Springs is interpreted by Mehringer *et al.* (1967) as representing exceptionally drier climate in southeastern Arizona, a trend that is apparently continuing today (Hastings and Turner, 1965).

The Gardner Spring alluvial pollen site (Fig. 9) occurs in the Chihuahuan Desert of southern New Mexico (Freeman, 1972). The pollen diagram is dominated by Chenopodiineae, short-spine Compositae, and Gramineae. The only major change in the pollen record is a decrease in Chenopodiineae concurrent with an increase in Gramineae about 4,000 yrs. B.P., probably representing a decline in *Atriplex* and greater abundance of grasses in response to slightly moister climate. Another alluvial pollen study in the Chihuahuan Desert grassland margin is the Garnsey Bison Kill site (Fig. 10) of southeastern New Mexico (Hall, 1984b). A steady decline in *Pinus* pollen from 10% to 1% over a 30 to 50 year period

indicates a strong but short-term drought in that area, occurring about 500 to 450 yrs. B.P.; the high degree of resolution is possible because of the rapid rate of alluvial deposition at the site. Owing to the short duration of the event, it is thought that the cause of the *Pinus* pollen decrease is due to diminished *Pinus* pollen production at a time when the pine trees are experiencing drought-related stress. The rapid increase of *Pinus* pollen percentages to pre-drought levels supports the contention that changes in *Pinus* pollen production, rather than fluctuations in pine abundance or range, may be responsible for much of the observed short-term record at the Garnsey Bison Kill site.

The Chaco Canyon pollen record (Fig. 11) spans the past 7,000 years but includes pollen spectra from two distinct facies of alluvium, one representing local side canyon sources and the other representing the upper reaches of the Chaco River drainage basin (Hall, 1977, 1983). The pollen succession indicates that the vegetation at Chaco Canyon has been a treeless arid shrub grassland throughout the past

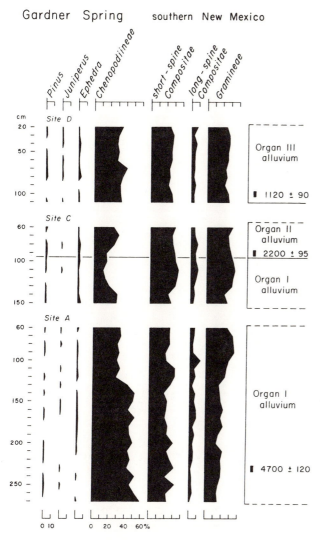

Figure 9. Summary pollen relative frequency diagram from Gardner Spring, southern New Mexico, documenting late-Holocene vegetation in the Chihuahuan Desert (modified from Freeman, 1972).

7,000 years although changes occurred in the distribution and abundance of pines in the regional montane forests. The greatest reduction in pine forests as seen from the Chaco Canyon pollen record occurred from at least 6,000 to 2,400 yrs. B.P.; pines increased in abundance at 2,400 yrs. B.P. and again at about 600 yrs. B.P. The discovery of pinyon pine and juniper macrofossils in prehistoric wood rat middens at Chaco Canyon led to the interpretation of a Holocene pinyon-juniper woodland vegetation (Betancourt and Van Devender, 1981) in contrast with the shrub grassland vegetation indicated by pollen from the canyon fill. Pollen analysis of matrix from the same series of wood rat middens from which the macrofossils were obtained verifies the alluvial pollen-based vegetation reconstructions. The

pinyon pine macrofossils likely represent isolated stands or individuals of trees along the canyon escarpments in numbers too small to significantly affect the pollen rain (Hall, 1981b, 1982). Just east of Chaco Canyon, Ashislepah Shelter shows a mid-Holocene decrease in *Pinus* pollen with increased Chenopodiineae, Compositae, and *Ephedra,* indicating the onset of drier conditions (Fredlund, 1984; Fredlund and Johnson, 1984). By 2,200 yrs. B.P., *Pinus* pollen frequencies increase, indicating expanded pine forests and moister climate, similar to the Chaco Canyon alluvial pollen record.

Alpine sites in southwestern Colorado (Lake Emma, Hurricane Basin, Twin Lakes) show a general pattern of high *Picea/Pinus* pollen ratios during mid-Holocene time followed by lower ratios beginning about 3,000 yrs. B.P., indicating lowered forest vegetation zones and a slightly cooler/moister climate. The record from Redrock Lake (Maher, 1972a, 1973) shows a chronology of elevational shifts of vegetation similar to the other alpine sites, although the apparent elevational position of timberline based on *Picea/Pinus* pollen ratios may be out of phase (Nichols, 1982). At lower elevation, the Alkalai Creek basin site shows a sharp decrease in *Pinus* pollen and an increase in *Artemisia* percentages about 4,000 yrs. B.P. (Markgraf and Scott, 1981), establishing the sagebrush grassland vegetation that characterizes the area today.

In Arizona, the pollen diagram from Laguna Salada shows a mid-Holocene (7,200 yrs. B.P.) decline in percentages of *Picea* and *Pinus* with a decline in pollen from aquatic plants accompanied by increases in Chenopodiineae percentages (Hevly, 1962, 1964). By about 3,500 yrs. B.P., *Pinus* and *Quercus* frequencies increase. In the White Mountains, a mid-Holocene dry period is followed by a cooler/moister episode lasting from 4,700 to 2,800 yrs. B.P.; the present-day vegetation was established by 2,800 yrs. B.P. (Batchelder and Merrill, 1976). The lacustrine pollen sequence from Walker Lake has a record of dessication during the mid-Holocene (Hevly, 1984). The Holocene pollen records from nearby Hay and Jacob lakes (Jacobs, 1983), Potato Lake (Whiteside, 1965), and San Agustin Plains, New Mexico (Markgraft *et al.,* 1983, 1984), are attenuated or stratigraphically shallow.

Only a few late-Holocene pollen sites have been reported from Utah. The pollen sequence from Snowbird Bog is complex and has been interpreted with the aid of *Picea/Pinus* and conifer/non-conifer pollen ratios. From about 8,000 to 6,000 yrs. B.P., pollen ratios indicate a period of warmer/drier

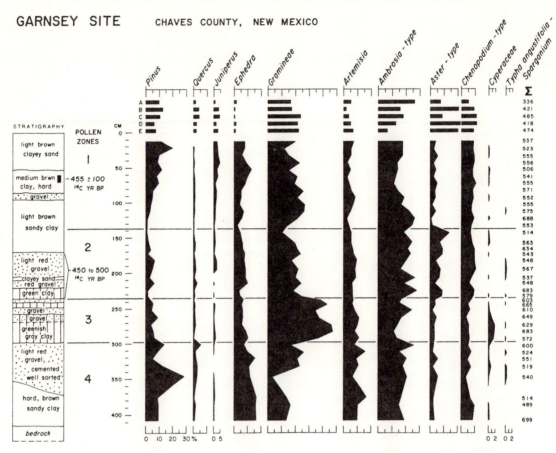

Figure 10. Summary pollen relative frequency diagram from Garnsey Bison Kill site, southeastern New Mexico, documenting short-term changes in late-Holocene vegetation (modified from Hall, 1984b).

climate (relative to the Holocene average) that resulted in expansion of conifers into former sagebrush areas at high elevation sites. From about 6,000 to 3,000 yrs. B.P. pollen ratios indicate generally warm/wet conditions with post-3,000 yrs. B.P. climate comparatively cool/dry (Madsen and Currey, 1979); the past 1,500 years is probably missing from the record. At lower elevation in southeastern Utah, a pollen diagram from 4,000-year-old Lake Pagahrit shows flucuations in *Pinus, Juniperus,* and Chenopodiineae pollen (Lipe *et al.,* 1975). However, the flucuations lack systematic trends and late-Holocene vegetation changes documented elsewhere in the Southwest do not appear in the Lake Pagahrit record. A Holocene pollen record from Lake Bonneville has not been completed but evidently shows mid-Holocene climatic drying (Madsen and Kay, 1982).

The Tule Springs pollen succession in southern Nevada suggests that the lower Mohave Desert vegetation of the Las Vegas Valley was established by 7,000 yrs. B.P. Mehringer (1967a) points out, however, that the Holocene pollen record at Tule Springs is incomplete and that the pollen site may insensitive to elevational shifts in vegetation of a magnitude less than 300 m. A late-Holocene pollen diagram from O'Malley Shelter, southeastern Nevada (Madsen, 1973), is dominated by Chenopodiineae and reflects a sagebrush vegetation with changing importance of juniper and pine. At the beginning of the shelter pollen sequence about 6,900 yrs. B.P., Gramineae pollen percentages are increasing and reach maximum abundance about 3,900 yrs. B.P.; *Juniperus* pollen decreases slightly during the period of increased grasses. At the change of decreasing Gramineae 3,900 yrs. B.P., *Juniperus* pollen increases and *Pinus* and *Ephedra* pollen begin to show up in the record. Approximately 2,000 yrs. B.P., *Pinus* pollen is well represented, indicating the expansion and development of a pinyon-juniper sagebrush vegetation like that of today near O'Malley Shelter.

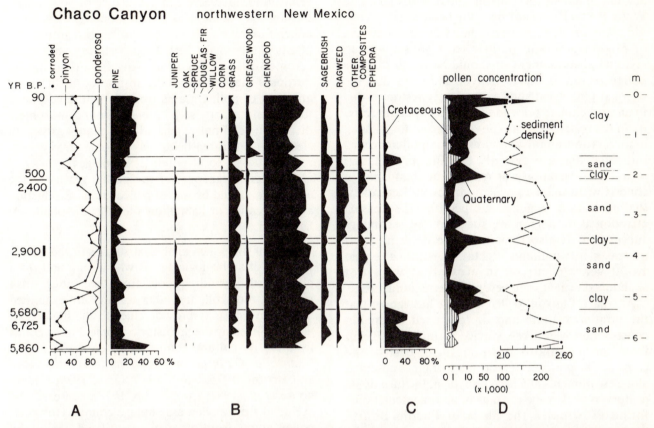

Figure 11. Summary pollen relative frequency diagram from Chaco Canyon, northwestern New Mexico, documenting late-Holocene vegetation (modified Gallo Wash II diagram from Hall, 1977; Hall, 1983). (A) Chronology at left of diagram includes radiocarbon dates from pollen section and age of correlated stratigraphy; relative percentages of pinyon *(Pinus edulis)* and ponderosa pine *(P. ponderosa)* and percentage of corroded pine pollen grains (all species); (B) pollen frequencies; (C) percentages of Cretaceous spores and pollen relative to all palynomorphs from late-Holocene alluvial section; (D) palynomorph concentrations, sediment density, and sediment texture.

ARCHEOLOGY AND THE VEGETATION RECORD

The Deforestation Problem

The decline in elm pollen percentages and other changes in the post-glacial pollen record of northwestern Europe were interpreted by Iversen (1941, 1956) as, instead of climatic in origin, a result of the clearance of forest trees for prehistoric farming. In the American Southwest, heavy land use activities associated with pueblo farming villages and towns, such as cutting timbers for construction, harvesting woody shrubs and small trees for firewood, and brush clearance for farm fields, may have had a sufficiently large impact on the vegetation to be visible in the pollen record.

A number of pollen analytical studies at Mesa Verde, a large concentration of Anasazi settlements in southwestern Colorado, has shown that since Anasazi puebloan abandonment about 650 years ago, *Pinus* and *Juniperus* pollen frequencies have steadily risen, giving support to the contention that secondary regrowth of pine and juniper woodland vegetation occurred after having been greatly reduced by the Anasazi (Martin and Byers, 1965; Wyckoff, 1977). At Navajo Reservoir, northwestern New Mexico, analyses of alluvial sections and room fill at sites show increases in *Pinus* and *Juniperus* pollen percentages beginning sometime after 1,000 yrs. B.P. (Schoenwetter, 1964). The Chaco Canyon alluvial pollen record also shows a regional increase in pine vegetation. Differentiation of the pollen grains indicates that the principal species involved in the increase is pinyon pine *(P. edulis)* with only a slight increase in abundance of ponderosa pine *(P. ponderosa)* (Hall, 1977, 1983). The period of greater *Pinus* pollen frequencies at Chaco Canyon is undated but occurs in Kirk Bryan's post-Bonito channel fill (Post-Bonito unit of Hall, 1977) which post-dates Anasazi abandonment of the canyon about 800 years ago. While the pollen record from the heart of the prehistoric Anasazi culture clearly shows increased growth of pinyon pine and juniper trees subsequent to regional abandonment, it should be noted that the pollen record at Chaco Canyon indicates that the regional pine woodlands were already restricted due to an arid climate prior to the time of Anasazi activity in the region. A more convincing case for Anasazi deforestation could be made if the pollen record indicated that a previously existing, pre-Anasazi woodland vegetation was reduced in abundance in concert with Anasazi population growth. Regardless, the increases in pine and juniper after the Anasazi abandoned the region are suggestive of secondary succession. It also remains to be determined if increases in woodland vegetation within the past 1,000 years occurred in areas that were not abandoned or that were beyond Anasazi influence.

At Chaco Canyon, pollen analysis has shown that the ponderosa pine and fir logs used in pueblo construction were likely carried as much as 70 km from the mountains west, south, or east of the canyon. The presence of dead snags of pines observed along escarpments at Chaco Canyon at the turn of the century as well as the remains of an ornamental pine found growing in the plaza of Pueblo Bonito suggested to earlier workers that the trees were the remnants of a pine forest in the canyon from which the Anasazi had harvested timbers for pueblo construction. However, the alluvial pollen record and the differentiation of fossil *Pinus* pollen grains to species indicate that ponderosa pines were not present in any large numbers during the late-Holocene (Hall, 1977).

Vegetation Reconstruction from Archeological Sites

Hundreds of pollen surface samples illustrate the relationship of pollen percentages to plant communities. Pollen analysis of a sedimentary deposit, ignoring depositional and post-depositional phenomena, will provide direct information on the relative abundance of plant species, genera, or families in the local and regional vegetation. In a pollen diagram, changes in the relative frequencies of pollen types correspond to changes in the abundance of those plants. At archeological sites where prehistoric inhabitants have disturbed the pre-existing vegetation, pollen abundances may be distorted from those that would have been produced by undisturbed vegetation: (a) the cutting of timber for construction and firewood and the clearing of brush for agriculture will result in locally decreased abundance of pollen from those woody plants and a corresponding increase in pollen from weeds that colonize disturbed ground; (b) plants gathered for fuel, food, weaving, medicine, ceremonies, and for processing for other special purposes may result in the introduction of larger amounts of pollen grains from the gathered plants than would otherwise be present from atmospheric dispersal; (c) finally, agricultural plants grown in cleared or uncleared fields may become greatly overrepresented in local pollen assemblages by way of atmospheric transport of the pollen grains to the adjacent area and by way of pollen grain introduction at a processing or habitation site after the plants are harvested.

Most of the pollen analytical studies in the Southwest are of archeological sites where sediment from pits, hearths, grinding tools, floor sweepings, trash middens, room fill, and other features are examined for pollen that may provide clues to the identity of plants utilized by the prehistoric people. Pollen grains from clearly cultural context, such as *Zea, Cleome, Cucurbita, Opuntia,* or *Phaseolus* (Williams-Dean and Bryant, 1975; Scott, 1979; Gish, 1979; Greenhouse *et al.,* 1981; Lytle-Webb, 1978), can easily be excluded from pollen assemblage counts. However, pollen grains from other plants utilized heavily by prehistoric inhabitants, such as species in the Chenopodiineae and Compositae, cannot be distinguished from pollen grains of individual plants of the same or similar species abundant in the native vegetation yet not used by prehistoric man. Also, since pollen percentages are relative, increases in one pollen type category result in decreases in the other categories, resulting in pollen frequencies that are biased towards man's activities and that misrepresent the extant vegetation. Analysts have attempted to bypass the problem of over-representation of pollen grains of cultural origin by using ratios of selected pollen types, such as (a) arboreal/nonarboreal, (b) *Pinus/Juniperus,* and (c) large pine/small pine pollen ratios (Hevly, 1968; Hevly *et al.,* 1979; Euler *et al.,* 1979). In the Southwest, the large and small pine grains are generally ponderosa and pinyon pine, respectively, although size alone can be misleading in species identification. Pollen data have also been manipulated through the use of adjusted pollen sums and adjusted arboreal pollen percentages (Schoenwetter, 1964, 1967). While pollen ratios and adjusted percentages have the advantage of highlighting one aspect of the pollen record, there is no way to be certain that those elements are without cultural biases and distortions. Ratios and adjusted percentages may be valuable supplements to a pollen diagram but should not be relied upon too strongly because they guarantee an interpretation with no surprises. Because of the many unresolved difficulties in using pollen from archeological sites for paleovegetation and paleoenvironmental interpretations, it seems advisable to investigate non-archeological deposits first.

The difficulty of environmental interpretation of pollen percentages from archeological sites is well documented by the Chuska Valley study in northwestern New Mexico. Schoenwetter (1967) analyzed pollen samples from thirty archeological sites, most of the sites represented by a single pollen sample collected from the floor of an excavated room. By adjusting the arboreal pollen percentages to the expected modern arboreal frequencies from a particular elevational zone (based on the analysis of 36 surface samples from the area), Schoenwetter put together a composite picture of changing vegetation and climate for the time period represented by the archeological sites. The resulting adjusted arboreal pollen curve indicated the occurrence of two cycles of rise and fall of vegetation zones, each by 600 meters, during the 300 year period between A.D. 1,000 and 1,300 (Schoenwetter, 1967, p. 100). Such rapid elevational shifts of vegetation zones have not been documented elsewhere, illustrating the unrealistic paleoenvironmental signal that can result from pollen counts obtained from archeological sites where pollen assemblages may have been greatly distorted by cultural activities.

Not all archeological sites result in distorted pollen records. The sites discussed above are Puebloan, characterized by year-long habitation and an agricultural economy. Rock shelters, cave sites, open sites that have been buried by alluvium, or other sites of seasonal or short-term occupation may yield pollen records that have not been biased by activities of prehistoric people.

CONCLUSIONS

The vegetational history of the Southwest is beginning to unfold. Baker's (1983) observation that paleoenvironmental reconstruction in the Western United States is 15 to 20 years behind that in the East applies to the Southwest as well. The primary reason for the lag in research lies in the fact that only a few pollen analysts are working in the Southwest and many of these are committed to studies of archeological sites.

A strong impression reinforced by this survey of Southwest Quaternary pollen analysis is the clear

need, not only for additional studies to fill the many geographic and temporal gaps, but for rigorous geomorphic and stratigraphic analyses in combination with the pollen work. Bogs, lakes, caves, rock shelters, alluvium, as well as archeological sites make sense only in the context of their contained stratigraphy. The pollen record is equally dependent upon careful, informed stratigraphic sampling. This is particularly important in sensitive Southwest landscapes where dry climates are intertwined with processes of erosion, deposition, and vegetational change. Adequate chronologic controls are also mandatory. Without a detailed chronology, the best pollen sequence is valueless for future correlations of vegetation history.

QUATERNARY VEGETATIONAL HISTORY: A SUMMARY

The early- and middle-Pleistocene vegetational history of the Southwest is unknown. Only a few pre-Wisconsinan sites have been reported. The available evidence suggests that middle-Pleistocene vegetation zones may have experienced a shift in elevation similar in magnitude to the change that occurred during the Wisconsinan.

Early- and full-glacial vegetation is known from only a few localities and details of the records are insufficient to recognize regional patterns. Several important studies are as yet unpublished. The early-glacial (>24,000 yrs. B.P.) and full-glacial (24,000 to 14,000 yrs. B.P.) records do not greatly differ from each other, although this apparent similarity may be due to the small number of sites. Overall, vegetation zones were 900 to 1,400 meters below present elevations. Sagebrush was more widespread during early- and full-glacial time than it is today, comprising an important proportion of the pollen influx at many glacial age sites.

The change from glacial to post-glacial vegetation with accompanying upward shift in elevation of vegetation zones occurred during the late-glacial period. A precise chronology for this major change in vegetation at any single locality is not well determined because of altered rates of sedimentation, drying of lakes, and too few radiocarbon dates from critical sections. From all the available evidence, however, it appears that the shift from glacial to post-glacial vegetation took place over a period of 2,000 years from about 14,000 to 12,000 yrs. B.P. At present our information on chronology is too imprecise to permit us to interpret, for example, time-transgressive changes with latitude.

The early-Holocene was generally cooler/moister than today and, in contrast to the dramatic vegetation changes that occurred during the glacial/post-glacial transition, was characterized by a gradual decrease in woodland vegetation and a warming and drying of the climate toward the extreme conditions of the middle-Holocene. Although the concept of a wet altithermal has greatly influenced paleoecological research in the past, pollen evidence from numerous bog, lake, and alluvial sites across the Southwest shows conclusively that the middle-Holocene was characterized by an extremely warm/dry climate, relative to either Holocene or historic averages. The principal timing of the maximum warm/dry climate is about 7,000 to 5,000 yrs. B.P., although several sites indicate the onset of warm/dry conditions as early as 9,000 yrs. B.P. while other sites indicate that the warm/dry episode continued until about 2,500 yrs. B.P. The middle-Holocene warm/dry climate is indicated by the reduction of woodland and forest vegetation at many sites. At alpine localities, the migration of krummholtz and treeline upslope to positions at least 70 meters above those of today also indicates warmer middle-Holocene climate. During the period of extreme aridity, alluvial valleys were eroded, resulting in truncation of many Southwestern alluvial pollen records.

The late-Holocene vegetation is characterized by a gradual increase in woodland and forest vegetation in response to the general lessening of the extreme conditions of the middle-Holocene. An exception may be southeastern Arizona, where desert vegetation is thought to have replaced broad areas of grassland during the past 1,000 years. Pollen data from many archeological sites and associated sediments provide brief glimpses of vegetation. However, it is difficult to reconcile these short records with the long-term trends of vegetation development seen at only a few localities. Further investigations into the late-Holocene pollen record are imperative if the important prehistoric cultural changes that occurred during the past 2,000 years are to be viewed within a broad paleoenvironmental framework.

ACKNOWLEDGEMENTS

Some of the material presented is based upon work supported by the National Science Foundation under Grant GA-35448 and Grant EAR-7911158 and by a North Texas State University faculty research grant. The preparation of the paper was also supported by North Texas State University. I thank Donna M. Argo, Jannifer W. Gish, and Linda J. Scott for special help.

References Cited

ADAM, D. P.
  1970   Some palynological applications of multivariate statistics. Ph.D. dissertation, University of Arizona, Tucson, 132 p.
ADAM, D. P., and MEHRINGER, P. J., JR.
  1975   Modern pollen surface samples—an analysis of subsamples. *Journal of Research U. S. Geological Survey,* 3: 733-736.
ANDREWS, J. T., CARRARA, P. E., KING, F. B., and STUCKENRATH, R.
  1975   Holocene environmental changes in the alpine zone, northern San Juan Mountains, Colorado: evidence from bog stratigraphy and palynology. *Quaternary Research,* 5: 173-197.
ANTEVS, E.
  1948   Climatic changes and pre-white man. *In:* The Great Basin, with emphasis on glacial and postglacial times. *University of Utah Bulletin,* 38 (20): 168-191.
  1955   Geologic-climatic dating in the west. *American Antiquity,* 20: 317-335.
  1962   Late Quaternary climates in Arizona. *American Antiquity,* 28: 193-198.

ARMS, B. C.
  1960   A silica-depressant method for concentrating fossil pollen and spores. *Micropaleontology,* 6: 327-328.
BACHHUBER, F. W.
  1971   Paleolimnology of Lake Estancia and the Quaternary history of the Estancia Valley, central New Mexico. Ph.D. dissertation, University of New Mexico, Albuquerque, 238 p.
BAKER, R. G.
  1983   Holocene vegetational history of the western United States. *In:* Wright, H. E., Jr. (ed.), *Late Quaternary Environments of the United States, Vol. 2. The Holocene.* University of Minnesota Press, Minneapolis, 109-127.
BATCHELDER, G. L., and MERRILL, R. K.
  1976   Late Quaternary environmental interpretations from palynological data, White Mountains, Arizona [abst.]. American Quaternary Association (AMQUA), 4th biennial mtg., Tempe, Arizona, *Abstracts,* 125.
BENSON, L., and DARROW, R. A.
  1981   *Trees and Shrubs of the Southwestern Deserts* (3rd edition). University of Arizona Press, 416 p.

BENT, A. M., and WRIGHT, H. E., JR.
1963 Pollen analyses of surface materials and lake sediments from the Chuska Mountains, New Mexico. *Geological Society of America Bulletin,* 74: 491-500.

BERRY, R. W., McCORMICK, C. W., and ADAM, D. P.
1982 Pollen data from a 5-meter upper Pleistocene lacustrine section from Walker Lake, Coconino County, Arizona. *U.S. Geological Survey Open-File Report* 82-0383, 108 p.

BETANCOURT, J. L., and VAN DEVENDER, T. R.
1981 Holocene vegetation in Chaco Canyon, New Mexico. *Science,* 214: 656-658.

BONNEFILLE, R.
1984 The origin of grassland and woodland in tropical East Africa. Sixth International Palynological Conference (IPC), Calgary, Canada. *Abstracts,* 11.

BROOKS, C. E. P.
1949 *Climate Through the Ages* (2nd revised edition), reprinted by Dover Publ., New York, 395 p.

BROWN, D. E. (ed.)
1982 Biotic communities of the American Southwest— United States and Mexico. *Desert Plants,* 4(1-4), 342 p.

BRUSH, G. S., and BRUSH, L. M., JR.
1972 Transport of pollen in a sediment-laden channel: a laboratory study. *American Journal of Science,* 272: 359-381.

BRYAN, K.
1925 Date of channel trenching (arroyo cutting) in the arid southwest. *Science,* 62: 338-344.

BULL, W. B.
1964 History and causes of channel trenching in western Fresno County, California. *American Journal of Science,* 262: 249-258.

CARRARA, P. E., MODE, W. N., RUBIN, M., and ROBINSON, S. W.
1984 Deglaciation and postglacial timberline in the San Juan Mountains, Colorado. *Quaternary Research,* 21: 42-55.

CLISBY, K. H., FOREMAN, F., and SEARS, P. B.
1957 Pleistocene climatic changes in New Mexico, U.S.A. *Veroff. Geobotanisches Inst. Rubel in Zurich,* 34: 21-26.

CLISBY, K. H., and SEARS, P. B.
1956 San Augustin Plains—Pleistocene climatic changes. *Science,* 124: 537-539.

COLE, K.
1982 Pleistocene packrat middens from the western Sierra Nevada, California [abst.]. American Quaternary Association (AMQUA), 7th biennial mtg., Seattle, Washington, *Program and Abstracts,* 83.
1983 Late Pleistocene vegetation of Kings Canyon, Sierra Nevada, California. *Quaternary Research,* 19: 117-129.

CROWDER, A. A., and CUDDY, D. G.
1973 Pollen in a small river basin: Wilton Creek, Ontario. *In:* Birks, H. J. B., and West, R. G. (eds.), *Quaternary Plant Ecology.* John Wiley & Sons, New York, 61-77.

CUSHING, E. J.
1967 Evidence for differential pollen preservation in late Quaternary sediments in Minnesota. *Review of Palaeobotany and Palynology,* 4: 87-101.

DEEVEY, E. S., JR., and FLINT, R. F.
1957 Postglacial hypsithermal interval. *Science,* 125: 182-184.

DELCOURT, P. A., and DELCOURT, H. R.
1980 Pollen preservation and Quaternary environmental history in the southeastern United States. *Palynology,* 4: 215-231.

DIXON, H. N.
1962 Vegetation, pollen rain, and pollen preservation, Sangre de Cristo Mountains, New Mexico. M.S. thesis, University of New Mexico, Alburquerque, 69 p.

EARDLEY, A. J., and GVOSDETSKY, V.
1960 Analysis of Pleistocene core from Great Salt Lake, Utah. *Bulletin of the Geological Society of America,* 71: 1323-1344.

EULER, R. C., GUMERMAN, G. J., KARLSTROM, T. N. V., DEAN, J. S., and HEVLY, R. H.
1979 The Colorado Plateaus; cultural dynamics and paleoenvironment. *Science,* 205: 1089-1101.

FAEGRI, K., and IVERSEN, J.
1975 *Textbook of Pollen Analysis* (3rd revised edition). Hafner Press, New York, 295 p.

FALL, P. L.
1981 Modern pollen spectra and their application to alluvial pollen sedimentology. M.S. thesis, University of Arizona, Tucson, 63 p.

FINLEY, R. B., JR.
1958 *The wood rats of Colorado: Distribution and ecology.* University of Kansas Publications, Museum of Natural History, 10(6): 213-552.

FREDLUND, G.
1984 Palynological analysis of sediments from Sheep Camp and Ashislepah shelters. *In:* Simmons, A. H. (ed.), *Archaic Prehistory and Paleoenvironments in the San Juan Basin, New Mexico: The Chaco Shelters Project.* University of Kansas, Museum of Anthropology, Project Report Series 53, 186-209.

FREDLUND, G., and JOHNSON, W. C.
1984 Palynological evidence for late Quaternary paleoenvironmental change in the San Juan Basin [abst.]. American Quaternary Association (AMQUA), 8th biennial mtg., Boulder, Colorado, *Program and Abstracts,* 46.

FREEMAN, C. E.
1972 Pollen study of some Holocene alluvial deposits in Dona Ana County, southern New Mexico. *Texas Journal of Science,* 24: 203-220.

GISH, J. W.
1979 Palynological research at Pueblo Grande Ruin. *The Kiva,* 44: 159-172.

GRAY, J.
1961 Early Pleistocene paleoclimatic record from Sonoran Desert. *Science,* 133: 38-39.

GREENHOUSE, R., GASSER, R. E., and GISH, J. W.
1981 Cholla bud roasting pits: an ethnoarchaeological example. *The Kiva,* 46: 227-242.

HAFSTEN, U.
1959 Bleaching + HF + acetolysis—a hazardous preparation process. *Pollen et Spores,* 1: 77-79.

HAGER, M. W.
1975 Late Pliocene and Pleistocene history of the Donnelly Ranch vertebrate site, southeastern Colorado. *University of Wyoming, Contributions to Geology, Special Paper* 2, 62 p.

HALL, S. A.

1977   Late Quaternary sedimentation and paleoecologic history of Chaco Canyon, New Mexico. *Geological Society of America Bulletin,* 88: 1593-1618.

1981a   Deteriorated pollen grains and the interpretation of Quaternary pollen diagrams. *Review of Palaeobotany and Palynology,* 32: 193-206.

1981b   Holocene vegetation at Chaco Canyon: pollen evidence from alluvium and pack rat middens [abst.]. Society for American Archaeology, 46th ann. mtg., San Diego, *Program and Abstracts,* p. 61.

1982   Reconstruction of local and regional Holocene vegetation in the arid Southwestern United States based on combined pollen analytical results from *Neotoma* middens and alluvium [abst.]. International Union for Quaternary Research (INQUA), XI Congress, Moscow, U.S.S.R., *Abstracts,* 1: 130.

1983   Holocene stratigraphy and paleoecology of Chaco Canyon. *In:* Wells, S. G., Love, D. W., and Gardner, T. W. (eds.), *Chaco Canyon Country.* American Geomorphological Field Group, 1983 Field Trip Guidebook, 219-226.

1984a   Pollen influx and distribution in the Southern Rockies and SW Plains (U.S.A.) [abst.]. Sixth International Palynological Conference (IPC), Calgary, Alberta, Canada. International Association for Aerobiology symposium, *Abstract issued separately.*

1984b   Pollen analysis of the Garnsey Bison Kill Site, southeastern New Mexico. *In:* Parry, W. J., and Speth, J. D., *The Garnsey Spring Campsite: Late Prehistoric Occupation in Southeastern New Mexico.* Museum of Anthropology University of Michigan Technical Reports 15, 85-108.

HANSEN, B. S., and CUSHING, E. J.

1973   Identification of pine pollen of late Quaternary age from the Chuska Mountains, New Mexico. *Geological Society of America Bulletin,* 84: 1181-1200.

HASTINGS, J. R., and TURNER, R. M.

1965   *The Changing Mile, An Ecological Study of Vegetation Change with Time in the Lower Mile of an Arid and Semiarid Region.* University of Arizona Press, Tucson, 317 p.

HAYNES, C. V., JR.

1968   Geochronology of late Quaternary alluvium. *In:* Morrison, R. B., and Wright, H. E., Jr. (eds.), *Means of Correlation of Quaternary Successions.* University of Utah Press, 591-631.

HEUSSER, C. J.

1977   A survey of Pleistocene pollen types of North America. *In:* Elsik, W. C. (ed.), Cenozoic Palynology. *American Association of Stratigraphic Palynologists Contribution Series 5A,* 111-129.

HEVLY, R. H.

1962   Pollen analysis of Laguna Salada. *In:* New Mexico Geological Society, 13th Field Conference, *Guidebook of the Mogollon Rim region, east-central Arizona,* 115-117.

1964   Paleoecology of Laguna Salada. *In:* Martin, P. S. *et al.,* Chapters in the prehistory of eastern Arizona, II. Chicago Natural History Museum. *Fieldiana: Anthropology,* 55: 171-187.

1968   Studies of the modern pollen rain in northern Arizona. *Journal of the Arizona Academy of Science,* 5: 116-127.

1984   Quaternary environments: Walker Lake, Coconino County, Arizona [abst.]. Sixth International Palynological Conference (IPC), Calgary, Alberta, Canada, *Abstracts,* 63.

HEVLY, R. H., KELLY, R. E., ANDERSON, G. A., and OLSEN, S. J.

1979   Comparative effects of climatic change, cultural impact, and volcanism in the paleoecology of Flagstaff, Arizona, A.D. 900-1300. *In:* Sheets, P. D., and Grayson, D. K. (eds.), *Volcanic Activity and Human Ecology.* Academic Press, N. Y., 487-523.

HEVLY, R. H., and MARTIN, P. S.

1961   Geochronology of Pluvial Lake Cochise, southern Arizona. I. Pollen analysis of shore deposits. *Journal of the Arizona Academy of Science,* 2: 24-31.

HEVLY, R. H., MEHRINGER, P. J., JR., and YOCUM, H. G.

1965   Modern pollen rain in the Sonoran Desert. *Journal of the Arizona Academy of Science,* 3: 123-135.

HOLLOWAY, R. G.

1981   Preservation and experimental diagenesis of the pollen exine. Ph.D. dissertation, Texas A & M University, College Station, 317 p.

IVERSEN, J.

1941   Landnam i Danmarks Stenalder. *Danm. Geol. Unders.,* Ser. II, 66: 1-68.

1956   Forest clearance in the Stone Age. *Scientific American,* 194: 36-41.

IZETTE, G. A., and WILCOX, R. E.

1982   Map showing localities and inferred distributions of the Huckleberry Ridge, Mesa Falls, and Lava Creek ash beds (Pearlette family ash beds) of Pleistocene age in the Western United States and southern Canada. U. S. Geological Survey, *Miscellaneous Investigations Series Maps,* I-1325.

JACOBS, B. F.

1983   Past vegetation and climate of the Mogollon Rim area, Arizona. Ph.D. dissertation, University of Arizona, Tucson, 166 p.

KAPP, R. O.

1965   Illinoian and Sangamon vegetation in southwestern Kansas and adjacent Oklahoma. University of Michigan, *Contributions from the Museum of Paleontology,* 19(14): 167-255.

1969   *How to Know Pollen and Spores.* Wm. C. Brown Co. Publishers, Dubuque, Iowa, 249 p.

KEARNEY, T. H., and PEEBLES, R. H.

1960   *Arizona Flora* (2nd edition with supplement). University of California Press, Berkeley, 1085 p.

KING, J. E.

1967   Modern pollen rain and fossil pollen in soils in the Sandia Mountains, New Mexico. *Papers of the Michigan Academy of Science, Arts, and Letters,* 52: 31-41.

KING, J. E., and SIGLEO, W. R.

1973   Modern pollen in the Grand Canyon, Arizona. *Geoscience and Man,* 7: 73-81.

KING, J. E., and VAN DEVENDER, T. R.

1977   Pollen analysis of fossil packrat middens from the Sonoran Desert. *Quaternary Research,* 8: 191-204.

KREMP, G. O. W.

1978   Pliocene palynological literature: five hundred implemented references. *Paleo Data Banks,* 9, 50 p.

KURTZ, E. B., JR., and TURNER, R. M.

1957   An oil-flotation method for the recovery of pollen from inorganic sediments. *Micropaleontology,* 3: 67-38.

LEGG, T. E., and BAKER, R. G.

1980   Palynology of Pinedale sediments in Devlins Park, Boulder County, Colorado. *Arctic and Alpine Research,* 12: 319-333.

LEOPOLD, L. B.

1951   Rainfall frequency: an aspect of climatic variation. *Transactions of the American Geophysical Union,* 32: 347-357.

LEOPOLD, L. B., EMMETT, W. W., and MYRICK, R. M.

1966   Channel and hillslope processes in a semiarid area, New Mexico. *U. S. Geological Survey Professional Paper* 352-G, 193-253.

LEWIS, W. H., VINAY, P., and ZENGER, V. E.

1983   *Airborne and Allergenic Pollen of North America.* Johns Hopkins University Press, Baltimore, 254 p.

LIPE, W. D., BREED, W. J., WEST, J., and BATCHELDER, G.

1975   Lake Pagahrit, southeastern Utah, a preliminary research report. *In:* Fassett, J. E. (ed.), *Canyonlands Country.* Four Corners Geological Society, 8th Field Conference, 103-110.

LYTLE-WEBB, J.

1978   Pollen analysis in Southwestern archaeology. *In:* Grebinger, P. (ed.), *Discovering Past Behavior, Experiments in the Archaeology of the Southwest.* Gordon and Breach Press, 13-28.

MADSEN, D. B.

1973   The pollen analysis of O'Malley Shelter. *In:* Fowler, D. D., Madsen, D. B., and Hattori, E. M., *Prehistory of Southeastern Nevada.* Desert Research Institute Publications in the Social Sciences 6, 137-142.

MADSEN, D. B., and CURREY, D. R.

1979   Late Quaternary glacial and vegetation changes, Little Cottonwood Canyon area, Wasatch Mountains, Utah. *Quaternary Research,* 12: 254-270.

MADSEN, D. B., and KAY, P. A.

1982   Late Quaternary pollen analysis in the Bonneville Basin [abst.]. American Quaternary Association (AMQUA), 7th conf., Seattle, Washington. *Program and Abstracts,* 128.

MAHER, L. J., JR.

1961   Pollen analysis and postglacial vegetation history in the Animas Valley region, southern San Juan Mountains, Colorado. Ph.D. dissertation, University of Minnesota, Minneapolis, 85 p.

1963   Pollen analyses of surface materials from the southern San Juan Mountains, Colorado. *Geological Society of America Bulletin,* 74: 1485-1504.

1972a   Absolute pollen diagram of Redrock Lake, Boulder County, Colorado. *Quaternary Research,* 2:531-553.

1972b   Nomograms for computing 0.95 confidence limits of pollen data. *Review of Palaeobotany and Palynology,* 13: 85-93.

1973   Pollen evidence suggests that climatic changes in the Colorado Rockies during last 5,000 years were out of phase with those in the northeastern United States [abst.]. International Union for Quaternary Research (INQUA), IX Congress, Christchurch, New Zealand, *Abstracts,* 227-228.

MARKGRAF, V.

1980   Pollen dispersal in a mountain area. *Grana,* 19: 127-146.

MARKGRAF, V., BRADBURY, J. P., FORESTER, R. M., McCOY, W., SINGH, G., and STERNBERG, R.

1983   Paleoenvironmental reassessment of the 1.6-million-year-old record from San Agustin Basin, New Mexico. *New Mexico Geological Society Guidebook, 34th Field Conference, Socorro Region,* 291-297.

MARKGRAF, V., BRADBURY, J. P., FORESTER, R. M., SINGH, G., and STERNBERG, R. S.

1984   San Agustin Plains, New Mexico: age and paleoenvironmental potential reassessed. *Quaternary Research,* 22: 336-343.

MARKGRAF, V., and SCOTT, L.

1981   Lower timberline in central Colorado during the past 15,000 years. *Geology,* 9: 231-234.

MARTIN, P. S.

1963a   *The Last 10,000 Years, A Fossil Pollen Record of the American Southwest.* University of Arizona Press, 87 p.

1963b   Geochronology of Pluvial Lake Cochise, southern Arizona. II. Pollen analysis of a 42-meter core. *Ecology,* 44:436-444.

1963c   Early man in Arizona, the pollen evidence. *American Antiquity,* 29: 67-73.

1964   Pollen analysis and the full glacial landscape. *In:* Hester, J. J., and Schoenwetter, J. (eds.), *The Reconstruction of Past Environments.* Fort Burgwin Research Center Publication 3, 66-75.

1969   Pollen analysis and the scanning electron microscope. *In:* Jahari, C. O. (ed.), *Scanning Electron Microscopy/ 1969; Proceeding of the 2nd Annual Scanning Electron Microscope Symposium.* IIT Research Institute, Chicago, 89-102.

1983   Pollen profile from the east bank of Cienega Creek. *In:* Eddy, F. W., and Cooley, M. E., Cultural and environmental history of Cienega Creek, southeastern Arizona. *Anthropological Papers of the University of Arizona* 43, 42-44.

MARTIN, P. S., and BYERS, W.

1965   Pollen and archaeology at Wetherill Mesa. *In:* Osborne, D., Contributions of the Wetherill Mesa Archaeological Project. *American Antiquity,* 31(2): 122-135 *(Society for American Archaeology Memoir* 19).

MARTIN, P. S., and DREW, C. M.

1969   Scanning electron photomicrographs of Southwestern pollen grains. *Journal of the Arizona Academy of Science,* 5: 147-176.

1970   Additional scanning electron photomicrographs of Southwestern pollen grains. *Journal of the Arizona Academy of Science,* 6: 140-161.

MARTIN, P. S., and GRAY, J.

1962   Pollen analysis and the Cenozoic. *Science,* 137: 103-111.

MARTIN, P. S., and MEHRINGER, P. J., JR.

1965   Pleistocene pollen analysis and biogeography of the Southwest. *In:* Wright, H. E., Jr., and Frey, D. G. (eds.), *The Quaternary of the United States.* Yale University Press, New Haven, 433-451.

MARTIN, P. S., and MOSIMANN, J. E.

1965   Geochronology of Pluvial Lake Cochise, southern Arizona. III. Pollen statistics and Pleistocene metastability. *American Journal of Science,* 263: 313-358.

MARTIN, W. C., and HUTCHINS, C. R.

1980   *A Flora of New Mexico.* 2 Volumes. A. R. Gantner Verlag, Vaduz, Lichtenstein, 2591 p.

McANDREWS, J. H., and KING, J. E.

1976   Pollen of the North American Quaternary: The top twenty. *Geoscience and Man,* 15: 41-49.

MEHRINGER, P. J., JR.

1965   Late Pleistocene vegetation in the Mohave Desert of southern Nevada. *Journal of the Arizona Academy of Science,* 3: 172-188.

1967a   Pollen analysis of the Tule Springs Site area, Nevada. *In:* Wormington, H. M., and Ellis, D. (eds.), *Pleistocene Studies in Southern Nevada.* Nevada State Museum Anthropological Papers 13, 129-200.

1967b   Pollen analysis and the alluvial chronology. *The Kiva,* 32: 96-101.

1977   Great Basin late Quaternary environments and chronology. *In:* Fowler, D. D. (ed.), *Models and Great Basin Prehistory: A Symposium.* Desert Research Institute Publications in the Social Sciences 12, 113-167.

MEHRINGER, P. J., JR., ADAM, D. P., and MARTIN, P. S.

1971   Pollen analysis at Lehner Ranch arroyo. *In:* American Association of Stratigraphic Palynologists, Field Trip Guide, *Lehner Early Man-Mammoth Site,* 10-26.

MEHRINGER, P. J., JR., and HAYNES, C. V., JR.

1965   The pollen evidence for the environment of early man and extinct mammals at the Lehner mammoth site, southeastern Arizona. *American Antiquity,* 31: 17-23.

MEHRINGER, P. J., JR., MARTIN, P. S., and HAYNES, C. V., JR.

1967   Murray Springs, a mid-postglacial pollen record from southern Arizona. *American Journal of Science,* 265: 786-797.

MILLINGTON, A. C.

1976   Late Quaternary paleo-environmental history of the Mary Jane Creek Valley, Grand County, Colorado. M. A. thesis, University of Colorado, Boulder, 194 p.

NELSON, A. R., MILLINGTON, A. C., ANDREWS, J. T., and NICHOLS, H.

1979   Radiocarbon-dated upper Pleistocene glacial sequence, Fraser Valley, Colorado Front Range. *Geology,* 7: 410-414.

NICHOLS, H.

1982   Review of late Quaternary history of vegetation and climate in the mountains of Colorado. *In:* Halfpenny, J. C. (ed.), *Ecological Studies in the Colorado Alpine, A Festschrift for John W. Marr.* University of Colorado, Institute of Arctic and Alpine Research, Occasional Paper 37, 27-33.

OLDFIELD, F.

1975   Pollen analytical results, Part II. *In:* Wendorf, F., and Hester, J. J. (eds.), *Late Pleistocene Environments of the Southern High Plains.* Publication of the Fort Burgwin Research Center 9, 121-147.

PECK, R. M.

1973   Pollen budget studies in a small Yorkshire catchment. *In:* Birks, H. J. B., and West, R. G. (eds.), *Quaternary Plant Ecology.* John Wiley & Sons, New York, 43-60.

PETERSEN, K. L., and MEHRINGER, P. J., JR.

1976   Postglacial timberline fluctuations, La Plata Mountains, southwestern Colorado. *Arctic and Alpine Research,* 8: 275-288.

POTTER, L. D.

1967   Differential pollen accumulation in water-tank sediments and adjacent soils. *Ecology,* 48: 1041-1043.

POTTER, L. D., and ROWLEY, J.

1960   Pollen rain and vegetation, San Augustin Plains, New Mexico. *Botanical Gazette,* 122: 1-25.

PRICE, C. R.

1971   Preliminary paleopalynological analysis of Alamosa Formation sediments. *In:* New Mexico Geological Society, 22nd Field Conference. James, H. L. (ed.), *Guidebook of the San Luis Basin, Colorado,* 219-220.

ROGERS, K. L.

1984   A paleontological analysis of the Alamosa Formation (south-central Colorado: Pleistocene: Irvingtonian). *In:* New Mexico Geological Society Guidebook, 35th Field Conference, *Rio Grande Rift: Northern New Mexico,* 151-155.

SANGSTER, A. G., and DALE, H. M.

1964   Pollen grain preservation of underrepresented species in fossil spectra. *Canadian Journal of Botany,* 42: 437-449.

SAYLES, E. B.

1965   Late Quaternary climate recorded by Cochise Culture. *American Antiquity,* 30: 476-480.

SCHOENWETTER, J.

1962   The pollen analysis of eighteen archaeological sites in Arizona and New Mexico. *In:* Martin, P. S., *et al.,* Chapters in the Prehistory of eastern New Mexico, I. Chicago Natural History Museum, *Fieldiana: Anthropology,* 53: 168-209.

1964   The palynological research. *In:* Schoenwetter, J., and Eddy, F. W., Alluvial and Palynological Reconstruction of Environments, Navajo Reservoir District. *Museum of New Mexico Papers in Anthropology* 13, 63-107.

1967   Pollen survey of the Chuska Valley. *In:* Harris, A. H., Schoenwetter, J., and Warren, A. H., An Archaeological Survey of the Chuska Valley and the Chaco Plateau, New Mexico. *Museum of New Mexico Research Records* 4 (Part I, Natural Science Studies), 72-103.

SCHOENWETTER, J., and DOERSCHLAG, L. A.

1971   Surficial pollen records for central Arizona. I. Sonoran Desert scrub. *Journal of the Arizona Academy of Science,* 6: 216-221.

SCOTT, L. J.

1979 Dietary inferences from Hoy House coprolites: a palynological interpretation. *The Kiva,* 44: 257-281.

1983 A model for the interpretation of pit structure activity areas at Anasazi sites (Basketmaker III-Pueblo I) through pollen analysis. M.A. thesis, University of Colorado, Boulder.

SEARS, P. B.

1937 Pollen analysis as an aid in dating cultural deposits in the United States. *In:* MacCurdy, G. G. (ed.), *Early Man.* J. B. Lippincott Co., London, 61-66.

1961 Palynology and the climatic record of the Southwest. *Annals of the New York Academy of Sciences,* 95: 632-641.

SEARS, P. B., and CLISBY, K. H.

1952 Two long climatic records. *Science,* 116: 176-178.

SOLOMON, A. M., BLASING, T. J., and SOLOMON, J. A.

1982 Interpretation of floodplain pollen in alluvial sediments from an arid region. *Quaternary Research,* 18: 52-71.

SOLOMON, A. M., and HARRINGTON, J. B.

1979 Palynology models. *In:* Edmonds, R. L. (ed.), *Aerobiology, The Ecological Systems Approach.* US/IBP Synthesis Series 10. Dowden, Hutchinson, and Ross, Inc., Stoudsburg, Pennsylvania, 338-371.

SOLOMON, A. M., KING, J. E., MARTIN, P. S., and THOMAS, J.

1973 Further scanning electron photomicrographs of Southwestern pollen grains. *Journal of the Arizona Academy of Science,* 8: 135-157.

SPAULDING, W. G., LEOPOLD, E. B., and VAN DEVENDER, T. R.

1983 Late Wisconsin paleoecology of the American Southwest. *In:* Wright, H. E., Jr. (ed.), *Late-Quaternary Environments of the United States. Vol. 1. The Late Pleistocene* (Porter, S. C., ed.). University of Minnesota Press, Minneapolis, 259-293.

VAN DEVENDER, T. R., and KING, J. E.

1971 Late Pleistocene vegetational records in western Arizona. *Journal of the Arizona Academy of Science,* 6: 240-244.

VAN DEVENDER, T. R., and TOOLIN, L. J.

1983 Late Quaternary vegetation of the San Andres Mountains, Sierra County, New Mexico. *In:* Eidenbach, P. L. (ed.), *The Prehistory of Rhodes Canyon, N.M. Survey and Mitigation.* Human Systems Research, Inc., Tularosa, New Mexico, 33-54.

VAN DEVENDER, T. R., and SPAULDING, W. G.

1979 Development of vegetation and climate in the Southwestern United States. *Science,* 204: 701-710.

VISHER, S. S.

1950 High temperatures and the seasonal distribution of precipitation and some ecological consequences or correlations. *The American Midland Naturalist,* 44: 478-487.

WATERS, M. R.

1984 Late Quaternary alluvial geology and early archaeology of Whitewater Draw, Cochise County, southeastern Arizona [abst.]. American Quaternary Association (AMQUA), 8th mtg., Boulder, Colorado, *Program and Abstracts,* 136.

WEBER, W. A.

1976 *Rocky Mountain Flora.* Colorado Associated University Press, Boulder, 484 p.

WELLS, P. V.

1970 Postglacial vegetational history of the Great Plains. *Science,* 167: 1574-1582.

WHITESIDE, M. C.

1965 Paleoecological studies of Potato Lake and its environs. *Ecology,* 46: 807-816.

WILLIAMS-DEAN, G., and BRYANT, V. M., JR.

1975 Pollen analysis of human coprolites from Antelope House. *The Kiva,* 41: 97-111.

WRIGHT, H. E., JR.

1971 Late Quaternary vegetational history of North America. *In:* Turekian, K. K. (ed.), *The Late Cenozoic Glacial Ages.* Yale University Press, New Haven, 425-464.

WRIGHT, H. E., JR., BENT, A. M., HANSEN, B. S., and MAHER, L. J., JR.

1973 Present and past vegetation of the Chuska Mountains, northwestern New Mexico. *Geological Society of America Bulletin,* 84: 1155-1180.

WYCKOFF, D. G.

1977 Secondary forest succession following abandonment of Mesa Verde. *The Kiva,* 42: 215-231.

QUATERNARY POLLEN RECORDS FROM CALIFORNIA

DAVID P. ADAM

United States Geological Survey, M.S.915
345 Middlefield Road
Menlo Park, California 94025

Abstract

Quaternary palynology in California is still in its formative stages, and many studies are unpublished. At least 109 sites have been studied; this paper presents a map of their locations and a brief description of the available data for each site. Although many sites have yielded only rather short records, a few records (Clear Lake, Tulelake, and Tulare Lake) extend back at least 100,000 years. Quaternary palynology has very high potential in California because of the wide range of depositional environments suitable for pollen preservation. Major problems include: (1) the size of the California flora and an inadequate understanding and documentation of the pollen types, (2) poor documentation of the work already done, and (3) the very wide range of topography and climate and biotic provinces within the state, which inhibits simple extrapolation of conclusions from individual studies into regional syntheses.

INTRODUCTION

The purpose of this paper is to assemble in one place a list of both published and unpublished pollen work from Quaternary stratigraphic sequences in California. The earliest paper using the techniques of modern pollen analysis for Quaternary sediments appeared in Sweden in 1916 (von Post, 1916); not until two decades later was any pollen work begun in western North America (Osvald, 1936). The earliest published pollen work in California was by Hansen (1942) but further work did not appear until much later (other work by Hansen reported in Frink and Kues, 1954; Roosma, 1958). Most reports on California Quaternary pollen studies have appeared since 1960.

Quaternary palynology in California is still in its formative stage, despite the studies enumerated below. Only a few parts of the state have yet produced published pollen records spanning most or all of Holocene time, and the size of the state and its great floristic and topographic diversity inhibit easy extrapolation of palynological results and experience from one region to another.

These difficulties notwithstanding, the opportunities for palynology within California are outstanding. The Sierra Nevada was heavily glaciated at various times during the Quaternary, and less extensive glacial advances also occurred in the northern Coast Ranges, the southern Cascade Range, the Transverse Ranges, and the higher ranges of the western Great Basin. These glacial events created closed depressions in many montane regions, and the deposits now found in these depressions offer the chance to develop pollen records at varying elevations, as well as the chance to evaluate pollen records in similar topographic situations but on opposite sides of the Sierra Nevada.

In much of the United States, such closed depressions of glacial origin are the primary source of continuous pollen records. In California, by contrast, other promising pollen sites include pluvial lake basins, actively-subsiding structural basins, volcanic basins, tectonic basins, closed basins impounded by or formed on top of landslides, and hillside swales.

The California pollen record is of particular interest because many of the basins mentioned above may well contain continous depositional sequences much longer than can be recovered from glaciated areas. The great subsiding basins of the southern San Joaquin Valley, for example, contain a Quaternary section well over 1 km thick in places, and the stratigraphy is reasonably well known because of the long history of exploration for oil and water resources (Davis *et al.*, 1959; Croft, 1972). East of the Sierra Nevada, thick Quaternary sequences have been recovered from the deposits of Searles Lake, China Lake, Owens Lake, and Long Valley, although very little pollen work has been done on the cores. The Modoc Plateau, in northeastern California, contains many valley-bottom deposits of complex origin. At least one of these deposits (Tulelake) contains a record which, although not continuous for pollen, extends back over 3 million years. In the northern Coast Ranges, the deposits of Clear Lake have yielded a detailed pollen record of the last complete glacial cycle (Adam, Sims, and Throckmorton, 1979; Adam and West, 1983; Adam, 1985a, 1985b, 1985c).

A major difficulty with California pollen work lies in the size and complexity of the flora, which includes over 5,000 species of native vascular plants in ten major floristic regions (Raven and Axelrod, 1978). No general key to the California pollen types is yet available, and the level of pollen taxonomy can undoubtedly be greatly improved.

A particular problem is that several major pollen types are at present identified at the generic level or above. The three main AP (arboreal pollen) types are pine (*Pinus*), oak (*Quercus*), and the TCT group, which includes the families Taxodiaceae, Cupress-aceae, and Taxaceae.

There are at least 20 species of *Pinus* native to California, 16 species of *Quercus,* and 22 species in the TCT group (Munz and Keck, 1973). The pollen grains of the various species in each group generally overlap in size and morphology, making it difficult or impossible to differentiate all species in a fossil assemblage. A major attempt to overcome these problems in the case of *Pinus* is the study by Ting (1966), who measured size distributions of saccus, colpus, and overall breadth of pollen grains in over 300 samples of reference pollen. Although he was able to identify single-species pollen slides correctly in many cases, he did not claim to be able to identify individual unknown grains to species, and I know of no California pollen studies that have attempted to apply his results to fossil material.

A more promising development is the claim by Margie Reed of the University of California at Davis that she can distinguish the pollen of *Quercus chrysolepis* from other species of oak in the pollen record from Dogwood Pond. Further studies of pollen morphology in California taxa will undoubt-edly lead to more precise and reliable work in the future.

DISTRIBUTION OF POLLEN STUDIES

Pollen studies have been done in most parts of the state (Fig. 1); however, the distances between sites are often large, and data for California Quaternary pollen studies are quite scattered. This review describes all

studies known to me, but is probably incomplete. Many of the studies are unpublished. The studies are grouped here into eight geographic areas: (1) the Sierra Nevada; (2) the western Great Basin; (3) the Modoc Plateau; (4) the northern Coast Ranges; (5) the Central Valley; (6) the southern Coast Ranges; (7) the Transverse Ranges; and (8) the coastal region. Offshore studies are not considered in this review; interested readers are referred to two papers by Linda Heusser (1978, 1981); see also Heusser and Shackleton, 1979, and the thesis by McGann (1985, in preparation).

The locations of the pollen sites are shown in Figure 1 and listed in Table 1. An earlier version of this paper (Adam, 1985d) includes a list of the addresses and affiliations of many of the researchers whose work is mentioned here.

Sierra Nevada

Published pollen studies in the Sierra Nevada deal with the Lake Tahoe Basin, the Yosemite area, and the Chagoopa Plateau. Unpublished studies range from Sierra County to just south of Walker Pass and are mostly on the western slope of the Sierra. Inter-pretation of much of the Sierra Nevada pollen work has been based on two transects of modern pollen surface samples across Carson and Tioga Passes (Adam, 1964b, 1967). More such transects are needed throughout California in order to interpret the fossil pollen record.

The northernmost pollen site in the Sierra Nevada is at Ross Relles Camp in Sierra County (Table 1, No. 13), which has been studied by G. J. West. The pollen record includes the past 2,000 years (G. J. West, oral communication, 1982). In the northern foothills, Kilbourne (1978) has studied several samples from the Auburn Dam site (Table 1, No. 21), and Matson (1972) has studied a 1.25-m section from the Spring Garden Ravine site (Table 1, No. 96).

The oldest site studied in the Tahoe Basin is near Tahoe City (Table 1, No. 17) and describes early-Pleistocene samples from beneath a basalt flow dated at 1.9 m.y. (Adam, 1973). The site is significant because the samples include significant amounts of *Picea* (spruce) pollen, which implies that a Mediterra-nean climate with its characteristic summer drought did not exist at that time.

A site in the Tahoe Basin that has received exten-sive study is Osgood Swamp (Table 1, No. 23), which has yielded a pollen record that extends from the present back to the late-Pleistocene (Adam, 1967,

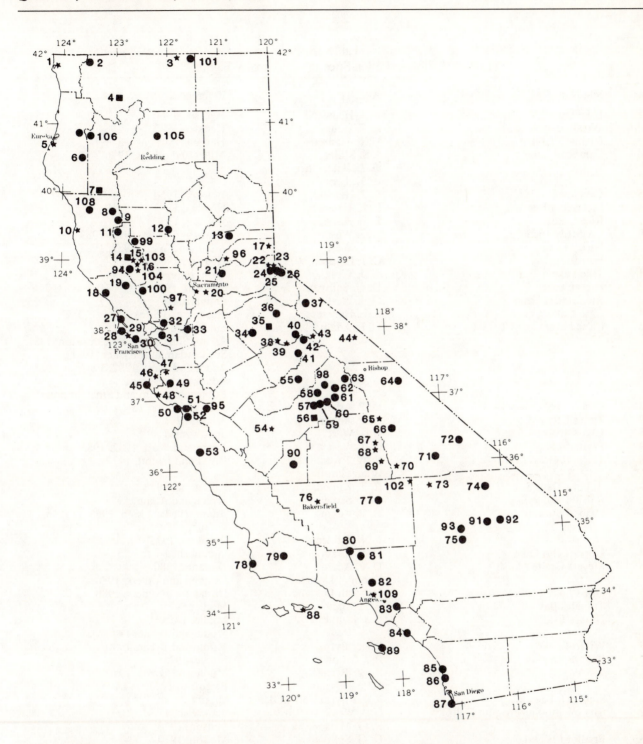

Figure 1. Map showing the location of pollen sites discussed in this report. Sites for which full pollen diagrams have been published are shown as stars; sites for which only abstracts have been published are shown as squares, and unpublished sites are shown as circles. The number given after each site name in the text is the reference number used in Figure 1 and Table 1 for that site.

1974); it has also been studied in detail by Zauderer (1973), whose study covered the past 3,000 years. The other published report for the area is that of Sercelj and Adam (1975) on Ralston Ridge Bog (Table 1, No. 22). The study spans about the last 3,000 years and includes only 11 pollen samples.

Table 1.
List of California Pollen Sites to Accompany Figure 1.

Map #	Site Name	Analyst	References
1	Lake Earl	C. J. Heusser	Heusser, 1960
2	Whisky Lake	L. E. Heusser	Heusser, 1982
3	Klamath Lake	H. P. Hansen	Hansen, 1942
4	Campbell Lake	J. S. Miller	Miller and others, 1976
		G. L. Batchelder	
		G. K. Lee	
5	Capetown	C. J. Heusser	Heusser, 1960
6	McClellan Bog	L. E. Heusser	Heusser, 1982
7	Rice Lake	L. E. Heusser	Heusser, 1982
8	CA-MEN-1633	G. J. West	Adam and West, 1982, 1983
9	Barley Lake	G. J. West	Adam and West, 1982, 1983
10	Fort Bragg	C. J. Heusser	Heusser, 1960
11	Tule Lake	G. J. West	West, in press
12	Packer Lake	D. G. Sullivan	Sullivan, 1982
13	Ross Relles Camp	G. J. West	unpublished
14	Clear Lake, Core CL-80-1	L. E. Heusser	Heusser and Sims, 1981
15	Clear Lake, Core CL-73-4	D. P. Adam	Sims, 1976
			Adam, Sims, and Throckmorton, 1981
			Sims, Adam and Rymer, 1981
			Adam and West, 1982, 1983
			Adam, in press a-c
			Robinson, Adam and Sims, in press
16	Clear Lake, Core CL-73-7	D. P. Adam	same as for site 15
17	Tahoe City	D. P. Adam	Adam, 1973
18	Lake Oliver	G. J. West	unpublished
19	Preston Lake	G. J. West	Adam and West, 1982, 1983
20	Teichert Gravel Pit	E. W. Ritter	Ritter and Hatoff, 1977
		Brian Hatoff	
21	Auburn Dam Site	R. T. Kilbourne	Kilbourne, 1978
22	Ralston Ridge Bog	Alojz Sercelj	Sercelj and Adam, 1975
23	Osgood Swamp	D. P. Adam	Adam, 1964a, 1964b, 1965a, 1965b,
			1967, 1974
		J. N. Zauderer	Zauderer, 1973
24	Upper Echo Lake	D. P. Adam	unpublished
25	Meyers Grade Marsh	D. P. Adam	Dorland, 1980
		Deborah Dorland	Dorland and others, 1980
26	Grass Lake	Deborah Dorland	Dorland and others, 1980
27	Tomales Bay	P. J. Mudie	Mudie, 1975
28	Drakes Estero	P. J. Mudie	Mudie, 1975
			Mudie and Byrne, 1980
29	Wildcat Lake	Roger Byrne	Mudie and Byrne, 1980
30	Bolinas Lagoon	P. J. Mudie	Mudie, 1975
		Roger Byrne	Mudie and Byrne, 1980
		J. R. Bergquist	Bergquist, 1977, 1978
31	Sindicich Lagoons	Roger Byrne and students	unpublished
32	Sacramento-San Joaquin Delta	G. J. West	unpublished
		R. G. Matson	Matson, 1970
33	Bradford Island	D. G. Sullivan	Matson, 1970
34	New Melones Dam Site	Diana Araki	unpublished
		W. G. Spaulding	Spaulding, in press
35	Swamp Lake	G. L. Batchelder	Batchelder, 1980
36	Catfish Lake	Roger Byrne and students	unpublished
37	Tule Lake	Roger Byrne and students	unpublished
38	Hodgdon Ranch	D. P. Adam	Adam, 1964b, 1967
39	Crane Flat	D. P. Adam	Adam, 1964b, 1967
40	Harden Lake	G. L. Batchelder	unpublished
		Roy Faverty	
41	Bridalveil Creek	D. P. Adam and students	unpublished

42	Polly Dome	John Batch	Batch, 1977
		G. L. Batchelder	
43	Soda Springs	D. P. Adam	Adam, 1964b, 1967
44	Black Lake	G. L. Batchelder	Batchelder, 1970a,b
			Mehringer, 1977
45	Año Nuevo Point	D. G. Sullivan	unpublished
46	Pearsons Pond	D. P. Adam	Adam, 1975
47	Mt. View Dump	D. P. Adam	Helley and others, 1972
48	Laguna de las Trancas	Edgar Luther	Adam, Byrne and Luther, 1981
	(McCrary's Marsh)		Heusser, 1982
49	Saratoga	D. P. Adam	Adam and others, 1983
50	Elkhorn Slough	P. J. Mudie	Mudie, 1975
51	Kelly Lake	Roger Byrne and students	unpublished
52	Warner Lake	Roger Byrne and students	unpublished
53	The Lakes	Sally Peterson	unpublished
54	Corcoran Clay	H. P. Hansen	Frink and Kues, 1954
55	Bass Lake (4-MAD-223)	James Schoenwetter	unpublished
56	Lower Kings Canyon	Kenneth Cole	Cole, 1983
57	CA-FRE-682	James Schoenwetter	unpublished
		Ella Stewart	
58	CA-FRE-741	James Schoenwetter	unpublished
		Ella Stewart	
59	CA-FRE-608	James Schoenwetter	unpublished
		Ella Stewart	
60	CA-FRE-756	James Schoenwetter	unpublished
		Ella Stewart	
61	CA-FRE-534	James Schoenwetter	unpublished
		Ella Stewart	
62	CA-FRE-661	James Schoenwetter	unpublished
		Ella Stewart	
63	Lake 11,100'	D. P. Adam	unpublished
64	Deep Springs Valley	W. Woolfenden	unpublished
65	Alabama Hills	W. L. Ting	Axelrod and Ting, 1961
66	Owens Lake	P. S. Martin	unpublished
67	Kennedy Meadows	W. L. Ting	Axelrod and Ting, 1961
68	Ramsay Meadows	W. L. Ting	Axelrod and Ting, 1961
69	Bakeoven Meadows	W. L. Ting	Axelrod and Ting, 1961
70	Little Lake	W. L. Ting	Axelrod and Ting, 1961
		P. J. Mehringer, Jr.	Mehringer and Sheppard, 1978
		J. C. Sheppard	
71	Warm Sulphur Springs	P. J. Mehringer, Jr.	unpublished
72	Badwater	P. J. Mehringer, Jr.	unpublished
73	Searles Lake	Aino Roosma	Roosma, 1958
		E. B. Leopold	Leopold, 1967
74	Saratoga Springs	P. J. Mehringer, Jr.	unpublished
75	East Rim site	James Schoenwetter	unpublished
		D. E. Buge	
76	Wasco site	L. E. Heusser	Davis and others, 1977
77	Bird Springs Pass	G. J. West	unpublished
78	Vandenberg AFB	G. L. Batchelder	unpublished
79	Zaca Lake	Roger Byrne	unpublished
80	R pond	G. J. West	unpublished
81	unnamed pond	K. Mossberg	unpublished
82	Van Norman Reservoir	D. P. Adam	unpublished
83	Slusher Estate	James Schoenwetter	unpublished
		Amie Limon	
84	R-ORA-64	D. E. Buge	unpublished
		James Schoenwetter	
85	Batiquitos Lagoon	Herbert Meyer	unpublished
			(U.C. Berkeley M.S. thesis?)
86	Los Penasquitos Lagoon	P. J. Mudie	Mudie, 1975
			Mudie and Byrne, 1980

87	Mission Bay	P. J. Mudie	Mudie and Byrne, 1980
	Santee Green	R. H. Hevly and	unpublished
	Pacific Bluffs	R. E. Diggs	
	Shadow Ridge	R. H. Hevly and	unpublished
	Rancho San Diego	R. E. Diggs	
88	Santa Cruz Island	R. H. Hevly	Hevly and Hill, 1970
89	Santa Catalina Island	R. G. Matson	unpublished
	(Toyon Bay)		
90	Tulare Lake	D. P. Adam	Atwater and others, in press
91	Cronese Lake	D. E. Buge	unpublished
92	Soda (Zzyzx) Springs	P. J. Mehringer, Jr.	unpublished
93	Calico site	Sally Peterson	unpublished
94	Kelseyville Formation	J. A. Wolfe	Rymer, 1981
95	Hollister	L. E. Heusser	McMasters and others, in press
96	Spring Garden Ravine site	R. G. Matson	Matson, 1970
97	Lagoon Valley	R. G. Matson	Matson, 1970
98	Balsam Meadow	O. K. Davis	unpublished
99	Lily Pond	G. J. West	unpublished
100	Dogwood Pond	Margie Reed	unpublished
101	Tulelake	D. P. Adam	unpublished
102	China Lake	P. S. Martin	Martin and Mehringer, 1965
103	High Valley colluvial wedge	W. E. Dietrich	Dietrich and Dorn, 1984
		Ronald Dorn	
104	Perini colluvial wedge	W. E. Dietrich	Dietrich and Dorn, 1984
		Ronald Dorn	
105	Potter Creek Cave	G. J. West	unpublished
	Dekkas Cave	G. J. West	unpublished
106	South Fork Mountain	G. J. West	unpublished
107	Pilot Ridge	G. J. West	unpublished
108	Big Lake	G. J. West	unpublished
109	Rancho La Brea	J. K. Warter	Warter, 1976

Unpublished work in the Tahoe Basin includes studies at Upper Grass Lake (Table 1, No. 26) by Deborah Dorland and D. P. Adam, Lower Grass Lake (Table 1, No. 26) by Adam, Meyers Grade Marsh (Table 1, No. 25) by Dorland and Adam, and work at Upper Echo Lake (Table 1, No. 24) by Adam. Parts of the Upper Grass Lake and Myers Grade Marsh work are described in a thesis by Dorland (1980), and in an abstract by Dorland and others (1980). All four sites span all of Holocene time, and the Upper Echo Lake core has yielded a radiocarbon age of 11,100±70 yrs. B.P. (USGS-1076) for lacustrine deposits that postdate the deglaciation of the valley.

South of the Tahoe Basin, Tule Lake (Table 1, No. 37), at Sonora Pass Junction on the east side of the Sierra in Mono County, has been studied by Roger Byrne and his students and spans the Holocene. Directly across the Sierra on the western slope are two sites with complete Holocene records, Catfish Lake (Table 1, No. 36) and Swamp Lake (Table 1, No. 35). Catfish Lake has been studied by Byrne and his students, and Swamp Lake has been studied by G. L. Batchelder. Only the transition from glacial to post-glacial conditions has been studied at Swamp

Lake, but a complete core of Holocene sediments is still preserved (G. L. Batchelder, oral communication, 1982). The Swamp Lake site has yielded a date of 15,565±820 yrs. B.P. for glacial outwash sediments overlying till that is the oldest date yet reported for the onset of Sierran deglaciation (Batchelder, 1980). At a lower elevation, both W. G. Spaulding (1984) and Diana Araki (Roger Byrne, oral communication, 1982) have studied short pollen records from archeological midden sites near the New Melones Dam (Table 1, No. 34).

Several sites have been studied in the Yosemite area. Harden Lake (Table 1, No. 40) has been studied by G. L. Batchelder and Roy Faverty, who encountered an anomalously young radiocarbon age of about 7,000 years for the lowest post-glacial sediments in their core from a small closed basin east of the lake. Another core from Harden Lake itself was collected by Adam and Byrne; it has not been studied for pollen but has yielded another very young radiocarbon age of about 8,000 yrs. B.P. for glacial clays which must have been deposited when the Grand Canyon of the Tuolumne was completely filled with ice below the site. I consider both of these ages to be in error, because they conflict with other data indicating a post-glacial climate was well established in California by 10,000 years ago (*e.g.,* Adam, 1967; Adam, Sims, and Throckmorton, 1981).

Several short pollen records along the Tioga Pass road were studied by Adam (1964b, 1967). These sites are Hodgdon Ranch (Table 1, No. 38), Crane Flat (Table 1, No. 39), and Soda Springs (Table 1, No. 43). The Hodgdon Ranch and Soda Springs records are undated but show sensitivity to climatic changes, whereas the Crane Flat record shows little sensitivity to climatic changes but is relatively well dated. Adam (unpublished data, 1970) also has studied a core from beside Bridalveil Creek on the Glacier Point road south of Yosemite Valley (Table 1, No. 41). That record is also undated but is of interest because it records the presence of *Typha* pollen (tetrads) above the maximum elevation at which it is presently found. A meadow site near Polly Dome (Table 1, No. 42) was studied by J. R. Batch and G. L. Batchelder (Batch, 1977).

Still farther south, a number of archeological sites on the western slope of the Sierra have been analyzed by James Schoenwetter and his students. One site is at Bass Lake (Table 1, No. 55) in Madera County; the other six sites (Table 1, Nos. 57-62) are between the Kings and San Joaquin Rivers in Fresno County (James Schoenwetter, written communication, 1982). A recent study at Balsam Meadow (Table 1, No. 98) near Shaver Lake, has yielded a pollen record that spans all of post-glacial time; the pollen diagram strongly resembles the Osgood Swamp diagram (Owen K. Davis, written communication, 1984).

An incomplete transect of modern pollen surface samples collected by Adam extends from Bishop to Auberry across the Sierra Nevada in northern Fresno County, and Adam has analyzed the pollen in a short core collected from meadow deposits at the edge of a lake at an elevation of 3,386 m (11,106 feet) (Table 1, No. 63) just south of the Glacier Divide in Kings Canyon National Park. A particularly interesting site is found just south of the Kings River in the lower Kings Canyon (Table 1, No. 56), where Cole (1983) has studied pollen from dated pack rat middens of glacial age.

Several pollen sites are on the Chagoopa Plateau, where Ting analyzed early-Pleistocene samples from Kennedy Meadows (Table 1, No. 67), Ramsay Meadows (Table 1, No. 68), and Bakeoven Meadows (Table 1, No. 69) (Axelrod and Ting, 1961). That study is remarkable for the degree of taxonomic precision claimed for the pollen identifications; most

pollen grains are identified to species, including all oak and conifer grains.

The southernmost pollen work in the Sierra Nevada is from an archeological site near Bird Springs Pass (Table 1, No. 77). The site has been studied by G. J. West and is no more than a few thousand years old (G. J. West, oral communication, 1982).

Western Great Basin

Several pollen studies have been undertaken in the western Great Basin area of California. Most of these are located in the Owens River drainage and the other drainages with which it coalesced during the pluvial conditions that prevailed during the last glacial cycle. These pollen sites are discussed here in a downstream direction along the pluvial drainage system.

The site farthest upstream is Black Lake (Table 1, No. 44), which lies in the southern end of the Adobe Valley between Mono Lake and Benton (Batchelder 1970a, 1970b). A summary pollen diagram has been published by Mehringer (1977, p. 127). A radiocarbon date of 11,350±350 yrs. B.P. near the base of Black Lake core 2 establishes that the core spans the entire Holocene; however, the pollen record does not reveal any major changes near the base that are of sufficient magnitude to be the result of the Pleistocene/Holocene warming. The Deep Springs Valley (Table 1, No. 64), located east of the White Mountains in northern Inyo County, has yielded a pollen record that spans the past 2,000 years as part of a study by W. Woolfenden.

Several sites have been studied in the Owens/Searles/Panamint/Death Valley drainage system. Early-Pleistocene records have been described by Axelrod and Ting (1961) at the Alabama Hills (Table 1, No. 65) and at Little Lake (Table 1, No. 70), and Little Lake has also provided a pollen record for the past 5,000 years (Mehringer and Sheppard, 1978).

Both E. B. Leopold and Allen Solomon have undertaken studies of the modern pollen rain in the southern Owens Valley, and a pollen record for the last 8,000 years has been recovered by P. S. Martin from Owens Lake (Table 1, No. 66). The Owens Lake core is of particular interest because Matthes (1941) based his original estimate of the age of the Altithermal period on the assumption that Owens Lake dried up during the Altithermal, and that the layer of salts precipitated during desiccation was then buried beneath a layer of mud. The age of the end of desiccation was then estimated by dividing the present salt content of Owens Lake by the estimated annual input of salt to the lake. The Altithermal was thus calculated to have ended about 4,000 years ago. However, the Owens Lake core showed that the only salt layer formed during the past 8,000 years was the one that formed in 1912 following the diversion of the Owens River to the Los Angeles water supply.

The long sedimentary record at Searles Lake (Table 1, No. 73) was among the first California sections studied for pollen (Roosma, 1958). Roosma's section was undated; subsequent work has established that the part of the section studied for pollen ranges in age from <10,000 to >32,000 years (Leopold, 1967; see also Spaulding, Leopold, and Van Devender, 1983). A very long Searles Lake core with a maximum age estimated at 3.2 million years extending to bedrock is being studied under the direction of G. I. Smith of the U.S. Geological Survey (Liddicoat and others, 1980), but pollen analyses have not yet been done. Unfortunately, the periodic desiccation of Searles Lake has prevented the preservation of a continuous pollen sequence, and the response of the pollen deposited in the lake as a function of the regional vegetation has varied as a result of changes in the size of the lake through time. Both of these factors make the Searles Lake pollen record difficult to interpret.

An older section has been studied by Martin from the China Lake basin (Table 1, No. 102), just west of Searles Lake, but only a pine pollen curve has been published (Martin and Mehringer, 1965, fig. 2). The base of the pollen-bearing section was interpreted as Illinoian in age by Martin and Mehringer (1965), but Spaulding and others (1983) believe that the oldest part of the China Lake section is of Early Wisconsin age.

Several pollen diagrams from playa-edge spring deposits in the Mojave Desert have been prepared by P. J. Mehringer, Jr. The vegetation at these sites has varied over the past several thousand years in response to variations in spring discharge and climate that have controlled moisture availability at the sites. Localities include Warm Sulphur Springs (Table 1, No. 71) in Panamint Valley, Badwater (Table 1, No. 72) in Death Valley, Saratoga Springs (Table 1, No. 74) at the south end of Death Valley, and Soda Springs (Table 1, No. 92) at Soda Lake.

Three other sites have been studied in the Mojave Desert, all in conjunction with archaeological work: a site at Cronese Lake (Table 1, No. 91) by D. Buge; the East Rim site (Table 1, No. 75) by Buge and J. Schoenwetter; and the Calico site (Table 1, No. 93) by

Sally Peterson Horn (Roger Byrne, oral communication, 1982).

Modoc Plateau

Although the area has great potential, almost no pollen work has been done in the Modoc Plateau area of northeastern California since the initial study of Klamath Lake (Table 1, No. 3) by Hansen (1942). His core included the sediments deposited since the eruption of Mt. Mazama about 7,000 years ago.

Interdisciplinary studies have just begun on a much longer record at Tulelake (Table 1, No. 101), where a 334-m core spans about the past 3.2 million years (Adam, unpublished data). Preliminary observations indicate pollen is present throughout the section, although some hiatuses in either deposition or preservation are likely. The site will be particularly important because the numerous tephra found in the core will permit precise correlations with other, shorter sequences.

The only other pollen work near the Modoc Plateau region is from Dekkas and Potter Creek Caves, near Shasta Lake (Table 1, No. 105). These records are undated, but are probably of Pleistocene age (G. J. West, written communication, 1984).

Northern Coast Ranges

Clear Lake, in Lake County, is at present the most important pollen site in California, and perhaps in North America, because of the length and continuity of the pollen records that have been recovered there. Eight cores were collected from Clear Lake in 1973 (Sims, 1976; Adam, Sims, and Throckmorton, 1981; Sims, Adam, and Rymer, 1981), and two more long cores were recovered in 1980 (Sims, Rymer, and Perkins, 1981; Sims, Rymer, Perkins, and Flora, 1981). The longest of the 1973 cores, CL-73-4 (Table 1, No. 15), is 115 m long and spans the time since the end of the Illinoian glaciation and deep-sea oxygen-isotopic Stage 6.

Selected variables from the Clear Lake core 4 pollen record are plotted against age in Figure 2. *Quercus, Pinus,* and TCT (= Taxodiaceae + Cupressaceae + Taxaceae) are the three dominant pollen types throughout the record, and always total at least 75% of the pollen. The three types vary widely in frequency, and show that major changes in the phytogeography of the northern Coast Ranges have occurred during the last glacial cycle. The curves for *Alnus, Chrysolepis* (=*Castanopsis*), and Rhamnaceae pollen also show stratigraphically significant changes in frequency, both within the Holocene (the past 10,000 years) and during the last interglacial interval.

The interpretation of the pollen record from Clear Lake is discussed by Adam, Sims, and Throckmorton (1981), who note that the frequency of oak pollen in the core appears to be a direct function of temperature. Adam and West (1982, 1983) have developed a transfer function based on modern pollen samples from lakes and marsh deposits in the Clear Lake area that relates the ratio oak/(oak + pine) to elevation, and then uses that relationship to reconstruct temperature and precipitation fluctuations for the past 130,000 years. The major warming that took place at the end of the Pleistocene began about 13,000 yrs. B.P. and reached its mid-point by about 11,500 yrs. B.P. in the Clear Lake area.

Further discussion of the pollen record for Clear Lake cores 4 and 7 is given in the Clear Lake symposium volume (Adam, in press a and b; and Robinson, Adam, and Sims, in press), and in Adam (in press c). The pollen counts for those cores are available as separate reports (Adam, 1979a, 1979b).

A preliminary description of pollen work on Clear Lake core CL-80-1 (Table 1, No. 14) has been published as an abstract by Heusser and Sims (1981), and the pollen data for that core have been documented by Heusser and Sims (1983). They suggest that the base of that core may be as old as 175,000 years. Older Pleistocene deposits of the Kelseyville Formation (Table 1, No. 94) have been studied for both plant macrofossils and pollen by J. A. Wolfe. The pollen data are unpublished but include samples of both glacial and interglacial age (Rymer, 1981).

Pollen analysis has been applied to the geomorphic problem of dating the onset of deposition of colluvial fill in hillside swales by Dietrich and Dorn (1984). They examined pollen profiles from the High Valley (Table 1, No. 103) and Perini (Table 1, No. 104) colluvial wedges near Clear Lake, and found a shift from high pine to high oak pollen frequencies in each section. They correlated the shift in each case with the major change observed at the end of the Pleistocene in Clear Lake cores 4 and 7, and were thus able to establish that the swales had not been emptied of colluvium during the Holocene.

Other work in the northern Coast Ranges includes studies by Linda Heusser at Whisky Lake (Table 1, No. 2), McClellan Bog (Table 1, No. 6), and Rice Lake (Table 1, No. 7). The McClellan Bog record is relatively short, spanning only the past 4,000 years, but the Whisky Lake record covers the past 10,000 years, and the Rice Lake record is at least 15,000 years long. In addition to the stratigraphic records named above, Linda Heusser (1983) has examined over 70 modern pollen surface samples from the northern Coast Ranges that are not shown on Figure 1.

Another study spanning the Holocene in the far northern Coast Ranges has been reported from Campbell Lake (Table 1, No. 4) in Siskiyou County by Miller and others (1976). Pine dominates their pollen record, but their diagram shows only the major pollen types in a second pollen sum that excludes pine. Other, shorter records in the northern Coast Ranges include a 5,000-year record from Pilot Ridge (Table 1, No. 107), undated spring mat records from Lemonade and McKay Springs on South Fork Mountain (Table 1, No. 106), and a short record from Big Lake (Table 1, No. 108) that probably records only the past few centuries (G. J. West, written communication, 1984).

West has also studied several sites in the southern part of the northern Coast Ranges. Tule Lake (West, in press) (Table 1, No. 11), Barley Lake (Table 1, No. 9), and a spring mat at California Archeological Site Survey Site CA-MEN-1633 (Table 1, No. 8) are montane sites located between Rice Lake and Clear Lake. The Tule Lake and Barley Lake records span the last 8,000 years, but the CA-MEN-1633 record is much shorter; it probably spans only the last 300-400 years (West, oral communication, 1982). Another site at Lily Pond, in the Lower Letts Valley northeast of Clear Lake, has produced a dated pollen record that spans about the last 9,700 years (West, written communication, 1984).

To the south, West has also studied two sites in Sonoma County. Preston Lake (Table 1, No. 19) is near the Russian River about 40 km from the coast, and the pollen record includes only the past 1,300 years (West, oral communication, 1982). The other site is Lake Oliver (Table 1, No. 18) at Plantation. The lake is a sag pond along the San Andreas fault only a few kilometers inland, and the deposits are probably not very old, perhaps a few thousand years. Further inland in the same general area, Dogwood Pond (Table 1, No. 100) has produced a pollen record with a date of 10,260±70 B.P. on basal peat (Margie Reed, oral communication, 1984).

Clear Lake, Core 4

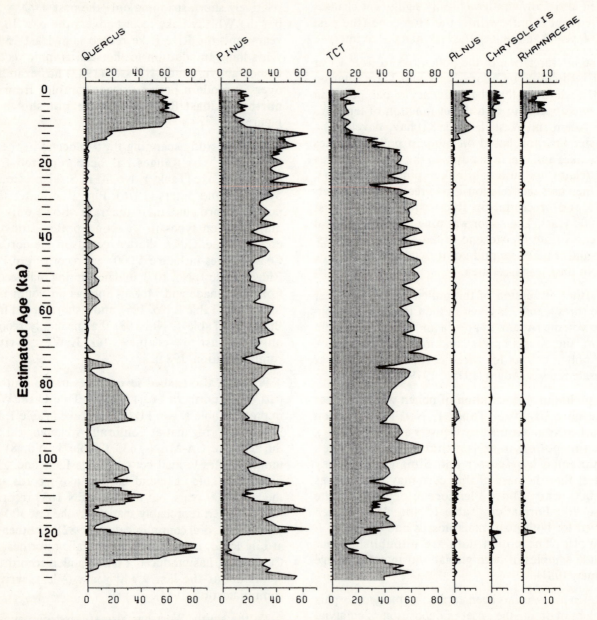

Figure 2. Diagram of selected pollen types from Clear Lake, California, core CL-73-4. Timescale is that of Robinson and others (in press); redrafted from Adam (in press c).

Along the eastern edge of the Coast Ranges, Matson (1970) has reported on a pollen diagram from a 0.5-m core from Lagoon Valley (Table 1, No. 97). G. J. West (oral communiation) reports that a 5-m core recovered from the same site did not yield any pollen.

The remaining sites in the northern Coast Ranges are so near the Pacific Ocean that their climates are strongly stabilized by the marine influence. These sites are mentioned below under "Coastal region."

Central Valley

Pollen work in the Central Valley has as yet received very little attention. In the north, Sullivan (1982) has studied a short record from Packer Lake (Table 1, No. 12) along the Sacramento River. The record appears to be no more than 1,000 years old. Sullivan has also examined samples with an estimated age of about 40,000 years from Bradford

Island (Table 1, No. 33) in the Delta area (Sullivan, oral communication, 1982). West (1981) has examined core samples from the past 5,000 years from the Delta area, but these are shown only schematically on the map (Table 1, No. 32), as is the 70-cm pollen profile from the midden site of California Archaeological Survey Site 4-SOL-35 reported by Matson (1970). Another archeological site in the Central Valley has also been studied for pollen (Germeshausen, 1970), but I have not actually seen the report, and the site is not shown on Figure 1.

East of Sacramento, Ritter and Hatoff (1977) have examined samples from the early part of the last glacial cycle from the Teichert gravel pit (Table 1, No. 20). Even older samples are reported in or stratigraphically near the Corcoran Clay Member of the Tulare and Turlock Lake Formations; Hansen studied a few samples from a core (Table 1, No. 54) in Fresno County (Frink and Kues, 1954) and Linda Heusser looked at a few samples from cores at the Wasco nuclear reactor site (Table 1, No. 76) (Davis and others, 1977).

A 35-m hole drilled by Brian Atwater of the USGS in Tulare Lake (Table 1, No. 90) has yielded a series of pollen samples that range in age from modern to at least 70,000 years (Atwater and others, in press). The samples establish that greasewood (*Sarcobatus*) grew in the Central Valley during the last full-glacial interval. The site has the potential to provide a pollen record for the entire Quaternary, but recovering a continuous core through the 1-km-thick Quaternary section will be expensive.

Southern Coast Ranges

Most of the pollen work in the southern Coast Ranges has been done in the region between Salinas and the Carquinez Straits (two sites in Santa Barbara County are treated here as being in the Transverse Ranges). Roger Byrne and his students have examined a short core from one of the Sindicich Lagoons (Table 1, No. 31), a series of landslide ponds in Contra Costa County, and from Kelly Lake (Table 1, No. 51) and Warner Lake (Table 1, No. 52) near Watsonville. The Sindicich Lagoons core is only a few hundred years old, and the Watsonville cores probably only a few thousand (Roger Byrne, oral communication, 1982).

The longest pollen record from the southern Coast Ranges is that of Adam, Byrne, and Luther (1981) from Laguna de las Trancas (Table 1, No. 48), a small landslide pond in northern coastal Santa Cruz

County. The site is referred to as McCrary's Marsh by Heusser (1982). That record extends back about 30,000 years and reveals the presence of fir (*Abies*) pollen during the last full-glacial interval.

Other pollen samples of Pleistocene age were recovered from excavations at the Mountain View Dump (Table 1, No. 47) on the shore of San Francisco Bay (Helley and others, 1972). Abundant wood provided a series of radiocarbon dates in the range 21,000-24,000 yrs. B.P., and plant macrofossils established the presence of incense cedar (*Calocedrus*), cypress (*Cupressus*), Douglas fir (*Pseudotsuga*), pine (*Pinus*), and possibly common juniper (*Juniperus communis*) at the site. No detailed pollen work was done, but results for several samples were noteworthy for the almost total absence of both redwood (*Sequoia*) and oak (*Quercus*) pollen (Adam, unpublished data).

A rather detailed pollen record for the past 3,000 years is available for Pearsons Pond (Table 1, No. 46), which is situated atop a landslide in southern San Mateo County (Adam, 1975). The algae in that pond (*Pediastrum* and *Botryococcus*) responded sensitively to hydrologic variations during the period from about 3,000 yrs. B.P. to 1,000 yrs. B.P. Since about 1,000 yrs. B.P. the site has been less sensitive.

A further site of passing interest here is found at Saratoga (Table 1, No. 49) in sediments of the Santa Clara Formation of latest Pliocene to early-Pleistocene age (Adam and others, 1983). Pollen was found in three of five samples analyzed. Although fossil leaves of willow (*Salix*) and oak (*Quercus*) were found in the deposit (J. A. Wolfe, oral communication, 1977), angiosperm pollen was very scarce, which suggests that differential pollen destruction has occurred. Nearly all of the pollen recovered was conifer pollen, which was of interest because it included from 50 to 64 percent spruce (*Picea*) pollen. Spruce pollen is very scarce in upper-Pleistocene samples from central California, and its abundance in the Saratoga samples suggests that the present Mediterranean summer-drought climate had not yet developed when the samples were deposited.

Other pollen work includes the study by Linda Heusser of several cores near Hollister (Table 1, No. 95) (McMasters and others, in press). Although the cores were long, the length of the record is only about 8,000 years because sedimentation rates were high. The only site south of Salinas that has been studied is The Lakes in the Santa Lucia Range (Table 1, No. 53), where Sally Peterson Horn has recovered a pollen record several thousand years in length.

Transverse Ranges

Little pollen work has been done in the Transverse Ranges. A core from Zaca Lake (Table 1, No. 79) in Santa Barbara County, obtained by Roger Byrne and studied by Sally Peterson Horn (1980), spans only the last few thousand years. A study by G. L. Batchelder at Vandenberg Air Force Base (Table 1, No. 78) encountered difficulty with poor pollen preservation. G. J. West and K. Mossberg have analyzed short cores from two ponds along the San Andreas fault near Gorman and Little Rock (Table 1, Nos. 80-81), and Adam (unpublished data) has produced a composite diagram from excavations for the Van Norman Reservoir (Table 1, No. 82). The Van Norman record includes dated materials both older than 40,000 years and within the past 1,000 years, but the pollen data are ambiguous and difficult to interpret. None of the pollen work on the Transverse Ranges has been published.

Coastal Region

The earliest pollen studies along the California coast were done by Calvin Heusser (1960), who analyzed cores from Lake Earl (Table 1, No. 1), Capetown (Table 1, No. 5), and Fort Bragg (Table 1, No. 10). Mudie and Byrne (1980) report partial pollen diagrams from Bolinas Lagoon (Table 1, No. 30) and Drakes Estero (Table 1, No. 28) in Marin County and Mission Bay (Table 1, No. 87) and Los Penasquitos lagoon (Table 1, No. 86) in San Diego County. Several archeological sites within the city of San Diego (Santee Green, Pacific Bluffs, Shadow Ridge, and Rancho San Diego) have been studied by R. H. Hevly and R. E. Diggs (R. H. Hevly, written communication, 1984).

In addition to estuarine deposits, Mudie and Byrne (1980) also report in passing on a core from Wildcat Lake (Table 1, No. 29) on the Point Reyes Peninsula, a landslide pond not directly affected by salt water. A later report by Russell (1983) presents a more complete discussion of the Wildcat Lake record. Mudie (1975) has also studied cores from Tomales Bay (Table 1, No. 27) and Elkhorn Slough (Table 1, No. 50), but no diagrams were published. Roger Byrne studied pollen in two cores taken from Bolinas Lagoon (Table 1, No. 30) (Bergquist, 1977, 1978). He found spruce (*Picea*) pollen at a depth of 43-44 m in his longest core, beneath a radiocarbon date of 8,400±100 yrs. B.P.; he also analyzed a 185-cm core from the lagoon. A core from Batiquitos Lagoon (Table 1, No. 85) was studied for a master's thesis by Herbert Meyer at the University of California at Berkeley. Betty Nybakken of Hartnell College in Salinas has examined several short sediment cores (not shown on Figure 1) from San Francisco Bay in an investigation of the effects of 19th-century hydraulic mining on sediments of the bay.

Two archeological sites in the Los Angeles area have been examined for pollen. Site 4-ORA-64 (Table 1, No. 84) is on the coast in Orange County and was studied by D. E. Buge and James Schoenwetter. The Slusher Estate (Table 1, No. 83) was the subject of pollen studies by Schoenwetter and Amie Limon (James Schoenwetter, written communication, 1982). Also in the Los Angeles area, Warter (1976) has done some pollen work in connection with her studies of plant macrofossils from the La Brea Tar Pits (Table 1, No. 109).

The Channel Islands have received only minimal study by palynologists. The only published work is by Hevly and Hill (1970), who investigated archeological middens on Santa Cruz Island (Table 1, No. 88). The only other work known to me was a small study of archeological samples from Toyon Bay (Table 1, No. 89) on Santa Catalina Island, which was done by R. G. Matson.

DISCUSSION

Although Figure 1 shows over 100 pollen sites that have been studied in California, the work has been done by numerous, very independent workers, and there has been little effort made to coordinate research or communicate results, in part because of a lack of appropriate channels. This paper represents an attempt to identify the available information, but more work is clearly needed.

Several lines of approach appear particularly promising. Detailed documentation of the rich California pollen flora will provide new insights into the taxonomy and evolution of the California flora, in addition to placing paleoecological pollen work on a sounder footing. California also offers unusual opportunities to develop transects of pollen profiles across large elevational ranges and across major biogeographic boundaries. Because biogeographic gradients are quite steep in many areas of the state, a much denser network of sites than now exists is needed before we can fully understand the behavior of the California flora under environmental changes.

The areas most in need of initial studies are: (1) in the southeastern part of the state, where almost nothing has been done except along the coast (Figure 1), (2) in the southern Coast Ranges, and (3) in the southern Cascades, Modoc Plateau, and northernmost Sierra Nevada. The northeastern part of the state appears particularly promising because of the probability that volcanic ashes will be fairly frequent and because long depositional sequences should be present.

In the long run, perhaps the greatest contribution California palynology can make to science is the description of many long and at least fairly continuous sequences from a wide range of latitudes. Several sites (Tulare Lake, Clear Lake, and Tulelake) have already yielded records extending back at least 100,000 years, and other sites such as Lake Elsinore, Tulare Lake, Searles Lake, China Lake, Owens Lake, Long Valley, Mono Lake, Lake Tahoe, Honey Lake, and various sites in the Modoc Plateau offer similar opportunities.

More immediate needs are for publication or formal archiving of the large number of unpublished studies described above, and for illustration of the California pollen types, particularly in light of the high degree of endemism in the state (Raven and Axelrod, 1978). Even if some work is difficult to interpret at our present state of knowledge, pollen diagrams should be described and published for most of the studies shown on Figure 1. The detailed integration of the California pollen record so clearly missing from this paper will then become feasible, and future work can be planned in a much broader context than is now possible.

ACKNOWLEDGMENTS

This paper would not be possible without the generous help of many workers who have shared their unpublished data and called my attention to various studies. The contributions of G. L. Batchelder, J. R. Bergquist, Roger Byrne, Kenneth Cole, L. E. Heusser, R. H. Hevly, S. P. Horn, M. McGann, P. J. Mehringer, Jr., Betty Nybakken, M. Reed, E. W. Ritter, James Schoenwetter, A. M. Solomon, G. J. West, and W. Woolfenden are gratefully acknowledged. Manuscript reviews by G. J. West, R. H. Hevly, and A. Sarna-Wojcicki improved the manuscript.

References Cited

ADAM, D. P.

1964a   Palynology of Osgood Swamp, Eldorado County, California (abstract). *Arizona Academy of Science, General Program, Eighth Annual Meeting, Tempe, Arizona, April 4, 1964*, p. 14-15.

1964b   Exploratory palynology in the Sierra Nevada, California. *Geochronology Laboratories, University of Arizona, Interim Research Report*, No. 4, 30 p.

1965a   Late Pleistocene and recent palynology in the central Sierra Nevada, California (abstract). *Abstracts of the VII INQUA Congress, Boulder, Colorado, August 30-September 5, 1965*, p. 2.

1965b   Exploratory palynology in the Sierra Nevada, California. Unpublished M.S. thesis, University of Arizona: 51 p.

1967   Late Pleistocene and Recent palynology in the central Sierra Nevada, California. *In:* E. J. Cushing and H. E. Wright, Jr. (eds.), *Quaternary Paleoecology.* Yale University Press, New Haven, p. 275-301.

1973   Early Pleistocene (?) pollen spectra from near Lake Tahoe, California. *United States Geological Survey Journal of Research*, 1 (6): 691-693.

1974   Palynological applications of principal component and cluster analyses. *United States Geological Survey Journal of Research*, 2 (6): 727-741.

1975   A late Holocene pollen record from Pearson's Pond, Weeks Creek Landslide, San Francisco Peninsula, California. *United States Geological Survey Journal of Research*, 3: 721-731.

1979a   Raw pollen counts from core 4, Clear Lake, Lake County, California. *United States Geological Survey Open-File Report* 79-663, 181 p.

1979b   Raw pollen counts from core 7, Clear Lake, Lake County, California. *United States Geological Survey Open-File Report* 79-1085, 92 p.

1985a   Pollen zonation and proposed informal climatic units for Clear Lake, California, cores CL-73-4 and CL-73-7. *Geological Society of America Special Paper,* Clear Lake Symposium Volume. In press.

1985b   Correlations of the Clear Lake core CL-73-4 pollen sequence with other long climatic records. *Geological Society of America Special Paper,* Clear Lake Symposium Volume. In press.

1985c   Palynology of two Upper Quaternary cores from Clear Lake, Lake County, California. *United States Geological Survey Professional Paper.* In press.

1985d   Quaternary pollen records from California. *Proceedings of the Conference on Holocene Climate and Archaeology of the California Coast and Desert, San Diego State University, February, 1982,* 19 p., 1 map, 2 tables. In press.

ADAM, D. P., ADAMS, D. B., FORESTER, R. M., McLAUGH-LIN, R. J., REPENNING, C. A., and SORG, D. H.
1983   An animal- and plant-fossil assemblage from the Santa Clara Formation (Pliocene and Pleistocene), Saratoga, California. *In:* Andersen, D. W., and Rymer, M. J. (eds.), *Tectonics and Sedimentation Along Faults of the San Andreas System.* Society of Economic Paleontologists and Mineralogists (Pacific Section), Los Angeles, p. 105-110.

ADAM, D. P., BYRNE, ROGER, and LUTHER, EDGAR
1981   A Late Pleistocene and Holocene pollen record from Laguna de Las Trancas, northern coastal Santa Cruz County, California. *Madroño,* 28 (4): 255-272.

ADAM, D. P., SIMS, J. D., and THROCKMORTON, C. K.
1981   A 130,000-year continuous pollen record from Clear Lake, Lake County, California. *Geology,* 9: 373-377.

ADAM, D. P., and WEST, G. J.
1982   Temperature and precipitation calibration of pollen records from Clear Lake, California, through the last full glacial cycle (abstract). *American Quaternary Association, Seventh Biennial Conference, Program and Abstracts,* p. 56.
1983   Temperature and precipitation estimates through the last glacial cycle from Clear Lake, California, pollen data. *Science,* 219: 168-170.

ATWATER, B. F., ADAM, D. P., BRADBURY, J. P., FORESTER, R. M., GOBALET, K. M., LETTIS, W. R., MARK, R. K., FISHER, G. R., and ROBINSON, S. W.
1985   A Tulare Lake, California, fan dam, and implications for the Wisconsin glacial history of the Sierra Nevada. *Geological Society of America Bulletin.* In press.

AXELROD, D. I., and TING, W. S.
1961   Early Pleistocene floras from the Chagoopa Surface, southern Sierra Nevada. *University of California Publications in Geological Sciences,* 39 (2); 119-194.

BATCH, J. R.
1977   A post-glacial pollen record from a subalpine valley in Yosemite National Park. Unpublished Masters thesis, San Francisco State University.

BATCHELDER, G. L.
1970a   Post-glacial ecology at Black Lake, Mono County, California. Unpublished Ph.D. dissertation, Arizona State University.
1970b   Post-glacial fluctuations of lake level in Adobe Valley, Mono County, California. *American Quaternary Association, Abstracts, First Meeting, August 28-September 1, 1970,* p. 7.
1980   A Late Wisconsinan and Early Holocene lacustrine stratigraphy and pollen record from the west slope of the Sierra Nevada, California. *American Quaternary Association Abstracts and Program, 6th Biennial Meeting,* p. 13.

BERGQUIST, J. R.
1977   Depositional history and fault-related studies, Bolinas Lagoon, California. Unpublished Ph.D. dissertation, Stanford University, 227 p.
1978   Depositional history and fault-related studies, Bolinas Lagoon, California. *United States Geological Survey Open-File Report* 78-802, 164 p.

COLE, KENNETH
1983   Late Pleistocene vegetation of Kings Canyon, Sierra Nevada, California. *Quaternary Research,* 19: 117-129.

CROFT, M. G.
1972   Subsurface geology of the Late Tertiary and Quaternary water-bearing deposits of the southern part of the San Joaquin Valley, California. *United States Geological Survey Water-Supply Paper,* 1999-H, 29 p. + 6 plates.

DAVIS, G. H., GREEN, J. H., OLMSTED, F. H., and BROWN, D. W.
1959   Ground-Water conditions and storage capacity in the San Joaquin Valley, California. *United States Geological Survey Water-Supply Paper,* 1469, 287 p. + 29 plates.

DAVIS, P., SMITH, J., KUKLA, G. J., and OPDYKE, N. D.
1977   Paleomagnetic study at a nuclear power plant site near Bakersfield, California. *Quaternary Research,* 7: 380-397.

DIETRICH, W. E., and DORN, RONALD
1984   Significance of thick deposits of colluvium on hillslopes: a case study involving the use of pollen analysis in the coastal mountains of northern California. *Journal of Geology,* 92: 147-158.

DORLAND, DEBORAH
1980   Two post-glacial pollen records from Meyers Grade Marsh and Grass Lake, El Dorado County, California. Unpublished M.A. thesis, San Francisco State University, 103 p.

DORLAND, DEBORAH, ADAM, D. P., and BATCHELDER, G. L.
1980   Two Holocene pollen records from Meyers Grade Marsh and Grass Lake, El Dorado County, California. *American Quaternary Association Abstracts and Program, 6th Biennial Meeting,* p. 64.

FRINK, J. W., and KUES, H. A.
1954   Corcoran Clay—A Pleistocene lacustrine deposit in San Joaquin Valley, California. *Bulletin of the American Association of Petroleum Geologists,* 38: 2357-2371.

GERMESHAUSEN, E.
1969   Pollen analysis of Sac-267. Manuscript on file with the Department of Anthropology, Sacramento State University.

HANSEN, H. P.
1942   A pollen study of peat profiles from Lower Klamath Lake of Oregon and California. *In:* Cressman, L. S., Archaeological Researches in the Northern Great Basin. *Carnegie Institution of Washington Publication* 538: 103-114, figs. 57-62.

HELLEY, E. J., ADAM, D. P., and BURKE, D. B.
1972   Late Quaternary stratigraphic and paleoecological investigations in the San Francisco Bay Area. *In:* Frizzell, Virgil (ed.), Unofficial Progress Report on the USGS Quaternary Studies in the San Francisco Bay Area, An Informal Collection of Preliminary Papers. *Guidebook for Friends of the Pleistocene, October 6, 7, 8, 1972,* p. 19-30.

HEUSSER, C. J.
1960   Late-Pleistocene environments of North Pacific North America. *American Geographical Society Special Publication* 35, 308 p.

HEUSSER, L. E.

1978   Marine pollen in Santa Barbara Basin, California: a 12,000-year record. *Geological Society of America Bulletin,* 89: 673-678.

1981   Pollen analysis of selected samples from Deep Sea Drilling Project Leg 63. *Initial Reports of the Deep Sea Drilling Project,* 63: 559-563.

1982   Quaternary paleoecology of northwest California and southwest Oregon. *American Quaternary Association Program and Abstracts, Seventh Biennial Conference,* June 28-30, 1982, p. 104.

1983   Contemporary pollen distribution in coastal California and Oregon. *Palynology,* 7: 19-42.

HEUSSER, L. E., and SHACKLETON, N. J.

1979   Direct marine-continental correlation: a 150,000-year oxygen isotope-pollen record from the North Pacific. *Science,* 204: 837-839.

HEUSSER, L. E., and SIMS, J. D.

1981   Palynology of core CL-80-1, Clear Lake, California. *Geological Society of America Abstracts with Programs,* 13 (2): 61.

1983   Pollen counts for core CL-80-1, Clear Lake, Lake County, California. *United States Geological Survey Open-File Report* 83-384, 28 p.

HEVLY, R. H., and HILL, J. N.

1970   Pollen from archaeological middens from Santa Cruz Island, California. *University of California Archaeological Survey Annual Report,* 12: 108-118.

HORN, SALLY PETERSON — see below under PETERSON.

KILBOURNE, R. T.

1978   Pollen studies in the Auburn Dam area. *California Division of Mines and Geology Open-File Report* 78-18 SF.

LEOPOLD, E. B.

1967   Summary of palynological data from Searles Lake. *In:* Pleistocene ecology and palynology, Searles Valley, California. *Guidebook for Friends of the Pleistocene, Pacific Coast Section, September 23-24, 1967,* p. 52-66.

LIDDICOAT, J. C., OPDYKE, N. D., and SMITH, G. I.

1980   Paleomagnetic polarity in a 930-m core from Searles Valley, California. *Nature,* 256: 22-25.

MARTIN, P. S., and MEHRINGER, P. J., Jr.

1965   Pleistocene pollen analysis and biogeography of the Southwest. *In:* Wright, H. E., Jr., and Frey, D. G. (eds.), *The Quaternary of the United States,* Princeton University Press, Princeton, p. 433-451.

MATSON, R. G.

1970   The pollen evidence for a recent arboreal transgression into grass lands in central California. *In:* Ritter, E. W., Shulz, P. D., and Kautz, Robert (eds.), Papers on California and Great Basin Prehistory. *Center for Archaeological Research at Davis Publication,* 2: 147-154.

1972   Pollen from the Spring Garden Ravine site (4-PLA-101). *In:* Ritter, E. W., and Shulz, P. D. (eds.), Papers on Nisenan Environment and Subsistence. *Center for Archaeological Research at Davis Publication,* 3: 24-27.

MATTHES, F. E.

1941   Rebirth of the glaciers of the Sierra Nevada during late Post-Pleistocene time (abstract). *Geological Society of America Bulletin,* 52 (12/2): 2030.

McGANN, MARY

1985   Quaternary Monterey submarine fan levee deposits: Foraminifers and pollen. M.A. thesis, Paleontology Department, University of California (Berkeley). In preparation.

McMASTERS, C. R., HERD, D. G., and THROCKMORTON, C. K.

1985   Subsurface stratigraphy of the eastern Hollister Valley, California, with pollen analysis by Linda E. Heusser. *United States Geological Survey, Miscellaneous Field Studies Map* MF-1461. In press.

MEHRINGER, P. J., Jr.

1977   Great Basin Late Quaternary environments and chronology. *Desert Research Institute Publications in the Social Sciences* 12: 113-167.

MEHRINGER, P. J., Jr., and SHEPPARD, J. C.

1978   Holocene history of Little Lake, Mojave Desert, California. *In:* Davis, E. L. (ed.), The Ancient Californians: Rancholabrean hunters of the Mojave Lakes country. *Los Angeles County Natural History Museum Science Series,* 29: 153-166.

MILLER, J. S., BATCHELDER, G. L., and LEE, G. K.

1976   A postglacial pollen record from Campbell Lake, Siskiyou County, California. *American Quaternary Association, Abstracts of the Fourth Biennial Meeting, October 9 and 10, 1976,* p. 151-152.

MUDIE, P. J.

1975   Palynology of Recent coastal lagoon sediments in southern and central California (abstract). *Annual Meeting of the Botanical Society of America, Paleobotany Section, Corvallis, Oregon, August 17-22, 1975.*

MUDIE, P. J., and BYRNE, ROGER

1980   Pollen evidence for historic sedimentation rates in California coastal marshes. *Estuarine and coastal Marine Science,* 10: 305-316.

MUNZ, P. A., and KECK, D. D.

1973   *A California Flora and Supplement.* University of California Press, Berkeley, 1681 + 224 p.

OSVALD, H.

1936   Stratigraphy and pollen flora of some bogs of the North Pacific coast of America. *Berichtungen des Schweizerisches Botanisches Gesellschaft,* 46.

PETERSON, SALLY

1980   Late Holocene sedimentation at Zaca Lake, Santa Barbara County, California. Undergraduate thesis on file at Geography Department, University of California (Berkeley), 61 p.

RAVEN, P. H., and AXELROD, D. I.

1978   Origin and relationships of the California Flora. *University of California Publications in Botany,* 72, 134 p.

RITTER, E. W., and HATOFF, B. W.

1977   Late Pleistocene pollen and sediments: an analysis of a central California locality. *Texas Journal of Science,* 29: 195-207.

ROBINSON, S. W., ADAM, D. P., and SIMS, J. D.

1985   Radiocarbon content, sedimentation rates, and a timescale for Clear Lake, Core CL-73-4. *Geological Society of America Special Paper,* Clear Lake Symposium Volume. In press.

ROOSMA, AINO

1958   A climatic record from Searles Lake, California. *Science,* 128: 716.

RUSSELL, E. W. B.

1983   Pollen analysis of past vegetation at Point Reyes National Seashore, California. *Madroño,* 30: 1-11.

RYMER, M. J.

1981   Stratigraphic revision of the Cache Formation (Pliocene and Pleistocene), Lake County, California. *United States Geological Survey Bulletin* 1502-C, 35 p.

ŠERCELJ, ALOJZ, and ADAM, D. P.

1975   A late Holocene pollen diagram from near Lake Tahoe, Eldorado County, California. *United States Geological Survey Journal of Research,* 3 (6): 737-745.

SIMS, J. D.

1976   Paleolimnology of Clear Lake, California, U.S.A. *In:* Horie, Shoji (ed.), *Paleolimnology of Lake Biwa and the Japanese Pleistocene,* 4: 658-702.

SIMS, J. D., ADAM, D. P., and RYMER, M. J.

1981   Late Pleistocene stratigraphy and palynology of Clear Lake, Lake County, California. *United States Geological Survey Professional Paper* 1141: 219-230.

SIMS, J. D., RYMER, M. J., and PERKINS, J. A.

1981   Description and preliminary interpretation of core CL-80-1, Clear Lake, Lake County, California. *United States Geological Survey Open-File Report* 81-751, 175 p.

SIMS, J. D., RYMER, M. J., PERKINS, J. A., and FLORA, L. A.

1981   Description and preliminary interpretation of core CL-80-2, Clear Lake, Lake County, California. *United States Geological Survey Open-File Report* 81-1323, 112 p.

SPAULDING, W. G.

1984   Archeobotanical and paleoecological investigations at archeological sites in the New Melones Reservoir area, Calaveras and Tuolumne Counties, California, *In:* Moratto, M. J., *et al., Final Report of the New Melones Archeological Project,* 4 (to National Park Service).

SPAULDING, W. G., LEOPOLD, E. B., and Van DEVENDER, T. R.

1983   Late Wisconsin paleoecology of the American Southwest. *In:* Wright, H. E., Jr. (editor), *Late Quaternary Environments of the United States, 1, The Late Pleistocene* (S. C. Porter, editor), University of Minnesota Press, Minneapolis, p. 259-293.

SULLIVAN, D. G.

1982   Prehistoric flooding in the Sacramento Valley: Stratigraphic evidence from Little Packer Lake, Glenn County, California. Unpublished M. A. thesis, Geography Department, University of California (Berkeley), 90 p. Reports pollen work in general terms, but gives no data or pollen diagram.

von POST, LENNART

1916   Om Skogstradpollen i sydsvenska torfmosslagerfoljder. *Geologiska Foreningens i Stockholm Forhandlingar,* 38: 384-394.

WAHRHAFTIG, CLYDE, and BIRMAN, J. H.

1965   The Quaternary of the Pacific mountain system in California, *In:* Wright, H. E., Jr., and Frey, D. G. (eds.), *The Quaternary of the United States.* Princeton University Press, Princeton, p. 299-340.

WARTER, J. K.

1976   Late Pleistocene plant communities—evidence from the Rancho La Brea Tar Pits, *In:* Ratting, June (ed.), Plant Communities of Southern California. *California Native Plant Society Special Publication,* 2: 32-39.

WEST, G. J.

1981   Walnut pollen in late-Holocene sediments of the Sacramento-San Joaquin Delta, California. *Madroño,* 28: 44-45.

1985   Pollen analysis of sediments from Tule Lake: a record of Holocene vegetation/climatic changes in the Mendocino National Forest, California. *Proceedings of the Conference on Holocene Climate and Archaeology of the California Coast and Desert, San Diego State University, February 1982.* In press.

ZAUDERER, J. N.

1973   A neoglacial pollen record from Osgood Swamp, California. Unpublished M.S. thesis, University of Arizona, 48 p.

QUATERNARY POLLEN RECORDS FROM THE PACIFIC NORTHWEST COAST: ALEUTIANS TO THE OREGON-CALIFORNIA BOUNDARY

CALVIN J. HEUSSER
Department of Biology
New York University
New York, NY 10003

Abstract

Quaternary sediments of lakes and mires in the glaciated portion of the North Pacific coast were deposited, for the most part, following Late Wisconsin wastage of the Cordilleran Glacier Complex. Some sequences in western Washington are older, covering time spans since the Early Wisconsin, and a few are of earlier interglaciations. Selected for geographic location and length of continuous sedimentation are 20 pollen records from sites distributed between Alaskan tundra and Pacific coastal forest of Washington and Oregon.

Pollen records dating back to the Early Wisconsin in Washington contrast assemblages of tundra, parkland, and closed coastal forest. Assemblages show response of vegetation to stadial and interstadial climatic and edaphic conditions. Tree line in the Pleistocene apparently fluctuated across southwestern Washington and during the Holocene advanced to higher latitudes in Alaska and to higher altitudes in the cordillera. Advance of forest northwestward along the coast was from unglaciated refugia located in Washington and possibly in the Queen Charlotte Islands of British Columbia and other places along the Pacific slope. Lodgepole pine and alder invaded deglaciated ground in British Columbia and southeast Alaska during the late-glacial and were followed in the Holocene by Sitka spruce, western hemlock, and mountain hemlock, which within approximately the past 2,000-3,000 years reached south-central Alaska. Advance of alder and spruce into tundra on Kodiak Island and southwestward is an apparent recent event.

Transfer functions calibrate temperature and precipitation for Quaternary pollen records from western Washington and British Columbia. Trends indicate relatively warm, wet intervals at >47,000 and around 30,000 yrs. B.P. and coldest and driest conditions between 28,000 and 13,000 yrs. B.P. Holocene warmth and dryness became pronounced about 8,000 years ago, after which climate was colder and more humid. During earlier interglaciations, temperatures comparable to the Holocene are evident in records of the Alderton and Whidbey Formations, whereas the record of temperature for the Puyallup Formation is about 2°C lower than for the Holocene.

INTRODUCTION

Stratigraphic records of pollen in Quaternary deposits of Pacific coastal North America were first investigated some 50 years ago in the peat bogs of Alaska, British Columbia, and Washington (Bowman, 1934; Osvald, 1936; Hansen, 1938). This pioneering work set forth the stage for an interpretation of Holocene vegetational and climatic changes since the wastage of ice-age glaciers. Previous studies connected with mapping of surficial deposits took into account peat deposits containing wood and other macrofossil remains which implied climatic events of Pleistocene glacial and interglacial ages (Willis, 1898; Lupton, 1914).

Peats and related biogenic deposits occur throughout the coastal region, where extensive glaciation has given rise to numerous basins of sedimentation and cool and wet oceanic climate contributes to the formation of ombrotrophic and soligenous mires (Rigg, 1925, 1958; Rigg and Richardson, 1938; Dachnowski-Stokes, 1930, 1936, 1941; Osvald, 1933, 1970). Pollen and spores found in north Pacific coastal peats and nonbiogenic deposits are described and illustrated in several publications (Mack, 1971; Moe, 1974; Bagnell, 1975; Mathewes, 1978, 1979a). The regional vegetational history of the coast during the Holocene and late-Pleistocene is included in recent reviews of the western conterminous United States and Alaska (Ager, 1982, 1983; Baker, 1983; Heusser, 1983a).

Quaternary palynology of Pacific coastal North America discussed in this account covers the maritime border between Oregon and Attu Island, the westernmost member of the Alaskan Aleutian Islands (Fig. 1). The stretch of coast, extending over a distance of some 6,000 km between 42° and 62° N, is in large part rugged and rocky (Fig. 2) with deep, glacially-scoured fiords and numerous islands. Coniferous evergreen forest of spruce, hemlock, and cedar clothes the Pacific slope between California and south-central Alaska. Beyond the forest, tundra forms the vegetation cover on the mainland of southwestern Alaska and on the islands of the Aleutian chain.

The Cordilleran Glacier Complex (Flint, 1971) spread over the region during the Late Wisconsin (25,000-10,000 yrs. B.P.). Except for a few places that appear to have been free of ice, glaciers mantled the coast as far south as northwestern Washington (Clague *et al.,* 1980, 1982a; Hamilton and Thorson, 1983; Waitt and Thorson, 1983). Today, glaciers remain in the higher parts of the cordillera; coalesced as ice fields in the north (Fig. 3), they feed numerous valley systems that first reach tidewater at 56°49′N in southeast Alaska. Northwestward, the huge piedmont lobes of the Malaspina and Bering Glaciers spread out along the Gulf of Alaska. Bering Glacier and its tributaries, the largest glacier-occupied area in northwestern North America, cover 5,700 km^2 (Meier and Post, 1962).

Most pollen records begin after the close of Late Wisconsin glaciation. Some extend beyond the limit of radiocarbon dating, while only a few date from interglacial ages. For this review, twenty records are selected of time spans of continuous deposition over the geographic extent of the region, and surface pollen data from a total of 180 stations apply in the calibration of temperature and precipitation from fossil records in western Washington and southwestern British Columbia.

THE NORTH PACIFIC COAST: VEGETATIONAL AND CLIMATIC SETTING

Vegetation

Pacific coastal forest (Fig. 1) extends for over 4,000 km between northern California and Kodiak Island in Alaska (Sudworth, 1908; Cooper, 1957; Hansen, 1947; Heusser, 1960; Küchler, 1964; Fowells, 1965; Krajina, 1965, 1969; Sigafoos, 1958; Daubenmire, 1969; Viereck and Little, 1972; Franklin and Dyrness, 1973; Barbour and Major, 1977). The forest, perhaps the grandest in temperate latitudes of the world (Waring and Franklin, 1979), coincides with the region of cool summers and moderately cold winters associated with high precipitation and cloud cover and low evaporation. At its northern limit, the forest grades into tundra or interior spruce-birch forest; attenuation in the south is largely at the edge of mixed evergreen forest.

The central locus of distribution of constituent trees is in western Washington and southwestern British Columbia (Fig. 4). Species reach varying distances north and south of this distribution center, while ranges of species from contiguous vegetation are in part coincident over the extent of coastal forest. The western cordillera, at altitudes well over timberline, limits distribution of coastal tree species inland, except along transverse valleys, for example, those of the Stikine, Taku, Alsek, and Copper Rivers in Alaska, which serve as avenues for migration into the interior. Names of species used in this chapter refer to Hultén (1968), Munz (1968), Hitchcock and Cronquist (1973), and Taylor and MacBryde (1977).

Three wide-ranging tree species, western hemlock *(Tsuga heterophylla),* western red cedar *(Thuja plicata),* and Sitka spruce *(Picea sitchensis),* form the unifying component of the Pacific coastal forest (Fig. 1). Below 600 m along the immediate coast and below 1,000 m on the western slope of inland cordillera, they make up the bulk of lowland forest. Douglas fir *(Pseudotsuga menziesii)* at lower latitudes is of importance where humidity and cloud cover decrease away from the coast. It is usually a seral species except under relatively dry conditions and in the absence of competition. In southwestern Oregon and California, redwood *(Sequoia sempervirens)* and Port Orford cedar *(Chamaecyparis lawsoniana)* are often prominent.

Conifers of less importance in the forest include grand fir *(Abies grandis)* and western yew *(Taxus brevifolia).* Lodgepole pine *(Pinus contorta)* and western white pine *(P. monticola)* achieve abundance locally under successional or edaphic conditions in parts of the forest; ponderosa pine *(P. ponderosa)* and sugar pine *(P. lambertiana),* as well as incense cedar *(Calocedrus decurrens),* occur occasionally toward the south, inland in Oregon and California; and closed-cone pine communities, distinguished by Bishop pine *(P. muricata),* Bolander pine *(P. contorta* var. *bolanderi),* and cypress *(Cupressus pygmaea* and *C. macrocarpa),* are of minor extent in California.

A broad-leaved element, both deciduous and evergreen, pervades the southern part of lowland forest. Most noteworthy among the deciduous trees is red alder *(Alnus rubra),* a successional species favored by human disturbance, which ranges northward to southeastern Alaska. It is common on floodplains and along streams, growing with big-leaf maple *(Acer macrophyllum),* vine maple *(A. circinatum),* black cottonwood *(Populus trichocarpa),* willow *(Salix scouleriana* and *S. lasiandra),* and, at times, Oregon ash *(Fraxinus latifolia).* Successional stands in valley bottoms and on slopes include cascara *(Rhamnus purshiana)* and dogwood *(Cornus nuttallii).*

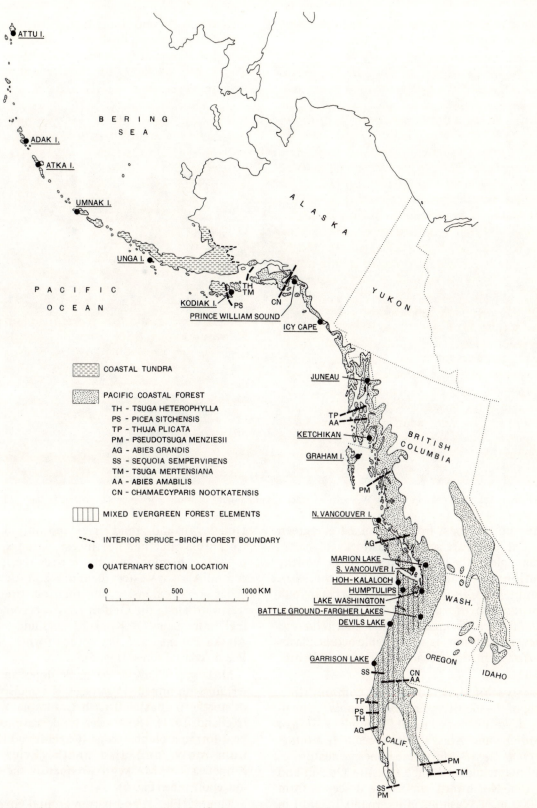

Figure 1. Location of Quaternary sections in North Pacific North America in relation to coastal vegetation and latitudinal range limits of major arboreal species.

Figure 2. Olympic Peninsula coast in western Washington. Seacliff is of flat-lying Pleistocene sediments overlying Tertiary bedrock.

Ranging up the coast from the mixed evergreen forest of California are several broad-leaved species that commingle in the coastal forest (Fig. 1). Worthy of attention are madrone *(Arbutus menziesii),* chinquapin *(Castanopsis chrysophylla),* tan oak *Lithocarpus densiflorus),* and California laurel *(Umbellularia californica).* Oregon Oak *(Quercus garryana),* in addition, is well established at places inland; oak and madrone reach as far north as British Columbia.

Distinctive high-altitude forest communities, consisting of Pacific silver fir *(Abies amabilis),* subalpine fir *(A. lasiocarpa),* mountain hemlock *(Tsuga mertensiana),* and Alaska yellow cedar *(Chamaecyparis nootkatensis),* form the cohesive feature of upper montane and subalpine forest in Oregon and Washington. Northward, timberline descends from about 2,000 m in Washington to around 1,000 m in southeasten Alaska; species reach sea level and intersperse with lowland forest species. Western red cedar, Pacific silver fir, and western yew reach their northern limits in southeastern Alaska, while the ranges of

Douglas fir and grand fir extend only into British Columbia. Farther northwest, Sitka spruce and mountain hemlock are major trees reaching south-central Alaska; western hemlock and Alaska yellow cedar are as extensive, but their occurrence north-westward along the Gulf of Alaska becomes increasingly infrequent. On Kodiak Island and the adjacent Alaska Peninsula, Sitka spruce forms the edge of Pacific coastal forest (Fig. 5).

Muskeg, a feature of low-angle slopes and flat-lying terrain with impeded drainage, is expansive in parts of northern coastal British Columbia and Alaska (Neiland, 1971). Its surfaces support communities of peat-forming plants, sedge (Cyperaceae) and sphagnum moss, and also heath (Ericaceae and Empetraceae). Muskeg is frequently the habitat of lodgepole pine (Fig. 6).

Tundra (Fig. 1) and barren ground cover most of southwestern Alaska (Hultén, 1960; Sigafoos, 1958; Viereck and Little, 1972). Communities consist of sedge, grass, heath, and a diverse selection of species. Tundra between the easternmost Aleutians and the

Figure 3. Glacier-covered Coast Mountains at the British Columbia boundary in southeast Alaska.

Figure 4. Pacific coast forest of western hemlock and western red cedar in western Washington.

Figure 5. Sitka spruce invading shrub tundra at the edge of Pacific coast forest, Kodiak Island, Alaska.

edge of the forest is a mosaic of herbaceous communities and dense thickets of willow *(Salix alaxensis* subsp. *alaxensis, S. barclayi, S. pulchra)* and Sitka alder *(Alnus crispa* subsp. *sinuata).* Sitka alder, common southward over much of coastal Alaska, ranges to the mountains of California. Where tundra communities come in contact with interior spruce-birch forest, white spruce *(Picea glauca),* black spruce *(P. mariana),* Kenai birch *(Betula kenaica),* balsam poplar *(Populus balsamifera),* and aspen *(P. tremuloides)* occur dispersed in an ecotone.

Climate

Cool rainy climate of the maritime northwest derives from geographical position and mountainous character of the coast which lies athwart tracks of storms that cross inland from off the Pacific Ocean (Bryson and Hare, 1974). Precipitation is pronounced in winter because of greater frequency of storms generated in the low-pressure Aleutian trough. Cloudiness and wind are a general feature of the coast; ocean fog is frequent in winter in the Gulf of Alaska and Aleutian sector and during summer between British Columbia and California.

The concentration of storm tracks across southeastern Alaska, British Columbia, and Washington is the reason for the greater average annual precipitation in this stretch of coast (Hare and Hay, 1974; Court, 1974). Mean values are close to 3,000 mm and vary locally, owing to topographic influence, from >5,000 to <2,000 mm. Precipitation progressively decreases away from this sector to <1,000 mm. Stations inland often receive half or less of amounts received at coastal stations.

Unlike precipitation, which registers peak values in southeastern Alaska and British Columbia, temperature increases more or less steadily at stations between the westernmost Aleutians and California (Heusser *et al.,* 1980). Averages for July are 8-10°C in tundra and 11-16°C in coastal forest. Summer temperatures are low because of moderating influence of cool ocean water; inland, conditions are warmer. Average July temperatures, for example, between west coast Vancouver Island and mainland British Columbia can differ by 4°C. In winter, the situation reverses; temperatures inland are lower than at the ocean.

Figure 6. Muskeg on Wrangell Island, Alexander Archipelago, Alaska.

Pacific coastal forest is limited in the south by high summer temperature and an unfavorable precipitation/evaporation ratio. Summer fog at this latitude is the key feature enabling redwood to grow on the California coast. In Alaska, high precipitation and low evaporation favor formation of muskeg. At the northern edge of the forest, excessive soil moisture, coupled with low summer temperature, limited sunshine, and much strong wind, restricts growth and migration of coastal tree species.

SURFACE POLLEN DATA

Modern pollen rain data of tundra and Pacific coastal forest are from 180 surface stations distributed between Attu Island in the Aleutians and the vicinity of Santa Cruz in California (Fig. 7; Table 1). Data are from several sources but mostly from Heusser (1960). Observations of pollen rain on the north Pacific coast are contained in studies by Hansen (1949), Florer (1972), Hebda (1977, 1983),

Alley (1979), Barnosky (1981), Peteet (1983), Heusser (1954, 1969, 1973a, 1978a, 1978b, 1978c, 1983b, 1983c), L. Heusser (1983a), and Warner (1984).

Pollen in Alaskan tundra (stations 1-44) is from herbs and shrubs belonging to the Gramineae, Cyperaceae, Umbelliferae, *Empetrum*-Ericaceae, and Compositae (Fig. 8). Also from shrubs, usually no more than 2-3 m in height, are most additional *Alnus, Betula,* and *Salix,* but some pollen comes from tree species in the forest-tundra ecotone, where *Alnus* increases in importance (stations 34-44). Small quantities of *Picea* are from spruce trees scattered in the ecotone. *Alnus* pollen found in the western Aleutians (stations 1-17) is transported there by wind.

At the edge of coastal forest on Kodiak Island, *Picea* is the exclusive coniferous pollen. The sharp drop in shrub and herb percentages (stations 44-45) closely coincides with extent of spruce stands. *Tsuga mertensiana* first increases at station 57 on the Kenai Peninsula and *T. heterophylla* at station 62 in Prince William Sound. These stations also approximate range limits of both species of hemlock. Amounts of

Table 1.
Locations of Modern Pollen Stations and Relevant Meterological Data[1]

Location	Composite Sites	Source Sites (Fig. 1)	Av. July Temperature (°C)	Av. Annual Precipitation (mm)	Location	Composite Sites	Source Sites (Fig. 1)	Av. July Temperature (°C)	Av. Annual Precipitation (mm)
ALASKA					**BRITISH COLUMBIA**				
Attu Is.	-	1	8.7	1020	Langara Is.	32	117-120	12.1	1780
Adak Is.	1	2-16	9.2	1400	Graham Is.	33	121-123	12.8	1500
Umnak Is.	2	17	9.7	1140	Graham Is.	34	124-125	12.8	1280
Chirikof Is.	3	18-20	11.2	1080	Prince Rupert	35	126-130	12.4	2500
Trinity Is.	4	21-25	11.2	1080	Pitt Is.	36	131	12.5	2000
Kodiak Is.	5	26-33	11.4	830	Susan Is.	37	132	12.6	2500
Alaska Pen.	6	34-35	11.5	1500	Namu	38	133	12.8	2800
Alaska Pen.	7	36	11.9	710	Cape Caution	39	134	12.8	2200
Alaska Pen.	8	37-38	12.1	580	Port Hardy-Hope Is.	40	135-137	13.0	2300
Kodiak Is.	9	39-50	11.9	1780	Harbledown Is.	41	138	14.0	1600
Kenai Pen.	10	52-56	11.8	700	Vancouver Is.	42	139	17.1	1190
Kenai Pen.	11	57-60	12.9	1050	Vancouver Is.	43	140	17.0	1100
Turnagain Arm	12	61	12.5	900	Vancouver Is.	44	141	16.5	900
Prince William Sd.	13	62-64	12.2	2200					
Cordova-Alaganik	14	65-69	12.5	2300	**WASHINGTON**				
Katalla	15	70-73	12.3	2350	Puget Sd.	45	142-145	16.3	900
Icy Cape	16	74	12.1	2800	Puget Sd.	46	146-154	17.9	1150
Lituya Bay	17	75-77	12.2	2800	Olympic Pen.	47	155-160	15.6	3140
Icy Point	18	78-80	12.2	2800	Olympic Pen.	48	161-166	15.2	3300
Haines	19	81	13.1	950	Seaview	49	167	15.0	2110
Excursion Inlet	20	82	12.9	1800					
Juneau	21	83-94	12.7	2000	**OREGON**				
Chichagof-Yakobi Is.	22	95-97	13.0	2200	Warrenton	50	168	15.0	2000
Baranof-Kruzof Is.	23	98-101	13.2	1850	Devils Lake	51	169	14.0	2200
Admiralty Is.	24	102-103	13.2	3200	Newport	52	170-171	13.8	2000
Kuiu-Kupreanof Is.	25	104-105	12.7	3500	Waldport	53	172	14.0	2100
Petersburg	26	106-107	13.2	3210	Yachats	54	173	14.0	2100
Wrangell Is.	27	108-109	12.6	1930	Tahkenitch Lake	55	174	16.0	2200
Prince of Wales Is.	28	110-111	12.9	2500	Garrison Lake	56	175	15.1	1930
Prince of Wales Is.	29	112	13.5	2600					
Ketchikan	30	113-115	13.7	4110	**CALIFORNIA**				
Prince of Wales Is.	31	116	13.0	2500	Lake Earl	57	176	14.5	1720
					Capetown	58	177	15.0	1250
					Fort Bragg	59	178	14.5	990
					San Francisco	-	179	14.8	470
					Santa Cruz	-	180	17.1	730

[1]Temperature and precipitation are estimated from data (1968-1977) at 43 weather stations (Heusser *et al.*, 1980).

Alnus are high in this sector of southern Alaska and southeastward along the coast until the vicinity of Alexander Archipelago (station 84). Pollen of *Picea* continues to be important in southeastern Alaska, while *Tsuga heterophylla* increases in quantity and *T. mertensiana* decreases.

Rise of *Pinus* (station 75) and its preeminence to the south at stations over the remainder of Alaska and in British Columbia result from local occurrence of lodgepole pine on muskeg. The drop in percentage of *Picea* in the south of British Columbia is probably because spruce trees are fewer in forests on the drier eastern side of Vancouver Island where surface sample stations are located. Pollen of *Tsuga mertensiana* also falls off in southern British Columbia, as mountain hemlock becomes distributed at montane and subalpine altitudes away from coastal stations.

Extensive heath barrens in the Queen Charlotte Islands account for increased percentages of *Empetrum*-Ericaceae pollen (stations 119-122).

Pollen of *Alnus* dominates most of remaining data. It is derived in large measure from red alder that has spread during the past century following logging and fire on the Pacific slope of the Cascade Range in Washington and Oregon. Quantities of *Pseudostuga menziesii* in Washington (stations 142-154) are an indication of Douglas fir importance in Puget Lowland communities, which are less mesic than communities on the west side of the Olympic Mountains. In the more humid, western Olympics (stations 155-166), higher amounts of *Picea* and *Tsuga heterophylla* pollen, by comparison, parallel increased proportions of Sitka spruce and western hemlock.

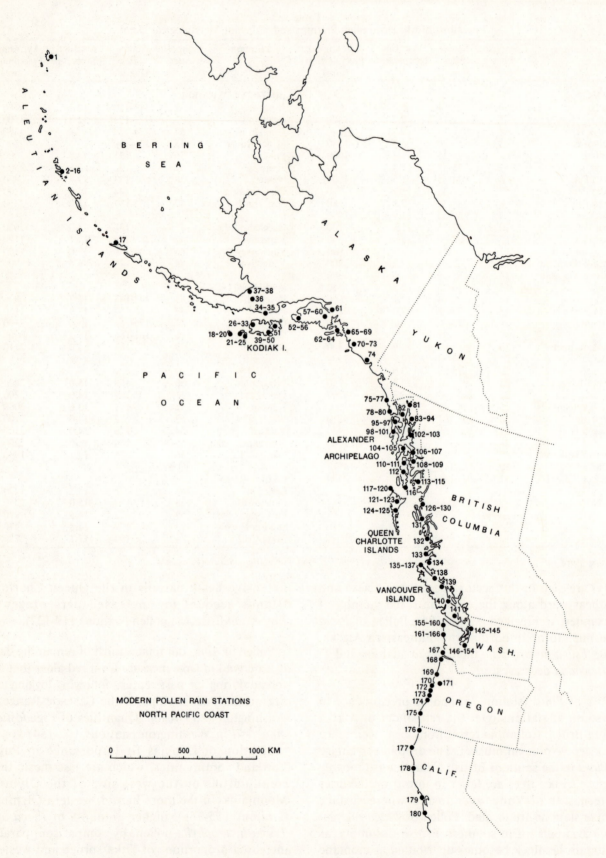

Figure 7. Distribution of modern pollen rain stations. Locations are given in Table 1.

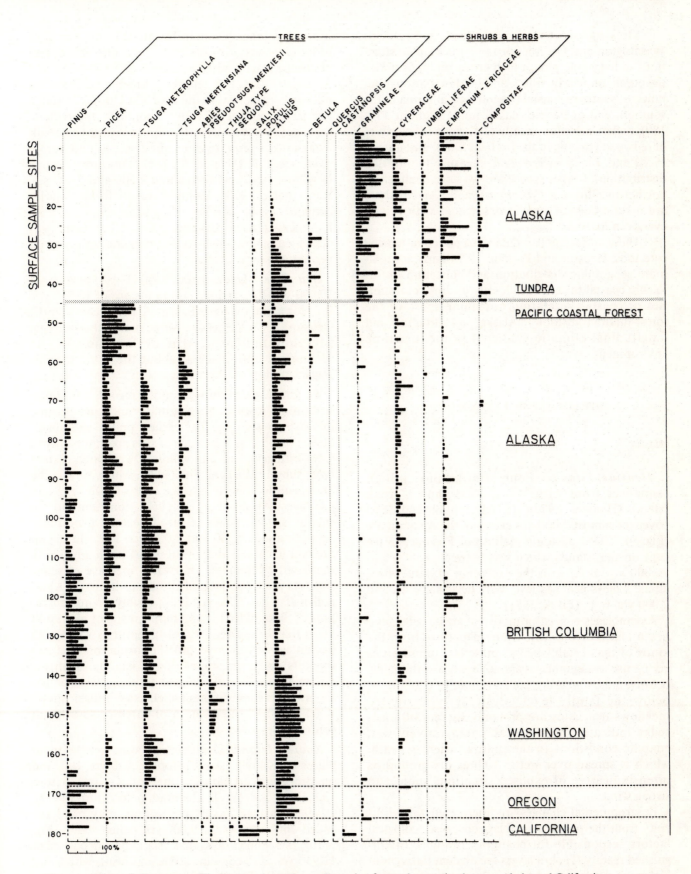

Figure 8. Frequency of leading taxa in modern pollen rain of coastal vegetation between Alaska and California.

Striking percentages of *Pinus* in southern Washington and Oregon portray local edaphic situations of lodgepole pine, which grows on bluffs above the ocean, on nearby mires, and on dunes throughout much of southern coastal Oregon. In California at the southern end of Pacific coastal forest (stations 179-180), pollen is predominantly of *Sequoia* from groves of redwood growing in the fog belt. Minor amounts of *Picea* and *Tsuga heterophylla* measure the limited distribution of Sitka spruce and western hemlock in northern California; *Castanopsis* and *Quercus* there are a reflection of species associated with mixed evergreen forest.

These surface pollen data (also pollen in marine core tops; Heusser and Balsam, 1977) fairly approximate geographic distribution and concentration of Pacific coastal taxa. They identify leading communities of coastal vegetation, including those of the forest-tundra transition in Alaska, and latitudinal and longitudinal climatic gradients in the maritime environment.

STRATIGRAPHIC PALYNOLOGY

Alaska

Aleutian Islands. Four stratigraphic pollen sequences from Attu, Atka, Adak, and Umnak Islands (Heusser, 1973b, 1978d, unpublished data) cover the almost 2,000-km extent of the Aleutian arc (Fig. 9). These postdate melting of Pleistocene ice caps on the islands which was between 12,000 and 10,000 yrs. B.P., with the exception of Attu where alpine valleys near sea level were deglaciated around 7,000 yrs. B. P. (Black, 1981).

Assemblages consist primarily of grass, sedge, and heath and indicate little if any departure from the kinds of taxa established at present. Communities giving rise to assemblages are not easily ascertainable, because grass and sedge pollen cannot be determined below the family level. Mesic herb-dominated meadows near sites are probably sources of grass; sedge indicates mostly local, comparatively wet, edaphic conditions in the tundra, whereas heath, which is spread over wet as well as dry ground, is often indicative of exposed situations subject to strong wind.

Environmental changes are interpreted with difficulty from the pollen assemblages because ecological factors responsible for changes cannot be distinguished readily. Tephra layers are evident throughout Holocene deposits, and their influence on vegetation in most instances is not differentiated from the effect of climatic variations. Conditions compatible with Holocene glacier advances dated at 7,500-5,500 and 3,500-2,000 yrs. B.P. (Black, 1981), for example, are not easily perceived in the data. On Umnak Island, a rise in percentage of willow, noted by the willow-grass-sedge assemblage (Fig. 9), is thought to result from milder climate after about 8,500 and before 3,000 yrs. B.P. (Heusser, 1973b). Transitory decline in willow, dated between about 7,500 and 5,000 yrs. B.P. (Black, 1974), seems to result from a cooler climate, possibly corresponding to an episode of mid-Holocene glacier advance. No tephra layer was deposited at this time, so that climatic influence on vegetation seems likely.

Shumagin Islands. Unga Island (Fig. 9) dates from 10,490 yrs. B.P. and is close to the age of 10,730 yrs. B.P. for a site on the adjacent Alaska Peninsula (C. Heusser, 1983b). Ages are for peat directly overlying till, thus indicating that deglaciation in the Shumagins took place at about the same time as deglaciation in the Aleutians.

Tundra, evident throughout the record, was apparently more open at the beginning and later became increasingly shrubby. The earliest pollen assemblage during the first millennium consists predominantly of grass with artemisia and willow. Dwarf birch (*Betula nana* subsp. *exilis*) and Sitka alder, which grow in the Shumagin Islands close to the ends of their ranges in southwest Alaska (Hultén, 1968), contribute to the shrub aspect. Pollen of birch became prominent between about 9,500 and 8,000 yrs. B.P., during an interval when the amount of dwarf birch apparently increased at the site; afterward, increase of sedge suggests a change toward impeded drainage. Alder pollen at about 7,000 yrs. B.P. possibly records migration of Sitka alder to the Shumagins. But migration more likely occurred in recent centuries when high percentages of alder record the formation of dense thickets on lower slopes of the islands. Multiple tephra layers directly related to changes in pollen assemblages show differential effects of vulcanism on vegetation and the creation of patchworks of plant communities.

Kodiak Island. Deglaciation was over with in places on Kodiak Island before 12,220 yrs. B.P. and on the Trinity Islands, south of Kodiak, before 13,800 yrs. B.P. (W. L. Coonrad, pers. comm., 1977). Glacier wastage locally on Kodiak may not have ended until later, as the 9,000 yrs. pollen record (Fig. 9) suggests. Pollen in Kodiak Island peat deposited at 11,930 yrs. B.P. contains sedge and heath with lesser amounts of willow, umbellifer, grass, and composite,

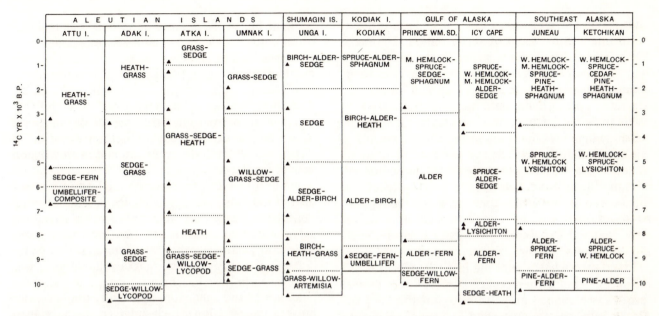

Figure 9. Correlation diagram of pollen assemblage zones of Quaternary sections in Alaska. Triangles indicate dated levels. See text for sources.

while Trinity Island peat dated at 13,800 yrs. B.P. is almost entirely of sedge.

The Kodiak sequence (Fig. 9) at first shows a sedge-fern-umbellifer pollen assemblage followed by alder-birch (Heusser, 1960). Sitka alder, the only species growing today on Kodiak, is presumed to be the pollen represented in this and later assemblages. Dwarf and Kenai birch, both established at the present time on the island, are not differentiated in the record. Dwarf birch ranges as far as northern arctic Alaska while Kenai birch is a subarctic arboreal species (Hultén, 1968). Under colder, early-Holocene climate, dwarf birch would be the species expected in the pollen assemblage, whereas later, in the next younger birch-alder-heath assemblage, the likelihood is that pollen includes Kenai birch. Sitka spruce apparently migrated to Kodiak Island within the past millennium, as shown by small quantities of its pollen below Mount Katmai tephra of A.D. 1912 in contrast to large quantities above the layer. The modern expansion of Sitka spruce was demonstrated by Griggs (1914, 1934) in his observations on distribution of first-generation forest.

Gulf of Alaska. The earliest pollen record in Prince William Sound (Heusser, 1983c) is dated at 10,000 yrs. B.P., four millennia later than the record on Kenai Peninsula to the west, in which a shrub birch tundra zone dated 13,730 yrs. B.P. follows an undated herb zone (Ager and Sims, 1982). East of Prince William Sound around the Copper River, plant invasion also appears to have taken place around 14,000 yrs. B.P. (Sirkin and Tuthill, 1969;

Connor, 1983). During the late-Pleistocene, heavy snowfall in the Chugach and Kenai Mountains evidently maintained the glacial lobe in Prince William Sound close to equilibrium, while glaciers wasted in neighboring regions.

Sedge, willow, and fern were principal pioneers at Prince William Sound (Fig. 9). Later, alder in association with fern became preponderant by 8,300 yrs. B.P. This succession, showing pollen influx ranging from 300 at the beginning to 2,700 grains cm^{-2} yr^{-1} later on, is interpreted to represent sedge tundra followed by shrub alder tundra. As alder thickets multiplied after 8,300 yrs. B.P., the proportion of fern decreased. At about 3,000 yrs. B.P., Sitka spruce and mountain hemlock entered the record, and by about 2,000 yrs. B.P., these trees rapidly displaced much of the alder; western hemlock is also evident but in minimal amounts.

A wetter and cooler climate is indicated during the past 2,000 years by rate of peat accumulation (contributed by sedge and sphagnum) which is much higher than during the previous 8,000 years. Establishment of stands of Pacific coastal forest in Prince William Sound followed migration of conifers northwestward along the coast from southeast Alaska. This event occurred later than arrival of conifers (white spruce and possibly black spruce) on the central Kenai Peninsula. Dated at about 8,000 yrs. B.P., migration is believed to have followed a route from interior Alaska (Ager and Sims, 1982).

At Icy Cape (Fig. 9), sedge-heath tundra is in evidence between 10,820 and about 10,000 yrs. B.P., after which alder, invading along with fern, formed the primary constituent of shrub tundra (Heusser, 1960; Peteet, 1983). An alder fen containing lysichiton *(Lysichiton americanum)* developed between about 7,600 and 7,500 years ago. During this interval, low rate of peat generation in the mire is ascribed to warmer and drier climate with greater periods of sunshine than prevail at present. After 7,500 yrs. B.P., regeneration at the surface, evident from deposition of sedge peat, is attributed to wetter and cloudier climate. Shrub tundra developed a park aspect, as Sitka spruce at about this time colonized the region. At about 3,800 yrs. B.P., when deposition of sedge peat reached excessive proportions for several hundred years, western hemlock reached Icy Cape, followed somewhat later by mountain hemlock. With the decline of alder in recent centuries, closed conifer forest has extended along this part of the Gulf of Alaska between the Bering and Malaspina Glaciers. Eastward bordering the gulf, the overall pattern of forest development over the past 10,000-11,000 yrs. is much the same as at Icy Cape (Peteet, 1983; Mann, 1983).

Southeast Alaska. Deglaciation about Juneau, based on dates of mollusks (McKenzie, 1970; Miller, 1973; Ackerman *et al.*, 1979), was earlier than 13,000-14,000 yrs. B.P. Tundra is not evident in younger pollen records (Heusser, 1952, 1960, 1965) at Juneau and Ketchikan (Fig. 9). The earliest assemblage at Juneau dated at 10,300 yrs. B.P. is composed of pine, alder, and fern. The species of pine is identified as lodgepole, and Sitka alder is probably the species with which it colonized deglaciated terrain of the early-Holocene. This alder has much the same role at present in Pacific coastal Alaska and has been the leading invader in recent centuries on drift deposited in the region during glacier recession (Lawrence, 1958). Lodgepole, on the other hand, is infrequent today and relegated to muskeg.

Spruce was established more or less contemporaneously in both records around 9,500 yrs. B.P. This was during the interval of alder predominance which lasted until close to 7,500 yrs. B.P. Western hemlock reached Alaska along with spruce but apparently gained importance first at Ketchikan. Significant is the fact that spruce is recorded in numbers at these sites some 2,000 years earlier than at Icy Cape. Duration of the alder interval is short, compared with its prolongation into the late-Holocene at Icy Cape and other sites to the west, and is possibly explained by availability of spruce and hemlock seed sources. Sitka alder is a seral species when competing with

conifers, but in the absence of arboreal competition, communities of alder can endure for millennia.

Coastal forest consisted of Sitka spruce and western hemlock until around 3,500 yrs. B.P. After about 7,800 until 3,500 yrs. B.P., lysichiton in the pollen assemblage was abundant in association with ligneous peat at the nongenerative surface of the mires. These data make evident a lengthier duration of warmer, drier mid-Holocene climate in this region than at Icy Cape.

Western hemlock has become the leading component of the coastal forest during the past 3,500 years. Pollen data imply that Sitka spruce was a secondary associate; also, mountain hemlock was more abundant at Juneau than at Ketchikan and cedar (western red cedar and/or Alaska yellow cedar) was established only at Ketchikan. Increase of lodgepole pine, heath, and sphagnum apparently resulted from muskeg regeneration which enlarged the habitat of these plants. Wetter and cooler climate shown by these data is also implied by major Neoglacial episodes dated regionally between 3,300-2,400 yrs. B.P. and during the last few centuries (Denton and Karlén, 1973).

British Columbia

Graham Island. Plant records from Graham Island, a member of the Queen Charlotte Islands in northern coastal British Columbia, date from between 45,700 and 27,500 yrs. B.P. during the Mid Wisconsin (Warner *et al.,* 1984) and from 16,000 yrs. B.P. to the present when ice-free conditions prevailed (Mathewes and Clague, 1982; Warner *et al.,* 1982, 1984; Warner, 1984). Between about 18,000-15,000 yrs. B.P., at the time Late Wisconsin glaciers reached their maximum in southern British Columbia and adjacent Washington, glaciers on the Queen Charlotte Islands were limited in extent (Clague *et al.,* 1980, 1982a).

The Mid Wisconsin sequence beginning with a sedge-herb assemblage, followed by spruce-mountain hemlock-sedge and mountain hemlock-spruce-herb assemblages (Fig. 10), implies mostly subalpine forest. Noteworthy is the pollen profile of *Abies,* a genus absent from the flora today. Macrofossils identified are of *Tsuga mertensiana, Salix, Hippuris vulgaris, Veronica scutellata, Potentilla palustris, Nuphar lutea, Menyanthes trifoliata, Carex,* and *Selaginella selaginoides.*

The Late Wisconsin at around 16,000 years ago shows a grass-herb pollen assemblage (Fig. 10)

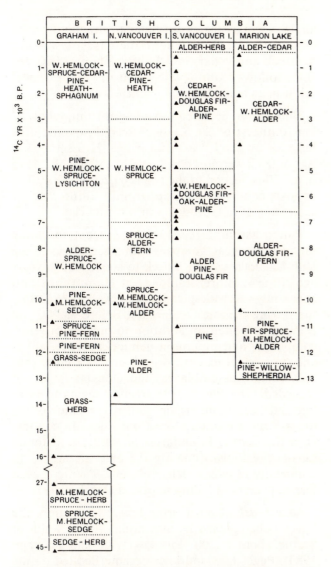

Figure 10. Correlation diagram of pollen assemblage zones of Quaternary sections in British Columbia. Triangles indicate dated levels. See text for sources.

supplemented by macrofossils of *Juncus,* Caryophyllaceae, *Rumex, Potamogeton filiformis, Callitriche, Ranunculus aquatilis, Salix reticulata,* and *Picea,* along with oospores of *Chara* and *Nitella.* Later, between 12,400-10,200 yrs. B.P., pollen assemblages are of grass and sedge with a variety of additional herb taxa, followed by lodgepole pine-fern, Sitka spruce-pine-fern, and finally, pine-mountain hemlock-sedge. As reconstructed, late-glacial vegetation included open, herb-dominated communities, which in part occupied a floodplain, sedge fen, and swamp forest. Ice-free parts of Graham Island evidently served as a locus for plant migration into adjacent coastal regions.

Holocene pollen sequences from the northern part of the Queen Charlotte Islands (Heusser, 1955, 1960;

Warner, 1984), in the main, resemble the sequence at Ketchikan (Fig. 9).

North Vancouver Island. A succession of pollen assemblages is described from a bog dated at 13,630 yrs. B.P. near Port Hardy (Hebda, 1983). The order and character of major taxa in assemblages (Fig. 10) compare with sequences described from other sites in the region (Heusser, 1960; Hebda, 1984), as well as from sites in the Queen Charlottes and southeastern Alaska.

Recognizable as an overall feature of the succession are lodgepole pine at the beginning, Sitka spruce later, and at the end, western hemlock. The manifestation of mountain hemlock in early assemblages in British Columbia and not in southeast Alaska seems to imply that Alaskan sites were covered by glaciers at the time. It is also possible that low-lying sites were beneath sea level. Shorelines in the Queen Charlottes were below their present positions between 13,700 and approximately 9,500-10,000 years ago (Clague *et al.,* 1982b).

South Vancouver Island. Pollen data in bog sections (Hansen, 1950; Heusser, 1960) and a marine core (L. Heusser, 1983b) from the drier part of southern Vancouver Island near Victoria are of the past approximately 12,000 year period (Fig. 10). Communities of lodgepole and to a lesser extent of western white pine during the late-glacial remained commonplace with alder (mostly red alder) and Douglas fir during the Holocene. Oregon oak, migrating northward via the Puget-Willamette Lowland of Washington and Oregon, multiplied after about 8,000 until 2,000-3,000 yrs. B.P. under a warmer, summer-dry climate (Hansen, 1947; Heusser *et al.,* 1980; Mathewes and Heusser, 1981). Communities afterward contained increased numbers of western hemlock and western red cedar in response to more humid and cooler climate. Effect of human settlement, beginning around A.D. 1900, is recorded by an increase of alder, weedy adventives, and charcoal.

Marion Lake. Located within the altitudinal limit of glacio-marine deposits, Marion Lake dates from 12,350 yrs. B.P. following coastal emergence (Mathewes, 1973). A detailed pollen stratigraphy (Fig. 10) matches the pattern described for southern Vancouver Island but with some modification owing to location of the site in a more humid zone in the Fraser River drainage.

The glacio-marine sediments, perhaps 13,000 years of age, contain pollen of lodgepole pine, willow, and shepherdia *(Shepherdia canadensis)* from semi-open communities at the lake. Lodgepole, with small amounts of spruce, fir, and mountain hemlock and with rising amounts of alder, also dominated the next younger assemblage zone (12,350-10,370 yrs. B.P.). Pine, later, was unimportant, whereas alder became dominant with fern (10,370-6,600 yrs. B.P.). Douglas fir is evident at the beginning of the interval, but the quantity fell off with the rise of western hemlock at about 7,600 yrs. B.P. The last 6,000 years trace the spread of western red cedar-Alaskan yellow cedar type. A sharp increase of alder most recently is attributed to human settlement.

Washington

Hoh-Kalaloch. This combined sequence (Fig. 11) brings together records of a seacliff section near Kalaloch, dating from the Early Wisconsin (>47,000 yrs. B.P.) until about 16,000 years ago, and a core from a bog in the Hoh River drainage, 23 km distant, covering the past approximately 16,000 years (Heusser, 1972, 1974). The site of the seacliff is unglaciated and was a Wisconsin-age refugium; the bog originated after a valley glacier from the Olympic Mountains, during the alpine phase of Fraser (Late Wisconsin) Glaciation (Porter *et al.,* 1983), withdrew from the lower Hoh.

Assemblages at Kalaloch frequently contain a high proportion of herbs. Grass, sedge, and composite are most abundant, along with subalpine-alpine indicators, for example, *Polygonum bistortoides* type, *Gentiana, Polemonium, Valeriana,* and *Artemisia.* These taxa, most distinctive of intervals estimated at >65,000 and between 40,000-34,000 and 28,000-16,000 yrs. B.P., are indicative of tundra or, with increased mountain hemlock at >65,000 and between 28,000-19,000 yrs. B.P., suggest park tundra. At other times at Kalaloch, especially <65,000-40,000 and 34,000-28,000 yrs. B.P., assemblages of western hemlock and spruce imply spread of coastal forest. Pine, often present in these assemblages, may be derived from trees restricted edaphically at the site. The western hemlock-spruce-alder assemblage dated >65,000 yrs. B.P. probably represents a noteworthy interstade.

An herb assemblage 15,600 years old is from earliest sediments in the Hoh bog and provides continuity with the topmost herb zone at Kalaloch which is <16,700 yrs. B.P. At about 13,000 yrs. B.P., increase in pollen influx contributed by pine, spruce, and mountain hemlock identifies an initial spread of forest communities. Between about 9,500-8,000 yrs. B.P., maximum influx of alder, Douglas fir, and bracken fern *(Pteridium aquilinum* var. *pubescens),*

all light-demanding, is attributed to structurally open communities. Subsequent increase of western hemlock and cedar, mostly after about 3,000 yrs. B.P., follows from development of closed coastal forest in an equable humid climate.

Humptulips. Sediments in a bog at Humptulips, 53 km southeast of the Kalaloch seacliff, overlie drift regarded as Early Wisconsin in age (Heusser, 1982). Age of the lowermost dated level at a depth of 2.45 m in a 7.70-m section of the bog is 63,200 yrs. B.P. The deposit consists of an apparently uninterrupted succession of interbedded peat, silt, and clay.

Pollen assemblages (Fig. 11), except for some minor differences, compare with assemblages in the Hoh-Kalaloch sequence back to about 65,000 years ago. Pine is more abundant at Humptulips and apparently was widespread in southwestern Washington during the late Quaternary. Before 65,000 yrs. B.P., most pollen, in addition to pine, is of grass, composite, and mountain hemlock; spruce, alder, and, to a lesser extent, western hemlock enter into only certain assemblages; and Douglas fir is not recorded. The Pleistocene treeline seems to have fluctuated across the region, so that vegetation often-times formed a tundra-forest transition. In contrast with fluctuating boundaries of tundra and forest during the Pleistocene are the extensiveness and uniformity of closed, late-Holocene, Pacific coastal forest apparent at Humptulips and Hoh-Kalaloch.

Lake Washington. During Fraser Glaciation, the Puget Lobe reached its maximum, 90 km south of Seattle, about 15,000 years ago (Waitt and Thorson, 1983). Puget Lowland later became flooded by the sea, but with isostatic rebound, tidal extent diminished. Pine and a small amount of spruce (Fig. 11) were deposited in freshwater lacustrine sediments in Lake Washington at Seattle from 13,430 until around 11,000 yrs. B.P., and along with these for about a millennium afterward, Douglas fir and true fir increased (Leopold *et al.,* 1982). Douglas fir, alder, and bracken fern subsequently reached maxima lasting until about 7,000 yrs. B.P., after which cedar type became dominant and alder secondary. Mt. Mazama tephra, deposited widely in the Puget Lowland, approximates the 7,000-year dated horizon.

Sediments in nearby Hall Lake (Tsukada *et al.,* 1981) cover about the same time span as Lake Washington and contain a significant amount of western hemlock (>20%) and cedar type during the last 7,000 years. In the pollen stratigraphy of Pangborn Lake (Hansen and Easterbrook, 1974), located near the Canadian border, trends of taxa are

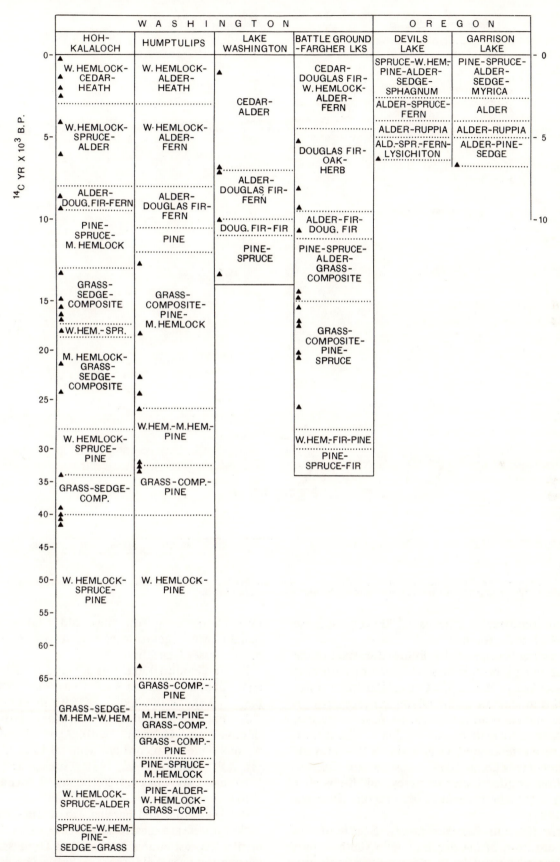

Figure 11. Correlation diagram of pollen assemblage zones of Quaternary sections in Washington and Oregon. Triangles indicate dated levels. See text for sources.

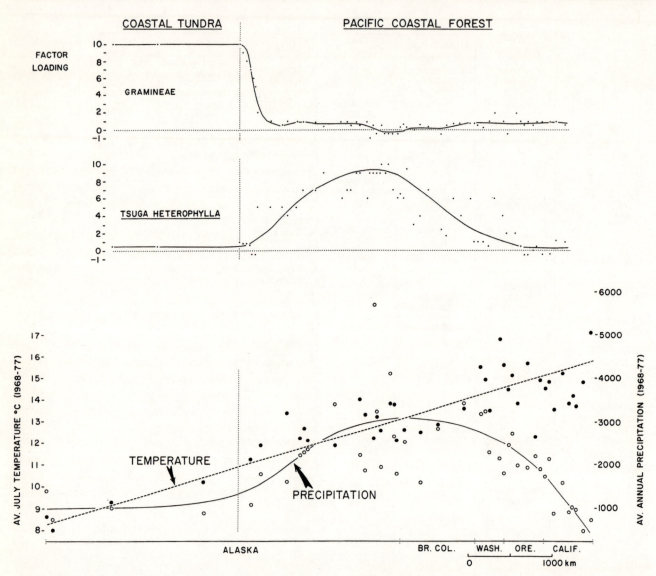

Figure 12. Temperature and precipitation trends in relation to factor loading of Gramineae (grass) and *Tsuga heterophylla* (western Hemlock) in modern pollen rain data from stations in coastal tundra and Pacific coastal forest.

generally comparable for the past 9,900 years to those established in the south.

Late-glacial vegetation that followed wastage of the Juan de Fuca lobe, which at the time of the Puget lobe extended to the Pacific Ocean, illustrates influence of rain shadow of the Olympic Mountains on successional communities. Near its western margin, forest species are evident after the Juan de Fuca Lobe began to waste around 14,500 yrs. B.P. (Heusser, 1973c), whereas about 100 km east in rain shadow, vegetation was dominated by herbs and shrubs after melting of the lobe at about 12,000 yrs. B.P. (Petersen *et al.,* 1983).

Battle Ground-Fargher Lakes. South of the maximum limit of the Puget Lobe in southwestern Washington, several lakes reach greater antiquity than lakes situated within the boundaries of the lobe.

Pollen sequences from these older lakes provide insight into vegetation of this region during and before Fraser Glaciation.

Battle Ground Lake was formed in a volcanic crater about 20,000 years ago (Barnosky, 1983), and Fargher Lake, estimated to be close to 34,000 years old, originated on drift of Cascade Range provenance (Heusser and Heusser, 1980). Ages of additional regional lakes of record are Mineral Lake at 17,500 yrs. B.P. (Tsukada *et al.,* 1981; Tsukada and Sugita, 1983) and Davis Lake at 26,100 yrs. B.P. (Barnosky, 1981).

Battle Ground and Fargher Lakes, approximately 10 km apart, are located at the edge of the western hemlock forest zone (Franklin and Dyrness, 1973). During the past 4,500 years at Battle Ground Lake (Fig. 11), western hemlock formed an assemblage

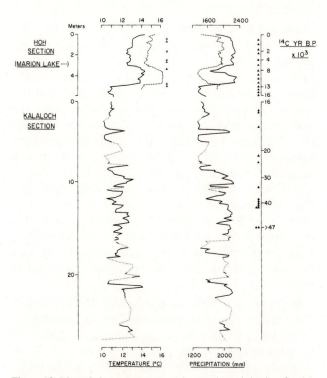

Figure 13. Mean July temperature and annual precipitation for the late-Quaternary of western Washington and southwestern British Columbia. Fluctuations are three-point moving averages of data. Solid lines apply to Hoh bog section (upper part) and Kalaloch seacliff section (lower part); dotted portions are intervals with low communalities. Dash lines relate to Marion Lake section. Triangles indicate dated levels (Hoh-Kalaloch on the right and Marion Lake in the middle).

Oregon

Devils Lake. Lakes along the coast appear to owe their origin to impoundment of streams by moving dunes at a time when the water table, being raised by Holocene eustatic rise of sea level, was approaching its present position. Devils Lake (4.5 m elev.), with an age of 6,300 yrs. B.P., is of freshwater origin (Heusser, 1960). A pollen assemblage from the dated horizon at -7 m (Fig. 11), possibly the level of the sea at the time, shows red alder, Sitka spruce, fern, and lysichiton. For a time later, when sea level may have reached its maximum, Devils Lake became estuarine, as indicated by sand interbedded with lake sediments and by ruppia of brackish provenance in the pollen assemblage. Devils Lake has since become fresh, surrounded by arboreal communities of Sitka spruce, western hemlock, western red cedar, Douglas fir, pine, and alder.

Garrison Lake. Garrison Lake (1.5 m elev.), 250 km south of Devils Lake, is also of mid-Holocene age and was originally fresh (Heusser, 1960). Limnic sediments dated at about 6,700 yrs. B.P. contain pollen of alder, pine, and sedge (Fig. 11). Subsequent marine incursion is evident from a prominence of ruppia and from interbedded sand. More recently, limnic deposition and a pollen assemblage of locally derived sedge and myrica and of regional pine, Sitka spruce, and alder identify restoration of the freshwater lake.

Overview

Vegetation in western Washington beyond the limit of Fraser Glaciation during the alpine phase between 22,000-18,500 yrs. B.P. (Porter *et al.,* 1983), graded from tundra, through parkland of spruce and other conifers, to pine woodland. At the time of the lowland phase between 17,000-14,000 yrs. B.P. (Waitt and Thorson, 1983), parkland of pine, spruce, and mountain hemlock prevailed, and tundra was limited to the colder valleys in the mountains.

Unglaciated western Washington was a Pleistocene refugium and important locus for late-glacial and Holocene plant migration northward into British Columbia. The Queen Charlotte Islands, unglaciated in part during the Late Wisconsin, supported another locus that supplied disseminules, enabling the spread of plants in British Columbia and southeast Alaska. Pine and alder were notable early pioneers that first invaded deglaciated coastal sectors, usually followed by open stands of conifers, which included Douglas

with cedar type and Douglas fir, suggesting proximity of hemlock forest. Earlier, western hemlock occurred only before an estimated 28,000 yrs. B.P. when true fir and pine were evidently associated in montane forest about Fargher Lake. Still earlier, forest communities consisted mostly of pine. Pine was also extensive in the northern part of Willamette Lowland in adjacent Oregon, as shown by quantities of its pollen in an undated late-Quaternary section from Onion Flats (Hansen, 1947).

After 28,000 until about 15,000 yrs. B.P., subalpine parkland of colonial pine and spruce growing in communities of grass and composite (including artemisia) is reconstructed. Open forest is later implied by an increase of montane-subalpine tree taxa and Sitka alder type until around 11,000 years ago when temperate trees (Douglas fir and red alder, for example) expanded. Douglas fir-alder woodland from 9,500-4,500 yrs. B.P., associated with oak communities, preceded development of the modern forest.

fir on Vancouver Island and principally Sitka spruce and western hemlock on the Queen Charlottes and in the Alexander Archipelago of southeast Alaska.

Pine migrated rapidly through southeast Alaska between 11,000-10,000 yrs. B.P. It was replaced soon after by spruce and western hemlock which continued to migrate northwestward along the Gulf of Alaska. Spruce reached Icy Cape about 7,600 yrs. B.P., Prince William Sound at 2,000-3,000 yrs. B.P., and its present-day limit on Kodiak Island probably within the past 1,000 years; western hemlock migrated to Icy Cape around 3,800 yrs. B.P. and Prince William Sound at about 2,000 yrs. B.P.; and mountain hemlock appeared at Icy Cape at about 3,500 yrs. B.P., later than spruce or western hemlock, and at Prince William Sound at the same time as spruce. Stands of conifer forest along the Gulf of Alaska displaced shrub tundra which was made up predominantly by alder. Beyond the forest in southwestern Alaska, alder has recently invaded herb tundra as far as the eastern Aleutian Islands.

Pollen records of Pleistocene age predating the Late Wisconsin in Oregon (Hansen and Allison, 1942; Gottesfeld *et al.,* 1981), Washington (Florer, 1972; Heusser, 1972, 1978e; Hansen and Easterbrook, 1974), and British Columbia (Mathewes, 1979b; Alley, 1979; Alley and Chatwin, 1979; Hebda *et al.,* 1983; Warner *et al.,* 1984) cover intervals during which plants adapted to a variety of glacial and nonglacial conditions. Development of coastal forest is especially manifest during past interglaciations in Washington (Hansen and Mackin, 1949; Leopold and Crandell, 1957; Heusser, 1977, 1982; Heusser and Heusser, 1981). Taxa forming steady-state communities and having a role in succession during interglaciations compare with modern taxa. Pollen data contain no evidence of Tertiary plants that stratigraphically range into the Pleistocene and become extinct. Remarkable are the stability and unity of Pacific coastal forest interglacial communities which seem to have undergone little change as a result of physical upheavals during ice ages.

LATE-QUATERNARY TEMPERATURE AND PRECIPITATION

Paleoclimatic data are derived from fossil pollen percentages at Hoh-Kalaloch using a pair of regression equations that relate surface pollen (Fig. 8) at 59 composite sites in coastal forest and tundra to mean July temperature and annual precipitation (Heusser *et al.,* 1980). Temperature and precipitation are estimated for sites from 10-year averages (1968-77) at 43 meteorological stations between the Aleutians and northern California (Table 1). Q-mode, rotated principal components analysis of 13 numerically important taxa, selected from a total of 67, identified four pollen factors, western hemlock, alder, grass, and pine, which account for 93% of the variance in the data. Stepwise regression of the factors with temperature and precipitation produced a pair of transfer functions. Western hemlock and pine proved to significantly predict precipitation, while these and grass predicted temperature; alder proved not to be a predictor. Standard errors in the equations are 1.1°C of temperature and 620 mm of precipitation. Relationships between grass and western hemlock factors and temperature-precipitation are shown in Figure 12.

Regression equations applied to Hoh-Kalaloch fossil pollen data yield a record of mean July temperature (10-15°C) and annual precipitation (1,300-2,400 mm) for western Washington dating back to the Early Wisconsin (Fig. 13). Climatic trends over the Pleistocene portion of the curves indicate relatively warm and wet intervals at >47,000 and around 30,000 yrs. B.P. with coldest and driest conditions in effect between 28,000 and 13,000 yrs. B.P. Mid Wisconsin temperature trends are corroborated by oxygen isotope measurements of dated speleothems from caves on Vancouver Island (Gascoyne *et al.,* 1981). Colder and drier climate during the Late Wisconsin after 30,000 yrs. B.P. is also indicated by calibrations of a Puget Lowland core (Heusser and Heusser, 1980). Temperature of about 10°C and precipitation around 1,300 mm during the Late Wisconsin glacial maximum compare with values measured near the present-day edge of coastal forest in Alaska (Fig. 12).

After 13,000 yrs. B.P., temperature increased 3°C to a maximum at about 8,000 yrs. B.P., while an initial increase of 1,100 mm of precipitation by about 10,000 yrs. B.P. was followed by a decrease amounting to some 900 mm when maximum temperature was reached. After 8,000 yrs. B.P., climate has been colder and wetter with temperature lowest and precipitation heaviest between about 5,000-2,000 yrs. B.P. These trends fundamentally conform with Hypsithermal and Neoglacial climatic changes implied by regional glacier fluctuations (Burke and Birkland, 1983).

Confirmation of trends after about 13,000 yrs. B.P. is found in quantification of temperature and precipitation at Marion Lake (Mathewes and Heusser, 1981); as a result of warmer and drier climate in southwestern British Columbia, curves appear offset

(Fig. 13). Drop in precipitation after 10,000 yrs. B.P. is pronounced and begins earlier at Marion Lake. Data show that during the Hypsithermal, temperature was 2°C warmer and precipitation 400-800 mm less than in western Washington.

Temperature calibrations of the last interglaciation made on the Whidbey Formation, exposed in the northern Puget Lowland, were found to be close to present-day temperature but also ranged several degrees lower (Heusser and Heusser, 1981). Included in the Whidbey, as in the Holocene or present interglaciation, are apparent intervals when temperatures were below maxima. In California, temperature maximum during the last interglaciation is estimated to be 1.5°C higher than today (Adam and West, 1983), whereas in western Washington, it is at most only 0.5°C higher. For earlier interglaciations in the Puget Lowland, which pre-date the zircon fission-track age of Salmon Springs drift at 0.84 ± 0.21 m.y. (Easterbrook *et al.*,1981), the Alderton Formation shows a temperature maximum much the same as the Whidbey, while Puyallup Formation temperature maxima are lower by approximately 2°C (Heusser and Heusser, 1981).

ACKNOWLEDGMENTS

I thank the late R. F. Black (Aleutian Islands), W. L. Coonrad (Chirikof Island, Trinity Islands, and parts of Kodiak Island), and L. E. Heusser (vicinity of San Francisco and Santa Cruz) for surface samples, and M. Stuiver for radiocarbon dates. Research supported by the National Science Foundation.

References Cited

ACKERMAN, R. E., HAMILTON, T.D., and STUCKENRATH, R.
1979   Early culture complexes on the northern northwest coast. *Canadian Journal of Archaeology,* 3:195-208.

ADAM, D. P., and WEST, G. J.
1983   Temperature and precipitation estimates through the last glacial cycle from Clear Lake, California, pollen data. *Science,* 219: 168-170.

AGER, T. A.
1982   Vegetational history of western Alaska during the Wisconsin glacial interval and the Holocene. *In:* Hopkins, D. M., Matthews, J. V., Jr., Schweger, C. E., and Young, S. B. (eds.), *Paleoecology of Beringia.* Academic Press, New York, pp. 75-93.
1983   Holocene vegetational history of Alaska. *In:* Wright, H. E., Jr. (ed.), *Late-Quaternary Environments of the United States. Volume 2. The Holocene.* University of Minnesota Press, Minneapolis, pp. 128-141.

AGER, T. A., and SIMS, J. D.
1982   Late Quaternary pollen record from Hidden Lake, Kenai Peninsula, Alaska. *Palynology,* 6: 271-272.

ALLEY, N. F.
1979   Middle Wisconsin stratigraphy and climatic reconstruction, southern Vancouver Island. *Quaternary Research,* 11: 213-237.

ALLEY, N. F., and CHATWIN, S. C.
1979   Late Pleistocene history and geomorphology, southwestern Vancouver Island, British Columbia. *Canadian Journal of Earth Sciences,* 16: 1645-1657.

BAGNELL, C. R., JR.
1975   Species distinction among pollen grains of *Abies, Picea,* and *Pinus* in the Rocky Mountain area (a scanning electron microscope study). *Review of Palaeobotany and Palynology,* 19: 203-220.

BAKER, R. G.
1983   Holocene vegetational history of the western United States. *In:* Wright, H. E., Jr., *Late-Quaternary Environments of the United States. Volume 2. The Holocene.* University of Minnesota Press, Minneapolis, pp. 109-127.

BARBOUR, M. G., and MAJOR, J.
1977   *Terrestrial Vegetation of California.* Wiley-Interscience, New York, 1,002 pp.

BARNOSKY, C. W.
1981   A record of late-Quaternary vegetation from Davis Lake, southern Puget Lowland, Washington. *Quaternary Research,* 16: 221-239.
1983   Late-Quaternary vegetational and climatic history of southwestern Washington. Ph. D. Thesis, University of Washington, Seattle, 201 pp.

BLACK, R. F.
1974   Late-Quaternary sea level changes, Umnak Island, Aleutians—their effect on ancient Aleuts and their causes. *Quaternary Research,* 4: 264-281.
1981   Late-Quaternary climatic changes in the Aleutian Islands, Alaska. *In:* Mahaney, W. C. (ed.), *Quaternary Paleoclimate.* Geoabstracts, Ltd., Norwich, England, pp. 47-62.

BOWMAN, P. W.
1934   Pollen analysis of Kodiak bogs. *Ecology,* 15: 694-708.

BRYSON, R. A., and HARE, F. K.
1974   The climates of North America. *In:* Bryson, R. A., and Hare, F. K. (eds.), *Climates of North America.* Elsevier, Amsterdam, pp. 1-47.

BURKE, R. M., and BIRKELAND, P. W.
1983   Holocene glaciation in the mountain ranges of the western United States. *In:* Wright, H. E., Jr. (ed.), *Late-Quaternary Environments of the United States. Volume 2. The Holocene.* University of Minnesota Press, Minneapolis, pp. 3-11.

CLAGUE, J. J., ARMSTRONG, J. E., and MATHEWS, W. H.
1980   Advance of the Late Wisconsin Cordilleran Ice Sheet in southern British Columbia since 22,000 yrs B.P. *Quaternary Research,* 13: 322-326.

CLAGUE, J. J., MATHEWES, R. W., and WARNER, B. G.
1982a   Late-Quaternary geology of eastern Graham Island, Queen Charlotte Islands, British Columbia. *Canadian Journal of Earth Sciences,* 19: 1786-1795.

CLAGUE, J. J., HARPER, J. R., HEBDA, R. J., and HOWES, D. E.
1982b   Late-Quaternary sea levels and crustal movement, coastal British Columbia. *Canadian Journal of Earth Sciences,* 19: 597-618.

CONNOR, C. L.
1983   Late Pleistocene paleoenvironmental history of the Copper River basin, south-central Alaska. *In:* Thorson, R. M., and Hamilton, T. D. (eds.). Glaciation in Alaska. *Alaskan Quaternary Center University of Alaska Museum Occasional Paper,* 2: 30-34.

COOPER, W. S.
1957   Vegetation of the Northwest-American Province, *Proceedings of the Eighth Pacific Science Congress,* 4: 133-138.

COURT, A.
1974   The climate of the conterminous United States. *In:* Bryson, R. A., and Hare, F. K. (eds.), *Climates of North America.* Elsevier, Amsterdam, pp. 193-343.

DACHNOWSKI-STOKES, A. P.
1930   Peat profiles in the Puget Sound basin of Washington. *Journal of the Washington Academy of Science,* 20: 193-209.

1936   Peat land in the Pacific coast states in relation to land and water resources. *United States Department of Agriculture Miscellaneous Publications,* 248, 68 pp.

1941   Peat resources in Alaska. *United States Department of Agriculture Technical Bulletin,* 769, 84 pp.

DAUBENMIRE, R. F.
1969   Ecologic plant geography of the Pacific Northwest. *Madrono,* 20: 111-128.

DENTON, G. H., and KARLEN, W. K.
1973   Holocene climatic variations—their patterns and possible cause. *Quaternary Research,* 3: 155-205.

EASTERBROOK, D. J., BRIGGS, N. D., WESTGATE, J. A., and GORDON, M. P.
1981   Age of the Salmon Springs Glaciation in Washington. *Geology,* 9: 87-93.

FLINT, R. F.
1971   *Glacial and Quaternary Geology.* Wiley, New York, 892 pp.

FLORER, L. E.
1972   Quaternary paleoecology and stratigraphy of the sea cliffs, western Olympic Peninsula, Washington. *Quaternary Research,* 2: 202-216.

FOWELLS, H. A. (ed.)
1965   Silvics of forest trees of the United States. *United States Department of Agriculture Forest Service Agriculture Handbook,* 271, 762 pp.

FRANKLIN, J. F., and DYRNESS, C. T.
1973   Natural vegetation of Oregon and Washington. *United States Department of Agriculture Forest Service General Technical Report,* PNW-8, 417 pp.

GASCOYNE, M., FORD, D. C., and SCHWARCZ, H. P.
1981   Late Pleistocene chronology and paleoclimate of Vancouver Island determined from cave deposits. *Canadian Journal of Earth Sciences,* 18: 1643-1652.

GOTTESFELD, A. S., SWANSON, F. J., and GOTTESFELD, L. M. J.
1981   A Pleistocene low-elevation subalpine forest in the western Cascades, Oregon. *Northwest Science,* 55: 157-167.

GRIGGS, R. F.
1914   Observations on the edge of the forest in the Kodiak region of Alaska. *Bulletin of the Torrey Botanical Club,* 41: 381-385.

1934   The edge of the forest in Alaska and the reasons for its position. *Ecology,* 15: 80-96.

HAMILTON, T. D., and THORSON, R. M.
1983   The Cordilleran Ice Sheet in Alaska. *In:* Porter, S. C. (ed.), *Late-Quaternary Environments of the United States. Volume 1. The Late Pleistocene.* University of Minnesota Press, Minneapolis, pp. 38-52.

HANSEN, B. S., and EASTERBROOK, D. J.
1974   Stratigraphy and palynology of late Quaternary sediments in the Puget lowland, Washington. *Bulletin of the Geological Society of America,* 85: 587-602.

HANSEN, H. P.
1938   Postglacial forest succession and climate in the Puget Sound region. *Ecology,* 19: 528-542.

1947   Postglacial forest succession, climate, and chronology in the Pacific Northwest. *Transactions of the American Philosophical Society,* 37: 1-130.

1949   Pollen content of moss polsters in relation to forest composition. *American Midland Naturalist,* 42: 473-479.

1950   Pollen analysis of three bogs on Vancouver Island, Canada. *Journal of Ecology,* 38: 270-276.

HANSEN, H. P., and ALLISON, I. S.
1942   A pollen study of a fossil peat deposit on the Oregon coast. *Northwest Science,* 16: 86-92.

HANSEN, H. P., and MACKIN, H.
1949   A pre-Wisconsin forest succession in the Puget Lowland, Washington. *American Journal of Science,* 247: 833-855.

HARE, F. K., and HAY, J. E.
1974   The climate of Canada and Alaska. *In:* Bryson, R. A., and Hare, F. K. (eds.), *Climates of North America.* Elsevier, Amsterdam, pp. 49-192.

HEBDA, R. J.
1977   The paleoecology of a raised bog and associated deltaic sediments of the Fraser River delta. Ph. D. thesis, University of British Columbia, Vancouver, 201 pp.

1983   Late-glacial and postglacial vegetation history at Bear Cove bog, northeast Vancouver Island, British Columbia. *Canadian Journal of Earth Sciences,* 61: 3172-3192.

1984   Postglacial vegetation history of Brooks Peninsula, Vancouver Island, British Columbia. *Abstracts Sixth International Palynological Conference,* 1984: 62.

HEBDA, R. J., HICOCK, S. R., MILLER, R. F., and ARM-
STRONG, J. E.
  1983   Paleoecology of Mid-Wisconsin sediments from Lynn
         Canyon, Fraser Lowland, British Columbia. *Geological
         Association of Canada Program with Abstracts,* 8: A31.
HEUSSER, C. J.
  1952   Pollen profiles from southeastern Alaska. *Ecological
         Monographs,* 22: 331-352.
  1954   Palynology of the Taku Glacier snow cover, Alaska and
         its significance in the determination of glacier regimen.
         American Journal of Science, 252: 291-308.
  1955   Pollen profiles from the Queen Charlotte Islands, British
         Columbia. *Canadian Journal of Botany,* 33: 429-449.
  1960   Late-Pleistocene environments of North Pacific North
         America. *American Geographical Society Special Publi-
         cation,* 35, 308 pp.
  1965   A Pleistocene phytogeographical sketch of the Pacific
         Northwest and Alaska. *In:* Wright, H. E., Jr., and Frey,
         D. G. (eds.), *The Quaternary of the United States.*
         Princeton University Press, Princeton, pp. 469-483.
  1969   Modern pollen spectra from the Olympic Peninsula,
         Washington. *Bulletin of the Torrey Botanical Club,* 96:
         407-417.
  1972   Palynology and phytogeographical significance of a late-
         Pleistocene refugium near Kalaloch, Washington.
         Quaternary Research, 2: 189-201.
  1973a Modern pollen spectra from Mount Rainier, Washing-
         ton. *Northwest Science,* 47: 1-8.
  1973b Postglacial vegetation on Umnak Island, Aleutian
         Islands, Alaska. *Review of Palaeobotany and Palynol-
         ogy,* 15: 277-285.
  1973c Environmental sequence following the Fraser advance of
         the Juan de Fuca Lobe, Washington. *Quaternary
         Research,* 3: 284-306.
  1974   Quaternary vegetation, climate, and glaciation of the
         Hoh River Valley, Washington. *Bulletin of the Geologi-
         cal Society of America,* 85: 1547-1560.
  1977   Quaternary palynology of the Pacific slope of Washing-
         ton. *Quaternary Research,* 8: 282-306.
  1978a Modern pollen rain in the Puget Lowland of Washing-
         ton. *Bulletin of the Torrey Botanical Club,* 105: 296-305.
  1978b Modern pollen rain of Washington. *Canadian Journal of
         Botany,* 56: 1510-1517.
  1978c Modern pollen spectra from western Oregon. *Bulletin of
         the Torrey Botanical Club,* 105: 14-17.
  1978d Postglacial vegetation on Adak Island, Aleutain Islands,
         Alaska. *Bulletin of the Torrey Botanical Club,* 105: 18-
         23.
  1978e Palynology of Quaternary deposits of the lower
         Bogachiel River area, Olympic Peninsula, Washington.
         *Canadian Journal of Earth Sciences,*15: 1568-1578.
  1982   Quaternary vegetation and environmental record of the
         western Olympic Peninsula. *Seventh Biennial Confer-
         ence of the American Quaternary Association, Guide for
         Day 2 of Field Trip G,* 23 pp.
  1983a Vegetational history of the northwestern United States
         including Alaska. *In:* Porter, S. C. (ed.), *Late-Quaternary
         Environments of the United States. Volume 2. The Late
         Pleistocene.* University of Minnesota Press, Minneapo-
         lis, pp. 239-258.
  1983b Pollen diagrams from the Shumagin Islands and
         adjacent Alaska Peninsula, southwestern Alaska.
         Boreas, 12: 279-295.

  1983c Holocene vegetation history of the Prince William
         Sound region, south-central Alaska. *Quaternary
         Research,* 19: 337-355.
HEUSSER, C. J., and HEUSSER, L. E.
  1980   Sequence of pumiceous tephra layers and the late
         Quaternary environmental record near Mount St.
         Helens. *Science,* 210: 1007-1009.
  1981   Palynology and paleotemperature analysis of the
         Whidbey Formation, Puget Lowland, Washington.
         Canadian Journal of Earth Sciences, 18: 136-149.
HEUSSER, C. J., HEUSSER, L. E., and STREETER, S. S.
  1980   Quaternary temperatures and precipitation for the
         north-west coast of North America. *Nature,* 286: 702-
         704.
HEUSSER, L. E.
  1983a Contemporary pollen distribution in coastal California
         and Oregon. *Palynology,* 7: 19-42.
  1983b Palynology and paleoecology of postglacial sediments in
         an anoxic basin, Saanich Inlet, British Columbia.
         Canadian Journal of Earth Sciences, 20: 873-885.
HEUSSER, L. E., and BALSAM, W. L.
  1977   Pollen distribution in the northeast Pacific Ocean.
         Quaternary Research, 7: 45-62.
HITCHCOCK, C. L., and CRONQUIST, A.
  1973   *Flora of the Pacific Northwest.* University of Washington
         Press, Seattle, 730 pp.
HULTEN, E.
  1960   *Flora of the Aleutian Islands.* Cramer, Weinheim, 376
         pp.
  1968   *Flora of Alaska and Neighboring Territories.* Stanford
         University Press, Stanford, 1008 pp.
KRAJINA, V. J.
  1965   Ecology of western North America. Volume 1. *Depart-
         ment of Botany, University of British Columbia, Vancou-
         ver,* 112 pp.
  1969   Ecology of western North America. Volume 2. *Depart-
         ment of Botany, University of British Columbia, Vancou-
         ver,* 147 pp.
KUCHLER, A. W.
  1964   Potential natural vegetation of the conterminous United
         States. *American Geographical Society Special Publica-
         tion,* 36, map and manual, 39 pp.
LAWRENCE, D. B.
  1958   Glaciers and vegetation in southeastern Alaska. *Ameri-
         can Scientist,* 46: 89-122.
LEOPOLD, E. B., and CRANDELL, D. R.
  1957   Pre-Wisconsin interglacial pollen spectra from Washing-
         ton State, USA. *Veröffentlichung Geobotanisches Institut
         Rübel in Zürich,* 34: 76-79.
LEOPOLD, E. B., NICKMANN, R., HEDGES, J. I., and ERTEL,
J. R.
  1982   Pollen and lignin records of late Quaternary vegetation,
         Lake Washington. *Science,* 218: 1305-1307.
LUPTON, C. T.
  1914   Oil and gas in the western part of the Olympic Penin-
         sula, Washington. *United States Geological Survey
         Bulletin,* 581: 23-81.
MACK, R. N.
  1971   Pollen size variation in some western North American
         pines as related to fossil pollen identification. *Northwest
         Science,* 45: 257-269.

MANN, D. H.
1983   The Quaternary history of the Lituya glacial refugium, Alaska. Ph.D. Thesis, University of Washington, Seattle, 268 pp.

MATHEWES, R. W.
1973   A palynological study of postglacial vegetation changes in the University Research Forest, southwestern British Columbia. *Canadian Journal of Botany,* 51: 2085-2103.
1978   Pollen morphology of some western Canadian *Myriophyllum* species in relation to taxonomy. *Canadian Journal of Botany,* 56: 1372-1380.
1979a Pollen morphology of Pacific Northwestern *Polemonium* species in relation to paleoecology and taxonomy. *Canadian Journal of Botany,* 57: 2428-2442.
1979b A paleoecological analysis of Quadra Sand at Point Grey, British Columbia, based on indicator pollen. *Canadian Journal of Earth Sciences,* 16: 847-858.

MATHEWES, R. W., and CLAGUE, J. J.
1982   Stratigraphic relationships and paleoecology of a late-glacial peat bed from the Queen Charlotte Islands, British Columbia. *Canadian Journal of Earth Sciences,* 19: 1185-1195.

MATHEWES, R. W., and HEUSSER, L. E.
1981   A 12,000 year palynological record of temperature and precipitation trends in southwestern British Columbia. *Canadian Journal of Botany,* 59: 707-710.

MC KENZIE, G. D.
1970   Glacial geology of Adams Inlet, southeastern Alaska. *Institute of Polar Studies Report,* 25: 1-121.

MEIER, M., and POST, A. S.
1962   Recent variations in mass net budgets of glaciers in western North America. *Union Géodésique et Géophysique Internationale. Association Internationale d'Hydrologie Scientifique. Commission des Neiges et Glaces. Colloque d'Obergurgl,* 63-77.

MILLER, R. D.
1973   Gastineau Channel Formation, a composite glaciomarine deposit near Juneau, Alaska. *United States Geological Survey Bulletin,* 1394-C, 20 pp.

MOE, D.
1974   Identification key for trilete microspores of Fennoscandian Pteridophyta. *Grana,* 14: 132-142.

MUNZ, P. A.
1968   *A California Flora.* University of California Press, Berkeley, 1,681 pp.

NEILAND, B. J.
1971   The forest-bog complex of southeast Alaska. *Vegetatio,* 22: 1-63. pp.

OSVALD, H.
1933   Vegetation of the Pacific coast bogs of North America. *Acta Phytogeographica Suecica,* 5, 32 pp.
1936   Stratigraphy and pollen flora of some bogs of the north Pacific Coast of America. *Bulletin de la Societe Botanique Suisse,* 46: 489-504.
1970   Vegetation and stratigraphy of peatlands in North America. *Nova Acta Regiae Societatis Scientiarum Upsaliensis,* 1, 96 pp.

PETEET, D. M.
1983   Holocene vegetational history of the Malaspina Glacier district, Alaska. Ph.D. Thesis, New York University, New York, 170 pp.

PETERSEN, K. L., MEHRINGER, P. J., and GUSTAFSON, C. E.
1983   Late-glacial vegetation and climate at the Manis Mastodon Site, Olympic Peninsula, Washington. *Quaternary Research,* 20: 215-231.

PORTER, S. C., PIERCE, K. L., and HAMILTON, T. D.
1983   Late Wisconsin mountain glaciation in the western United States. *In:* Porter, S. C. (ed.), *Late-Quaternary Environments of the United States. Volume 2. The Late Pleistocene.* University of Minnesota Press, Minneapolis, pp. 71-111.

RIGG, G. B.
1925   Some sphagnum bogs of the north Pacific coast of North America. *Ecology,* 6: 259-278.
1958   Peat resources of Washington. *State of Washington Division of Mines and Resources Bulletin,* 44, 272 pp.

RIGG, G. B., and RICHARDSON, C. T.
1938   Profiles of some sphagnum bogs of the Pacific coast of North America. *Ecology,* 19: 408-434.

SIGAFOOS, R. S.
1958   Vegetation of northwestern North America, as an aid in interpretation of geologic data. *United States Geological Survey Bulletin,* 1061-E: 165-183.

SIRKIN, L. A., and TUTHILL, S.
1969   Late-Pleistocene palynology and stratigraphy of Controller Bay region, Gulf of Alaska. *Études sur le Quaternaire dans le Monde. VIII Congrès International Quaternary Association, Paris:* 197-208.

SUDWORTH, G. B.
1908   *Forest Trees of the Pacific Slope.* United States Government Printing Office, Washington, 441 pp.

TAYLOR, T. N., and MAC BRYDE, B.
1977   *Vascular Plants of British Columbia, A Descriptive Resource Inventory.* University of British Columbia Press, Vancouver, 754 pp.

TSUKADA, M., and SUGITA, S.
1983   Late Quaternary dynamics of pollen influx at Mineral Lake, Washington. *Botanical Magazine of Tokyo,* 95: 401-418.

TSUKADA, M., SUGITA, S., and HIBBERT, D. M.
1981   Paleoecology in the Pacific Northwest. I. Late Quaternary vegetation and climate. *Verhandlungen International Verein Limnologie,* 21: 730-737.

VIERECK, L. A., and LITTLE, E. L., JR.
1972   Alaska trees and shrubs. *United States Department of Agriculture Forest Service Agriculture Handbook,* 410, 265 pp.

WAITT, R. B., JR., and THORSON, R. M.
1983   The Cordilleran Ice Sheet in Washington, Idaho, and Montana. *In:* Porter, S. C. (ed.), *Late-Quaternary Environments of the United States. Volume 1. The Late Pleistocene.* University of Minnesota Press, Minneapolis, pp. 53-70.

WARING, R. H., and FRANKLIN, J. F.
1979   Evergreen coniferous forests of the Pacific Northwest. *Science,* 204: 1380-1386.

WARNER, B. G.
1984   Late Quaternary paleoecology of eastern Graham Island, Queen Charlotte Islands, British Columbia, Canada. Ph. D. Thesis, Simon Fraser University, Burnaby, 190 pp.

WARNER, B. G., CLAGUE, J. J., and MATHEWES, R. W.
1984   Geology and paleoecology of a Mid-Wisconsin peat from the Queen Charlotte Islands, British Columbia, Canada. *Quaternary Research,* 21: 337-350.

WARNER, B. G., MATHEWES, R. W., and CLAGUE, J. J.
  1982   Ice-free conditions on the Queen Charlotte Islands,
          British Columbia, at the height of late Wisconsin glacia-
          tion. *Science,* 218: 675-677.

WILLIS, B.
  1898   Drift phenomena of Puget Sound. *Bulletin of the Geolog-
          ical Society of America,* 9: 111-162.

LATE-QUATERNARY POLLEN RECORDS FROM THE INTERIOR PACIFIC NORTHWEST AND NORTHERN GREAT BASIN OF THE UNITED STATES

PETER J. MEHRINGER, JR.
Departments of Anthropology and Geology
Washington State University
Pullman, Washington 99164

Abstract

Nearly 50 years ago Henry P. Hansen began fossil pollen studies of lakes and bogs in the interior Pacific Northwest and northern Great Basin. He foresaw the potential of palynology for tracing vegetation history as influenced by climatic change and the natural catastrophes that characterize this region.

Although most pollen profiles date from within the last 12,500 radiocarbon years, a few sequences detail Late Wisconsin spread of cold steppe vegetation throughout the region and montane conifers in the Great Basin. With shrinking lakes, wasting glaciers and catastrophic flooding, pollen profiles show initial success of pioneer species as early-Holocene forest expanded toward the north and to higher elevations. To the south, relative decline of conifer pollen, and increase of grass and sagebrush pollen signal retreat of montane trees and expanding warm steppe.

Before 9,000 yrs. B.P. pollen records reflect Holocene vegetation and rising temperatures over the entire region. By 7,000 yrs. B.P. shadscale and sagebrush communities had expanded at the expense of grass, and conifers lost ground to grass and sagebrush. By 5,400 yrs. B.P. this trend had slowed; by 4,000 yrs. B.P. it had reversed with the return of climatic patterns resulting in apparently more effective moisture. The moist maritime forests of northern Idaho and adjacent states may date from less than 2,500 yrs. B.P. Charcoal/pollen ratios mark changing Holocene fire regimes.

Specific details, not usually available from study of lake and bog deposits, come from fossil pollen of caves, ancient woodrat middens, volcanic ashes and archaeological sites. These studies have revealed season and duration of ashfalls, details of human diet, and progress of northward expanding juniper-pinyon woodland.

INTRODUCTION

Herein I present a bibliography and brief review of pollen records from the interior Pacific Northwest and northern Great Basin beginning with the papers of pioneering palynologist Henry P. Hansen (Fig. 1, Table 1). Hansen investigated Holocene vegetation history from semi-arid steppe (1941a), to the high Cascades and Glacier National Park (1946a, 1948), to the moist maritime forests of northern Idaho (1939a). He used pollen, especially in association with volcanic ashes, for dating (Hansen, 1944) and employed modern analogues to interpret fossil assemblages (Hansen, 1949). He recognized the importance of the details of vegetation history as related to climate, to eruptions of Cascade volcanos (Hansen, 1942a, 1942b), and to fire in forest and steppe (Hansen, 1939b, 1943b). Before the advent of radiocarbon dating, and despite tenuous conclusions drawn from specific identification of conifer pollen (Hansen, 1947a; Mack, 1971), he illustrated the instability of Holocene vegetation in response to local conditions and regional climatic trends.

Hansen broke ground as the first palynologist to join interdisciplinary paleoecological teams in western North America (Merriam, 1941). With colleagues he illustrated the importance of climatic fluctuations, and corresponding histories of lakes and marshes, to changing human populations (Hansen, 1942c). Hansen's (1946b, 1947a, 1951) climatic interpretations and alliances with archaeologists lent credence to Ernst Antevs' (1955) geographically more encompassing scheme—a three part post-glacial climatic sequence, with extreme mid-Holocene drought—that, with few exceptions, proved gospel to two generations of western North American archaeologists.

References to palynological investigations (Fig. 2, Table 2), reflect the interests and contributions of palynologists who have studied this region since the time of Hansen and are briefly reviewed as they apply to:

1. general vegetation history reconstructed from pollen of late-Pleistocene, late-glacial and

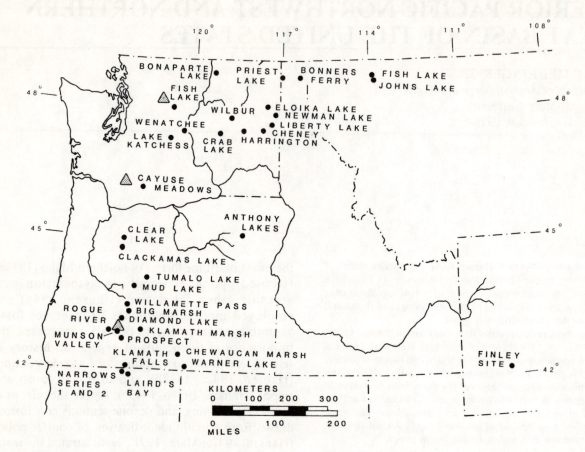

Figure 1. Late-Quaternary pollen localities investigated by H. P. Hansen (Table 1).

Holocene sedimentary sequences, and from pollen analysis of ancient woodrat middens;

2. modern pollen rain and historic pollen records;
3. pollen analysis of archaeological sites including human coprolites;
4. pollen content of tephra as an indicator of season and duration of ashfalls;
5. history of fire as reflected in pollen and microscopic charcoal.

Discussion of these topics for adjacent and overlapping areas are found in this volume and in general reviews by Baker (1983), Heusser (1983), Mehringer (in press), and Spaulding, Leopold and Van Devender (1983). My object is not to provide a complete or critical review of vegetation history, but rather to use examples that generalize findings and approaches as a guide to the literature of Quaternary palynology of the interior northwest of the U.S.

VEGETATION HISTORY

For the last 1,000,000 years and more the earth was held in the grip of glacial climates. Vegetation

between the Sierra Nevada-Cascades and the Rocky Mountains (Fig. 3) responded to varying intensity of numerous glacial-interglacial cycles in which brief warmth punctuated long, cold intervals. Each cycle brought the growth and wasting of glaciers (Waitt and Thorson, 1983; Porter, Pierce and Hamilton, 1983), rise and fall of lakes (Smith and Street-Perrott, 1983; Currey, Atwood and Mabey, 1983), frozen ground (Péwé, 1983), and catastrophic floods (Bretz, 1969; Baker and Nummedal, 1978). Plants responded as vagaries of climate, dispersal potential, competition, selection, soils, topography, volcanic eruptions, fire, man and chance dictated. A hazy outline of these responses has been traced through pictures painted with pollen extracted from yesterday's mud. Sketchy as the images may be, they prove the dynamic nature of this area's vegetation.

Within the shadow of the Cascade-Sierra Nevada axis lies a land of elevational and seasonal contrasts. In the northern Great Basin scant rainfall and searing summer heat produce barren salt deserts at the foot of mountains, ringed by woodlands and capped by perennial snow (Billings, 1951; Davis, 1981). On the Columbia Plateau dry gray sagebrush and rolling green grasslands (Daubenmire, 1970, 1975a, 1982)

Table 1.
References to sites investigated by H. P. Hansen (Figure 1).

SITE	REFERENCE	SITE	REFERENCE
CALIFORNIA		Warner Lake	1947b
Laird's Bay	1942c	Willamette Pass	1942b
Narrows Series 1 and 2	1942c	**MONTANA**	
IDAHO		Fish Lake	1948
Bonners Ferry	1943b	Johns Lake	1948
Priest Lake (Hagar Pond)	1939a	**WASHINGTON** (eastern)	
OREGON (eastern)		Bonaparte Lake (Bonaparte Meadows)	1940
Anthony Lakes	1943a	Cayuse Meadows	1947a
Big Marsh	1942b	Crab Lake	1941a
Clackamas Lake	1946a	Eloika Lake	1944
Clear Lake	1947a	Fish Lake (near Cheney)	1943c
Chewaucan Marsh	1947b	Fish Lake (near Wenatchee)	1941b
Diamond Lake	1946a	Harrington Bog	1944
Klamath Falls	1942c	Lake Katchess	1947a
Klamath Marsh	1947b	Liberty Lake	1944
Mud Lake	1942b	Newman Lake	1939b
Munson Valley	1942b	Wenatchee	1939c
Prospect	1946a	Wilbur (Creston Bog)	1944
Rogue River	1946a	**WYOMING**	
Tumalo Lake	1942a	Finley Site	1951

merge beyond the eastern horizon with a seemingly endless blanket of dark green forests penetrated by jagged snow-white peaks (Arno, 1979; Daubenmire and Daubenmire, 1968; Steele *et al.,* 1981, 1983). Although the northwestern U.S. is noted for its remarkably diverse forests, vast expanses of saltbush, sagebrush and grass communities predominate, even in the semi-arid interiors of Oregon and Washington (Franklin and Dyrness, 1973).

That steppe characterized considerably larger areas during most of the late-Quaternary is perhaps the most significant paleoclimatic signal—cold continental conditions—to come from pollen records of this region. Rapid expansion of steppe at the expense of forest or woodland, or increasing importance of warmth-requiring conifers during the mid-Holocene are consistently recognized events. The full-glacial fate and Holocene history of northern Idaho's moist maritime hemlock and cedar forests—unrecorded in fossil pollen sequences before 2,500 yrs. B.P.—remain a puzzle.

Pollen records from present forests east of the Cascades consistently reveal that the first invading conifers flourished on what had been glacier- or lake-covered terrain, flood tracts, or frozen ground supporting cold steppe during the last full-glacial episode. In some cases these conifers persisted; in others they gave way to steppe shrubs that remain to the present day or, after a few thousand years and changing climate, were in turn overrun by forest that burned repeatedly but held its ground. Finally—with

the coming of cows and agriculture—alien weeds and wheat, and different fire regimes forever changed the landscape. The fossil pollen localities most important in deciphering this story and that from east of the Sierra Nevada to the south will be reviewed next according to their age—Pleistocene, late-glacial, and Holocene.

Pleistocene

The few localities containing late-Quaternary pollen records older than 13,000 years or so share two characteristics:

1. Interstadial assemblages, though sometimes similar in certain aspects, usually bear only superficial resemblance to Holocene pollen spectra from the same areas.
2. Pollen assemblages dating from Wisconsin-age stadial periods may show increased arboreal percentages in the northern Great Basin. Yet, even there, the abundance of sagebrush *(Artemisia)* pollen highlights its ice-age importance. In sites to the north and at higher elevations, sagebrush pollen dominates overwhelmingly.

South of continental ice and below the mountain glaciers cold steppe typified much of the northwestern interior of the U.S. Details of variation in vegetation of the last interglacial and interstadials is best seen in sites in the Yellowstone National Park area

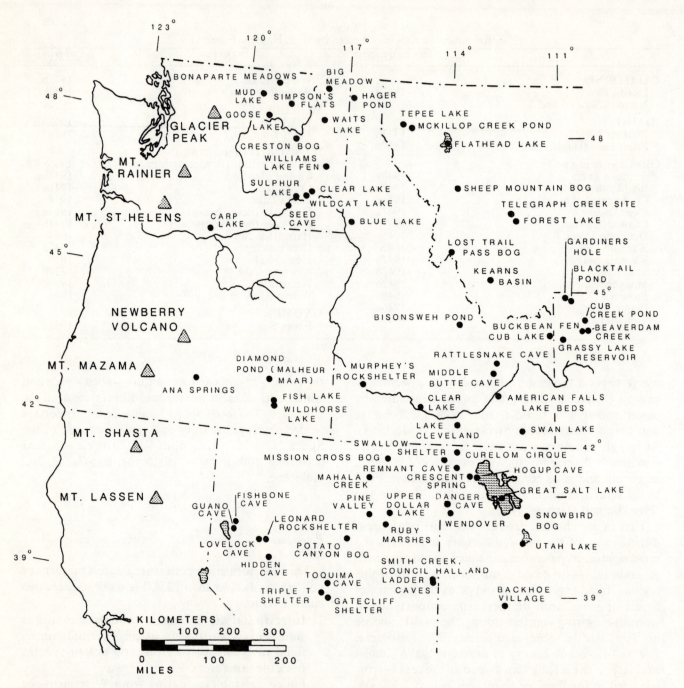

Figure 2. Sites of late-Quaternary pollen studies since the time of H. P. Hansen (Table 2).

(Grassy Lake Reservoir and Beaverdam Creek), Wyoming (Baker and Waln, this volume).

Carp Lake, southwestern Columbia Basin, Washington, holds sediments spanning parts of the past 33,000 years. Sagebrush and grass pollen dominate this record. Slightly more conifer pollen before 23,000 yrs. B.P. suggests temperate steppe, whereas full-glacial temperatures too cold for trees apparently produced periglacial steppe or tundra. Spruce pollen is conspicuous by its continued presence in small percentages. Palynological investigations of pluvial lake deposits at Ana Springs, Oregon, promise to provide a sequence chronologically bridging those from Carp Lake and Lake Bonneville.

Pollen profiles from the Lake Bonneville Basin, Utah (Crescent Spring, Great Salt Lake and Wendover), show the period of deepest pluvial lake dominated by conifer pollen, whereas interstadial-age sediment contains relatively more non-arboreal

Table 2.
Locations and references to sites of Quaternary pollen studies (Figure 2).

SITE	REFERENCE
IDAHO	
American Falls Lake Beds (~1344 m)	Bright, 1982
Bisonsweh Pond (2220 m)	Chatters, 1982
Blue Lake, near Lewiston (1035 m)	Mehringer, unpubl.; Smith, 1983
Clear Lake, near Buhl (915 m)	Davis, pers. commun.
Cub Lake (1840 m)	Baker, 1983
Grays Lake Marsh*	Beiswenger, pers. commun.
Hager Pond (860 m)	Mack, Rutter, Bryant, and Valastro, 1978a
Lake Cleveland (2519 m)	Davis, 1981, 1984b
Middle Butte Cave (1593 m)	Davis and Bright, 1983; Davis, 1984b
Murphey's Rockshelter (~808 m)	Henry, 1984
Rattlesnake Cave (1596 m)	Davis, 1981; Bright and Davis, 1982
Swan Lake (1450 m)	Bright, 1966
MONTANA (western)	
Flathead Lake (880 m)	Onken, 1984
Forest Lake (1895 m)	Brant, 1980
Kearns Basin (2200 m)	Mehringer, unpubl.
Lost Trail Pass Bog (2152 m)	Mehringer, Arno, and Petersen 1977; Mehringer, Blinman, and Petersen 1977
McKillop Creek Pond (920 m)	Mack, Rutter, and Valastro, 1983
Sheep Mountain Bog (1920 m)	Hemphill, 1983; Mehringer, Sheppard, and Foit 1984; Mehringer, unpubl., herein
Telegraph Creek Site (2130 m)	Brant, 1980, 1982
Tepee Lake (1270 m)	Mack, Rutter, and Valastro, 1983
NEVADA (northern)	
Council Hall Cave (2040 m)	Thompson, 1984
Fishbone Cave (1237 m)	Sears and Roosma, 1961
Gatecliff Shelter (2319 m)	Thompson and Kautz, 1983
Guano Cave (1237 m)	Sears and Roosma, 1961
Hidden Cave (1251 m)	Wigand and Mehringer, 1985
Ladder Cave (~2040 m)	Thompson, 1984
Leonard Rockshelter (1273 m)	Byrne, Busby, and Heizer, 1979
Lovelock Cave (1372 m)	Napton and Kelso, 1969
Mahala Creek (1950 m)	Madsen, unpubl.
Mission Cross Bog (2424 m)	Thompson, 1984
Pine Valley (1555 m)	Thompson, 1984
Potato Canyon Bog (1725 m)	Madsen, unpubl.
Ruby Marshes (1818 m)	Thompson, 1984
Smith Creek Cave (~2040 m)	Thompson, 1984, 1979
Toquima Cave (2432 m)	Kautz and Thomas, 1972
Triple T Shelter (2024 m)	Thompson and Kautz, 1983
Upper Dollar Lake (2990 m)	Thompson, 1984
OREGON (eastern)	
Ana Springs (1250 m)	Pippin, unpubl.
Fish Lake (2250 m)	Mehringer, unpubl., herein; Verosub and Mehringer, 1984
Diamond Pond, Malheur Maar (1265 m)	Mehringer and Wigand, 1985b; Wigand, 1985
Wildhorse Lake (2565 m)	Mehringer, unpubl., herein
UTAH (western)	
Backhoe Village (1645 m)	Madsen and Lindsay, 1977
Crescent Spring (~1300 m)	Mehringer, 1977, herein
Curelom Cirque (2835 m)	Mehringer, Nash, and Fuller, 1971; Mehringer, 1977, unpubl.
Danger Cave (1318 m)	Kelso, 1970
Great Salt Lake, Bird Island (1280 m)	Mehringer, herein
Great Salt Lake, South Arm (1280 m)	Spencer *et al.*, 1984
Hogup Cave (1433 m)	Kelso, 1970
Remnant Cave (1477 m)	Hull, 1976
Snowbird Bog (2470 m)	Madsen and Currey, 1979
Swallow Shelter (1768 m)	Dalley, 1976
Utah Lake (1368 m)	Bushman, 1980
Wendover (~1295 m)	Martin and Mehringer, 1965

* Southeastern Idaho; not shown on Figure 2.

Table 2. (continued)
Locations and references to sites of Quaternary pollen studies (Figure 2).

SITE	REFERENCE
WASHINGTON (eastern)	
Big Meadow (1040 m)	Mack, Rutter, Bryant, and Valastro, 1978b
Bonaparte Meadows (1021 m)	Mack, Rutter, and Valastro, 1979
Carp Lake (714 m)	Barnosky, 1984, 1985
Clear Lake (396 m)	Bartholomew, 1982; Mehringer, unpubl.
Creston Bog (Wilbur) (~710 m)	Mack, Bryant and Fryxell, 1976
Goose Lake (373 m)	Nickmann, pers. commun.; Nickmann and Leopold, 1984
Seed Cave (~137 m)	Thompson, 1985
Mud Lake (655 m)	Mack, Rutter, and Valastro, 1979
Simpson's Flats (535 m)	Mack, Rutter, and Valastro, 1978
Sulphur Lake (210 m)	Mehringer, unpubl.
Waits Lake (540 m)	Mack, Rutter, Valastro, and Bryant, 1978
Wildcat Lake (342 m)	Blinman 1978; Blinman, Mehringer and Sheppard, 1979; David, Kolva and Mehringer, 1977; Mehringer, unpubl., herein
Williams Lake Fen (635 m)	Nickmann, 1979
WYOMING (western)	
Beaverdam Creek (2466 m)	Baker and Richmond, 1978
Blacktail Pond (2018 m)	Gennett, 1977
Buckbean Fen (2367 m)	Baker, 1970, 1976
Cub Creek Pond (2485 m)	Waddington and Wright, 1974
Grassy Lake Reservoir (2200 m)	Baker and Richmond, 1978
Gardiners Hole (2215 m)	Baker, 1983

pollen, especially sagebrush. Spruce pollen is important in deposits of full-glacial age (Fig. 4).

A pollen sequence from Council Hall Cave in the east-central Great Basin spans 40,000 years. Pine pollen dominates Middle Wisconsin deposits and remains the most abundant type through the Late Wisconsin. Minor but clear peaks in spruce and then fir pollen preceded the decline of pine and increase in Cupressaceae pollen that identify Holocene deposits. Sagebrush pollen is common throughout the cave fill.

In the Ruby Valley, Nevada, fossil algae and pollen of aquatic plants indicate that a shallow brackish lake, dating from about 40,000 to 23,000 yrs. B.P., became deeper and fresher then shallower again by 10,000 yrs. B.P. However, unlike the Bonneville Basin, this change in lake size was poorly reflected in pollen of terrestrial species including conifers; sagebrush pollen dominated the entire 30,000-year sequence. Similar abundance of sagebrush pollen at times between about 70,000 and 10,000 yrs. B.P. (American Falls Lake Beds, Middle Butte Cave and Grays Lake Marsh) indicates persistence of cold steppe on the Snake River Plain as well.

Late-glacial

Rapid wasting of glaciers, shrinking of vast lakes and final catastrophic flooding attended the onset of post-glacial conditions. Diverse pollen spectra of this age retain aspects of full-glacial vegetation and reflect initial successes of pioneer invaders on newly available terrains. They share several characteristics separating them from later Holocene samples from the same sites, including:

1. an initial treeless interlude indicated by importance of non-arboreal pollen dominated by *Artemisia* and often accompanied by abundant grass pollen;

2. common occurrence of pollen of *Shepherdia canadensis, Juniperus* (probably *J. communis*) and small percentages of *Picea* sometimes accompanied or followed by *Abies;*

3. combinations of pollen types such as *Rumex-Oxyria, Bistorta, Polemonium, Eriogonum,* and *Koenegia;*

4. unusual abundance of *Selaginella densa*-type along with other spores such as *Selaginella selaginoides, Botrychium* and *Lycopodium annotinum.*

The alpine and subalpine character of these assemblages is similar over much of the interior Northwest, including mountains of the northern Great Basin and adjacent Snake River Plain. Nonetheless, regions and individual sites may show considerable variation as illustrated by the late-glacial distribution of *Betula* pollen.

Birch may or may not be locally important in the late-glacial of Yellowstone Park as indicated by abundant pollen and by macrofossils of *Betula glandulosa* from Buckbean Fen, and by its insignificant representation at Blacktail Pond. Birch pollen is rare at Lost Trail Pass, Montana, whereas the Kearns Basin, Forest Lake and Telegraph Creek sites, Montana, all show a distinct zone of abundant birch pollen directly above Glacier Peak tephra. Likewise, in northern Washington a late-glacial or early-Holocene birch pollen episode occurs in sediments already dominated by pine pollen (Goose Lake, Mud Lake and Simpson's Flats), but is not apparent in the Waits Lake diagram.

At lower elevations in unglaciated terrain rapid environmental changes may be obvious from pollen records where conifer pollen was important and/or where changing depositional environments *(e.g.,* shrinking lakes) must themselves contribute to varying pollen spectra (Ruby Marshes, Lake Bonneville-Great Salt Lake). Drastic decline in full- and late-glacial conifer pollen at Swan Lake, Idaho (Fig. 5), and Great Salt Lake, Utah, reflects expanding steppe at this elevation (1,580 meters). Increasing

conifer pollen in the Wasatch (Snowbird Bog), Raft River (Curelom Cirque) and Albion (Lake Cleveland) ranges indicates wasting of ice and passing of cold, nearly treeless sagebrush dominated communities as conifers crept upslope or arrived from distant ice-age homes.

At Fish Lake, Steens Mountain, Oregon, sagebrush steppe followed retreating glaciers to 2,300 meters by 12,000 yrs. B.P. where it now persists. There, late-glacial pollen spectra, dominated by sagebrush and grass, are distinguished from those of the Holocene by larger values of juniper pollen (probably *Juniperus communis).*

Holocene

Pollen sequences of the last 10,000 years often have been characterized in three parts—a cool-moist early-Holocene, a xeric middle period (Altithermal), and a

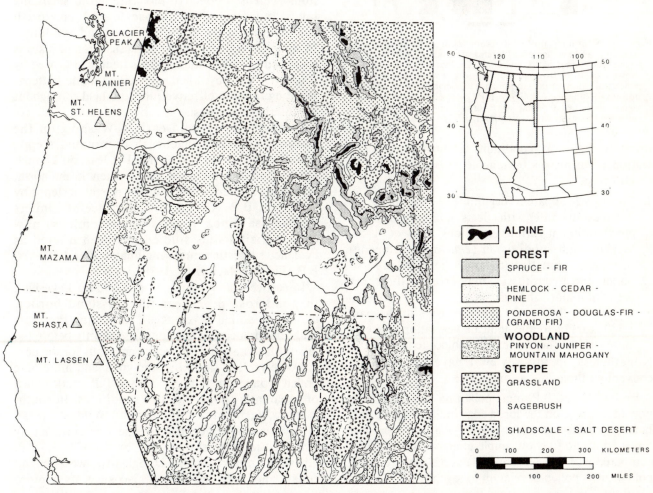

Figure 3. Generalized vegetation map of the interior Pacific Northwest and Northern Great Basin of the U.S. (adapted from Kuchler, 1964).

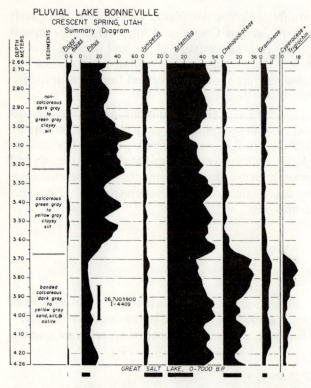

Figure 4. Summary pollen diagram from Lake Bonneville sediments and average post-Mazama ash pollen percentages, of a core from Great Salt Lake, Utah (see Fig. 10). The diagram shows transition from shallow to deeper water and conifer pollen (including spruce) importance that marks the rise of Lake Bonneville about 24,000 yrs. B.P. Note continuously large sagebrush values and that larger juniper pollen values distinguish the Holocene (from Mehringer, 1977).

return to cool-moist conditions. Though this generalization often proves reasonable, it is too simple in several ways:

1. Timing and apparent magnitude of specific vegetational events differ with effects of temperature and precipitation at various elevations.
2. Within each subdivision (*e.g.,* mid-Holocene) punctuated climatic variation, as reflected in fossil pollen and macrofossils, may exceed differences between major subdivisions.
3. At least in the north, pollen records of the past 2,500 years or so may indicate vegetational assemblages unlike those of earlier Holocene times.

Nonetheless, in this review I have followed the convenient three-fold arrangement.

By 10,000 yrs. B.P. most pollen records from sites now in forest or on the forest edge indicate the importance of coniferous trees. Those lying below present forest continue to show ample effective moisture (as compared with the present) in large values of conifer pollen, grass pollen in relation to sagebrush pollen, or

sagebrush in relation to pollen of chenopods, including greasewood *(Sarcobatus)*. At Carp Lake, Washington, warm steppe is in evidence by 10,000 B.P.

Douglas-fir pollen and macrofossils occur at Sheep Mountain Bog, Montana, by 10,000 yrs. B.P. (Fig. 6) and the vegetational events of the Holocene, already set in motion, are apparent over the entire region by 9,000 yrs. B.P. Responses to rising temperature are indicated by decreasing conifer pollen percentages in northern Washington (Goose Lake, Williams Lake Fen, Bonaparte Meadows), declining Haploxylon (primarily whitebark pine) in relation to Diploxylon (primarily lodgepole pine) pollen in western Montana and Wyoming (Sheep Mountain Bog, Lost Trail Pass Bog, Buckbean Fen, Cub Creek Pond), and increasing sagebrush or chenopod pollen in southern Idaho, western Utah and Nevada (Murphey's Rockshelter, Swan Lake, Council Hall Cave).

Although details may vary, by 7,000 yrs. B.P. pollen sequences from this region, without exception, indicate apparent warming. At least at lower elevations decreased effective moisture led to shrinking lakes and expanding shadscale and sagebrush communities at the expense of grassland and forest. In some cases (Ruby Marshes) desiccation of lakes is suggested by pollen studies. This trend had slowed by 5,400 yrs. B.P.; by 4,000 yrs. B.P. climatic patterns resulting in more effective moisture held throughout the interior Northwest.

In north-central and eastern Washington the Columbia Basin's sagebrush dominated steppe (Daubenmire, 1956, 1970) stood at least 50 kilometers beyond its present perimeter. Then, in the north, increasing conifer pollen indicates retreat of steppe by 4,000 yrs. B.P. Arrival or dominance of conifers (ponderosa pine, Douglas-fir or hemlock) now characterizing sites studied has been offered as evidence suggesting modern climates after 2,500 yrs. B.P. (Fig. 7) (Mack *et al.,* 1978a, 1978b).

In east-central Washington at Williams Lake Fen, in ponderosa pine near the grassland border, sagebrush pollen percentages, indicating the most arid mid-Holocene conditions, rise shortly before the fall of Mazama ash then decline gradually for the next 3,000 years. Less sagebrush and grass, and more pine pollen (ponderosa?) after 4,000 yrs. B.P. mark return of apparently cooler, moister conditions. In southeastern Washington (Sulphur and Wildcat lakes), mid-Holocene sagebrush steppe was contracting toward the state's arid core by 4,400 yrs. B.P. Growing importance of Douglas-fir, western red cedar, and hemlock pollen also suggests a more

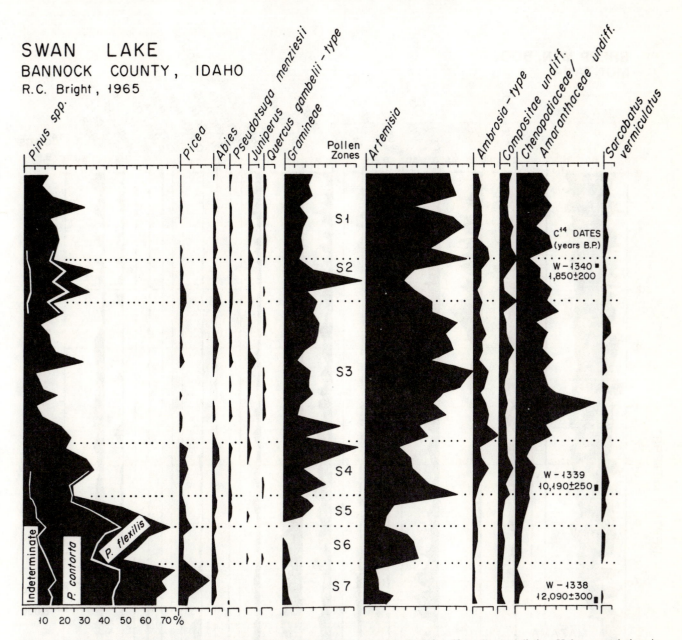

Figure 5. Summary pollen diagram from Swan Lake, Idaho (Bright, 1966), shows decline of conifer pollen and distinct Holocene variations in the relative abundance of grass, sagebrush and conifer pollen (from Wright, 1971).

humid, cooler phase after 4,000 yrs. B.P. at Carp Lake, southwestern Columbia Basin.

Sites in mountains to the east show similar trends. Early Holocene whitebark and lodgepole pine forests of Yellowstone Park gave way to more dense lodgepole forests. Mixed forests of lodgepole pine, spruce, fir, and some Douglas-fir emerged after 5,000 yrs. B.P.

Before the fall of Glacier Peak ash (11,250 yrs. B.P.; Mehringer, Sheppard and Foit, 1984), sub-alpine conifers began to fill higher elevations of the Bitterroot Range near Lost Trail Pass where they grow

today. Then, shortly before the fall of Mazama ash (6,800 yrs. B.P.) (Bacon, 1983), grass and sagebrush pollen percentages along with pollen of the warmth-requiring Douglas-fir, and lodgepole pine replaced the more cold-tolerant whitebark pine. Up-slope migration and persistence of Douglas-fir marks a period of undoubted warming; yet, an unbroken sedimentary record indicates that the small pond at Lost Trail Pass did not go dry.

About 4,000 yrs. B.P. Douglas-fir lost dominance to whitebark and lodgepole pine, perhaps for the first time in over 3,500 years, and retreated to warmer

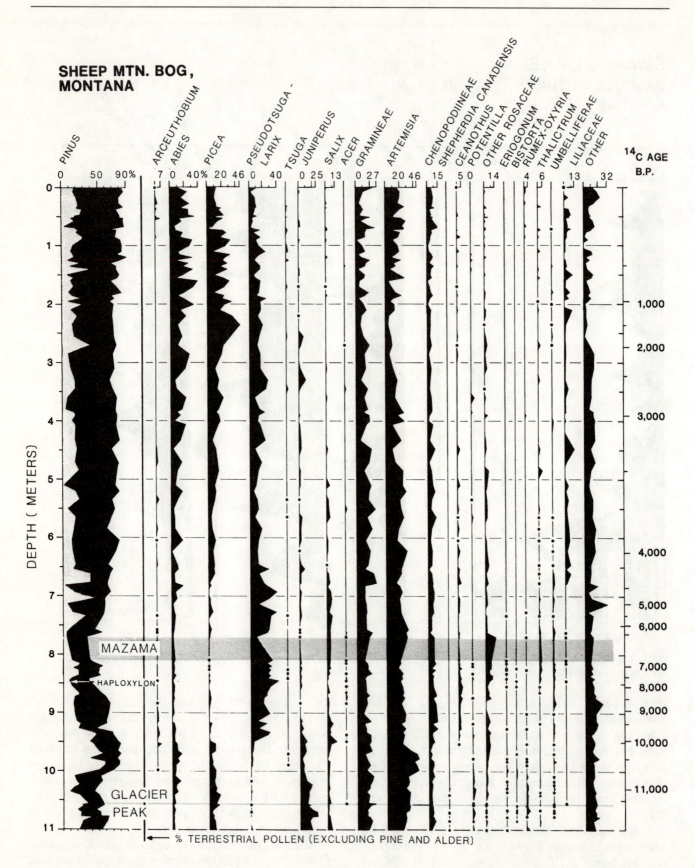

Figure 6.

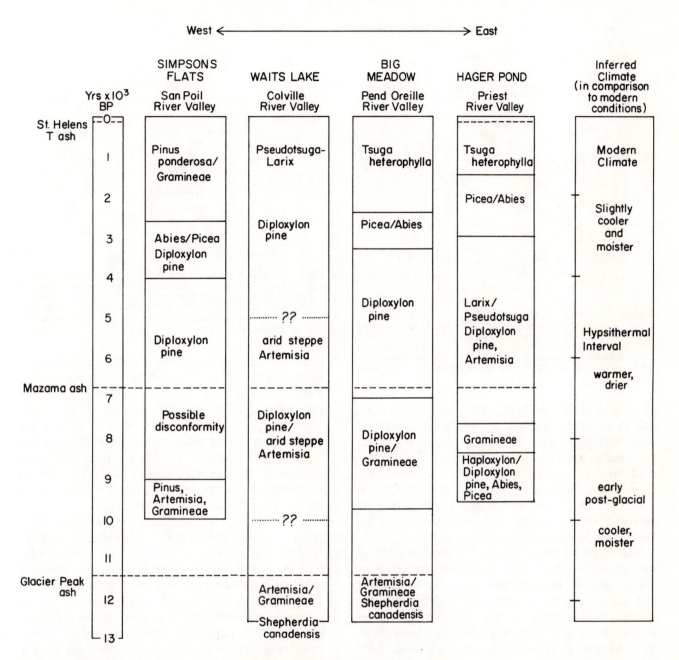

Figure 7. Summary of vegetation dominants and inferred climatic changes for sites in northeastern Washington and northern Idaho since recession of the Late Wisconsin ice sheet (from Mack, Rutter, Valastro, and Bryant, 1978).

down-slope positions. Since that time and especially after 1,750 yrs. B.P., the vegetation was apparently similar to the present with at least one unusually cool (and wet?) climatic episode around 3,600 yrs. B.P. This later event is perhaps the same one noted elsewhere (*e.g.,* Blue Lake near Lewiston, Idaho; Carp and Wildcat lakes, Washington).

Blue Lake, Idaho, is now surrounded by Douglas-fir and ponderosa pine; varying abundances of the pollen and macrofossils of these two trees suggest

Figure 6. Late-glacial and Holocene summary pollen diagram from Sheep Mountain Bog, 18 km northeast of Missoula, Montana. At 1,920 m, this site lies at the upper limit of ponderosa pine *(Pinus ponderosa)* and larch *(Larix occidentalis)* on south-facing slopes, yet neither of these trees is represented by macrofossils. According to distribution of fossil conifer needles and cones, Douglas-fir *(Pseudotsuga menziesii)* has grown near this site since at least 10,000 yrs. B.P., and the Haploxylon and Diploxylon pines are primarily whitebark *(Pinus albicaulis)* and lodgepole *(Pinus contorta)*. In addition to these conifers, mixed forests with subalpine fir *(Abies lasiocarpa)* and Engelmann spruce *(Picea engelmannii)* surround the bog.

TERNARY PLOTS OF FOSSIL POLLEN (0-5700 B.P.)

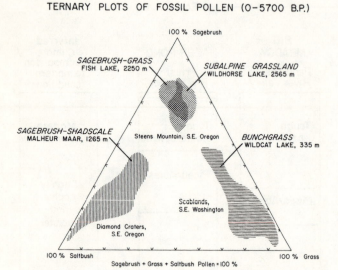

Figure 8. Ternary plots of fossil pollen from four steppe communities in Oregon and Washington. Relative abundances of saltbush (including *Sarcobatus*), sagebrush and grass pollen distinguish steppe associations that most often produce monotonous Holocene pollen profiles (see Fig. 10) dominated by nonarboreal pollen and by pine pollen transported from long-distances.

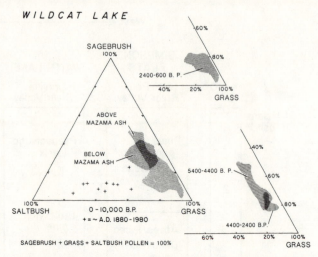

Figure 9. Ternary plots of saltbush, sagebrush and grass pollen from selected periods of the last 10,000 years from Wildcat Lake, Washington. Note that the relatively largest grass values occur within the last 2,400 years and that samples of the past 100 years (+) are distinctive as compared with those of the last 10,000 radiocarbon years.

important variations in effective moisture over the past 4,300 years. A relatively warm-moist interval, dominated by Douglas-fir, from 4,300-4,000 yrs. B.P. preceded a cooler period from 4,000 to 3,000 yrs. B.P. with mixed Douglas-fir lodgepole and ponderosa pine. Then, conifer pollen decreased from 3,000 to 1,700 yrs. B.P. and dry, open ponderosa park persisted until 1,000 yrs. B.P. when vegetation like the present emerged.

The moist hemlock forests *(Tsuga heterophylla)* with grand fir *(Abies grandis)* and western red cedar, *(Thuja plicata)* so characteristic of northern Idaho and adjacent states may be quite young as a major vegetation unit east of the Cascades. At least pollen profiles from this region (Big Meadow, Hager Pond) show no clear indication of its presence before 2,500 years ago. Likewise, grass pollen, in relation to sagebrush and chenopods, reaches its largest values near the western Palouse-Scabland border (Wildcat Lake) in the last 2,400 years (Figs. 8, 9).

Sequences from the northeastern Great Basin reveal details that, like those further north, show vegetational responses to decreasing effective moisture in the mid-Holocene (Snowbird Bog, Swan Lake, Curelom Cirque). Pollen analysis of cores from the above sites and from Cleveland and Great Salt (Fig. 10) lakes, Crescent Spring (Fig. 11), and of stratigraphic sequences from the caves and rockshelters of the northern Great Basin and Snake River Plain all show changes attending the return of greater effective

moisture apparent by 4,000 yrs. B.P. and climatic fluctuations thereafter (Mehringer, in press). For example, the abundant grass pollen from Hogup Cave since about 1,500 yrs. B.P. is unmatched in the preceding 7,000 radiocarbon years. It parallels a relative increase in grass pollen at Swallow Shelter (about 2,000-1,500 yrs. B.P.) where increasing pine, juniper and sagebrush pollen, relative to those of saltbushes and composites, suggests more effective moisture starting about 3,500 yrs. B.P.

A three-part pollen sequence (pine-saltbush-pine) from Leonard Rockshelter, Nevada, is said to indicate the mid-Holocene desiccation of Humboldt Sink. At nearby Hidden Cave, Carson Sink, pine and sagebrush pollen percentages declined between 15,000 and 10,000 yrs. B.P., and sagebrush pollen values decreased further as saltbush pollen increased about 7,000 yrs. B.P. Cat-tails and sedges—represented by both pollen and macrofossils in cave fill and human coprolites—indicate re-expansion of lakes and marshes in Carson Sink about 3,800-3,600 yrs. B.P. in response to greatly increased effective moisture in the northern Great Basin.

Steens Mountain, southeastern Oregon, is unusual in lacking a montane coniferous forest zone (McKenzie, 1982) and is, therefore, ideal for study of changing steppe vegetation. There, as elsewhere at lower elevations in the Great Basin (Crescent Spring, Great Salt Lake), on the Snake River Plain (Middle Butte and Rattlesnake caves, and Murphey's Rockshelter), and Columbia Basin (Clear, Sulphur and Wildcat lakes), varying amounts of pollen produced by grasses, saltbushes and sagebrush are the primary

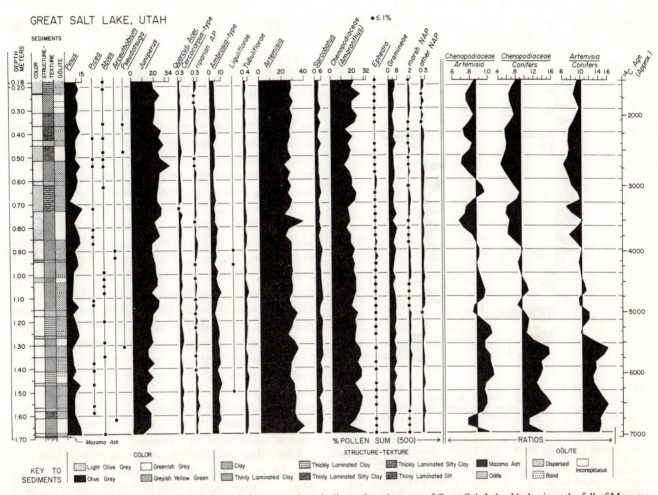

Figure 10. Smoothed pollen ratios, plotted about their mean values, indicate changing area of Great Salt Lake, Utah, since the fall of Mazama ash. Chronology for this core is derived from radiocarbon dates whose apparent ages were corrected to bring the core age at the Mazama ash level to 6,700 yrs. B.P. (Mehringer *et al.*, 1971). Decreasing saltbush as compared to sagebrush pollen indicates loss of salt flat halophytes and shadscale genera to rising water and to competition with sagebrush under the influence of greater effective moisture. The loss of sagebrush to even higher lake stands and/or the general expansion or increasing density of conifers, especially juniper, would account for smaller sagebrush/conifer values. The upper 18 cm was disturbed in coring. This core provided samples for the isotope studies of Grey and Bennett (1972).

clues to relative abundance of major steppe genera as influenced by climate (Fig. 8).

Cores from mountain lakes (Fish and Wildhorse lakes) and from Diamond Pond, on the sagebrush-shadscale desert ecotone, are precisely correlated by six volcanic ashes deposited over the last 7,000 radio-carbon years. Fossil pollen exhibits a general three-part division of Holocene vegetational change with variations in the timing of specific events at each site. These differences resulted from effects of temperature and precipitation at various elevations.

Greater relative abundance of sagebrush pollen in relation to grass pollen indicates relatively low effective moisture at Fish Lake (2,250 meters) between about 8,700 and 4,700 yrs. B.P. The mid-Holocene episode of sagebrush pollen abundance began 1,500 years earlier than the temperature-controlled upward expansion of sagebrush to Wildhorse Lake (2,565

meters) about 7,200 yrs. B.P. Also, it ended at least 1,000 years before grass again assumed dominance at Wildhorse Lake (about 3,800 yrs. B.P.) (Fig. 12), marking the end of this prolonged but variable period of relatively higher temperatures and reduced snowpack.

About the same time, at Diamond Pond (Malheur Maar, 1,265 meters), juniper and grass pollen percentages increase with declining values of cheno-pod pollen in relation to sagebrush pollen. Radiocarbon-dated macrofossils of western juniper *(Juniperus occidentalis)* from woodrat middens in lava tubes and caves of the Diamond Craters flows further suggest relatively mesic conditions at lower elevations. They leave little doubt that sagebrush and juniper grass-lands replaced xeric shadscale vegetation as suggested by the pollen sequence.

CRESCENT SPRING, UTAH

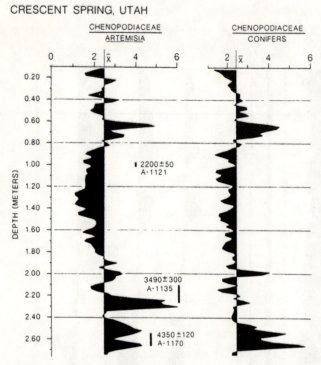

Figure 11. Smoothed pollen ratios, plotted about their mean values, indicate changing relative abundance of salt desert halophytes and shadscale genera in relation to sagebrush pollen and to conifer pollen transported from long-distances. Crescent Spring is a small salt marsh at the playa edge below Hogup Cave, Utah. These sediments uncomformably overlie pluvial-age deposits (see Fig. 4) and date from the time that isostatic rebound slowed and/or water tables rose to allow accumulation rather than erosion of playa-edge marsh sediments. Though interpretation of this record is complicated by rebound and flooding of the Great Salt Desert, larger sagebrush values to 2,000 yrs. B.P. follow mid-Holocene chenopod pollen importance characteristic of this region (see Figs. 5 and 10). At higher elevations and to the north grass replaces sagebrush pollen during this same late-Holocene period (see Figs. 9 and 12).

Pollen and Woodrat Middens

Studies of radiocarbon-dated plant macrofossils from ancient woodrat (*Neotoma* spp.) middens (Van Devender, 1983) are revealing the responses of desert shrubs and forest trees (Spaulding, 1984) to late-Quaternary climatic variation in detail seldom achieved solely through fossil pollen. In combination the two complimentary methods give greater resolution than either alone by:

1. adding fossil species represented only by pollen recovered from woodrat dung and from dust trapped in the urine-coated dens;
2. confirming species represented by non-distinctive pollen types;
3. confirming regional vegetational trends suggested by fossil pollen;
4. calibrating pollen frequencies.

Examples from the northern Great Basin illustrate all of these advantages.

Both dung and dust trapped within woodrat middens hold abundant pollen. Woodrat dung may harbor pollen types that, by their animal selection and concentration, reveal a segment of vegetation unlikely to occur as wind-born pollen and perhaps unrecognized or absent as macrofossils. Dust trapped within the den is more likely to contain pollen types representing regional and very local pollen rain, as well as the collecting activities of woodrats.

In east-central Nevada late-Pleistocene pollen of spruce with abundant pine, from the fill of Council Hall Cave and middens from Ladder Cave, complement the macrofossils of bristlecone (*Pinus longaeva*) and limber (*Pinus flexilis*) pines, and spruce (*Picea engelmannii*) from woodrat middens of these caves and from nearby Smith Creek Cave. Decline in pine pollen and importance of juniper pollen over the past 10,000 years is mirrored by abundant macrofossils of Utah juniper (*Juniperus osteosperma*) in Holocene middens. Both pollen content of cave fill and lakes, and first appearances of Utah juniper and pinyon (*Pinus monophylla*) seeds and twigs in woodrat middens illustrate the Holocene northward passage of pinyon and juniper in Nevada.

The full-glacial northern perimeter of juniper-pinyon woodlands stretched across the northern Mohave Desert below 1800 meters elevation at about 37° north latitude (Spaulding, Leopold and Van Devender, 1983). When released from the chilling grip of glacial climates, woodland species streamed northward and upward into areas relinquished by pluvial lakes, cold steppe species, and montane conifers. Man the seed carrier must have played a role in northward advance of woodland and can hardly be doubted as an accidental or intentional propagator of a food so desirable as the pinyon nut (Mehringer, in press).

Increased values of pine and juniper pollen in Gatecliff Shelter about 6,000 yrs. B.P. herald arrival of Utah juniper and single-needle pinyon in the Toquima Range, central Nevada. Macrofossils from cave fill and woodrat middens confirm their presence by 5,000 yrs. B.P. (Thompson and Hattori, 1983). Similarly, after 4,000 yrs. B.P. the deposits of Gatecliff and Triple T shelters show larger percentages of joint-fir (*Ephedra* cf. *viridis*) pollen, in accord with recovery of joint-fir macrofossils from woodrat middens.

Increasing juniper and pine pollen values, suggestive of juniper-pinyon woodland by 4,000 yrs. B.P.,

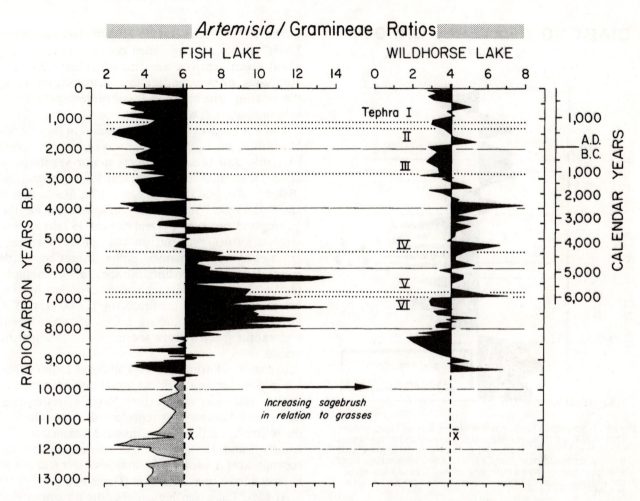

Figure 12. The ratios of sagebrush/grass pollen plotted about their means of the last 9,700 and 9,300 radiocarbon years reveal details of vegetation and climatic change for the Steens Mountain, Oregon. Increasing sagebrush in relation to grass at Fish Lake (2,250 m) indicates less effective moisture. The same variations at Wildhorse Lake, which lies 315 m higher, at the current upper elevational limit of sagebrush, suggests upward advance of sagebrush owing to warmer conditions with fewer snow patches lasting into the summer.

indicate advance of woodland to the Ruby Range, Nevada. Woodrat middens from limestone shelters of the southern Ruby Mountains are within 3 kilometers of Ruby Marsh coring sites. As may have been predicted, woodrat midden macrofossils confirm presence of pinyon pine and Utah juniper since at least 3,000 yrs. B.P. (Thompson, 1984).

Studies at Diamond Pond (Malheur Maar), Oregon, illustrate calibration of pollen frequencies (Mehringer and Wigand, 1985b; Wigand, 1985). Juniper pollen is present in deposits spanning the last 6,000 years and its percentages (of terrestrial plant pollen) fluctuate suggesting periods of varying juniper abundance. However, larger fossil juniper pollen percentages might indicate either expansion or increasing density of junipers, or both.

Sites 100 meters to 1 kilometer from the nearest living junipers contain woodrat middens dominated by twigs and seeds of western juniper *(Juniperus occidentalis).* Radiocarbon dates of these juniper macrofossils correspond to the ages of larger or increasing juniper pollen percentages (Fig. 13). Thus, juniper value differences of 5% or so accurately reflect downslope movement or retreat of juniper woodland near Diamond Pond equivalent to at least 30 to 100 meters elevation as compared with historic distribution of junipers at Diamond Craters. Juniper pollen percentages in modern surface samples suggest similar conclusions.

HISTORIC AND MODERN POLLEN SPECTRA

Soil surface, moss polsters, and lake and pond bottoms from representative vegetation types hold recently deposited pollen often used as analogues of past pollen rain. Because depositional environment influences pollen accumulation, ideal comparative samples would be collected from areas resembling

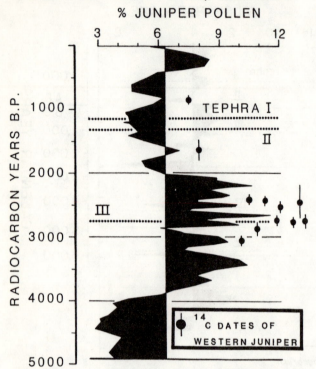

DIAMOND CRATERS, OREGON
% JUNIPER POLLEN

Figure 13. Smoothed juniper pollen percentages of samples from Diamond Pond (Malheur Marr) cores, plotted about their mean of the last 5,000 ¹⁴C years, and radiocarbon dates associated with western juniper *(Juniperus occidentalis)* remains from woodrat middens from nearby lava tubes 100 m to 1 km outside the present range of juniper (from Mehringer and Wigand, 1985b).

sites being investigated. For example, soil surface pollen might best approximate pollen deposited in alluvium (Adam and Mehringer, 1975) or washed into caves, whereas analogs for pollen spectra contained within lake cores would be sought from other lakes of similar size and bathymetry, in comparable topographic situations and various vegetation types. It is usually impractical or impossible to meet these conditions locally; so most surface sample studies offer only a sketchy, but nonetheless important outline of specific relationships between current vegetation and pollen contributed to appropriate depositional environments.

In the Great Basin it is nearly a tradition to collect modern surface or moss polster samples along elevational and vegetational gradients and from selected communities (Adam, 1967; Bright, 1966; Mehringer, 1967) to aid in the interpretation of fossil records from settings as diverse as cave fill, pluvial lakes, desert marshes and alpine bogs (Bright and Davis, 1982; Kautz, 1983; Thompson, 1984). Statistical comparisons of modern surface counts with fossil pollen spectra have recently been applied on the

Snake River Plain (Davis, 1984a; Henry, 1984). Davis' studies relating pollen percentages to existing plant communities are the most detailed and ambitious, even extending to conclusions regarding the relationship between fossil pollen spectra and solar insolation (Davis, 1984b).

The few modern pollen rain studies in the interior Northwest have outlined the ways in which pollen currently and broadly reflects major vegetational units (Heusser, 1978a, 1978b) and local communities (Baker, 1976; Mack and Bryant, 1974; Mack, Bryant and Pell, 1978). Unfortunately, in some areas logging, grazing and agriculture (Daubenmire, 1975b) have so altered natural vegetation that modern surface samples as well as historic pollen values bear little resemblance to pre-disturbance spectra. For example, in the Columbia Basin chenopod pollen attains values unrecorded in the preceding 11,000 years (Fig. 9). Additionally, distinctive pollen of native weeds and exotic species mark sediments of the historic period.

In eastern Washington and adjacent Idaho (Blue, Clear, Sulphur and Wildcat lakes) pollen spectra of the last 100 years or so exhibit larger percentages of native and exotic weeds represented by members of the saltbush, sunflower and mustard families, and by *Erodium* and *Triticum*. Fungal spores, notably the coprophilous *Sporormiella,* may also increase where livestock have been abundant (Davis *et al.,* 1977).

At Clear Lake, southern Idaho, the historic period is distinguished by pollen of *Populus* and *Elaeagnus* attributed to the introduced Lombardy poplar and Russian olive. Juniper pollen percentages also increase, perhaps reflecting recent expansion of woodlands. Dandelion *(Taraxacum officinale)* marks historic levels of Utah Lake. Lastly, lack of a routine and distinctive fall in grass pollen percentages in profiles from present steppe or sites near the steppe-forest border (Williams Lake Fen) may reflect replacement of over-grazed native grasses by the exotic and ubiquitous cheat grass *(Bromus tectorum),* by other grasses encouraged by disturbance such as shrub and forest clearing, and by cereals.

POLLEN OF ARCHAEOLOGICAL SITES

Many sites that harbour human leavings contain pollen indicative of the natural environment, disturbance and subsistence of prehistoric peoples (Seed and Toquima caves, and Triple T Shelter). Also, nearby lakes and bogs, studied in conjunction with archaeological excavations and surveys have revealed details of vegetational changes corresponding to events recorded within archaeological sites. Pollen studies of Bisonsweh Pond and Murphey's Rockshelter, Idaho, Gatecliff Shelter and Hidden Cave, Nevada, and Hogup Cave, Utah, can all be related to kinds of small mammals recovered from these or nearby sites. Analyses of lake cores from southeastern Oregon, were initiated with archaeological explorations to provide paleoenvironmental basis for understanding human occupation of caves and dunes (Mehringer and Wigand, 1985a).

Pollen from cave fill and coprolites of northwestern Utah (Danger, Hogup, and Remnant caves, and Swallow Shelter) reveal a period of unusual abundance of grass pollen and macrofossils beginning around 1,500 yrs. B.P. It is probably not by chance that this event corresponds to increased importance of bison and the Fremont Culture.

Human coprolites contain clues to ancient diet, preparation and storage of food, environments, and seasons of occupation. Some of over 5,000 human coprolites recovered from Lovelock Cave, western Nevada, are composed wholly of charred cat-tail pollen. Pollen and macrofossils from cave fill and coprolites of nearby Hidden Cave confirm prehistoric importance of cat-tail pollen and other marsh species available from mid-summer through early autumn. By comparison, desert resources like the halophytic pickleweed *(Allenrolfea occidentalis)* and chenopod pollen dominate coprolites of Utah's Archaic hunter-gatherers of the Great Salt Desert region.

At Backhoe Village, west-central Utah, sediments collected from room fill, house floors, metates, and vessels held abundant chenopod-amaranth pollen and some corn. Samples from near fire basins of pithouses contained relatively large percentages of cat-tail pollen. Apparently, these Fremont horticulturalists supplemented a diet of marsh and upland wild plant foods with cultivated crops and field weeds.

POLLEN, FIRE AND TEPHRA

It is reasonable to conclude that late-Quaternary volcanic eruptions (Luedke *et al.,* 1983; Sarna-Wojcicki *et al.,* 1983), wildfires (Arno, 1980) and fires set by prehistoric inhabitants (Barrett and Arno, 1982) influenced vegetation history. Yet, the immediate effects, and long-term consequences of any prehistoric catastrophe, even the eruption of Mount Mazama, remain uncertain.

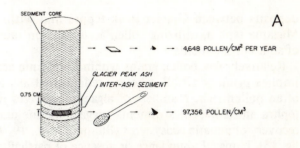

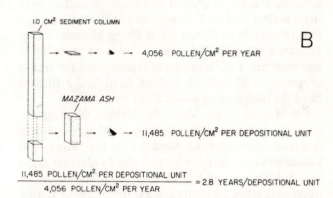

Figure 14. Calculation of duration of deposition within and between volcanic ashes by sampling with a spoon of known volume (A) or from a column (B), and comparing the pollen content with pollen accumulation rates of adjacent sediments (after Mehringer, Blinman and Petersen, 1977). The pollen types within a tephra layer may reflect the season of ashfall, whereas the number of pollen may indicate its duration.

Explosive destruction and collapse of a great mountain into the caldera now occupied by Crater Lake, Oregon, followed recurrent eruptions that sent Mazama tephra beyond the region of this review. Yet, one cannot say how vegetation of the northwestern U.S. would be different if Mount Mazama, Mount St. Helens or Glacier Peak had not been active in the last 12,000 years, nor in what ways the magnitude and timing of their eruptions determined or altered the course of Holocene vegetational "development." Although these questions remain unanswered, pollen associated with tephra layers has furnished some clues to the timing, magnitude and vegetational responses to past volcanic eruptions.

Effects of an ashfall on terrestrial or aquatic ecosystems depend on thickness of primary and secondary tephra, season and duration of the ashfall, and time separating recurrent eruptions. Therefore, estimates of the depositional chronology of ashfalls are essential to evaluating their influences (Mehringer, Blinman and Petersen, 1977). Figure 14 illustrates methods employed to estimate time represented by deposits between Glacier Peak tephra and within Mazama tephra, utilizing pollen accumulation rates of adjacent sediments.

Relatively few pollen grains within a volcanic ash implies primary airfall, whereas redeposited ash is often pollen-rich. Numbers of algae and *Isoetes* in tephra and adjacent sediments suggest decline and recovery of aquatic ecosystems (Blinman *et al.*, 1979, fig. 15). Unusual abundance or absence of particular pollen types may also indicate the season of a prehistoric eruption. Sagebrush pollen from Mazama tephra at Lost Trail Pass, Montana, suggested an autumn ashfall. A graded bed of clean tephra with little pollen and lack of seasonal indicators implies that Glacier Peak tephra fell at Sheep Mountain Bog, Montana, when the lake was ice-free and probably in late summer. Pollen counts of tephra from the recent eruption of Mount St. Helens furnished a further basis for interpreting pollen from volcanic ash.

On May 18, 1980, Mount St. Helens, Washington, erupted sending volcanic ash eastward across North America. As the cloud passed over my home, 400 kilometers downwind, it turned day to blackest night that put the birds to bed. Ash, which accumulated in traps and lakes of eastern Washington, held pollen spectra easily distinguished from those of other late-Quaternary volcanic ashes and lake sediments of the region. The relatively few pollen grains consisting of uniquely large percentages of hemlock (*Tsuga heterophylla*) and alder, with many charred grains, were accompanied by exceptionally large numbers of charcoal fragments. This distinctive combination resulted from the violent eruption that drew pollen and charcoal from scorched and devastated forests, and surrounding areas, skyward into the hot, rising ash cloud.

That Mount St. Helens erupted during the flowering season of the Cascade hemlock forests is recorded wherever the initial airfall tephra lies preserved in lakes east of the Cascades. In addition, that it occurred on a forested mountain is suggested by its abundant charcoal. I expect that successive eruptions, within the time necessary for recovery of forests, could be distinguished from the first eruption by their lesser charcoal. Varying charcoal accumulation rates and charcoal/pollen ratios may also reveal the history of fire in forest and steppe.

The observation that charcoal was unusually abundant over the past 2,000 years or so near Lost Trail Pass, Bitterroot Mountains, Montana, led to three studies of the late-Holocene relationships between reconstructed vegetation, charcoal bands, and charcoal and pollen abundance in lake sediments

(Table 3). At Blue Lake, near Lewiston, Idaho, severe fires, resulting in distinctive bands of washed-in charcoal, burned through Douglas-fir forests about once every hundred years between 4,300 and 4,000 yrs. B.P. Only two severe fires in mixed conifer forests left charcoal bands between 4,000 and 3,100 yrs. B.P., but such fires were nearly twice as common between 3,100 and 1,700 yrs. B.P. From 1,700-1,000 yrs. B.P. severe fires averaged only one in 175 years. Light surface fires characterized the last 700 years.

Table 3. Comparison of charcoal/pollen ratios from three sites for the periods of pollen zones established for Lost Trail Pass Bog, Montana.

^{14}C Age B.P.	Charcoal/Pollen Ratios		
	Lost Trail	Sheep Mt.	Blue Lake
600- 0	0.60	0.67	1.04
1,750- 600	0.31	0.37	0.87
4,000-1,750	0.17	0.24	0.69
4,800-4,000	0.19	0.36	
7,000-4,800	0.34	0.39	
11,500-7,000	0.11	0.19	

Fire frequency, as evidenced by the charcoal/pollen ratio at Sheep Mountain Bog, Montana, declined gradually between 7,000 and 2,000 yrs. B.P. and then increased to produce a superabundance of microscopic charcoal during the past 1,000 years. Fires known from tree fire-scar studies and historic records were correlated with peaks of charcoal abundance in bog sediments. Pollen and charcoal values of cores from Flathead Lake, Montana, apparently reveal no indications of significant changes in either vegetation or frequency of fire over the past 4,000 years. Current studies in eastern Oregon and Washington indicate that fire has played an important role in steppe throughout the Holocene.

SUMMARY AND CONCLUSIONS

Henry P. Hansen pioneered pollen analysis of the interior Pacific Northwest and northern Great Basin nearly 50 years ago. Although some questions posed by Hansen remain unanswered, shadowy shapes of vegetation history are slowly appearing. In addition, pollen has emerged as a tool for deciphering the history and effects of prehistoric fires and volcanic eruptions. Pollen studies aid in investigating ancient woodrat middens and pollen from archaeological sites tells of past environments, human subsistence and seasons of occupation.

During the last full-glacial episode much of this area was covered by continental ice sheet, mountain glaciers, and pluvial lakes. Montane conifers grew at lower elevations in the Great Basin, but all in all, open, cold steppe-like conditions characterized vast areas of the interior as indicated by non-arboreal pollen, especially sagebrush. Distinctive late-glacial pollen assemblages reflect success of pioneer species *(e.g., Shepherdia canadensis)* and persistence of established vegetation.

Holocene pollen records are often and conveniently characterized in three parts. Mazama tephra fell during the middle and warmest period. However, fossil pollen studies reveal sufficient variation that, except for broad regional generalizations, the three-part division often disguises details resulting from elevational differences and the relatively recent emergence of some vegetation associations. Additionally, Holocene pollen sequences show unquestionable fluctuations in vegetation wrought by short, sharp climatic episodes, by fire and by volcanic eruptions. Interpretations of these pollen records may be aided by charcoal and macrofossils in lake deposits, and by macrofossils from woodrat middens.

To progress beyond these facts and speculations will require approaches that take full advantage of correspondences between "modern" comparative pollen spectra and climatic parameters, and between fossil pollen and climatic sequences reconstructed from tree-rings, and the rare opportunity for precise regional correlation through both tephra and paleomagnetic relationships. With more data and improved methods this region offers many opportunities for vegetational and climatic reconstruction afforded by studies of its fossil pollen in conjuction with plant macrofossils.

ACKNOWLEDGMENTS

Previously unpublished pollen studies reported herein were partially supported by a contract from Battelle Pacific Northwest Laboratories, USDA Forest Service Cooperative Agreements through the Fire Effects and Use Research and Development Program of the Intermountain Forest and Range Experiment Station, Ogden, Utah, and by NSF Grants (BNS-77-12556 and BNS-80-06277) in conjunction with the Steens Mountain Prehistory Project.

References Cited

ADAM, D. P.
  1967   Late-Pleistocene and Recent palynology in the central Sierra Nevada, California. *In:* Cushing, E. J., and Wright, H. E., Jr. (eds.), *Quaternary Paleoecology.* Yale University Press, New Haven, p. 275-301.

ADAM, D. P., and MEHRINGER, P. J., JR.
  1975   Modern pollen surface samples—an analysis of subsamples. *Journal of Research of the U. S. Geological Survey,* 3(6): 733-736.

ANTEVS, E.
  1955   Geologic-climatic dating in the west. *American Antiquity,* 20(4): 317-333.

ARNO, S. F.
  1979   Forest regions of Montana. *USDA Forest Service Research Paper,* INT-218. Intermountain Forest and Range experiment Station, Ogden. 39 p.
  1980   Forest fire history in the northern Rockies. *Journal of Forestry,* 78(8): 460-465.

BACON, C. R.
  1983   Eruptive history of Mount Mazama and Crater Lake Caldera, Cascade Range, U.S.A. *Journal of Volcanology and Geothermal Research,* 18(1-4): 57-115.

BAKER, R. G.
  1970   Pollen sequence from Late Quaternary sediments in Yellowstone Park. *Science,* 168(3938): 1449-1450.
  1976   Late Quaternary vegetation history of the Yellowstone Lake Basin, Wyoming. *United States Geological Survey Professional Paper 729-E,* 43 p.
  1983   Holocene vegetational history of the western United States. *In:* Wright, H. E., Jr., (ed.), *Late Quaternary Environments of the United States, Vol 2., The Holocene.* University of Minnesota Press, Minneapolis, p. 109-127.

BAKER, R. G., and RICHMOND, G. M.
  1978   Geology, palynology, and climatic significance of two pre-Pinedale lake sediment sequences in and near Yellowstone National Park. *Quaternary Research,* 10(2): 226-240.

BAKER, V. R., and NUMMEDAL, D. (eds)
  1978   *The Channeled Scabland, a guide to the geomorphology of the Columbia Basin, Washington:* Planetary Geology Program, Office of Space Sciences, National Aeronautics and Space Administration, Washington, D.C., 186 p.

BARNOSKY, C. W.
1984   Late Pleistocene and early Holocene environmental history of southwestern Washington State, U.S.A. *Canadian Journal of Earth Science,* 21(6): 619-629.
1985   A record of late-Quaternary vegetation from the southwestern Columbia Basin, Washington. *Quaternary Research,* 23(1): 109-122.

BARRETT, S. W., and ARNO, S. F.
1982   Indian fires as an ecological influence in the northern Rockies. *Journal of Forestry,* 80(10): 647-651.

BARTHOLOMEW, M. J.
1982   Pollen and sediment analyses of Clear Lake, Whitman County, Washington: the last 600 years. Washington State University, M.A. Thesis, 81 p.

BILLINGS, W. D.
1951   Vegetational zonation in the Great Basin of western North America. *In:* Les Bases Ecologiques de la regeneration de la vegetation des zones arides. *Internationale Union Societe Biologique, Series B,* (9): 101-122.

BLINMAN, E.
1978   Pollen analysis of Glacier Peak and Mazama volcanic ashes. Washington State University, M.A. Thesis, 49 p.

BLINMAN, E., MEHRINGER, P. J., JR., and SHEPPARD, J. C.
1979   Pollen influx and the deposition of Mazama and Glacier Peak tephra. *In:* Sheets, P. D., and Grayson, D. K. (eds.), *Volcanic Activity and Human Ecology.* Academic Press, New York, p. 393-425.

BRANT, L. A.
1980   A palynological investigation of postglacial sediments at two locations along the continental divide near Helena, Montana. Pennsylvania State University, Ph.D. Thesis, 162 p.
1982   A trail back through time. *Montana Outdoors,* 13(1): 20-22.

BRETZ, J. H.
1969   The Lake Missoula floods and the Channeled Scabland. *Journal of Geology,* 77(5): 505-543.

BRIGHT, R. C.
1966   Pollen and seed stratigraphy of Swan Lake, southeastern Idaho: it's relation to regional vegetational history and to Lake Bonneville history. *Tebiwa,* 9(2): 1-47.
1982   Paleontology of the lacustrine member of the American Falls Lake Beds, southeastern Idaho. *In:* Bonnichsen, B., and Breckenridge, R. M. (eds.), Cenozoic Geology of Idaho *Idaho Bureau of Mines and Geology Bulletin,* 26: 597-614.

BRIGHT, R. C., and DAVIS, O. K.
1982   Quaternary paleoecology of Idaho National Engineering Laboratory, Snake River Plain, Idaho. *American Midland Naturalist,* 108(1): 21-33.

BUSHMAN, J. R.
1980   The rate of sedimentation in Utah Lake and the use of pollen as an indicator of time in the sediments. *Brigham Young University Geology Studies,* 27(3): 35-43.

BYRNE, R., BUSBY, C., and HEIZER, R. F.
1979   The altithermal revisited: pollen evidence from the Leonard Rockshelter. *Journal of California and Great Basin Anthropology,* 1(2): 280-294.

CHATTERS, J. C.
1982   Evolutionary human paleoecology: climatic change and human adaptation in the Pahsimeroi Valley, Idaho, 2500 BP to the present. University of Washington, Ph.D. Thesis. 452 p.

CURRY, D. R., ATWOOD, G., and MABEY, D. R.
1983   Major levels of Great Salt Lake and Lake Bonneville. *Utah Geological and Mineral Survey, Utah Department of Natural Resources,* Map 73.

DALLEY, G. F.
1976   Palynology of the Swallow Shelter deposits. *In:* Dalley, G. F., Swallow Shelter and Associated Sites. *University of Utah Anthropological Papers,* 96: 171-174.

DAUBENMIRE, R.
1956   Climate as a determinant of vegetation distribution in eastern Washington and northern Idaho. *Ecological Monographs,* 26(2): 131-154.
1970   Steppe vegetation of Washington. *Washington Agricultural Experiment Station, Technical Bulletin,* 62: 131 p.
1975a   Ecology of *Artemisia tridentata* subsp. *tridentata* in the state of Washington. *Northwest Science,* 49(1): 24-35.
1975b   Plant succession on abandoned fields, and fire infuences, in a steppe area in southeastern Washington. *Northwest Science,* 49(1): 36-48.
1982   The distribution of *Artemisia rigida* in Washington: a challenge to ecology and geology. *Northwest Science,* 56(3): 162-164.

DAUBENMIRE, R., and DAUBENMIRE, J. B.
1968   Forest vegetation of eastern Washington and northern Idaho. *Washington Agricultural Experiment Station, Technical Bulletin,* 60: 104 p.

DAVIS, O. K.
1981   Vegetation migration in southern Idaho during the late-Quaternary and Holocene. University of Minnesota, Ph.D. Thesis, 252 p.
1984a   Pollen frequencies reflect vegetation patterns in a Great Basin (U.S.A.) mountain range. *Review of Palaeobotany and Palynology,* 40(4): 295-315.
1984b   Multiple thermal maxima during the Holocene. *Science,* 225(4662): 617-619.

DAVIS, O. K., and BRIGHT, R. C.
1983   Late-Pleistocene vegetation history of the Idaho National Engineering Laboratory. *In:* Markam, O. D. (ed.), *Idaho National Engineering Laboratory Radioecology and Ecology Programs 1983 Progress Report. National Technical Information Service, ID-12098: 162-171.*

DAVIS, O. K., KOLVA, D. A., and MEHRINGER, P. J., JR.
1977   Pollen analysis of Wildcat Lake, Whitman County, Washington: the last 1000 years. *Northwest Science,* 51(1): 13-30.

FRANKLIN, J. F., and DYRNESS, C. T.
1973   Natural vegetation of Oregon and Washington. *USDA Forest Service General Technical Report,* PNW-8. Pacific Northwest Forest and Range Experiment Station, Portland, 417 p.

GENNETT, J. A.
1977   Palynology and paleoecology of sediments from Blacktail Pond, northern Yellowstone Park, Wyoming. University of Iowa, M.S. Thesis, 74 p.

GREY, D. C., and BENNETT, R.
1972   A preliminary limnological history of Great Salt Lake. *In:* The Great Salt Lake and Utah's Water Resources. *Proceedings of the First Annual Conference, Utah Section of the American Water Resources Association.* Salt Lake City, p. 3-18.

HANSEN, H. P.

1939a Pollen analysis of a bog in northern Idaho. *American Journal of Botany,* 26(4): 225-228.

1939b Pollen analysis of a bog near Spokane, Washington. *Bulletin of the Torrey Botanical Club,* 66(4): 215-220.

1939c Paleoecology of a central Washington bog. *Ecology,* 20(4): 563-568.

1940 Paleoecology of a montane peat deposit at Bonaparte Lake, Washington. *Northwest Science,* 14(3): 60-68.

1941a A pollen study of post-Pleistocene lake sediments in the upper sonoran life zone of Washington. *American Journal of Science,* 239(7): 503-522.

1941b Paleoecology of a montane peat deposit near Wenatchee, Washington. *Northwest Science,* 15(3): 53-65.

1942a The influence of volcanic eruptions upon post-Pleistocene forest succession in central Oregon. *American Journal of Botany,* 29(2): 214-219.

1942b Post-Mount Mazama forest succession on the east slope of the central Cascades of Oregon. *American Midland Naturalist,* 27(2): 523-534.

1942c A pollen study of peat profiles from lower Klamath Lake of Oregon and California. *In:* Cressman, L. S., *Archaeological Researches in the Northern Great Basin.* Carnegie Institution of Washington Publication, 538: 103-114.

1943a A pollen study of a subalpine bog in the Blue Mountains of northeastern Oregon. *Ecology,* 24(1): 70-78.

1943b Post-Pleistocene forest succession in northern Idaho. *American Midland Naturalist,* 30(3): 796-802.

1943c Paleoecology of a peat deposit in east central Washington. *Northwest Science,* 17(2): 35-40.

1944 Postglacial vegetation of eastern Washington. *Northwest Science,* 18(4): 79-87.

1946a Postglacial forest succession and climate in the Oregon Cascades. *American Journal of Science,* 244(10): 710-734.

1946b Early Man in Oregon. *Scientific Monthly,* 62(1): 52-62.

1947a Postglacial forest succession, climate and chronlogy in the Pacific Northwest. *American Philosophical Society,* 37(1): 1-130.

1947b Postglacial vegetation of the northern Great Basin. *American Journal of Botany,* 34(3): 164-171.

1948 Postglacial forests of the Glacier National Park region. *Ecology,* 29(2): 146-152.

1949 Pollen content of moss polsters in relation to forest composition. *American Midland Naturalist,* 42(4): 473-479.

1951 Pollen analysis of peat sections from near the Finley site, Wyoming. *In:* Moss, J. H., *Early Man in the Eden Valley,* University of Pennsylvania Museum Monographs, p. 113-118.

HENRY, C.

1984 Holocene paleoecology of the Western Snake River Plain, Idaho. University of Michigan, M.S. Thesis, 171 p.

HEMPHILL, M. L.

1983 Fire, vegetation, and people—charcoal and pollen analyses of Sheep Mountain Bog, Montana: the last 2800 years. Washington State University, M.A. Thesis, 70 p.

HEUSSER, C. J.

1978a Modern pollen spectra from western Oregon. *Bulletin of the Torrey Botanical Club,* 105(1): 14-17.

1978b Modern pollen rain of Washington. *Canadian Journal of Botany,* 56(13): 1510-1517.

1983 Vegetational history of the northwestern United States including Alaska. *In:* Wright, H. E., Jr., and Porter, S. C. (eds.), *Late-Quaternary Environments of the United States, Vol. 1, The Late Pleistocene.* University of Minnesota Press, Minneapolis, p. 239-258.

HULL, F. W.

1976 Comparative pollen sampling techniques at Remnant Cave. *In:* Dalley, G. F., Swallow Shelter and Associated Sites, *University of Utah Anthropological Papers,* 96: 175-179.

KAUTZ, R. R.

1983 Contemporary pollen rain in Monitor Valley. *In:* Thomas, D. H., The Archaelogy of Monitor Valley, 1. Epistomology. *The Anthropological Papers of the American Museum of Natural History,* New York, 58(1): 106-117.

KAUTZ, R. R., and THOMAS, D. H.

1972 Palynological investigations of two prehistoric cave middens in central Nevada. *Tebiwa,* 15(2): 43-54.

KELSO, G.

1970 Hogup Cave, Utah: comparative pollen analysis of human coprolites and cave fill. *In:* Aikens, C. M., Hogup Cave, *University of Utah Anthropological Paper,* 93: 251-262.

KUCHLER, A. W.

1964 Potential natural vegetation of the coterminous United States (map). *American Geographical Society Special Publication,* 36.

LUEDKE R. G., SMITH, R. L., and RUSSELL-ROBINSON, S. L.

1983 Map showing distribution, composition, and age of late Cenozoic volcanoes and volcanic rocks of the Cascade Range and vicinity, northwestern United States. *Miscellaneous Investigations Series, U.S. Geological Survey,* Map I-1507.

MACK, R. N.

1971 Pollen size variation in some western North American pines as related to fossil pollen identification. *Northwest Science,* 45(4): 257-269.

MACK, R. N., and BRYANT, V. M., JR.

1974 Modern pollen spectra from the Columbia Basin, Washington. *Northwest Science,* 48(3): 183-194.

MACK, R. N., BRYANT, V. M., JR., and FRYXELL, R.

1976 Pollen sequence from the Columbia Basin, Washington: reappraisal of postglacial vegetation. *American Midland Naturalist,* 95(2): 390-397.

MACK, R. N., BRYANT, V. M., JR., and PELL, W.

1978 Modern forest pollen spectra from eastern Washington and northern Idaho. *Botanical Gazette,* 139(2): 249-255.

MACK, R. N., RUTTER, N. W., BRYANT, V. M., JR., and VALASTRO, S.

1978a Reexamination of postglacial vegetation history in northern Idaho: Hager Pond, Bonner Co. *Quaternary Research,* 10(2): 241-255.

1978b Late Quaternary pollen record from Big Meadow, Pend Oreille County, Washington. *Ecology,* 59(5): 956-965.

MACK, R. N., RUTTER, N. W., and VALASTRO, S.
1978 Late Quaternary pollen record from the Sanpoil River Valley, Washington. *Canadian Journal of Botany,* 56(14): 1642-1650.
1979 Holocene vegetation history of the Okanogan Valley, Washington. *Quaternary Research,* 12(2): 212-225.
1983 Holocene vegetational history of the Kootenai River Valley, Montana. *Quaternary Research,* 20(2): 177-193.

MACK, R. N., RUTTER, N. W., VALASTRO, S., and BRYANT, V. M., JR.
1978 Late Quaternary vegetation history at Waits Lake, Colville River Valley, Washington. *Botanical Gazette,* 139(4): 499-506.

MADSEN, D. B., and CURREY, D. R.
1979 Late Quaternary glacial and vegetation changes, Little Cottonwood Canyon area, Wasatch Mountains, Utah. *Quaternary Research,* 12(2): 254-270.

MADSEN, D. B., and LINDSAY, L. M.
1977 Backhoe Village. *Antiquities Section Selected Papers,* Utah Department of Development Service, Division of State History, 4(12): 119 p.

MARTIN, P. S., and MEHRINGER, P. J., JR.
1965 Pleistocene pollen analysis and biogeography of the Southwest. *In:* Wright, H. E., and Frey, D. G. (eds.), *The Quaternary of the United States,* Princeton University Press, Princeton, New Jersey, p. 433-451.

McKENZIE, D.
1982 The northern Great Basin. *In:* Bender, G. L., (ed.), *Reference Handbook on the Deserts of North America.* Greenwood Press, Westport, Connecticut, p. 67-82.

MEHRINGER, P. J., JR.
1967 Pollen analysis of the Tule Springs area, Nevada. *In:* Wormington, H. M. and Ellis, D. (eds.), *Nevada State Museum Anthropological Paper,* No. 13. Carson City, p. 129-200.
1977 Great Basin late Quaternary environments and chronology. *In:* Fowler, D. D. (ed.), *Models and Great Basin Prehistory,* Desert Research Institute Publications in Social Science, University of Nevada, Reno, 12: 113-167.
1985 Prehistoric environment, Chapter 3. *In:* d'Azevedo, W. L. (ed.), *Handbook of North American Indians, Volume 11: The Great Basin.* Smithsonian Institution, Washington, D.C. In press.

MEHRINGER, P. J., JR., ARNO, S. F., and PETERSEN, K. L.
1977 Postglacial history of Lost Trail Pass Bog, Bitterroot Mountains, Montana. *Arctic and Alpine Research,* 9(4): 345-368.

MEHRINGER, P. J., JR., BLINMAN, E., and PETERSEN, K. L.
1977 Pollen influx and volcanic ash. *Science,* 198(4314): 257-261.

MEHRINGER, P. J., JR., NASH, W. P., and FULLER, R. H.
1971 A Holocene volcanic ash from northwestern Utah. *Proceedings of the Utah Academy of Sciences, Arts, and Letters,* 48(1): 46-51.

MEHRINGER, P. J., JR., SHEPPARD, J. C., and FOIT, F. F., JR.
1984 The age of Glacier Peak tephra in west-central Montana. *Quaternary Research,* 21(1): 36-41.

MEHRINGER, P. J., JR., and WIGAND, P. E.
1985a Holocene history of Skull Creek Dunes, Catlow Valley, Oregon, U.S.A., *Journal of Arid Environments.* In press.
1985b Prehistoric distribution of western juniper. *In:* Proceedings of the Western Juniper Management Short Course, Bend, Oregon, October 15-16, 1984. Oregon State University Extension Service, Corvallis, p. 1-9.

MERRIAM, J. C.
1941 Paleontology, Early Man and Historical Geology. *Carnegie Institution of Washington Year Book,* 40: 316-333.

NAPTON, L. K., and KELSO, G. K.
1969 Part III: Preliminary palynological analysis of human coprolites from Lovelock Cave, Nevada. *In:* Kemper, R. V. (ed.), *Kroeber Anthropological Society Papers,* 2: 1-98.

NICKMANN, R.
1979 The palynology of Williams Lake Fen, Spokane County, Washington. Eastern Washington University, M.S. Thesis, 57 p.

NICKMANN, R. J., and LEOPOLD, E.
1984 A postglacial pollen record from Goose Lake, Okanogan County, Washington: evidence for an early Holocene cooling. *Chief Joseph Summary Report.* Office of Public Archaeology, University of Washington, Seattle.

ONKEN, T. L.
1984 Prehistoric fire activity and vegetation near Flathead Lake, Montana. University of Montana, M.S. Thesis, 59 p.

PÉWÉ, T. E.
1983 The periglacial environment in North America during Wisconsin time. *In:* Wright, H. E., Jr., and Porter, S. C., (eds.), *Late-Quaternary Environments of the United States, Vol. 1, The Late Pleistocene.* University of Minnesota Press, Minneapolis, p. 157-189.

PORTER, S. C., PIERCE, K. L., and HAMILTON, T. D.
1983 Late Wisconsin mountain glaciation in the western United States. *In:* Wright, H. E., Jr., and Porter, S. C. (eds.), *Late-Quaternary Environments of the United States, Vol. 1, The Late Pleistocene.* University of Minnesota Press, Minneapolis, p. 71-111.

SARNA-WOJCICKI, A. M., CHAMPION, D. E., and DAVIS, J. O.
1983 Holocene volcanism in the conterminous United States and the role of silicic volcanic ash layers in correlation of latest-Pleistocene and Holocene deposits. *In:* Wright, H. E., Jr. (ed.), *Late-Quaternary Environments of the United States, Vol. 2, The Holocene.* University of Minnesota Press, Minneapolis, p. 52-77.

SEARS, P. B., and ROOSMA, A.
1961 A climatic sequence from two Nevada caves. *American Journal of Science,* 259(9): 669-678.

SMITH, C. S.
1983 A 4300 year history of vegetation, climate, and fire from Blue Lake, Nez Perce County, Idaho. Washington State M.A. Thesis, 86 p.

SMITH, G. I., and STREET-PERROTT, F. A.
1983 Pluvial lakes of the western United States. *In:* Wright, H. E., Jr., and Porter, S. C., (eds.), *Late-Quaternary Environments of the United States, Vol. 1, The Late Pleistocene.* University of Minnesota Press, Minneapolis, p. 190-212.

SPAULDING, W. G.
1984   The last glacial-interglacial climatic cycle: its effects on woodlands and forests in the American West. *In:* Lanner, R. M. (ed.), *Proceedings of the Eighth North American Forest Biology Workshop.* Utah State University, Logan, p. 42-69.

SPAULDING, W. G., LEOPOLD, E. B., and VAN DEVENDER, T. R.
1983   Late Wisconsin paleoecology of the American Southwest. *In:* Wright, H. E., Jr., and Porter, S. C. (eds.), *Late-Quaternary Environments of the United States, Vol. 1, The Late Pleistocene.* University of Minnesota Press, Minneapolis, p. 259-293.

SPENCER, R. J., BAEDECKER, M. J., EUGSTER, H. P., FORESTER, R. M., GOLDHABER, M. B., JONES, B. F., KELTS, K., MCKENZIE, J., MADSEN, D. B., RETTIG, S. L., RUBIN, M., and BOWSER, C. J.
1984   Great Salt Lake, and precursors, Utah: the last 30,000 years. *Contributions to Mineralogy and Petrology,* 86(4): 321-334.

STEELE, R., COOPER, S. V., ONDOV, D. M., ROBERTS, D. W., and PFISTER, R. D.
1983   Forest habitat types of eastern Idaho-western Wyoming. *USDA Forest Service General Technical Report,* INT-144. Intermountain Forest and Range Experiment Station, Ogden, 122 p.

STEELE, R., PFISTER, R. D., RYKER, R. A., and KITTAMS, J. A.
1981   Forest habitat types of central Idaho. *USDA Forest Service General Technical Report,* INT-114. Intermountain Forest and Range Experiment Station, Ogden, 138 p.

THOMPSON, R. S.
1979   Late Pleistocene and Holocene packrat middens from Smith Creek Canyon, White Pine County, Nevada. *In:* Tuohy, D. R., and Rendall, D. L. (eds.), The Archaeology of Smith Creek Canyon, Eastern Nevada. *Nevada State Museum Anthropological Papers,* 17: 362-380.

1984   Late Pleistocene and Holocene environments in the Great Basin. University of Arizona, Ph.D. Thesis, 256 p.

1985   Paleoenvironmental investigations at Seed Cave (Windust Cave H,45FR46). Contract Report 100-41, *Archaeological and Historical Services, Eastern Washington University, Cheney.*

THOMPSON, R. S., and KAUTZ, R. R.
1983   Paleobotany of Gatecliff Shelter. *In:* Thomas, D. H., The Archaeology of Monitor Valley, 2. Gatecliff Shelter. *The Anthropological Papers of the American Museum of Natural History, New York,* 59(1): 136-157.

THOMPSON, R. S., and HATTORI, E. M.
1983   Packrat *(Neotoma)* middens from Gatecliff Shelter and Holocene migrations of woodland plants. *In:* Thomas, D. H., The Archaeology of Monitor Valley, 2. Gatecliff Shelter. *The Anthropological Papers of the American Museum of Natural History, New York,* 59(1): 157-167.

VAN DEVENDER, T. R.
1983   Our first curators. *Sonorensis: Arizona-Sonoran Desert Museum Newsletter,* 5(3): 5-10.

VEROSUB, K. L., and MEHRINGER, P. J., JR.
1984   Congruent paleomagnetic and archeomagnetic records from the western United States: A.D. 750 to 1450. *Science,* 224(4697): 387-389.

WADDINGTON, J. C. B., and WRIGHT, H. E., JR.
1974   Late Quaternary vegetational changes on the east side of Yellowstone Park, Wyoming. *Quaternary Research,* 4(2): 175-184.

WAITT, R. B., JR., and THORSON, R. M.
1983   The Cordilleran ice sheet in Washington, Idaho, and Montana. *In:* Wright, H. E., Jr., and Porter, S. C., (eds.), *Late-Quaternary Environments of the United States, Vol. 1, The Late Pleistocene.* University of Minnesota Press, Minneapolis, p. 53-70.

WIGAND, P. E.
1985   Diamond Pond, Harney County, Oregon: man and marsh in the eastern Oregon desert. Ph.D. Thesis, Washington State University.

WIGAND, P. E., and MEHRINGER, P. J., JR.
1985   Pollen and seed analyses, Chapter 6. *In:* Thomas, D. H. (ed.), *Archaeology of Hidden Cave Nevada,* Vol. 61, pt. 1, *Anthropological Papers of the American Museum of Natural History, New York.* p. 108-124.

WRIGHT, H. E., JR.
1971   Late Quaternary vegetational history of North America. *In:* Turekian, K. K. (ed.), *Late Cenozoic Glacial Ages.* Yale University Press, New Haven, p. 425-464.

QUATERNARY POLLEN RECORDS FROM THE GREAT PLAINS AND CENTRAL UNITED STATES

RICHARD G. BAKER
Department of Geology
University of Iowa
Iowa City, Iowa 52242

KIRK A. WALN
Department of Geology
University of Iowa
Iowa City, Iowa 52242

Abstract

Tertiary plant macrofossil and faunal records indicate a gradual shift from subtropical forest to savannas and grasslands in Oligocene and Miocene time on the Great Plains, while Rocky Mountain floras were nearly modern in character. Subtropical conditions apparently disappeared sometime in middle Quaternary time; vertebrate faunas suggest that southwestern Kansas and northwestern Oklahoma experienced repeated faunal shifts, indicating an increasingly continental climate. No direct vegetational record has been recovered for most of the Great Plains during much of the Pleistocene, but sparse records of purported Illinoian, Sangamon, and Early Wisconsin pollen and vertebrate sequences in southwestern Kansas suggest that grasslands were interspersed in a pine-parkland.

Probably Sangamon to recent pollen records indicate that Yellowstone National Park underwent substantial vegetational changes ranging from *Pseudotsuga* forests representing climates warmer than those at present, to montane and subalpine forests indicating cooler-than-present conditions, to tundra and icecap environments indicating much colder conditions. Post-glacial *Pinus contorta* forests imply a mid-Holocene climate warmer than present.

In Farmadalian time grasslands with a few conifers and deciduous trees prevailed in western Iowa, whereas *Picea-Pinus* forests covered eastern Iowa. As Late Wisconsin glaciers advanced, tundra environments spread southward across ice-free areas in the northern half of Iowa.

Previous reviews have shown that *Picea* and *Larix* both grew across the Northern Great Plains during late-glacial time. Molluscan and small-mammalian faunas suggest that grasslands prevailed in the Late Wisconsin of western Kansas and Nebraska at this same time. An ecotone must have separated forests or parklands on the northeastern plains from grasslands on the western and central plains. The late-glacial *Picea* forests gave way directly to prairie on the northern plains, and only along the eastern plains border did deciduous forest intervene in this transition. Most of the Great Plains has apparently remained a grassland throughout the Holocene.

INTRODUCTION

The purpose of this paper is to summarize the Quaternary vegetational history of the Great Plains, Missouri, Iowa, and Wyoming. The paper is short because several reviews cover the Quaternary environments of parts or all of the Great Plains and adjacent areas (Wells, 1970a; Dort and Jones, 1970; Bradbury, 1980; Watts, 1983; Graham, 1981; Spaulding, *et al.*, 1983; Baker, 1983), and we will stress the areas not well covered in previous reviews. It is organized into three sections: the southern plains including Missouri, Oklahoma and Kansas; the northern plains including Iowa, Nebraska, North and South Dakota; and Wyoming (Fig. 1). Although the primary sources of information are supposed to be palynology, in this paper collateral information from plant macrofossils, molluscs, and small mammals is also used, because for large areas in our region no palynological information is available. A list of the Latin names used in the text is provided in Table 1.

GEOGRAPHY AND GEOLOGY

The Great Plains is both a geomorphic and vegetational province. This large surface generally slopes eastward from the Rocky Mountains to the Mississippi or Missouri Rivers. It is underlain in many places by Eocene through Pliocene fluvial sediments, but Cretaceous marine rocks and Paleocene non-marine rocks are exposed in places in the Dakotas, and late-Paleozoic marine rocks are found in southeastern Kansas and Oklahoma. A thin cover of glaciogenic sediments overlies the northern and eastern parts of the Great Plains. Wyoming is characterized

Table 1.
Latin and common names of plants used in the text.

Abies lasiocarpa	subalpine fir	*Panicum*	panic grass
Acer negundo	boxelder	*Platanus*	sycamore
Alnus	alder	*Picea engelmannii*	Engelmann spruce
Ambrosia	ragweed	*Pinus albicaulis*	whitebark pine
Andropogon	bluestem	*Pinus banksiana*	jack pine
Artemisia	sagebrush, wormwood	*Pinus cembroides*	Mexican pinyon pine
Atriplex	saltbush	*Pinus contorta*	lodgepole pine
Betula glandulosa	dwarf birch	*Pinus edulis*	Colorado pinyon pine
Betula papyrifera	paper birch	*Pinus flexilis*	limber pine
Bouteloua	grama grass	*Pinus monophylla*	single-needle pinyon
Buchloë dactyloides	buffalo grass	*Pinus ponderosa*	pine
Carpinus	blue beech	*Populus balsamifera*	ponderosa pine
Carya	hickory	*Populus deltoides*	balsam poplar
Celtis occidentalis	hackberry	*Populus tremuloides*	eastern cottonwood
Chenopodiaceae	goosefoot family	*Pseudotsuga menziesii*	trembling aspen
Compositae	daisy family	*Quercus macrocarpa*	Douglas fir
Corylus cornuta	beaked hazel	*Quercus marilandica*	bur oak
Cupressaceae	cypress family	*Quercus rubra*	blackjack oak
Cyperaceae	sedge family	*Quercus stellata*	northern red oak
Dryas intergrifolia	mountain avens	*Salix*	post oak
Empetrum	crowberry	*Sarcobatus*	willow
Fraxinus nigra	black ash	*Saxifraga aizoides*	greasewood
Gramineae	grass family	*Selaginella selaginoides*	yellow mountain
Juglans	walnut	*Shepherdia canadensis*	saxifrage
Juniperus scopulorum	Rocky Mountain juniper	*Sorghastrum nutans*	spikemoss
Larix	larch	*Tilia americana*	buffalo-berry
Leguminosae	pea family	*Ulmus americana*	indian grass
Ostrya	ironwood	*Zelkova*	basswood
			American elm

by high mountain ranges, many with Precambrian cores, and intervening basins with a cover of mid-Tertiary sediments. Western Wyoming is a part of a thrust belt, and the Tetons represent the eastern edge of Great Basin faulting. Yellowstone National Park is a large plateau of Quaternary rhyolites bordered on its eastern margin by Eocene volcanic and volcaniclastic rocks of the Absaroka Range. Iowa and Missouri are both underlain by early to late-Paleozoic rocks of marine origin.

The climate of the Great Plains is controlled largely by the rain shadow of the Rocky Mountains which intercept a westerly flow of Pacific air (Borchert, 1950; Bryson, *et al.*, 1970). This dry Pacific air forms a wedge extending eastward through Iowa (Bryson, *et al.*, 1970). The plains are drier along the western edge and become progressively moister eastward as Gulf of Mexico airmasses play increasingly important roles in causing precipitation. In Wyoming, the

mountainous areas are much cooler and moister than the semi-arid basins.

The Great Plains was a vast prairie prior to cultivation, and it was dominated by short grasses (*Buchloë dactyloides, Bouteloua* ssp.) in the west, and tall grasses *(Andropogon* spp., *Panicum* spp., *Sorghastrum nutans)* in the east. The Compositae and Leguminosae are other dominant families, and there are numerous other taxa that are locally important. Trees are generally limited to river valleys (and farmyards). Predominantly eastern trees such as *Acer negundo, Betula papyrifera, Celtis occidentalis, Ostrya virginiana, Quercus macrocarpa, Quercus stellata, Quercus marilandica, Tilia americana,* and *Ulmus americana* extend westward along these valleys or escarpments and a few Rocky Mountain species like *Pinus ponderosa, P. flexilis* and *Juniperus scopulorum* extend eastward. In Wyoming the vegetation of the basins is a steppe dominated by shrubs like *Atriplex* spp. and other chenopods, *Artemisia*

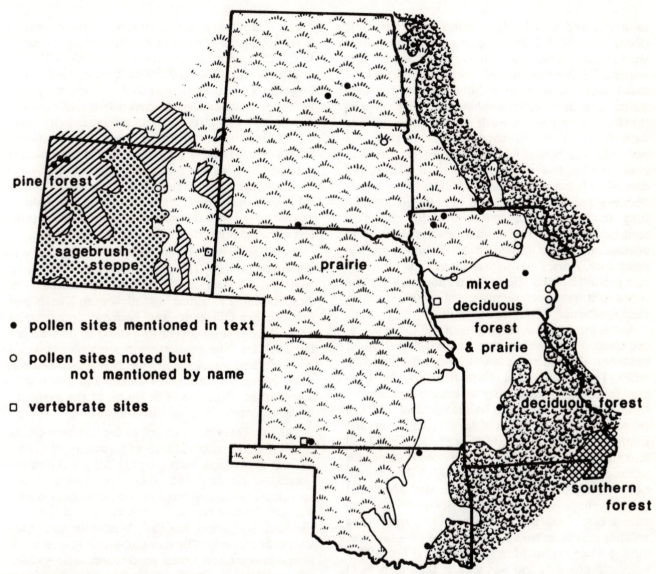

Figure 1. Map showing the area reviewed. Location of pollen and small mammal sites are shown by dots and squares respectively. Map generalized from Küchler, 1964.

spp., and various grasses. The mountains are generally forested with altitudinally-controlled vegetation belts of *Pinus ponderosa* and/or *P. flexilis* (foothills forest), *Pseudotsuga menziesii,* and *Pinus contorta,*(montane forests) and *Picea engelmannii, Abies lasiocarpa* and *Pinus albicaulis* (subalpine forests). Tundra is present atop the high mountains.

LATE-TERTIARY STUDIES

The history of the Great Plains grasslands is controversial: did grasslands exist since the middle-Tertiary, or did they develop only after full-glacial forests retreated? Evidence for the establishment of grasslands in the Tertiary comes from vertebrate remains and plant macrofossils (Thomasson, 1979).

No pollen studies of late-Tertiary or early-Quaternary deposits are published from this area. Elias (1942) hypothesizes that treeless grasslands have persisted since late-Miocene. Others believe that these plains were wooded during glacial maxima, and that present grasslands have existed only since early-Holocene times (MacGinitie, 1962; Wells, 1970b). Evidence from fossil hares (family Leporidae) (Dawson, 1967), tortoises (Brattstrom, 1961), horses (Shotwell, 1961), and opalized grass anthoecia and other angiosperm reproductive structures (Thomasson, 1979) indicates that the environment of the Great Plains during late-Miocene and Pliocene was a subhumid and subtropical savanna or savanna-parkland (Thomasson, 1979, 1980). Paleosols from Badlands National Park also indicate a change from humid subtropical forests to savannas and grasslands in middle-Tertiary time (Rettalack, 1983). Clearly, open grass-covered areas were present in the late-Tertiary, but trees were also present. For example, modern tortoises similar to the fossil ones do not burrow, but they require shade during periods of warm sunshine, and hence suggest that trees were present. Such tortoises also indicate frost free winters (Hibbard, 1970). Miocene and early-Pliocene rabbits were adapted for short runs to cover and had less cement in their teeth, indicating a diet of soft herbage; these gave way in the late-Pliocene and early-Pleistocene to hares adapted for long, straight-line runs and with highly cemented teeth adapted for eating abrasive grasses. Both these adaptations suggest a change from savanna to treeless conditions. Horses that were adapted for savanna life also prevailed in the Miocene and Pliocene, and these types of horses became restricted to the south in early-Quaternary (Thomasson, 1979).

As the Pliocene was coming to an end in the southern Great Plains a flora that had developed under subtropical conditions was dominant (Hibbard, 1970). The interpretation of this flora, and the floras that followed through early to middle-Quaternary time, are based upon the habitat preferences and related climatic tolerances of fossil mammalian and molluscan faunas. The late-Pliocene Rexroad local fauna in southwestern Kansas provides evidence for subtropical conditions (Hibbard, 1970). The vegetation associated with this fauna consisted of wooded areas, perhaps gallery forests along stream systems, with areas of marshland nearby. The surrounding uplands may have been a tall grass prairie—savanna, with isolated stands of trees (Hibbard, 1938, 1970). Effective precipitation was greater than at present with frost-free winters and milder summer temperatures (Hibbard, 1970).

SOUTHERN PLAINS

By the beginning of the Quaternary Epoch faunas of the southern plains were responding to cyclical climatic variations. Repeated shifts of both the ranges of species, and the composition of the faunas through time, can be correlated with glacial-interglacial intervals. The change in the composition of faunas through time suggests that there was progressive deterioration of the climate through the Quaternary (Hibbard, 1970; Taylor, 1960).

Although changed from the generally stable climate of the Pliocene, early- to middle-Quaternary climates were less severe than modern climatic regimes in the region. Mammalian and molluscan evidence suggests that subtropical conditions existed during the early- to middle-Quaternary at least as far north as Kansas and Nebraska (Hibbard, 1970; Taylor, 1960). The large land tortoise *Geochelone* and thermophilic genera of molluscs suggest a generally equable climate during at least parts of these intervals, with mild winter conditions and greater effective precipitation than received today (Hibbard, 1970). Prominant caliche horizons are found within many early- to middle-Quarternary interglacial deposits. Whether these deposits are indicative of prolonged drought or altered yearly rainfall patterns, the caliche implies less effective precipitation than in previous times.

Pollen analytical reconstruction of the floras was attempted by Kapp (1965, 1970). However, he found that early-Quaternary sediments associated with fossil mammal localities in southwestern Kansas and northwestern Oklahoma were barren or contained so little pollen that no conclusions could be drawn from the analyses. The Pearlette Ashes (then considered one unit) and associated sediments in South Dakota, Iowa, Nebraska, Kansas, Oklahoma and Texas were also barren of pollen. The floras of the early- to middle-Quaternary remain little known.

The late-Quaternary record in the southern plains is better than that of the earlier Quaternary, but there are still many gaps, both geographically and chronologically. Kapp (1965) recovered pollen from sediments associated with vertebrate sites in northwestern Oklahoma and southwestern Kansas of Illinoian, Sangamon, and Early Wisconsin age. Three sites believed to be Illinoian (full-glacial) based upon faunal content and faunal progression, have yielded pollen diagrams. The diagrams show abundant *Picea* (5-8%) and *Pinus* (15-80%) pollen with some *Pseudotsuga* (.7-1.3%) and varying amounts of *Ambrosia* (0-2%) *Artemisia* (5-28%) and Gramineae and Compositae (5-62%) (Kapp, 1970). Pollen of the arboreal deciduous element was of minor importance, although *Quercus, Carya, Juglans, Alnus, Betula, Fraxinus,* and *Salix* were all present. Kapp (1970) interprets these data as representing a pine savanna in the region, with *Picea* and *Pseudotsuga* as minor constitutents of the arboreal coniferous element. Grasses and composites presumably were abundant on the upland savanna while deciduous trees were restricted to stream margins and other mesic habitats.

By Late Illinoian time *Pinus* pollen percentages begin to drop to low values (about 21%), *Picea* disappears from most sites and there is a concomitant rise in Gramineae/Compositae pollen (about 28%). *Juniperus* pollen appears at some sites (Kapp, 1970,

1965). These changes, interpreted as a general trend toward less effective precipitation, due either to increased temperatures, decreased rainfall, or both, continued in northwestern Oklahoma and southwestern Kansas through the Late Sangamon culminating in the development of caliche horizons (Kapp, 1965). Such a climatic change should result in expansion of grasslands.

At the Jinglebob site, also in southwestern Kansas, palynological evidence from latest Sangamon or Early Wisconsin time suggests the redevelopment of a flora requiring more mesic conditions. *Pinus* pollen percentages reach 10 to 64%, *Picea* reaches up to 8%, and Gramineae/Compositae pollen values are high, ranging from 24 to 80% (Kapp, 1970). Faunal evidence from the site suggests a climate with greater effective precipitation and milder summer temperatures which produced a generally equable climate during the period (Hibbard, 1955).

In west-central Missouri a Middle to Late Wisconsin pollen record provides our only information on interstadial and full-glacial vegetation in this area (King, 1973). The record, recovered from spring deposits, is a composite of profiles from five separate springs. The oldest deposits dated at 40,000 yrs. B.P. to approximately 20,000 yrs. B.P. are characterized by high percentages of *Pinus* pollen (5-70%), and NAP (40-80%) (comprised mostly of Cyperaceae but also containing Gramineae, *Ambrosia*-type and other Compositae). Low percentages of *Picea* pollen (about 1%) were recorded (King, 1973). Pollen of deciduous trees including *Betula, Quercus, Salix, Alnus, Fraxinus, Carya,* and *Ostrya/Carpinus* was present but of minor importance. King (1973) interprets this flora as being an open *Pinus*-parkland. *Picea* was probably not present in this part of Missouri and the absence of *Picea* macrofossils and the low levels of *Picea* pollen support this conclusion (King, 1973). Macrofossils of *Pinus banksiana* indicate that this pine grew near the spring in Middle Wisconsin time (King, 1973).

The vegetation of southwestern Kansas was quite unlike the *Pinus*-parkland in west-central Missouri. Vertebrate remains from the Jones local fauna (about 26,000 yrs. B.P.) suggest that moist grassland was present and few, if any, trees grew near the site (Davis, 1975).

In northeastern Kansas at Muscotah and Arrington marshes, Grüger (1973) reports that pollen of *Picea,* Gramineae, Compositae, and *Artemisia* along with Cyperaceae (to 85%) reached high values prior to approximately 24,000 yrs. B.P. Deciduous trees pollen include *Quercus, Fraxinus, Ostrya/Carpinus,*

Carya, and *Platanus,* each composing less than 2% of the pollen spectra. *Alnus* (25%), *Betula* (to 9.4%) and *Salix* (to 14.3%) were important members of the local flora, perhaps growing on floodplains (Grüger, 1973). *Picea* and *Pinus* were present in northeast Kansas at this time, yet the relatively low pollen percentages suggest that there were not extensive stands of these trees. Perhaps the *Pinus*-parkland of west-central Missouri extended into northeastern Kansas with *Picea* only in favorable localities.

By approximately 20,000 yrs. B.P. the *Pinus*-parkland of west-central Missouri was declining and *Picea* forest had replaced the parkland. *Picea* pollen values from the spring deposits in Missouri reach 73 to 92% with very low values of arboreal deciduous pollen types. King (1973) concludes that the western Ozark highlands were covered by a closed spruce forest at this time.

In northeastern Kansas at Muscotah and Arrington marshes, the shift to a *Picea* forest is dated at approximately 24,000 yrs. B.P. (Grüger, 1973). Although not dated, Grüger (1973) places the end of *Picea* dominance at approximately 11,500 yrs. B.P. By approximately 9,900 yrs. B.P. *Picea* was gone and prairie, marked by an increase in *Ambrosia* pollen, became dominant in northeastern Kansas (Grüger, 1973). Beginning at approximately 5,100 yrs. B.P. *Ambrosia* and *Artemisia* percentages declined and pollen of deciduous forest trees increased, signaling the end of true prairie in northeastern Kansas and the development of the modern *Quercus-Carya* prairie woodland (Grüger, 1973).

In Missouri the *Picea* forest declined and was invaded by arboreal deciduous elements by approximately 16,500 yrs. B.P. and by 13,600 yrs. B.P. a mixed *Picea* deciduous forest with *Fraxinus, Alnus, Salix, Quercus, Ulmus, Ostrya, Juglans,* and *Acer* was established (King, 1973). The vegetational changes associated with the change to the present prairie-deciduous forest are not recorded in the spring deposits in Missouri but faunal evidence from Rodgers Cave, Missouri (McMillan, 1976) near the spring sites indicates that the prairie border moved eastward approximately 8,500 yrs. B.P.

In southern Oklahoma a series of cores taken from Ferndale Bog record a *Quercus*-savanna prior to approximately 5,200 yrs. B.P. The *Quercus*-savanna was replaced by an essentially modern *Quercus-Carya* forest approximately 2,700 yrs. B.P. (Albert, 1981). After this time *Pinus* pollen values increase, followed by repeated fluctuations in dominance by *Quercus-Carya* and *Pinus* that continued to the recent (Albert, 1981). In northeastern Oklahoma a record recovered from several rock-shelter sites, spanning approximately the last 2,000 years, suggests that an essentially modern *Quercus-Carya* forest has been dominate throughout this period (Hall, 1982).

NORTHERN PLAINS

No early or middle-Quaternary pollen sequences are known from this area. The late-Quaternary environments of the northern plains have been thoroughly reviewed (Wright, 1970, 1981; Wells, 1970a, 1970b), and not much new work has been done since those reviews. A number of pollen sites along the eastern plains border are late- and post-glacial in age, but a few sequences are known that predate the Late Wisconsin.

A peat bed exposed along Waterman Creek in northwestern Iowa and dated at greater than 40,000 yrs. B.P. contains over 80 percent *Picea* pollen throughout its thickness (Van Zant, 1983). The deposit lies between Tazewell till (20,000 to 25,000 years old in two dated sections in Iowa; Ruhe, 1969; Kemmis *et al.*, 1981) and pre-Illinoian till, and it probably is Early Wisconsin in age. It likely represents a closed *Picea* forest either during an early stage or near the end of a glacial advance.

The sequence at Wolf Creek in Minnesota, is outside our area, but is briefly described here because of its age, location near the prairie border, and significant flora. It lies in the conifer/hardwood forest about 80 km from the prairie border and dates from 20,500 to about 9,000 yrs. B.P. (Birks, 1976). Pollen and plant macrofossils of a diverse assemblage of tundra plants indicates that a tundra-like environment existed in central Minnesota from 20,500 to 14,700 yrs. B.P. A transitional shrubland with *Shepherdia canadensis, Empetrum, Betula* (probably *B. glandulosa), Salix,* and *Alnus* developed between 14,600 and 13,600 yrs. B.P. This assemblage gave way to a *Picea*-dominated woodland from 13,600 to 10,000 yrs. B.P. From 10,000 to about 9,000 yrs. B.P., *Pinus banksiana* replaced *Picea* to form a mixed forest with a variety of deciduous trees (Birks, 1976). Similar tundra-like environments are indicated for two eastern Iowa sites. An assemblage of small mammals dated at 20,530±130 yrs. B.P. (BETA-2748) in northeastern Iowa contains tundra species and has a sympatry in the central and northern Yukon Territory (Woodman, 1982). The Conklin Quarry site in southeastern Iowa apparently represents a short depositional interval and has a date of 17,170±205 (DIC-240); it contains an insect,

mammal, and molluscan fauna and a flora that indicate tundra conditions at or close to treeline (Baker, *et al.,* 1982). The pollen diagram is dominated by *Picea, Pinus* and Cyperaceae, and is not very informative. The floral macrofossils include tundra species such as *Dryas integrifolia, Betula glandulosa, Saxifraga aizoides* and *Selaginella selaginoides,* along with *Picea* needles. Both sites are close to the present prairie-forest border. It is not known whether areas further out on the Great Plains ever supported any areas of tundra.

Pollen diagrams of Farmdalian (about 28,000 to 22,000 yrs. B.P.) interstadial deposits in eastern Iowa show a pattern of change similar to Kings (1973) Missouri sites. From about 34,000 to 28,000 yrs. B.P. *Picea* and *Pinus* were co-dominants in pollen profiles, and Gramineae and Cyperaceae were also present in small amounts (Fig. 2). The vegetation postulated at this time is an open *Picea-Pinus* forest in eastern Iowa. *Picea* and *Pinus* were even more abundant from about 28,000 to 22,000 yrs. B.P., and herb pollen was almost completely absent, suggesting closed forest conditions. At the end of the Farmdalian interstadial, *Pinus* pollen percentages decline sharply, and herb percentages rise again. *Picea* pollen percentages rise in some diagrams and remain constant in others (Fig. 2). Apparently this period is the time of expansion of Wisconsin ice sheets, and the vegetation again became more open in eastern Iowa (Mundt and Baker, 1979; Hallberg, *et al.,* 1980; Van Zant, *et al.,* 1980).

In the loess hills of Western Iowa, the only record of Farmdalian environments is the Craigmile local fauna dated at 23,240 yrs. B.P. (DIC-1369) (Rhodes, 1984). The modern habitats of 31 taxa of small mammals recovered in an alluvial fill indicate that a predominantly grassland habitat prevailed. Some conifer and deciduous trees were apparently present, because a few of the mammals found require them. These trees probably inhabited favorable exposures on steep slopes.

The later Waubonsie local fauna (14,830+1060-1220 (DIC-1688) and 14,430±1030 (I-7496) yrs. B.P.) from the same area yielded 23 taxa of small mammals that indicated a conifer forest with hardwoods present locally and areas of open northern grasslands (Rhodes, 1984). Apparently the environments of western Iowa were more open in both Farmalian and late-glacial times than those in eastern Iowa.

Closed *Picea* forests with *Fraxinus nigra* and *Larix* apparently dominated the early late-glacial in Iowa shortly after 14,000 yrs. B.P. (Van Zant, 1979; Van

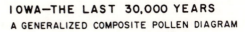

IOWA—THE LAST 30,000 YEARS

A GENERALIZED COMPOSITE POLLEN DIAGRAM

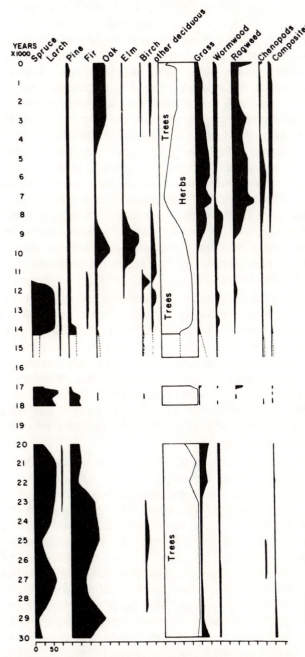

Figure 2. A generalized and composite pollen diagram for the last 30,000 years in Iowa. The period from 20,000 to 30,000 years is derived from several Farmdalian pollen sequences in eastern Iowa. The 17,000 to 18,000 section comes from two short sequences in eastern Iowa. The upper 14,000 years is principally derived from Lake West Okoboji (Van Zant, 1979).

Zant and Hallberg, 1976; Kim, 1982; Szabo, 1980). By about 12,500 years ago deciduous trees such as *Quercus rubra* and *Acer negundo,* and *Corylus cornuta* were present within the *Picea-Larix* forests (Baker, *et al.,* 1980). *Picea* and *Larix* disappeared about 11,000 yrs. B.P., and deciduous trees, especially

Quercus and *Ulmus* were dominant in west-central Iowa until about 9,000 yrs. B.P. (Van Zant, 1979; Kim, 1982, Baker and Van Zant 1980; Hudak, 1984). The deciduous forest declined as prairie taxa began to encroach about 9,000 yrs. B.P., and by 7,700 yrs. B.P. the prairie was dominant. Prairie continued to cover most of Iowa throughout the Holocene (Kim, 1982), but *Quercus* did return to selected sites about 3,000 yrs. B.P. (Fig. 2) (Van Zant, 1979).

Further west, the Rosebud site in southernmost South Dakota indicates that *Picea* was dominant in the vegetation about 12,500 yrs. B.P. in the northern Nebraska Sand Hills (Watts and Wright, 1966). Both *Picea* and *Larix* were present in the basal sediments from Woodworth Pond in central South Dakota (McAndrews, *et al.,* 1967.) *Picea* wood was also found in basal pond sediments at Tappen in central North Dakota (Moir, 1958), dated at 11,480 yrs. B.P. (W-542). At Hafichuk in south-central Saskatchewan, dated from about 11,600 to 10,000 yrs. B.P. (Ritchie and de Vries, 1964) macrofossils indicate that *Picea* was accompanied by *Populus (P. balsamifera, P. tremuloides,* and *P. deltoides).* Clearly the northern plains were covered by *Picea* and in places *Larix* and *Populus* spp. during the late-glacial.

The southern limit of *Picea* on the Great Plains during Wisconsin time is unclear. Bryant (1977) suggested that the southern boundary may have been in south-central Texas but low *Picea* pollen percentages on the Texas high plains indicate that it was not abundant. Many sites in southeastern United States indicate that *Picea* was widespread at the latitude of Texas and may have reached the Gulf Coast (Watts, 1983). The southern limit of *Larix, Populus,* and other trees on the Great Plains in Wisconsin time is not known.

In the central plains of Nebraska, Kansas, and Oklahoma terrestrial molluscs from Wisconsin loess are dominated by prairie and woodland-border taxa (Leonard and Frye, 1954). These faunas imply open prairies with riparian woodlands along streams. Graham (1981) and Lundelius, *et al.,* (1983) summarize western Great Plains mammalian faunas, and conclude that the Late Wisconsin vegetation was prairie despite a cooler climate with greater effective moisture. Graham (1981) suggests that an ecotone existed on the central plains between grasslands on the west and *Picea* forests on the north and east during the Late Wisconsin. Several faunas from southwestern Wyoming and northeastern Colorado indicate that these areas became forested between about 11,000 and 8,000 yrs. B.P. (Roberts, 1970; Graham, 1981). Perhaps remnants of these forests remained to form the *Pinus* woodland at 4,000 yrs. B.P., as indicated by macrofossils of *Pinus ponderosa* and *Juniperus scopulorum* in pack rat (genus *Neotoma)* middens of the Laramie Basin in southern Wyoming (Wells, 1970a).

WYOMING

The records of past vegetation are also scarce in Wyoming. Pollen studies from Miocene sediments in Jackson Hole, northwestern Wyoming, indicate that the pollen rain was dominated by *Pinus,* with lower percentages of *Picea, Abies* Cupressaceae, *Artemisia,* other Compositae, *Sarcobatus* and other Chenopodiaceae (Barnosky, 1984). Except for the presence of *Ulmus-Zelkova* and other Teritiary relicts, this pollen assemblage could be modern. Other indications of the early modernization of the Rocky Mountain flora are detailed in Leopold (1967), who shows that 90% of the genera in Miocene megafossil floras from Wyoming and Colorado are now native to these states. Late-Pliocene pollen samples from Yellowstone National Park also indicate virtually modern conditions (Richmond, *et al.,* 1978). These samples also are dominated by *Pinus,* with lower percentages of *Picea, Abies, Artemisia,* Chenopodiaceae, and Gramineae.

Yellowstone National Park also contains the most complete Quaternary record in the Rocky Mountains. A number of unpublished pollen diagrams from early and middle-Pleistocene deposits by E. B. Leopold and William Mullenders show fluctuations from *Pinus*-dominated to *Artemisia*-dominated pollen spectra. Based on late-Pinedale (Wisconsin) to present pollen sequences (Baker, 1976, Waddington and Wright, 1974), the *Pinus*-dominated intervals represent conditions comparable to the Holocene, and *Artemisia*-dominated periods represent colder conditions analogous to late-glacial environments. A series of pollen diagrams from interstadial and interglacial sequences (Baker and Richmond, 1978; Baker, 1981; Baker in preparation; Richmond and Bradbury, 1982) indicate similar fluctuating conditions during probable Sangamon and Early Wisconsin time (Fig. 3). Three of the sections (EP-5, EP-6, and EP-8) are found in stratigraphic succession in the southeastern quarter of Yellowstone Park. Another section is at Grassy Lake Reservoir adjacent to the southwest corner of the park. Radiocarbon dates and potassium-argon dates provide some chronologic control for these sections (Baker and Richmond, 1978). The oldest of the sections (EP-6) contains a pollen sequence that begins at the base with a zone of *Artemisia* dominance. This zone is considered to

represent cold conditions and an open, tundra-like environment (Fig. 3). *Picea, Abies,* and *Pinus albicaulis* dominate in the overlying zone, and cool, subalpine forests like those below the modern treeline were prevalent. *Picea* and *Abies* are less important in the third zone above the base, and *Pinus* was dominant as it is in the modern forest. The uppermost zone in section EP-6 has a peak of *Pseudotsuga* pollen percentages, and numerous *Pseudotsuga* needles occur at this level. The vegetation was apparently a *Pseudotsuga* forest with accompanying *Pinus ponderosa* and *Pinus flexilis.* This foothills forest is now present at much lower altitudes than site EP-6, and such an assemblage indicates a climate warmer than any in the Holocene. Based on its stratigraphic position and the warm climatic reconstructions this section is believed to be Sangamon and correlative with Oxygen Isotope stage 5e (Shackleton and Opdyke, 1973).

Slightly higher in the same sequence, a small stream cut exposes a short sequence (EP-5) dominated by *Picea, Abies,* and *Pinus.* This assemblage indicates a return to cooler conditions characteristic of the modern subalpine environment (Fig. 3). These two pollen sequences (EP-6 and EP-5) occur below two gravel units that underlie a peat radiocarbon-dated at 68,000 +2,200 –1700 yrs. B.P. (GrN-7332) (Grootes, 1977). This date supports a probable Sangamon age.

The Grassy Lakes Reservoir pollen sequence shows changes similar to those in the lower zones at EP-6. A lower *Artemisia*-dominated zone is interpreted to represent tundra, followed by a zone dominated by *Picea, Abies,* and *Pinus albicaulis*-type pollen, indicating subalpine forest. This subalpine assemblage is replaced upward by pollen spectra dominated by *Pinus contorta* type. This third zone indicates that a *Pinus contorta* forest similar to the modern forest grew near the site. However, the percentages of *Picea* and *Abies,* indicative of subalpine conditions, remain higher than they are in the modern pollen rain at Grassy Lake Reservoir (Baker and Richmond, 1978). Thus conditions are interpreted as cooler than the present for the *Pinus contorta* zone. The uppermost zone shows a return to dominance of *Artemisia* and other non-tree pollen types (Baker and Richmond, 1978). This change is believed to indicate a return to cold, open, tundra-like conditions (Fig. 3).

At section EP-8, most of the thick section of lake sediments is characterized by an *Artemisia*-dominated pollen assemblage interrupted only by a short segment containing low peaks of *Pinus albicaulis, Picea,* and *Abies* (Baker and Richmond, 1978).

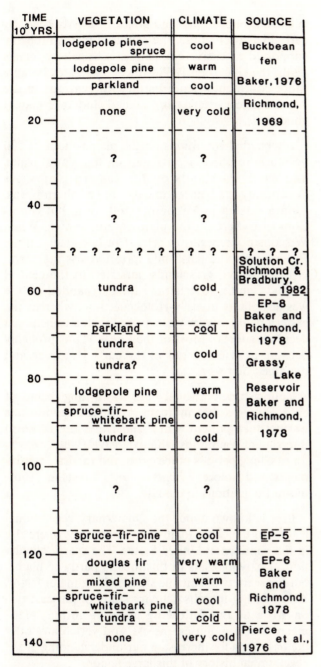

TIME 10³ YRS.	VEGETATION	CLIMATE	SOURCE
	lodgepole pine-spruce	cool	Buckbean fen Baker, 1976
	lodgepole pine	warm	
	parkland	cool	
20	none	very cold	Richmond, 1969
	?	?	
40	?	?	
	? – ? – ? – ? – ?	? – ?	? – ? – ? Solution Cr. Richmond & Bradbury, 1982
60	tundra	cold	EP-8 Baker and Richmond, 1978
	parkland	cool	
	tundra	cold	
	tundra?	cold	Grassy Lake Reservoir Baker and Richmond, 1978
80	lodgepole pine	warm	
	spruce-fir-whitebark pine	cool	
	tundra	cold	
100	?	?	
120	spruce-fir-pine	cool	EP-5
	douglas fir	very warm	EP-6 Baker and Richmond, 1978
	mixed pine	warm	
	spruce-fir-whitebark pine	cool	
	tundra	cold	
140	none	very cold	Pierce et al., 1976

Figure 3. Tentative sequence of environmental changes for the last 135,000 years in Yellowstone National Park. Derived from Baker, 1976; Baker and Richmond, 1978; Baker, 1981; Richmond and Bradbury, 1982; Baker, unpublished manuscript.

This long sequence indicates that very cold, tundra-like conditions were dominant through much of the Early Wisconsin. These conditions were interupted, apparently briefly, by a short period of subalpine forest (Fig. 3). The 68,000 year date reported above comes from peat near the base of this section.

Richmond and Bradbury (1982) report an *Artemisia*-dominated assemblage throughout a small section

at Solution Creek, radiocarbon-dated at about 54,000 ±500 yrs. B.P. (GrN-8140). Apparently cold tundra-like conditions continued to prevail in Early Wisconsin time (Fig. 3).

No known pollen records date between 54,000 and 14,000 in Yellowstone Park. Pierce *et al.,* (1976), Pierce (1979), and Richmond (1969) indicate that Pinedale glaciers covered much of Yellowstone National park from about 30,000 or more to 15,000 yrs. B.P. Glaciers retreated and sediment began to accumulate in kettle ponds about 14,000 yrs. B.P. Pollen spectra in this lowermost zone are dominated by *Artemisia* along with peak pollen percentages of *Abies, Picea, Pinus albicaulis* and Gramineae (Baker, 1976). Apparently a cool parkland became established as the ice retreated. *Pinus* pollen percentages rise to dominance about 12,000 yrs. B.P., and a short period of *Pinus albicaulis* dominance was superceded by high *Pinus contorta* pollen percentages. It seems clear that *Pinus contorta* forests covered much of the Yellowstone Plateau during the entire Holocene. However, a slight increase in *Picea* and *Abies* pollen percentages at sites above 2,250 m during the last 5,000 years indicates that a slight cooling occurred, causing these two taxa to move down the mountain slopes and onto the plateau (Baker, 1970, 1976; Waddington and Wright, 1974). At sites near the lower forest border (below 2,250 m), *Pinus contorta* and *Artemisia* remained at relatively high pollen percentages throughout the Holocene, and *Pinus* pollen reached peak levels within the last 1,500 years or so. Such a late peak suggests that the warmest interval of the Holocene may have lasted until about 1,500 years ago at low elevations, and then cooler climates resulted in a shift towards closed *Pinus contorta* forests.

Three sites in the Bighorn Mountains, all located in the middle of large belts of *Pinus contorta* forest, show no indication of Holocene vegetational changes. All are dominated by *Pinus contorta*-type pollen throughout post-glacial time (Burkart, 1976). However, a 24,000 year pollen record from moraine-lake sediments at low elevations along the south side of the Wind River Range (Bright, written communication, 1968) shows fluctuations from *Pinus* to *Artemisia* dominance like those in the Yellowstone Park sequences. The absence of more sites at low elevations precludes reconstruction of past vegetation in the foothills and basins during Wisconsin time. Many important questions can not be answered directly until such sites are found and investigated. For example, when tundra expanded down the mountainsides to cover present montane and subalpine sites, did all the forest zones migrate downward an equal amount? Did forests expand across basin areas in Wyoming during glacial episodes? Two lines of inference can be used to attempt to address these questions. One is the evidence from pack rat middens in southwestern U.S.; this evidence indicates that woodlands of *Juniperus* along with *Pinus edulis, P. cembroides,* and *P. monophylla* (pinyon pines) dominated in desert basins now supporting desert scrub species (Spaulding, *et al.,* 1983). However, the composition of these glacial-age woodlands was not the same as that in modern woodlands; many desert species not presently associated with these trees grew together with them at that time. The ecological response to changing climate was thus individualistic, rather than community-wide. If similar lowering and mixing of vegetational units occurred in Wyoming, woodlands could have spread across the Wyoming basins as well, and may have been mixed with steppe taxa in parkland habitats. We are unaware of any studies of pack rat middens in the Wyoming basins.

A second line of evidence is the vertebrate record in Wyoming. A number of recent investigations indicates that the Wyoming basins were a relatively open, cold environment (Walker, 1982). Terms like steppe-savannah or steppe tundra have been used to describe this environment. It is unclear whether plants presently restricted to tundra lived on the plains, or whether the plains were cold grasslands, parklands, or shrublands. If these open environments represent a downward shift in treeline and the level of the basin floors was above climatic treeline, the dislocation of treeline would be much greater than previously predicted. Such a shift implies that trees would be sparse over the whole state, and forests probably non-existent. However, if the basin vegetation was a biome no longer existent (steppe-savanna or steppe-tundra?), forests may have grown above this biome on the lower mountain slopes, with alpine tundra above the forest. These questions probably won't be answered until paleobotanical sites in the basins are located.

CONCLUSIONS

The Great Plains environment changed from subtropical forests in the Oligocene to savannas with substantial open grassland areas in the Miocene. By early-Pleistocene time, the southern plains probably supported extensive grasslands, with gallery forests along the rivers. Early-Pleistocene climates became more continental, and distinct fluctuations in the biota suggest alternating glacial and interglacial conditions.

Interstadial conditions on the plains about 25,000 years ago resulted in grasslands on the central plains and *Pinus*-parklands on the eastern plains. As Wisconsin ice advanced to cover the northern plains, tundra existed in Minnesota and Iowa, but conditions elsewhere on the plains are unknown. When Wisconsin ice began to retreat in late-glacial time, *Picea* forests and parklands spread across the northern plains, but grasslands persisted in the central plains. Conifer forests may have expanded on the western plains in early-Holocene time, whereas the eastern plains border was covered with a transitional deciduous forest. Most of the Great Plains probably became prairie-covered in the early-Holocene and remained that way.

Yellowstone National Park has a long record of environments that indicate interglacial conditions warmer than present with *Pseudotsuga* forests; warm interstadial conditions with extensive *Pinus contorta* forests; long periods of ice cover and tundra environments; and Holocene *Pinus contorta* forests that show minor fluctuations in climate.

Less is known about the Quaternary vegetational history of the Great Plains than that of anywhere else in the continental United States. The sparse population of palynologists in this area is partly to blame, but the scarcity of suitable sites is also a major factor. Sites offering interdisciplinary study of (at least two or more of the following groups) pollen, plant macrofossils, molluscs, insects, and vertebrates offer the best hope for increasing our understanding of the vegetational history of this large region.

ACKNOWLEDGEMENTS

We thank Rebecca Bush for typing the manuscript.

References Cited

ALBERT, L. E.

1981 Ferndale Bog and Natural Lake: five thousand years of environmental change in southeastern Oklahoma. *Oklahoma Archeological Survey Studies in Oklahoma's Past,* 7: 99 p.

BAKER, R. G.

1970 Pollen sequence from late Quaternary sediments in Yellowstone Park. *Science,* 168, 1449-1450.

1976 Late Quaternary vegetation history of the Yellowstone Lake Basin, Wyoming. *United States Geological Survey Professional Paper,* 729-E 48 p.

1981 Interglacial and interstadial environments in Yellowstone National Park. *In:* Mahaney, W. C., (ed), *Quaternary Paleoclimate.* Geo Books, University of East Anglia, Norwich, p. 361-375.

1983 Holocene vegetational history of western United States. *In:* Wright, H. E., Jr., (ed.), *Late Quaternary Environments of the United States, Volume II.* University of Minnesota Press, Minneapolis, p. 109-127.

1985 Paleoenvironmental records from Yellowstone National Park during the last glacial-interglacial cycle. In prep.

BAKER, R. G., RHODES, R. S., FREST, T. J., SCHWERT, D. P., and ASHWORTH, A. C.

1982 A full-glacial biota from Iowa, Midwestern USA. *XI INQUA Congress,* (abs.) 2: 18.

BAKER, R. G., and RICHMOND, G. M.

1978 Geology, Palynology, and Climatic significance of two pre-Pinedale lake sediment sequences in and near Yellowstone National Park. *Quaternary Research,* 10: 226-240.

BAKER, R. G., and VAN ZANT, K. L.

1980 Holocene vegetational reconstruction in northwestern Iowa. *In:* Anderson, D. C. and Semken, H. A. Jr., (eds.), *The Cherokee Excavations: Holocene Ecology and Human Adaptations in Northwestern Iowa.* Academic Press, New York, p. 123-138.

BAKER, R. G., VAN ZANT, K. L., and DULIAN, J. J.

1980 Three late-glacial pollen and plant macrofossil assemblages from Iowa. *Palynology,* 4: 197-203.

BARNOSKY, C. W.

1984 Late Miocene vegetational and climatic variations inferred from a pollen record in northwest Wyoming. *Science,* 223: 49-51.

BIRKS, H. J. B.

1976 Late-Wisconsinan vegetational history at Wolf Creek, central Minnesota. *Ecological Monographs,* 46: 395-429.

BORCHERT, J. R.

1950 Climate of the central North American grassland. *Annals of the Association of American Geographers,* 40: 1-39.

BRADBURY, J. P.

1980 Late Quaternary vegetation history of the central Great Plains and its relationship to eolian processes in the Nebraska Sand Hills. *In:* Geologic and paleoecologic studies of the Nebraska Sand Hills, *United States Geological Survey Professional Paper,* 1120-C, p. 29-38.

BRATTSTROM, B. H.

1961 Some new fossil tortoises from western North America with remarks on the zoogeographic and paleoecology of tortoises. *Journal of Paleontology,* 35: 543-560.

BRYANT, V. M., JR.

1977 A 16,000 year pollen record of vegetation change in central Texas. *Palynology,* 1, 143-156.

BRYSON, R. A., BAERREIS, D. A., and WENDLAND, W. M.

1970 The character of late-glacial and post-glacial climatic changes. *In:* Dort, W., Jr., and Jones, J. K., Jr., (eds.), *Pleistocene and Recent Environments of the Central Great Plains.* University of Kansas Press, Lawrence, p. 53-74.

BURKART, M. R.

1976 Pollen biostratigraphy and late Quaternary vegetation history of the Bighorn Mountains, Wyoming. University of Iowa Ph.D. Dissertation, 70 p.

DAVIS, L. C.

1975 Late Pleistocene geology and paleoecology of the Spring Valley Basin Meade County, Kansas. University of Iowa Ph.D. Dissertation, 170 p.

DAWSON, M. R.

1967 Lagomorph history and the stratigraphic record. *In:* Teichert, C. and Yochelson, E. L. (eds.), *Essays in Paleontology and Stratigraphy.* University of Kansas Department of Geology Special Publication 2. University of Kansas Press, Lawrence, p. 286-316.

DORT, W. R., JR., and JONES, J. K., JR., (eds.)

1970 *Pleistocene and Recent Environments of the Central Great Plains.* University of Kansas Press, Lawrence, 433 p.

ELIAS, M. K.

1942 Tertiary prairie grasses and other herbs from the High Plains. *Geological Society of America Special Paper,* 41: 1-176.

GRAHAM, R. W.

1981 Preliminary report on late Pleistocene vertebrates from the Selby and Dutton archeological/paleontological sites, Yuma County, Colorado. *Contributions to Geology, The University of Wyoming,* 20: 33-56.

GROOTES, P. M.

1977 Thermal Diffusion isotopic enrichment and radiocarbon dating beyond 50,000 yrs. B.P. University of Groningen Ph.D. Dissertation.

GRÜGER, J.
1973 Studies on the Late Quaternary vegetation history of northeastern Kansas. *Geological Society of America Bulletin,* 84: 239-250.

HALL, S. A.
1982 Late Holocene paleoecology of the southern plains. *Quaternary Research,* 17: 391-407.

HALLBERG, G. R., BAKER, R. G., and LEGG, T.
1980 A mid-Wisconsinan pollen diagram from Des Moines County, Iowa. *Proceedings of the Iowa Academy of Science,* 87: 41-44.

HIBBARD, C. W.
1938 An Upper Pliocene fauna from Meade County, Kansas. *Transactions of the Kansas Academy of Science,* 40: 239-265.

1955 The Jinglebob interglacial (Sangamon?) fauna from Kansas and its climatic significance. *Contributions of the Museum of Paleontology, University of Michigan,* 12: 179-228.

1970 Pleistocene mammalian local faunas from the Great Plains and Central Lowland Provinces of the United States. *In:* Dort, W., Jr., and Jones, J. K., Jr. (eds.), *Pleistocene and Recent Environments of the Central Great Plains.* University of Kansas Press, Lawrence, p. 395-433.

HUDAK, C. M.
1984 Paleoecology of an early Holocene faunal and floral assemblage from the Dows local biota of north-central Iowa. *Quaternary Research,* 21: 351-368.

KAPP, R. O.
1965 Illinoian and Sangamon vegetation in southwestern Kansas and adjacent Oklahoma. *Contributions from the Museum of Paleontology,* University of Michigan 19: 167-255.

1970 Pollen analysis of pre-Wisconsin sediments. *In:* Dort, W., Jr. and J. K. Jones, Jr. (eds.), *Pleistocene and Recent Environments of the Central Great Plains,* University of Kansas Press, Lawrence, Kansas, p. 143-155.

KEMMIS, T. J., HALLBERG, G. R., and LUTENEGGER, A. J.
1981 Depositional environments of glacial sediments and landforms on the Des Moines Lobe, Iowa. *Iowa Geological Survey Guidebook Series* no. 6, 132 p.

KIM, H. K.
1982 Late-glacial and postglacial pollen studies from the Zuehl Farm site, north-central Iowa and the Cattail channel bog, northwestern Illinois. University of Iowa M.S. thesis, 57 p.

KING, J. E.
1973 Late Pleistocene palynology and biogeography of the western Missouri Ozarks. *Ecological Monographs,* 43: 539-565.

KÜCHLER, A. W.
1964 Potential natural vegetation of the conterminous United States. *American Geographical Society, Special Publication,* 36: 38 p.

LEOPOLD, E. B.
1967 Late-Cenozoic patterns of plant extinction. *In:* Martin, P. S. and Wright, H. E., Jr. (eds.), *Pleistocene Extinctions: The Search for a Cause.* Yale University Press, New Haven, p. 203-245.

LEONARD, A. B., and FRYE, J. C.
1954 Ecological Conditions accompanying loess deposition in the Great Plains region of the United States. *The Journal of Geology,* 62: 399-404.

LUNDELIUS, E. L., JR., GRAHAM, R. W., ANDERSON, E., GUILDAY, J., HOLMAN, J. A., STEADMAN, D. W., and WEBB, S. D.
1983 Terrestrial vertebrate faunas. *In:* Porter, S. C. (ed.), *Late Quaternary Environments of the United States.* University of Minnesota Press, Minneapolis, p. 311-353.

MacGINITIE, H. D.
1962 The Kilgore flora, a late Miocene flora from northern Nebraska. *University of California Publications in Geological Sciences,* 35: 67-158.

McANDREWS, J. H., STEWART, R. E., JR., and BRIGHT, R. C.
1967 Paleoecology of a prairie pothole: A preliminary report. *In:* Clayton, L., and Frees, T. F. (eds.), Glacial geology of the Missouri Coteau and adjacent areas. *North Dakota Geological Survey Miscellaneous Series* 30, p. 101-113.

McMILLAN, R. B.
1976 Rodgers Shelter: a record of environmental and cultural change. *In:* Wood, W. R., and McMillan, R. B., (eds.) *Prehistoric Man and His Environment: A Case Study in the Ozark Highland.* Academic Press, New York, p. 111-122.

MOIR, D. R.
1958 Occurrence and radiocarbon date of coniferous wood in Kidder County, North Dakota. *In:* Laird, W. M., Lemke, R. W., and Hanson, M. (eds.) Mid-Western Friends of the Pleistocene Guidebook, 9th Annual Field Conference, *North Dakota Geological Survey, Miscellaneous Series,* 10. p. 108-114.

MUNDT, S., and BAKER, R. G.
1979 A mid-Wisconsinan pollen diagram from Black Hawk County, Iowa. *Proceedings of the Iowa Academy of Science,* 86: 32-34.

PIERCE, K. L.
1979 History and dynamics of glaciation in the northern Yellowstone National park area. *United States Geological Survey Professional Paper,* 729-F, 90 p.

PIERCE, K. L., OBRADOVICH, J. D., and FRIEDMAN, I.
1976 Obsidian hydration dating and correlation of Bull Lake and Pinedale glaciations near West Yellowstone, Montana. *Geological Society of America Bulletin,* 87: 703-710.

RETALLACK, G. J.
1983 Late Eocene and Oligocene paleosols from Badlands National Park, South Dakota. *Geological Society of America Special Paper,* 193, 82 p.

RHODES, R. S. II
1984 Paleoecology and regional paleoclimatic implications of the Farmdalian Craigmile and Woodfordian Waubonsie mammalian local faunas, southwestern Iowa. *Illinois State Museum Reports of Investigations,* 40, 51 p.

RICHMOND, G. M.
1969 Development and stagnation of the last Pleistocene icecap in the Yellowstone Lake Basin, Yellowstone National Park, USA. *Eiszeitalter und Gegenwart,* 20: 196-203.

RICHMOND, G. M., MULLENDERS, W., and COREMANS, M.
1978 Climatic implications of two pollen analyses from newly recognized rocks of latest Pliocene age in the Washburn Range, Yellowstone National Park. *United States Geological Survey Bulletin* 1455, 13 p.

RICHMOND, G. M., and BRADBURY, J. P.
1982 Environmental amelioration at beginning of early/middle Wisconsin Interstadial about 54,000 B.P. in Yellowstone National Park. *American Quaternary Association Program and Abstracts, 7th Biennial Conference,* University of Washington, Seattle, p. 156.

RITCHIE, J. D., and DE VRIES, B.
1964 Contributions to the Holocene paleoecology of west central Canada. A late-glacial deposit from the Missouri Coteau. *Canadian Journal of Botany,* 42: 677-692.

ROBERTS, M. F.
1970 Late glacial and postglacial environments in southeastern Wyoming. *Palaeogeography, Palaeoclimatology, Palaeoecology,* 8: 5-17.

RUHE, R. V.
1969 *Quaternary Landscapes in Iowa.* Iowa State University Press, Ames, 255 p.

SHACKLETON, N. J., and OPDYKE, N. D.
1973 Oxygen isotope and palaeomagnetic stratigraphy of equatorial Pacific core V28-238: Oxygen isotope temperatures and ice volumes on a 10^5 year and 10^6 year scale. *Quaternary Research,* 3: 39-55.

SHOTWELL, J. A.
1961 Late Tertiary biogeography of horses in the northern Great Basin. *Journal of Paleontology,* 35: 203-217.

SPAULDING, G. W., LEOPOLD, E. B., and VAN DEVENDER, T. R.
1983 Late Wisconsin paleoecology of the American Southwest. *In:* Porter, S. C. (ed.), *Late Quaternary Environments of the United States, Volume I.* University of Minnesota Press, Minneapolis, p. 259-293.

SZABO, J. P.
1980 Two pollen diagrams from Quaternary deposits in east central Iowa, U.S.A. *Canadian Journal of Earth Sciences,* 17: 453-458.

TAYLOR, D. W.
1960 Late Cenozoic molluscan faunas from the High Plains. *United States Geological Survey Professional Paper* 337: 94 p.

THOMASSON, J. R.
1979 Late Cenozoic grasses and other angiosperms from Kansas, Nebraska, and Colorado: Biostratigraphy and relationships to living taxa. *Kansas Geological Survey Bulletin,* 218, 68 p.

1980 *Archaeoleersia nebraskensis* Gen. et sp. nov. (Gramineae-Oryzeae), a new fossil grass from the late Tertiary of Nebraska. *American Journal of Botany,* 67: 876-882.

VAN ZANT, K. L.
1979 Late glacial and postglacial pollen and plant macrofossils from Lake West Okoboji, northwestern Iowa. *Quaternary Research,* 12: 358-380.

1983 Lithological and palynological analyses of a Quaternary exposure, Waterman Creek, O'Brien County, Iowa. *Proceedings of the Iowa Academy of Science,* 90: 125-127.

VAN ZANT, K. L., and HALLBERG, G. R.
1976 A late-glacial pollen sequence from northeastern Iowa: Sumner Bog revisited. *Iowa Geological Survey Technical Information Series* 3: 17 p.

VAN ZANT, K. L., HALLBERG, G. R., and BAKER, R. G.
1980 A Farmadalian pollen diagram from east-central Iowa. *Proceedings of the Iowa Academy of Science,* 87: 52-55.

WADDINGTON, J. C. B., and WRIGHT, H. E., JR.
1974 Late Quaternary vegetational changes on the east side of Yellowstone National Park, Wyoming. *Quaternary Research,* 4: 175-184.

WALKER, D. N.
1982 Early Holocene vertebrate fauna. *In:* Frison, G. C. and Stanford, D. J. (eds.), *The Agate Basin Site: A Record of the Paleoindian Occupation of the Northwestern High Plains.* Academic Press, New York, New York, p. 274-309.

WATTS, W. A.
1983 Vegetational history of the eastern United States 25,000 to 10,000 years ago. *In:* Porter, S. C. (ed.), *Late Quaternary Environments of the United States, Volume I.* University of Minnesota Press, Minneapolis, p. 294-310.

WATTS, W. A., and WRIGHT, H. E., JR.
1966 Late Wisconsin pollen and seed analysis from the Nebraska sandhills. *Ecology,* 47: 202-210.

WELLS, P. V.
1970a Postglacial vegetational history of the Great Plains. *Science,* 167: 1574-1582.

1970b Vegetational history of the Great Plains: A post-glacial record of coniferous woodland in southeastern Wyoming. *In:* Dort, W., Jr., and Jones, J. K., Jr. (eds.), *Pleistocene and Recent Environments of the Central Great Plains.* University of Kansas Press, Lawrence, p. 185-202.

WOODMAN, N.
1982 A subarctic fauna from the Late Wisconsinan Elkader Site, Clayton County, Iowa. University of Iowa MS thesis, 56 p.

WRIGHT, H. E., JR.
1970 A vegetational history of the central Plains. *In:* Dort, W., Jr., and Jones, J. K., Jr. (eds.) *Pleistocene and Recent Environments of the Central Great Plains.* University of Kansas Press, Lawrence, p. 157-172.

1981 Vegetation east of the Rocky Mountains 18,000 years ago. *Quaternary Research,* 15: 113-125.

LATE-QUATERNARY POLLEN RECORDS AND VEGETATIONAL HISTORY OF THE GREAT LAKES REGION: UNITED STATES AND CANADA

RICHARD G. HOLLOWAY
Palynology Laboratory
Department of Anthropology
Texas A&M University
College Station, TX 77843

VAUGHN M. BRYANT JR.
Department of Anthropology and
Department of Biology
Texas A&M University
College Station, TX 77843

Abstract

The Great Lakes Region is extremely important palynologically. Historically, this region was one of the first examined for fossil pollen in hopes that it would provide answers to questions of changing vegetational composition. Numerous reports are available from this region extending back to the late 1920's. Since 1960, however, a continually increasing data base consisting of radiocarbon dated pollen sequences has been accumulated. Much of this work has been conducted in an attempt to quantify the relationship between the pollen spectra and physical parameters such as climate and vegetational composition. Quantification of surface pollen assemblages has also been utilized to map existing boundaries and this approach has been extended to fossil pollen assemblages in an attempt to understand the vegetational response to changing environmental conditions.

We have presented an overview of the pollen analytical reports which are available from this region. We have described the major radiocarbon dated pollen sequences from this area and using these data have summarized the general trends of the vegetational history of this region during the last 20,000 years.

INTRODUCTION

The Great Lakes Region is a vast area extending throughout a large portion of the North American mid-continent. It encompasses entirely, or in part, the states of New York, Pennsylvania, Ohio, Indiana, Illinois, Michigan, Wisconsin, Minnesota, Iowa, North and South Dakota, and the Canadian Province of Ontario. We have selected to ignore the present geopolitical boundaries in favor of defining the region based on similar physiographic and/or environmental parameters. In this sense we follow Whittaker's (1967) concept of a gradient distribution of vegetation. As such, some overlap of site areas necessarily must occur. For example, Gaudreau and Webb (this volume) have included sites in New York and Pennsylvania, while we have also included data from the western portions of those states. The same type of overlap also occurs between this chapter and the ones on Eastern Canada (Anderson, this volume), the Great Plains (Baker and Waln, this volume), and the Southeastern United States (Delcourt and Delcourt, this volume).

Historically, the Great Lakes Region has been of great importance in the development of Quaternary palynology and paleoenvironmental studies. Even as early as the 1920's this importance was recognized by Auer (1927, 1928) who conducted the first Quaternary age pollen analytical study in North America from deposits located within this area. The importance of this geographical region is based in part on the physiognomy and topography of the landscape, which formed as a direct result of successive periods of glacial activity. This geologic activity resulted in a landscape literally dotted with numerous lakes and bogs (Figures 1 and 2), most of which contain sediments which are capable of providing detailed histories of past environmental conditions. Pollen and plant macrofossil analyses have been one of the primary tools utilized to reconstruct the paleoenvironmental record of the region. Sites previously reported and discussed in this chapter are presented in Table 1.

The long history of Quaternary palynological research conducted in this region is, unfortunately, in part responsible for much of the confusion in the North American Quaternary pollen record. Following Auer's (1927, 1928) analysis, a number of additional

TABLE 1: PALEOENVIRONMENTAL SITES FROM THE GREAT LAKES REGION

Site Number	Site Name	Reference
NEW YORK SITES		
1	Nichols Brook	Calkin and McAndrews, 1980
2	Wintergulf	Calkin and McAndrews, 1980
3	Winter Springs	Schwert and Morgan, 1982
4	Lockport Gulf	Miller and Morgan, 1982
		Miller, 1973
5	Belmont Bog	Spear and Miller, 1976
6	Allenberg Bog	Miller, 1973b
7	Houghton Bog	Miller, 1973b
8	Genesee Valley Peat Works	Miller, 1973b
9	Unnamed	Miller, 1973b
10	Protection Bog	Miller, 1973b
PENNSYLVANIA SITES		
11	Rose Lake	Cotter and Crowl, 1981
12	Cory Bog	Droste et al., 1959
13	Crystal Lake	Walker and Hartman, 1960
14	Mud Lake	Walker and Hartman, 1960
15	Dollar Lake	Walker and Hartman, 1960
16	Old Log Road Bog	Walker and Hartman, 1960
17	Mercer Bog	Walker and Hartman, 1960
18	Columbus Bog	Walker and Hartman, 1960
OHIO SITES		
19	Battaglia Bog	Shane, 1975
20	Silver Lake	Ogden, 1966
21	Browns Lake	Sanger and Crowl, 1979
22	Bucyrus Bog	Sears, 1930b
23	Sunbeam Prairie Bog	Kapp and Gooding, 1964a
24	Canton Bog	Potter, 1947
25	Mogadore Bog	Potter, 1947
26	Mud Lake	Sears, 1931
27	Long Lake Bog	Potter, 1947
28	Luna Lake Bog	Potter, 1947
29	Orrville Bog	Potter, 1947
30	Tiro Bog	Potter, 1947
31	Savanna Lake Bog	Potter, 1947
32	Lodi Bog	Potter, 1947
33	Granger Bog	Potter, 1947
34	New Haven Bog	Potter, 1947
35	Peru Bog	Potter, 1947
36	New London Bog	Potter, 1947
37	Hartland Center Bog	Potter, 1947
38	Pittsfield Bog	Potter, 1947
39	Camden Lake Bog	Potter, 1947
40	Oberlin Bog	Potter, 1947
41	Birmingham Bog	Potter, 1947
42	Eaton Bog	Potter, 1947
43	Norwalk Bog	Potter, 1947
44	Castalia Bog	Potter, 1947
INDIANA SITES		
45	Bacon's Swamp	Engelhardt, 1959, 1962
		Otto, 1937
		Potzger, 1946
46	Fox Prairie	Engelhardt, 1959, 1962
47	Hendricks Lake	Potzger, 1943b
48	Myers Lake	Frey, 1959
49	Christensen Mastodon Site	Whitehead et al., 1972
50	Lake Cicott Bog	Smith, 1937
51	Cabin Creek Bog	Friesner and Potzger, 1946
52	Pretty Lake	Williams, 1974
53	Reed Bog	Griffin, 1950
54	Kokomo Bog	Howell, 1937
		Potzger, 1946
55	Cranberry Bog	Barnett, 1937
56	Otterbein Bog	Richards, 1937
57	Mill Creek Bog	Swickard, 1941
58	Yountsville Bog	Swickard, 1941
59	Hovey Lake	Crisman and Whitehead, 1975
60	Jeff Bog	Potzger, 1946
61	Tippecanoe Lake	Potzger, 1946
62	Townsend Farm	Kapp and Gooding, 1974
63	Hendley Farm	Kapp and Gooding, 1974
64	Whitewater Basin	Kapp and Gooding, 1964b
65	Unnamed	Potzger, 1946
ILLINOIS SITES		
66	Seminary School Basin	Grüger, 1972a
67	Pittsburg Basin	Grüger, 1972a
68	Hickory Ridge Basin	Grüger, 1972a
69	Macon County	Grüger, 1972a
70	Richland Creek Section	Grüger, 1972a
71	Turtle Pond	Griffin, 1951
72	Chatsworth Bog	Voss, 1937
		King, 1981
73	Volo Bog	King, 1981

Site Number	Site Name	Reference
74	North Manlius Bog	Voss, 1937
75	South Manlius Bog	Voss, 1937
76	Langley Bog	Voss, 1937
77	Buda Deposit	Voss, 1937
78	North Sheffield Bog	Voss, 1937
79	Canal Bog	Voss, 1937
80	Arlington Bog	Voss, 1937
81	Lily Lake	Voss, 1937
82	Voegelli Farm	Whitecar and Davis, 1982
83	Canton Bog	Smith and Kapp, 1964
MICHIGAN SITES		
84	Cheboygan County	Farrand et al., 1968
85	Vestaburg Bog	Gilliam et al., 1967
86	George Reserve	Andersen, 1954
87	Isle Royale	Potzger, 1954
		Raymond et al., 1975
88	Crystal Marsh	McMurray et al., 1978
89	Garden Peninsula	Kapp and Means, 1977
90	Pontiac	Stoutamire and Benninghoff, 1964
91	Douglas Lake	Wilson and Potzger, 1943a
92	Middle Fish Lake	Wilson and Potzger, 1943a
93	Tobico Marsh	Jones and Kapp, 1972
94	Montcalm County	Hushen et al., 1966
95	Algonquin Peat Bog	Kapp et al., 1969
96	Barney Lake	Kapp et al., 1969
97	South Haven	Zumberge and Potzger, 1955
98	South Haven	Zumberge and Potzger, 1956
99	Chippewa Bog	Bailey and Ahearn, 1981
100	Hartford Bog	Zumberge and Potzger, 1956
101	Upper Twin Lakes	Saarnisto, 1974
102	Prince Lake	Saarnisto, 1974
103	Thaller Mastodon Site	Held and Kapp, 1969
104	Smith Farm Site	Oltz and Kapp, 1963
105	Pitt Farm Site	Oltz and Kapp, 1963
106	Clinton County Sites	Miller, 1973a
107	Unnamed, Westenow CO	Potzger, 1946
108	Unnamed, Gratiot CO	Potzger, 1946
109	Unnamed, Midland CO	Potzger, 1946
110	Unnamed, Cheboygan CO	Potzger, 1946
111	Gilbert Bog	Potzger, 1946
112	Birge Bog	Potzger, 1946
113	Casnovia	Kapp, 1978
114	Wintergreen Lake	Manny et al., 1978
115	Lawrence Lake	Wetzel and Manny, 1977
116	Marl Pond	Kapp et al., 1977
117	Demont Lake	Kapp et al., 1977
118	Mud Lake	Kapp et al., 1977
119	Crystal Lake	Kapp et al., 1977
WISCONSIN SITES		
120	Lake Mary	Webb, 1974a
121	Two Creeks	Morgan and Morgan, 1978
		West, 1961
122	Tamarack Creek	Hansen, 1939
		Davis, 1979
123	Iola Bog	Schweger, 1969
124	Duck Creek Ridge	Schweger, 1969
125	Peters Quarry	Schweger, 1969
126	Sheep Ranch Road Bog	Potzger, 1943a
127	Gravel Pit Bog	Potzger, 1943a
128	Sunken Highway Bog	Potzger, 1943a
129	Four-mile Lake Bog	Potzger, 1943a
130	Draper Bog	Potzger, 1943a, 1946
131	Hell's Kitchen Lake	Swain, 1978
132	Dell's Bog	Hansen, 1937
133	Baraboo Bog	Hansen, 1937
134	Hub City Bog	Hansen, 1939
		Davis, 1977
135	Mormon Coulee Bog	Hansen, 1939
136	Douglas County	Wilson, 1938
137	Ranch Road Bog	Potzger, 1946
138	Forestry Bog Lake	Potzger, 1946
139	Allequah Lake	Potzger, 1946
140	Forest Lake Bog	Potzger, 1946
141	Blue Mounds Creek	Davis, 1977
142	Disterhaft Farm Bog	West, 1961
		Baker, 1970
143	Seidel Lake	West, 1961
144	Bog A	Potzger and Richards, 1941
145	Bog B	Potzger and Richards, 1941
146	Bog C	Potzger and Richards, 1941
147	Bog D	Potzger and Richards, 1941
148	Grassy Lake Mat	Potzger and Richards, 1941
149	Wood Lake	Heide, 1984
150	Kelly's Hollow	Heide, 1984
151	Stewart's Dark Lake	Heide, 1984
		Peters and Webb, 1979

Site Number	Site Name	Reference
MINNESOTA SITES		
152	Kirchner Marsh	Wright *et al.,* 1963
		Winter, 1962
153	Madelia	Jelgersma, 1962
		Winter, 1962
154	Lake Carlson	Wright *et al.,* 1963
155	Aitkin	Farnham *et al.,* 1964
156	Crooked Lake	Wilson and Potzger, 1943b
		Potzger, 1946
157	Ham Lake	Wilson and Potzger, 1943a
158	Bethel Bog	Wilson and Potzger, 1943a
159	Island Lake	Wilson and Potzger, 1943a
160	Tamarack Lake	Wilson and Potzger, 1943a
161	Big Woods	McAndrews, 1968
162	Stevens Pond	Janssen, 1967
163	Red Lake Bog	Griffin, 1977, 1978
164	Lake Sallie	Birks *et al.,* 1976
165	St. Clair Lake	Birks *et al.,* 1976
166	Elk Lake	Birks *et al.,* 1976
		Dean *et al.,* 1984
167	Wolf Creek	Birks, 1976
168	Glatsch Lake	Wright and Watts, 1969
		Norris and McAndrews, 1970
169	Martin Pond	McAndrews, 1966
170	Cindy Pond	McAndrews, 1966
171	Bog D Pond	McAndrews, 1966
172	Bog A Pond	McAndrews, 1966
173	Bad Medicine	McAndrews, 1966
174	McCraney Pond	McAndrews, 1966
175	Reichow Pond	McAndrews, 1966
176	Terhell Pond	McAndrews, 1966
177	Horse Pond	McAndrews, 1966
178	Faith Pond	McAndrews, 1966
179	Thompson Pond	McAndrews, 1966
180	Kylen Lake	Birks, 1981
181	Lake of the Clouds	Swain, 1972, 1973
		Craig, 1972
182	Norwood	Ashworth *et al.,* 1981
183	Rice Lake	McAndrews, 1969
184	Fox Pond	McAndrews, 1969
185	Rutz Lake	Waddington, 1969
186	North Branch	Fries *et al.,* 1961
187	Weber Lake	Fries, 1962
188	Horseshoe Lake	Cushing, 1967
189	White Lily Lake	Cushing, 1967
190	Hubbard Lake	Potzger, 1946
191	Henzel Lake	Potzger, 1946
192	Lake Superior	Maher, 1977
193	Gervais Farm	Ashworth, 1980
194	Koturanta Lake	Wright and Watts, 1969
195	Myrtle Lake	Janssen, 1968
IOWA SITES		
196	West Lake Okoboji	Van Zant, 1979
197	Dows	Hudak, 1984
198	Hughes Peat Bed	Hall, 1971
199	Colo Bog	Brush, 1967
200	McCulloch Bog	Brush, 1967
201	Jewell Bog	Brush, 1967
202	Nichols Silts	Baker *et al.,* 1980
203	Sumner Bog	Baker *et al.,* 1980
204	Brayton Gravel Pit	Baker *et al.,* 1980

Site Number	Site Name	Reference
305	Pioneer Creek	Szabo, 1980
206	Des Moines CO	Szabo, 1980
207	Woden Bog	Durkee, 1971
208	Butler Farm	Van Zant *et al.,* 1980
209	Black Hawk CO	Mundt and Baker, 1980
ONTARIO SITES		
210	Wasaga Beach	Terasmae 1979
211	Kenora	McAndrews, 1982
212	Lake Algonquin	Cwynar, 1977
213	Alfies Lake	Saarnisto, 1974
214	Antoine Lake	Saarnisto, 1974
215	Prince Lake	Saarnisto, 1974
216	Kincardine Bog	Karrow *et al.,* 1975
217	Eighteen Mile River	Karrow *et al.,* 1975
218	Cookstown Bog	Karrow *et al.,* 1975
219	Nicholston Cut	Karrow *et al.,* 1975
220	St. Mary's	Sigleo, Karrow, 1977
221	unnamed	Janson, Halfert, 1936
222	English River Bog	Wilson, Webster, 1943
223	Beckworth Bog	Wilson, Webster, 1943
224	Hawk Lake	Wilson, Webster, 1943
225	Little Savanna Bog	Wilson, Webster, 1943
226	Toronto	Churcher, Peterson, 1982
		Williams *et al.,* 1981
		Radforth, Terasmae, 1960
227	Greenleaf Lake	Cwynar, 1978
228	Mer Bleue Peat Bog	Mott, Camfield, 1969
229	Dows Lake Bogs (5)	Mott, Camfield, 1969
230	St. David's Gorge	Hobson, Terasmae, 1968
231	Harrowsmith Bog	Terasmae, 1968; Karrow, 1963
232	Victoria Road Bog	Terasmae, 1968; Karrow, 1963
233	Grieff Kettle Bog	Terasmae, 1968; Karrow, 1963
234	North Bay Bog	Terasmae, 1968
235	Wood Lake	Terasmae, 1968
236	Blind River Bog	Terasmae, 1968
237	Alderdale Bog	Terasmae, 1968
238	Attawapiskat Lake	Terasmae, 1968
239	Brampton	Terasmae, Matthews, 1980
240	Newington Bog	Potzger, Courtemanche, 1956
241	Alfred Bog	Potzger, Courtemanche, 1956
242	Guelph	Karrow *et al.,* 1982
243	Kitchener	Anderson, 1982
244	Don Beds	Radford, Terasmae, 1960
		Terasmae, 1960
245	Scarborough	Radford, Terasmae, 1960
		Terasmae, 1960
246	Maplehurst Lake	Mott, Farley-Gill, 1978
247	Sutton Ridge	McAndrews *et al.,* 1982
248	Pond Mills Road	McAndrews, 1981
249	Edward Lake	McAndrews, 1981
250	Found Lake	McAndrews, 1981
251	Lake Louis	McAndrews, 1981
252	Lac Yelle	McAndrews, 1981
253	Lake Ontario	McAndrews, Power, 1973
254	Ridgetown Island	Dreimanis, 1966
255	Mud Lake	Terasmae, Mott, 1964
256	Meach Lake	Terasmae, Mott, 1964
257	Meach Lake Bog	Terasmae, Mott, 1964
258	Mer Bleue Bog	Terasmae, Mott, 1964
259	McKay Lake	Terasmae, Mott, 1964
260	Manitoulin Island	Warner *et al.,* 1984

pollen studies from this region were undertaken. As was common during the early development period of pollen analysis, different methods of pollen extraction and statistical analysis of the resulting data were commonplace. Auer (1930), Voss (1933, 1937) and Sears (1930a, 1930b), among others, routinely examined relatively few pollen grains (100-150) per sample. This procedure continued throughout this early period even though Barkley (1934) demonstrated the utility of standardizing a 200 grain pollen count. A second major problem during this early period concerned the selection of pollen types used to construct a basic pollen sum. Many of these early reports analyzed only tree pollen types, often completely disregarding all other pollen grains. Later researchers, notably Hansen (1937, 1939) and Potzger and his students (Potzger, 1943a, 1943b, 1946, 1954;

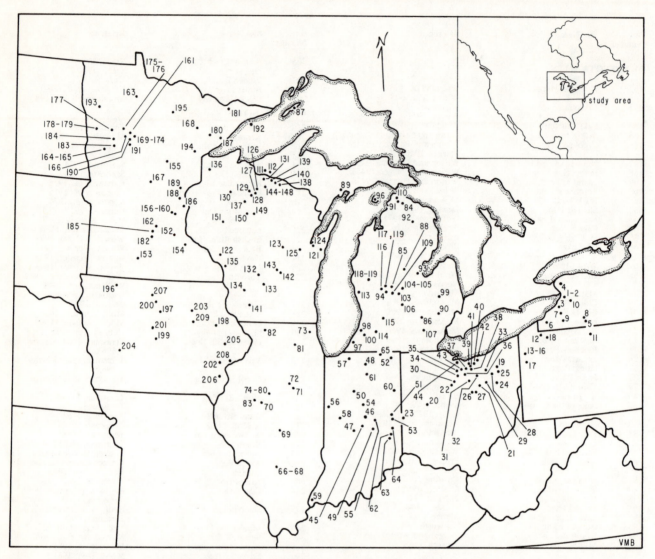

Figure 1. Map of Upper Midwest States showing pollen sampling site localities. Numbers refer to Table 1.

and his students (Potzger, 1943a, 1943b, 1946, 1954; Potzger and Richards, 1941; Barnett, 1937; Griffin, 1951; Howell, 1937; Otto, 1937; Richards, 1937) did record non-arboreal pollen (NAP) in their analyses, but at best these types were recorded as a percentage of the arboreal pollen (AP) and still were not included within the pollen sum. Thus, the majority of these early analyses are not statistically comparable to later studies. Unfortunately, some researchers continued this procedure through the 1960's and it has been only recently that palynologists working in the area have more or less standardized their methods.

Beginning with the development and increased usage of radiocarbon dating, a growing number of sites were added to the earlier data base. The establishment of well-dated radiocarbon sequences beginning in the 1960's for the first time allowed

researchers to investigate the timing and spatial analyses of the inferred changes.

Modern Pollen Rain Studies

For any geographical region it is essential to understand the modern pollen rain and to utilize these data to analyze the actual changes which have occurred in the fossil pollen record. Several previous studies have attempted to provide these data for the upper midwest (McAndrews, 1966; Janssen, 1967; Davis *et al.* 1971; Webb, 1973, 1974a, 1974b, 1974c; Webb and McAndrews, 1976; Webb *et al.* 1981; Peterson, 1978) and from adjacent Canada (Lichti-Federovich and Ritchie, 1968; King and Kapp, 1963). Later, Webb and McAndrews (1976) used these surface pollen spectra to construct pollen isopoll maps and

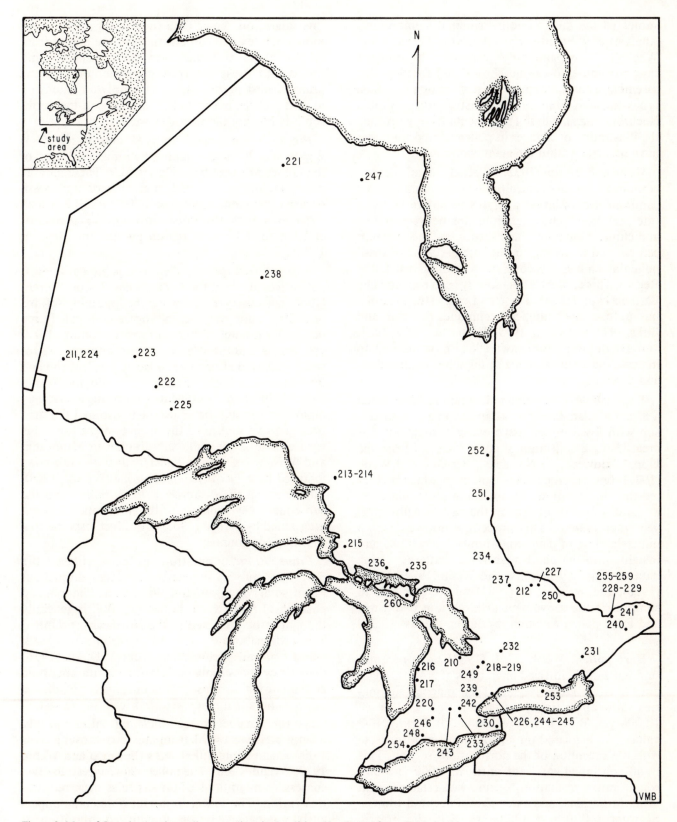

Figure 2. Map of Ontario showing pollen sampling site localities. Numbers refer to Table 1.

subjected the data to both Principal Components Analysis and Trend-surface Analysis. Bernabo and Webb (1977) expanded the techniques by applying this procedure to radiocarbon dated fossil pollen assemblages and mapping the distribution of pollen types through time by the construction of pollen isochrone maps. These refinements have permitted the illustration of changes in pollen composition that have occurred within the plant community.

Webb and Bryson (1972) collected climatic data in conjunction with surface pollen samples and used cannonical correlation analyses to obtain climatic interpretations. Once relationships between pollen and climatic data are established, these relationships can be used as analogs in the interpretation of fossil pollen assemblages recovered from the Great Lakes Region. This approach has recently been expanded by Bartlein et al. (1984). These and other studies utilizing multivariate statistical techniques (Gordon and Birks, 1974; Birks et al. 1975; Webb et al. 1978, 1981; and Heide and Bradshaw, 1982) have helped to increase our understanding of the forest composition and dynamics.

These studies of modern pollen rain have provided a necessary data base for extrapolation and comparison with fossil pollen assemblages throughout this area. Using a collection of cores obtained from the Upper Midwest and Northeast, Bernabo and Webb (1977) constructed pollen isopoll maps which permitted the identification of plant community migration at various periods of time in the past (usually 1,000 year increments). This procedure allowed for an understanding of plant community interactions and insights into the plant migrational patterns. Using data provided by Bernabo and Webb (1977), Davis (1976), and Webb et al. (1983) we are now able to summarize the movement of important plant taxa and fossil pollen assemblages during the last 15,000 years.

Herb pollen. Webb et al. (1983) have recently summarized the inferred migrations of NAP throughout the Great Lakes Region. Their mapped summaries show an initial low frequency of NAP at *ca.* 10,000 yrs. B.P., but by 9,000 yrs. B.P., the NAP values had increased throughout most of Minnesota with the exception of the northeastern section. This trend continued eastward until *ca.* 8,000 yrs. B.P. when prairie vegetation became well established in Iowa, Minnesota, and extended into central Illinois. Between 6,000 and 3,000 yrs. B.P. the prairie extended eastward but at the same time was decreasing westward across Minnesota. Webb et al. (1983) showed that the 20% contour for herb pollen reflected

this migration. Their data further suggested much more complex plant interactions in the period from 7,000-500 yrs. B.P. than previously suspected and noted that these interpreted movements in the prairie/forest border were not simply an extension toward the east but included variations in the north-south borders of the prairie as well.

Picea. Pollen isopoll maps (Webb et al., 1983) suggest that Picea populations had migrated north of the Upper Midwest by 8,000 yrs. B.P. These authors note that the most rapid migration of this taxon occurred between 11,000 and 10,000 yrs. B.P. which correlates with the most pronounced period of climate shift for that region (Webb and Bryson, 1972).

A generalized spruce pollen zone is present in most pollen records from the Great Lakes Region. Wright (1968) has observed that while the beginnings of this period are quite variable, the spruce dominated time period terminates rather uniformly around 11,000 yrs. B.P. revealing the very rapid climate change which occurred at this time (Ogden, 1967; Webb and Bryson, 1972). Later research (Barnabo and Webb, 1977; Webb et al., 1983) while confirming a relatively rapid climatic shift have also demonstrated the time transgressive nature of the spruce migration. This spruce zone is generally sub-divided in Minnesota and Wisconsin into a *Picea/Fraxinus* sub-zone followed by a *Picea/Betula* sub-zone (Wright, 1968). Furthermore, the decrease in *Picea* pollen and the subsequent development of the pine zone has been interpreted by Wright (1968) as a direct response to a warming environment.

Pinus. Pinus populations quickly replaced the northward moving populations of *Picea* throughout Minnesota, Michigan, and Wisconsin. The mapped position of *Pinus* at *ca.* 11,000 yrs. B.P. is generally thought to be restricted to the Northeastern United States (Bernabo and Webb, 1977; Webb et al., 1983; Gaudreau and Webb, this volume). Davis (1976) does show a few isolated localities containing pine pollen but for the most part is in agreement with the other authors that pines were restricted to the east.

As the spruce pollen declined, pine quickly immigrated westward from the east moving over 1,000 km in about 1,000 years (Bernabo and Webb, 1977; Wright, 1968). The pollen data suggest that two successive migrations of pine species occurred into this region. The initial immigration of pine apparently was caused by the northward migration of spruce brought on by the ameliorating climate. The spruce was quickly replaced by either *Pinus banksiana* or *P. resinosa,* both members of the subgenus haploxylon. Initially, the speciation of pine pollen grains was based solely on size measurements (Cain, 1940). Later investigators, utilizing the criteria of Ueno (1958, 1960) separated the taxa into two major groups: haploxylon and diploxylon. In the Great Lakes Region the initial migration of pines consisted of the haploxylon types with the diploxylon species *Pinus strobus* migrating into the area at a somewhat later time period.

The later arrival of *Pinus strobus* is recorded in sediments dating: to 10,500 yrs. B.P. at Volo Bog, Illinois (King, 1981); to 9,800 yrs. B.P. at Pretty Lake, Indiana (Williams, 1974); to 8,500 yrs. B.P. at Lake Mary, Wisconsin (Webb, 1974a); to 8,000 yrs. B.P. at Cedar Bog Lake, Minnesota (Cushing, 1963); and to 7,800 yrs B.P. in northwestern Michigan (Brubaker, 1975). According to Jacobson (1979) the white pine migration occurred in two successive periods. The initial migration of white pine into areas of Minnesota was interrupted by the mid-Holocene expansion of the prairie. With the decrease in prairie areas in Minnesota at *ca.* 4,000 yrs. B.P. the westward migration of white pine was accelerated, resulting in a second immigration of this species during this more recent period. The expansion of this species appears to have occurred from northeastern Minnesota in a southerly direction (Jacobson, 1979). Jacobson (1979) suggested that *Pinus strobus* migrated into open, mixed deciduous forests dominated by *Quercus* in central and southern Minnesota.

Quercus. In the eastern sections of the Midwest, *Quercus* shows a gradual trend to increasing values between 10,000 and 6,000 yrs. B.P. on a north to south gradient (Webb *et al.*, 1983). Post-6,000 year old sites reflect a southward migration of the conifer-hardwood/deciduous forest ecotone. In the western section, however, the patterns show more complex interactions. Webb *et al.* (1983) have interpreted the isochrones of *Quercus* as representing the location of its population relative to both the conifer-hardwood forest to the north, and the prairie vegetation to the west.

According to Jacobson (1979) *Quercus* migrated into Minnesota from either a south-west, or southerly direction (Jacobson, 1979). Watts and Bright (1968) reported its occurrence in sediments from Pickeral Lake, South Dakota, dated at 10,670 yrs. B.P., from Kirchner Marsh, Minnesota (Winter, 1962) and Cedar Bog Lake, Minnesota sediments (Cushing, 1963) at 10,000 yrs. B.P., and from Bog D (McAndrews, 1966) at 8,560 yrs. B.P. The migration occurred slightly later (*ca.* 7,000 yrs. B.P.) at Nelson Pond (Jacobson, 1979) and Jacobson Lake (Wright

and Watts, 1969). The eastward expansion of *Quercus* appears to have been stopped by the westward migration of *Pinus strobus* (Jacobson, 1979). He further suggested that it was *Quercus macrocarpa* which was initially replaced by *Pinus strobus* during this time period.

Fraxinus. *Fraxinus* and *Ulmus* were well established in the Great Lakes Region prior to 10,000 yrs. B.P. Webb *et al.* (1983) suggested that population expansions of these taxa were in response to larger scale environmental shifts. The high *Fraxinus* peaks tend to remain somewhat later in Illinois through Ohio than those occurring in southern Michigan and Minnesota. In both areas, however, the succession is from *Fraxinus* to *Alnus* and generally, *Fraxinus* follows immediately after the spruce peak, or co-occurs with *Picea*.

Tsuga. Davis (1976) estimated that populations of *Tsuga* also moved northward at a rapid rate of 200-300 m per year. It followed *Pinus strobus* by approximately 500-1,000 years. In many areas of the Upper Midwest, *Tsuga* appears to become dominant during the early-Holocene. At approximately 4,600 yrs. B.P. there is an almost synchronous decline in the percentages of this pollen type in sediments from New England through Minnesota (Davis, 1981; Webb, 1982; Gaudreau and Webb, this volume). The decline in *Tsuga* has been suggested as a chronostratigraphic marker for this region. Davis (1981) has suggested that this decline was the result of a pathogen attacking the immense hemlock forests throughout the northeastern United States.

PRE-WISCONSIN SITES

Indiana

Townsend Farm Section, Indiana. Kapp and Gooding (1974) examined an exposed black humic unit from Fayette County, Indiana. The authors designated two units, 4 and 5. Unit 4 (the lower) was a black humic unit. At the contact between Unit 4 and 5, a twig layer was present, thought to be Illinoian in age. The pollen from Unit 4 was badly degraded and consisted primarily of conifer types, but the pollen content was too sparse for interpretation. Based on the macrofossil data, the authors inferred that Unit 4 represented a rather cold period. The authors interpret this sequence as representing the Early Illinoian period rather than the Yarmouthian Interglacial as previously reported by Gooding (1966).

Hendley Farm Section, Indiana. The Hendley farm section is a stream bank exposure of lacustrine clay sediments which are thought to date to the Yarmouth Interglacial (Kapp and Gooding, 1974b). Eight meters of sediment were examined and the resulting pollen samples were divided into three zones. Zone I (oldest) consisted of high percentages of *Ostrya-Carpinus* (20-35%) pollen and other temperate species such as *Tilia, Corylus, Populus, Betula,* and *Pinus.* The upper portion of this zone shows the first appearance of *Carya* and decreasing percentages of *Pinus* pollen. In Zone II, *Ostrya-Carpinus* pollen decreases initially to 15% (2A) and finally to 5% (2B). There is a concurrent increase in the pollen frequencies of *Carya, Fraxinus, Acer,* and *Celtis. Planera* and *Liquidambar* pollen are also consistently present. In Zone III the pollen is partially degraded and is characterized as a zone of soil weathering, characterized by high *Carya, Populus, Pinus,* and Poaceae pollen, with decreases in the pollen percentages of other arboreal taxa. NAP increased to 25% in the uppermost sample. Kapp and Gooding (1974) proposed that the pollen record of this Yarmouthian exposure represented a time when temperatures were warmer than during either the Sangamon or the Holocene.

Illinois

Darrah Farm Sangamon Soil. This section is located less than one mile from the Smith Farm Section. The pollen profile (Kapp and Gooding, 1964b) is essentially similar to the Smith Farm Section. Englehardt (1962) also examined this section and his pollen sequences are very similar to those obtained by Kapp and Gooding (1964b), from near Canton, Illinois. Smith and Kapp (1964) reported on the analyses of an organic deposit from the Illinoian interstadial. The deposit consisted primarily of several *Larix* wood pieces directly above the basal till. Only two of the six sediment samples yielded close to a statistical pollen count. Of these, *Pinus* and *Picea* pollen were dominant but with deciduous pollen in low percentages. Two size ranges of *Picea* were also reported suggesting the presence of different species of *Picea.* Overall, these assemblages appear too weak for proper interpretation.

Iowa

Pioneer Creek, Iowa. A sedimentary deposit dated at >39,000 yrs. B.P. and identified as Pre-Illinoian in age was analyzed by Szabo (1980). The assemblage contained both *Picea* and *Pinus* pollen in about equal percentages with minor amounts of *Abies, Larix,* and some *Alnus, Salix, Betula,* and *Quercus.* The NAP component was dominated by Cyperaceae pollen with small amounts of *Ambrosia,* Cheno-am, and Poaceae pollen. Szabo (1980) interpreted these assemblages as representing a coniferous forest with small open areas present. The larch and fir was probably more local than the small percentages suggested.

EARLY- AND MIDDLE-WISCONSIN SITES

Indiana

Smith Farm Section, Indiana. Kapp and Gooding (1964b) reported on the pollen analysis of a Sangamon age soil from the Smith Farm. This soil was radiocarbon dated to greater than 37,500 yrs. B.P. Zone I, at the base of these deposits, was dominated by pine and spruce pollen. Zone II marked a decline of spruce pollen with *Pinus* reaching a maximum of 90%. Zone III reflected a decrease in pine pollen and a replacement by pollen of hardwoods, shrubs, and some herbs. This association was replaced in Zone IV by an oak-hickory forest containing elements of *Acer, Fagus, Fraxinus, Juglans, Liquidambar, Liriodendron,* and *Ostrya-Carpinus. Ulmus, Tsuga,* and *Larix* pollen also appear for the first time. The upper portion of Zone V is transitional in that the hardwood forest gives way to boreal conifers. Zone VI had higher percentages of *Picea* and *Abies* pollen than were present in the other levels. Abington Interstadial deposits of Illinoian age typically contained high percentages of spruce and pine pollen and were of short duration. Pollen analysis suggested that this Abington interstadial was of short duration. Deposits of New Paris (>38,000-40,000 yrs. B.P.) and Connersville (*ca.* 20,000 yrs. B.P.) Interstadial deposits from the Smith Farm section are also reported by Kapp and Gooding (1964b). These deposits contained abundant mollusk remains, mosses, wood fragments, and are very similar palynologically. They are both interpreted as representing forests dominated largely by conifers but again, with a small amount of hardwood pollen present.

Illinois

Voegelli Farm, Illinois. A 2.5 M peat deposit was reported by Whittecar and Davis (1982) with a basal radiocarbon age of 40,500 ± 1700 yrs. B.P. Below the peat are diamicton deposits which are believed to be Early or Middle Wisconsin in age. Zone I was dominated by *Pinus* (40-50%) and *Picea* (50%) with NAP less than 7%. In Zone II, *Picea* pollen ranges between 50-60%, *Pinus* between 25 and 30%, with *Betula, Alnus, Salix*, present throughout. This is interpreted to represent an increase in *Picea* at the expense of pine throughout Altonian and Farmdalian Periods which may indicate a southward and westward expansion of the spruce boreal forest. Peat accumulation stops before Late Wisconsin (Woodfordian) times and may be correlated with an increase in loess fall and colluviation.

Richland Creek Station, Illinois. Two profiles were examined by Grüger (1972) from these sites. The peat sample was clearly Farmdalian in age. *Pinus* and *Picea* pollen were dominant with small values of *Abies, Larix,* and Cupressaceae, and correspondingly lower values of NAP. Grüger (1972) interpreted this region as a boreal conifer forest. The Morton Loess profile was dated between 22,000 and 20,000 yrs. B.P. and presented an essentially similar pollen assemblage, Grüger (1972) apparently could not be sure if the *Pinus* and *Picea* pollen was the result of long distance transport or local in origin. Hence, he interpreted the assemblage as reflecting either a coniferous boreal forest or tundra depending on the actual source of pine and spruce pollen.

Michigan

Casnovia, Michigan. Kapp (1978) reported on the presence of both *Larix* and *Picea* wood dated from 25,050 ± 700 yrs. B.P. The wood and pollen evidence suggest an aboreal assemblage. It is suggested that the Casnovia deposits correlate with the Plum Point Interstadial assemblage from Ontario as reported by Dreimanis (1964).

Iowa

Des Moines County, Iowa. Hallberg *et al.* (1980) reported on a buried peat deposit which was radiocarbon dated between 28,720 and 24,900 yrs. B.P. These authors recognized two distinct pollen zones. Zone B was older than 24,900 yrs. B.P. and was characterized by high percentages of *Pinus* (37-67%), high *Picea*, and a peak in *Betula* and *Alnus* pollen in the middle of the zone. The upper Zone A (24,900 yrs. B.P.) was dominated by *Picea* with pine present at 37%. This is interpreted as an area covered by a pine/spruce forest during early Farmdale times.

Butler Farm, Iowa. The radiocarbon ages of 28,800-22,750 yrs. B.P. place this deposit within the Farmdalian substage of the Wisconsin Glacial Period. Van Zant *et al.* (1980) report the dominance of *Picea* and *Pinus* pollen, with *Abies,* and *Larix* pollen generally low. Pollen of thermophilous taxa such as *Quercus, Fraxinus,* and *Platanus* occur sporadically in low percentages throughout the deposit. NAP ranged between 20-30%. Van Zant *et al.* (1980) suggested an interpretation of this deposit as reflecting cool and moist conditions. They saw no evidence of a postulated warming trend (Harmon *et al.* 1979) as reported elsewhere.

Black Hawk County, Iowa. Mundt and Baker (1979) reported on a Farmdalian deposit recovered from a cut bank along the Wapsipinian River. This deposit was radiocarbon dated at 20,850 + 2,000 − 2,700 yrs. B.P. The deposit was clearly dominated by *Picea, Pinus,* and Cyperaceae. Three zones were indentified showing a succession from pine, to spruce, to spruce-NAP assemblages.

Ontario

St. David's Gorge, Ontario. Hobson and Terasmae (1969) reported on the pollen analysis of interstadial beds near Niagra Falls. All of these pollen assemblages were dominated by pine and spruce with occasional grains of hardwood taxa such as *Quercus, Ulmus,* and *Carya*. They were interpreted as being the result of long distance transport. This assemblage also compares with that of the Plum Point Interstadial (Dreimanis, 1964). The St. David's Gorge deposits are dated at 22,800 ± 450 yrs. B.P. and the abundance of arboreal pollen argues for a forested condition near the site. The presence of *Artemisia, Ambrosia, Sheperdia canadensis,* Poaceae, and Cyperaceae pollen all suggest much colder conditions for this area than at present.

Guelph, Ontario. The discovery of a well developed paleosol containing fossil pollen was reported by Karrow *et al.* (1982). The arboreal pollen component of the paleosol was dominated by *Pinus* (<66%), *Picea* (<33%), and some thermophilous tree pollen composed of *Quercus, Carya, Acer,* and *Tilia* (6%). Karrow *et al.* (1982) interpreted this assemblage as representing a pine/spruce forest in xeric and mesic habitats. The vegetational reconstruction inferred drier and cooler conditions than are available at the present time. A minimum radiocarbon age of 45,000 yrs. B.P. was determined for this assemblage on the basis of two separate analyses.

Waterloo, Ontario. Karrow and Warner (1984) analyzed an interstadial deposit recovered from Waterloo, Ontario. These authors identified two pollen zones. The older zone was characterized by high *Picea* and *Juniperus* type pollen and interpreted as representing a lowland forest. The increase in Cyperaceae pollen, Pediastrum (algae) colonies, and aquatic plant macrofossils suggested that the area underwent a transition to a wetland associated with more open water. The deposit was accelerator radiocarbon dated at 40,080 ± 1,200 yrs. B.P.

Missinaibi River Sections, Ontario. A number of non-glacial peat deposits were described from exposures along the Missinaibi River, Ontario (Terasmae, 1958), and one has been cited as representative. This pollen assemblage was essentially dominated by *Picea* and *Abies*. The basal portion of the assemblage showed a very high percentage of NAP and fern spores. Although Terasmae (1958) did not explain these percentages, this assemblage quite likely shows succession from a herb tundra to a spruce dominated vegetation which is common throughout the region. Radiocarbon dates place this deposit at >38,000 years B.P.

LATE WISCONSIN AND HOLOCENE AGE DEPOSITS

New York State.

Bellmont Bog, New York. Spear and Miller (1976) reported on the pollen analysis of a 7.1 m sediment column from Bellmont Bog, New York. Four zones and four sub-zones were identified by their analyses. Their lowest zone, Zone T, dated between 16,400 and 12,500 yrs. B.P. This zone was marked by high NAP (50-70%) with pine and spruce dominating the AP. They interpreted this zone as representing a boreal forest/tundra type assemblage. Between 12,500 and 11,000 yrs. B.P. (Zone A) they identified a vegetation shift to a open spruce woodland. Pine pollen increased and NAP was still high (25-35%) but not as high as in Zone T. Zone B (11,000-8,500 yrs. B.P.) was dominated by pine (60%), with a decrease in NAP to 10-25%. The pine maximum was estimated to occur at 10,100 yrs. B.P. The sub-zones of Zone C were delimited based on the fluctuating percentages of *Tsuga* pollen. Zone C1 (8,500 - 4,400 yrs. B.P.) was characterized by high *Tsuga* pollen with some *Fagus*, *Acer*, and *Betula* pollen. The NAP is low, less than 5%. Zone C2 starts with a *Tsuga* decline which is dated at 4,000 yrs. B.P. but by 1,700 yrs. B.P. (Zone C3) *Tsuga* began increasing, continuing this trend until *ca.* 175 B.P. This upper portion (Zone C4)

reflected the post settlement history of the area which is characterized by high NAP, especially members of the Poaceae and *Ambrosia*.

Lockport, New York. Miller and Morgan (1982) reported on the analysis of a Coleopterous assemblage. They recovered 1.9 meters of organic rich marls with a basal date of 10,920 yrs. B.P. From these deposits the authors recovered over 780 individual fossil insects. These assemblages have no modern analog but the majority of the species are found within the boreal forest with a mean July temperature range from 11°C - 17°C. Between 10,920 ± 160 − 9,145 ± 110 yrs. B.P., the insect data suggest a fairly stable climate. There is no evidence of the spruce-pine transition which is thought to occur at *ca.* 10,600 − 10,500 yrs. B.P. The earlier levels from this site were dominated by taxa associated with a boreal forest. A number of the insect species were identified as bark beetles indicating the close proximity of *Picea* trees. A clear change in the vegetation from a boreal forest to one characterized by taxa associated with the present-day Great Lakes-St. Lawrence Forest Region is indicated by 9,700 yrs. B.P.

Winter Gulf, New York. Schwert and Morgan (1980) analyzed an 80 cm organic deposit dated at 12,700 yrs. B.P. and identified three fossil insect zones. Zone W1 is the oldest and dates > 12,700 yrs. B.P. It indicates an open shoreline mire, composed primarily of sedges, with no indication of nearby forest cover. At *ca.* 12,700 yrs. B.P. (Zone W2) the area was still fairly open but a moist, meadow-marsh environment is indicated. The trees in the area were probably dominated by *Picea*. By 12,600 yrs. B.P. (Zone W3) a shoreline environment is indicated. Also, by this time the data suggest a coniferous forest dominated by spruce was in close proximity to the shoreline.

Pennsylvania

Hartstown, Bog Area, Pennsylvania. Four postglacial sites were examined by Walker and Hartman (1960). Six-inch segments in the core were mixed and then sub-sampled for pollen analyses. Only AP was used in the calculation of the pollen sum (200 grain counts) and the NAP was calculated outside the sum as a percentage of the AP. In spite of the gross sampling procedures these cores show a forest history compatable with much of the Great Lakes Region. This sequence shows an initial fir-spruce-oak-pine forest followed by pine which was succeeded by an oak-hemlock association and finally a beech-oak assemblage.

Rose Lake, Pennsylvania. Cotter and Crowl (1981) reported on the analysis of a 13 m core. Unfortunately, only two radiocarbon dates were taken and these were located 10 meters apart. Their analysis revealed an initial late-glacial closed boreal forest dominated mainly by *Picea, Pinus,* and *Alnus.* This was succeeded in the post-glacial by a mixed pine-oak forest which was later replaced by an oak-hemlock-beech forest for the remainder of the Holocene. Their uppermost zone was further sub-divided into a hemlock sub-zone, a maple-beech-birch sub-zone, and a hemlock-chestnut-NAP sub-zone which recorded the reestablishment of *Tsuga.* No pollen evidence of the European settlement period was noted from this column.

Ohio

Battaglia Bog, Ohio. Shane (1975) reported on the analysis of a 3.8 m sediment core radiocarbon dated at 16,500 yrs. B.P. The basal date was estimated by extrapolation from the established sedimentation rate within the three pollen zones which were identified. Zone 1A was dated 16,500-13,600 yrs. B.P. and is thought to document the first stage of the ice retreat of the most recent ice advance. There was a small percentage of NAP (less than 10%) with *Picea* (greater than 50%), clearly dominate. This assemblage, with the presence of *Abies,* was interpreted as representing a cool to cold climate. Zone 1B was dated 13,600-11,000 yrs. B.P. and shows a slow but steady increase in the diversity of temperate deciduous taxa, including the first appearance of Nymphaceae pollen which suggests a warming environment. *Picea* pollen percentages declined and were replaced by increasing deciduous pollen. Zone 2 (11,000-9,000 yrs. B.P.) was clearly dominated by *Pinus* and suggested a fairly sudden change in the environment. A trend to increased drying was suggested by maximum values of Nymphaceae followed by *Sphagnum* and the presence of *Ditrema,* a *Sphagnum* parasite. Zone 3 (9,000-*ca.* 6,000 yrs. B.P.) was virtually destroyed by plowing and draining of the bog. The pollen counts from this uppermost zone were too low for a statistically valid count but Shane (1975) suggested that an increase in deciduous tree pollen did occur.

Bucyrus Bog, Ohio. A 14 ft sediment core extending through marl and into "quicksand" was reported and analyzed by Sears (1930b). Preservation was poor as he noted that "when possible" 100 grain counts were attempted. Samples were taken at 1 ft intervals, and only AP pollen constituted the pollen sum. Sears (1930b) proposed a forest sequence from an initial *Picea-Abies* forest with spruce replacing fir and then replaced by pine. The pine forest in turn was replaced first by an oak-pine assemblage and finally by an oak-hickory forest.

Browns Lake, Ohio. A 14 meter core composed of limnetic sediments with a basal date of 16,000 yrs. B.P. was reported by Sanger and Crowl (1979). These authors analyzed the fossil pigments which revealed that the Browns Lake area evolved rapidly after ice-retreat to a nutrient-rich lake which remained meromictic for a period of several thousand years. The shore line eventually supported a reed-swamp habitat which provided a source of organic nutrients for the development of a fen-forest. The pond eventually stabilized as a mesotrophic area until the present time.

Silver Lake, Ohio. Ogden (1966) reported on the analyses of an 8.9 m sediment core with a radiocarbon date of 10,778 ± 210 yrs. B.P. and recognized three pollen zones (Fig. 3). Zone 1 indicated periglacial conditions existed in the region with the pollen assemblage dominated by Poaceae, Cyperaceae, *Polygonum,* and *Saxifraga.* Ogden (1966) could not absolutely identify the existence of an arctic-type tundra but suggested that an increase in forest cover occurred throughout this zone. Zone 2 terminated at 10,778 ± 210 yrs. B.P. and was characterized by an abrupt decline in spruce pollen, a sharp rise in NAP, and a rise in oak pollen, similar to other Great Lakes Region sites. NAP reaches a maximum during Zone 2 and Ogden (1966) correlates these with the Two Creeks Interval. The Valders readvance may likewise be indicated by a slight rise in spruce at the top of this zone. Zone 3a reveals the virtual extinction of boreal conifers and an abrupt rise in *Quercus,* and *Carya.* Zone 3 showed a maximum in *Fagus* and *Juglans* pollen and suggests a warm-moist interval. Zone 3b was characterized by maximum values of *Fagus,* and an increase in NAP. This sub-zone is radiocarbon dated at 4,479 ± 212 yrs. B.P. A maximum in *Carya* pollen suggested that the climate was drier and warmer. Zone 3c was the uppermost zone and records the impact of forest clearance and European agriculture.

Indiana

Bacon's Swamp, Indiana. Englehardt (1959, 1962, 1965) reported on pollen analyses of an undated sediment column from Bacon Swamp. Initially a *Picea-Abies* climax forest developed after the retreat

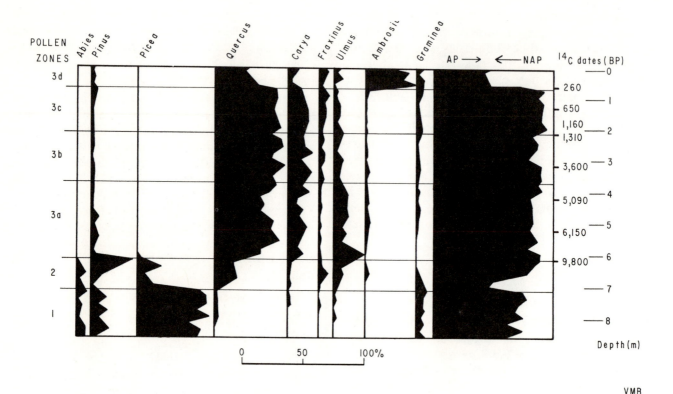

Figure 3. Pollen diagram from Silver Lake, Ohio. Redrawn from Ogden, 1966.

of the Late Wisconsin glacial advance. The establishment of this forest suggested that cool-moist conditions existed until approximately 12,500 yrs. B.P.
Pine increased in percentages as the spruce-fir pollen
declined, becoming dominate for a short interval.
Pines were replaced by increased percentages of the
deciduous taxa which suggested warmer and drier
conditions. Englehardt (1959, 1962, 1965) interpreted
the slight rise in conifer pollen as correlating with the
Valders age readvance. Following these increases in
conifer pollen, Englehardt (1959) reported a shift to
warmer and drier conditions as evidenced from the
rapid increase of deciduous pollen percentages.

Fox's Prairie, Indiana. Englehardt (1959, 1962,
1965) examined a 13 m core from a site in close
proximity to Bacon Swamp. The pollen analysis
reflected the same general forest sucessional pattern
as described above.

Hendrick's Lake, Indiana. Potzger (1943b) reported
on the analysis of an undated shallow peat deposit
from Hendrick's Lake. Potzger (1943b) noted that the
basal sediments were dominated by *Picea* pollen
which was then replaced by pollen of broad leaved
genera, especially *Quercus.* Shortly after this the lake
drained, sedimentation ceased, and no later pollen
records were preserved.

Christensen Mastodon Site, Indiana. Faunal
remains located within lake and bog sediments were
reported by Whitehead *et al.* (1972). These sediments
were dated greater than 14,000 to 13,000 yrs. B.P. and
pollen analyses suggested an open, white spruce
boreal forest was present in this region. Immigration
of hardwood taxa such as *Fraxinus, Quercus,* and
Ulmus began by 13,000 yrs. B.P. as evidenced from
the fossil pollen record.

Cabin Creek Bog, Indiana. Freisner and Potzger
(1946) sampled an undated 31 ft deposit at one-foot
intervals. They recorded changes in forest composition from a *Picea* to *Pinus-Quercus* and finally to an
oak-broad leaved forest.

Pretty Lake, Indiana. Williams (1974) identified 3
major pollen zones and several subzones from this
lake area (Fig. 4). Sedimentation was initiated
between 14,300-13,800 yrs. B.P. in Zone 1a. The
pollen influx was extremely low (less than 700 grains/
cm²) and was dominated by high percentages of NAP.
The presence of *Eleagnus* and *Saxifraga oppositifolia*
pollen suggested tundra conditions. Zone 1b was
deposited between 13,300 - 12,978 yrs. B.P. *Picea*
reached a maximum of 90% with *Abies* comprising
8%. This was interpreted as an open park-like
woodland dominated by black spruce. Zone 1c

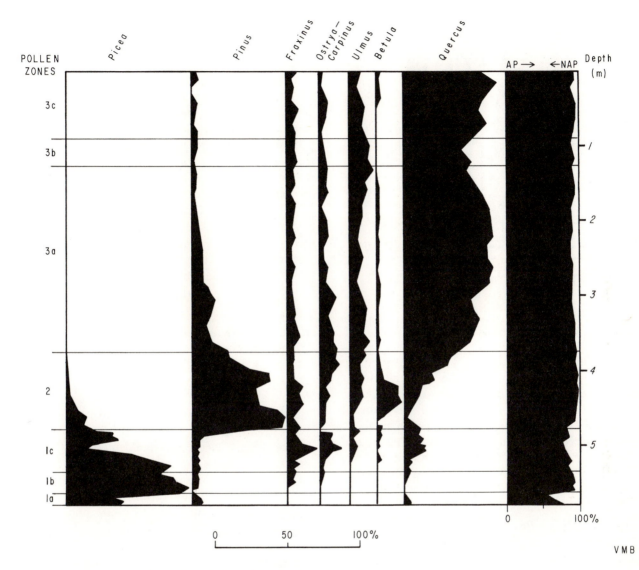

Figure 4. Pollen diagram from Pretty, Lake, Indiana. Redrawn from Williams, 1974.

(12,978 - 10,652 yrs. B.P.) reflected a decrease in *Picea* and peaks in *Fraxinus, Ostrya-Carpinus,* and *Quercus.* The lower portion of this zone was correlated with the Port Huron glaciation. Zone 2 (10,652 - 9,588 yrs. B.P.) shows maximum development of both pines and birches. *Picea* and *Abies* decrease rapidly and AP is at a minimum. In Zone 3a (9,588 - 6,100 yrs. B.P.) *Quercus* was dominant, *Pinus* decreased and an increase of NAP was noted. Zone 3b (6,100 - 4,436 yrs B.P.) showed a decrease in *Quercus* with a concurrent maximum in *Fraxinus* and *Ulmus.* NAP decreased during this zone and was interpreted as representing a mixed mesophytic forest. Between 4,436 - 1,685 yrs. B.P. (Zone 3C), oak is dominant in an open oak-hickory forest. This forest assemblage is replaced by a *Fagus-Acer* forest

between 1,685 - 1,670 yrs. B.P. (Zone 3d). An oak-hickory forest returned for a short period of about 500 years (670-150 yrs. B.P.) until historic times when the pollen record gives evidence of forest clearance (150 yrs. B.P.- present) dominated by *Ambrosia, Plantago,* and *Rumex.*

Illinois

Pittsburg Basin, Illinois. Grüger (1972a, 1972b) examined only the Late Wisconsin age sediments which included his Zone 4 and parts of Zones 3 and 5. Zone 3 (oldest) was characterized by high values of NAP, (> 70%) composed of such types as Cheno-ams, (30%), Poaceae, (12-20%), and Cyperaceae (8%). The upper portion of Zone 3 decreased in the percentage

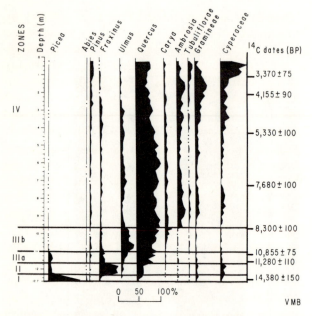

Figure 5. Pollen diagram from Chatsworth Bog, Illinois. Redrawn from King, 1981.

of NAP. Grüger (1972a) defined 3 sub-zones for Zone 4. In Zone 4a *Pinus* pollen reached a maximum at approximately 34%, with some *Ulmus, Betula,* and *Corylus,* also present. Zone 4b shows a maximum of *Artemisia, Ulmus, Betula,* and *Corylus* pollen and a decrease in NAP to 40%. In Zone 4c, *Picea* rose to 19% while *Pinus, Abies, Betula, Ulmus,* and *Corylus* pollen decreased, and were of no further importance in the core. In Zone 5 the vegetation was dominated by deciduous tree pollen up to 50%.

Hickory Ridge Basin, Illinois. Grüger (1972a, 1972b) reported on the analyses of an undated 5.45 m sediment column which probably dates from the same time period as the deposit from Pittsburgh Basin. The same 3 zones are identified. Zone 3 saw the initiation of sedimentation in the Hickory Ridge Basin. Zone 4 differs from Zone 4 at Pittsburgh Basin only in the percentages, and not the overall characteristics of the zones. In Zone 5, the NAP increased sharply and *Quercus* was dominant at between 28 and 50%.

Niantic and Macon County, Illinois. Grüger (1972a, 1972b) also reported on the analysis of two, very similar sites. Only two zones were represented at each site. The lower zone was dominated by NAP chiefly by Chenopodiineae, Poaceae and Cyperaceae. Zone 2 was dominated by boreal conifers: *Pinus* (30%) and *Picea* (60%).

Turtle Pond, Illinois. An undated, 24-foot deposit was recorded by Griffin (1951) and was sampled at one-foot intervals. A minimum of 150 grains of

arboreal pollen grains per sample were counted. The lower levels indicated a cool-moist climate, dominated by spruce. The lowest most NAP zone, common in late-glacial sites in Illinois, was missing. *Picea* pollen decreased and was replaced by *Quercus, Salix, Ulmus,* with small amounts of *Carya,* and *Betula* pollen. Pine did not excede 5% until near the top of this column. The mid-section of the core was dominated by *Quercus* followed at the 5 ft level by pine which reached its maximum at that point. Next, oak pollen decreases, and in the upper sediments pine pollen was replaced by *Picea* and *Larix.* Griffin (1951) gives no indication that NAP was identified, nor even tabulated, and this may partially explain the differences in the interpretation of forest secession between Griffin's (1951) data and records from other, nearby fossil pollen localities.

Chatsworth Bog, Illinois. This site was originally investigated by Voss, (1937) and more recently by King (1981). King recovered a 12.7 meter core and identified four pollen zones (Fig. 5). Zone 1 dated 14,700-13,800 yrs. B.P. and was dominated by spruce (76%) and small amount of *Abies, Larix, Ulmus, Betula,* and *Quercus.* This zone exhibited a rather low species diversity and because of low pollen influx values, King (1981) interpreted this assemblage as representing a spruce woodland mixed with tundra, or possibly a forest-tundra transition. Zone 2 (13,800-11,600 yrs. B.P.) showed a decrease in *Picea* pollen with concomitant increases in *Ostrya-Carpinus, Ulmus,* and *Quercus* pollen. *Fraxinus* pollen reached a maximum of 44% while pine remained very low, never above 2%. The low percentages of pine pollen suggest a period of gradually increasing temperatures with ash invading the lowland regions. Zone 3a (11,600-10,600 yrs. B.P.) showed a decrease in *Fraxinus* pollen with a peak in *Alnus* and an increase in *Ulmus* and *Quercus* pollen. Zone 3b (10,600-8,300 yrs. B.P.) showed a decrease in *Alnus* pollen with maximum percentages of *Ulmus,* and *Ostrya-Carpinus. Carya* and *Quercus* pollen increased and King (1981) interpreted this subzone as the culmination of the transition from a boreal forest to an oak dominated forest. Zone 3 showed a shift from a cool-temperate to warm temperate deciduous forest. Cyperaceae pollen decreased which may suggest a reduction of the lake area. Zone 4, dated at 8,300 yrs. B.P. to the present, was characterized by high percentages of NAP. Once the prairie vegetation was established on the uplands as indicated by the Zone 4 pollen assemblage, it remained dominant until the present time.

Volo Bog Illinois. This site was previously investigated first by Artist (1936) and later by King (1981). King recovered 8.7 meters of sediments and indentified 5 pollen zones beginning with Zone 1 which was radiocarbon dated at the base at 11,070 ± 210 yrs. B.P. *Picea* (60%) was dominant with *Abies, Larix, Pinus, Alnus, Betula, Ulmus, Quercus, Fraxinus,* and *Ostrya-Carpinus,* all low. *Picea* pollen begins decreasing at the top of this zone. King (1981) has interpreted this assemblage as representing a spruce woodland. Zone 2 ended at 10,800 yrs. B.P. and this assemblage shows a decrease in *Picea,* increases in *Pinus* and *Fraxinus,* and high percentages of *Betula* and Alder. *Abies* pollen increased slightly but by the top of Zone 2, deciduous elements are dominant. Zone 3 (10,300-7,900 yrs. B.P.) was dominated predominently by deciduous taxa such as *Quercus* and *Ostrya-Carpinus* pollen that showed an increase, with *Carya, Acer,* and *Tilia* pollen present. *Ambrosia* and Poaceae pollen also increased during Zone 3. King (1981) interpreted Zone 3 as reflecting a mixture of upland wet-mesic lowland types and some scattered pine. Zone 4a (7,900-900 yrs. B.P.) was dominated by *Quercus* pollen but with the complete absence of *Betula* and *Alnus* pollen with decreases in the pollen of *Fraxinus* and *Ulmus.* In Zone 4b (900-400 yrs. B.P.) there is a birch pollen peak which King (1981) has used as the prime characteristic to define this period. Essentially, King (1981) feels that Zone 4a represented an oak dominated forest with some boreal expansion related to Neoglacial cooling. Zone 5 records the *Ambrosia* peak reflecting local forest clearing which occurred around 140 years ago. This was accompanied by an abrupt increase in *Rhus* pollen.

Michigan

Pontiac, Michigan. A 10 cm diameter clay plug, was removed from the right tusk socket of an American Mastodon found in deposits near Pontiac, Michigan (Stoutamire and Benninghoff, 1964). Twigs and wood of *Larix;* twigs, wood, seeds, cone-scales, and needles of *Picea;* and twigs of *Salix* were recovered from this deposit. Moss remains of *Scorpidium scorpiodes, Campylium stellatum,* and *Drepanoclaudus revolvens;* and molluscs of *Lymnaea parva, L. humilif,* and *Gyraulus parvus* were also recovered. The material was dated at 11,900 ± 350 yrs. B.P. and five sediment samples were analyzed for pollen. *Picea, Potomogeton, Cyperaceae,* and *Salix* were recovered, all of which are characteristic of late-glacial vegetational assemblages.

Douglas Lake and Middle Fish Lake, Michigan. Both profiles were reported by Wilson and Potzger (1943). The fossil pollen records an initial spruce pollen zone containing some pine pollen which was succeeded first by a pine forest, and then by an immigration of broad leaved genera. Since these deposits were not dated, it is difficult to correlate these cores with dated cores from the same region.

Tobico Marsh, Michigan. A one-meter sediment core was recovered and analyzed by Jones and Kapp (1972). A radiocarbon date of 1,805 ± 180 yrs. B.P. was recorded from the 10-13 cm level. Fossil pollen analyses of these sediments reflected a regional vegetational shift which occurred *ca.* 1,600 yrs. B.P. The lower portion of the core was believed to correlate with the Sub-Atlantic Period.

Barney Lake, Beaver Island, Michigan. Kapp *et al.* (1969) analyzed a sediment section between 25 - 450 cm in depth for pollen. Zone 2 was dominated by *Pinus* (>60%) pollen. Zone 3 was dominated by a mixture of pine (<60%) and hardwood pollen. *Picea* pollen was present in low percentages and finally disappeared from the record between 80 - 230 cm which may correlate with the Hypsithermal Period. Pollen of *Tilia, Carya,* and *Fagus* first appeared near the base of Zone 3 (*ca.* 7,000 yrs. B.P.) and these continue in importance throughout the core. One radiocarbon date of 7,280 ± 160 yrs. B.P. was taken from the top of Zone 2. After the retreat of the last glacial (Valders) advance at 11,000 yrs. B.P. the initial forest cover was probably a spruce woodland. Because of the island's distance (20 miles) from mainland Michigan, plant succession probably lagged somewhat.

Algonquin Peat Bog, Beaver Island, Michigan. Kapp *et al.* (1969) also reported on the analysis of this 3.85 m section. Zone 1 was characterized by high spruce and pine with high percentages of NAP. In the lower subzone (Zone 1a) *Picea* pollen was greater than 60% and the presence of abundant *Pediastrum* (a colonial green alga) suggest open water conditions. In Zone 1b *Picea* pollen decreased to 10-30%, *Pinus* pollen increased to 30-50%, *Ulmus* pollen is present at 16%, and *Betula* and *Quercus* increased at the expense of NAP. Zone 2 showed a pine maximum between 50 and 80%. *Tsuga* pollen first appears, and *Quercus* and *Betula* pollen percentages are low. In Zone 3 pine varies between 30 - 55%, and the zone marks the first occurence of *Fagus* pollen which Kapp *et al.* (1969) used to identify the base of this zone. *Sphagnum* was present at 42% and the disappearance of Nymphaceae pollen suggested the formation of the bog during this time period.

South Haven, Michigan. Zumberge and Potzger (1955) reported on the analysis of a 30 inch thick peat deposit which was underlain first by sand and then by blue silt. The peat deposits record a succession from a spruce/fir forest through pine to a pine/oak/chestnut association. This is succeeded by an oak/pine/hemlock broadleaved forest and the upper layer probably reflected the beginning of the Hypsithermal Period. The spruce-fir forest was present in the area by 8,000 yrs. B.P.; the pine by 6,000 yrs. B.P.; and peat deposition terminated by 4,000 yrs. B.P. A radiocarbon date on wood underlying blue silt revealed a date of 11,000 yrs. B.P.

Chippewa Bog, Michigan. Bailey and Ahearn (1981) identified four pollen zones from this site. Zone 1 (10,280 - 9,600 yrs. B.P.) fossil pollen data suggested the area was covered by a spruce/fir forest. The fossil pollen assemblage compared most favorably with the present surface pollen spectra from the area near James Bay, Ontario. Zone 2, which lasted from 9,600 - 8,100 yrs. B.P., was dominated by pine pollen and there was a slight increase in both *Quercus* and *Ulmus* and the first appearance of thermophilous types such as *Carya, Fagus, Acer,* and *Tilia*. Zone 3 was subdivided into three subzones. Zone 3a (8,100 - 4,500 yrs. B.P.) was dominated by an oak/elm/beech forest. Zone 3b (4,500 - 1,900 yrs. B.P.) showed a decrease in *Ulmus* pollen and the establishment of an oak/beech/ash forest. Zone 3c (1,900 - 1,150 yrs. B.P.) is identified as the beech maximum with a gradual increase in pine at the expense of oak. Zone 4 (1,150 - 550 yrs. B.P.) showed a pine hardwood pollen assemblage which was probably maintained by periodic fires.

Hartford Bog, Michigan. Zumberge and Potzger (1956) analyzed a 32-foot deposit which yielded a basal radiocarbon date of 13,000 yrs. B.P. The fossil pollen record revealed that following the retreat of the Cary Ice lobe between 13,000 and 10,000 yrs. B.P., the area was first dominated by a spruce-fir forest. Later, between 10,000 and 8,500 yrs. B.P. this forest type disappeared and was replaced by Jackpines until they dominated by 8,000 yrs. B.P. The Jackpine forest was succeeded by a white and red pine forest by 6,000 yrs. B.P., and by 5,000 yrs. B.P. an oak/pine forest was dominant which was in turn replaced by an oak/hickory/pine assemblage by 3,500 yrs. B.P. The upper 2,500 years of pollen deposition indicated the area was dominated by an oak/pine forest. Unfortunately, only 150 aboreal pollen grains per sample were counted and no record was made of either shrub or NAP types.

Upper Twin Lakes and Prince Lake, Michigan. Saarnisto (1974) analyzed a series of end-moraines across northern Ontario and Michigan and analyzed two pollen cores from the east and north shore of Lake Superior. However, Saarnisto (1974) only reported on the records from late-glacial and early-Holocene age sediments. The basal zones of both areas were dominated by NAP (30%) while pine and spruce were the dominant arboreal types. The spruce maximum (50%) occurred between 9,940 ± 210 and 8,760 ± 270 yrs. B.P. Spruce was replaced by birch (40%) with significant levels of pine pollen by 8,640 ± 140 yrs. B.P. and this assemblage was succeeded by a pine forest, probably consisting primarily of *Pinus strobus*.

Thaller Mastodon site, Gratiot County, Michigan. Bones of *Mammut americanum* were located immediately above glacial till in gray sand (Held, Kapp, 1969). The tusks dated to 9,910 ± 350 yrs. B.P. yet this date was questioned since it was older than a date on the underlying marl. Karrow *et al.* (1984) have recently demonstrated the problem of dating marls accurately, thus the 9,910 yrs. B.P. date may actually be valid. Pollen analysis revealed a spruce dominated assemblage in association with the Mastodon bones.

Pitt Mastodon Site, Michigan. Deposits associated with the Pitt Mastodon Site revealed a pollen record dominated by pine (59%), and Picea (27%) (Oltz, Kapp, 1963). This assemblage was interpreted as a spruce dominated site. The presence of *Typha, Salix,* and *Cyperaceae* pollen in this assemblage indicated a pond margin environment with open water in close proximity. This is further supported by the presence of high percentages of *Pediastrum,* a colonial alga.

Smith Mastodon Site, Michigan. Oltz and Kapp (1963) found pollen evidence of a similar spruce dominated forest and the same open water indicators as found at the Pitt Mastodon Site.

Flint Mammoth Site, Michigan. Oltz and Kapp's (1963) fossil pollen study at this Mammoth Site reported somewhat higher percentages of oak than were found at either of the above mastodon sites. Also no *Tilia, Ulmus,* or *Juglans* pollen was found at this locale but higher percentages of Poaceae, Asteraceae, and *Artemisia* were recorded. Based on these data, Oltz and Kapp (1963) suggested an earlier date for this site than for the Pitt and Smith Mastodon locales.

Cheboygan County, Michigan. Farrand *et. al.* (1968) reported on the pollen analysis of a 13,000-12,500 yrs. B.P. deposit. Their study recovered remains of Bryophytes such as *Calliergon turgescens* and *Dripanocladus ingermedius* and fossil pollen suggesting that the area was treeless during this late-glacial period.

Vestaburg Bog, Michigan. Gilliam *et al.* (1967) analyzed nine meters of sediment and identified four pollen zones. Zone 1 was divided into two sub-zones: Zone 1a (basal) was comprised mainly of sandy sediments containing *Picea* (<40%) and *Pinus* (>30%) pollen with minor amounts of *Quercus*, *Betula*, and *Fraxinus* pollen. Zone 1b was dated at 10,328 ± 436 yrs. B.P. and was dominated by spruce pollen (60%), while both pine and hardwood pollen increased slightly. Zone 2 marked the pine maximum (65%) which increased at the expense of *Picea*. There are also slight increases in *Ulmus*, *Acer*, *Carya*, and *Ostrya-Carpinus* pollen. Zone 3 was characterized by an oak/mixed-hardwood assemblage and was further divided into two sub-zones. Zone 3a records the period prior to the immigration of *Tsuga* while Zone 3b (7,982 ± 250 to 3,146 ±237 yrs. B.P.) showed large increases in *Tsuga* pollen. Zone 4 pollen records suggest a decrease in all hardwood taxa and increases in *Ambrosia*, Poaceae, and agricultural weeds including Cheno-ams and *Rumex*. Zone 4 is presumed to be historic as indicated by this particular pollen assemblage elsewhere in the Upper Midwest.

George Reserve, Michigan. Anderson (1954) reported on the analyses of a 14 foot sediment section obtained from a small pond in Livingstone County, Michigan. The entire assemblage was dominated by *Picea* until replaced by pine in the upper levels. Anderson (1954) thought the presence of deciduous pollen in association with pollen of boreal conifers was the result of contamination with sediments of inter-glacial age but this profile is consistent with other pollen records from this region. The basal section, dominated by *Picea* with low NAP, correlated with the Two Creeks Interval while the later *Picea* pollen decrease and NAP increase was limited to the Mankato readvance. The upper portion of the sediment column reflected the increasing dominance of pine in the local vegetation.

Isle Royale, Michigan. Potzger (1954) reported on the analyses of nine bog deposits from this island in Lake Superior. No individual discussion of the pollen stratigraphy was provided. Potzger (1954) conducted 200 grain counts of arboreal pollen at 1-foot intervals. Herbaceous pollen was counted but not used in the analyses. Potzger (1954) interpreted his results to reflect an initial forest of pine or pine/spruce assemblage that could be correlated with similar records from the mainland. He also felt that there was a pine rise at the initiation of the Xerothermic Period and

noted that the forest history of Isle Royale was very similar to the record from southern Quebec. Potzger (1954) believed that the rise of *Picea* pollen in the upper one-third of the profile reflected a cooling trend in recent times.

More recently Raymond *et al.* (1975) investigated 12 sites from Isle Royale to determine their sedimentary history. The basal sediments were composed of a reddish brown silty clay which probably related to the Valders Ice retreat. These were overlain by homogeneous green clays which, depending on their presence or absence, denotes the degree to which the basin was affected by post-glacial lobes of the Minong-Nipissing interval.

Crystal Marsh, Michigan. A 5.7 M sediment core with polliniferous sediments to 5.0 M was reported by McMurrey *et al.* (1978). Zone 1 (basal) was dominated by *Pinus* pollen and was correlated with the pine maximum from other nearby areas. This was immediately followed in Zone 1 by increases in *Ostrya-Carpinus* pollen. Zone 2a showed increasing percentages of *Ulmus*, high *Quercus*, and the presence of *Ostrya-Carpinus* pollen. The estimated date of Zone 2a, based on average rates of sedimentation, was thought to be between 9,000 - 7,150 yrs. B.P. Zone 2b was dated between 7,150 - 5,000 yrs. B.P. and revealed an increase in *Fagus* pollen, accompanied by high amounts of *Acer* and *Tilia* pollen. Zone 2c (5,000 - 125 yrs. B.P.) showed an increase in conifer pollen, especially pine, and was interpreted to reflect a conifer/mixed-hardwood forest. Zone 3 is historic in age and shows the effect of forest removal as indicated by high NAP percentages.

Garden Peninsula, Michigan. Kapp and Means (1977) analyzed a small peat deposit from a sink hole on the Garden Peninsula and identified two pollen zones. Zone 1 (lowest), reflected the shift from a spruce/fir forest to one composed of *Alnus/Betula* with some *Tsuga* pollen present. Zone 2 showed significant amounts of hardwood pollen, and increased amounts of NAP. *Juglans* pollen was high in the topmost sample. The top portion of the core was dated 560 ± 130 yrs. B.P., which is a minimum date for the immigration of butternut into the region. Based on percentages of *Tsuga* pollen it is believed that the basal sediments were deposited prior to 7,000 yrs. B.P.

Wisconsin

Lake Mary, Wisconsin. Webb (1974a) analyzed a 2.45 m sediment core and delimited four distinct pollen zones. Zone 1, dated > 10,000 to 9,800 yrs.

B.P., and was dominated by *Picea* and *Pinus*. *Quercus, Ulmus, Fraxinus, Larix, Betula,* and *Artemisia* were present but in low quantities, probably due to the over representation of pine percentages. The basal date of this zone is estimated to be close to 10,000 yrs. B.P. since the accepted date range for the immigration of pine into the area was between 10,000 - 10,500 yrs. B.P. (Wright, 1968). At the top of Zone 1 *Picea* declines very rapidly, and is followed by a rise in *Pinus* pollen which is the dominant constituant of Zone 2 (9,800 - 8,400 yrs. B.P.). *Betula* and *Alnus* pollen increased slightly in Zone 2 as compared with Zone 1 levels. Zone 3 (8,400 - 3,500 yrs. B.P.) is dominated by white pine with a slight decrease in *Betula* pollen while *Ostrya-Carpinus* pollen is high and *Quercus* shows a gradual increase. Zone 4 represents the last 3,500 years and was characterized by reduced frequencies of pine pollen and increased percentages of both *Betula* and *Tsuga* pollen. The Zone 4 vegetation was interpreted as essentially a pine-northern hardwood forest.

Two Creeks, Wisconsin. Morgan and Morgan (1979) described the fossil insect analysis of a forest bed dated as being deposited at 1,000 ± 850 yrs. B.P. They analyzed a 21 kg soil sample and recovered insects of Coleoptera, Diptera, and Hymonoptera. The majority of the Carabidae recovered suggest the presence of moderately moist to dry, open ground, with little arboreal vegetation present. However, at least some *Picea* was present in the area since most of the bark beetles recovered are associated with *Picea.*

Tamarack Creek, Wisconsin. Davis (1979) reported the fossil pollen analysis of a 3.6 meter core with a record of sedimentation extending back to 4,400 yrs. B.P. Zone 1 was interpreted as representing a local environment composed primarily of emergent aquatics and a sedge meadow surrounding the site. AP and NAP influx was highest in Zone 1 than in any other portion of the core. The regional vegetational component was dominated during this time by *Quercus* and *Pinus* yet the pollen data suggest that *Picea* and *Abies* were probably present only a short distance from the site. The lower portion of the core (2.25 - 3.6 m) contained little pollen. In Zone 2 the local vegetation was probably dominated by a sedge meadow/bog shrub community. The observed increase in *Quercus* pollen during the Zone 2 period is interpreted as reflecting local encroachment rather than a regional shift. The pollen record from Zone 3 indicated that by this time the area was covered by a shrub and *Larix* community with *Larix* pollen increasing after 1,100 yrs. B.P. to 15%.

Iola Bog, Wisconsin. Iola Bog is a steep sided kettle bog which produced an 810 cm sediment core (Schweger, 1969). Pollen was recovered to the 793 cm level and Schweger (1969) identified 9 different pollen zones. Zone 1 (lowest) is characterized by high percentages of NAP (32-73%), high amounts of *Picea* pollen and the presence of *Fraxinus* and *Salix* pollen representing an open habitat and possibly correlated with the Late Cary (late-glacial) advance. In Zone 2, there is a sharp rise in AP to 77%, composed primarily of *Picea* and *Fraxinus* which suggests a mixed-hardwood spruce forest with open areas. Zone 3 showed a rise in NAP to 42% which is interpreted to reflect a deterioration of the Zone 2 period forest. Schweger (1969) correlated the forest decline with the Valders readvance. AP reaches 90% in Zone 4 and this is interpreted as the institution of a closed forest of pine, spruce, and hardwood taxa. The early postglacial period is reflected in Zones 5 and 6 by the dominance of pine pollen (72%), with some *Quercus* and *Ulmus* pollen present. Zone 7 shows a slight pine pollen decrease with concommitant increases in *Quercus, Ulmus,* and *Ostrya - Carpinus* pollen. The increase in NAP to 32% signified an opening in the forest cover which Schweger (1969) correlated with the Hypsithermal Interval. In Zone 8, pine pollen increased, *Quercus* pollen was high and *Betula* pollen reached a maximum which was interpreted as representing a conifer/hardwood forest developed after the dry Hypsithermal period. Zone 9 represented forest clearance during historic times and was characterized by a sharp increase in the NAP.

Duck Creek Ridge, Wisconsin. Schweger (1969) reported on the pollen analyses of a paleosol formed on lacustrine clays and silts which were deposited on Cary till and were overlain by a Pre-Valders age clay. A *Larix* log from the paleosol was radiocarbon dated at 11,600 ± 350 yrs. B.P. and a mat of mosses, needles, twigs and cones was recovered. Cyperaceae pollen was high at the base of the paleosol and was replaced by *Picea* at the top. *Larix* was an important forest component and reached its maximum percentages during the spruce maximum. This assemblage reflected the transition from a herb dominated assemblage to a forest habitat.

Peters Quarry, Wisconsin. A second paleosol was also reported on by Schweger (1969) and was based on the pollen analysis of a single composite sample. The fossil pollen assemblage was similar to that obtained from the Duck Creek site. *Picea* and Cyperaceae were dominant and NAP reached 52%. This was interpreted as representing an assemblage dominated by open vegetation with wetland pioneers.

Because of the similarity of this assemblage to the one from Duck Creek assemblage, the paleosol from Peters Quary is thought to date from the same period (*ca.* 10,500 yrs. B.P.).

Hells Kitchen Lake, Wisconsin. Swain (1978) recovered a 2,000 year record of pollen, charcoal, and seeds from this site. Six zones were identified and Swain (1978) recognized the effects of both long-term and short-term changes which he felt were caused by the fire history of the area. Swain recognized two moist intervals; one from 2,000 - 1,700 yrs B.P. and the other between 600 - 100 yrs. B.P. These moist intervals were characterized by increases in pine and hemlock pollen with corresponding decreases in the charcoal influx.

Dells Bog, and Baraboo Bog, Wisconsin. Hansen (1937) reported on the pollen analyses of two bog sites. In each bog a six-inch sampling interval was used throughout the deposits. In Baraboo Bog the basal zone reflected a spruce/fir forest that was soon replaced by a forest dominated by oak and pine. It was during this oak/pine period that sedimentation began in Dells Bog. Later, spruce returned to the area as indicated by the fossil pollen data, and was correlated to the Mankato sub-stage. Hansen (1937) also noted the presence of a Xerothermic Interval which showed decreases in the percentages of pollen from both *Pinus* and Cyperaceae and increases in overall NAP percentages.

Hub City, Wisconsin, Blue Mounds Creek, Wisconsin. These two sites were located in the Driftless Region of Wisconsin and were reported by Davis (1977). The decline in *Picea* pollen, usually dated at 10,500 yrs. B.P. from sites in this region, did not occur at these sites until 9,500 yrs. B.P. This suggests that a local black spruce/tamarack community existed for a longer period of time than was found in nearby regions. Davis (1977) suggested that the disappearance of the boreal forest was possible due to fire. In addition, there were no indications of the Hypsithermal Prairie Period expansion evident from these cores. *Pinus* and *Quercus,* once established are dominant throughout the column. Davis (1977) suggested that at least locally, the entire Holocene was characterized by a mosaic distribution of the vegetation.

Disterhaft Farm Bog, Wisconsin. Baker (1970) reported on an 8.78 m core from this site. The fossil pollen chronology was basically the same as that reported earlier by West (1961) but the radiocarbon dates alter West's (1961) correlations with glacial events. Baker's (1970) late-glacial period pollen record was dominated by spruce and suggests that

Larix was also locally available. Deglaciation appears to have been completed by 15,000 yrs. B.P. The variations in *Picea* and *Artemisia* that West (1961) correlated with the Valders and Two Creeks (late-glacial) intervals appear to have occurred about 1,000 years earlier suggesting that they may correlate with the Cary-Port Huron (late-glacial) Interstadial instead. *Pinus* pollen replaced *Picea* by 11,000 yrs. B.P. and remained dominant until *ca.* 8,500 yrs. B.P. The Hypsithermal Period (8,500 - 5,300 yrs. B.P.) was dominated by *Quercus, Ulmus,* and *Ostrya/Carpinus.* The prairie invasion, normally associated with this interval, does not appear to have occured in Wisconsin. From 5,400 - 2,850 yrs. B.P. the vegetation was dominated by an oak forest or oak savanna and during the last 2,850 years was dominated by an oak-savanna with local populations of *Betula* and *Tsuga* as reflected by the fossil pollen record.

Minnesota

Kirchner Marsh, Minnesota. Wright *et al.* (1963) reported on the analyses of a 12.5 M core. The basal portion of this core was composed of till and both the till and 50 cm of overlying clays were essentially devoid of pollen. Zone K, the earliest pollen bearing zone was deposited 13,270 yrs. B.P. and was characterized as an open plant community with *Picea* and Cyperaceae dominant. Zone Aa dated between 13,270 - 12,050 yrs. B.P. and consisted primarily of *Picea* and *Fraxinus* pollen. Wright et al. (1963) correlated this sub-zone with the Two Creeks Interval. Valders age sediments were reflected in zone Ab which was dominated by *Picea* and *Artemisia.* Zone B was deposited slightly before 10,230 yrs. B.P. and was initially a *Betula-Alnus* community followed by a shift to a pine dominated assemblage near the top of this zone. Zone Ca was dominated by *Ulmus* and *Quercus* and terminated prior to 7,120 yrs. B.P. Zone Cb reflected a drying trend in which the *Quercus* dominance was replaced by NAP which continued to *ca.* 5,450 yrs. B.P. Zone Cc represented the period from 5,100 yrs. B.P. to present. This zone reflected the re-establishment of oak dominance which implied a return to moister conditions and in the upper portion of this core, *Quercus* pollen declined with concurrent increases first in *Betula* and finally by *Pinus.* The pine rise is estimated to have occurred around 1,660 years ago.

Madelia, Minnesota. A 3.2 m section analyzed by Jelgersma (1962) reported the presence of 5 pollen zones. Zone 1 consisted of pioneer vegetation shortly after the ice retreat and was characterized by high

NAP percentages which were interpreted to represent a park tundra with components of *Picea* and *Salix*. Zone 2 dated between 12,650 - 11,200 yrs. B.P. and represented the development of a spruce forest, correlated with the Two Creeks (late-glacial) Interval. Zone 3 (11,200 - *ca.* 10,300 yrs. B.P.) represented the Valders (late-glacial) stage and showed a decline in *Picea* pollen, a slight increase in NAP, and an abundance of shrubs of *Betula, Alnus, Salix,* and *Corylus*. Zone 4 marked a decline in *Picea, Alnus,* and NAP with increases in *Betula* pollen to a maximum level. Zone 5 was characterized mostly by deciduous pollen taxa with no boreal conifers present.

Lake Carlson, Minnesota. Wright *et al.* (1963) sampled a 14.68 m core from Lake Carlson but no pollen was recovered below the 13.2 m level. Only the upper portions of Zone A (see Wright *et al.* 1963) sediments were recovered at the base of this core. No radiocarbon dates were reported but based on pollen zonation, sedimentation was estimated to have begun sometime after 12,050 yrs. B.P. Wright *et al.* (1963) suggested that the initial *Betula-Artemisia* community was replaced first by a seral community of *Betula-Alnus* (prior to the establishment of pine) and then pine forests that were replaced by an elm/oak/hardwood forest during the early-Holocene. A drier period, indicated by the increase in NAP percentages, suggested the opening of the forest. When moister conditions returned, oak again dominated.

Aitkin, Minnesota. Farnham *et al.* (1964) reported on the analyses of a buried soil horizon developed on sand and overlain by marl. This soil was radiocarbon dated at 11,675 yrs. B.P. Fossil pollen percentages from this soil horizon were based on a pollen sum consisting of AP plus *Ambrosia, Artemisia,* Cheno-ams and most wind pollinated NAP, excluding Poaceae and Cyperaceae. Thus, the resulting pollen percentages may be slightly inflated but still provide data for a meaningful comparison. Three pollen zones were identified from this buried soil. Zone 1 was identified as the Al Horizon of a paleosol and was dominated by *Picea* and pollen of *Juniperus, Larix,* and *Salix. Fraxinus* (7%) was present and implied a temperate climate comparable to the southern boreal forests at the present time. This zone was correlated with the Two Creeks Interval. In Zone 2, *Picea* pollen decreased to 30% with accompanying decreases in both *Artemisia* and *Juniperus. Betula* pollen increased during this time and continued through Zone 3 times. *Pinus* abruptly increased during Zone 3, eventually dominating the assemblage. *Alnus, Ulmus* and *Quercus* increased and the

presence of *Carya, Tilia, Juglans,* and *Acer* was also noted. Poaceae increased dramatically in Zone 3.

Anoka County Sites. Minnesota. Wilson and Potzger (1943a) reported on the analyses of five bog and lake deposits from Anoka County. All five pollen profiles were very similar in their major features. Either *Picea* or *Picea/Abies* were dominate in each of the earliest levels. *Pinus* dominance then replaced spruce and *Ulmus* was common at this time. The pine forest is superceded by an oak/pine association. Poaceae pollen increased towards the top of these cores but according to Wilson and Potzger (1943a) may represent lake filling and the contribution of pollen from aquatic grasses and not necessarily a regional shift.

Big Woods, Minnesota. McAndrews (1968) investigated the vegetational succession which occurred in the Big Woods region since the time of European settlement. Pre-settlement "Big Woods" were dominated by *Ulmus* and *Quercus*. A succession to more mesic conditions with a forest dominated by *Acer* and *Tilia* was initiated several hundred years prior to the settlement. The present oak forest was the result of a seral succession of a oak-savanna *after* European settlement in the area.

Stevens Pond, Minnesota. Janssen (1967) reported five pollen zones from this site. Zone 1 (basal) was completely dominated by pine (80-90%) pollen and contained high levels of *Pteridium* spores suggesting the presence of an open forest. This assemblage was then replaced by *Quercus* /Poaceae/*Artemisia* and *Quercus/Ostrya* communities. Next, *Pinus strobus* succeeded the oak/ironwood zone and was itself succeeded by the *Ambrosia* Zone indicating land clearance and European settlement.

Red Lake Bog, Minnesota. Griffen (1977) identified three pollen zones from Red Lake Bog. Zone 1 was characterized by oak, elm, ash, *Ostrya-Carpinus, Populus,* and *Corylus* pollen. Prairie indicators such as *Petalostemum* and *Amphora* were also present. *Pinus* was the dominant arboreal type. Zone 2 showed a definite increase in deciduous pollen types and prairie indicators were almost completely absent, *Larix* pollen was present in small amounts throughout this zone. Zone 3 documented a rise in *Ambrosia* pollen reflecting the European settlement. The Red Lakes were formed by the retreat of Glacial Lake Agassiz *ca.* 11,000 yrs. B.P. yet the ^{14}C chronology of these Bog sediments reflected only the last 3,000 years. Griffen's (1977) interpretation that a dry prairie was established prior to the start of peat formation (greater than 3,100 yrs. B.P.) was based on a well-developed A$_1$ horizon in the mineral substrate of the bog. Griffen (1977) also believed that a prairie bog - prairie marsh existed between 3,100 - 2,700 yrs. B.P. and was followed by a transition to a sedge marsh pond around 2,700 years ago.

Lake Sallie, Minnesota. Birks *et al.,* (1976) reported on the recent forest history of the Lake Sallie area based on a series of short cores. During the deposition of sediments between 70 - 75 cm the region was surrounded by an oak savanna. Sediments deposited between 70 - 48 cm represented the period from AD 1850 - 1870. The vegetation during this time was still an oak-savanna but contained mesic species suggesting a rising water level. Subsequent vegetational changes after AD 1890 were related to problems of eutrification of the lake basin.

Elk Lake, Minnesota. Dean *et al.* (1984) recently examined Elk Lake using a variety of information sources. They found that between 10,400 - 8,500 yrs. B.P. the lake varves were simple, composed of light and dark laminae. Between 8,500 - 4,000 yrs. B.P. the lake became shallower and more saline with a higher influx of clastic material which altered the characteristics of the varves. From 4,000 yrs. B.P. to the present, sediments suggest that the lake became deeper, less saline, and accumulated little detrital material. Of particular interest is their characterization of the prairie period (8,500 - 4,000 yrs. B.P.) deposits showing evidence of two dry intervals separated by a period of moister conditions.

Wolf Creek, Minnesota. Birks (1976) analyzed over 500 cm of material at this site (Fig. 6). Zone WO-1 dated between 20,500 - 13,600 yrs. B.P. NAP was in excess of 60% which was dominated primarily by Asteraceae and Cyperaceae pollen. *Picea* dominated the AP (10 - 25%) and the vegetation compared most favorably to present arctic treeless areas. The upper portion of Zone WO-1 was dated 14,700 - 13,600 yrs. B.P. and was more comparable to a forest/tundra transition. Zone WO-2 dated between 13,600 - 10,000 yrs. B.P. and was dominated by *Picea* (40-70%) and *Larix* pollen with moderate frequencies of NAP. The closest modern analogue is the mixed conifer-hardwood forest yet the moderate frequencies of NAP suggest that the forest was not closed. The period 12,250 - 10,000 yrs. B.P. was masked by increases in *Betula, Populus,* and *Abies* pollen. Zone WO-3 was dated 10,000 - 9,150 yrs. B.P. and was dominated by *Pinus* (72%), *Pteridium,* and increases in deciduous pollen taxa. NAP also increased in Zone WO-3 and the high amounts of *Pteridium* spores suggests that the area was covered by an open pine forest. The upper-Holocene levels were not recorded.

Itasca, Minnesota. McAndrews (1966, 1967) investigated a number of sites along a transect west of Lake Itasca, Minnesota. McAndrews (1967) compared the present vegetation with surface samples he examined from 11 short cores he obtained from this transect. In addition, four longer cores were taken to analyze the post-glacial history of the region. The basal zones in the deep cores were dominated by *Picea/Populus* with low NAP (15%) and dated between 12,300 - 11,000 yrs. B.P. Wood remains of both *Populus* and *Picea* were recovered along with low percentages of *Artemisia* pollen suggesting an open quality to this forest. This zone was succeeded by a *Pinus banksiana/resinosa -Pteridium* zone between 11,000 - 8,560 yrs. B.P. and was characterized by low amounts of NAP and significant quantities of boreal conifers.

In the western portion of the transect area the above zones were succeeded by a Poaceae/*Artemisia* zone. The NAP frequencies were greater than 25% with Poaceae the dominant type. Apparently these communities formed the source of the prairie elements within this region. An *Ambrosia* peak is recorded between 8,000 - 7,000 yrs. B.P. At Thompson Pond the *Ambrosia* peak reached 60% but at Terhill Pond, presently in the mesic deciduous forest area, it reached only 35%. This zone was then replaced by *Ostrya-Carpinus* and *Ulmus* pollen. McAndrews (1966) viewed these arboreal pollen types as representing long distant transport of these pollens from the east.

In the eastern portion of the Lake Itasca transect the *Pinus banksiania/resinosa - Pteridium* zone was replaced by a zone dominated by *Quercus,* Poaceae and *Artemisia* between 8,560 - 4,000 yrs. B.P. This zone was vegetationally equivalent to the establishment of the oak-savanna which was succeeded by a zone dominated by *Quercus* and *Ostrya.* The NAP decreased to less than 25% which suggests a closed forest. Oak and *Betula* increased and McAndrews (1966) interpreted this assemblage as representing the dominance of a mixed deciduous forest which replaced the existing oak savanna around 4,000 yrs. B.P. *Pinus strobus* migrated into this area by 2,700 yrs. B.P. and by 2,000 yrs. B.P. dominated the forest.

Kylen Lake, Minnesota. Birks (1981b) reported on the analysis of a 15,850 year record of sedimentation which he divided into four pollen zones (Fig. 7). Zone 1a dated between 15,850 - 14,300 yrs. B.P. The pollen preservation in this zone was extremely poor which led Birks (1981b) to suggest that at this late-glacial time, the area was an open tundra barren with a discontinuous cover of stunted grasses, sedges, and dwarf willows. Zone 1b (14,300 - 13,600 yrs. B.P.)

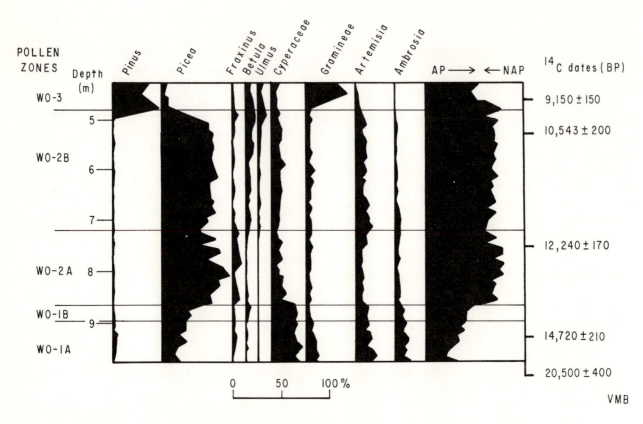

Figure 6. Pollen diagram from Wolf Creek, Minnesota. Redrawn from Birks, 1976.

was characterized by significant increases in *Artemisia* pollen. In Zone 1c (13,600 -12,000 yrs. B.P.) *Artemisia* pollen exceeded 25% and oak pollen concentration values and pollen influx values reflected an extensive vegetational cover during this period. Zone 1b and 1c reflected a gradual transition to more heavily forested conditions. Zone 2 (12,000 - 10,700 yrs. B.P.) was dominated by high percentages of *Betula,* some *Picea,* and high values of *Artemisia* pollen. Zone 3a (10,700 - 10,400 yrs. B.P.) showed a decrease in *Betula* pollen with concurrent increases in *Picea* pollen. The Zone 3a assemblage resembled the mixed coniferous - deciduous forest such as currently found in southern Manitoba. Zone 3d dated between 10,400 - 9,250 yrs. B.P. and noted a dramatic increase in pine with a pine maximum at 10,400 yrs. B.P. which predates similar peaks in central Minnesota suggesting an immigration of pine from the east. Zone 4, dated 9,250 - 8,400 yrs. B.P., showed a pine dominance with decreases in both *Picea* and *Populus* pollen. The high *Pteridium* percentages suggested the presence of an open pine forest. The upper-Holocene sediments were not reported.

Norwood, Minnesota. Ashworth *et al.* (1981) investigated a sediment deposit which consisted of 1.6 m of silt deposited on till and overlain by peat. The peat

was radiocarbon dated between 12,400 ± 60 and 11,200 ± 250 yrs. B.P. The lowest portion of the silt contained very little pollen and most of the recovered grains were Cretaceous in age. The middle silt contained insect remains of types which were predominately aquatic or semi-aquatic in habitat. Sedge pollen was high, *Picea* pollen was greater than 20%, and fragments of *Artemisia* were also recovered. This middle silt assemblage suggested that there was a colonization of the water margin area. The upper silt contained several insect taxa which are no longer found in Minnesota. A shallow, weedy pond environment was inferred from the accompanying beetle assemblage. Thick growths of vegetation during the deposition of the upper silts were indicated. Pollen and plant macrofossil remains suggesting an open environment with some boreal conifers (as noted from the presence of *Picea* and *Larix* needles) were present. The peat deposits above the silts contained twigs and cones of *Picea.* There is no modern analogue for this assemblage, for a while the pollen records suggest a tundra vegetation, the insect assemblage is at best boreal.

Rice Lake Minnesota. McAndrews (1969) analyzed a 360 cm core from this site. The base of the core shows the *Quercus* /Poaceae/*Artemisia* assemblage

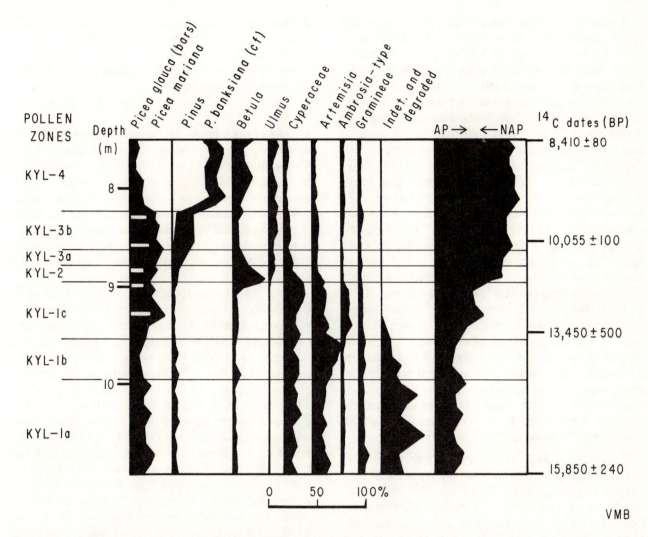

Figure 7. Pollen diagram from Kylen Lake, Minnesota. Redrawn from Birks, 1981.

similar to the one found at Pond D in the Itasca transect (McAndrews 1966). McAndrews (1969) noted that in general the sequence of vegetational succession was similar to that identified from other cores in the Itasca region.

Rutz Lake, Minnesota. Waddington (1969) identified four pollen zones in her analysis of Rutz Lake sediments. Zone 1 (12,000-10,500 yrs. B.P.) contained high percentages of *Picea, Populus, Alnus,* and *Larix* pollen with a basal zone of plant detritus consisting of needles, seeds, and cones suggesting the deposit formed from a melting ice block. Zone 2 (10,500-*ca.* 8,500 yrs. B.P.) showed an increase in NAP (60%), and an AP count consisting initially of *Betula* and *Alnus* pollen which was quickly replaced by *Pinus, Quercus, Ulmus,* and *Ostrya-Carpinus* pollen suggesting an apparent abrupt vegetational change from Zone 1 to Zone 2. By Zone 3 (8,800-4,200 yrs. B.P.) the AP values decreased except for

pine and oak while both *Artemisia* and *Ambrosia* pollen remained high suggesting that a prairie replaced the previous parkland vegetation. An increase in charcoal in Zone 3 deposits suggests fire was an important factor during this time. Zone 4 (4,240 yrs. B.P. - present) noted a gradual re-forestation of the region to an open parkland composed of grasses and deciduous trees. The upper portion of this zone (Zone 4b) showed a pollen record indicating the development of the "Big Woods" consisting of *Ulmus, Ostrya-Carpinus,* and *Tilia.*

Northbranch, Minnesota. Fries *et al.* (1961) reported on a radiocarbon dated deposit of between 12,700 ± 250 and 12,030 ±200 yrs. B.P. The pollen assemblage consisted of high percentages of *Picea,* NAP, and significant amounts of *Quercus* and *Fraxinus. Larix, Picea mariana, Sparganium,* and *Potomogeton* also were present in the immediate area as indicated by the plant macrofossils.

Weber Lake, Minnesota. Fries (1962) identified seven pollen zones from Weber Lake, Minnesota. Zone 1 was dominated by NAP and was interpreted to represent an open, treeless area, possible a tundra. Radiocarbon dates of 10,550 ± 300 and 10,180 ± 160 yrs. B.P. were taken just above the boundary between Zones 1 and 2. Zone 2 showed a maximum in *Betula* pollen which Fries (1962) identified as *Betula nana* based on cone scales recovered from these levels. Zone 3 is characterized by high percentages of *Ulmus* pollen with lesser percentages of *Quercus* and *Fagus* pollen. Zone 3 terminated by 9,150 ± 130 yrs. B.P. and was replaced by an alder zone (Zone 4), which also showed a higher frequency of *Pinus* pollen accompanied by a decline in thermophilous trees pollen. This suggests a depauperation of the forest in Zone 4 by climatic deterioration. The top of Zone 4 showed an increase in NAP and was dated at 7,300 ± 140 yrs. B.P. Zone 5 showed a high *Pinus* pollen maximum and an accompanying pollen record suggesting an undergrowth of deciduous taxa which formed the mixed conifer-hardwood forest. In Zone 6, the pine dominance continued but with higher percentages of *Betula* and *Picea* pollen. Zone 7 is not clearly separated from Zone 6 but does show a slight decrease in *Pinus* and *Betula* with increases in *Picea*, *Larix*, and *Abies* pollen.

Horseshoe Lake, Minnesota. Cushing (1967) analyzed the basal section of a core from Horseshoe Lake dating from 14,000-18,000 yrs. B.P. and recognized four pollen assemblage zones. The lowest zone, Zone HL-1 was characterized by high Cyperaceae (> 15%) pollen with total NAP values exceeding 40% while *Picea* dominated the AP percentages. The basal sediments also contained a layer of plant detritus consisting of *Dryas integrifolia* leaves, Cyperaceae fruits, fungal hyphe and *Tofiedia glutinosa* (3%) pollen. In Zone HL-2 *Picea* dominated the deposit and was associated with *Larix, Populus, Fraxinus, Artemisia, Ambrosia,* and *Urtica* pollen. *Fraxinus* pollen is greater than 5% in the lower shrub zone (HL-1) and *Betula* pollen is greater than 5% in the upper sub-zone (HL-2). In Zone HL-3, pine replaced *Picea* pollen as the dominant type, *Pteridium* spores are conspicuous, and *Quercus* and *Ulmus* pollen increased in importance. In Zone HL-4 there was a high percentage of NAP pollen, dominated by Poacae and Cyperaceae.

White Lilly Lake, Minnesota. Cushing (1967) analyzed the basal section of a core from this site and found a similar sequence to that obtained from Horseshoe Lake. In Zone WLL-1 Cyperaceae pollen was greater than 15% and there was a very high percentage of indeterminate pollen. Zone WLL-2, 3, and 4 were comparable to Zones HL-2, 3, and 4 (Horseshoe Lake) respectively. Based on these and other reports from the area, Cushing (1967) proposed the identification and use of several biostratigraphic zones which are completely described in the 1967 report.

Lake of the Clouds, Minnesota. Swain (1972, 1973) examined the fire history of the Boundary Waters Canoe Area (BWCA) by examination of charcoal and pollen influx. He found that there had been no major changes in the major pollen types during the last 1,000 years. Secondly, Swain (1972, 1973) was able to demonstrate that fire had been an important factor in determining forest composition yet he was unable to substantiate the hypothesis that Europeans had substantially altered the vegetation.

Craig (1972) analyzed a long core from Lake of the Clouds and recognized six pollen zones. Zone LC-1 was dominated by Asteraceae and Cyperaceae with overall high NAP (40%), suggesting tundra conditions between 11,000-10,300 yrs. B.P. Zone LC-2 was dominated by *Picea* (>30%) pollen and high amounts of *Larix, Populus, Ulmus, Quercus,* and *Carya* pollen. This zone was deposited between 10,300-9,200 yrs. B.P. and was predominantly a boreal spruce forest with some *Abies* present. Zone LC=3 (9,200 = 8,300 yrs. B.P.) was dominated by *Pinus* (> 50%) and very low NAP. Zone LC-4 dated between 8,300-6,500 yrs. B.P. and was recognized by a *Pinus/Betula/Alnus* pollen assemblage. Except for the presence of alder pollen, this zone was comparable to Zone LC-3. Zone LC-5 (6,500-3,000 yrs. B.P.) contained a pine pollen increase to 60-70% suggesting the migration of *Pinus banksiania* into the region. Zone LC-6 (3,000-200 yrs. B.P.) showed a decrease in pine pollen and an increase in *Picea* pollen which signified a return to more moist conditions.

Iowa

West Lake Okoboji, Iowa. Van Zant (1979) recovered and analyzed eight pollen zones from an 11.68 meter sediment core above glacial till from Little Millers Bay (Fig. 8). The basal *Picea-Larix* zone was dominated by *Picea* (35-70%), *Larix* (1-4%), low NAP which decreased towards the top of this zone and some thermophilous tree pollen in low percentages. This record represented a closed, conifer forest that existed in northwest Iowa before 13,990 ±135 yrs. B.P. The *Picea/Fraxinus nigra* Zone was characterized by a high spruce (36-45%), *Fraxinus* (6-15%) and *Larix* (2-3%) pollen. This zone was then succeeded by

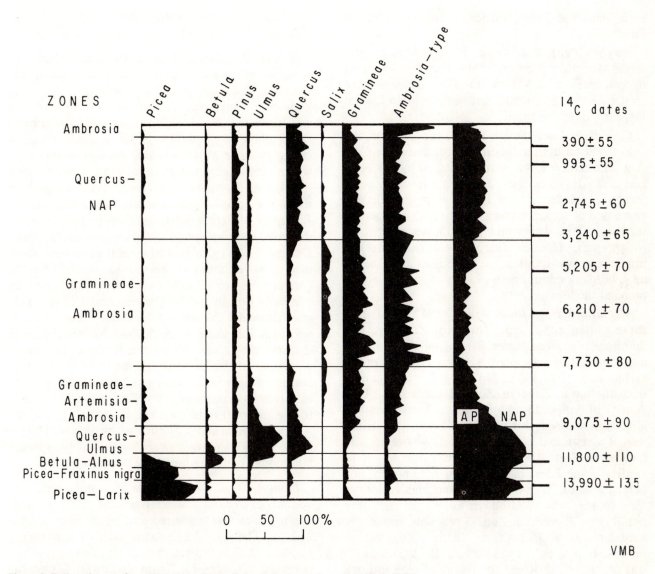

Figure 8. Pollen diagram from West Lake Okoboji, Iowa. Redrawn from Van Zant, 1979.

a *Betula/Alnus* Zone which saw the rapid rise and fall of *Betula* pollen which peaked at 11,800 ± 110 yrs. B.P. *Picea* steadily decreased during this period marking the end of late-glacial times. Next, the zone dominated by *Quercus* (30%) and *Ulmus* (40%) pollen occurred between 11,000-9,075 yrs. B.P. and was interpreted to reflect a deciduous forest composed of additional taxa such as *Carya, Tilia,* and some *Fraxinus.* This zone was succeeded by an assemblage dominated by Poaceae/*Artemisia/Ambrosia* which showed a rapid decrease in AP percentages and macrofossil remains indicating a decrease in water depth, characteristic of drier conditions. A shift to a Poaceae/*Ambrosia* assemblage was completed by 7,730 yrs. B.P. and remained until 3,240 yrs. B.P. During this period *Artemisia* pollen decreased and

there was a rapid increase in *Ambrosia* pollen and AP pollen reached its lowest levels. *Lemna* pollen was restricted to this level and this taxon, in association with a sand lens, suggests the lake almost completely dried up. After 3,240 yrs. B.P. the vegetation was characterized by a *Quercus*/ NAP Zone consisting of 12-28% oak and the constant presence of NAP. As seen elsewhere in the Upper Midwest, post-settlement assemblages are dominated by *Ambrosia.*

Dows, Iowa. Hudak, (1984) examined a fossiliferous bed dated at 9,380 ± 130 yrs. B.P. The overlapping ranges of the flora and fauna identified from this early period at Dows, Iowa are similar to present day Central Wisconsin. Furthermore, the data imply cooler and moister conditions with greater effective precipitation and the presence of a moist deciduous forest.

Hughes Peat Bed, Iowa. Hall (1971) briefly reported on the pollen and mollusk analysis of a peat deposit dated at 11,800 yrs. B.P. His data suggested a spruce dominated boreal forest was present in Iowa at this early time period.

Colo Bog, Iowa. Brush (1967) reported on the analysis of an undated sediment column from Colo Bog. The earliest levels were dominated by low NAP and high AP percentages suggesting a spruce/pine/fir forest. The intermediate zone was predominately composed of deciduous taxa with fir pollen present at the base. *Ulmus* was dominant and was accompanied by *Quercus, Acer, Betula,* and *Alnus* pollen. The upper zone was dominanted by grasses and Cheno-ams with oak common and increases in pine pollen noted at the top.

McCulloch Bog, Iowa. Bruch (1967) identified three pollen zones from this bog. Zone 1A was dominated by *Picea, Abies, Betula,* and *Alnus* pollen accompanied by very low NAP percentages. A radio-carbon date of 14,500 ± 340 yrs. B.P. was obtained from the top of Zone 1a. Zone 1b contained a high density of pollen. In addition to the boreal conifers already mentioned, this zone also contained pollen of *Acer, Quercus,* and *Juglans.* Zone 1c was essentially similar to Zone 1b in composition but lower in density. Zone 2 was dominated by pine pollen, low percentages of *Picea* and *Abies,* and high counts of fern spores, which, although not identified, may possibly be *Pteridium.* Zone 3 was split into three subzones. Zone 3a contained little pollen yet an increase in pine pollen was noted at the top. Zone 3b was characterized by the dominance of pine and oak accompanied by some *Acer, Alnus, Betula Cornus* and *Corylus* pollen. The NAP pollen is generally low in Zone 3b but grass pollen percentages were increasing. The upper zone (3c) was dominated by pine and oak pollen with increased amounts of both types at the top and high NAP values throughout the zone.

Jewell Bog, Iowa. Brush (1967) examined sediments from Jewell Bog and identified three pollen zones from the site. Zone 1 was the *Picea/Abies* zone which also contained abundant *Quercus, Betula, Ulmus, Alnus,* Cheno-ams and *Ambrosia* pollen. Zone 2 was characterized by the dominance of pollen of deciduous taxa, and the absence of spruce and fir pollen. *Quercus* pollen was high, and *Acer, Ulmus, Alnus, Tilia, Juglans,* and *Corylus* pollen was present in small amounts. *Quercus* was abundant in Zone 3 as were the grasses, Cheno-ams, and *Ambrosia* pollen. No elm pollen was present but other deciduous taxa such as *Acer, Betula, Alnus, Juglans, Tilia* and *Corylus* were present in some quantity.

Woden Bog, Iowa. Durkee (1971) analyzed a 10 meter sediment core which extended into glacial till. He identified three pollen assemblage zones along with several subzones. Zone 1 which had AP of 90% was identified as a *Picea/Abies* zone and had a basal radiocarbon date of 11,570 ± 330 yrs. B.P. Wood of either *Picea* or *Larix* was also present in this zone which was interpreted as reflecting a spruce boreal forest. Both *Picea* and *Larix* pollen decline rapidly at the top of Zone 1. Zone 1a was dominated by *Betula/Pinus/Abies. Betula* pollen reached its maximum in this zone succeeding a pine maximum of 30%, which Durkee (1971) calculated as consisting of 95% *Pinus banksiana/resinosa* type. Zone 2 was identified as the *Quercus/Carya/Tilia/Ulmus* pollen zone and was primarily dominated by *Quercus* and *Ulmus .* This zone showed a change to a deciduous forest at about 9,300 yrs. B.P. Zone 3a (*ca.* 9,000-7,200 yrs. B.P.) was dominated by *Quercus* /Asteraceae/Cheno-am pollen and showed the progressive decline of AP. Pine reached some of this taxon's highest percentages and *Acer* is at its maximum of 9%. Zone 3 (7,200 yrs. B.P. to the present) is termed the Poaceae/Cheno-am/Asteraceae zone and is dominated by high NAP. This zone reflects a shift to a drier climate which no longer favored the continuation of the deciduous forest.

Pioneer Creek, Iowa. A Late Wisconsin sediment column of 518 cm was analyzed for pollen (Szabo, 1980). Pollen was recovered only in the interval 213-442 cm. The assemblage dated older than 14,050 ± 180 to 10,810 ± 540 yrs. B.P. The pollen assemblage revealed a succession from an initial (Zone 1a) spruce/pine assemblage with small amounts of decid-uous tree pollen. The NAP was composed of Cyper-aceae, Asteraceae, and Cheno-am. In Zone 1b, the pine/spruce frequencies remain the same but decreases in deciduous pollen were noted, while the NAP increased. Spruce pollen dominated in Zone 2 (40%) and other boreal conifers, notable *Abies* and *Larix,* were present. Zone 3 showed a decrease in *Picea* pollen with concurrent increases in *Pinus* and a rise in deciduous tree pollen. Szabo (1980) inter-preted this assemblage as indicating a succession from a transitional vegetational component to one consisting of an upland community composed of thermophilous taxa and a floodplain community dominated by *Alnus, Fraxinus, Salix,* and *Larix.* This was succeeded by a spruce dominated forest with some fir. By Zone 3 times, pine was beginning to invade into the region.

Ontario

Kenora, Ontario. McAndrews (1982) analyzed sediments recovered within a buried bison skull for pollen. The skull was dated at 4,850 ± 60 yrs. B.P. and the pollen analyses indicated that the bison died in a shallow pond surrounded by a pine-poplar woodland. The Kenora deposit correlated to the mid-Holocene Period which, based on data from the Hays Lake Core, extended from 9,200-3,600 yrs B.P.

Greenleaf Lake, Ontario. Cwynar (1978) reported on Greenleaf Lake located within Algonquin Park, Ontario. His analyses reported a 1,200 year pollen record and identified two zones based on McAndrews' (1972) earlier work. Cwynar's (1978) Zone 7 was dominated by pine (50-67%) with *Pinus strobus* as the primary contributor, *Betula* in quantities of 10 - 22%, *Ulmus* represented as the most abundant shrub and weak herb pollen. Zone 8, the uppermost section, recorded an increase in herb pollen and reflected a mixed conifer-hardwood forest showing the effects of land clearance. The environment seems to have been essentially stable during this 1,200 year period.

Alfies and Antoine Lakes, Ontario. Saarnisto (1974) reported on the early vegetational history of this region. The lowest zone was dominated by *Picea* (55%), *Populus* was common, and the NAP was insignificant. This lowest zone ended by 9,210 ± 100 yrs. B.P. The spruce zone was followed by a *Betula* maximum of 40% and a strong pine component was present. This zone dated 7,650 ± 180 yrs. B.P. near the top, and was in turn succeeded by a pine dominated assemblage. Based on these and other records from the Sault Ste Marie area, Saarnisto (1974) documented a major climate shift that occurred in this area *ca.* 10,000 yrs. B.P. Based on pollen and geological data Saarnisto (1974) showed that a steep climate gradient existed in this region which produced a very narrow, if any, initial tundra zone.

Upper Thane Lake, Ontario. Saarnisto (1974) reported on the lower sediments of several small lake basins on the eastern and northern shores of Lake Superior. The lowest zone in these deposits dated 10,650 ± 265 yrs. B.P. and contained NAP in excess of 30% with Cyperaceae, *Artemisia,* and *Ambrosia* pollen percentages all declining in the upper portion of the zone. The upper boundary of that initial zone is marked by a rise in *Picea* pollen. The following spruce zone contained *Picea* pollen up to 50%, yet the upper sediments show a decrease in *Picea* pollen and an increase in *Betula* pollen. The lower sediments were radiocarbon dated at 9,940 ± 210 yrs. B.P. and birch increases to 40%. This zone is dated between 8,460 ± 140 yrs. B.P. - *ca.* 8,000 yrs. B.P. The *Betula* zone is replaced by a zone of maximum pine and analyses of the pine pollen suggests an immigration *Pinus strobus* into the area.

Prince Lake, Ontario. The lowest zone was dominated by high NAP values (Saarnisto, 1974) and the upper portion of this zone was radiocarbon dated at 10,800 ± 360 yrs. B.P. Saarnisto (1974) questioned the age of this date since the pollen assemblage suggested a slightly later emergence of this area from the glacial lake stage. This initial zone was succeeded by a spruce period which terminated at 9,050 ± 110 yrs. B.P. which was again followed by a zone of *Betula* maximum.

Kincardine Bog, Ontario. Karrow *et al.* (1975) analysed a 5 m section from this site. Zone 2b (lowest) recorded low amounts of *Picea* and high amounts of *Juniperus/Thuja* pollen. Also present were *Quercus, Ostrya-Carpinus, Fraxinus, Salix,* and *Artemisia.* Karrow *et al.* (1975) interpreted this lowest zone as representing a retreating spruce forest that occurred at a time earlier than the one ^{14}C date of 11,200 ± 170 yrs. B.P. removed from a meter above. Their Zone 2c was reflected by a maximum of *Picea* pollen with low percentages of all other taxa. Zone 2d represented a transition from the spruce forest to one dominated by pine, indicating a slightly warmer climate. Zone 3 was the pine maximum. Zones 4-7 were treated together and reflect a bi-modal *Tsuga* curve showing the *Tsuga* invasion of the area at *ca.* 7,620 yrs. B.P. that was accompanied by pollen of *Quercus, Ulmus, Acer,* and *Fraxinus.*

Eighteen Mile River, Ontario. Two exposures dated respectively at 10,600 ± 160 and 10,500 ± 150 yrs. B.P. were sampled by Karrow *et al.* (1975). Wood of *Picea* was found in each exposure and was interpreted as evidence of the existence of a large eutrophic lake at this early period.

Cooks Town Bog, Ontario. Karrow *et al.* (1975) analyzed a sediment core for both pollen and diatoms. The same zonation used for Kincardine Bog was used for these sediments. Zone 2b contained low percentages of *Picea* pollen but had thermophilous pollen taxa suggesting the presence of an open spruce forest. Zone 2c reflected a dominant *Picea* forest with some *Betula* present. Zone 2d showed the transition between the spruce and pine maximum. Zone 3 was completely dominated by *Pinus* pollen with low frequencies of all other taxa. In Zones 4-7, pine percentages decline and are succeeded by a *Tsuga/Acer* pollen association. Zones 2c/2d were analyzed for diatoms and showed that shallow water was present throughout this time period with an increasing trend to shallower conditions accompanied by lush macrophytic vegetation. Alkline and eutrophic conditions also were indicated in the upper diatom levels.

Nicholston Cut, Ontario. Karrow *et al.* (1975) reported on the analyses of a 0.5 m section of Lake Algonquin sediments. Only in this 0.5 m section was good pollen preservation found. Based on pollen percentages, parts of Zones 2c and 2d were recovered. Zone 2c again is a spruce dominated assemblage. Zone 2d shows a deteriorating spruce forest and the replacement pine. Based on comparative data from New York and Ohio, this transition probably occurred at 10,600 yrs. B.P.

St. Marie, Ontario. Sigleo and Karrow (1977) reported on the analyses of a 16 M core through the Wildwood Silts which were correlated with the Erie Interstadial at 16,000 yrs. B.P. Three pollen zones were recognized. Zone 1 contains high pine (12-29%), *Picea,* and *Quercus* (7.5%). Zone 2 showed a reduction of pine with increases in all other AP and NAP. Zone 3 showed an initial decrease in *Pinus* pollen with increases in most deciduous taxa and higher NAP percentages. Sigleo and Karrow (1977) interpreted the pine pollen as most likely the result of long distance transport. The presence of *Artemisia* and other Asteraceae in significant amounts in Zone 3 suggest an open or disturbed area. In addition, they felt that *Picea* was locally present and that the combined record was indicative of a forest/tundra transition.

English River, Bedworth, and Hawk Lake Bog, Ontario. Wilson and Webster (1943) reported on three shallow, undated peat deposits from southwestern Ontario. The pollen records suggested that the early forest were spruce dominated with considerable Jackpine present. Later, a recent shift from birch to spruce is indicated. These sites are completely post-glacial in age.

Little Savanna Bog, Ontario. Wilson and Webster (1943) recovered only a 36 inch deposit from this locale. *Picea* showed a gradual decrease, pine had a bi-modal distribution and *Tsuga* was present in the pollen record even though it is not present in the area today.

Toronto, Ontario. A radiocarbon date of 11,350 ± 325 yrs. B.P. was obtained from the soil associated with the antler of a new genus of fossil deer *(Torontoceros hypogaeus)* (Churcher, Petterson, 1982). The pollen assemblage for this deposit suggested that *Picea, Pinus,* and *Cyperaceae* were

major components of the vegetation implying that it represented a typical interstadial or post-glacial climate with mixed forest.

Mer Bleu Peat Bog, Dowells Lake Bog, Ontario. Mott and Camfield (1969) reported on the pollen history of several bog sites in this region. The Dowells Lake region contained five bogs whose basal sediments dated between 8,200 - 7,600 yrs. B.P. Jackpine was high at the bottom with low *Picea* and NAP percentages. Pine begins to decrease with concurrent increases in *Quercus* and *Betula* pollen, accompanied by an abrupt rise in *Tsuga, Ulmus,* and other broad leaf genera. NAP pollen sharply increased near the top.

Wasaga Beach, Ontario. Terasmae (1979) reported on the combined palynological and radiocarbon studies of peat deposits in this area. Sedimentation in this area began about 5,700 yrs. B.P. and terminated by *ca.* 4,500 yrs. B.P. The results of Terasmae's (1979) analyses of a 6 m core suggested that, in general, the pollen assemblages conformed to the established palynostratigraphic sequence for southern Ontario as previously described. The base of the underlying sediments of marl correspond to the upper boundary of the pine zone which is estimated at 7,500 yrs. B.P.

Brampton, Ontario. Terasmae and Matthews (1980) analyzed a small peat and pond sediment deposit on the top of the Brampton esker, near Toronto. This deposit was dated at 12,320 ± 360 yrs. B.P. Palynological studies on those sediments clearly indicated the assemblage was dominated by *Picea* pollen, with some *Pinus* and a consistent presence of the Chlorophycean alga *Pediastrum.* Towards the upper sediments, *Picea* pollen decreases with the consistent presence of *Abies, Juniperus, Quercus, Fraxinus,* and other hardwood species. Cones of *Picea glauca,* in addition to needles and twigs of this species, were also recovered throughout the deposit. The pollen assemblage is mostly analagous to a northern boreal forest type vegetation with spruce being locally available. The radiocarbon date of 12,320 yrs. B.P. provides a minimum date for the base of the spruce zone from this area.

Kitchener, Ontario. Anderson (1982) reported on the analysis of a peat covered marl deposit located at Kitchener, Ontario. These deposits range from 22,000 to 6,900 yrs. B.P.; however, the basal dates may be inaccurate due to problems associated with the radiocarbon dating of marl (Schwert, 1978; Karrow *et al.* 1984). Anderson (1982) identified four pollen assemblage zones and recorded a succession from a herb pollen zone dominated by *Salix* and

Artemisia, through a *Picea* pollen zone, a *Pinus* pollen zone, and finally to a zone dominated by *Tsuga* and the hardwoods *Acer* and *Fagus.*

Manitoulin Island, Ontario. Warner *et al.* (1984) analyzed sediments from Manitoulin Island for pollen, plant macrofossils, and Cladoceran remains. That portion of the deposit which dated older than 10,500 yrs. B.P. was dominated by *Picea, Artemisia,* and Cyperaceae pollen, indicating a cool, dry, climate. Between 10,500-10,000 yrs. B.P. the assemblage was dominated by *Pinus* pollen which was interpreted as representing more open pine parklands.

Maplehurst Lake, Ontario. Pollen analysis of a 765 cm organic deposit was reported by Mott and Farley-Gill (1978) from southwestern Ontario. They obtained a basal radiocarbon date of 12,400 ± 180 yrs. B.P. and identified eight pollen assemblage zones. Pollen influx calculations were made from five radiocarbon dates. The basal zone (M-8) was dominated by pollen of *Artemisia* and Cyperaceae with *Picea* >30%. *Picea* dominated in Zone M-7 with percentages reaching 70%. Zone M-6 records a transition to a *Pinus* dominated assemblage which occured at 9,650 ± 110 yrs. B.P. Pine pollen was represented by two types with *Pinus strobus* type increasing towards the top of the zone at 7,690 ± 170 yrs. B.P. Pine pollen generally declines and is replaced by pollen of *Tsuga canadensis* (Zone M-5). Zone M-4 is characterized by increases in *Fagus, Ulmus,* and *Acer* pollen but with *Tsuga* pollen continually present. Zone M-3 is characterized by minimum values of *Tsuga* pollen and increases in *Quercus* pollen. Generally, pollen of hardwood taxa are more abundant in this zone. Zone M-2 reflects increases in *Tsuga, Fagus, Ulmus,* and *Acer* pollen. The uppermost zone (M-1) reflects the impact of European forest clearance and is dominated by *Ambrosia* and Poaceae pollen.

Lake R, Ontario. McAndrews *et al.* (1982) reported on the pollen analysis of a 450 cm core recovered from the Hudson Bay Lowland. Based on radiocarbon dates, a basal date of 8,200 yrs. B.P. was estimated for the emergence of the area by isostatic rebound. McAndrews *et al.* (1982) identified two pollen zones and further sub-divided the lower zone into two sub-units. The fossil pollen and plant macrofossil data revealed a succession from tundra to shrub tundra with the modern spruce woodland present in the area from 6,490 yrs. B.P. until the present.

SUMMARY

Full-Glacial Period

In the western section of this region, few study sites provide a record extending back to the Wisconsin full-glacial. In part, this is due to the present environment of the Great Plains which is generally not conducive to long-term pollen preservation. Based on those sites from which data are available, *Picea* probably covered much of this region during the Late Wisconsin. In fact, it appears that the treeless prairie was not established in this region until just prior to the beginning of the Holocene.

By Farmdalian times, Van Zant *et al.* (1980) showed that the forest cover of Iowa was dominated by both *Picea* and *Pinus.* By 22,000 yrs. B.P. *Picea* had replaced *Pinus* at least in central and eastern Iowa.

At Boney Springs, Missouri, *Pinus banksiana,* accompanied by a large NAP component, was present in that area from before 27,410 to *ca.* 21,000 yrs. B.P. (King, 1973). This community was replaced by an assemblage competely dominated by *Picea* pollen, which after 16,500 yrs. B.P. became more diverse with the addition of several broadleaved genera. King (1973) also noted that *Picea* populations remained in the area until at least 13,500 yrs. B.P. when poor pollen preservation terminated the pollen record.

Picea forests were also present in Illinois in areas outside the glacial limits (Grüger, 1972a, 1972b). At the Seminary School Basin site, *Picea* pollen was present between 24,200 to after 21,370 yrs. B.P. and at nearby Pittsburg Basin, while *Picea* pollen was still dominant, *Quercus* trees were also an important component of the forest. Watts (1983) has recently argued that, by interpolation, Grüger's (1972a, 1972b) data may represent Farmdale deposits. These sections are dominated by NAP and Watts (1983) suggested that the vegetation during Farmdale times may have been treeless, the arboreal pollen being the result of long distance transport. Recently, King obtained additional cores from this site covering the time period from 20,000 yrs. B.P. to the present. King's (personal communication) interpretation is similar to Watts' (1983) in that King proposes an open vegetation with arboreal pollen deposited as the result of long-distance transport. Thus the reported occurrence of forests in close proximity to the ice front during this full-glacial period cannot be substantiated.

The Wisconsin full-glacial at more northerly locales is much more complex. This is caused in part by the expansion and contraction of lobes of the Laurentide Ice Sheet (Watts, 1983). At Wolf Creek, Minnesota (Birks, 1976), which may contain a record in excess of 20,000 yrs. B.P., the lower, full-glacial deposits contained evidence of many dwarf shrubs and herbs such as *Dryas, Antennaria, Silene,* and *Vaccinium* which are typical of a tundra environment. By 14,700 yrs. B.P. the diversity of the plant community increased and *Picea* trees became dominant by 13,600 yrs. B.P. Tundra floras are known from a number of sites in this region including Lake Kotiranta (Watts and Wright, 1969), Kylen Lake (Birks, 1981b) and White Lily Lake (Cushing, 1967.)

The northward movement of the tree-line following the Wisconsin full-glacial was time transgressive as shown by both Cushing (1967) and Wright (1971) and is denoted by the arrival of *Picea* dominated assemblages at these sites. A transitional community from tundra to the *Picea* zone, consisting of *Dryas, Shepherdia, Eleagnus, Populus, Juniperus,* and *Salix* was reported from Norwood Minnesota by Ashworth *et al.* (1981). While the pollen records suggest a tundra environment, the insect remains reported by Ashworth *et al.* (1981) suggest primarily boreal conditions were present at this time. Thus the evidence for tundra vegetation at this site is at best equivocal. The plant succession appears continuous and there is no evidence of dramatic fluctuations during this period (Watts, 1983).

The mixture of boreal conifers with deciduous taxa in these late-glacial communities has been difficult to explain. Various explanations have been offered including long-distance transport, and redeposition (Anderson, 1954). Baker *et al.'s.* (1980) report of a *Quercus* acorn cup provides the first solid evidence for the existence of this taxon in association with boreal conifers within a single vegetational community.

Late-Glacial Period.

Wright (1981) has suggested that only at Wolf Creek (Birks, 1976) was the existence of a tundra-like vegetation growing between 20,500 and 14,500 yrs. B.P. conclusively demonstrated, although several authors (Williams, 1974; Watts and Wright, 1969; Birks, 1976, 1981a; Cushing, 1967; King, 1981; and others) had suggested the existence of a herb tundra in northern sites. Williams (1974) data from northern Indiana was based primarily on the high pollen percentages of NAP, principally Cyperaceae, Poaceae, and *Artemisia.* However, at Madelia, Jelgersma

(1962) reported a boreal assemblage dominated by NAP and Watts (cited in Cushing, 1965) reported the recovery of *Dryas integrifolia* leaves from Norwood, Minnesota. King's (1981) inference, unlike the others, of tundra-like conditions at Chatsworth Bog, Illinois, was based on very low pollen influx values.

Van Zant (1979) interpreted the region of northwestern Iowa as being covered by a *Picea/Larix* forest prior to 13,990 yrs. B.P. Earlier studies from the state (Brush, 1967; Durkee, 1971) had suggested that *Abies* and *Pinus* were also a component of this early forest. Recent investigations such as those by Van Zant (1979), Szabo (1980), and Baker *et al.* (1980) have all failed to confirm the presence of *Abies* in the late-glacial record from Iowa.

Additionally, *Tsuga* wood dated at 12,000 to 16,500 yrs. B.P. had previously been reported from central Iowa (Ruhe and Scholtes, 1956). However, *Tsuga* pollen is not represented in significant amounts from any deposit within this area (Ruhe *et al.,* 1957a, 1957b; Durkee, 1971; Van Zant, 1979).

During this same time period, pollen evidence from Wolf Creek (Birks, 1976) and Kylen Lake (Birks, 1981b), Minnesota suggest that this area was comparable to the forest-tundra vegetation of present day northern Canada. In southern Minnesota, a *Picea/Fraxinus* community was dominant (Wright *et al.* 1963). Schweger (1969) reports the same type of forest composition from Iola Bog, Wisconsin at about the same time. Unfortunately, because of the lack of radiocarbon dates from Iola Bog, this time correlation is at best tenuous.

At Pittsburg Basin, Illinois (Grüger, 1972a, 1972b) the presence of forest in close proximity to the ice sheet was demonstrated. The vegetation was dominated by *Picea* and *Quercus* and the NAP may be suggestive of openings within the forest. No direct evidence of tundra plants were observed (Grüger, 1972a, 1972b). King (1981) has recently questioned Grüger's (1972a, 1972b) interpretation. King (1981) views the area as essentially open with the vegetation composed primarily of very low pollen producers. The high percentages of arboreal pollen are interpreted as the result of long distance transport. Grüger's (1972a, 1972b) lack of tundra plants would tend to confirm King's (1981) interpretation. Wright (1981) has suggested for this same time period that while pine may have been present throughout the upper Midwest prior to 18,000 yrs. B.P., by this date it had migrated east and south. Wright (1981) cites the high pine pollen percentages from Battaglia Bog, Ohio (Shane, 1975) as evidence for the movement of this species.

Following the basal herb pollen assemblage, the pollen data from most sites throughout the Midwest suggest that a widespread spruce dominated forest or woodland covered most of the area. By 14,000 yrs. B.P. the spruce dominated forest were moving northward as indicated by the pollen record from Chatsworth Bog (King, 1981). While high values of spruce pollen were recorded at most sites throughout this region, pine was noticeably absent (Webb, 1981). Sites with high percentages of pine pollen were restricted to the eastern portion of North America. At the eastern margin of the Great Lakes Region, there are relatively few sites dating to the late-glacial period. Spear and Miller (1976) interpreted the pollen assemblage from Belmont Bog, New York as representing a boreal forest/tundra transition. The main difference between this site and those in northern Minnesota is reflected by the higher percentages of pine pollen in the eastern locales. Additional open environments are indicated from Winter Gulf, New York (Schwert and Morgan, 1980). In Ohio, however, the area was dominated by a spruce-fir forest (Shane, 1975).

Many of the reports from this area record exceptionally high frequencies of *Picea* pollen during late-glacial times. However, the presence of *Picea* pollen in high amounts in more recent investigations (King, 1981; Lawrenz, 1975; Williams, 1974; Van Zant, 1979) suggest that this phenomenon was real and not a factor of the pollen sum in use.

As the climate began to ameliorate and as the ice sheets were retreating northward, the *Picea* forests likewise moved north. *Picea* pollen was rapidly replaced by pollen of other taxa. In the central portion of the Midwest, *Picea* pollen is first replaced by *Fraxinus* pollen and pollen from a few additional hardwood genera (King, 1981). This succession is not quite as noticeable at either the eastern or western extremes of the Great Lakes Region. Because of the location of the retreating ice-front, the central Great Lakes Region would be most susceptable to minor oscillations in climate. This can be seen in a second *Picea* pollen increase which is recorded at approximately 11,600 yrs. B.P. from both Chatsworth Bog (King, 1981) and from Pretty Lake (Williams, 1974). In fact, both authors attributed this *Picea* increase to a cooler climate associated with the Valders (GreatLakean) readvance. Most sites record a date between 10,000 and 11,000 yrs. B.P. for the rapid retreat of *Picea* populations, denoting the termination of the spruce period. For example Kapp and Gooding (1964a) report an age of 10,600 ± 150 yrs. B.P. at Sunbeam Prairie Bog and Ogden (1966)

reports an age 10,500 yrs. B.P. from Silver Lake, Ohio. In Iowa at *ca.* 11,000 yrs. B.P. the *Picea/Larix* forests were replaced initially by a seral *Betula/Alnus* zone quickly followed by a forest dominated by *Quercus* and *Ulmus* (Van Zant, 1979). This replacement was quite rapid in the pollen records from Iowa (Hall, 1971; Szabo, 1980; Hudak, 1984).

Following the Valders (late-glacial) readvance the percentages of *Picea* pollen declined. At most localities throughout the Midwest this spruce decline occurs at approximately the same time, or very shortly before, the immigration of pines into the region. The pine zone is then generally replaced by pollen of hardwood genera, notably *Quercus* and *Ulmus*.

Tundra like conditions were still evident at *ca.* 11,000 yrs. B.P. in sites from northern Minnesota such as Lake of the Clouds (Craig, 1972), and Weber Lake (Fries, 1962). At Wolf Creek, the forest was primarily composed of *Picea/Larix* trees (Birks, 1981) while at Kylen Lake, Birks (1981) identified a *Betula/Artemisia* zone. The western portion of Minnesota was covered by an open *Picea/Populus* forest (McAndrews, 1965, 1967). However, the typical *Picea/Fraxinus* zone is well represented in sediments from Kirchner Marsh (Wright *et al.* 1963).

The northward migration of *Picea* was interrupted briefly by the readvance of ice lobes identified primarily from Wisconsin. The beginnings of the spruce decline occurred at approximately 11,000 yrs. B.P. at the type locality of Two Creeks, Wisconsin (Morgan and Morgan, 1979). The Two Creeks interval is well documented from sites in both Wisconsin (West, 1961; Schweger, 1969; Baker, 1970), and Minnesota (Wright *et al.* 1963; Jelgersma, 1962). The readvance of the Valders Lobe in this area resulted in increased *Picea* percentages from these deposits. Recently however, Everson *et al.* (1976) have suggested a revision to the Cary-Valders glacial sequence in which they have proposed a gradual, climatically controlled deglaciation (GreatLakean) in preference to the existing model recognizing a Post-TwoCreekian readvance. According to these authors, this type of model would explain the lack of a "Two Creeks/Valders" sequence in the other areas of the Great Lakes Region.

By 11,000 yrs. B.P. pine populations were migrating into the area of western Minnesota (McAndrews, 1965, 1967) but apparently did not encroach into Iowa (Van Zant, 1979). This early immigration of pine was generally confined to *Pinus banksiana* or *P. resinosa*. In the western region, the pine advance was also generally accompanied by a *Pteridium* under-story component. The early post-glacial pine peak is also evident in sites from Wisconsin (Schweger, 1969; Webb, 1974a, 1974c; Heide, 1984) although it arrives somewhat later. Jackpine was also dominant in Michigan from 10,000 to 8,500 yrs. B.P. (Zumberge and Potzger, 1956). As proposed by Shay (1971), the radiocarbon date from Bog D Pond (McAndrews, 1966) may be 1,000 years too old. If this date is adjusted, the anamolously early appearance of pine in this region disappears and the sequence shows a steady westward expansion from pine refugia in the east (Webb *et al.,* 1983).

The spruce dominated assemblage in Illinois is replaced by a *Pinus/Fraxinus* community by about 10,800-10,300 yrs. B.P. at Volo Bog (King, 1981). Pines are scarce at Chatsworth Bog (King, 1981) located further to the south but attain maximum percentages during the period 10,600-9,580 yrs. B.P. at Pretty Lake, Indiana (Williams, 1974).

Early-Holocene

The early-Holocene forests in Iowa and Minnesota contained higher percentages of *Ulmus, Ostrya/Carpinus,* and other thermophilous deciduous trees than are common in many deciduous forest communities at the present time. The elm dominated woodlands were spreading rapidly eastward as can be seen in the pollen assemblages from Iowa (Van Zant, 1979), and Pickerel Lake, South Dakota (Watts and Bright, 1968). Webb, *et al.* (1983) have implied that this reflected early-Holocene conditions which were cooler and moister than at the present. The trend throughout the early-Holocene, therefore, is a replacement of boreal coniferous forest with decidu-ous forests and this trend continued for several thousands of years.

The sites in Iowa reveal that the deciduous woodland was short-lived. The establishment of the prairie occurred early in the Holocene as indicated by Van Zant (1979), Szabo (1980), Durkee (1971), and McAndrews (1965) from Minnesota. The prairie spread rapidly eastward reaching its maximum eastward extension between 6,000 and 7,000 yrs. B.P.

The early deciduous forest remained somewhat longer in the eastern portion of this area than in the west. At Kirchner Marsh (Wright *et al.,* 1963) the decline of the *Ulmus/Quercus* zone did not occur until around 7,120 yrs. B.P. By this time in western Minnesota, the vegetation had shifted to a Poaceae/*Artemisia* assemblage with a pronounced *Ambrosia* peak occurring between 8,000 and 7,000 yrs. B.P.

(McAndrews, 1966, 1967). During this same time at the eastern end of the Itasca transect, a *Pinus banksiana/Pteridium* forest had become established by 8,000 yrs. B.P. (McAndrews, 1965).

In Michigan, much of the area was still covered by a boreal conifer forest. At Chippewa Bog (Bailey and Ahearn, 1981) this boreal forest was not replaced by deciduous taxa until at least 8,100 yrs. B.P. At more northerly sites in Michigan, the early-Holocene period was dominated by pine forests (Zumberge and Potzger, 1955, 1956; Kapp *et al.,* 1969; Saarnisto, 1974).

To the east, the early dominance of pine pollen is succeeded generally by *Tsuga. Tsuga* pollen becomes dominant by *ca.* 8,500 yrs. B.P. at Bellmont Bog, New York (Spear and Miller, 1976). This same general succession is also recorded elsewhere at Rose Lake, Pennsylvania (Cotter and Crowl, 1981).

The region immediately north of Lake Superior and Lake Huron was still covered with ice at least as recently as the Valders readvance (Terasmae, 1967). Thus, palynologically this area produces a record of vegetation extending at best, only 11,000 yrs. B.P. Terasmae (1980) used a sequence of 20 radiocarbon dated lakes and bogs to infer that deglaciation had occurred between 12,500 and 11,500 yrs. B.P. at this southern Ontario location.

Immediately following deglaciation and the subse-quent draining of the large glacial lakes which occupied the area (Terasmae, 1967, 1968) the pollen record shows little evidence for tundra growth. The basal fossil pollen assemblage from southern Ontario (Terasmae, 1968) suggest an almost immediate colonization by the boreal forest. The character of this forest differed from that of the modern one by the absence of pine, a feature consistent within the Great Lakes Region (Wright, 1968; Webb *et al.* 1983; Davis, 1976; Kapp, 1977). This forest type existed for several thousand years terminating by *ca* 9,000 yrs. B.P. This closely approximates the changes which occurred north of Lake Huron 9,000-8,500 yrs. B.P., which Terasmae (1967) interprets as a replacement by the Great Lakes-St. Lawrence Forest type (mixed conifer-hardwood taxa, see Rowe, 1972).

Middle and Late-Holocene

The mid-Holocene Period records a characteristic time-transgressive expansion of the prairie from west to east. The prairie was well established in Iowa and western Minnesota by 8,000 yrs. B.P. (McAndrews, 1966, 1967; Van Zant, 1979), yet it is only through

Indiana and Illinois that we have evidence of vegetational responses during this period. The extension of the prairie is clearly shown in the assemblages from Chatsworth and Volo Bogs, Illinois (King, 1981). However, as Webb *et al.* (1983) have noted, the lack of radiocarbon dated sediments to the south of the "prairie peninsula" precludes the identification of just how extensive was the replacement of forested regions by the prairie.

The response to a progressively drying climate is not universally recorded in sediments from the Great Lakes Region. No shift is noted within the pollen assemblages recorded at Bellmont Bog (Spear and Miller, 1976), Rose Lake (Cotter and Crowl, 1981), nor at Silver Lake, Ohio (Ogden, 1966). Thus, while certainly dramatic changes were occurring throughout most of this region, the eastern woodlands section may have been unaffected.

By 5,000 yrs. B.P., the prairie was again migrating westward. The retreat of the prairie was accompanied by an expansion of *Picea* in both a north and south direction (Webb, 1981). The westward retreat of the prairie continued until about 2,000 yrs. B.P. when the prairie/forest ecotone reached its present position (Webb, 1981).

The general trends in plant succession are thought to be climatically controlled. The replacement of the spruce forests by pine species alternatively could have occurred as a response to increased incidence of fire. Based on reports from many different sites in this region, the first group of pines to invade the Great Lakes Region were *Pinus banksiana/resinosa* type which are fire adapted. Amundson and Wright (1979) examined the charcoal evidence from a few selected sites from Minnesota and concluded that there was no demonstrable change in charcoal influx at the end of the spruce period. Thus, they interpreted the forest changes as having occurred primarily as the result of climate.

In addition to climatically controlled changes, other factors may have significantly contributed to the development of the broadscale vegetational patterns. For example, Davis (1981) has argued convincingly for the role of disease in explaining the abrupt synchronous decline in *Tsuga* pollen throughout the northeastern United States around 4,800 yrs. B.P. Further, while discussing these plant migrations, we must remember that individual taxa migrate at varying rates, creating new plant associations that are at best, ephemeral (Davis, 1976).

Southeastern Ontario revealed a change from boreal forest to the Great Lakes-St. Lawrence Forest type *ca.* 8,000 yrs. B.P. (Potzger and Courtemanche,

1956). This forest has been in existence since that time even though minor fluctuations in the position of the forest elements did occur. For example, evidence for the effects of the Hypsithermal Period is shown by the northward migration of *Pinus strobus* prior to 5,500 yrs. B.P.

Terasmae (1968) further proposed a mechanism for the rapid and time synchronous displacement of boreal forest by the Great Lakes-St. Lawrence Forest type. He envisioned a widespread boreal spruce muskeg whose acidic conditions prevented the immigration of less tolerant species. This muskeg complex is essentially self-perpetuating and developed initially on the exposed floodplain of recently drained glacial lakes. The amelioration of the climate caused these areas to dry, and through the impact of fire, enabled them to be very rapidly colonized by other forest tree taxa. Terasmae (1968) also proposed this explanation in conjunction with the extinction of mastodon, adapted to boreal conditions, which occurred around 9,000 yrs. B.P. Terasmae (1968) speculated that the mastodon became extinct when the Boreal Forest disappeared rather rapidly from southern Ontario.

The basal, herb dominated assemblage zones commonly found in pollen sites from the Upper Midwest, are absent from sites in Ontario. In its place, the initial post-glacial vegetation was dominated by spruce. McAndrews (1981) suggested that the absence of a herb dominated zone was caused by a lag between deglaciation and the melting of buried ice to form kettle bogs. The subsequent *Picea* decline occurred in southern Ontario *ca.* 10,500 yrs. B.P. (Liu, 1981). This decline has been attributed to a major climatic shift occurring in this area (Ogden, 1967; Moran, 1973). At sites in central Minnesota and Wisconsin, Webb and Bryson (1972) attributed the cause to an initial shift in the air mass position. Webb and Bryson (1972) further postulated that a second shift, occurring about 9,500 yrs. B.P., was caused by further retreat of the ice front.

As spruce decreased in dominance after 10,000 yrs. B.P., pine invaded southern Ontario in two waves, initially by *Pinus banksiana,* and secondly by *P. strobus.* Between 8,000 and 7,000 yrs. B.P., populations of thermophilous hardwood taxa such as *Acer* and *Fagus* began to appear in the area (Davis, 1976: McAndrews, 1981).

In southern Ontario, the climatic warming trend which characterized much of the early- and middle-Holocene, began earlier than in the northern part of the province and is first represented by the spruce decline. Terasmae (1968) postulated that the expansion of *Pinus strobus* northward indicated the effects of the Hypsithermal Period. Alternatively, McAndrews (1981) suggested that the Hypsithermal, if it occurred in southern Ontario, was not of sufficient magnitude to alter the existing zonal vegetation.

There are two major trends apparent in the late-Holocene. The first reflects cooler and moister conditions returning to areas after the Hypsithermal (Liu, 1981). This trend is reflected in the migration of arboreal pollen frequencies of boreal conifers in a southward direction. The second major trend is the recognition in the pollen assemblages of both the aboriginal and later European agricultural practices

and the associated forest clearance (McAndrews, 1976).

ACKNOWLEDGMENTS

We would like to express our sincere thanks and appreciation to Dr. Thompson Webb III for his invaluable criticisms and comments at various stages of the development of this chapter. We similarly wish to express our thanks to Dr. Ronald O. Kapp and Dr. James E. King for providing comments on an earlier draft of this manuscript. We are also indebted to Mrs. Celinda A. Stephens for typing many drafts of the manuscript.

References Cited

AMUNDSON, D. C., and WRIGHT, H.E.
1979   Forest changes in Minnesota at the end of the Pleistocene. *Ecological Monographs,* 49:1-16.

ANDERSEN, S.T.
1954   A late-glacial pollen diagram from southern Michigan USA. *Denmarks Geologiske Undersogelse,* 2(80):140-155.

ANDERSON, T.W.
1974   The Chestnut pollen decline as a time horizon in lake sediments in eastern north America. *Canadian Journal of Earth Sciences,* 11:678-685.

1982   Pollen and plant macrofossil analyses of Late Quaternary sediments at Litchener, Ontario. *In: Current Research, Part A, Geological Survey of Canada, Paper* 82-1A:131-136.

1985   Late-Quaternary pollen records from Eastern Ontario, Quebec, and Atlantic Canada. *In:* Bryant, V.M. and Holloway, R.G. (eds) *Pollen Records of Late-Quaternary North American Sediments.* American Association of Stratigraphic Palynologists. Dallas, Texas.

ARTIST, R.C.
1936   Stratigraphy and preliminary pollen analysis of a Lake County, Illinois bog. *Butler University Botanical Series,* 3:191-198.

ASHWORTH, A.C.
1980   Environmental implications of a bettle assemblage from the Gervais Formation (Early Wisconsinan?), Minnesota. *Quaternary Research,* 13:200-213.

ASHWORTH, A.C., SCHWERT, D.P., WATTS, W.A., and WRIGHT, H.E.
1981   Plant and insect fossils at Norwood in south central Minnesota: a record of late-glacial succession. *Quaternary Research,* 16:66-79.

AUER, V.
1927   Botany of the interglacial peat beds of Moose River Basin. *Geological Survey Canada Summary Report 1926, part C:* 45-47.

1928   Some problems of peat bog investigations in Canada. *Geological Survey Canada Summary Report 1927 part C:* 96-111.

1930   Peat bogs in southeastern Canada. *Geological Survey Canada Memoir* 162:1-32.

BAILEY R.E., and AHEARN, P.J.
1981   A late and postglacial pollen record from Chippewa Bog, Lapeer Co, MI: further examination of white pine and beech immigration into the central Great Lakes Region. *In:* Romans, R.C. (ed) *Geobotany II.* :53-74 Plenum Publishing Company, New York.

BAKER, R.G.
1970   A radiocarbon dated pollen chronology for Wisconsin: Disterhaft farm bog revisited. *Geological Society of America Abstracts,* 2:488.

BAKER, R.G., VAN ZANT, K.L., and DULIAN, J.J.
1980   Three late-glacial pollen and plant macrofossil assemblages from Iowa. *Palynology,* 4:197-203.

BAKER, R.G., and WALN, K.
1985   Pollen records from the Great Plains and central United States. *In:* Bryant, V.M. and Holloway, R.G. (eds) *Pollen records of Late-Quaternary North American Sediments.* American Association of Stratigraphic Palynologists. Dallas, Texas.

BARKLEY, F.A.
1934   The statistical theory of pollen analysis. *Ecology,* 15:283-289.

BARNETT, J.
1937   A pollen study of Cranberry Pond near Emporia, Madison County, Indiana. *Butler University Botanical Studies,* 4:55-64.

BARTLEIN, P.J., WEBB, T., III, and FLERI, E.
1984   Holocene climatic change in the northern Midwest: Pollen derived estimates. *Quaternary Research,* 22:361-375.

BERNABO, J.C., and WEBB, T., III.

1977   Changing patterns in the Holocene pollen record of northeastern North America: a mapped summary. *Quaternary Research,* 8:69-96.

BIRKS, H.H., WHITESIDE, M.C., STARK, D.M., and BRIGHT, R.C.

1976   Recent paleolimnology of three lakes in northwestern Minnesota. *Quaternary Research,* 6:249-272.

BIRKS, H.J.B.

1976   Late-Wisconsin vegetational history at Wolf Creek central Minnesota. *Ecological Monographs,* 46:395-429.

1981a   Long-distance pollen in late Wisconsin sediments of Minnesota, U.S.A.: a quantitative analysis. *New Phytologist,* 87:630-661.

1981b   Late Wisconsin vegetational and climatic history at Kylen Lake, northeastern Minnesota. *Quaternary Research,* 16:322-355.

BIRKS, H.J.B., WEBB, T., III, and BERTI, A.A.

1975   Numerical analysis of pollen samples from central Canada: a comparison of methods. *Review of Paleobotany and Palynology,* 20:133-169.

BRUBAKER, L.B.

1975   Postglacial forest patterns associated with till and outwash in northcentral upper Michigan. *Quaternary Research,* 5:488-527.

BRUSH, G.S.

1967   Pollen analyses of late-glacial and post-glacial sediment in Iowa. *In:* Cushing, E.J. and Wright, H.E. (eds) *Quaternary Paleoecology,* pp. 99-105, Yale University Press, New Haven, Connecticut.

CAIN, S.A.

1940   The identification of species in fossil pollen of *Pinus* by size frequency distribution. *American Journal of Botany,* 27:301-308.

CALKIN, P.E. and McANDREWS, J.H.

1980   Geology and paleoontology of two late Wisconsin sites in western New York state. *Geological Society of America Bulletin,* 91:295-306.

COTTER, J.F.P. and CROWL, G.H.

1981   The paleolimnology of Rose Lake, Potter Co, Pennsylvania a comparison of palynologic and paleopigment studies. *In:* Romans, R.C. (ed). *Geobotany II:* 91-122, Plenum Publishing Company, New York.

CRAIG, A.J.

1972   Pollen influx to laminated sediments: a pollen diagram from northeastern Minnesota. *Ecology,* 53:46-57.

CRISMAN, T.L. and WHITEHEAD, D.R.

1975   Environmental history of Hovey Lake, Southwestern Indiana. *American Midland Naturalist,* 93:198-204.

CHURCHER, C.S., and PETERSON, R.L.

1982   Chronologic and environmental implications of a new genus of fossil deet from late Wisconsin deposits at Toronto, Canada. *Quaternary Research,* 18:184-195.

CURRIER, P.J. and KAPP, R.O.

1974   Local and regional pollen rain components at Davis Lake, Montcalm County, Michigan. *Michigan Academician,* 7:211-225.

CUSHING, E.J.

1963   Late-Wisconsin pollen stratigraphy in east-central Minnesota. Ph.D. dissertation, University of Minnesota.

1965   Problems in the Quaternary phytogeography of the Great Lakes Region. *In:* Wright, H.E. and Frey, D.M. (eds) *The Quaternary of the United States.* : 403-417. Princeton University Press, Princeton.

1967   Late Wisconsin pollen stratigraphy and the glacial sequence in Minnesota. *In:* Cushing, E.J. and Wright, H.E. (eds). *Quaternary Paleoecology.* :59-88. Yale University Press, New Haven Connecticut.

CWYNAR, L.C.

1978   Recent history of fire and vegetation from laminated sediments of Greenleaf lake, Algonquin Park, Ontario. *Canadian Journal of Botany,* 56:10-21.

DAVIS, A.M.

1977   The prairie-deciduous forest ecotone in the upper middle west. *Annals of the Association of American Geographers,* 67:204-213.

1979   Wetland succession, fire, and the pollen record: a midwestern example. *American Midland Naturalist,* 102:86-94.

DAVIS, M.B.

1976   Pleistocene Biogeography of temperate deciduous forests. *Geoscience and Man,* 13:13-26.

1981   Outbreaks of forest pathogens in Quaternary history. *In: Proceedings of the IV International Palynological Conference, Lucknow (1976-79),* 3:216-227.

DAVIS, M.B., BRUBAKER, L.B., and BEISWENGER, J.

1971   Pollen grains in lake sediments: pollen percentages in surface sediments from southern Michigan. *Quaternary Research,* 1:450-467.

DEAN, W.E., BRADBURY, J.P., ANDERSON, R.Y., and BARNOSKY, C.W.

1984   The variability of Holocene climatic change: evidence from varved lake sediments. *Science,* 226:1191-1194.

DELCOURT, H.R. and DELCOURT, P.A.

1985   Quaternary palynology and vegetational history of the southeastern United States. *In:* Bryant, V.M. and Holloway, R.G. (eds) *Pollen records of Late-Quaternary North American Sediments.* American Association of Stratigraphic Palynologists. Dallas, Texas.

DREIMANIS, A.

1958   Wisconsin stratigraphy at Port Talbot on the north shore of Lake Erie, Ontario. *Ohio Journal of Science,* 58:65-84.

1964   Notes on the Pleistocene time-scale in Canada. *In:* Osborne, F. (ed) *Geomorphology in Canada:* 139-156. University of Toronto Press, Toronto, Ontario.

DROSTE, J.B., RUBIN, M., and WHITE, G.W.

1959   Age of marginal Wisconsin drift at Corry, northwestern Pennsylvania. *Science,* 130.1760.

DURKEE, L.H.

1971   A pollen profile from Woden bog in northcentral Iowa. *Ecology,* 52:837-844.

ENGLEHARDT, D.W.

1959   A comparative pollen study of two early Wisconsin bogs in Indiana. *Proceedings of the Indiana Academy of Science,* 69:110-118.

1962   A palynological study of post-glacial and interglacial deposits in Indiana. Unpublished Ph.D. dissertation, Indiana University.

1965   A late-glacial-postglacial pollen chronology for Indiana. *American Journal of Science,* 263:410-415.

EVERSON, E.B., FARRAND, W.R., ESCHMAN, D.F., MICHELSON, D.M., and MAHER, L.J.
1976   Great lakean substage: a replacement for Valderan substage in the Lake Michigan basin. *Quaternary Research,* 6:411-424.

FARNHAM, R.S., McANDREWS, J.H., and WRIGHT, H.E.
1964   A late Wisconsin buried soil near Aikin Minnesota and its paleobotanical setting. *American Journal of Science,* 262:393-412.

FARRAND, W.R., AAHNER, R., and BENNINGHOFF, W.S.
1968   Buried Bryophte bed, Cheboygan County, Michigan. *Abstract, paper presented at North Central Section Geological Society of America meetings,* Iowa City.

FREY, D.G.
1959   The Two Creeks Interval in Indiana pollen diagrams. *Investigations of Indiana Lakes and Streams,* 4:131-139.

FRIES, M.
1962   Pollen profiles of late Pleistocene and recent sediments at Weber Lake, northeastern Minnesota. *Ecology,* 43:295-308.

FRIES, M., WRIGHT, H.E., and RUBIN, M.
1961   A late Wisconsin buried peat at North Branch, Minnesota. *American Journal of Science,* 259:679-693.

FRIESNER, R.C., and POTZGER, J.E.
1946   The Cabin Creek raised bog, Randolph County, Indiana. *Butler University Botanical Studies,* 8:24-43.

GAUDREAU, D.C., and WEBB, T., III.
1985   Late-Quaternary pollen stratigraphy and isochrone maps for the northeastern United States. *In:* Bryant, V.M. and Holloway, R. G. (eds) *Pollen records of Late-Quaternary North American Sediments.* American Association of Stratigraphic Palynologists, Dallas, Texas.

GILLIAM, J.A., KAPP, R.O. and BOGUE, R.D.
1967   A post-Wisconsin pollen sequence from Vestaburg bog, Montcalrim County, Michigan. *Michigan Academy of Science, Arts, and Letters,* 52:3-17.

GOODING, A.M.
1966   The Kansan glaciation in southeastern Indiana. *Ohio Journal of Science,* 4:426-433.

GORDON, A.D., and BIRKS, H.J.B.
1974   Numberical Methods in Quaternary Paleoecology II. *New Phytologist,* 73:221-249.

GRIFFIN, C.D.
1950   A pollen profile from Reed Bog, Randolph County, Indiana. *Butler University Botanical Studies,* 9:131-139.
1951   Pollen analysis of a peat deposit in Livingston County, Illinois. *Butler University Botanical Studies,* 10:90-99.

GRIFFIN, K.O.
1975   Vegetation studies and modern pollen spectra from Red Lake peatland, northern Minnesota. *Ecology,* 56:531-546.
1977   Paleoecological aspects of the Redlake peatland, northern Minnesota. *Canadian Journal of Botany,* 55:172-192.

GRUEGER, E.
1970   The development of the vegetation of southern Illinois since late Illinoian time (preliminary report). *Revue de Geographie Physique et de Geologie Dynamique,* 12:143-148.
1972a   Late Quaternary vegetation development in south-central Illinois. *Quaternary Research,* 2:217-231.
1972b   Pollen and seed studies of Wisconsin vegetation in Illinois. *Geological Society of America Bulletin,* 83:2715-2734.

HALL, S.A.
1971   A postglacial mollusk and pollen succession associated with *Bison occidentalis* in east-central Iowa. *Michigan Academy of Science, Art. and Letters, Annual Meeting Program with Abstracts:* 19.
1985   Quaternary pollen analysis and vegetational history of the southwest. *In:* Bryant, V.M. and Holloway, R.G. (eds) *Pollen Records of Late-Quaternary North American Sediments.* American Association of Stratigraphic Palynologists. Dallas, Texas.

HALLBERG, G.R., BAKER, R.G., and LEGG, T.
1980   A mid-Wisconsinan pollen diagram from Des Moines County, Iowa. *Proceedings of the Iowa Academy of Science,* 87:41-44.

HANSEN, H.P.
1937   Pollen analysis of two Wisconsin bogs of different ages. *Ecology,* 18:136-148.
1939   Postglacial vegetation of the Driftless area of Wisconsin. *American Midland Naturalist,* 21:752-762.

HARMON, R.S., SCHWARCZ, H.P., FORD, D.C., and KOCH, D.L.
1979   An isotopic paleotemperature record for late Wisconsinian time in northeast Iowa. *Geology,* 7:430-433.

HEIDE, K.M.
1984   Holocene pollen stratigraphy from a lake and small hollow in north-central Wisconsin, U.S.A. *Palynology,* 8:3-21.

HEIDE, K.M., and BRADSHAW, R.
1982   The pollen-tree relationship within forests of Wisconsin and Upper Michigan U.S.A. *Review of Paleobotany and Palynology,* 36:1-23.

HELD, E.R., and KAPP, R.O.
1969   Pollen analysis at the Thaller Mastodon site, Gratiot County, Michigan. *The Michigan Botanist,* 8:3-10.

HOBSON, G.D., and TERASMAE, J.
1969   Pleistocene geology of the buried St. Davids Gorge Niagra Falls, Ontario: Geophysical and palynological studies. *Geological Survey Canada Paper,* 68-67:1-16.

HOWELL, J.W.
1937   A fossil pollen study of Kokomo Bog, Howard County, Indiana. *Butler University Botanical Studies,* 4:117-127.

HUDAK, C.W.
1984   Paleoecology of an early Holocene faunal and floral assemblage from the Dows Local Biota of north-central Iowa. *Quaternary Research,* 21:351-368.

HUSHEN, T.W., KAPP, R.O., BOGUE, R.D., and WORTHINGTON, J.T.
1966   Presettlement forest patterns in Montcalm County Michigan. *Michigan Botanist,* 5:192-211.

JACOBSON, G.L.
1979   The paleoecology of white pine in Minnesota. *Journal of Ecology,* 67:697-726.

JANSON, E., and HALFERT, E.
1936   A pollen analysis of a bog in northern Ontario. *Papers Michigan Academy of Science, Arts, and Letters,* 22:95-98.

JANSSEN, C.R.
1967   A postglacial pollen diagram from a small Typha swamp in northwestern Minnesota, interpreted from pollen indicators and surface samples. *Ecological Monographs,* 37:145-172.
1968   Myrtle Lake: a late and postglacial pollen diagram from northern Minnesota. *Canadian Journal of Botany,* 46:1397-1410.

JELGERSMA, S.
1962 A late-glacial pollen diagram from Madelia, south-central Minnesota. *American Journal of Science,* 260:522-529.

JONES, C.L., and KAPP, R.O.
1972 Relationship of Bay County Michigan presettlement forest patterns to Indian cultures. *Michigan Academician,* 5:17-28.

KAPP, R.O.
1977 Late Pleistocene and postglacial plant communities of the Great Lakes Region. *In:* Romans, R.C. (ed) *Geobotany:* 1-26. Plenum Publishing Co. New York.

1978 Plant remains from a Wisconsin Interstadial dated 25,000 B.P. Muskegon County, Michigan. *American Midland Naturalist,* 100:506-509.

KAPP, R.O., AHEARN, P. and KLOUS, G.
1977 Pollen analysis at Demont and Crystal Lakes. *In: Paleoecology Field Trip. Paleoecology Section,*

KAPP, R.O., BUSHOUSE, S., and FOSTER, B.
1969 A contribution to the geology and forest history of Beaver Island, Michigan. *Proceedings 12th Conference Great Lakes Research:* 225-236.

KAPP, R.O., and GOODING, A.M.
1964 A radiocarbon-dated pollen profile from Sunbeam prairie bog, Darke County, Ohio. *American Journal of Science,* 262:259-266.

1964b Pleistocene vegetational studies in the Whitewater basin, southeastern Indiana. *The Journal of Geology,* 72:307-326.

1974 Stratigraphy and pollen analysis of Yarmouthian interglacial deposits in Southeastern Indiana. *The Ohio Journal of Science,* 74:226-238.

KAPP, R.O., OGG, J. and LEACH, J.
1977 Pollen analysis of post glacial and presettlement sediments in Newaygo County. *In: Paleoecology Field Trip Paleoecology Section of Ecological Society of America:* 42-44.

KAPP, R.O., and MEANS, T.P.
1977 Pollen analysis of peat from a sinkhole in the garden peninsula, Delta County, Michigan. *Michigan Botanist,* 16:55-62.

KARROW, P.F., ANDERSON, T.W., CLARKE, A.H., DELORME, L.D., and NIVASA, M.R.
1975 Stratigraphy, paleontology, and age of Lake Algonquin sediments in southwestern Ontario, Canada. *Quaternary Research,* 5:49-87.

KARROW, P.F., CLARKE, A.H., and HERINGTON, H.B.
1972 Pleistocene molluscs from Lake Iroquois deposits in Ontario. *Canadian Journal of Earth Sciences,* 9:589-595.

KARROW, P.F., HEBDA, R.J., PRESANT, E.W., and ROSS, G.J.
1982 Late Quaternary inter-toll paleosol and biota at Guelph, Ontario. *Canadian Journal of Earth Sciences,* 19:1857-1872.

KARROW, P.F., and WARNER, B.G.
1984 A subsurface middle Wisconsin interstadial site at Waterloo, Ontario. *Boreas,* 13:67-85.

KARROW, P.F., WARNER, B.G., and FRITZ, P.
1984 Corry Bog, Pennsylvania: a case study of the radiocarbon dating of marl. *Quaternary Research,* 21:326-336.

KING, J.E.
1973 Late Pleistocene palynology and biogeography of the western Ozarks. *Ecological Monographs,* 43:539-565.

1981 Late Quaternary vegetational history of Illinois. *Ecological Monographs,* 51:43-62.

KING, J.E., and KAPP, R.O.
1963 Modern pollen rain studies in eastern Ontario. *Canadian Journal of Botany,* 41:243-252.

KING, J.E., LINEBACK, J.A., and GROSS, D.L.
1976 Palynology and sedimentology of Holocene deposits in southern Lake Michigan. *Illinois State Geological Survey Circular* 496:1-24.

LAWRENZ, R.
1975 Biostratigraphic study of Green Lake, Michigan. M.S. Thesis, Central Michigan University, Mt. Pleasant, Michigan.

LEWIS, C.F.M., and ANDERSON, T.W.
1976 Basin deposits, a record of longterm coastal evolution (recession) in the Great Lakes? *In:* Rukavina, N.A. (ed) *Proceedings of the Workshop on Great Lakes Coastal erosion and sedimentation. Environment Canada. Canadian Centre Inland Waters, Burlington:* 109-115.

LICHTI-FEDEROVICH, S., and RITCHIE, J.C.
1968 Recent pollen assemblages from the western interior of Canada. *Review of Paleobotany and Palynology,* 7:297-344.

LIU, K.B.
1981 Pollen evidence of Late-Quaternary climatic changes in Canada; a review Part II: eastern Canada. *Ontario Geography,* 17:61-82.

MAHER, L.J.
1977 Palynological studies in the western arm of Lake Superior. *Quaternary Research,* 7:14-44.

MANNY, B.A., WETZEL, R.G. and BAILEY, R.E.
1978 Paleolimnological sedimentation of organic carbon, nitrogen, phosphorous, fossil pigments, pollen and diatoms in a hypereutrophic hardwater lake: a case story of eutrophication. *Polskie Archiwum hydrobiologi,* 25:243-267.

McANDREWS, J.H.
1966 Postglacial history of prairie, savanna, and forest in northwestern Minnesota. *Bulletin of the Torrey Botanical Club,* 22:1-72.

1967 Pollen analysis and vegetational history of the Itasca region, Minnesota. *In:* Cushing, E.J., Wright, H.E. (ed) *Quaternary Paleoecology.* Yale University Press: 219-236. New Haven, Connecticut.

1968 Pollen evidence for the protohistoric development of the "Big Woods" in Minnesota (U.S.A.). *Review of Paleobotany and Palynology,* 7:201-211.

1969 Paleobotany of a wild rice lake in Minnesota. *Canadian Journal of Botany,* 47:1671-1679.

1976 Fossil history of man's impact on the Canadian flora: an example from southern Ontario. *Canadian Botanical Association Bulletin Supplement to Vol. 9:1-6.*

1981 Late Quaternary climate of Ontario: temperature trends from the fossil pollen record. *In:* W.C. Mahaney (ed). *Quaternary Paleoclimate.* Geo Abstracts Lts. University of East Anglia Norwich, England: 319-333.

1982 Holocene environment of a fossil bison from Kenora Ontario. *Ontario Archaeology,* 37:41-51.

McANDREWS, J.H., and POWER, D.M.
1973 Palynology of the Great Lakes: the surface sediments of Lake Ontario. *Canadian Journal of Earth Science,* 10:777-792.

McANDREWS, J.H., RILEY, J.L., and DAVIS, A.M.
1982 Vegetation history of the Hudson Bay Lowland: a postglacial pollen diagram from the Sutton Ridge. *Naturaliste Canadien,* 109:597-608.

McMURRAY, M., KLOOS, G., KAPP, R.O., and SULLVIAN, K.
  1978   Paleoecology of Crystal Marsh, Montcalm County,
          based on macrofossil and pollen analysis. *Michigan
          Academician,* 10:403-417.

MILLER, N.G.
  1973a   Pollen analysis of deeply buried Quaternary sediments
          from southern Michigan. *American Midland Naturalist,*
          89:217-223.
  1973b   Late-glacial and postglacial vegetation change in south-
          western New York State. *New York State Museum and
          Science Service Bulletin,* 420:102 p.

MILLER, R.F., and MORGAN, A.V.
  1982   A postglacial Coleopterous assemblage from Lockport
          Gulf, New York. *Quaternary Research,* 17:258-275.

MOTT, R.J., and CAMFIELD, M.
  1969   Palynological studies in the Ottawa area. *Geological
          Survey Canada Paper,* 69-38:1-16.

MOTT, R.J., and FARLEY-GILL, L.D.
  1978   A Late-Quaternary pollen profile from Woodstock
          Ontario. *Canadian Journal of Earth Sciences,* 15:1101-
          1111.

MORAN, J.M.
  1973   Late-glacial retreat of 'Arctic Air' as suggested by onset
          of *Picea* decline. *Professional Geographer,* 25:373-376.

MORGAN, A.V., and MORGAN A.
  1979   The fossil Coleoptera of the Two Creeks Forest Bed,
          Wisconsin. *Quaternary Research,* 12:226-241.

NORRIS, G. and McANDREWS, J.H.
  1970   Dinoflagellate cysts from post-glacial lake muds, Minne-
          sota (U.S.A.). *Review of Paleobotany and Palynology,*
          10:131-156.

OGDEN, J.G. III
  1966   Forest history of Ohio I. radiocarbon dates and pollen
          stratigraphy of silver lake, Logan County, Ohio. *Ohio
          Journal of Science,* 66:387-400.
  1967   Radiocarbon and pollen evidence for a sudden change in
          climate in the Great Lakes Region, approximately
          10,000 years ago. *In:* E.J. Cushing, Wright, H. E. (eds).
          Quaternary Paleoecology, Yale University Press, New
          Haven, Connecticut:117-127.

OLTZ, D.F., JR.; KAPP, R.O.
  1963   Plant remains associated with Mastodon and Mammoth
          remains in central Michigan. *American Midland
          Naturalist,* 70:339-346.

OTTO, J.H.
  1937   Forest succession of the southern limits of early Wiscon-
          sin glaciation as indicated by a pollen spectrum from
          Bacon's Swamp, Marion County, Indiana. *Butler
          University Botanical Studies,* 4:93-116.

PETERS, A. and WEBB, T., III
  1979   A radiocarbon dated pollen diagram from west-central
          Wisconsin. *Bulletin of the Ecological Society of America,*
          60:102.

PETERSON, G.M.
  1978   Pollen spectra from surface sediments of lakes and
          ponds in Kentucky, Illinois, and Missouri. *American
          Midland Naturalist,* 100:333-340.

POTTER, L.D.
  1947   Postglacial forest sequence in Ohio. *Ecology,* 28:396-
          417.

POTZGER, J.E.
  1943   Pollen studies of five bogs in Price and Sawyer Counties,
          Wisconsin. *Butler University Botanical Studies,* 6:54-64.

  1943b   Pollen profile from sediments of an extinct lake in
          Hendricks county Indiana marks time of drainage.
          Proceedings Indiana Academy Science, 52:83-86.
  1946   Phytosociology of the primeval forests in central-north-
          ern Wisconsin and upper Michigan and a brief postgla-
          cial history of the Lake Forest formation. *Ecological
          Monographs,* 16:211-250.
  1954   Post-Algonquin and Post-Nipissing forest history of Isle
          Royale, Michigan. *Butler University Botanical Studies,*
          11:200-209.

POTZGER, J.E. and COURTEMANCHE, A.
  1956   A series of bogs across Quebec from the St. Lawrence
          valley to James Bay. *Canadian Journal of Botany,*
          34:473-500.

POTZGER, J.E. and RICHARDS, R.R.
  1941   Forest succession in the Trout Lake, Vilas County,
          Wisconsin area: a pollen study. *Butler University Botan-
          ical Studies,* 5:179-189.

RADFORTH, N.W. and TERASMAE, J.
  1960   Palynological study relating to the Pleistocene Toronto
          formation. *Canadian Journal of Botany,* 38:571-580.

RAYMOND, R.E., KAPP, R.O. and JAMKE, R.A.
  1975   Postglacial and recent sediments of inland lakes of Isle
          Royale National Park, Michigan. *Michigan Academi-
          cian,* 7:453-465.

RICHARDS, R.R.
  1937   A pollen profile of Otterbein Bog, Warren County,
          Indiana. *Butler University Botanical Studies,* 4:128-140.

ROWE, J.S.
  1972   *Forest Regions of Canada.* Department of the Environ-
          ment. Canadian Forestry Service, Publication No. 1300.,
          Ottawa, Ontario.

RUHE, R.V., RUBEIN, M. and SCHOLTES, W.H.
  1957   Late Pleistocene radiocarbon chronology in Iowa.
          American Journal of Science, 255:671-689.
  1957   Ages and development of soil landscapes in relation to
          climatic and vegetational changes in Iowa. *Soil Science
          Society of America Proceedings,* 20:264-273.

RUHE, R.V., and SCHOLTES, W.H.
  1956   Ages and development of soil landscapes in relation to
          climate and vegetational change in Iowa. *Soil Science
          Society of America Proceedings,* 20:204-273.

SAARNISTO, M.
  1974   The deglaciation history of the Lake Superior Region
          and its climatic implications. *Quaternary Research,*
          4:316-339.

SANGER, J.E. and CROWL, G.H.
  1979   Fossil Pigments as a guide to the paleolimnology of
          Browns Lake, Ohio. *Quaternary Research,* 11:342-353.

SCHWEGER, C.E.
  1969   Pollen analysis of Iola Bog and paleoecology of the Two
          Creeks forest bed, Wisconsin. *Ecology,* 50:859-868.

SCHWERT, D.P.
  1978   Paleoentomological analyses of two postglacial sites in
          eastern North America. Unpublished Ph.D. thesis,
          University of Waterloo, Ontario.

SCHWERT, D.P. and MORGAN, A.V.
  1980   Paleoenvironmental implications of a late glacial insect
          assemblage from Northwestern New York. *Quaternary
          Research,* 13:93-111.

SEARS, P.B.
  1930   Common fossil pollens of the Erie Basin. *Botanical
          Gazetter,* 89:95-106.

1930b A record of post-glacial climate in northern Ohio. *Ohio Journal of Science,* 30:205-217.

SHANE, L.C.K.

1975 Palynology and radiocarbon chronology of Battaglia Bog, Portage County, Ohio. *Ohio Journal of Science,* 75:96-102.

SIGLEO, W.R. and KARROW, P.F.

1977 Pollen-bearing Erie interstadial sediments from near St. Mary's Ontario. *Canadian Journal of Earth Science,* 14:1888-1896.

SMITH, J.G. and KAPP, R.O.

1964 Pollen analysis of some Pleistocene sediments from Illinois. *Transactions of the Illinois State Academy of Science,* 57:158-162.

SMITH, W.M.

1937 Pollen spectrum of Lake Cicott bog, Cass County, Indiana. *Butler University Botanical Studies,* 4:43-54.

SPEAR, R.W. and MILLER, N.G.

1976 A radiocarbon dated pollen diagram from the Allegheny Plateau of New York State. *Journal of the Arnold Arboretum,* 57:369-403.

STOUTAMIRE, W.P., BENNINGHOFF, W.S.

1964 Biotic assemblage associated with a mastodon skull from Oakland County, Michigan. *Papers of the Michigan Academy of Science, Arts, and Letters,* 49:47-60.

SWAIN, A.M.

1972 A fire history of the Boundary Waters Canoe Area as recorded in lake sediments. *Naturalist,* 1972:24-41.

1973 A history of fire and vegetation in northeastern Minnesota as recorded in lake sediments. *Quaternary Research,* 3:383-396.

1978 Environmental changes during the last 2,000 years in north-central Wisconsin: analysis of pollen, charcoal, and seeds from varved lake sediments. *Quaternary Research,* 10:55-68.

SWICKARD, D.A.

1941 Comparison of pollen spectra from bogs of early and late Wisconsin glaciation in Indiana. *Butler University Botanical Studies,* 5:67-84.

SZABO, J.P.

1980 Two pollen diagrams from Quaternary deposits in east-central Iowa, USA. *Canadian Journal of Earth Sciences,* 17:453-458.

TERASMAE, J.

1955 A palynological study relating to the Toronto formation (Ontario) and the Pleistocene deposits in the St. Lawrence Lowland (Quebec). Ph.D. thesis, McMaster University, Hamilton, Ontario.

1958 Non-glacial deposits along Missinaibi River, Ontario. *Geological Survey of Canada Bulletin,* 46:29-34.

1960 A palynological study of Pleistocene Interglacial beds at Toronto, Canada. *Geological Survey Canada Bulletin,* 56.

1967 Postglacial chronology and forest history in the northern Lake Huron and Lake Superior regions. *In:* Cushing, E.J., and Wright, H.E. (eds) *Quaternary Paleoecology,* Yale University Press, New Haven, Connecticut: 45-58.

1968 A discussion of deglaciation and the boreal forest history in the northern Great Lakes Region. *In: Proceedings of the Entomological Society of Ontario,* 99:31-43.

1980 Some problems of late Wisconsin history and geochronology in southwestern Ontario. *Canadian Journal of Earth Science,* 17:361-381.

1981a Radiocarbon dating and palynology of glacial Lake Nipissing deposits at Wasaga Beach, Ontario. *Journal of Great Lakes Research,* 5:292-300.

1981b Some problems of late Wisconsin history and geochronology in southeastern Ontario. *Canadian Journal of Earth Sciences,* 17:361-381.

TERASMAE, J. and MATTHEWS, H.L.

1980 Late Wisconsin white spruce (Picea glauca (Moench) Voss) at Brampton, Ontario. *Canadian Journal Earth Science,* 17:1087-1095.

TERASMAE, J. and MOTT, R.J.

1964 Pollen deposition in lakes and bogs near Ottawa Canada. *Canadian Journal Botany,* 42:1355-1363.

TURNER, J.V., FRITZ, P., KARROW, P.F. and WARNER, B.

1983 Isotopic and geochemical composition of marl lake waters and implications for radiocarbon dating of marl lake sediments. *Canadian Journal of Earth Sciences,* 20:599-615.

UENO, J.

1958 Some Palynological observations of pinaceae *Journal of the Institute of Polytechnics, Osaka City University Series D.,* Vol. 9:163-187.

1960 Studies on Pollen grains of Gymnospermae. *Journal of the Institute of Polytechics, Osaka City University,* D-11:109-136.

VAN ZANT, K.

1979 Late glacial and postglacial pollen and plant macrofossils from Lake West Okoboji, Northwestern Iowa. *Quaternary Research,* 12:358-380.

VAN ZANT, K. and HALLBERG, G.R.

1976 A late-glacial pollen sequence from northeastern Iowa: Sumner bog revisited. *Iowa Geological Survey Technical Information Series,* 3:1-17.

VAN ZANT, K.L., HALLBERG, G.R.M. and BAKER, R.G.

1980 A farmdalian pollen diagram from east-central Iowa. *Proceedings of the Iowa Academy of Science,* 87:52-55.

VOSS, J.

1933 Pleistocene forests of central Illinois. *Botanical Gazette,* 94:808-814.

1937 Comparative study of bog on Cary and Tazewell drift in Illinois. *Ecology,* 18:119-135.

WADDINGTON, J.C.B.

1969 A stratigraphic record of the pollen influx to a lake in the Big Woods of Minnesota. *Geological Society of America Special Paper,* 123:263-282.

WALKER, P.C. and HARTMAN, R.T.

1960 The forest sequence of the Hartstown Bog area in western Pennsylvania. *Ecology,* 41:461-469.

WARNER, B.G., HEBDA, R.J. and HANN, R.J.

1984 Postglacial, paleoecological history of a cedar swamp, Manitoulin Island, Ontario, Canada. *Paleogeography, Paleoclimatology, Paleoecology,* 45:301-345.

WATTS, W.A.

1983 Vegetational history of the Eastern United States 25,000 to 10,000 years ago. *In:* Wright, H.E., Porter, S. (eds). *Late Quaternary Environments of the United States, Vol. 1.,* University of Minnesota Press. Minneapolis, Minnesota: 294-310.

WATTS, W.A. and BRIGHT, R.C.

1968 Pollen, seed, and mollusk analysis of a sediment core from Pickeral Lake, northeastern South Dakota. *Geological Society of America Bulletin,* 74:855-876.

WEBB, T., III

1973   A comparison of modern and presettlement pollen from southern Michigan. *Review of paleobotany and palynology,* 16:137-156.

1974   A vegetational history from northern Wisconsin: evidence from modern and fossil pollen. *American Midland Naturalist,* 92:12-34.

1974   The pollen-vegetation relationship in southern Michigan: an application of isopolls and principal components analysis. *Geoscience and Man,* 9:7-14.

1981   The past 11,000 years of Vegetational change in eastern North America. *Bioscience,* 31:501-506.

WEBB, T., III, and BRYSON, R.A.

1972   Late and postglacial climatic change in the northern midwest U.S.A.: quantitative estimates derived from fossil pollen spectra by multivariate statistical analysis. *Quaternary Research,* 2:70-115.

WEBB, T., III, CUSHING, E.J. and WRIGHT, H.E.

1983   Holocene changes in the vegetation of the midwest. *In:* Wright, H.E. (ed) *Late-Quaternary Environments of the United States Vol. 2: the Holocene:* 142-166. University of Minnesota Press, Minneapolis, Minnesota.

WEBB, T., III, HOWE, S., BRADSHAW, R.H.W. and HEIDE, K.M.

1981   Estimating plant abundances from pollen percentages: the use of regression analysis. *Review of Paleobotany and Palynology,* 34:269-300.

WEBB, T., III, LASESKI, R.A. and BERNABO, J.G.

1978   Sensing vegetational patterns with pollen dates choosing the data. *Ecology,* 59:1151-1163.

WEBB, T., III, and McANDREWS, J.H.

1976   Corresponding patterns of contemporary pollen and vegetation in central North America. *Geological Society of America Memoir,* 145:267-299.

WEST, R.G.

1961   Late and postglacial vegetational history in Wisconsin particularly changes associated with the Valders readvance. *American Journal of Science,* 259:766-783.

WETZEL, R.G. and MANNY, B.A.

1977   Postglacial rates of sedimentation, nutrient and fossil pigment deposition in a hardwater marl lake of Michigan. *In: Second International Symposium of Paleolimnology Mikolajki, Poland Mitt. Int. Ver. Limnol.*

WHITEHEAD, D.R., JACKSON, S.T., SHEEHAN, M.C. and LEYDEN, B.W.

1972   Late-glacial vegetational associated with Caribou and Mastodon in Central Indiana. *Quaternary Research,* 17:241-258.

WHITECAR, G.R. and DAVIS, A.M.

1982   Sedimentology and palynology of Middle Wisconsinan deposits in the Pectaonica river valley, Wisconsin and Illinois. *Quaternary Research,* 17:228-241.

WHITTAKER, R.H.

1967   Gradient analysis of vegetation. *Biological Review,* 49:207-264.

WILLIAMS, A.S.

1974   Late-glacial-postglacial vegetational history of the Pretty Lake region, northeastern Indiana. *United States Geological Survey Professional Paper,* 686-B 23p.

WILLIAMS, N.E., WESTGATE, J.A., WILLIAMS, D.D., MORGAN, A. and MORGAN, A.V.

1981   Invertebrate fossils Insecta: Trichoptra, Diptera, Coleoptera) from the Pleistocene Scarborough Formation at Toronto, Ontario and their paleoenvironmental significance. *Quaternary Research,* 16:146-167.

WILSON, I.T., POTZGER, J.E.

1943a   Pollen records from lakes in Anoka County, Minnesota: a study on methods of sampling. *Ecology,* 24:382-392.

1943b   Pollen study of sediments from Douglas Lake, Cheboygan County, and Middle Fish Lake, Montmorecy County, Michigan. *Proceedings Indiana Academy Science,* 52:87-92.

WILSON, L.R.

1938   The postglacial history of vegetation in northwestern Wisconsin. *Rhodora,* 40:137-175.

WILSON, L.R. and WEBSTER, R.M.

1943   Microfossil studies of four southwestern Ontario bogs. *Proceedings Iowa Academy Science,* 50:261-272.

WINTER, T.C.

1962   Pollen sequence at Kirchner Marsh, Minnesota. *Science,* 138:526-528.

WRIGHT, H.E.

1964   Aspects of the early postglacial forest succession in the Great Lakes Region. *Ecology,* 45:439-448.

1971   Late Quaternary vegetational history of North America. *In:* Turekian, K.K. (ed). *The Late Cenozoic glacial ages:* 425-464. Yale University Press, New Haven, Connecticut.

1972   Interglacial and postglacial climates: the pollen record. *Quaternary Research,* 2:274-282.

1976   Ice retreat and revegetation in the western Great Lakes Region. *In:* Mahaney, W.C. (ed) *Quaternary stratigraphy of North America:* 119-132. Dowden, Hutchinson, Ross, Stroudsburg, PA.

1981   Vegetation east of the rocky mountains 18,000 years ago. *Quaternary Research,* 15:113-125.

WRIGHT, H.E. and WATTS, W.A.

1969   Glacial and vegetational history of North-eastern Minnesota. *Minnesota Geological Survey Special Paper,* SP-11:1-59.

WRIGHT, H.E., WINTER, T.C. and PATTEN, H.L.

1963   Two pollen diagrams from southeastern Minnesota

problems in the late and postglacial vegetational history. *Geological Society of America Bulletin,* 74:1371-1396.

ZUMBERGE, J.H. and POTZGER, J.E.

1955   Pollen profiles, radiocarbon dating, and geologic chronology of the Lake Michigan Basin. *Science,* 121:309-311.

1956   Late Wisconsin chronology of the Lake Michigan Basin correlated with pollen studies. *Geological Society of America Bulletin, 67:271-288.*

LATE-QUATERNARY POLLEN STRATIGRAPHY AND ISOCHRONE MAPS FOR THE NORTHEASTERN UNITED STATES

DENISE C. GAUDREAU
Department of Geology
Southampton College
Southampton, New York 11968

THOMPSON WEBB, III
Department of Geological Sciences
Brown University
Providence, Rhode Island 02912

Abstract

Pollen data from sixty-three sites illustrate the regional variation in the late-Quaternary pollen stratigraphy across the northeastern United States. Many of the pollen diagrams from this region contain a sequence of four major pollen zones but the timing and composition of the zones vary with latitude and elevation. In the southern and central area, for example, the four zones are characterized by increased percentages of: 1) herb, 2) spruce *(Picea)*, 3) pine *(Pinus)*, and 4) oak *(Quercus)* pollen, but to the north hemlock *(Tsuga)*, birch *(Betula)*, and beech *(Fagus)* pollen are the dominant pollen types in the fourth zone. The only regionally synchronous (± 300 yr.) stratigraphic events are the mid-Holocene decline in hemlock pollen percentages and the recent human-induced increase in ragweed pollen percentages, but the magnitude of these events varies regionally with the hemlock decline most pronounced in the north and the ragweed increase more marked in the west and southern lowlands. Isopoll and isochrone maps illustrate how the changing regional vegetational patterns of the past 14,000 years resulted in the different local stratigraphies. Open spruce woodland had moved into the southern part of the region by 14,000 yrs. B.P. and was dominant over most of the area at 12,000 yrs. B.P. From 10,000 to 8,000 yrs. B.P., pine forests predominated and moved northward, being replaced in the south by oak-dominated forests and in the north by forests with abundant birch, hemlock, and beech trees. During the past 8,000 years, the ecotone between the deciduous and mixed conifer/hardwood forests has remained fairly stable in its location from central Pennsylvania to central New England. South of the ecotone both hickory *(Carya)* and chestnut *(Castanea)* populations expanded after 8,000 yrs. B.P., and north of the ecotone hemlock populations declined about 4,800 yrs. B.P. and then increased again by 3,000 yrs. B.P. Within the past 4,000 years, spruce populations have increased in the northern highlands. The mapped patterns and interpreted vegetational changes illustrate the limitations of using formal stratigraphic zones for the full interpretation of late-Quaternary pollen sequences.

INTRODUCTION

The first pollen studies in the Northeast revealed regionally similar pollen stratigraphies in southern and northern New England (Deevey, 1939, 1943; Leopold, 1956a; Deevey, 1951). From these studies came a widely recognized and still used (Miller, 1973a, Newman, 1977) sequence of pollen zones that summarize the main stratigraphic changes in pollen percentages. These zones were labeled the herb (T or L), spruce (A), pine (B), and oak (C) pollen zones. Deevey (1939, 1943, 1951) originally referred to them as a "climatic chronology", and correlated the zones and their subzones to bio- and climatostratigraphic units from Europe (*e.g.*, Allerod and Blytt-Sernander sequences) and the Midwest (*e.g.*, the Two Creeks Interval and the pollen zones defined by Sears, 1932).

In southern New England, the classic set of pollen zones (T, A, B, C) record regional vegetational changes from tundra to spruce parkland, pine forest, and oak forest. To the north, the vegetation associated with the L, A, B, and C zones differed somewhat from the interpreted sequence in the south, and Deevey (1951) recognized this difference in his original definition of the zones in northern Maine. For example, the increase and abundance of hemlock pollen played a more important role in designating the beginning of the C zone in Maine than in Connecticut where the increase in oak pollen was more pronounced. This difference in the composition of the pollen zones is regionally consistent with the vegetational differences between the two areas. The initial radiocarbon dates for the pollen zones (Flint and Deevey, 1951; Deevey, 1958) allowed the zones to be correlated chronostratigraphically within New England and elsewhere. The first dates showed that similar pollen zones were time transgressive within

Pollen Records of Late-Quaternary North American Sediments

New England, and Deevey (1958) concluded that the herb zone (L) in northern Maine postdated the herb zone (T) in southern New England.

In her comprehensive review, M. B. Davis (1965), listed the available radiocarbon dates and discussed the history of the classic zones as climatic, vegetational and temporal units. She noted Cushing's (1964; 1967) suggestion that pollen zones be considered "pollen assemblage zones," which are contiguous bodies of sediment with a similar pollen content. This definition of pollen zones has led to the recent practice of defining local pollen zones within each diagram before seeking biostratigraphic correlations with zones at other sites (Wright, 1976). M. B. Davis (1965, p. 387) also noted that pollen zones as "stratigraphic units of correlation" are used because of the often "apparent simplification" that they provide but that the concept of zones can be "too typological to be useful in all cases." As an alternative way to display and summarize these data, maps of pollen data can permit new interpretations of these data and illustrate important features in the spatial and temporal vegetational changes that created the local sequence of pollen zones at each site. Previous studies that mapped fossil pollen data from the Northeast include Sears (1942), Leopold (1958), M. B. Davis (1976, 1981b, 1983), Bernabo and Webb (1977), Webb and Bernabo (1977), Webb (1981), Delcourt and Delcourt (1981), and R. B. Davis and Jacobson (in press).

Our paper re-examines the classic pollen zones in terms of the radiocarbon-dated studies available from the Northeast and presents maps that allow inferences about the changes in vegetational structure and composition over the past 14,000 years and show features of vegetational distribution such as the position of ecotones. We have included summary pollen diagrams for eleven sites and isochrone maps for nine pollen types. The nine types include those used to define the classic pollen zones, and the maps illustrate many of the major changes in the vegetation of the Northeast during the past 14,000 years. We have also illustrated and described certain details of the late-glacial pollen stratigraphy by including summary pollen diagrams from nine sites from 16,000 to 8,000 yrs. B.P.

Many of the questions raised by the early pollen research in the Northeast led to several innovations for palynology. Among these were: (1) the study of how pollen and plant abundances are related, including (a) plots of scatter diagrams (Leopold, 1964) and determination of "correction factors" (M. B. Davis

and Goodlett, 1960; M. B. Davis, 1963; Livingstone, 1968; Miller, 1973a), (b) compilations and maps of modern pollen data (Leopold, 1958; M. B. Davis, 1967b, R. B. Davis and Webb, 1975), and (c) the calculation of pollen accumulation rates (M. B. Davis and Deevey, 1964; M. B. Davis, 1969a, 1969b); (2) the study of sedimentation processes, including (a) the reworking of sediments by benthic fauna in lake basins (R. B. Davis, 1967, 1974), and (b) sediment focusing (Lehman, 1975; M. B. Davis and Ford, 1982); (3) the development of equipment for sediment recovery (Livingstone, 1955; R. B. Davis and Doyle, 1969); and (4) the study of the ecological implications of pollen stratigraphic changes, including (a) the individualistic behavior of plant taxa (Deevey, 1949; M. B. Davis, 1965, 1976), (b) the biotic control of plant dispersal (M. B. Davis, 1976, 1978, 1981b), and (c) the role of plant pathogens (Anderson, 1974; Brugam, 1978; M. B. Davis, 1981a).

DATA AND METHODS

Our data base consists of 63 sites of which 51 have radiocarbon dates and three are annually laminated (Figs. 1,2; Table 1). Pollen data are available from other sites (Florer, 1972; Sirkin, 1967, 1971, 1972, 1976, 1977; Connally and Sirkin, 1970, 1971; Sirkin and Minard, 1972; Sirkin et al., 1970; Carmichael, 1980; see Ogden, 1965, and M. B. Davis, 1965 for references to earlier work), but many lack radiocarbon dates, are from isolated organic lenses, or are from specialized habitats, such as salt marshes, and have not been included in our data base. Six of the 63 sites had records that extended no earlier than 4,000 yrs. B.P., and three of these were from varved-sediment lakes (Table 1). The sites with the oldest dates (Table 1; Fig. 1) are Ninepin 24, Maryland (20,490 yrs. B.P.); Buckles Bog, Maryland (18,550 yrs. B.P.); Rogers Lake, Connecticut (14,910 yrs. B.P.); Rose Lake, Pennsylvania (14,170 yrs. B.P.); Hawley Bog, Massachusetts (14,000 yrs. B.P.); Winneconnet Pond, Massachusetts (13,360 yrs. B.P.); and Tannersville Bog, Pennsylvania (13,300 yrs. B.P.). The coverage of our data set is therefore restricted to the past 20,000 years and our main description begins with the data at 16,000 yrs. B.P. (See Donner, 1964; Lougee, 1957; and Sirkin, 1977, for descriptions of pre-Wisconsin pollen data and Newman, 1977, for a list of the oldest late-glacial radiocarbon dates in the Northeast.)

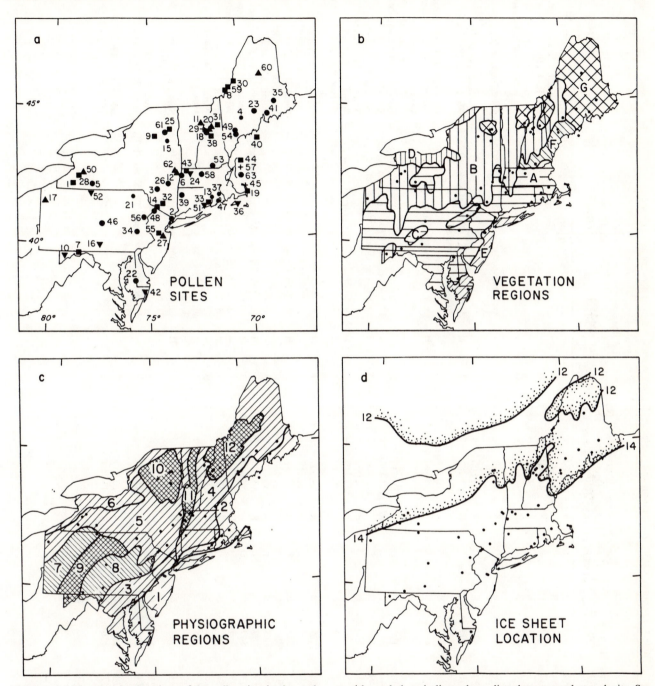

Figure 1. (a) Map showing locations of the pollen sites in the study area with symbols to indicate the earliest data mapped at each site. See Table 1 for a key to the identifying number and dating information for each site. The earliest date mapped at each site was determined by extrapolation of up to 500 years from the oldest radiocarbon date at the site, and is indicated as follows: >14,000 yrs. B.P. (▼); 13,000 to 12,000 yrs. B.P. (●); 11,000 to 10,000 yrs. B.P. (■); 9,000 to 8,000 yrs. B.P. (▲); 7,000 to 6,000 yrs. B.P. (+); <4,000 yrs. B.P. (●).

(b) Map of the major vegetation regions of the Northeast (after Eyre, 1980): oak-hickory forest (A); maple-beech-birch forest (B); oak-pine forest (C); elm-ash-cottonwood forest (D); loblolly-shortleaf pine forest (E); red-jack-white pine forest (F); spruce-fir forest (G).

(c) Map of the major physiographic regions of the Northeast (after Lull, 1968; see also Denny, 1982): Coastal Plain (1); New England Seaboard (2); Piedmont (3); New England Upland (4); Glaciated Allegheny Plateau (5); Great Lakes Section (6); Unglaciated Allegheny Plateau (7); Valley and Ridge Province (8); Allegheny Mountains (9); Adirondacks (10); Green Mountains (11); White Mountains (12). In general, the elevations of the pollen sites correspond with these physiographic regions as follows: regions 1, 2: <100 m; regions 3, 4, 5, 6: 100-400 m; regions 7, 8: 300-600 m; regions 9, 10, 11, 12: >600 m.

(d) Map with isochrones (in 10^3 radiocarbon years) showing the approximate location of the Laurentide ice sheet (after Denton and Hughes, 1981; R. B. Davis and Jacobson, in press) for the time intervals mapped in the isochrone maps (Figs. 4 to 13).

Table 1: Location, dating information and references for pollen sites. Site numbers are as in Figure 1.

Sitename	State	Latitude (°N)	Longitude (°W)	Elevation (m)	Site Type (L=lake, B=bog)	Top Date (yrs BP)	Ragweed Rise (R)	Oldest Date[1] (yrs BP)	No. 14C Dates	No. Dates Used	No. Stratig. Dates Used	Hemlock Decline Date Inserted (H)	Modification of Dates[2]	Source
1. Allenberg Bog	NY	42°15'	78°53'	494	B	0	R	(10,000)	0	3	3	H	ad: 9K, 12K (from Belmont and Protection)	Miller, N. G., 1973a
2. Alpine Peat Bog	NJ	40°57'	73°54'	30	B	0	R	12,840	5	5	0	–		Nickmann, R. J., unpubl.
3. Balsam Lake	NY	42° 2'	74°37'	762	L	0	R	12,600	8	8	0	–		Ibe, R. A., 1982
4. Basin Pond	ME	44°28'	70° 3'	124	L	0	R	1,580	varves	varves	0	–		Swain, A. M., unpubl.
5. Belmont Bog	NY	42°15'	77°55'	497	B	0	R	12,565	6	6	0	–		Spear, R. W. & N. G. Miller, 1976
6. Berry Pond	MA	42°30'	73°19'	600	L	0	R	12,679	16	8	0	–	av: 5 date pairs; de: 2,665, 4,050, 5,070 yrs. B.P.	Whitehead, D. R., 1979
7. Big Pond	PA	39°46'	78°33'	634	L	0	R	11,000	1	2	1	H		Watts, W. A., 1979
8. Boundary Lake	ME	45°34'	70°41'	603	L	0	R	11,200	5	2	2	H	de: 3 dates; corr: 1,000 yr at 5.7K, 10.2K (to match Dufresne)	Mott, R. J., 1977
9. Brandreth Lake Bog	NY	43°55'	74°41'	583	B	0	R	10,360	5	4	0	–		Overpeck, J. T., 1985
10. Buckle's Bog	PA	39°34'	79°16'	814	B	0	R	18,550	6	7	1	H	av: 1 date pair at 3,185 yrs. B.P.	Maxwell, J. & M. B. Davis, 1972
11. Bugbee Bog	VT	44°22'	72° 9'	398	B	0	R	8,510	4	4	0	–		McDowell, L. L. et al., 1971
12. Burden Lake	NY	42°36'	73°34'	192	L	500	–	8,730	4	4	0	–		Gaudreau, D. C., unpubl.
13. Carbuncle Pond	RI	41°42'	71°47'	342	L	0	R	(1,100)	0	1	1	–	ad: basal date of 1,100 yrs. B.P. a guess	Bernabo, J. C., 1977
14. Cedar Bog	NJ	41°40'	74°40'	137	B	0	–	(10,000)	0	3	3	H	ad: 9K, 10K (from Tannersville)	Niering, W. A., 1953
15. Clear Pond	NY	43°45'	74° 1'	510	L	0	R	2,000	varves	varves	0	–		Swain, A. M., unpubl.
16. Crider's Pond	PA	39°58'	77°33'	290	L	0	R	15,210	4	3	0	–		Watts, W. A., 1979
17. Crystal Lake	PA	41°33'	80°22'	313	L	500	–	9,310	1	3	2	H	de: 10,915 yrs. B.P. (reversal)	Walker, P. C. & R. I. Hartman, 1960
18. Deer Lake Bog	NH	44° 2'	71°50'	1325	B	0	R	13,000	6	7	1	H		Spear, R. W., 1981
19. Duck Pond (core b)	MA	41°50'	70° 0'	2	L	0	R	11,710	6	6	0	–		Winkler, M. G., 1982
20. Eagle Lake Bog	NH	44°10'	71°40'	1275	B	0	R	9,155	3	5	2	H	ad: 10.5K for spruce decline (from Deer)	Spear, R. W., 1981
21. Ely Lake	PA	41°46'	75°50'	385	L	0	R	2,320	varves	varves	0	–		Swain, A. M., unpubl.
22. Federalsburg	MD	38°42'	75°46'	15	B	13K	–	13,420	1	1	0	–		Sirkin, L. A. et al., 1977
23. Gould Pond	ME	44°44'	69°19'	89	L	0	R	13,280	13	11	1	H	de: 5,565, 7,410, 8,480 yrs. B.P. (reversals)	Jacobson, G. L., unpubl.
24. Hawley Bog Pond	MA	42°34'	72°53'	549	L	500	–	14,000	8	6	0	H		Patterson, W. A. III, unpubl.
25. Heart Lake	NY	44°11'	73°58'	664	L	0	–	10,475	8	8	1	H	ad: 12.4 K (from Upper Wallface)	Whitehead; D. R., unpubl.
26. Heart's Content Bog	PA	42°14'	73°58'	107	B	0	–	(12,600)	0	5	5	H	ad: 7.3K, 9K, 9.6K, 12.6K (from Balsam)	Ibe, R. A., 1982
27. Helmetta Bog	NJ	40°23'	74°26'	15	B	0	R	9,640	1	1	0	–		Watts, W. A., 1979
28. Houghton Bog	NY	42°33'	78°40'	428	L	0	R	11,880	1	5	4	H	ad: 1.3K, 9K, 11.2K (from Protection, Belmont)	Miller, N. G., 1973a
29. Kinsman Pond	NH	44° 8'	71°44'	1140	L	0	R	8,830	3	3	2	H	ad: 10.5K (from Deer); de: 225 and 4K basal dates	Spear, R. W., 1981
30. Lac Dufresne	PQ	45°51'	70°21'	650	L	0	R	11,200	6	6	0	–		Mott, R. J., 1977
31. Lake of the Clouds	NH	44°16'	71°19'	1542	L	0	–	11,530	5	5	3	–	de: 7K, 10.3K (reversal), 11.5K; ad: 10.7K, 13K (from Deer)	Spear, R. W., 1981
32. Lake Rogerine	NJ	41°35'	74°20'	137	L	500	–	10,500	2	2	0	–		Nicholas, J., 1968
33. Lantern Hill Pond-3	CT	41°28'	71°57'	136	L	0	R	11,160	5	5	0	–		Trent, K. M., unpubl.

Table 1: Location, dating information and references for pollen sites. Site numbers are as in Figure 1. (Continued)

Sitename	State	Latitude (°N)	Longitude (°W)	Elevation (m)	Site Type (L=lake B=bog)	Top Date[1] (yrs BP)	Ragweed Rise (R)	Oldest[1] Date (yrs BP)	No. 14C Dates	No. Dates Used	No. Stratig. Dates Used	Hemlock Decline Date Inserted (H)	Modification of Dates[2]	Source
34. Long Swamp	PA	40°29'	75°40'	192	B	0	R	12,540	6	6	2	H	de: 2 dates at 12K (reversal)	Watts, W. A., 1979
35. Loon Pond	ME	45° 2'	68°12'	110	L	0	R	12,615	10	9	0	–	corr: 500 yr at all dates (based on spruce and hemlock declines); av: 2 equal dates	Jacobson, G. L., unpubl.
36. Martha's Vineyard (MV-7A)	MA	41°25'	70°30	24	B	9K	–	(15,000)	0	3	3	–	ad: 9.7K, 12K, 15K (from Rogers)	Ogden, J. G. III, 1959
37. Mashapaug Pond	RI	41°47'	71°26'	12	L	0	R	3,820	1	1	0	–		Bernabo, J. C., 1977
38. Mirror Lake	NH	43°57'	71°42'	200	L	500	–	11,300	9	8	0	–		Likens, G. E. & M. B. Davis, 1975
39. Mohawk Pond	CT	41°49'	73°17'	360	L	200	–	12,460	6	6	0	–		Gaudreau, D. C., unpubl.
40. Monhegan Island Meadow	ME	43°46'	69°18'	3	B	0	R	10,745	17	9	0	–	choose 10,745 yrs. B.P. as basal date; de: 5 dates below basal date, 2 dates at 10K (reversal), 995 yrs. B.P.	Bostwick, L. K., 1978
41. Moulton Pond	ME	44°38'	68°38'	143	L	0	R	13,510	16	11	0	–	de: 3 top dates, and 3,075, 5,560 yrs. B.P. (reversal)	Davis, R. B. et al., 1975
42. Ninepin 24	MD	38°18'	75°17'	15	B	16K	–	20,490	2	2	0	–		Sirkin, L. A. et al., 1977
43. North Pond	MA	42°39'	73° 3'	586	L	0	R	11,600	13	8	0	–	av: date pairs at 3K, 4K, 7K, 8K, 9K	Whitehead, D. R. & T. Crisman, 1978
44. Nunkets Pond	MA	41°58'	71° 3'	18	L	0	R	11,500	4	4	0	–		Bradshaw, R. H. W., unpubl.
45. Pamet Cranberry Bog	MA	42° 0'	70° 2'	3	B	0	R	6,980	3	3	0	–		Patterson, W. A. III, unpubl.
46. Panther Run Pond	PA	40°48'	77°25'	634	L	0	R	12,610	2	3	1	H	de: 625 yrs. B.P. at greatest ragweed rise	Watts, W. A., 1979
47. Pasacaco Pond	RI	41°31'	71°27'	3	L	0	R	1,280	2	2	0	–		Bernabo, J. C., 1977
48. Pine Bog	NJ	41°17'	74°45'	137	B	0	–	(13,000)	0	4	4	H	ad: 9K, 10K, 13K (from Tannersville)	Niering, W. A., 1953
49. Poland Spring Pond	ME	44° 2'	70°21'	94	L	0	R	12,860	11	4	0	–	de: 135 yr. B.P.	Jacobson, G. L., unpubl.
50. Protection Bog	NY	42°37'	78°28'	430	B	0	R	9,030	3	4	1	–	ad: 12.4K (from Belmont)	Miller, N. G., 1973a
51. Rogers Lake	CT	41°22'	72° 7'	91	L	0	R	14,910	53	30	0	–	de: 23 dates (minor reversals)	Davis, M. B., 1969a
52. Rose Lake	PA	41°55'	77°56'	631	L	0	R	14,170	2	5	3	H	ad: 9K, 11.2K (from Belmont)	Cotter, J. F. P. & Crowl, G. H., 1981
53. Sandogardy Pond	NH	42°50'	71°40'	100	L	0	–	12,500	4	4	0	–		Davis, M. B., 1978
54. Sinkhole Pond	ME	43°58'	70°21'	95	L	0	R	12,790	12	11	0	–	de: 8,375 yrs. B.P. (reversal)	Jacobson, G. L., unpubl.
55. Szabo Pond	NJ	40°24'	74°29'	29	L	0	R	11,400	3	2	0	–	de: 11,950 yrs. B.P. (reversal)	Watts, W. A., 1979
56. Tannersville Bog	PA	41° 2'	75°16'	277	B	0	R	13,330	5	5	0	–		Watts, W. A., 1979
57. Titicut Swamp	MA	41°58'	71° 3'	18	B	0	R	7,250	5	5	2	–	ad: 9.8K and 11.5K (from Nunkets)	Nelson, S. N., 1984
58. Tom Swamp	MA	42°31'	72°13'	229	B	200	–	12,830	8	8	0	–		Gaudreau, D. C., unpubl.
59. Unknown Lake	ME	45°37'	70°38'	489	L	0	R	(11,500)	4	3	3	H	de: 4 dates (14.8K too old); ad: 9.8K, 11.5K (from Dufresne)	Mott, R. J., 1977
60. Upper South Branch Pond	ME	46° 5'	68°54'	300	L	0	R	9,970	7	4	2	H		Anderson, R. S., 1979
61. Upper Wallface Pond	NY	44° 9'	74° 3'	945	L	200	–	12,390	6	5	0	–	de: 2,105 yrs. B.P.	Whitehead, D. R., 1985
62. West Sand Lake Peat Bog	NY	42°38'	73°36'	170	B	2K	–	9,260	1	5	4	H	ad: 7.3K, 8.7K (from Burden), 12.8K (from Connally and Sirkin, 1970)	Gaudreau, D. C., unpubl.
63. Winneconnet Pond	MA	41°58'	71° 7'	20	L	0	R	13,600	7	7	0	–		Suter, S. M., 1985

[1]Parentheses around the oldest date used for a site indicate a stratigraphic date applied from a nearby site.

[2]For sites where the dates used differ from the published radiocarbon dates, the notes on date modifications list deletions (de), additions (ad), corrections (corr) and averaged (av) dates, and list all stratigraphic dates used, besides those for the ragweed (R) rise and hemlock (H) decline both of which are listed in separate columns. "K" indicates 10³ years.

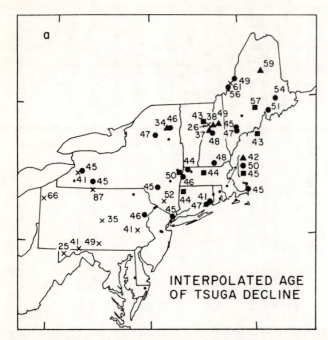

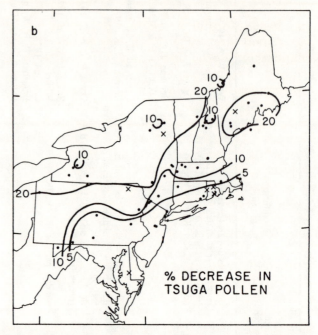

Figure 2.(a) Age in 10³ radiocarbon years of the mid-Holocene decrease in hemlock percentages that occurred throughout the Northeast. Age was interpolated for the midpoint of the interval of maximum decrease in hemlock pollen percentages. The reliability of the assigned date at each site is ranked as follows: (i) radiocarbon date within the interval of the decline in pollen percentages (●); (ii) bracketing radiocarbon dates both within ±2,000 yrs. (■); one within ±2,000 and other ±4,000 yrs. (▲), greater than ±4,000 yrs. (×); (iii) no radiocarbon dates or no clear decline in hemlock pollen percentages or no pollen data during this stratigraphic interval (●).
(b) Map with isopolls (isofrequency contours) showing the percentage decrease in hemlock pollen across the Northeast during the hemlock decline interval. "×" indicates that a site has no pollen data for this interval.

Two sets of pollen diagrams are included for a representative group of sites (Figs. 3A, 3B, 3C and 15A, 15B) in order to illustrate the general pollen stratigraphy of the past 16,000 years and to show details of the late-glacial stratigraphy. We also produced isochrone maps for selected isopolls of nine pollen types and drew contours for each 2,000-year interval from 14,000 yrs. B.P. to present (Figs. 4 to 13). The nine pollen types (sedge, spruce, pine, birch, oak, hemlock, beech, hickory and chestnut) are the principal ones that were used to define the classic pollen zones for the Northeast. We also included an isopoll map showing the distribution of ragweed pollen today (Fig. 14). The percentages of all pollen types illustrated in our paper are based on a sum of total tree, shrub, and herb pollen with spores and aquatic pollen types deleted. The percentages used for contouring were linearly interpolated values from the two nearest samples bracketing the date mapped (Webb *et al.,* 1983a, 1983b). Contours at the edges of our mapped region are constrained by values at nearby sites outside of the mapped region (Gaudreau, 1985).

The percentages mapped for each pollen type were chosen to indicate either significant populations of the plants producing the pollen within the pollen-source area of each site or high abundances of the

plants. In the latter case, the plants are likely to be dominant members of the vegetation. In the text, percentages meeting the first criterion are often referred to as indicating "significant" occurrences of a plant taxon, and those meeting the second, as indicating "high" abundances of a plant taxon. For five of the pollen types (spruce, pine, oak, beech and hemlock) both percentage levels were mapped. For the other types, the mapped value represents significant populations. In choosing the pollen percentages to be mapped, we used information from modern data about how pollen percentages are related to tree percentages (Webb *et al.,* 1981; Delcourt *et al.,*1984; Gaudreau, 1985). The relationships between pollen percentages and tree percentages are less well-understood for pollen samples from before 10,000 yrs. B.P., when the vegetational structure differed from the forested conditions that developed over most of the region after that time. Inferences about the size of plant populations during the late-glacial are therefore less certain than those for the Holocene.

The mapped pollen percentages show the distributions of well-established plant populations and not the initial appearance of a plant taxon in a region. Immigration of individual trees and establishment of small populations may significantly predate the increase in population levels as recorded by the appearance of significant amounts of a pollen type at individual sites (Gaudreau, 1984, 1985). The relationship between immigration of a taxon and an increase in its populations to significant levels can vary among the sites and will depend upon the prevailing climatic and edaphic conditions and such biotic factors as seed dispersal and competition. R. B. Davis and Jacobson (in press) have recently defined a new set of palynological criteria for plant immigration and produced maps that differ from those of M. B. Davis (1981b).

The mapped patterns in the pollen percentages often reveal vegetational features such as the areal extent of a vegetation region through time, the shape of the gradients in plant distributions, and the locations of boundaries and ecotones between vegetation regions (Bernabo and Webb, 1977; Webb *et al.*, 1983a, 1983b). Pollen grains often can be identified to the genus level, and the individual species within each genus have different ecological requirements and may occur in different vegetation regions today. A general sense of which species in a given genus occur at a particular time and place can sometimes be gained by careful study of the temporal and spatial variations in the mapped isochrones of a pollen type in relation to the mapped patterns of other pollen types (Webb *et al.*, 1983b). The mapped patterns aid the interpretation of pollen stratigraphies at individual locations by illustrating local vs. regional stratigraphic changes and by locating each site with respect to boundaries between vegetation regions, and gradients within these regions.

Establishment of the Chronology at Each Site

All pollen data and chronostratigraphic information are stored as computer files at Brown University, and the data were analyzed via a series of computer programs described by Webb *et al.*, (1983b). At most sites, linear interpolation between the available radiocarbon or stratigraphic dates was used to estimate the age for each pollen sample in the core. Weighted cubic regression was used at three sites (Heart Lake, Monhegan Island, and Moulton Pond) whose several dates showed too much scatter to justify linear interpolation (Table 1). We attempted to use the radiocarbon dates as published at each site and to add just the standard stratigraphic dates of 0 ± 30 yrs. B.P. for the top of most cores, and 150 to 300 ± 50 yrs. B.P. for the date of the "ragweed rise", *i.e.*, the increase in the percentages of *Ambrosia* (ragweed) pollen in the uppermost sediments (Solomon and Kroener, 1971;

Brugam, 1978; Russell, 1980; M. B. Davis, 1983). At twelve sites, the date of the topmost sample in the computer file was earlier than 0 yrs. B.P., and this fact is shown on Table 1 along with designation of whether the ragweed rise was present and whether various corrections, additions, or deletions were made to the available dates at each site.

Experience in mapping pollen data has revealed a need to check the stratigraphic and radiocarbon dates used at each site. The general philosophy is to use the available dates as much as possible, but each site must be checked for dating biases, errors, or reversals. Some of these problems were evident as unusual changes in sedimentation on depth-versus-date scatter plots for each site. Dating reversals also occurred, and for sites whose consistent pollen stratigraphies were "normal" for their geographic location, *e.g.*, Criders Pond and Moulton Pond, the choice of deleting reversed dates seemed reasonable (Table 1). The over-abundance of closely spaced dates at Rogers Lake made deletion of reversed dates a reasonable choice there, but we could also have used curvilinear regression or splines to fit a curve to the dates at Rogers Lake (Overpeck and Fleri, 1982).

Isopoll maps of the data at selected dates provided another means of checking for anomalous dates. In general, the vegetation varies in consistent patterns along environmental gradients that are monotonic with latitude and elevation. Trends in isopolls that are nonmonotonic over short distances can signal potential dating problems. For example, if just the one basal radiocarbon date of 11,880 yrs. B.P. were used at Houghton Bog, then the mapped values there would be unusual from 10,000 to 8,000 yrs. B.P. because of inadequate time control. Supplemental stratigraphic dates were therefore needed from nearby sites (Miller, 1973a)(Table 1). Biases in the radiocarbon dates due to old carbon (Deevey *et al.*, 1954) were also evident at certain sites *e.g.*, Loon Lake (Table 1), but the originally published correction for dates from Rogers Lake (Stuiver, 1967) was not used (M. B. Davis, 1983).

The Date for the "Hemlock Decline"

The regional decline in the percentages of hemlock pollen is a key indicator of possible dating problems (Table 1), because both M. B. Davis (1978, 1981a, 1983) and Webb (1982) have produced evidence that this pollen stratigraphic event was synchronous within radiocarbon error. M. B. Davis (1981a) proposed that the decline was caused by disease.

Webb (1982) plotted a histogram of the interpolated dates for this event and showed that they are normally distributed with a mean of 4,650 yrs. B.P. and a standard deviation of 300 years. Because this date for the "hemlock decline" was used to supplement the dates at 20 sites (Table 1), we have provided a brief review of its dating in the Northeast.

Fifty of the 63 sites record the "hemlock decline," and 45 of these had radiocarbon dates that could be used to date it. Twenty-one sites have radiocarbon dates within the stratigraphic interval in which the hemlock percentages decreased, and, of these, 19 sites have interpolated ages for the decline within two standard deviations of 4,650 yrs. B.P. (Fig. 2). Six other sites with dates tightly bracketing the decline and seven other sites with dates loosely bracketing the decline also yielded interpolated ages for the decline within two standard deviations of the mean.

Dates differing by more than two standard deviations from 4,650 yrs. B.P., exist at three well-dated sites in Maine and at ten less well-dated sites in Maine, New Hampshire, New York, Pennsylvania and Maryland (Fig. 2). No geographic pattern is evident in these dates although all of the anomalous dates in Maine are too old. At one of these sites, Loon Lake, a carbonate bias existed, and at three of the other Maine sites, dates for other events (*e.g.,* the late-glacial spruce pollen decline and the early-Holocene hemlock pollen increase) were also too old (Table 1). Our judgment was to correct these early dates and to bring the dates at these Maine sites into line with dates at surrounding sites in Maine, Quebec, Prince Edward Island, Nova Scotia, and New Brunswick.

VEGETATION AND PHYSIOGRAPHY OF THE NORTHEAST

The vegetation of the northeastern U.S. is usually described as being composed of the deciduous forest formation (Braun, 1950) or of variously defined deciduous forest regions to the south and mixed conifer/northern-hardwood forest regions to the north (Nichols, 1935; Küchler, 1964; Lull, 1968; Eyre, 1980). Boreal forest elements extend into the Northeast in the north and at high elevations, and elements of the southern conifer forest extend northward along the coast (Fig. 1). This macroscale distribution of the vegetation reflects the broad latitudinal control of climate. At the meso-scale, the vegetation reflects the influence of topographic and edaphic differences associated with the major physiographic regions of the Northeast (Fig. 1). These regions include the southern Appalachian Mountains, the Adirondack,

Green and White Mountains and their associated uplands, and the lowlands of the Atlantic coastal plain, the Great Lakes and major rivers. Evidence for the strong imprint of physiography on meso-scale vegetational distributions shows up in the general correspondence in the geographic distribution of the northeastern forest regions described by Braun (1950), Küchler (1964), and Lull (1968), who each used somewhat different criteria to define their vegetational regions.

The Society of American Foresters (Eyre, 1980) classifies the existing vegetation of the study area into seven forest types (Fig. 1). The deciduous forest region is composed of oak-hickory forest (Fig. 1; Table 2). Predominant are various species of oak (white, northern red, black, chestnut, scarlet) and of hickory (mockernut, pignut, shagbark). Yellow-poplar and sweetgum are more important southward as are species of oaks with southern distributions (*e.g.,* pin, swamp, willow, chestnut oaks). Before the blight that originated in New York in A.D. 1904, chestnut was an important tree in the oak forests of the Appalachian Highland region (Braun, 1950; Anderson, 1974).

The mixed conifer/hardwood forest is composed primarily of maple-beech-birch forest and white-red-jack pine forest (Fig. 1). In the former, the predominant trees are northern hardwoods, which include mainly sugar maple, American beech, and yellow birch, and also sweet birch, basswood, white ash, northern red oak, and eastern hemlock, a conifer. The white-red-jack pine forest borders the coast of northern New England and includes taxa of the maple-beech-birch forest.

The loblolly-shortleaf pine forest, extensive along the lowlands of the southern U.S., grows in sandy coastal plain areas of the Northeast (Fig. 1). Besides the two pine species for which the forest type is named, other important species include pitch, sand, Virginia and pond pines, and blackgum, and the hardwoods of the southern oak-hickory forest.

Three other forest types have isolated distributions in the Northeast, and two include species of either the southern or northern forest types described above. The oak-pine forest (Fig. 1), found in the Appalachian plateau, includes southern pines and southern hardwoods. The more northern, elm-ash-cottonwood forest (Fig. 1), found in moist soils along the Great Lakes, includes lowland hardwoods, such as black, green and white ash, balsam poplar and red maple. Boreal forest elements predominate in the spruce-fir forests (Fig. 1) in the north. Red, white and black

Table 2.
Common and Latin names (as in Eyre, 1980) for plant taxa referred to in the text.

COMMON NAME	LATIN NAME	COMMON NAME	LATIN NAME
Alder	*Alnus spp.*	Oak	*Quercus spp.*
speckled	*A. rugosa*	black	*Q. velutina*
Ash	*Fraxinus spp.*	chestnut	*Q. prinus*
black	*F. nigra*	northern red	*Q. rubra*
green	*F. pennsylvanica*	pin	*Q. palustris*
white	*F. americana*	scarlet	*Q. coccinea*
Basswood, American	*Tilia americana*	swamp white	*Q. bicolor*
Beech, American	*Fagus grandifolia*	white	*Q. alba*
Birch	*Betula spp.*	willow	*Q. phellos*
sweet	*B. lenta*	Pine	*Pinus spp.*
shrub	*B. glandulosa*	jack	*P. banksiana*
white (paper)	*B. papyrifera*	loblolly	*P. taeda*
yellow	*B. lutea*	pitch	*P. rigida*
Blackgum (tupelo)	*Nyssa sylvatica*	pond	*P. serotina*
Chestnut	*Castanea dentata*	red	*P. resinosa*
Composite (daisy) family	Compositae	sand	*P. clausa*
Cottonwood	*Populus spp.*	shortleaf	*P. echinata*
eastern	*P. deltoides*	Virginia	*P. virginiana*
Elm, American	*Ulmus americana*	white	*P. strobus*
Fir, balsam	*Abies balsamea*	Poplar	*Populus spp.*
Grass family	Gramineae	balsam	*P. balsamifera*
Hemlock, eastern	*Tsuga canadensis*	Ragweed	*Ambrosia spp.*
Hickory	*Carya spp.*	Sedge family	Cyperaceae
mockernut	*C. tomentosa*	Spruce	*Picea spp.*
pignut	*C. glabra*	black	*P. mariana*
shagbark	*C. ovata*	red	*P. rubens*
Ironwood/hornbeam	*Ostrya/Carpinus*	white	*P. glauca*
Maple	*Acer spp.*	Sweetgum	*Liquidambar styraciflua*
red	*A. rubrum*	Tamarack	*Larix spp.*
sugar	*A. saccharum*	Willow	*Salix spp.*
Myrtle family	Myricaceae	Yellow-poplar	*Liriodendron tulipifera*

spruce, balsam fir, balsam poplar, white and yellow birch are important species in this forest region.

GENERALIZED POLLEN STRATIGRAPHY AND ISOCHRONE MAPS: 16,000 yrs. B.P. to PRESENT

The eleven pollen diagrams in Figures 3A, 3B, and 3C summarize much of the general pollen stratigraphy for the last 16,000 years in the Northeast. The diagrams are from sites representative of the range of latitude and physiography in the study area (Fig. 1; Table 1). These pollen diagrams show a generally similar sequence of pollen changes from high basal percentages of sedge, other herb (composites, grass, "other herbs") and pine pollen, to high percentages of spruce, then pine, and finally oak pollen. These percentage patterns correspond with pollen zones T, A, B, and C, respectively, (Deevey 1958). Of the eleven pollen diagrams in Figures 3A, 3B, and 3C, four were published with pollen zonations labeled by the Deevey (1958) scheme. These were Rogers Lake (M. B. Davis, 1969a), Berry Pond (Whitehead, 1979), Belmont Bog (Spear and Miller, 1976), and Moulton

Pond (R. B. Davis *et al.,* 1975). At each of these sites, local zones were identified and then either the Deevey designations (T, A, B, C, with subzones) were used as labels or the local zones were correlated with the Deevey zones. Except for Belmont Bog, these sites are within the areas in which Deevey defined his zones: Connecticut (Deevey, 1939), southern New England (Deevey, 1943), and Maine (Deevey, 1951).

Though the eleven pollen diagrams show the same general pollen stratigraphic sequence, they differ in the percentage maxima reached by each pollen type, and they also differ in the relative timing of similar stratigraphic features. The isochrone maps show the patterns in these differences and how these patterns reflect the vegetational differences associated with latitude and with the physiography of the Northeast.

Mapped Patterns: 14,000-8,000 yrs. B.P.

From 14,000 to 8,000 yrs. B.P., the isochrone maps for sedge, spruce, pine, and birch pollen (Figs. 4 to 7) suggest vegetational changes in physiognomy (from tundra, to open parkland, to closed forest) and in composition within and between vegetational regions. The mapped patterns of sedge, spruce, and pine pollen translate into the time-transgressive pollen stratigraphies of the classic T, A, and B pollen zones that span late-glacial and early-Holocene time. (The pollen stratigraphy of these zones is discussed in a later section.)

After the southern edge of the ice sheet retreated northward (Fig. 1), the initial vegetation contained significant populations of sedges (Fig. 4) and other herbs or shrubs. High values of the pollen from these plants were used to characterize the L (Deevey, 1951) or T (Leopold, 1956a) zone. Sedge pollen had values above 5% at the few sites with records at 14,000 yrs. B.P. (Figs. 1 and 4). Sedge pollen was regionally significant at 12,000 yrs. B.P., but by 10,000 yrs. B.P. its distribution was reduced to areas in the southern Appalachian and White Mountains and in New Jersey and southern New York. At 8,000 yrs. B.P., significant values of sedge pollen continued to be confined to isolated areas with suitable local conditions and this patchy distribution has continued to present.

High values of spruce pollen (>20%) appeared first in the south at 14,000 yrs. B.P., were prevalent at most northeastern sites at 12,000 yrs. B.P., and then became patchy in distribution by 10,000 yrs. B.P. (Fig. 5). All sites in the Northeast had >5% spruce pollen at 12,000 yrs. B.P., but by 10,000 yrs. B.P. spruce populations began to decline throughout the Northeast. By 8,000 yrs. B.P., spruce pollen had a patchy distribution and in the south, only sites in the Appalachians had >5% spruce pollen. The spruce maps show the progression northward of the major populations of spruce trees. The combined evidence from both the sedge and spruce maps is that the vegetation associated with the spruce pollen zone (A) of Deevey (1939) was primarily a spruce parkland (M. B. Davis, 1967a). Only in the north, and after 12,000 yrs. B.P., did spruce forests develop (M. B. Davis et al., 1980; Jackson, 1983; Overpeck, 1985; Webb et al., 1983b).

At 12,000 yrs. B.P., pine populations initially expanded and >20% pine pollen occurred at coastal sites and centrally in New England (Fig. 6). By 10,000 yrs. B.P., pine populations re-expanded primarily in the south, as seen in the first regional distribution of high values (>40%) of pine pollen. At this time, the 40% and 20% isochrones mark the northern edge, across New York and New England, of significant populations of pine (probably white pine; Fig. 10.9 in M. B. Davis, 1981b). By 8,000 yrs. B.P., pine had a more northern distribution and the 40% and 20%

isochrones mark the southern edge of the significant pine populations. These isochrones show the history of the pine pollen zone (B), which regionally was most important between 10,000 and 8,000 yrs. B.P., and appeared at the southern sites earlier than at the northern sites.

Birch is an important mid- to late-Holocene pollen type at sites in northern New England, but peaks in percentages of birch pollen occur regionally in the late-glacial herb-dominated pollen assemblages and also in the early-Holocene, ca. 10,000 yrs. B.P. (Figs. 3A, 3B, 3C). While none of the classic zones were designated a "birch zone," subzone designations show the importance of birch in the pollen stratigraphy of the Northeast from the late-glacial to present (Deevey, 1951). Significant values of birch pollen (>10%) first appeared in northern sites at 12,000 yrs. B.P., were observed regionally in the east by 10,000 yrs. B.P., and then were confined to northern sites in the east and high elevation sites elsewhere in the Northeast by 8,000 yrs. B.P. (Fig. 7). From 8,000 to 4,000 yrs. B.P., the region of sites with >10% birch pollen expanded westward, as forests with significant numbers of birch trees developed in the north.

The mapped patterns of pine and birch pollen percentages show a sequence of initial appearance, widespread occurrence, reduced distribution and then re-expansion (Figs. 6 and 7). This sequence agrees with the time-transgressive population increases of different species of pine and birch. The maps of pine pollen probably indicate the regional occurrence of red and/or jack pine (Pinus resinosa, P. banksiana) until 12,000 yrs. B.P. The expanded populations of pine which appeared in the south at 10,000 yrs. B.P., and then shifted northward by 8,000 yrs. B.P., probably represent populations of white pine (P. strobus). Identification of pine pollen at individual sites supports this interpretation (M. B. Davis, 1958, 1969a; Whitehead, 1979; R. B. Davis et al., 1975; Mott, 1977; Webb et al., 1983b; Jackson, 1983; M. B. Davis et al., 1980). The birch pollen appearing in northern sites at 12,000 yrs. B.P. probably represents populations of shrub birch (Betula glandulosa). The expanded distribution of birch pollen in the south and west at 10,000 yrs. B.P. (after the decline in sedge and spruce pollen percentages), probably indicates white birch (B. papyrifera) populations, which declined across the region by 8,000 yrs. B.P., and now are important in the boreal forest (Richard, 1977). The increase in birch pollen percentages by 6,000 yrs. B.P. probably represents the expansion of yellow

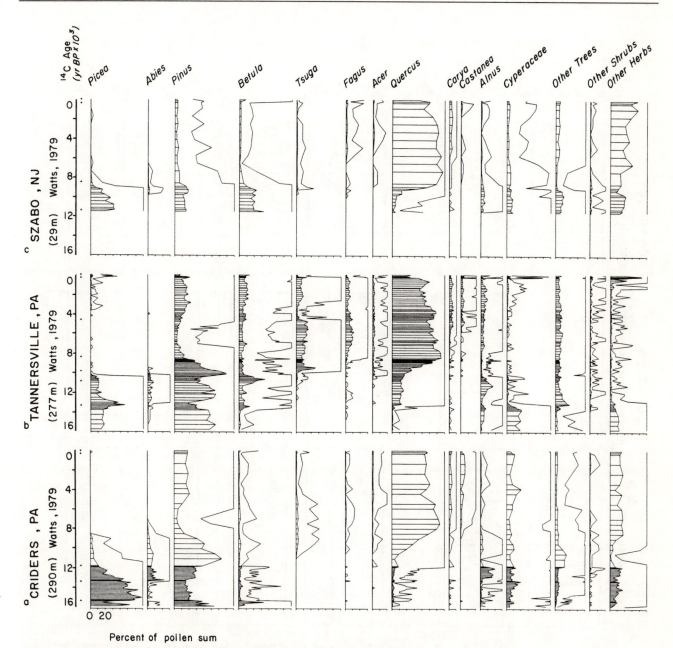

Figure 3A. Summary pollen diagrams from 16,000 to 0 yrs. B.P. for three of eleven sites in the Northeast. Pollen types were chosen to illustrate the general features of the pollen stratigraphy across the region. Chronologies are estimated according to the information in Table 1. The pollen diagrams are ordered by latitude (S to N) and physiography (inland highlands to coastal lowlands). Plus signs indicate the downcore location of radiocarbon and stratigraphic dates and a 10x magnification is shown for all pollen types.

birch *(B. lutea)* populations; this species is an element of the mixed conifer/hardwood forest. Measurements of birch grains, which allow differentiation of shrub from tree birch *(e.g.,*Leopold, 1956b; M. B. Davis, 1958; R. B. Davis *et al.,* 1975), and the identification of plant macrofossils (Jackson, 1983; Watts, 1979) support our interpretation of the species distributions for birch.

Mapped Patterns: 8,000 - 0 yrs. B.P.

Isochrone maps of pine, birch, oak, hemlock, beech, hickory and chestnut pollen (Figs. 6 to 13) and an isopoll map of ragweed pollen (Fig. 14) summarize the regional patterns for the numerically important pollen types from the early-Holocene to the present and record many of the changes within the classic C

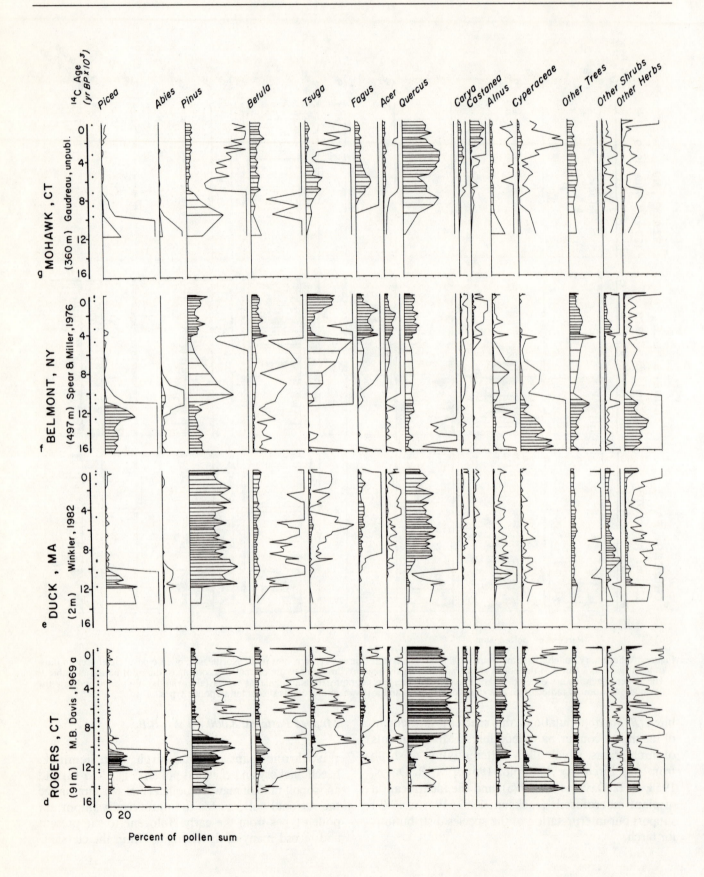

Percent of pollen sum

pollen zone. The pollen stratigraphies (Figs. 3A, 3B, 3C) show a regionwide increase in oak pollen percentages but differ in a regionally consistent way in the timing of certain changes. These differences in pollen composition correspond with regional differences in the C-subzones.

From 8,000 to 6,000 yrs. B.P., the regional occurrence of birch pollen (>10%) expanded westward (Fig. 7), representing the expansion of populations of yellow birch, now important in the mixed conifer/hardwood forests of northern New England and New York. With the birch expansion, populations of pine (primarily white pine) contracted (Fig. 6). The region with significant populations of birch trees continued to increase until 4,000 yrs. B.P., but has changed little in area since then. The southern edge of this region marks the southern border of the mixed conifer/hardwood forests. At 8,000 yrs. B.P., the 20% isopoll for pine marked this border. After 8,000 yrs. B.P., pine decreased in regional distribution, with its largest populations remaining in sandy coastal areas and in the northernmost latitudes (Fig. 6).

The 8,000 yrs. B.P. isochrones for oak show the first regional distributions for both significant (>20%) and high (>40%) amounts of oak pollen in southern and east-central sites (Figs. 8 and 9). The increase in oak pollen percentages, defining the beginning of the C zone (Deevey, 1939), occurred regionally but is most pronounced in the southern part of the study area (Figs. 3A, 3B, 3C). The isochrones for oak pollen show that its populations maintained a fairly stable distribution southward of central New England after 8,000 yrs. B.P.

The isochrones for oak pollen mark the northern ecotone of the oak-dominated deciduous forest. The position of this ecotone has been fairly stable over the past 8,000 years, during which time the dominant pollen type to the north changed from pine to birch (Figs. 6 and 7) and the composition of the forests to both the north and the south of this ecotone changed with the expansion of populations of hemlock, beech, hickory, and chestnut.

Hemlock populations first had a broad regional distribution along with white pine at 8,000 yrs. B.P. (Figs. 6, 10, 11). Hemlock populations had expanded further by 6,000 yrs. B.P. The principal region for the hemlock populations is north of the ecotone evident in the maps for pine, birch, and oak pollen (Figs. 6, 7,

8, 9). The beginning of the C zone in northern New England was defined by a maximum in hemlock pollen percentages (Deevey, 1951), while in southern New England the C-1 subzone was defined by increased oak and hemlock pollen percentages, with the former more pronounced (Deevey, 1939, 1943). At 4,000 yrs. B.P., the patchy distribution of hemlock values above 10% and the contraction of the isochrones for 5% hemlock pollen record the regionwide "hemlock decline," at approximately 4,700 yrs. B.P. (Fig. 2). Deevey (1939, 1943, 1951) used this decrease in hemlock pollen percentages to designate the beginning of the C-2 subzone throughout the Northeast. Hemlock populations re-expanded between 4,000 and 2,000 yrs. B.P. (Figs. 10 and 11).

The regional expansion of beech populations followed that of hemlock by 1,000 years or more. Significant values of beech pollen (>5%) occurred at a few scattered sites at 8,000 yrs. B.P. (Figs. 10 and 12), and first showed a regional distribution at 6,000 yrs. B.P., when its distribution was similar to that of hemlock at 8,000 yrs. B.P. Beech populations expanded further by 4,000 yrs. B.P., then were stable until 2,000 yrs. B.P., when the isopolls for both beech and hemlock coincided. Like hemlock, the main center for beech populations is north of the deciduous forest/mixed forest ecotone.

During the period of expansion for hemlock and beech, both hickory and chestnut also expanded their distributions (Fig. 13). Hickory populations occurred primarily in the south at low elevations, and chestnut pollen had its highest values in the southern Appalachians until its populations became significant in forests of central and southwestern New England. The late population increases of both of these taxa affected the composition of the forests in the deciduous forest region, but neither taxon occurred in significant populations in northern New England.

Within southern New England, the increased percentages in hickory pollen defined the C-2 subzone (Deevey, 1939, 1943), which began with the decline in hemlock percentages. The "hemlock decline" was more pronounced in the north, where hickory was not as important in the pollen stratigraphies (Figs. 2 and 3A, 3B, 3C). The C-1/C-2 boundary was defined, then, both by the time-transgressive and regionally restricted increase in hickory pollen percentages, and the synchronous "hemlock decline."

Figure 3B. Summary pollen diagrams from 16,000 to 0 yrs. B.P. for four of eleven sites in the Northeast. Pollen types were chosen to illustrate the general features of the pollen stratigraphy across the region. Chronologies are estimated according to the information in Table 1. The pollen diagrams are ordered by latitude (S to N) and physiography (inland highlands to coastal lowlands). Plus signs indicate the downcore location of radiocarbon and stratigraphic dates and a 10x magnification is shown for all pollen types.

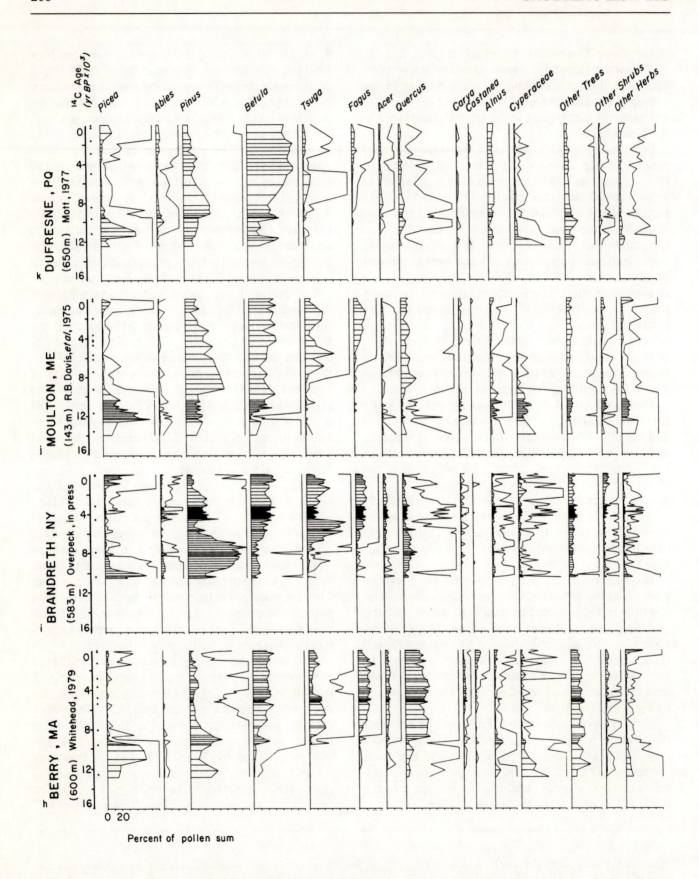

Percent of pollen sum

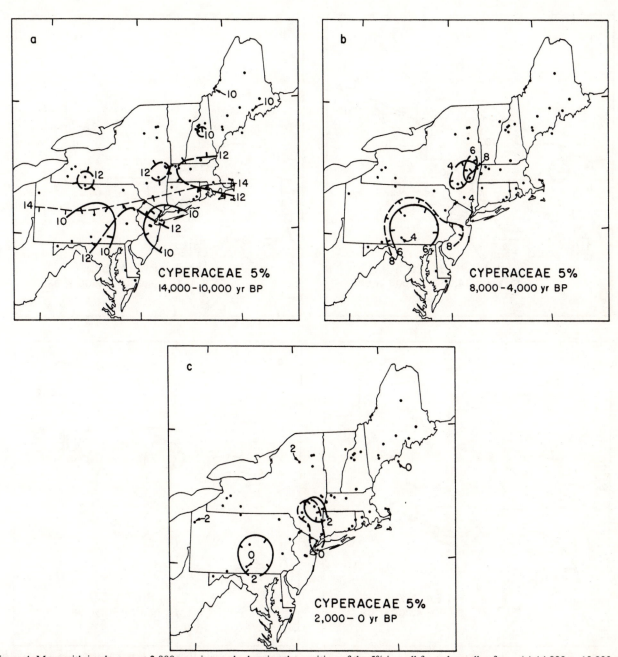

Figure 4. Maps with isochrones at 2,000-year intervals showing the position of the 5% isopoll for sedge pollen from: (a) 14,000 to 10,000 yrs. B.P., (b) 8,000 to 4,000 yrs. B.P., and (c) 2,000 to 0 yrs. B.P. (Contours are in 10^3 years on each map; isolated numbers indicate the presence of pollen at the mapped abundance level for single sites. Percentages of each pollen type increase in the direction of the hatchures on each isochrone.)

Within southern New England, the late-Holocene increase in chestnut pollen, and also increases in spruce and hemlock pollen percentages, were used to define the C-3 subzone (Deevey, 1939, 1943). The greatest increases in chestnut pollen percentages were in highland sites in the southern Appalachians and in southwestern New England, primarily in the highlands of western Massachusetts and Connecticut

Figure 3C. Summary pollen diagrams from 16,000 to 0 yrs. B.P. for four of eleven sites in the Northeast. Pollen types were chosen to illustrate the general features of the pollen stratigraphy across the region. Chronologies are estimated according to the information in Table 1. The pollen diagrams are ordered by latitude (S to N) and physiography (inland highlands to coastal lowlands). Plus signs indicate the downcore location of radiocarbon and stratigraphic dates and a 10x magnification is shown for all pollen types.

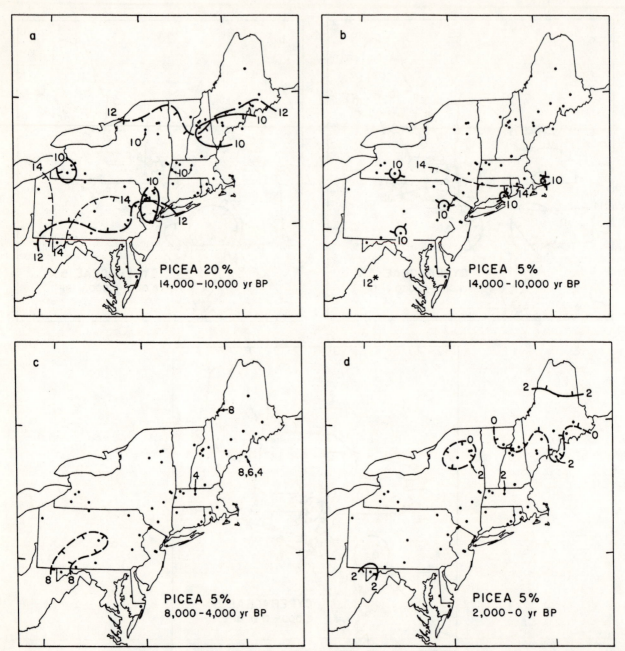

Figure 5. Isochrone maps of the 20% or 5% isopolls for spruce. (a) 20% isopolls from 14,000 to 10,000 yrs. B.P. After this time, spruce pollen did not occur regionally at this abundance level. (b) 5% isopolls from 14,000 to 10,000 yrs. B.P. At 12,000 yrs. B.P., all sites had >5% spruce pollen, as indicated by "12*". (c) 5% isopolls from 8,000 to 4,000 yrs. B.P., and (d) 5% isopolls from 2,000 to 0 yrs. B.P.

(Figs. 3A, a, b; 3B, d, g; 3C, h; Fig. 13). A new radio-carbon-dated diagram from Mohawk Pond in western Connecticut shows the C-3 oak-chestnut subzone in the region where Deevey (1943) defined it. Paillet (1982) has recently described some of the ecological factors that influenced the late-Holocene expansion of chestnut in New England.

In northern New England, Deevey (1951) used the late-Holocene increases in both spruce and hemlock pollen percentages to define subzone C-3. Spruce

pollen was a minor component of the regional pollen assemblages from the early-Holocene until 2,000 yrs. B.P., when percentages of spruce pollen >5% appeared at sites at high elevations and in the north (Fig. 5). Since then, spruce populations have extended their distribution farther south.

Between 2,000 and 0 yrs. B.P., most arboreal pollen types shrank in distribution and the distributions for some types became patchy (Figs. 5 to 13). In the north and at high elevations, spruce populations expanded and the populations of beech, hemlock and

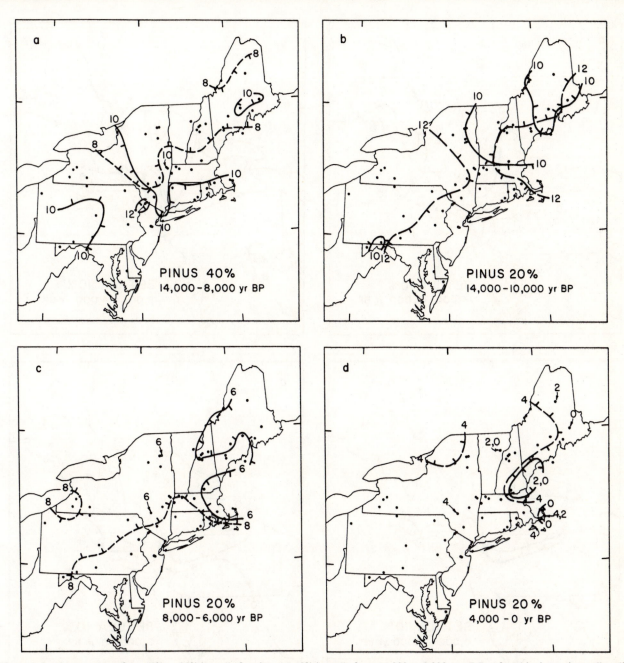

Figure 6. Isochrone maps of the 40% or 20% isopolls for pine. (a) 40% isopolls from 14,000 to 8,000 yrs. B.P. After this time, pine pollen did not occur regionally at this abundance level. (b) 20% isopolls from 14,000 to 10,000 yrs. B.P., (c) 20% isopolls from 8,000 to 6,000 yrs. B.P., and (d) 20% isopolls from 4,000 to 0 yrs. B.P.

other types declined (Figs. 5 to 13). At 0 yrs. B.P., ragweed pollen percentages are high in western New York, northern Pennsylvania, New Jersey, and southern New England (Fig. 14). The increased percentages of ragweed pollen record the recent logging and land clearance by Euro-Americans, and was recognized as subzone C-3b at Rogers Lake in Connecticut (M. B. Davis, 1969a). The recent changes in the pollen distributions reflect various influences, including

climatic and biotic factors. The expansion of spruce populations reflects continued climatic cooling in the Northeast since 4,000 yrs. B.P. (Webb *et al.*, 1983b; Bartlein and Webb, in press). Anthropogenic activity caused the decrease in pollen percentages of most southern trees, and a disease caused the "chestnut decline," which began in New England about A.D. 1915 (Anderson, 1974; Brugam, 1978; M. B. Davis, 1983).

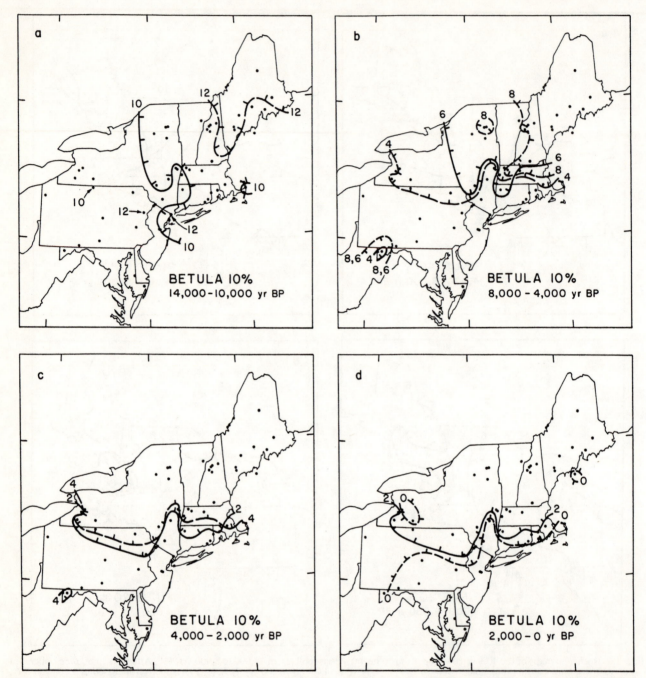

Figure 7. Isochrone maps of the 10% isopolls for birch from: (a) 14,000 to 10,000 yrs. B.P., (b) 8,000 to 4,000 yrs. B.P., (c) 4,000 to 2,000 yrs. B.P., and (d) 2,000 to 0 yrs. B.P.

Summary of Vegetation History 14,000 - 0 yrs. B.P.

The maps suggest that the vegetation was initially open with sedges and other herbaceous plants present, and that spruce trees moved in from the south from 14,000 to 12,000 yrs. B.P. An open spruce parkland in which sedges and shrub birch grew predominated at 12,000 yrs. B.P., and changed to closed forest as pine trees moved in from the south

between 12,000 and 10,000 yrs. B.P. Populations of red and/or jack pine were followed by those of white pine. Oak populations increased in the south and by 8,000 yrs. B.P. had established an ecotone across New England and Pennsylvania. The last 8,000 years show compositional changes in the development of deciduous forest and mixed conifer/hardwood forest which dominate in the region at present. The maps show that the ecotone between these two forest regions remained stable during the middle- and late-

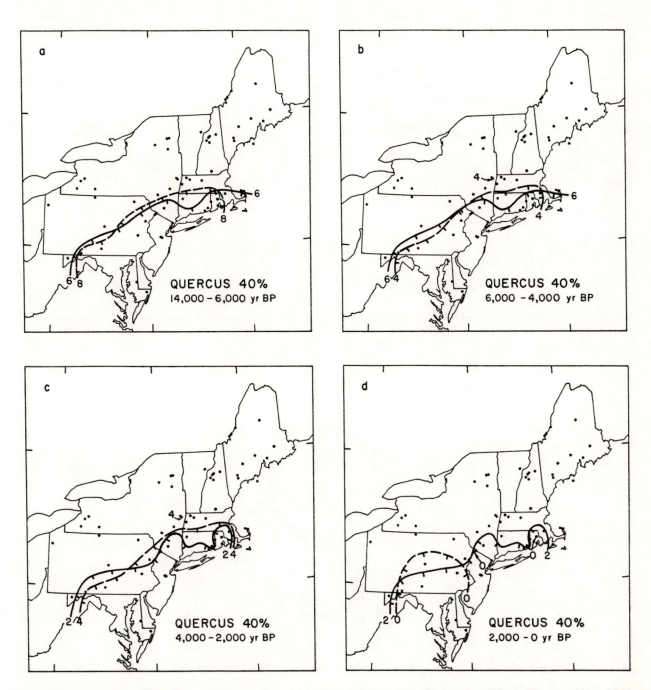

Figure 8. Isochrone maps of the 40% isopolls for oak from: (a) 14,000 to 6,000 yrs. B.P., (b) 6,000 to 4,000 yrs. B.P., (c) 4,000 to 2,000 yrs. B.P., and (d) 2,000 to 0 yrs. B.P.

Holocene. First, major populations of hemlock and beech became established to the north of this ecotone and later, populations of hickory and chestnut expanded within the deciduous forest.

Anomalies in this regional vegetational history of the Northeast are associated with elevational and edaphic influences. The extremes most clearly are represented by the highest mountain elevations and the sandy coastal plain. Davis *et al.,* (1980), Spear (1981), and Jackson (1983) have described the vegetational histories and special problems of obtaining vegetational interpretations in areas with sharp elevational gradients, and Winkler (in press) has described the development of the pine-dominated vegetation of Cape Cod, Massachusetts.

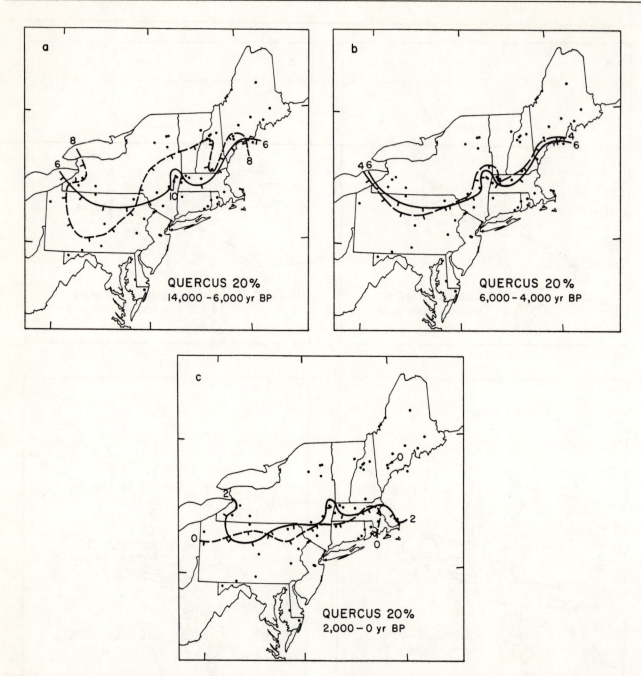

Figure 9. Isochrone maps of the 20% isopolls for oak from: (a) 14,000 to 6,000 yrs. B.P., (b) 6,000 to 4,000 yrs. B.P., and (c) 2,000 to 0 yrs. B.P.

LATE-GLACIAL POLLEN STRATIGRAPHY

Nine pollen diagrams show details of the late-glacial pollen stratigraphy across the Northeast (Figs. 15A, 15B). The southernmost site is Criders Pond (Watts, 1979), dating from 16,000 yrs. B.P., and located in unglaciated southeastern Pennsylvania. The seven other sites are from glaciated areas at different elevations and physiographies; most have sediments dating from 14,000 yrs. B.P. Certain strati-

graphic features (e.g., peak pollen values or abrupt declines in value) appear in each of these diagrams, but their relative brevity compared to the 2,000-year intervals on the isochrone maps meant that several of these features were incompletely represented in the isochrone maps. We review here the main stratigraphic features of the pollen assemblages with high percentages of herb pollen (the classic T zone) and of spruce pollen (the classic A zone).

The paleoecological and paleoenvironmental inter-

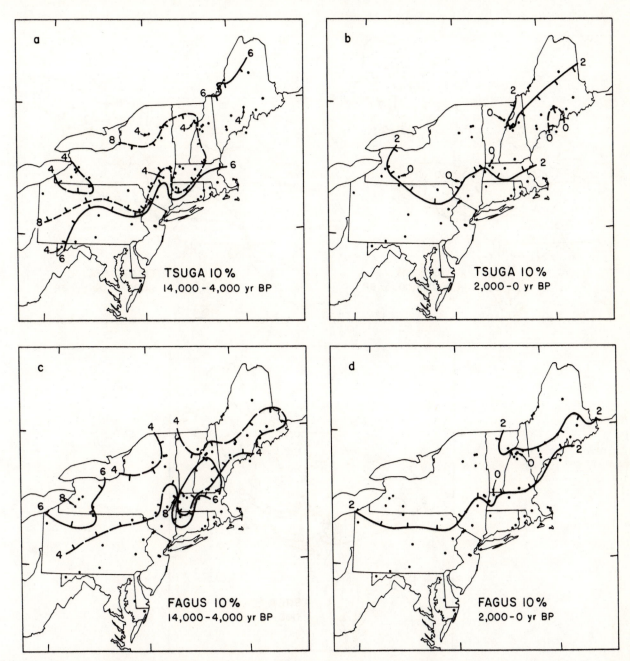

Figure 10. Isochrone maps of the 10% isopolls for hemlock and beech. (a) hemlock, from 14,000 to 4,000 yrs. B.P., and (b) hemlock, from 2,000 to 0 yrs. B.P. (c) beech, from 14,000 to 4,000 yrs. B.P., and (d) beech, from 2,000 to 0 yrs. B.P.

pretations of the late-glacial pollen stratigraphy in the Northeast are especially complex. We used plant macrofossils and pollen accumulation rates as sources of additional information to aid interpretations, as have M. B. Davis (1983) and Watts (1983) in recent reviews of late-Quaternary vegetational history of the eastern United States. We compiled Table 3 to show the dates and magnitude of the major pollen stratigraphic events at the eight sites in Figures 15A and 15B for which pollen accumulation rate data were available.

Deevey (1951) published the first record of treeless conditions in the Northeast as recorded by non-arboreal pollen from late-glacial sediments recovered in northern Maine and cited Potzger and Friesner (1948) as having a record of similar conditions in southern Maine (Deevey, 1949). Late-glacial sediments were soon recovered from several sites in southern New England (Leopold, 1956a; Ogden, 1959; M. B. Davis, 1958, 1960, 1961). Attention at all these sites was focused on pollen evidence for climatic oscillations that could be correlated with

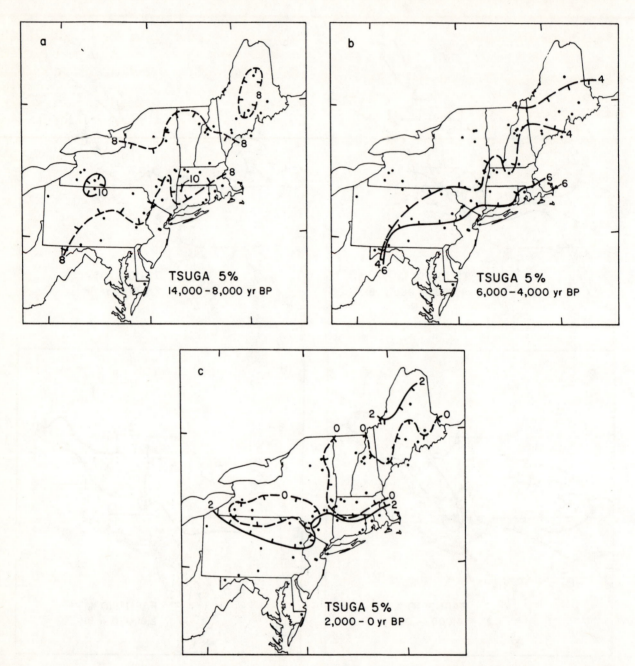

Figure 11. Isochrone maps of the 5% isopolls for hemlock from: (a) 14,000 to 8,000 yrs. B.P., (b) 6,000 to 4,000 yrs. B.P., and (c) 2,000 to 0 yrs. B.P.

events in the Midwest and in Europe. Odgen (1965) reviewed this early emphasis on climatic correlations. Rogers Lake (M. B. Davis and Deevey, 1964; M. B. Davis, 1967a, 1969a), recorded many of these late-glacial oscillations in pollen abundance. With its 53 radiocarbon dates and pollen accumulation rates, the Rogers Lake pollen stratigraphy became a standard for southern New England, but its chronology is less certain due to the question of whether to use a 700-year correction to its radiocarbon dates (M. B. Davis,

1983). We have included two new pollen diagrams from southern New England for comparison (Fig. 15A d, e). One of these diagrams (Winneconnet Pond) is located near the undated late-glacial diagram by M. B. Davis (1960) near Taunton, Massachusetts, and the other diagram is from a new core at Tom Swamp, Petersham, Massachusetts, and provides dates for M. B. Davis' (1958) late-glacial diagram there.

Herb-rich Pollen Assemblages

Percentages of sedge pollen and other nonarboreal pollen (composites, grasses and "other herbs") were highest from before *ca.* 14,000 yrs. B.P. in glaciated Pennsylvania (Fig. 15A, b) to before 12,000 yrs. B.P. in northern sites (Figs. 15A, 15B; Table 3 a). Percentages of sedge pollen declined below 10% by 11,000 yrs. B.P. (Table 3, b; Fig. 4). The percentages of herb pollen remained high at high elevation sites in the White Mountains until *ca.* 10,000 yrs. B.P. (M. B. Davis *et al.,* 1980) as well as at Moulton Pond, in coastal Maine (Fig. 15B, h). (The interpretation of the late-glacial stratigraphy at Moulton Pond is complicated because the site was located on an island during a marine transgression until about 12,400 yrs. B.P., and because of poor core recovery of sediments dated at 10,000 yrs. B.P. [R. B. Davis, *et al.,* 1975].)

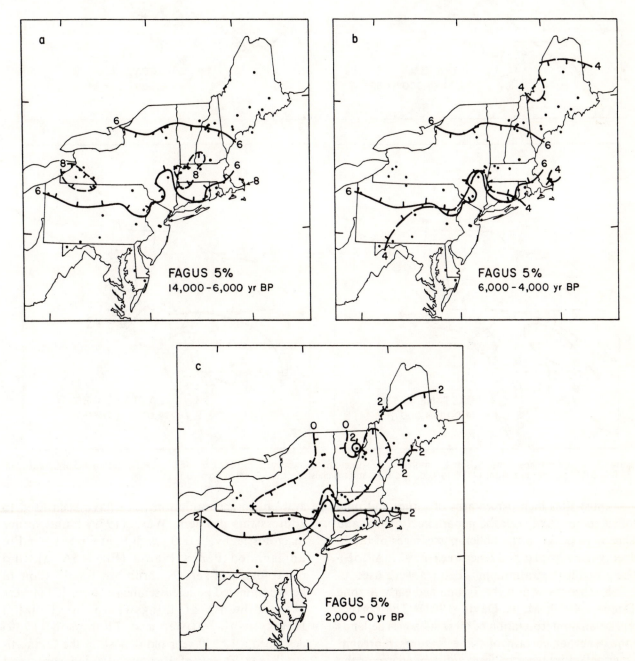

Figure 12. Isochrone maps of the 5% isopolls for beech from: (a) 14,000 to 6,000 yrs. B.P., (b) 6,000 to 4,000 yrs. B.P., and (c) 2,000 to 0 yrs. B.P.

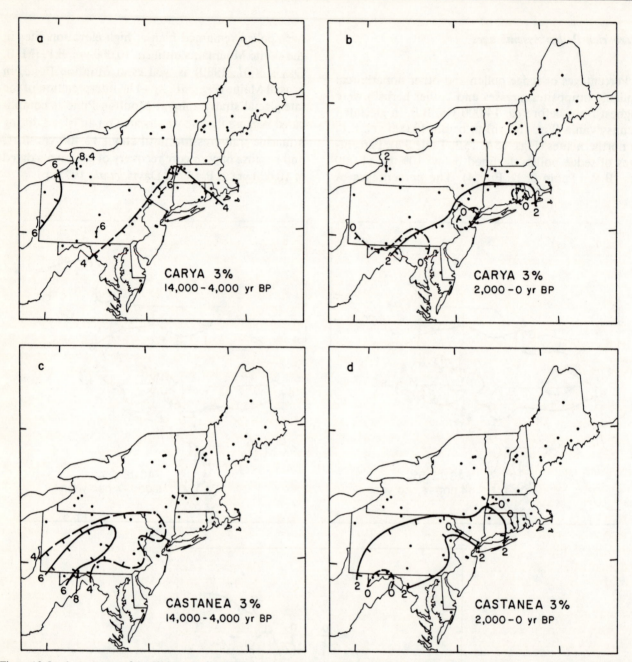

Figure 13. Isochrone maps of the 3% isopolls for hickory and chestnut. (a) hickory, from 14,000 to 4,000 yrs. B.P., and (b) hickory, from 2,000 to 0 yrs. B.P. (c) chestnut, from 14,000 to 4,000 yrs. B.P., and (d) chestnut, from 2,000 to 0 yrs. B.P.

At most sites high percentages of spruce and pine pollen co-occurred with the nonarboreal pollen, and there were peaks in the pollen percentages of birch, alder, willow, poplar and other trees (Figs. 15A, 15B). Many of these stratigraphic features were used to define subzones within the T zone and early A zone (Deevey, 1958; M. B. Davis, 1961). The correct vegetational interpretation of these subzones depends upon whether certain of these features represent either pollen transported from a distant source or the local occurrences of a few trees. Plant macrofossils

and pollen-accumulation rates have been used to address this problem. Watts (1979) found spruce macrofossils by 15,000 yrs. B.P. at Criders Pond in unglaciated Pennsylvania (Fig. 15A, a), and macrofossils of spruce, shrub birch and willow in herb-dominated pollen assemblages from 12,000 year old sediments at Longswamp in glaciated Pennsylvania. Miller and Thompson (1979) described a 12,500 year old deposit in the Connecticut River Valley of Vermont, which contained macrofossils of spruce, balsam poplar, willow, and

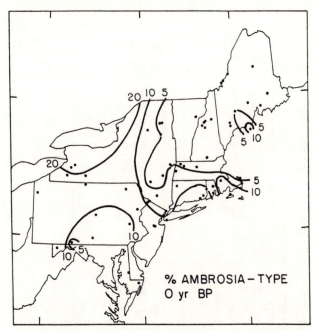

Figure 14. Map with isopolls for ragweed pollen (*Ambrosia*-type) for 0 yrs. B.P.

shrub birch along with those of herbs that grow in the tundra today. From an organic deposit dated at 12,100 yrs. B.P. in northwestern New York, Miller (1973b) recovered black and white spruce and tamarack macrofossils and those of many herbs, though not tundra indicators. These macrofossils from Vermont and New York were found with pollen assemblages having high percentages of herbs (with sedge pollen >10%), spruce and pine, similar to the herb-rich pollen assemblages found in late-glacial sediments throughout the Northeast (Figs. 15A, 15B). At sites in the White Mountains (Fig. 1), M. B. Davis *et al.*, (1980) analyzed the macrofossils in samples having high percentages of herb pollen. They found abundant macrofossils of spruce by 11,000 yrs. B.P. at low elevation sites, but by 10,000 yrs. B.P., spruce macrofossils were present at sites from all elevations and poplar macrofossils had also appeared.

During the period with high percentages of herb pollen, pollen accumulation rates for spruce and other tree types, as well as for herbs, were low. Total pollen accumulation rates in the Northeast range from a few hundred to a few thousand grains/cm^2/ year (Table 3 c). These pollen accumulation rates are regionally consistent and are similar to pollen accumulation rates in existing tundra vegetation and in late-glacial sediments elsewhere (M. B. Davis *et al.*, 1973). The macrofossils, pollen accumulation rates, and mapped patterns of the pollen percentage data suggest that although trees were found on the late-glacial landscape, the vegetation was open. Better

understanding of the processes of pollen accumulation in minerogenic sediments, of macrofossil taphonomy, and of pollen dispersal in a treeless landscape, is necessary to complete the vegetational interpretations of the pollen data from the herb-rich pollen assemblages.

Spruce-dominated Pollen Assemblages

The decline in nonarboreal pollen percentages and a rise in spruce pollen percentages defined the A zone of Deevey (1951). In Maine, Deevey (1951) named subzones either for the interpreted vegetation or the pollen assemblages: A-1 ("birch woodland"), A-2 ("spruce forest"), and A-3 ("birch-fir-alder"). In southern New England, Deevey (1943) defined an undifferentiated A zone ("spruce-fir-pine") in northwestern sites; and in southern and northeastern sites, he defined two subzones based on the occurrence of two spruce pollen peaks: A-1 ("spruce-fir; with pine due to overrepresentation"), A-2 ("spruce-fir with birch"). Leopold (1956a) also found two maxima for spruce in the A zone in southern Connecticut and she called it the "Durham Spruce Zone," which was widely referred to in the literature for the region (M. B. Davis, 1958, 1960, 1961; Ogden, 1965). The two spruce peaks were separated by a pine peak (which Leopold, 1956a, correlated with the Two Creeks Forest Bed). M. B. Davis (1960, 1961) also found a lower and an upper "Durham" spruce peak in diagrams from central Massachusetts, Taunton and Cambridge.

Two spruce peaks occurred at Rogers Lake in southern New England and at two new radiocarbon-dated sites, Tom Swamp and Winneconnet Pond in Massachusetts (Fig. 15A, c, d, e; Table 3, d, e); while a single spruce peak occurred at sites from other areas in the Northeast (Figs. 15A, 15B; Table 3, d, e). The earliest spruce peak is at Criders Pond at about 15,000 yrs. B.P. (Fig. 15A, a). Southern and central glaciated sites (Figs. 15A, b, c, d, e; 15B, f) had a spruce pollen peak between 13,000 and 12,000 yrs. B.P., and a second peak occurred about 11,000 yrs. B.P. in the southern New England sites (Fig. 15A, c, d, e). Only a single peak, about 11,000 yrs. B.P., is seen in the pollen stratigraphies of sites in the Berkshires and in the north (Fig. 15B, g, h, i; Table 3, d, e). A peak in pine pollen percentages preceded the 11,000 yrs. B.P. spruce peak; this pine peak did not occur at the northern sites (Fig. 15B, h, i) where spruce pollen had not reached a maximum before 11,000 yrs. B.P.

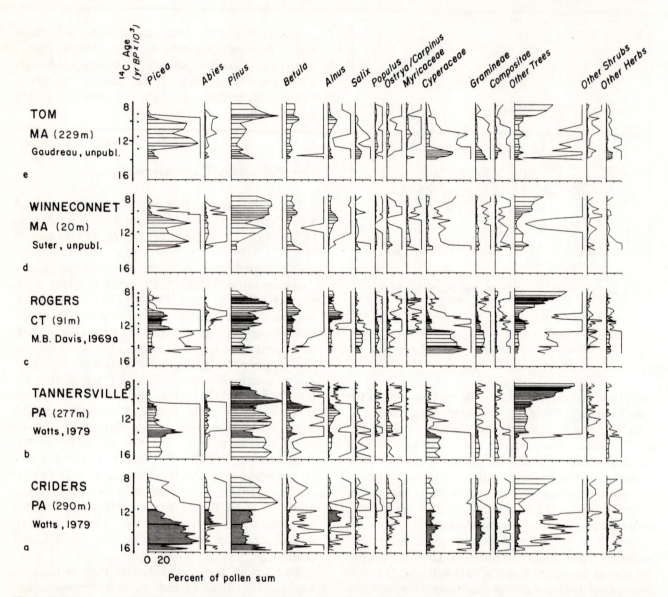

Figure 15A. Summary pollen percentage diagrams from 16,000 to 8,000 yrs. B.P. for five of nine sites in the Northeast. Pollen types show the main features of the late-glacial pollen stratigraphy. Chronologies are estimated as in Figures 3A, 3B, and 3C. The pollen diagrams are ordered by latitude (S to N) and physiography (inland highlands to coastal lowlands). Plus signs indicate the downcore location of radiocarbon and stratigraphic dates and a 10x magnification is shown for all pollen types.

The increased percentages of spruce pollen (that define the A zone) are associated with increased pollen accumulation rates for spruce and for the other tree taxa which also showed increased pollen percentages during this time (e.g., fir, birch, alder, poplar, hornbeam in Figs. 15A, 15B). Within the A zone, the trends in the pollen percentage and accumulation-rate data correspond more closely than they did in the T zone. Pollen accumulation rates for this interval are similar throughout the Northeast (Table 3, f), and range from a few thousand to tens of thousands of grains/cm²/year, with values declining with latitude. These rates may reflect increased numbers of trees within the pollen source area of a site. Among

these trees were spruce and other taxa such as pine which is a prolific pollen producer (Webb et al., 1981). The increased pollen-accumulation rates may also reflect the initiation of organic sedimentation in the lake basins and the beginning of sediment focusing (Lehman, 1975; M. B. Davis and Ford, 1982). Trees were more numerous during the A zone than during the T zone, but open parkland vegetation probably grew around most sites, and closed forest formed during the A zone only at northern sites after 11,000 yrs. B.P.

Peaks in the percentages of alder and birch pollen occurred regionally from 11,000 to 9,000 yrs. B.P. (Figs. 15A, 15B; Table 3, g, h). These percentage

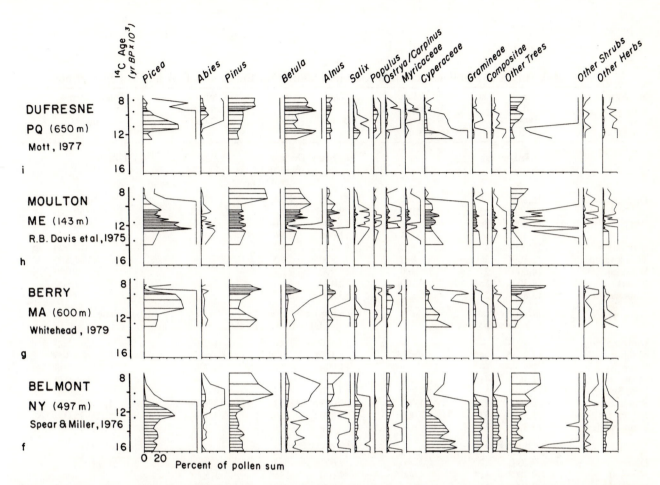

Figure 15B. Summary pollen percentage diagrams from 16,000 to 8,000 yrs. B.P. for four of nine sites in the Northeast. Pollen types show the main features of the late-glacial pollen stratigraphy. Chronologies are estimated as in Figures 3A, 3B, and 3C. The pollen diagrams are ordered by latitude (S to N) and physiography (inland highlands to coastal lowlands). Plus signs indicate the downcore location of radiocarbon and stratigraphic dates and a 10x magnification is shown for all pollen types.

increases were associated with a decline in the percentages of spruce pollen. The birch pollen probably represents white birch which is abundant in the southern boreal forest region today (Richard, 1977). Myricaceous (myrtle family) pollen was regionally important during this time, and had its highest percentages at sites nearer the coast and in the north (Fig. 15A, c, d, e; 15B, h, i). The regional increases in alder and Myricaceae pollen are of interest because of the nitrogen-fixing ability of these taxa (Sprent *et al.,* 1978). Percentages of alder and birch pollen had peak values in many herb-dominated pollen assemblages, but the percentage increases in birch and alder pollen after the spruce-decline are also associated with increased pollen accumulation rates for these types (Table 3, g, h), and seem to indicate increases in plant numbers within the pollen source areas of the sites. Macrofossils of speckled alder were found in 13,000 year old sediments at Criders Pond (Watts, 1979); these correspond with the earliest peak in alder pollen percentages after the sedge-decline at this site.

Jackson (1983) found macrofossils of white birch in the Adirondacks by 10,200 yrs. B.P., predating the peaks in birch pollen percentages at northern sites (Table 3, h).

The spruce-dominated pollen assemblages (the classic A zone) ended when the percentages of pine pollen increased about 10,000 yrs. B.P. and peaked by 9,000 yrs. B.P. (Table 3, i). Total pollen accumulation rates increased to Holocene highs throughout the Northeast (Table 3, j). The percentages of "other trees" (Figs. 15A, 15B) also increased, some slightly later. These changes marked the beginning of the development of forest regions structurally like those that occur in the Northeast today.

DISCUSSION

As informal biostratigraphic units, the classic pollen assemblage zones (T, A, B, C, of Deevey, 1939,

Table 3.
Dates, peak values and pollen accumulation rate maxima for the classic T, A and B pollen zones.

Site Name	Date of Event (yrs. B.P.)	%	Pollen Accumulation Rate (grains/cm²/yr)	Date of Event	%	Pollen Accumulation Rate	Total Pollen Accumulation Rate
	(a) Sedge Peak			**(b) Sedge Decline** % at and after decline			**(c) T Zone**
Dufresne	12,500	39	<400	11,700	21-5		100-1,000
Moulton	13,900	25	<100	9,800*	10-0		150-920
Berry	12,800	36	n.a.¹	12,800	36-10		(ave) 200
				11,000	10-4		
Belmont	13,000	49	500	12,400	10-3		1,200-3,000
Tom Swamp	13,100	36	860	12,900	24-5		425-2,500
Winneconnet	14,000	21	400	14,000	21-4		5,000
Rogers	14,300	55	<500	12,400	31-5		<1,000
Tannersville	13,800	21	n.a.	13,300	13-7		4,000
	(d) Spruce Peak before 12,000 yrs. B.P.			**(e) Spruce Peak** 12,000 - 10,000 yrs. B.P.			**(f) A Zone**
Dufresne	–	–	–	11,400	47	1,000	2,000
Moulton	12,300*	64*	<500*	11,300*	34*	<500*	3,800*
Berry	–	–	–	11,000	53	3,900	400-16,000
Belmont	12,400	41	1,400	–	–	–	3,000-6,000
Tom Swamp	12,300	67	10,000	10,200	52	10,000	12,500-18,500
Winneconnet	12,800	55	4,000	11,000	57	13,000	5,000-20,000
Rogers	12,300	33	2,000	10,900	26	3,000	1,000-20,000
Tannersville	13,300	46	n.a.	–	–	–	8,000-17,000
	(g) Birch Peak			**(h) Alder Peak**			
Dufresne	9,200	7	2,000	9,600	43	16,000	
Moulton	9,800*	11*	<600*	9,800*	35*	1,500*	
Berry	9,500	11	1,500	9,200	22	4,400	
Belmont	8,000	22	1,400	9,000	5	600	
Tom Swamp	9,800	18	6,300	9,800	12	4,300	
Winneconnet	10,700	11	5,500	10,600	23	4,500	
Rogers	11,000	20	1,500	10,400	19	1,500	
Tannersville	11,300	17	n.a.	10,600	34	n.a.	
	(i) Pine Peak						**(j) B Zone**
Dufresne	9,200	36	12,000				20,000
Moulton	9,200	52	8,000				13,300-24,900
Berry	9,000	43	12,300				16,000-34,000
Belmont	10,200	60	9,600				n.a.
Tom Swamp	9,400	65	50,000				35,600-76,900
Winneconnet	9,200	55	25,000				45,000
Rogers	10,100	57	>25,000				20,000-50,000
Tannersville	10,100	70	n.a.				10,000-22,000

¹n.a. indicates that pollen accumulation rate data are not available for a site.

*Percentage values for spruce, pine, birch and alder are for peaks occurring after the decline in sedge pollen percentages at each site, except at Moulton for which values with an asterisk are pre-sedge decline. The distinction of the T and the A zones is not clear at Moulton.

1943, 1958) remain in use and are evident in southern New England pollen diagrams (Figs. 3A, 3B, 3C and 15A, 15B). Variations in vegetation history across the Northeast mean that the definition and pollen composition of such zones as the C zone must vary if a C zone is to be recognized in both southern and northern New England. Current practice is to define local zones for each pollen diagram and to use labels unique to the site *(e.g.,* Cushing, 1967; Wright, 1976). These local zones can then be pollen-stratigraphically correlated with the classic zones or with local zones from other sites. This practice leads to a critical definition of the zones at each site.

The classic zones were originally defined by Deevey (1939, 1943) before radiocarbon dates were available. At this time, the pollen assemblage zones were often thought to have chronostratigraphic and climatostratigraphic meaning. The linkage of the original zones to climate was not simple, and Deevey's (1939) view of this linkage led him to discuss why certain zones might be time-transgressive. Our maps confirm the time-transgressiveness of the zones, but we would interpret their differences in space and time as directly reflecting the time-transgressiveness of regionally varying climatic change rather than reflecting the time-transgressive vegetational response to a globally synchronous sequence of climatic changes.

Deevey's (1939) underlying climate model included a globally (or perhaps only a northern hemisphere mid-latitude) synchronous hypsithermal interval, and this view led Deevey and Flint (1957) to define the "long, warm" hypsithermal interval as a chronostratigraphic unit with a local biostratigraphic expression *(i.e.,* "The time represented by four pollen zones, V through VIII in the Danish system," p. 182). In light of current understanding of past climates, this definition makes "hypsithermal" an almost useless term, because its name says high or peak temperature conditions, but these conditions did not prevail in all regions during this time interval (Bartlein *et al.,* 1984; Bartlein and Webb, in press). Wright (1976) noted this problem and proposed that the hypsithermal, "...a mid-Holocene episode of warmer (and/or) drier climate with indefinite time boundaries..." be redefined from being "...a time stratigraphic unit with time-parallel terminations..." to being "...a climatic episode with time-transgressive boundaries" (pp. 592-593). Geological evidence for the hypsithermal episode would be any Holocene data indicating local or regional temperatures higher than those today.

Global networks of such data are needed before we can understand the full spatial and temporal variations of the hypsithermal episode.

As biostratigraphic units, the classic pollen zones of Deevey (1958) can not be chronostratigraphic units (Watson and Wright, 1980). Their use as climatostratigraphic units is also questionable in light of the hemlock decline defining the C-1/C-2 boundary. The classic zones may yield different vegetational and climatic interpretations at different dates and different sites, especially if the A, B, and C zones are used at sites from New Jersey to northern Maine. Differences in scale contribute to the confusion about the stratigraphic implications of pollen zones because relationships change with scale. Local pollen zones, which by definition are biostratigraphic units, can often serve locally as both chronostratigraphic and climatostratigraphic units. They can even serve regionally *(e.g.,* within southern Connecticut) as biostratigraphic units with possible chronostratigraphic and climatostratigraphic uses. As the geographic region of correlation enlarges, however, significant differences in dating and climate can arise.

The classic zones continue to have value in southern New England, and their original definition allows them to be correlated with local zones in new diagrams from this area. The availability of radiocarbon dates, however, has decreased the need to use any pollen zones as chronostratigraphic units. Certain stratigraphic features, *e.g.,* the hemlock decine, may be synchronous over a large area, but pollen assemblage zones are not. The direct use of zones in climatic interpretations is also diminished because of our current understanding of how climate varies spatially and temporally (Bartlein *et al.,*1984), and because of our methods for climatic interpretation of pollen data (Howe and Webb, 1983).

For late-Quaternary studies, the definition of pollen zones will continue to be useful mainly as an initial method for simplifying the description of pollen diagrams. Research in line with what Cushing (1967) proposed for pollen assemblage zones in Minnesota can lead to new insights, but other methods of data analysis and display are preferable if vegetational and climatic interpretations are the main research goals. The definition and location of spatial gradients are key in these interpretations. Our maps of the northeastern pollen data show some of what can be gained by this method of describing pollen data.

CONCLUSIONS

We have summarized the late-Quaternary pollen data in the Northeast and critically examined the classic pollen zonation sequence for this region (Deevey, 1939, 1943, 1951). We have also mapped the percentage data of individual pollen types chosen to show the important features of the regional pollen stratigraphy. The maps provide an understanding of the time and space variations in pollen zones across the Northeast and show many of the vegetational features and changes that created the zonal sequence.

ACKNOWLEDGEMENTS

This research was supported by the NSF Climate Dynamics Program. R. H. W. Bradshaw, G. L. Jacobson, R. J. Nickmann, W. A. Patterson III, A. M. Swain, and D. R. Whitehead kindly provided unpublished data. J. C. Bernabo, R. A. Ibe, J. Nicholas and R. W. Spear provided data from doctoral dissertations. R. S. Anderson, L. (Bostwick) Bjerk, and M. G. Winkler provided data from Masters theses, and K. M. Trent provided data from a senior thesis at Brown University. We also thank the many people who provided pollen data from their published studies. We thank R. Arigo, J. Avizinis, E. Doro, S. Klinkman, B. Mellor, J. Overpeck, M. Ryall and S. Suter for technical assistance.

References Cited

ANDERSON, R. S.

1979   A Holocene record of vegetation and fire at Upper South Branch Pond in northern Maine. M.S. thesis, University of Maine, Orono.

ANDERSON, T. W.

1974   The chestnut pollen decline as a time horizon in lake sediments in eastern North America. *Canadian Journal of Earth Sciences,* 11: 678-685.

BARTLEIN, P. J., WEBB, T. III, and FLERI, E. C.

1984   Holocene climatic change in the northern Midwest: pollen-derived estimates. *Quaternary Research,* 22: 361-374.

BARTLEIN, P. J., and WEBB, T. III

1985   Mean July temperature for eastern North America at 6,000 yrs. B.P.: regression equations for estimates based on fossil-pollen data. *Syllogeus.* In press.

BERNABO, J. C.

1977   Sensing climatically and culturally induced environmental changes with palynological data. Ph.D. dissertation, Brown University.

BERNABO, J. C., and WEBB, T. III

1977   Changing patterns in the Holocene pollen record from northeastern North America: a mapped summary. *Quaternary Research,* 8: 64-96.

BOSTWICK, L. K.

1978   An environmental framework for cultural change in Maine: pollen influx and percentage diagrams from Monhegan Island. M.S. thesis, University of Maine, Orono.

BRAUN, E. L.

1950   *Deciduous Forests of Eastern North America.* The Blakiston Company, Philadelphia, 596 p.

BRUGAM, R. B.

1978   Pollen indicators of land-use change in southern Connecticut. *Quaternary Research,* 9: 349-362.

CARMICHAEL, D. P.

1980   A record of environmental change during recent milennia in the Hackensack tidal marsh, New Jersey. *Bulletin of the Torrey Botanical Club,* 107: 514-524.

CONNALLY, G. G., and SIRKIN, L. A.

1970   Late glacial history of the upper Wallkill Valley, New York. *Geological Society of America Bulletin,* 81: 3297-3306.

1971   The Luzerne readvance near Glens Falls, New York. *Geological Society of America Bulletin,* 82: 989-1008.

COTTER, J. F. P., and CROWL, G. H.

1981   The paleolimnology of Rose Lake, Potter Co., Pennsylvania: a comparison of palynologic and paleopigment studies. *In:* Romans, R. C. (ed.), *Geobotany II.* Plenum Press, New York, p. 91-116.

CUSHING, E. J.

1964   Application of the code of stratigraphic nomenclature to pollen stratigraphy. (Unpublished manuscript).

1967   Late-Wisconsin pollen stratigraphy and the glacial sequence in Minnesota. *In:* Cushing, E. J., and Wright, H. E. Jr., *Quaternary Paleoecology.* Yale University Press, New Haven, p. 59-88.

DAVIS, M. B.

1958 Three pollen diagrams from central Massachusetts. *American Journal of Science*, 256: 540-570.

1960 A late-glacial pollen diagram from Taunton, Massachusetts. *Bulletin of the Torrey Botanical Club*, 87: 258-270.

1961 Pollen diagrams as evidence of late-glacial climatic change in southern New England. *Annals of the New York Academy of Sciences*, 95: 623-631.

1963 On the theory of pollen analysis. *American Journal of Science*, 261: 897-912.

1965 Phytogeography and palynology of northeastern United States. *In:* Wright, H. E. Jr., and Frey, D. G. (eds.), *The Quaternary of the United States*. Princeton University Press, Princeton, p. 377-401.

1967a Pollen accumulation rates at Rogers Lake, Connecticut, during late- and postglacial time. *Review of Paleobotany and Palynology*, 2: 219-230.

1967b Late-glacial climate in northern United States: a comparison of New England and the Great Lakes region. *In:* Cushing, E. J., and Wright, H. E. Jr. (eds.), *Quaternary Paleoecology*. Yale University Press, New Haven, p. 11-44.

1969a Climatic changes in southern Connecticut recorded by pollen deposition at Rogers Lake. *Ecology*, 50: 409-422.

1969b Palynology and environmental history during the Quaternary period. *American Scientist*, 57: 317-332.

1976 Pleistocene biogeography of temperate deciduous forests. *Geoscience and Man*, XIII: 13-26.

1978 Climatic interpretation of pollen in Quaternary sediments. *In:* Walker, D., and Guppy, J. C. (eds.), *Biology and Quaternary Environments*. Australian Academy of Science, Canberra, p. 35-51.

1981a Outbreaks of forest pathogens in Quaternary history. *In: Proceedings of the Fourth International Palynological Conference*, Vol. 3. Lucknow, India (1976-1977), p. 216-227.

1981b Quaternary history and the stability of forest communities. *In:* West, D. C., Shugart, H. H., and Botkin, D. B. (eds.), *Forest Succession*. Springer-Verlag, New York, p. 132-177.

1983 Holocene vegetational history of the eastern United States. *In:* Wright, H. E. Jr. (ed.), *Late Quaternary Environments of the United States, Vol. 2. The Holocene*. University of Minnesota Press, Minneapolis, p. 166-181.

DAVIS, M. B., BRUBAKER, L. B., and WEBB, T. III

1973 Calibration of absolute pollen influx. *In:* Birks, H. J. B., and West, R. G. (eds.), *Quaternary Plant Ecology*. John Wiley and Sons, New York, p. 9-25.

DAVIS, M. B., and DEEVEY, E. S. JR.

1964 Pollen accumulation rates: estimates from late-glacial sediment of Rogers Lake. *Science*, 145: 1293-1295.

DAVIS, M. B., and FORD, M. S.

1982 Sediment focusing in Mirror Lake, New Hampshire. *Limnology Oceanography*, 27: 137-150.

DAVIS, M. B., and GOODLETT, J. C.

1960 Comparison of the present vegetation with pollen-spectra in surface samples from Brownington Pond, Vermont. *Ecology*, 41: 346-357.

DAVIS, M. B., SPEAR, R. W., and SHANE, L. C. K.

1980 Holocene climate of New England. *Quaternary Research*, 14: 240-250.

DAVIS, R. B.

1967 Pollen studies of near-surface sediments in Maine lakes. *In:* Cushing, E. J., and Wright, H. E. Jr. (eds.), *Quaternary Paleoecology*. Yale University Press, New Haven, p. 143-173.

1974 Stratigraphic effects of tubificids in profundal lake sediments. *Limnology Oceanography*, 19: 466-488.

DAVIS, R. B., BRADSTREET, T. E., STUCKENRATH, R. JR., and BORNS, H. W. JR.

1975 Vegetation and associated environments during the past 14,000 years near Moulton Pond, Maine. *Quaternary Research*, 5: 436-465.

DAVIS, R. B., and DOYLE, R. W.

1969 A piston corer for upper sediment in lakes. *Limnology Oceanography*, 14: 643-648.

DAVIS, R. B., and JACOBSON, G. L. JR.

1985 Lateglacial and early Holocene landscapes in northern New England and adjacent areas of Canada. *Quaternary Research*. In press.

DAVIS, R. B., and WEBB, T. III

1975 The contemporary distribution of pollen from eastern North America: a comparison with the vegetation. *Quaternary Research*, 5: 395-434.

DEEVEY, E. S. JR.

1939 Studies on Connecticut lake sediments. I. A postglacial climatic chronology for southern New England. *American Journal of Science*, 237: 691-723.

1943 Additional pollen analyses from southern New England. *American Journal of Science*, 241: 717-752.

1949 Biogeography of the Pleistocene Part I: Europe and North America. *Geological Society of America Bulletin*, 60: 1315-1416.

1951 Late-glacial and postglacial pollen diagrams from Maine. *American Journal of Science*, 249: 177-207.

1958 Radiocarbon-dated pollen sequences in eastern North America. *Veroffentlichungen des Geobotanischen Institutes Rubel in Zürich*, 34: 30-37.

DEEVEY, E. S. JR., GROSS, M. S., HUTCHINSON, G. E., and KRAYBILL, H. L.

1954 The natural C-14 content of materials from hard-water lakes. *Proceedings of the National Academy of Sciences*, 40: 285-288.

DEEVEY, E. S. JR., and FLINT, R. F.

1957 Postglacial hypsithermal interval. *Science*, 125: 182-184.

DELCOURT, P. A., and DELCOURT, H. R.

1981 Vegetation maps for eastern North America. *In:* Romans, R. C. (ed.), *Geobotany II*. Plenum Press, New York, p. 123-165.

DELCOURT, P. A., DELCOURT, H. R., and WEBB, T. III

1984 Atlas of mapped distributions of dominance and modern pollen percentages for important tree taxa of eastern North America. *American Association of Stratigraphic Palynologists Contributions Series*, no. 14, 131 p.

DENNY, C. S.

1982 Geomorphology of New England. *United States Geological Survey Professional Paper*, 1208, 18 p.

DENTON, G. H., and HUGHES, T. J. (eds.)

1981 *The Late Great Ice Sheets*. Wiley Interscience, New York.

DONNER, J. J.

1964 Pleistocene geology of eastern Long Island, New York. *American Journal of Science*, 262: 355-376.

EYRE, F. H. (ed.)

1980 *Forest Cover Types of the United States and Canada*. Society of American Foresters, Washington, D. C., 148 p.

FLINT, R. F., and DEEVEY, E. S. JR.
   1951   Radiocarbon dating of late-Pleistocene events. *American Journal of Science,* 249: 257-300.

FLORER, L. E.
   1972   Palynology of a postglacial bog in the New Jersey Pine Barrens. *Bulletin of the Torrey Botanical Club,* 99: 135-138.

GAUDREAU, D. C.
   1984   Features of Holocene plant population dynamics inferred from mapped pollen data from northeastern North America. *Bulletin of the Ecological Society of America,* 65: 211.
   1985   Holocene vegetational history of central New England: the use of elevational contrasts and latitudinal gradients in pollen data for paleoecological interpretation. Ph.D. dissertation, Yale University.

HOWE, S., and WEBB, T. III
   1983   Calibrating pollen data in climatic terms: improving the methods. *Quaternary Science Reviews,* 2: 17-51.

IBE, R. A.
   1982   Quaternary palynology of five lacustrine deposits in the Catskill Mountain region of New York. Ph.D. dissertation, New York University.

JACKSON, S. T.
   1983   Late-glacial and Holocene vegetational changes in the Adirondack Mountains (New York): a macrofossil study. Ph.D. dissertation, Indiana University.

KÜCHLER, A. W.
   1964   Potential natural vegetation of the conterminous United States (Map and Manual). *American Geographical Society, Special Publication* 36.

LEHMAN, J. T.
   1975   Reconstructing the rate of accumulation of lake sediment: the effect of sediment focusing. *Quaternary Research,* 5: 541-550.

LEOPOLD, E. B.
   1956a Two late-glacial deposits in southern Connecticut. *Proceedings of the National Academy of Sciences,* 52: 863-867.
   1956b Pollen-size frequency in New England species of the genus *Betula. Grana Palynologica,* 1: 140-147.
   1958   Some aspects of late-glacial climate in eastern United States. *Veroffentlichungen des Geobotanischen Institutes Rubel in Zürich,* 34: 80-85.
   1964   Reconstruction of Quaternary environments using palynology. *In: Reconstruction of Past Environments. Proceedings of the Fort Burgwin Conference on Paleoecology,* pub. no. 3. Taos, New Mexico, p. 43-50.

LIKENS, G. E., and DAVIS, M. B.
   1975   Post-glacial history of Mirror Lake and its watershed in New Hampshire, U.S.A.: an initial report. *Verhandlungen der Internationalen Vereinigung für theoretische und angewandte Limnologie,* 19: 982-993.

LIVINGSTONE, D. A.
   1955   A lightweight piston sampler for lake deposits. *Ecology,* 36: 137-139.
   1968   Some interstadial and postglacial pollen diagrams from eastern Canada. *Ecological Monographs,* 38: 87-125.

LOUGEE, R. J.
   1957   Pre-Wisconsin peat in Millbury, Massachusetts (abst.). *Geological Society of America Bulletin,* 68: 1896.

LULL, H. W.
   1968   *A Forest Atlas of the Northeast.* Northeastern Forest Experiment Station, Upper Darby, Pennsylvania, 46 p.

MC DOWELL, L. L., DOLE, R. M. JR., HOWARD, M. JR., and FARRINGTON, R. A.
   1971   Palynology and radiocarbon chronology of Bugbee Wildflower Sanctuary and Natural Area, Caledonia County, Vermont. *Pollen et Spores,* XIII: 73-91.

MAXWELL, J. A., and DAVIS, M. B.
   1972   Pollen evidence of Pleistocene and Holocene vegetation on the Allegheny Plateau, Maryland. *Quaternary Research,* 2: 506-530.

MILLER, N. G.
   1973a Late-glacial and Postglacial vegetation change in southwestern New York State. *Bulletin of the New York State Museum and Science Service,* 420, 102 p.
   1973b Lateglacial plants and plant communities in northwestern New York State. *Journal of the Arnold Arboretum,* 54: 123-159.

MILLER, N. G., and THOMPSON, G. G.
   1979   Boreal and western North American plants in the late Pleistocene of Vermont. *Journal of the Arnold Arboretum,* 60: 167-218.

MOTT, R. J.
   1977   Late-Pleistocene and Holocene palynology in southeastern Quebec. *Géographie Physique et Quaternaire,* XXXI: 139-149.

NELSON, S.
   1984   Upland and wetland vegetational changes in southeastern Massachusetts: a 12,000 year record. *Northeastern Geology,* 6: 181-191.

NEWMAN, W. S.
   1977   Late Quaternary paleoenvironmental reconstruction: some contradictions from northwestern Long Island, New York. *Annals of the New York Academy of Sciences,* 288: 545-570.

NICHOLAS, J.
   1968   Late Pleistocene palynology of southeastern New York and northern New Jersey. Ph.D. dissertation, New York University.

NICHOLS, G. E.
   1935   The hemlock-white pine-northern hardwood region of eastern North America. *Ecology,* 16: 403-422.

NIERING, W. A.
   1953   The past and present vegetation of High Point State Park, New Jersey. *Ecological Monographs,* 23: 127-148.

OGDEN, J. G. III
   1959   A late-glacial pollen sequence from Martha's Vineyard, Massachusetts. *American Journal of Science,* 257: 366-381.
   1965   Pleistocene pollen records from eastern North America. *The Botanical Review,* 31: 481-504.

OVERPECK, J. T.
   1985   A pollen study of a late-Quaternary peat bog: south-central Adirondack Mountains, New York. *Geological Society of America Bulletin,* 96: 145-154.

OVERPECK, J. T., and FLERI, E. C.
   1982   The development of age models for Holocene sediment cores: northeast North American examples. *American Quaternary Association, Abstracts,* p. 152.

PAILLET, F. L.
   1982   The ecological significance of American chestnut (*Castanea dentata* (Marsh.) Borkh.) in the Holocene forests of Connecticut. *Bulletin of the Torrey Botanical Club,* 109: 457-473.

POTZGER, J. E., and FRIESNER, R. C.

1948   Forests of the past along the coast of southern Maine. *Butler University Botanical Studies*, 8: 178-203.

RICHARD, P.

1977   *Histoire post-wisconsinienne de la végétation du Québec méridional par l'analyse pollinique*. Service de la Recherche, Direction Générale des Forêts, Ministère des Terres et Forêts, Québec, tome 1, 312 p.; tome 2, 142 p.

RUSSELL, E. W. B.

1980   Vegetational change in northern New Jersey from precolonization to the present: a palynological interpretation. *Bulletin of the Torrey Botanical Club*, 107: 432-446.

SEARS, P. B.

1932   The archaeology of environment in eastern North America. *American Anthropology*, 34: 610-622.

1942   Postglacial migration of five forest genera. *American Journal of Botany*, 29: 684-691.

SIRKIN, L. A.

1967   Late-Pleistocene pollen stratigraphy of western Long Island and eastern Staten Island, New York. *In:* Cushing, E. J., and Wright, H. E. Jr. (eds.), *Quaternary Paleoecology*. Yale University Press, New Haven, p. 249-274.

1971   Surficial glacial deposits and postglacial pollen stratigraphy in central Long Island, New York. *Pollen et Spores*, XIII: 93-100.

1972   Origin and history of Maple Bog in the Sunken Forest, Fire Island, New York. *Bulletin of the Torrey Botanical Club*, 99: 131-135.

1976   Block Island, Rhode Island: evidence of fluctuation of the late Pleistocene ice margin. *Geological Society of America Bulletin*, 87: 574-580.

1977   Late Pleistocene vegetation and environments in the Middle Atlantic region. *Annals of the New York Academy of Sciences*, 288: 206-217.

SIRKIN, L. A., DENNY, C. S., and RUBIN, M.

1977   Late Pleistocene environment of the central Delmarva Peninsula, Delaware-Maryland. *Geological Society of America Bulletin*, 88: 139-142.

SIRKIN, L. A., and MINARD, J. P.

1972   Late Pleistocene glaciation and pollen stratigraphy in northwestern New Jersey. *United States Geological Survey Professional Paper*, 800-D: 51-56.

SIRKIN, L. A., OWENS, J. P., MINARD, J. P., and RUBIN, M.

1970   Palynology of some upper Quaternary peat samples from the New Jersey coastal plain. *United States Geological Survey Professional Paper*, 700-D: 77-87.

SOLOMON, A. M., and KROENER, D. F.

1971   Suburban replacement of rural land uses reflected in the pollen rain of northeastern New Jersey. *New Jersey Academy of Science "Bulletin"*, 16: 30-44.

SPEAR, R. W.

1981   The history of high-elevation vegetation in the White Mountains of New Hampshire, Ph.D. dissertation, University of Minnesota.

SPEAR, R. W., and MILLER, N. G.

1976   A radiocarbon dated pollen diagram from the Allegheny Plateau of New York State. *Journal of the Arnold Arboretum*, 57: 369-403.

SPRENT, J. I., SCOTT, R., and PERRY, K. M.

1978   The nitrogen economy of *Myrica gale* in the field. *Journal of Ecology*, 66: 657-668.

STUIVER, M.

1967   Origin and extent of atmospheric C-14 variations during the past 10,000 years. *In: Radioactive Dating and Methods of Low Level Counting*. International Atomic Energy Agency, Vienna, p. 27-40.

SUTER, S.M.

1985   Late-glacial and Holocene vegetational history in southeastern Massachusetts: a 14,000 year pollen record. *Current Research*. In press.

WALKER, P. C., and HARTMAN, R. T.

1960   The forest sequence of the Hartstown Bog area in western Pennsylvania. *Ecology*, 41: 461-474.

WATSON, R. A., and WRIGHT, H. E. JR.

1980   The end of the Pleistocene: a general critique of chronostratigraphic classification. *Boreas*, 9: 153-163.

WATTS, W. A.

1979   Late Quaternary vegetation of central Appalachia and the New Jersey coastal plain. *Ecological Monographs*, 49: 427-469.

1983   Vegetational history of the eastern United States 25,000 to 10,000 years ago. *In:* Porter, S. C. (ed.), *Late-Quaternary Environments of the United States, Vol. 1, The Late Pleistocene*. University of Minnesota Press, Minneapolis, p. 294-310.

WEBB, T. III

1981   The past 11,000 years of vegetational change in eastern North America. *Bioscience*, 31: 501-506.

1982   Temporal resolution in Holocene pollen data. *Third North American Paleontological Convention, Proceedings*, 2: 569-572.

WEBB, T. III, and BERNABO, J. C.

1977   The contemporary distribution and Holocene stratigraphy of pollen types in eastern North America. *In:* Elsik, W. C. (ed.), *Contributions of Stratigraphic Palynology, Vol. 1. Cenozoic Palynology. American Association of Stratigraphic Palynologists, Contribution Series 5A:* 130-146.

WEBB, T. III, CUSHING, E. J., and WRIGHT, H. E. JR.

1983a   Holocene changes in the vegetation of the Midwest. *In:* Wright, H. E. Jr. (ed.), *Late Quaternary Environments of the United States, Vol. 2. The Holocene,* University of Minnesota Press, Minneapolis, p. 142-165.

WEBB, T. III, HOWE, S. E., BRADSHAW, R. H. W., and HEIDE, K. M.

1981   Estimating plant abundances from pollen percentages: the use of regression analysis. *Review of Paleobotany and Palynology*, 34: 269-300.

WEBB, T. III, RICHARD, P. J. H., and MOTT, R. J.

1983b   A mapped history of Holocene vegetation in southern Quebec. *Syllogeus*, 49: 273-336.

WHITEHEAD, D. R.

1979   Late-glacial and Postglacial vegetational history of the Berkshires, western Massachusetts. *Quaternary Research,* 12: 333-357.

WHITEHEAD, D. R., and CRISMAN, T. L.

1978   Paleolimnological studies of small New England (U.S.A.) ponds. Part I. Late-glacial and postglacial trophic oscillations. *Polskie Archiwum Hydrobiologii*, 25: 471-481.

WINKLER, M. G.

1982   Late-glacial and postglacial vegetation history of Cape
        Cod and the paleolimnology of Duck Pond, South Well-
        fleet, Massachusetts. M.S. thesis, University of Wiscon-
        sin, Madison.

1985   A 12,000-year history of vegetation and climate for Cape
        Cod, Massachusetts. *Quaternary Research.* In press.

WRIGHT, H. E. JR.

1976   The dynamic nature of Holocene vegetation. *Quaternary
        Research,* 6: 581-596.

LATE-QUATERNARY POLLEN RECORDS FROM EASTERN ONTARIO, QUEBEC, AND ATLANTIC CANADA

THANE. W. ANDERSON
Terrain Sciences Division
Geological Survey of Canada
601 Booth St.,
Ottawa, Ontario K1A 0E8

Abstract

This treatment is a compilation of all pollen stratigraphic sites in eastern Canada published to the end of 1983. Fifteen sites date to the Early Wisconsin and 292 sites are Late Wisconsin and Holocene in age. Pre-Late Wisconsin sites occur mainly in Nova Scotia, Magdalen Islands and southern Quebec except for an isolated occurrence in southwestern Newfoundland. The heaviest concentration of Late Wisconsin and Holocene sites by far is in southern Quebec. Areas clearly deficient in core sites are central Quebec between Gulf of St. Lawrence and James Bay, Nouveau-Québec, and mainland Newfoundland. On-going studies in these areas will help alleviate this deficiency.

Pre-Late Wisconsin records from southwest Newfoundland, the Magdalen Islands, Quebec, and at seven sites in Nova Scotia contain pollen assemblages that are indicative of interglacial (Sangamon) environments. Records at two other sites (Bay St. Lawrence in Nova Scotia and Pointe-Fortune in the Ottawa Valley) have been assigned interstadial status and possibly correspond with the St. Pierre Interstade of the St. Lawrence Lowlands.

Chrono-palynostratigraphic transects from the Atlantic Provinces to Newfoundland and from eastern Ontario to northernmost Nouveau-Québec and interior Labrador show that the Late Wisconsin-Holocene pollen stratigraphy is time-transgressive from south to north and from the coastal regions towards the interior. Pollen zones indicative of a mixed forest throughout much of the Holocene characterize sites in the Atlantic Provinces-eastern Ontario-south-central Quebec region. The Holocene pollen zonation of Newfoundland, Labrador and Nouveau-Québec is boreal in character whereas in northernmost Ungava herb zones imply tundra persisted throughout all of Holocene time. The Holocene pollen stratigraphy of south-central Quebec, in particular, reflects north-south movements of the forest in response to Hypsithermal and post-Hypsithermal climatic change.

INTRODUCTION

Pollen stratigraphic studies constitute an integral part in understanding the late-Quaternary stratigraphy of eastern Canada. Although the majority of the pollen studies are concerned with Late Wisconsin and Holocene vegetation and climatic history, there are well-known localities in eastern Canada where the pollen records relate to the last interglacial period. Pollen analysis of sediments which are beyond finite radiocarbon dating are essential in establishing a chronological framework for these sediments, for correlating purposes, and for providing data on the environmental conditions at the time.

The earliest and most extensive palynological analyses undertaken in eastern Canada were the surveys by Auer (1930). Later studies included those by Potzger (1953), Potzger and Courtemanche (1954a, b; 1956a, b), Ignatius (1956), Terasmae (1958, 1960), Terasmae and LaSalle (1968), and LaSalle (1966) in Quebec; Wenner (1947), Grayson (1956) and Morrison (1970) in northern Quebec and Labrador; Livingstone and Livingstone (1958), Livingstone and Estes (1967), Livingstone (1968), Ogden (1960) and Terasmae (1963) in the Atlantic Provinces and Newfoundland, and Mott and Camfield (1969) in eastern Ontario. The earlier pollen diagrams, i.e., those published prior to 1954, lack chronological control whereas those published since about 1960 generally contain at least a basal radiocarbon date.

The more recent investigations include those by Terasmae (1980) in southeastern Ontario; Terasmae and Anderson (1970), Mott (1976, 1977), Mott and Farley-Gill (1981), Mott et al. (1981), Richard (1971, 1973a, b, c, 1975, 1977, 1978, 1979, 1980), Labelle and Richard (1981a) and Savoie and Richard (1979) in southern and central Quebec; Labelle and Richard (1981b) in Gaspé; Bartley and Matthews (1969), McAndrews and Samson (1977), Richard (1981a), Richard et al. (1982), Short (1978), Short and Nichols (1977), Stravers (1981) and Lamb (1980) in Nouveau-

Québec and Labrador; and Hadden (1975), Railton (1972), Mott (1971, 1975a), Green (1976, 1981), Macpherson (1982), Mott *et al.* (1982), Anderson (1980, 1983), and Walker and Paterson (1983) in the Atlantic Provinces and Newfoundland. Some of the more recent studied sites may contain up to several radiocarbon dates but the number of dates is still low in terms of a regional synthesis of eastern Canada.

Comprehensive reports have been prepared on the history of vegetation and climate for certain regions, *i.e.*, southern Quebec by Richard (1981b) and Webb *et al.* (1983), Nouveau-Québec and Labrador by Short (1980), Stravers (1980) and Richard (1981a), and Newfoundland and Labrador by Macpherson (1981). Hills and Sangster (1980) compiled a listing of pollen and plant macrofossil sites known to 1977 supplemented by comments on paleoecology and geochronology.

This paper is a compilation of all known localities in the Atlantic Provinces (Nova Scotia, New Brunswick, Prince Edward Island), Newfoundland-Labrador, eastern Ontario and Quebec where pollen stratigraphic data is available. An attempt is made at integrating the data to develop representative pollen zonations for certain regions, to develop chrono-palynostratigrahic transects, and to infer the vegetation and climate for specific time intervals of the late-Quaternary.

The chrono-palynostratigraphy can only be as reliable as the quantity and quality of the fossil pollen data produced and of the radiocarbon dates made available. This synthesis is based on carefully selected reference pollen stratigraphies for certain areas, and the pollen records are supported by at least one, in most cases several, radiocarbon dates that are reliable and correlative with other dates in the area.

REGIONAL PHYSICAL CHARACTERISITCS

Physiography

Eastern Canada is divided into five physiographic regions as outlined in Bostock (1970). Boundaries of the regions and of smaller units within the regions relate to bedrock geology, the nature and distribution of surface materials and topographic relief (Fig. 1).

The Appalachian Region includes almost all of Newfoundland, the Atlantic Provinces and the highlands of Quebec east of the St. Lawrence. The physiography is dominated by the Cretaceous-aged peneplain which is highest in the northwest and slopes gently towards the Atlantic Ocean. Lowlands,

highlands and uplands have evolved as a result of differential erosion of the underlying bedrock. Elevations reach as high as 970 m in southwest Quebec, 1220 m in the Shickshock Mountains in the Gaspé and just over 600 and 900 m in the highlands of northwest New Brunswick and Newfoundland, respectively. Areas of lowest relief are the Newfoundland Central Lowland and Maritime Plain where elevations average 150 m to 200 m.

Inland from the Appalachian Region is the St. Lawrence Lowlands Physiographic Region. It comprises the St. Lawrence and lower Ottawa River valleys and the inner portion of the Gulf of St. Lawrence. The Region is underlain by flat-lying Paleozoic rocks giving origin to the St. Lawrence Lowland plain. The plain reaches no more than 150 m elevation except for the intrusive rocks of the Monteregian Hills.

Bordering the St. Lawrence Lowland Region on the west is the Shield which is generally believed to represent an exhumed pre-Paleozoic peneplane surface (Bostock, 1970). The easternmost physiographic region of the Shield, the Laurentian Region, is generally eastward sloping. It has a mountainous-like relief with elevations of 300 to 600 m in the southeast and up to 915 m, and at places to 1220 m in the interior. In the northeast, the region becomes deeply incised forming plateaus and uplands in the higher elevations with undulating lowlands in the intervening areas.

The largest part of mainland Quebec lies within the James Physiographic Region which extends from the Ontario-Quebec boundary to Hudson Strait and Labrador. Much of this Region slopes westward to Hudson and James Bays; in the north, drainage is northeastward into Ungava Bay and Hudson Strait. The surface terrain resembles a rolling plain with elevations ranging from 275 m to 365 m in the southwest to more than 915 m in the uplands of the interior. Elevations range from sea level to 150 m in the lowlands around James Bay.

Northeastern Quebec and northernmost Labrador fall within the southern part of the Davis Physiographic Region. This Region has the greatest relief in eastern Canada. It is characterized by deeply incised plateaus and highlands where elevations commonly reach 1065 m and occasionally 1525 m in the Torngat Mountains of northernmost Labrador. Inland, the old peneplain surface is less incised and undulating and elevations range between 300 and 600 m. Bedrock predominates; in places it has been glacially scoured into cirques and deep U-shaped valleys and fiords along the coast.

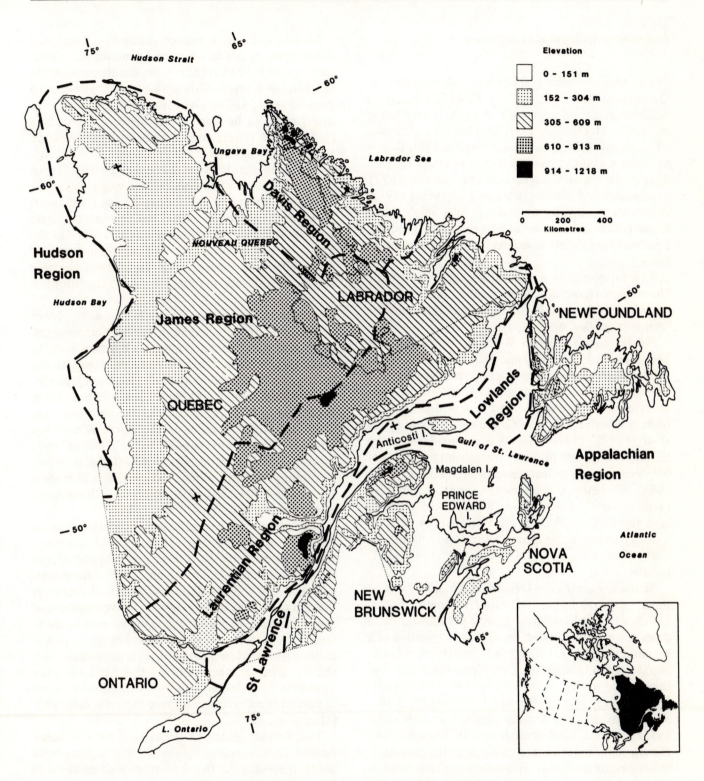

Figure 1. Physiographic regions and topographic relief of eastern Canada.

Quaternary Stratigraphy and Events

The stratigraphic framework and sequence of Quaternary events in eastern Canada have been outlined in several papers (Prest 1970, 1977; Prest *et al.,* 1972; McDonald, 1971, McDonald and Shilts, 1971; Grant, 1977; Denton and Hughes, 1981; Rogerson, 1982; and Occhietti, 1982). The entire Quaternary record is summarized in Prest (1970, 1977). McDonald and Shilts (1971) and Occhietti (1982) deal with the Wisconsin interval as a whole while Grant (1977) and Denton and Hughes (1981) concentrate more on events associated with Late Wisconsin history.

The following is a generalized account of the Quaternary record and major events. More detailed descriptions for certain time intervals follow in the section on pollen stratigraphy.

Pre-Wisconsin. The pre-Wisconsin stratigraphy is best exposed in Nova Scotia, Magdalen Islands and southeastern Quebec. The oldest Quaternary deposits in eastern Canada are thought to be the Bridgewater Conglomerates in southeast Nova Scotia and the Mabou Conglomerates on Cape Breton Island (Prest *et al.,* 1972). These deposits are considered to be Illinoian in age and probably correspond to Stage 6 of the deep-sea oxygen isotope record (Grant and King, 1984). Evidence of pre-Wisconsin stratigraphy also exists in the Appalachians in Quebec. Deeply weathered gravels thought to be pre-Wisconsin or possibly Tertiary in age occur beneath the Johnville till, an Early Wisconsin till (McDonald and Shilts, 1971, LaSalle, 1984).

Sangamonian-Early Wisconsin. Organic beds at three localities in Nova Scotia (Addington Forks, Leitches Creek, East Milford) have produced infinite dates and pollen assemblages that are suggestive of a warm interglacial interval, most likely the Sangamonian Interglacial. Fossiliferous-rich peat and organic clay at East Milford (Mott *et al.,* 1982) overlie gravelly, grey clay rubble which is probably a pre-Wisconsin till. The organic deposits in part show evidence of a hardwood forest in the area and climatic conditions at least as warm as the present.

Sangamonian-Early Wisconsin stratigraphy is probably best revealed in the Magdalen Islands, Quebec. The fossil record in a 2 m section of silt over peat over silt and sand denotes a climate that was warmer than present in the area (Prest *et al.,* 1976). The peat interval and inferred climate are tentatively correlated with Stage 5e of the deep-sea record (Grant and King, 1984).

Also in Quebec, a section along the Harricana River in the James Bay Lowland contains peat dated at >42,000 B.P. (Y-1165) in rhythmites which are correlative with the Missinaibi beds of Ontario (Stuiver *et al.,* 1963). The Missinaibi beds, orignally interpreted to be Early Wisconsin by Terasmae (1958), were re-interpreted as being Sangamonian equivalent by Skinner (1973).

The only "old" site discovered thus far in Newfoundland is a pond-sediment sequence exposed at Woody cove near Codroy, southwestern Newfoundland, where wood has produced a date >40,000 B.P. (I-10203) (Brookes *et al.,* 1982). On the basis of pollen stratigraphy and analysis of foraminifera the sequence is given interglacial status and is tentatively assigned to the Sangamonian interglacial.

Sangamonian-correlative organic beds are overlain by till or colluvium in Nova Scotia, Cape Breton Island and Newfoundland and by a diamicton in the Magdalen Islands. These overlying deposits are believed to be related to an early glaciation either in Late Sangamonian (Grant and King, 1984) or Early Wisconsin time (Prest, 1977). The equivalent interval in southeastern Quebec is the Early Wisconsin Nicolet Stade represented by the Bécancour, Johnville and Pointe St-Nicolas tills (LaSalle, 1984). This period of early widespread glaciation in eastern Canada is correlated with Stage 4 of the oxygen isotope record (Grant and King, 1984).

Inter-till or sub-till organic sequences are present at localities in Nova Scotia (Miller Creek) and Cape Breton Island (Bay St. Lawrence, Whycocomagh, River Inhabitants, Hillsborough). Peat at these sites has produced infinite dates and pollen assemblages that are more indicative of boreal-like vegetation and cooler climatic conditions than at present in these areas. Mott (1971, 1973) and Mott and Prest (1967) tentatively correlated some of these sequences with the St. Pierre Interstade of the St. Lawrence Lowlands. Correlation with the Sangamonian record at East Milford now seems more likely for these sites (Mott *et al.,* 1982).

The Bécancour till of the central St. Lawrence Lowlands is overlain by the St. Pierre peat beds which represent the best known exposures of Early Wisconsin non-glacial sediments in eastern Canada (Terasmae, 1958, Gadd, 1971). The equivalent beds in the Appalachians and Quebec City area are the Massawippi and L'Anse-aux-Hirondelles formations overlying the Johnville and Pointe St-Nicolas tills, respectively (LaSalle, 1984). The St. Pierre beds date 74,700 $^{+2700}$ (QL-198) (Stuiver *et al.,* 1978) and are characterized by a boreal-like pollen assemblage and cooler-than-present climatic conditions (Terasmae, 1958). They are correlated with the transition from oxygen isotope Stage 5a to Stage 4 and represent the onset of climatic deterioration after the previous Sangamonian interglacial (Stuiver *et al.,* 1978).

The St. Pierre peat beds in the type area of the central St. Lawrence Lowlands were buried first by glaciolacustrine sediments of Lac Deschaillons and subsequently by a single till, the Gentilly Till. Elsewhere in southeastern Quebec they are overlain by a complex of one or more tills and intervening glaciolacustrine sediments. This period of renewed glacial activity is referred to as the Gentilly Stade (LaSalle, 1984) when ice from a Maritime ice cap centered over northern Maine, the Gaspé and New Brunswick coalesced with the Laurentide ice off the Shield (Shilts, oral commun., 1983). Grant and King (1984) refer to the equivalent interval of generally widespread glaciation in the Atlantic Provinces as the Acadian Glaciation and as the Terra Novan Glaciation in Newfoundland.

Middle Wisconsin. The Gentilly Glaciation occupied the entire St. Lawrence Lowlands from approximately 75,000 to 12,000 B.P., but during this interval, the ice front retreated from the Appalachians, readvanced, and retreated again on one or more occasions. One of the interstades, the Gayhurst Interstade of McDonald and Shilts (1971), is considered to be Middle Wisconsin in age and it is tentatively correlated with the Port Talbot Interstade in the Great Lakes region (McDonald and Shilts, 1971; LaSalle, 1984).

Apparently there were also periods in Middle Wisconsin time in certain parts of the Atlantic Provinces when glaciers were possibly less widespread than at other times. Prest (1977) has alluded to the existence of non-glacial environments on Cape Breton Island from 31,800 to 32,100 B.P. A Middle Wisconsin age is also inferred for a foraminifera assemblage in a coastal section in Burin Peninsula, Newfoundland (Tucker and McCann, 1980).

Contradictory evidence in favour of an extensive glaciation over Nova Scotia at least during Middle Wisconsin time is the presence of end moraines and associated proglacial silts on the Scotian Shelf about 30 km off the Atlantic coast of Nova Scotia (King, 1969). The end moraine complex, previously interpreted as a Late Wisconsin event (King, 1969), is now believed to have been deposited by a grounded ice sheet at approximately 26,000 B.P. (Wightman, 1980).

Late Wisconsin. Late Wisconsin ice was at its maximum limit in eastern Canada for much of the time between 14,000 and 21,000 B.P. (Denton and Hughes, 1981). Prest (1983) shows two models of ice extent, a maximum model in which ice may have extended onto the continental shelf but not to the extent of previous glaciations and a minimum model of restricted ice caps. The minimum model shows Laurentide Ice covered all of Quebec except Anticosti Island and parts of Gaspé Peninsula. Appalachian Ice consisting of "satellitic ice caps" (Hillaire-Marcel *et al.,*1980) over Newfoundland, Cape Breton Island, Nova Scotia, Prince Edward Island, New Brunswick and Gaspé apparently coalesced with Laurentide Ice in southwestern New Brunswick, northern New Brunswick–Gaspé and northwest Gulf of St. Lawrence area. Parts of the coastal regions of Newfoundland, Labrador, southwestern Nova Scotia, northern New Brunswick and Gaspé, the highlands of Cape Breton Island, and the Magdalen and Anticosti Islands had remained ice free (Grant, 1977, Prest, 1983).

Retreat of Late Wisconsin Ice which deposited the correlative Gentilly, Fort Covington, St. Jacques, Lennoxville, Thedford Mines, Quebec City tills in the central St. Lawrence–Quebec City region and one or more equivalent tills in west-central Gaspé Peninsula gave rise to a glaciolacustrine phase which covered large segments of this region (LaSalle, 1984). This glacial lake phase extended up the St. Lawrence River valley to Lake Ontario and is likely diachronous with part of the Glacial Lake Iroquois–post-Iroquois glacial lake phase in the Lake Ontario basin. This is based on the conclusion that Lennoxville Ice, at least, retreated from the Appalachians by about 12,500 B.P. (McDonald and Shilts, 1971) and that Lake Iroquois formed prior to about 12,200 B.P. (Calkin, 1982) and drained to lower "post-Iroquois" levels shortly after 12,000 B.P. but before 11,000 B.P. (Karrow, *et al.,* 1975).

Retreat of Laurentide Ice out of the lower St. Lawrence–Gulf area allowed marine submergence of the isostatically depressed lowlands first by the Goldthwait Sea and later by the Champlain Sea (Hillaire-Marcel and Occhietti, 1980). Marine clay and sand are widespread throughout the Ottawa and upper St. Lawrence River valleys to Lake Ontario; they constitute the Champlain Sea Episode in the stratigraphic subdivision of LaSalle (1984). Radiocarbon dates obtained on marine shells show that Champlain Sea formed as early as 12,800 B.P. in the Ottawa Valley; it terminated shortly after 10,000 B.P. Pollen and radiocarbon evidence from sites in eastern

Ontario indicate, however, that the marine shell chronology may be too old by as much as 1,000 years.

The early stages in the deglaciation of the Atlantic Provinces and Newfoundland involved ice drawdown and formation of calving embayments that migrated rapidly inland from the offshore and coastal regions. Ice centres became isolated over Newfoundland, Nova Scotia, Cape Breton Island, New Brunswick and offshore Northumberland Strait. Several deglacial phases characterized by radially-spreading ice masses record the pattern of ice recession starting possibly as early as 14,000 B.P. (Denton and Hughes, 1981, Grant and King, 1984).

Organic sediments resting on till and overlain by a diamicton or other mineral sediment date from 13,400 B.P. in Newfoundland and from 12,500 to 11,000 B.P. in Nova Scotia; dates on gyttja above the mineral layers are as old as 10,700 B.P. in Newfoundland and 10,000 B.P. or slightly less in Nova Scotia. Pollen evidence (Mott, 1983; Anderson, 1983) indicates that climatic conditions improved gradually from the time of deglaciation to almost 11,000 B.P. followed by climatic deterioration between 10,000 and 11,000 B.P. and by dramatic warming after 10,000 B.P. Supporting evidence for climatic deterioration between 10,000 and 11,000 B.P. is a late-ice readvance dating to 10,900 B.P. in Newfoundland (Grant, 1969).

Recession of Laurentide Ice proceeded gradually northward in mainland Quebec except for a stillstand or possibly minor readvance at approximately 11,000 B.P. to form the St. Narcisse Moraine. LaSalle and Elson (1975) interpret the moraine as a climatic event and there may be some relationship between this event and the 10,000 to 11,000 B.P. climatic oscillation detected in the Atlantic region. Hillaire-Marcel and Occhietti (1980), on the other hand, maintain that the advance was unrelated to any climatic change at the time and invoke a concept of re-equilibrium of the ice sheet margin to explain the St. Narcisse advance into the Champlain Sea.

Continued northward retreat gave rise to proglacial Lake Ojibway in the James Bay Lowland. At approximately 8,300 B.P. ice out of Hudson Bay readvanced (Cochrane surges) into Glacial Lake Ojibway which drained shortly thereafter (7,900 B.P.), and marine waters of the Tyrrell Sea invaded the lowlands (Hillaire-Marcel et al., 1980; Denton and Hughes, 1981). By 7,900 to 8,000 B.P. Laurentide Ice (Nouveau-Quebec Ice) stood at the position of the Sakami moraine in west-central Nouveau-Québec (Hardy, 1982).

The Torngat Mountains of northernmost Labrador apparently were overriden in pre-Late Wisconsin time but they were nunataks during the Late Wisconsin maximum. During the waning stages of the Late Wisconsin ice, late readvances occurred at several areas in the mountains but an active ice front was maintained throughout (Prest, 1970). Rapid retreat of the Nouveau-Québec ice-dome took place after about 8,000 B.P. (Hillaire-Marcel et al., 1980). The final remnants of Nouveau-Québec ice dissipated between 5,500 and 6,500 B.P. (Richard et al., 1982). Ice persisted in western Labrador until about 6,000 B.P. according to Rogerson (1982).

Climate

The present-day climatic details of eastern Canada are derived from several sources, Hare and Thomas (1974) and Putnam (1940) on the Atlantic Provinces, Banfield (1981) for Newfoundland, and Wilson (1971, 1973) and Houde (1978) on Quebec. Bryson (1966), Bryson and Hare (1974) and Wendland and Bryson (1981) deal with climatic data on a regional scale showing the relationship of eastern Canada in the North American and northern hemisphere pattern of airstream circulation.

Hare and Thomas (1974) divide eastern Canada into four climatic regions, Atlantic Canada, Great Lakes–St. Lawrence, Boreal, and Arctic. The boundaries are defined on the basis of airstreams and confluences between airstreams and on physiography, relief, vegetation and distance from the open sea.

Eastern Canada is affected more by the continental winds, air masses and weather systems moving generally eastward and northward than by maritime air masses even though a large part of the region is close to the sea. It is influenced by four main air masses, a southern airstream, a western (Pacific) system and trajectories off the Arctic and Atlantic Oceans (Bryson and Hare, 1974). The main storm tract is north or south of the Great Lakes down the St. Lawrence and eventually northeast via the Strait of Belle Isle. Frontal storms characterize the Pacific and Arctic airstreams in winter bringing alternating cold and warm weather with accompanying precipitation and high westerly winds. The southerly airstream often brings heavy precipitation if it is not lost over the Appalachians. The Atlantic airmasses can be intense and produce moderate or heavy rainfall along the St. Lawrence and heavy falls and intense fog along the seaboard. The open water of the Atlantic Ocean has a moderating effect in the Atlantic

Provinces and Newfoundland. Easterly winds off the Atlantic in winter often allow high temperatures and high humidities to penetrate inland, whereas in spring and summer, onshore winds often bring about a cooling effect along the coast.

Figures 2 and 3 show temperature variations as well as the effects of physiography, relief, and distance from the open ocean. Mean July temperatures range from a maximum of 20°C in southeastern Quebec and eastern Ontario to a minimum of 7.5°C in northern Nouveau-Québec and Labrador. The cooling effect of the Atlantic Ocean is indicated by lower mean July temperatures along the outer coasts of Nova Scotia and Newfoundland and higher mean July temperatures in inland Newfoundland, west-central Nova Scotia, New Brunswick and Prince Edward Island. Mean annual temperatures range from a high of 7.0°C in southwestern Nova Scotia to a low of −7.5°C in northern Nouveau-Québec. Both mean annual and monthly temperatures do not vary more than 1°C in the more protected parts of the Gulf of St. Lawrence (Prince Edward Island) on account of being near to the sea but not directly exposed to the open ocean (Putnam, 1940).

Average annual days with precipitation and mean annual precipitation (Fig. 4) are highest (>200 days and >140 cm) in eastern Nova Scotia and Newfoundland and lowest (<120 days and <40 cm) in northernmost Nouveau-Québec. Most of the precipitation occurs as rainfall along the seaboard, but in the highlands of east-central Quebec it falls as snow and as much as 400 cm can occur annually (Hare and Thomas, 1974).

The frost-free period decreases inland and with increasing elevation (Canada Department of Energy, Mines and Resources, 1969). It varies from a high of 140 days in the central St. Lawrence to a low of <60 days in central Quebec giving a growing season of about 200 days maximum in southwestern Nova Scotia and about 140 days minimum in northern Nouveau-Québec.

Present-day Flora and Vegetation

The vegetation subdivision of eastern Canada (Fig. 5) basically follows that in Rowe (1977). The southernmost ecoregion in a north-south transect across eastern Canada is the Great Lakes-St. Lawrence Forest Region. This forest region is confined to the lower Ottawa River valley and central St. Lawrence Lowlands and is characterized by a mixed forest dominated by sugar maple *(Acer saccharum)* and beech *(Fagus grandifolia)*, with oak *(Quercus* spp), hemlock *(Tsuga canadensis)* and white pine *(Pinus strobus)*. Local occurrences include white oak *(Quercus alba)*, red ash *(Fraxinus pennsylvanica)*, grey birch *(Betula populifolia)*, rock elm *(Ulmus thomasii)*, blue-beech *(Carpinus caroliniana)* and bitternut hickory *(Carya cordiformis)*. Butternut *(Juglans cincera)*, eastern cottonwood *(Populus deltoides)* and slippery elm *(Ulmus rubra)* are sporadically distributed in river valleys and black ash *(Fraxinus nigra)* and eastern white cedar *(Thuja occidentalis)* in poorly drained depressions. This formation passes northward into a forest dominated by sugar maple, yellow birch *(Betula lutea)* and white pine, the northernmost forest of Rowe's Great Lakes-St. Lawrence Forest Region. The hardwoods tend to segregate out on the finer-textured soils of south-facing slopes; the conifers, particularly black spruce, eastern white cedar and tamarack *(Larix laricina)*, accompanied by black ash, prevail in poorly-drained uplands and lowland bog depressions.

Variations of these formations characterize the Acadian Forest Region of Rowe (1977) and the forest classification of the Atlantic Provinces by Loucks (1962). The coastal forest of eastern Nova Scotia and highlands of Cape Breton Island consist mainly of fir *(Abies balsamea)*, white and black spruce, white birch *(Betula papyrifera)* and yellow birch. Red maple *(Acer rubrum)* and white birch are the most common hardwoods although beech and sugar maple occur on the higher hills. Red spruce *(Picea rubens)* is absent along the Atlantic shore but is common along the Fundy Bay coast. In the interior of Nova Scotia and throughout the Maritime Plain of central New Brunswick and Prince Edward Island the mixed forest is dominated by the spruces (white, red and black), balsam fir, hemlock, red and white pine, red and sugar maple, and white and yellow birch. The hardwoods beech, sugar maple and yellow birch and the conifers red spruce, hemlock and white pine are typical present-day colonizers of the well-drained uplands. Red and white spruce are conspicuous along streams, fence rows and on abandoned farmlands in Prince Edward Island.

The next major forest region north of the Great Lakes-St. Lawrence Forest Region is the Boreal Forest. The southernmost formation which characterizes south-central Quebec consists of balsam fir, black and white spruce and white pine often in combination with birch. In addition to pure stands of balsam fir and black spruce, mixed stands of fir-spruce, often with eastern white cedar, are common. White pine and white and yellow birch decrease in frequency

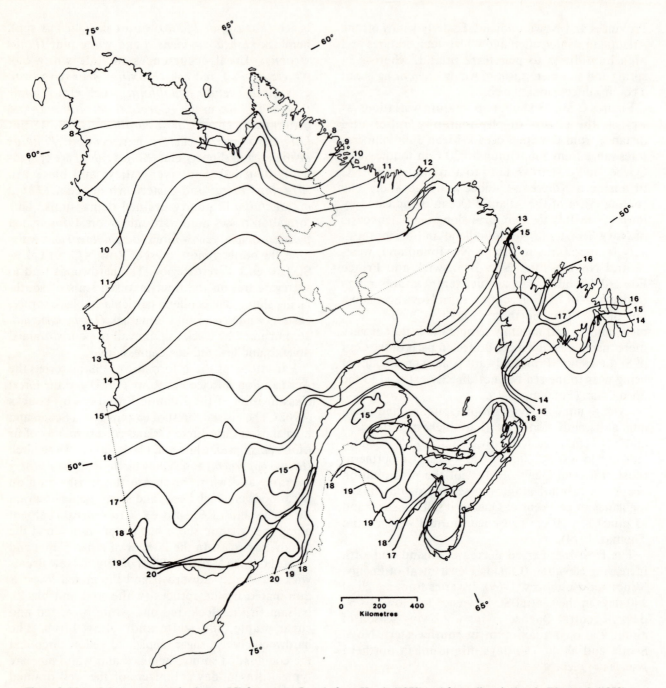

Figure 2. Mean July temperature isotherms (°C) for eastern Canada from Houde (1978) and from climatic data in Wernstedt (1972).

northwards. Outliers of this Boreal Forest formation occur in northern New Brunswick, Gaspé and Anticosti Island.

The main part of the Boreal Forest Region is a productive forest belt which extends throughout much of central Quebec east to southern Labrador. Black spruce is the most abundant species forming closed spruce stands in the peaty lowlands and on well-drained upland sites. White birch, trembling aspen *(Populas tremuloides)* and balsam poplar

(Populus balsamifera) are less conspicuous except in the immediate vicinity of rivers and lake shores.

The black spruce forest passes northward into a woodland and barrens region of lakes, rivers, bogs and muskeg with areas of upland heath barrens, lichens and open park-like woodland interspersed with black spruce. Closed spruce stands give way to isolated spruce groves in the wet lowlands. Jack pine *(Pinus banksiana)*, trembling aspen and white birch are uncommon except along the southern boundary

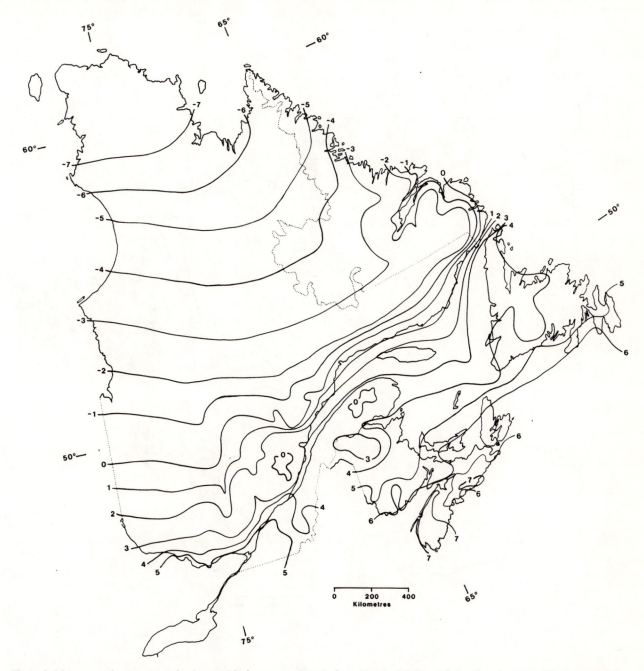

Figure 3. Mean annual temperature isotherms (°C) for eastern Canada from Houde (1978) and from climatic data in Wernstedt (1972).

and balsam poplar occurs infrequently along streams. Sparsely forested heath-and-moss barrens in southeast Labrador and south-central and western Newfoundland represent the easternmost extension of the Boreal Forest Region. The vegetation is a stunted, open and patchy or sometimes continuous cover of black spruce and balsam fir alternating with moss and heath barrens.

The woodland and barrens region passes northward into a transitional zone separating the park-like woodland in the south from tundra to the north. The primary plants of this forest-tundra ecotone are lichens *(Cladonia* spp), shrub birch *(Betula glandulosa),* shrub willow *(Salix cordifolia),* black spruce and tamarack. Balsam fir, white birch, trembling aspen and balsam poplar are infrequent.

Tundra vegetation is widespread in northern Nouveau-Québec and Labrador and characterizes a narrow strip along the coast of Labrador southeastward to the northern tip of western Newfoundland.

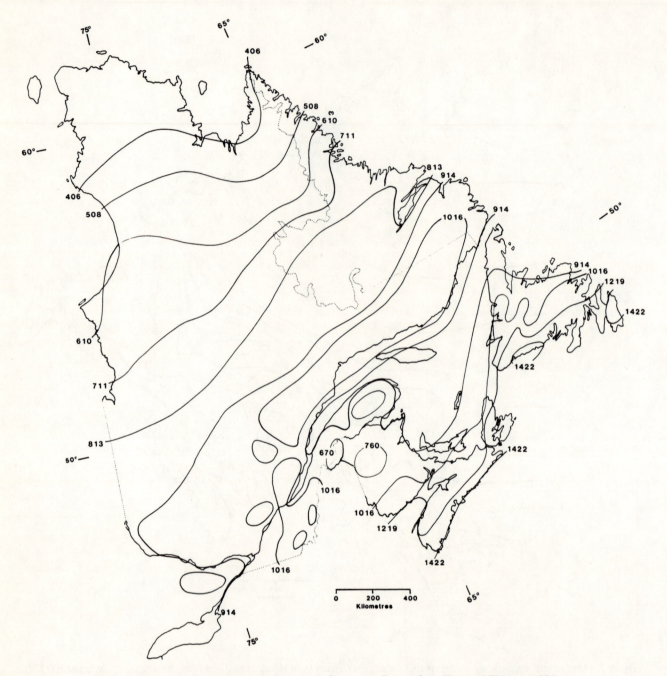

Figure 4. Mean annual precipitation isohyets (mm) for eastern Canada from Hare and Thomas (1974).

Shrub birch and willow, lichens, sedges (Cyperaceae) and heath plants (Ericaceae) predominate. Shrub birch and willow are less common in the narrow, lichen-heath physiognomic region along Hudson Strait. Here lowland heathy communities are mainly restricted to occasional small areas that offer unusually favourable conditions except for small late-snow areas of *Cassiope*-or *Salix herbacea*-heath (Polunin, 1948).

THE DATA AND METHODOLOGY

The Data Base

Pollen diagrams of all known lake, peat bog and section sites in eastern Canada published to the end of 1983 were compiled (Table 1) and the site locations illustrated (Fig. 6). The western boundary of

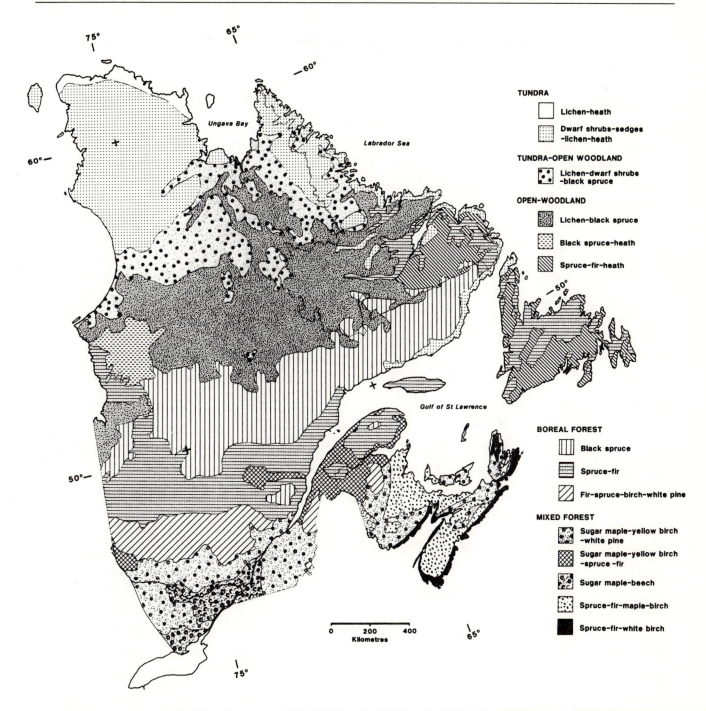

Figure 5. Vegetation regions of eastern Canada modified from National Atlas of Canada, Canada Department of Energy, Mines and Resources (1973). Modifications in northwest Noveau-Québec are based on recent studies by Payette (1983).

the compilation lies along a line from the Ganonoque area in eastern Ontario northwest to Pembroke on the Ottawa River and hence to James Bay (Fig. 6). About 300 pollen stratigraphic sites now exist in eastern Canada. The unpublished diagram for Lake Irene (Ignatius, 1956) is also included in this compilation as this is the only radiocarbon-dated site from mid-south-central Quebec between Gaspé and James

Bay. Several investigators have offered unpublished data for certain areas. Figure 6 also shows those areas where current studies are in progress by various investigators. Current studies include sites which have been analyzed or are presently being investigated but are not yet published, as well as sites for which core and sample data are available but which are not yet analyzed.

TABLE 1 Locational information and radiocarbon data of pollen stratigraphic sites in eastern Canada. Latitude and longitude are shown to the nearest minute.

SITE NUMBER AND NAME	LATITUDE (N)	LONGITUDE (W)	ALTITUDE (M)	SITE TYPE LAKE = L BOG = B SECTION = S	NUMBER OF RADIOCARBON DATES	OLDEST RADIOCARBON DATE (YR B.P.)	POLLEN DIAGRAM (THIS PAPER)	REFERENCE(S)
1. Perch	46°02'	77°32'	145	L	1	9830 ± 250		Terasmae (1980)
2. Dows Lake	45°23'	75°42'	61	B				Mott and Camfield (1969)
3. Site 3	45°24'	75°42'	61	S	1	8830 ± 190		Mott and Camfield (1969)
4. Site 4	45°25'	75°42'	68.6	S	2	7870 ± 160		Mott and Camfield (1969)
5. Site 5	45°26'	75°39'	61	S				Mott and Camfield (1969)
6. Site 6	45°21'	75°48'	68.6	S	1	8220 ± 150		Mott and Camfield (1969)
7. Mer Bleue	45°24'	75°30'	69	B	1	7650 ± 210		Mott and Camfield (1969)
8. Alfred	45°30'	74°48'		B				Potzger and Courtemanche (1956a); Auer (1930)
9. Northfield (Newington)	45°08'	74°56'	99	B	1	9430 ± 140		Terasmae (1965); Auer (1930)
10. Waterton	44°25'	75°58'	88.2	B	3	10,500 ± 140	★	Anderson (this paper)
11. Lambs Pond	44°39'	75°48'	113.6	L	4	12,300 ± 230	★	Anderson (this paper)
12. Atkins	44°45'	75°51'	116	L	2	11,100 ± 270		Terasmae (1980)
13. Pink	45°28'	75°49'	162	L	3	10,600 ± 150		Mott and Farley-Gill (1981)
14. Ramsay	45°36'	76°06'	200	L	5	10,800 ± 180	★	Mott and Farley-Gill (1981)
15. Pointe-Fortune	45°32'	74°23'	45	S	2	>40,000	☆	Gadd et al. (1981)
16. Jack	51°59'	78°04'		B				Potzger and Courtemanche (1956a)
17. Smoky Hill Falls	51°27'	78°32'		B	1			Potzger and Courtemanche (1965a; 1954a)
18. Smoky Hill Rapids	51°28'	78°45'		B	1	2350 ± 200		Potzger and Courtemanche (1954b)
19. Rupert House	51°28'	78°45'		B				Potzger and Courtemanche (1956a)
20. Lac Horden	50°54'	77°54'		B				Potzger and Courtemanche (1956a)
21. Iroquois Falls	50°39'	78°02'		B				Potzger and Courtemanche (1956a)
22. Lac Soscumica	50°24'	77°46'		B				Potzger and Courtemanche (1956a)
23. Bachelor Lake	49°32'	76°08'		B				Potzger and Courtemanche (1956a)
24. Lac Lacroix	49°02'	75°23'		B				Potzger and Courtemanche (1956a)
25. Val St. Gilles	49°01'	79°05'	289.5	S	2	6460 ± 140		Terasmae and Anderson (1970)
26. Clo	48°29'	79°21'	280	L	3	8310 ± 80		Richard (1980)
27. Clova	48°07'	75°22'		B				Potzger and Courtemanche (1956a)
28. Louis	47°17'	79°07'	300	L	3	9090 ± 240		Vincent (1973)
29. Lac des Loups	47°02'	76°50'		B				Potzger (1953)
30. Lac Caupal	46°46'	76°00'		B				Potzger (1953)
31. Kazabazua	45°57'	76°04'		B				Potzger and Courtemanche (1956b)
32. Lacroix	46°16'	76°00'		B				Potzger and Courtemanche (1956b)
33. Hobblety Creek	46°11'	76°21'		B				Potzger and Courtemanche (1956b)

TABLE 1 (continued)

SITE NUMBER AND NAME	LATITUDE (N)	LONGITUDE (W)	ALTITUDE (M)	SITE TYPE LAKE = L BOG = B SECTION = S	NUMBER OF RADIOCARBON DATES	OLDEST RADIOCARBON DATE (YR B.P.)	POLLEN DIAGRAM (THIS PAPER)	REFERENCE(S)
34. Cleaver Lake	46°13'	76°27'		B				Potzger and Courteinanche (1956b)
35. Brock Lake	46°17'	76°21'		B				Potzger and Courteinanche (1956b)
36. Lac Rouge	46°56'	74°40'		B				Potzger and Courteinanche (1956a)
37. Lac Mazanaskwa	47°07'	74°32'		B				Potzger and Courteinanche (1956a)
38. Creek Savanne	46°26'	74°12'		B				Potzger and Courteinanche (1956a)
39. Aux Quenoilles	46°10'	74°24'	403	L	3	10,820 ± 160		Savoie et Richard (1979); Webb et al. (1983)
40. Bellerive	46°10'	74°59'		B				Potzger 1953
41. Nominingue	46°10'	75°10'		B				Potzger 1953
42. Mont Tremblant Park	46°30'	74°30'		B				Potzger and Courteinanche (1954b)
43. Lac Shaw	46°19'	74°32'		B				Potzger and Courteinanche (1956a)
44. Mont Tremblant	46°15'	74°34'		B				Potzger and Courteinanche (1956a)
45. Lac à Pit	46°11'	74°29'		B				Potzger and Courteinanche (1956a)
46. Lac des Plages	46°00'	74°51'		B				Potzger and Courteinanche (1956a)
47. St. Agathe	46°03'	74°28'	454	L	3	10,170 ± 530		Savoie et Richard (1979) Webb et al. (1983)
48. Borne	46°00'	74°22'	425	L	7	8620 ± 165		Richard (1977)
49. St. Germain	45°56'	74°22'	473	L	6	10,420 ± 430		Savoie et Richard (1979)
50. St.-Michel	45°30'	74°30'		B				Potzger (1953)
51. Nantel	45°52'	74°20'		B				Potzger (1953)
52. St.-Lin	45°55'	73°47'		B				Potzger and Courteinanche (1956a)
53. Ste.-Agathe	45°50'	74°10'		B				Potzger (1953)
54. St. Janvier	45°20'	73°50'		B				Potzger (1953)
55. Large Tea Field	45°07'	74°15'		B				Auer (1930)
56. Hertel	45°33'	73°09'	168	L	1	10,880 ± 260		LaSalle (1966) Terasmae and LaSalle (1968)
57. St. Hilaire	45°33'	73°09'	259	B	1	12,570 ± 220		LaSalle (1966) Terasmae and LaSalle (1968)
58. St. Hilaire	45°31'	73°08'	43	S	3	11,000 ± 100		Mott, et al. (1981)
59. St. Bruno	45°32'	73°18'	130	B				LaSalle (1966)
60. Ste. Victoire	45°45'	73°00'		B				Potzger (1953)
61. St. Bonaventure	45°58'	72°20'	50	B				Terasmae (1960)
62. Shefford Farnham	45°21' 45°15'	72°35' 73°00'	282	B B	10	11,400 ± 340	★	Richard (1978) Potzger (1953)
63. Waterloo	45°20'	72°30'	208	L				Ouellet et Poulin (1976)
64. Barnston	45°07'	71°53'	415	L	1	11,020 ± 330		Mott (1977)
65. St. Germain	45°50'	72°35'	91	B	2	9550 ± 600		Terasmae (1960)
66. Birch	45°47'	72°34'	79	B				Terasmae (1960)
67. St. Eugène	45°50'	72°36'	65	B				Terasmae (1960)

TABLE 1 (continued)

SITE NUMBER AND NAME	LATITUDE (N)	LONGITUDE (W)	ALTITUDE (M)	SITE TYPE LAKE = L BOG = B SECTION = S	NUMBER OF RADIOCARBON DATES	OLDEST RADIOCARBON DATE (YR B.P.)	POLLEN DIAGRAM (THIS PAPER)	REFERENCE(S)
68. St. Antoine	45°48'	73°18'	21	S				LaSalle (1966)
69. St. Henri	45°59'	73°18'	18	B	6	5960 ± 130		Comtois (1982)
70. St. Joseph	45°59'	73°18'	18	B	3	4790 ± 140		Comtois (1982)
71. Coteau-Jaune	45°57'	73°20'	18	B	5	6490 ± 110		Comtois (1982)
72. Romer	45°57'	73°19'	18	L	4	6920 ± 80		Comtois (1982)
73. St. Jean	46°00'	73°13'	20	B	4	4730 ± 95		Comtois (1982)
Lanoraie	46°00'	73°13'		B				Potzger (1953)
74. Pierreville a.	46°05'	72°50'	15	S	1	>29,630		Terasmae (1958)
b.	46°05'	72°50'		S				Terasamae (1960)
Section 11 c.	46°05'	72°45'		S				Terasmae (1960)
75. Gabriel	46°16'	73°28'	250	L	5	9105 ± 175		Richard (1977)
76. Baie des Onze Îles	46°44'	73°07'	395	B				Richard (1977)
77. Wapizagonke	46°43'	73°01'	230	B	2	9730 ± 140		Richard (1977)
78. Dauphinais (Mauricie)	46°48'	72°59'	405	B				Richard (1977) Webb et al. (1983)
79. Sud du Lac du Noyer	46°47'	72°50'	270	L	5	9670 ± 190		Richard (1977)
80. Les Vielles Forges	46°22'	72°40'		S	2	>30,840	☆	Terasmae (1958; 1960)
81. Ste. Monique-de-Nicolet	46°08'	72°30'		S				Terasmae (1958)
82. Ste. Brigitte	46°01'	72°28'		S				Terasmae (1958)
83. Albion	45°40	71°19°	320	L	4	10,880 ± 160		Richard (1975a;1977)
84. Weedon	45°41'	71°25'	245	B				Richard (1975a;1977)
85. Princeville	46°08'	71°56'	135	L				Richard (1975a;1977)
86. St. Pierre	46°30'	72°10'	3	S	3	>40,000		Terasmae (1958)
87. a) St. Paulin	47°25'	73°01'	152	B				Terasmae (1960)
b) Patterson Lake	46°30'	73°00'	183	B				Terasmae (1960)
88. a) St. Boniface	46°30'	72°25'	106	B				Terasmae (1960)
b) St. Etienne	46°26'	72°23'	88	B				Terasmae (1960)
c) Marchand	46°25'	72°21'	61	B				Terasmae (1960)
89. Grondines	46°37'	72°04'	33	B				Terasmae (1960)
90. St. Alban	46°39'	72°03'	67	B				Terasmae (1960)
91. St. Adelphe	46°43'	72°23'	128	B				Terasmae (1960)
92. St. Raymond	46°53'	71°48'	160	B	2	7970 ± 140		Richard (1973a;1977)
93. Lotbinière	46°36'	71°46'	70	B				Richard (1975a;1977)
94. Boundary Pond	45°35'	70°41'	603	L	3	11,200 ± 200		Mott (1977)
95. Dufresne	45°51'	70°21'	650	L	2	11,200 ± 160		Mott (1977)
96. St.-Benjamin	46°16'	70°35'	330	L	1	9100 ± 150		Richard (1973c;1977)
97. Colin	46°43'	70°18'	658	L	8	11,100 ± 180	★	Mott (1977)
98. Petite Lac Terrien	46°35'	70°37'	404	L	1	12,640 ± 190		Mott (1977)
99. Dosquet	46°27'	71°30'	140	B	1	8835 ± 145		Terasmae (1960) Richard (1973c;1977)
100. Clair	47°00'	71°00'		B				Auer (1930)
101. St. Jean, Île D'Orléans	46°56'	70°56'	68	B	3	6100 ± 160		Richard (1971;1977)
102. Sagamité	47°15'	71°07'		B				Auer (1930)
103. Marcotte	47°04'	71°25'	503	L	5	9885 ± 230		Labelle et Richard (1981a)
104. Joncas	47°15'	71°09'	747	B	2	7140 ± 130		Richard (1971;1977)

TABLE 1 (continued)

SITE NUMBER AND NAME	LATITUDE (N)	LONGITUDE (W)	ALTITUDE (M)	SITE TYPE LAKE = L BOG = B SECTION = S	NUMBER OF RADIOCARBON DATES	OLDEST RADIOCARBON DATE (YR B.P.)	POLLEN DIAGRAM (THIS PAPER)	REFERENCE(S)
105. À l'Ange	47°28'	70°41'	640	L	7	10,710 ± 215		Labelle et Richard (1981a)
106. St. Eugene	47°04'	70°19'	144	S		11,050 ± 130		Mott, et al. (1981)
107. Mimi	47°29'	70°22'	411	L	6	11,050 ± 460		Richard (1977) Richard et Poulin (1976)
108. À la Fourche	47°36'	70°36'	305	L	1	9490 ± 230		Labelle et Richard (1981a)
109. Malbaie	47°35'	70°58'	800	B	2	8095 ± 155		Richard (1977)
110. Montagnais	47°54'	71°10'	800	B	1	8510 ± 140		Richard (1973b;1977)
111. Caribou	47°38'	71°14'	790	B	2	5145 ± 105		Richard (1977)
112. Kénogami	48°21'	71°34'	166	B	5	7630 ± 120		Richard (1973b;1977) LaSalle (1966) Radforth (1945)
113. Shipshaw	48°25'	71°15'		S				
114. Rivière Eternité I, II	48°09'	70°20'		B				Potzger (1953)
115. St.-Siméon	47°47'	69°49'		B				Potzger (1953)
116. R. Ouelle	47°30'	70°00'		B				Auer (1930)
117. R. du Loup	47°40'	69°25'		B				Potzger (1953)
118. St. Fabien	48°10'	68°45'		B				Potzger (1953)
119. Amqui I, II	48°19'	67°30'		B				Potzger (1953)
120. St. Omer	48°00'	66°20'		B				Potzger (1953)
121. Harriman	48°14'	65°50'	71	L				Livingstone (1968)
122. Côté	48°58'	65°57'	961	L	1	9810 ± 360		Labelle and Richard (1981b)
123. Turcotte	·49°09'	65°45'	456	L	2	10,360 ± 170		Labelle and Richard (1981b)
124. À Léonard	49°12'	65°48'	17	L	5	8970 ± 140		Labelle and Richard (1981b)
125. Sept-Îles("LD")	50°08'	67°07'	122	L	2	6960 ± 300	★	Mott (1976)
126. Matamek River	50°24'	65°47'		B				Bowman (1931)
127. Grand Falls	47°30'	67°44'	122	B	1	9830 ± 160		Terasmae (1973) Mott (1975b)
128. Upper Kent	46°36'	67°39'	152	B				Mott (1975b)
129. Hartland	46°20'	67°38'	122	B				Mott (1975b) Tersmae (1973)
130. Fredericton	45°56'	66°41'	110	B				Terasmae (1973)
131. Basswood Road	45°15'	67°19'	106	L	6	12,600 ± 270	★	Mott (1975a,b)
132. Little	45°08'	66°43'	64	L	4	16,500 ± 350		Mott (1975a,b)
133. Spruce Lake	45°20'	65°45'		B				Osvald (1970)
134. Hicks	46°15'	65°00'		B				Auer (1930)
135. Escuminac	47°00'	64°52'		B				Auer (1930)
136. Wood's Pond	45°54'	64°22'	25	L	1	9930 ± 350		Walker and Paterson (1983)
137. Portey Pond	45°51'	64°25'	45	L				Walker and Paterson (1983)
138. Portage	46°40'	64°04'	8	B	3	9880 ± 150		Anderson (1980)
139. East Bideford	46°38'	63°54'	8	B	6	8070 ± 80		Anderson (1980)
140. Mermaid Lake	46°15'	63°01'	15	B	1	8630 ± 180		Anderson (1980)
141. MacLaughlin	46°22'	62°50'	24	L	5	9670 ± 130	★	Anderson (this paper)
142. East Baltic	46°24'	62°09'	45	B	3	8430 ± 150		Anderson (1980)

TABLE 1 (continued)

SITE NUMBER AND NAME	LATITUDE (N)	LONGITUDE (W)	ALTITUDE (M)	SITE TYPE LAKE = L BOG = B SECTION = S	NUMBER OF RADIOCARBON DATES	OLDEST RADIOCARBON DATE (YR B.P.)	POLLEN DIAGRAM (THIS PAPER)	REFERENCE(S)
143. Portage-du-Cap	47°14'	61°54'	13	S	1	>35,000		Prest, et al. (1976)
144. Iris Station	45°56'	62°39'		B	1	6600 ± 270		Frankel and Crowl (1961)
145. Nicholas Point	45°54'	62°48'		S	1	915 ± 90		Frankel and Crowl (1961)
146. Bay St. Lawrence	47°01'	60°27'	10	S	1	>38,300	☆	Mott and Prest (1967) de Vernal, Richard and Occhietti (1983)
147. Wreck Cove	46°32'	60°26'		L	1	9030 ± 170		Livingstone and Estes (1967)
148. Whycocomagh	45°58'	61°07'		S	1	>44,000		Mott and Prest (1967)
149. Hillsborough	46°04'	61°22'	3	S	1	>51,000	☆	Livingstone (1968) Mott and Prest (1967)
150. Port Hood Section I	46°01'	61°34'	45	S	1	11,000 ± 170		Terasmae (1974)
151. Port Hood Section 2	46°01'	61°34'	45	S	1	7140 ± 140		Terasmae (1974)
152. McDougal	46°03'	60°25'	152	L				Livingstone (1968)
153. Leitches Creek	46°09'	60°23'		Drill Core	1	>52,000		Mott et al. (1982) Mott (1973)
154. Salmon River	45°38'	60°46'		L	2	8770 ± 150		Livingstone (1968)
155. Gillis	45°39'	60°46'		L	1	10,240 ± 220		Livingstone and Livingstone (1958)
156. River Inhabitants	45°41'	61°20'		S	2	>49,000		Mott (1971); Mott et al. (1982)
157. Glasgow Head	45°20'	61°10'		B				Osvald (1970)
158. Addington Forks	45°34'	62°06'		S	2	>42,000.		MacNeill (1969); Prest (1977); Mott et al. (1982)
159. Mulgrave	45°30'	61°22'		B				Auer (1930)
160. Folly	45°33'	63°33'		B				Livingstone (1968)
161. East Milford	45°00'	63°25'	30	S	1	>50,000	☆	Mott et al. (1982)
162. Shaws	45°01'	64°11'	30	B	4	9180 ± 255		Hadden (1975)
163. Caribou	45°02'	64°46'		B				Auer (1930); Ogden (1960)
164. Silver	44°33'	63°38'	69	L	3	9650 ± 150	★	Livingstone (1968)
165. Bluff	44°33'	63°38'	69	L				Livingstone (1968)
166. Miller Creek	45°30'	64°04'		S		>52,000		MacNeill (1969); Stea and Hemsworth (1979); Mott et al. (1982)
167. Cherryfield	44°20'	64°45'		B				Auer (1930)
168. Makoke	44°00'	66°00'		B				Auer (1930)
169. Fusket	44°00'	66°00'		B				Auer (1930)
170. Everitt	44°27'	65°52'		L				Green (1981)
171. Heath	43°50'	65°45'		B				Auer (1930)
172. Salmon River	44°10'	66°00'		B				Auer (1930)
173. Sable Island	43°58'	59°56'		Drill Core	5	10,900 ± 160		Terasmae and and Mott (1971)
174. Woody Cove	47°51'	59°22'		S	2	>40,000	☆	Brookes, et al. (1982)
175. Burin Peninsula	46°55'	55°36'	114	L	2	13,400 ± 140	★	Anderson (1983)
176. Whitbourne	47°25'	53°32'		B	1	8420 ± 300		Terasmae (1963)
177. Hawke Hills (Kettle) Hawke Hills (Col Gully)	47°20' 47°20'	53°08' 53°08'	220 190	L L	3	7290 ± 150		Macpherson (1982) Macpherson (1982)
178. Goulds	47°29'	52°46'		B	1	7400 ± 150		Terasmae (1963)

TABLE 1 (continued)

SITE NUMBER AND NAME	LATITUDE (N)	LONGITUDE (W)	ALTITUDE (M)	SITE TYPE LAKE = L BOG = B SECTION = S	NUMBER OF RADIOCARBON DATES	OLDEST RADIOCARBON DATE (YR B.P.)	POLLEN DIAGRAM (THIS PAPER)	REFERENCE(S)
179. Sugarloaf	47°37'	52°40'	100	L	2	9270 ± 150	★	Macpherson (1982)
180. Clarenville	48°02'	53°48'		B	1	3610 ± 100		Terasmae (1963)
181. L'Anse aux Meadows	51°36'	55°32'	3.4	S	3	2150 ± 60		Mott (1975c)
182. St. Anthony	51°22'	55°38'	30	B				Wenner (1947)
183. Muskrat Falls	53°10'	60°50'		L				Wenner (1947)
184. Hopedale	50°25'	60°20'	63	B				Wenner (1947)
185. Whitney's Gulch	51°31'	57°18'	98	L	6	9820 ± 110		Lamb (1980)
186. Paradise	53°03'	57°45'	180	L	3	9810 ± 120		Lamb (1980)
187. Eagle	53°14'	58°33'	400	L	5	10,550 ± 290	★	Lamb (1980)
188. Cartwright	53°40'	57°00'	4	B				Wenner (1947)
189. South Stag Island	54°07'	57°12'	14	B				Wenner (1947)
190. Indian Harbour	54°28'	57°15'	6	B				Wenner (1947)
191. Holton Island	54°37'	57°17'	16	B				Wenner (1947)
192. Cape Harrison	54°52'	58°01'	40	B				Wenner (1947)
193. Aliuk Pond	54°34'	57°22'	25	L	2	7170 ± 180		Jordon (1975)
194. Sand Cove Pond	54°24'	57°43'	100	L	4	4555 ± 145		Jordon (1975)
195. Rigolet	54°07'	58°20'	30	B				Wenner (1947)
196. St. John Pond	53°57'	58°55'	137	L	2	10,240 ± 1240		Jordon (1975)
197. North West River Pond	53°51'	60°10'	29	L	1	4805 ± 53		Jordon (1975)
198. Suzan River	53°40'	61°03'	73	B				Wenner (1947)
199. North West River 2	53°30'	60°15'	90	B				Wenner (1947)
200. North West River 3	53°30'	60°15'	4	B				Wenner (1947)
201. Alexander	53°20'	60°35'	143	L	3	5985 ± 140		Jordon (1975)
202. Landing	53°37'	63°22'	512	L				Morrison (1970)
203. Pole	53°30'	63°20'	439	L				Morrison (1970)
204. Sona West	53°35'	63°57'	429	B	1	5575 ± 250		Morrison (1970)
205. Churchill Falls North	53°36'	64°19'	398	B	1	5255 ± 200		Morrison (1970)
206. Churchill Falls South	53°35'	64°18'	398	B	1	5450 ± 200	★	Morrison (1970)
207. Churchill Falls East	53°36'	63°18'	384	B				Morrison (1970)
208. Grand Falls	53°35'	64°30'	500	B				Wenner (1947)
209. Esker	53°52'	66°25'	491	B				Morrison (1970)
210. Daumont	54°53'	69°24'	607	L	4	5490 ± 80		Richard et al. (1982)
211. Delorme 1	54°25'	69°55'	513	L	4	5330 ± 120		Richard et al. (1982)
212. Delorme 2	54°25'	69°55'	538	L	5	6320 ± 180	★	Richard et al. (1982)
213. Brisay 2	54°21'	70°21'	595	L	3	5980 ± 240		Richard et al. (1982)
214a. Lac Romanet	56°23'	67°51'	305	B				Terasmae et al. (1966)
214b. Track	55°46'	65°10'	442	L	7	4755 ± 85		Short and Nichols (1977)
215. Kogaluk Plateau	56°04'	63°45'	535	L	5	8610 ± 925		Short and Nichols (1977)
216. Second Rapid's Lake	54°45'	60°07'	66	B				Wenner (1947)
217. Kaipokok River	54°45'	60°07'	5	B				Wenner (1947)
218. Cape Aillik	55°10'	59°17'	24	B				Wenner (1947)

TABLE 1 (continued)

SITE NUMBER AND NAME	LATITUDE (N)	LONGITUDE (W)	ALTITUDE (M)	SITE TYPE LAKE = L BOG = B SECTION = S	NUMBER OF RADIOCARBON DATES	OLDEST RADIOCARBON DATE (YR B.P.)	POLLEN DIAGRAM (THIS PAPER)	REFERENCE(S)
219. Cape Aillik	55°10'	59°17'	17	B				Wenner (1947)
220. Cape Aillik	55°10'	59°17'		B				Wenner (1947)
221. Hopedale Pond	55°28'	60°17'	76	L	5	5440 ± 150		Short and Nichols (1977)
222. Windy Tickle	55°45'	60°22'	42	B				Wenner (1947)
223. Windy Tickle	55°45'	60°22'	83	B				Wenner (1947)
224. Davis Inlet	55°52'	60°53'	67	B				Wenner (1947)
225. Webb's Bay	56°45'	61°45'	8	B				Wenner (1947)
226. Webb's Bay	56°45'	61°45'	8	B				Wenner (1947)
227. Webb's Bay	56°45'	61°45'	8	B				Wenner (1947)
228. Webb's Bay	56°45'	61°45'	206	B				Wenner (1947)
229. Webb's Bay	56°45'	61°45'	287	B				Wenner (1947)
230. Nain Pond	56°32'	61°49'	92	L	7	12,235 ± 780		Short and Nichols (1977)
231. Njakungaratsuk	56°38'	62°00'	5	B				Wenner (1947)
232. Ukalik	56°39'	61°18'	66	B				Wenner (1947)
233. First Lake, Fraser River	56°37'	62°20'	90	B				Wenner (1947)
234. Kasigiatsite	56°39'	62°23'	10	B				Wenner (1947)
235. Kasigiatsite	56°39'	62°23'	10	B				Wenner (1947)
236. Fourth Lake, Fraser River	56°32'	62°25'	3	B				Wenner (1947)
237. Kangitlotanak	56°39'	62°25'	145	B				Wenner (1947)
238. Kangitlotanak	56°39'	62°25'	117	B				Wenner (1947)
239. Kankitlotanak	56°39'	62°25'	300	B				Wenner (1947)
240. Kasigiatsite	56°43'	62°22'	7	B				Wenner (1947)
241. Itiplik	56°41'	61°52'	160	B				Wenner (1947)
242. Itiplik	56°41'	61°52'	334	B				Wenner (1947)
243. Itiplik	56°41'	61°52'	334	B				Wenner (1947)
244. Lac des Roches Moutonnées	56°46'	64°49'	410	L	4	4090 ± 250		McAndrews and Samson (1977)
245. Pyramid Hills	57°38'	65°10'	381	L	7	6815 ± 125		Short and Nichols (1977)
246. Ublik Pond	57°23'	62°03'	122	L	4	10,260 ± 360	★	Short and Nichols (1977)
247. Lindburg Island	57°22'	61°30'	36	B				Wenner (1947)
248. House Island	57°15'	61°40'	8	B				Wenner (1947)
249. House Island	57°15'	61°40'	31	B				Wenner (1947)
250. Port Manvers	57°02'	61°25'	16	B				Wenner (1947)
251. Port Manvers	57°02'	61°25'	7	B				Wenner (1947)
252. Slambang Bay Aulatsivik Is.	56°45'	61°30'	176	B				Wenner (1947)
253. Salmon Is. (Killialuk)	56°52'	61°00'	35	B				Wenner (1947)
254. Salmon Is. (Killialuk)	56°52'	61°00'	10	B				Wenner (1947)
255. Jonathon Is. (Tessiujalialuk)	56°48'	61°03'	25	B				Wenner (1947)
256. Scolpin Is. (Kanayoktok)	56°47'	61°09'	7	B				Wenner (1947)
257. Big Black Is.	56°47'	61°09'	11	B				Wenner (1947)

TABLE 1 (continued)

SITE NUMBER AND NAME	LATITUDE (N)	LONGITUDE (W)	ALTITUDE (M)	SITE TYPE LAKE = L BOG = B SECTION = S	NUMBER OF RADIOCARBON DATES	OLDEST RADIOCARBON DATE (YR B.P.)	POLLEN DIAGRAM (THIS PAPER)	REFERENCE(S)
258. Big Black Is.	56°47'	61°09'	17	B				Wenner (1947)
259. Napaktok	57°55'	62°34'	143	L	3	8735 ± 235		Short (1980)
260. Hebron	58°12'	63°04'	168	L	3	10,075 ± 255		Short (1980)
261. Moraine	58°22'	63°33'	550	L	1	11,160 ± 520		Short (1980)
262. Lac Nakvak	58°40'	63°41'	400	S	3	2450 ± 380		de Vernal, Mathieu and Gangloff (1983)
263. Miriam	59°33'	63°53'	0	L	3	1630 ± 80		Short (1980)
264. Vieux-Port-Burwell A	60°25'	64°49'	6.7	B	1	4980 ± 100		Savoie and Gangloff (1980)
265. Vieux-Port-Burwell B	60°25'	64°49'	6.7	B	3	5070 ± 185	★	Savoie and Gangloff (1980)
266. Vieux-Port-Burwell C	60°25'	64°49'	24.4	B	1	2440 ± 85		Savoie and Gangloff (1980)
267. Nedlouc	57°39'	71°39'	330	L	7	3950 ± 150		Richard (1981a)
268. Rivière aux Feuilles-T	58°14'	72°04'	200	B	4	4960 ± 120		Richard (1981a)
269. Rivière aux Feuilles-1	58°14'	72°04'	200	L	3	5235 ± 185	★	Richard (1981a)
270. Rivière aux Feuilles-2	58°13'	71°57'	235	L	7	4880 ± 170		Richard (1981a)
271. Faribault	58°52'	71°43'	215	L	3	3885 ± 185		Richard (1981a)
272. Aupaluk	59°12'	70°14'	92	S	3	7350 ± 320		Lauriol, et al. (1979)
273. Diana-VHC	60°47'	69°50'	50	L	1	6460 ± 160		Richard (1977;1981a)
274. Diana-1	60°56'	69°58'	15	B				Richard (1981a)
275. Iglou – 1	60°51'	69°53'	25	B				Richard (1981a)
276. Iglou – 2	60°51'	69°53'	25	B	2	2095 ± 95		Richard (1981a)
277. Diana 375	60°59'	69°57'	114	L	3	6820 ± 155		Richard (1981a)
278. Diana 500	60°58'	69°59'	153	L				Richard (1981a)
279. Riviere Saule	61°31'	74°05'	87	S	1	525 ± 100		Bartley and Matthews (1969)
280. R. aux Poissons	62°06'	75°45'	39	S	1	3990 ± 140		Bartley and Matthews (1969)
281. R. aux Roches R. Renard-Noir	62°02' 62°07'	74°32' 74°38'		S S				Bartley and Matthews (1969) Bartley and Matthews (1969)
282. R. de l'Airelle (Vaccinium Valley) Sugluk	62°11' 62°12'	75°53' 75°39'	116	S S	1 1	1625 ± 175 2840 ± 160		Bartley and Matthews (1969) Terasmae et al. (1966) Bartley and Matthews (1969)
283. R. Tourbe 1	62°10'	75°55'	11.3	S	1	1600 ± 140		Bartley and Matthews (1969)
284. R.Tourbe 2	62°10'	75°55'	11.3	S	1	670 ± 120		Bartley and Matthews (1969)
285. R.Tourbe 3	62°10'	75°54'	16.5	S	1	5230 ± 130		Bartley and Matthews (1969)
286. Kugluk Cove	62°19'	76°02'		S				Bartley and Matthews (1969)
287. Lac Faucon	62°29'	76°20'	107	S				Bartley and Matthews (1969)
288. Chism-1	54°47'	76°08'	340	L	1	5465 ± 140		Richard (1979)
289. St. Joseph	45°59'	73°18'	18	B	3	4790 ± 140		Comtois (1979)
290. Bereziuk	54°03'	76°07'	205	L	3	6630 ± 170	★	Richard (1979)
291. Kanaaupscow	54°01'	76°38'	200	L	1	6450 ± 190		Richard (1979)
292. Lac Desaulniers	53°30'	77°23'	155	B	1	9410 ± 580		Ouellet et Poulin (1975)
293. Chism-2	53°05'	76°20'	273	L	1	6430 ± 160		Richard (1979)
294. Lake Irene	49°32'	74°46'		B	1	6960 ± 90	★	Ignatius (1956); Barendsen et al. (1957)

Sites in Figure 6 are differentiated on the basis of age. Fifteen sites comprise the pre-Late Wisconsin group and range in age from the Sangamonian Interglacial to the Early Wisconsin. The remainder and majority of the sites date from the time of retreat of the last ice; these sites make up the Late Wisconsin-Holocene group.

It is only feasible in this paper to discuss the pollen stratigraphy of a selected number of sites which are considered to be representative of a region. Figure 6 shows the locations of sites where pollen diagrams have been reproduced or modified in abbreviated form. Lake and peat bog sites were preferred over section sites with respect to Late Wisconsin and Holocene pollen records. Sections and drill cores, on the other hand, provide the only record of pre-Late Wisconsin stratigraphy.

The Late Wisconsin-Holocene pollen stratigraphy, the late-glacial of south-central Quebec in particular, contains a significant representation of herb and shrub pollen and consequently sub-zones within dominant pollen assemblages can be recognized on a fine scale (Richard 1981b; Labelle and Richard, 1981a). However, because of the vastness of the area and inter-regional variabilities of the pollen records, it is only feasible in this treatment to consider the palynostratigraphy as a whole on a somewhat simplified basis, i.e. at the dominant pollen assemblage level.

Methods

For the most part the field and laboratory procedures used in obtaining cores and sample material and for extracting pollen and spores are consistent from study to study. Pollen counts are presented in the form of abbreviated percentage diagrams which are correlative from site to site and several such diagrams provide a network or transect of pollen stratigraphic sites for eastern Canada. Diagnostic pollen profiles have been redrawn in an effort to maintain consistency for a particular region. Diagrams which were unzoned originally were zoned for ease of discussion and others were rezoned. Pollen influx estimates provide useful information in interpreting the pollen records; consequently total pollen influx (grains/cm^2/year) is shown on the percentage diagrams for several sites.

Sites with several radiocarbon dates were preferred over sites having only single datings or where ^{14}C dating was not available. The most reliable basal dates were used in preference to dates which are considered to be anomalous. The chronology for sites characterized by hemlock pollen profiles was adjusted with respect to the age of the middle-Holocene, hemlock pollen decline similar to the method used by Webb *et al.* (1983). The decline in hemlock is attributed to widespread wipeout of mature stands of hemlock in northeastern North America by a forest pathogen at about 4,800 B.P. (Davis, 1981).

POLLEN STRATIGRAPHY AND VEGETATION RECONSTRUCTION

Sangamonian—Early Wisconsin Records, Atlantic Provinces and Newfoundland

Buried nonglacial deposits containing peat overlain by till or by a diamicton occur at several localities throughout eastern Canada. One of the better documented sites discovered thus far is the East Milford Quarry site in Nova Scotia where wood in organic clay was dated at >50,000 B.P. (GSC-1642) (Mott *et al.,* 1982). An abbreviated pollen diagram from this site (Fig. 7) shows a basal assemblage (Zone 1) dominated by beech and basswood *(Tilia)* along with smaller amounts of oak, hickory, elm and maple. The hardwoods decline and are replaced by substantial increases in birch, spruce and fir in Zone 2. In Zone 3 birch declines and fir, followed by spruce, increases to maximum percentages; hardwoods decline to minor occurrences. The uppermost zone, Zone 4, is dominated first by alder *(Alnus)* and fir and later almost exclusively by spruce.

The pollen stratigraphy is interpreted to be representative of a late phase of an interglacial interval, more than likely the Sangamon Interglacial. Hardwood pollen supported by macrofossils of beech nuts, birch seeds and fruits of associated understory shrubs and herbs indicate a mature hardwood forest was present early in the interglacial. Climatic conditions at the time were at least as warm as the present. The dominance of birch followed by a succession to fir, alder and spruce is interpreted to indicate the hardwood forest was replaced by a mixed forest, which in turn, gave way to a coniferous forest. Ericaceae (heaths), Cyperaceae (sedges) and *Sphagnum* in association with spruce and fir suggest a landscape similar in many respects to the northernmost Boreal Forest today. The climate had obviously cooled in Late Sangamonian time before the onset of glaciation which eventually deposited the overlying till.

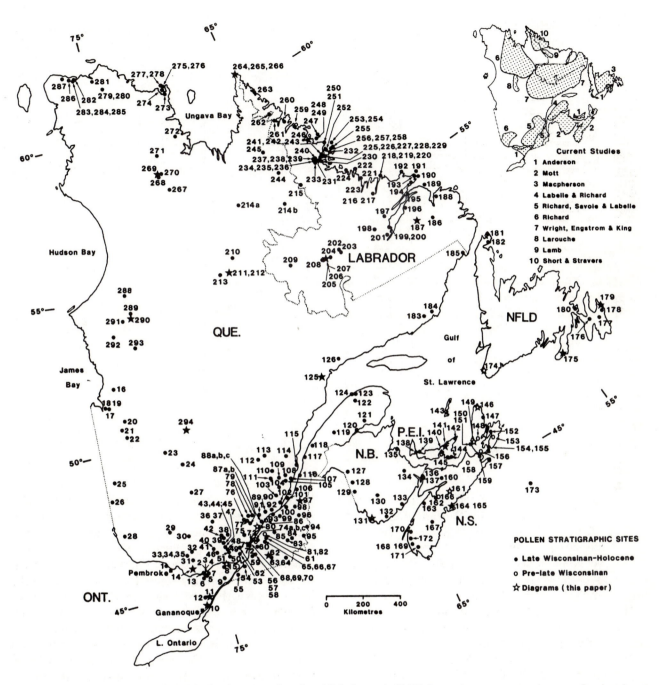

Figure 6. Locations of pollen stratigraphic sites in eastern Canada published to end of 1983. Insert shows current study areas of various investigators. Table 1 provides a key to the name and number of each site.

Because the East Milford Quarry shows a reasonably complete and continous sequence from an interglacial environment through to glacial conditions, this site serves as the reference section to which other sites in eastern Canada are correlated. Most of these sites contain buried organic beds and, as stated in Mott *et al.* (1982), the organics probably accumulated under similar environmental conditions over a broad region but they may not relate temporally.

Peat overlying silt and clay and underlying till at the Hillsborough site (Fig. 8) Cape Breton Island, contains a pollen sequence dominated by high alder and fir grading upward to high spruce (Mott and Prest, 1967; Livingstone, 1968). Peaks in pollen of Ericaceae and Cyperaceae within the spruce period (Livingstone, 1968) indicate Boreal Forest conditions prevailed in the area. The pollen sequence and >38,000 (W-157) and >51,000 (GSC-570) B.P. dates

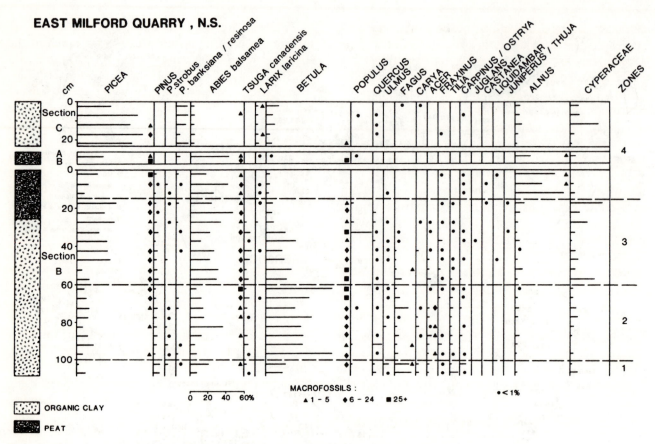

Figure 7. Abbreviated pollen diagram of the East Milford Quarry, Nova Scotia, modified from Mott *et al.* (1982).

(Rubin and Suess, 1955; Mott and Prest, 1967) on the peat suggest correlation with the Sangamonian record at East Milford. The close correspondence between the spruce, pine, fir, birch and alder percentages in the Whycocomagh section (Mott and Prest, 1967) and in the >39,000 B.P. (GSC-1406) section at River Inhabitants (Mott, 1971; Mott *et al.*, 1982) with those at Hillsborough and East Milford indicates the plant-bearing beds at Whycocomagh and River Inhabitants are similarly Late Sangamonian in age.

Recent studies by de Vernal, Richard and Occhietti (1983) on the >38,300 (GSC-283) B.P. sequence at Bay St. Lawrence reveal a pollen assemblage dominated by herbs with an intervening interval characterized by spruce, birch and alder (Fig. 9). A complete cycle of vegetation change from tundra, to boreal conditions, to tundra again is invoked for the basal part of the sequence. An Early to Middle Wisconsin interstade correlative with Stages 3, 5a or 5b of the oxygen isotope record is suggested.

Pond sediments beneath till at Woody Cove (Fig. 10), southwest Newfoundland, contain wood dating >40,000 (I-10203) B.P. (Brookes *et al.*, 1982). The sediment sequence records a pollen succession from birch, pine and herbs at the base to an intervening interval of spruce, fir, birch and pine to pine and herbs at the top. The corresponding inferred vegetation was tundra – boreal forest – tundra. Climatic conditions during the boreal interval were interpreted to have been warm if not warmer than present. The authors give the pond sequence full interglacial (Sangamonian) status partly on the basis of the pollen record and partly from foraminiferal evidence.

The similarities between the cycles of vegetation change at Woody Cove and at Bay St. Lawrence, Cape Breton Island, are quite apparent; the major difference between the two sites are the higher values of spruce and fir at Woody Cove. The modern spectra are not too dissimilar which is expected since the sites are only about 125 km apart. It might be argued whether the spruce and fir differences are pronounced enough to warrant interglacial status at one site (Woody Cove) and an interstadial one at the other (Bay St. Lawrence). Additional pollen analyses presently being carried out on other recently discovered sites in Nova Scotia (Mott, oral commun., 1983) might clarify the regional bio- and chronostratigraphy.

HILLSBOROUGH SECTION , N.S.

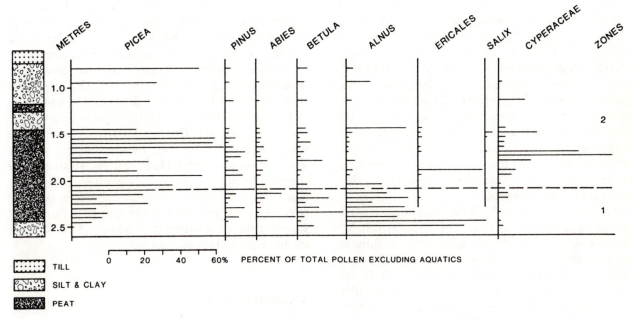

Figure 8. Pollen diagram of the Hillsborough site, Cape Breton Island, modified from Livingstone (1968). Aquatics = aquatic plants.

BAY ST. LAWRENCE SECTION, N.S.

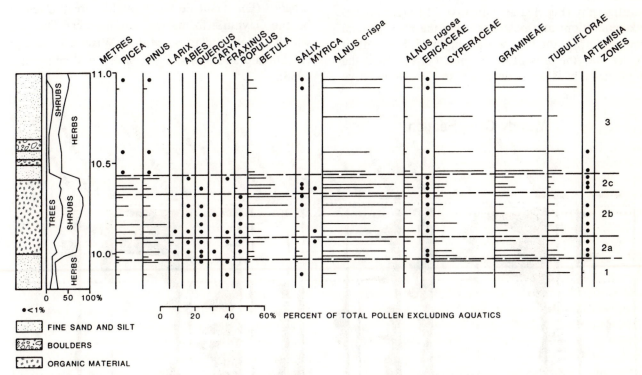

Figure 9. Pollen diagram of the Bay St. Lawrence section, Cape Breton Island, modified from de Vernal, Richard and Occhietti (1983). Aquatics = aquatic plants.

Sangamonian—Early Wisconsin Records in Quebec.

Few sites of this age range occur in Quebec but some of these are well known classical sites. A buried peat deposit in a gravel pit exposure at Portage-du-Cap, Magdalen Islands, Quebec, has produced ages >35,000 (BGS-259) and >38,000 (GSC=2313) B.P. (Prest *et al.,* 1976). Pollen analyses carried out on the peat show spruce, birch, pine and fir pollen dominate and oak and beech percentages are significantly higher than those presently occuring in the area. The pollen evidence indicates climatic conditions during deposition of the plant detritus were more favourable than the present and an interglacial period (probably the Sangamonian) is implied.

The best known Early Wisconsin non-glacial deposits in eastern Canada are the >75,000 B.P. St. Pierre beds, an inter-till peat and varved sediment sequence in the St. Lawrence lowlands, Quebec. Pollen profiles from five localities show the St. Pierre beds are characterized by spruce (up to 80% maximum) and lesser amounts of pine (50% maximum), fir (to about 10% maximum), birch (30% maximum), alder (almost 25% maximum) and thermophilous hardwoods (2 to 5% maximum). Terasmae (1958) has interpreted the pollen stratigraphy (Fig. 11) as indicative of Boreal Forest-like conditions and postulated that the mean annual temperature at the time was about 3 to 5° Farhrenheit lower than the present in the area. On this basis the non-glacial beds were considered to be representative of an interstadial interval and were collectively referred to as the St. Pierre Interstade.

Gadd *et al.* (1981) have related buried stratified and cross-bedded sand and silt over organic-rich sand at Pointe-Fortune, Quebec to the St. Pierre Interstade and the overlying till to the Gentilly Glaciation of the St. Lawrence lowlands. Three ^{14}C ages on wood in the sand sequence are >42,000 (GSC-2932) B.P., >40,000 (GSC-3459) B.P. and >38,000 (GSC-3444) B.P. Pollen studies carried out on the organic interval and underlying silt and fine sand show an assemblage which is clearly dominated by spruce, pine, willow, alder, grass and sedge (Fig. 12). Pollen of fir and thermophilous hardwoods is minor. The pollen assemblage indicates cooler-than-present climatic conditions had prevailed in the area at the time and supports the supposition by Gadd *et al.* (1981) that the interval is St. Pierre equivalent.

Middle Wisconsin

To the author's knowledge pollen evidence for the Middle Wisconsin does not exist at the present time in eastern Canada. Pollen studies, however, are in progress (Université de Montréal, Montréal, Québec) on the Gayhurst formation, southeastern Quebec, which is believed to be Middle Wisconsin in age (McDonald and Shilts, 1971; Shilts, oral commun., 1983).

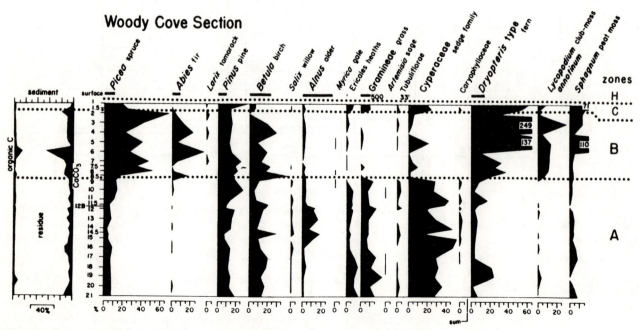

Figure 10. Percentage pollen diagram of the Woody Cove section, Newfoundland from Brookes *et al.* (1982).

LES VIELLES FORGES SECTION, QUEBEC

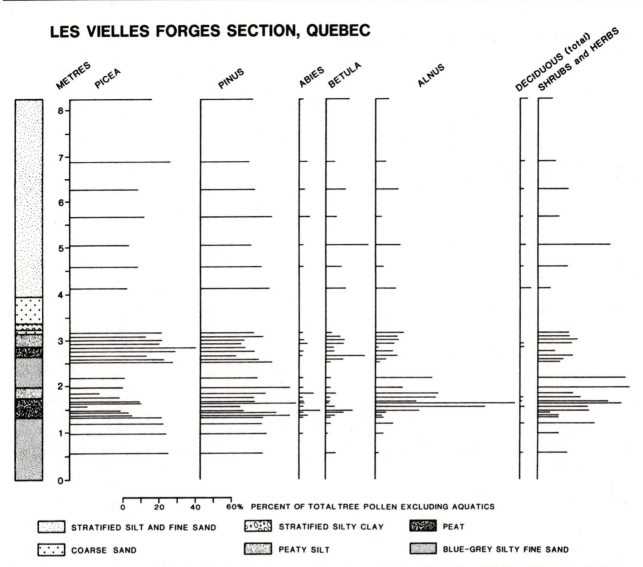

Figure 11. Percentage pollen diagram of Les Vieilles Forges section, Quebec, modified from Terasmae (1958). Aquatics = aquatic plants.

Late Wisconsin—Holocene Records, Atlantic Provinces and Newfoundland

The earliest Late Wisconsin pollen zones recorded thus far in the Atlantic Provinces and Newfoundland are dominated by tundra-like assemblages but the composition of these assemblages vary from region to region. Shrub tundra consisting of *Salix,* Ericaceae and *Myrica* was present in Burin Peninsula, Newfoundland (Fig. 13) possibly as early as 13,400 B.P. (Anderson, 1983). Herb tundra characterized by high percentages of *Salix, Artemisia,* Gramineae and Cyperaceae existed at Leak Lake, Nova Scotia at 12,900 ±160 B.P. (GSC-2728) (R. J. Mott, unpubl. GSC Palynological Report 79-13, in Wightman,

1980), if date is valid, and at Basswood Road Lake, New Brunswick (Fig. 14) at 12,600 ± 270 B.P. (GSC-1067) (Mott, 1975a). Prior to about 12,000 B.P. tundra vegetation was probably limited to areas which had been deglaciated about 1,000 years earlier such as southwest New Brunswick and parts of Bay of Fundy (Prest, 1970) and areas which had not been covered by Late Wisconsin ice such as the coastal regions of southeast Newfoundland. An assemblage dominated by *Betula* (shrub birch, *B. glandulosa*) succeeded tundra at some sites (*e.g.* Leak Lake), but at others (Basswood Road Lake), tundra was replaced by *Betula* and *Populus* or by *Betula* and *Picea.*

The southern Gulf of St. Lawrence region is believed to have been largely ice-covered or submerged by the sea during this period of restricted tundra vegetation. Marine shell dates of 12,410 ± 170 B.P. (GSC-101) and 12,670 ± 340 B.P. (GSC-160)

POINTE FORTUNE SECTION, QUEBEC

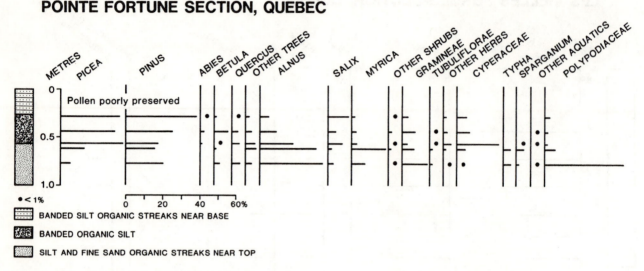

Figure 12. Pollen diagram of the Pointe-Fortune Section, Quebec.

LAKE SITE SOUTH BURIN PENINSULA

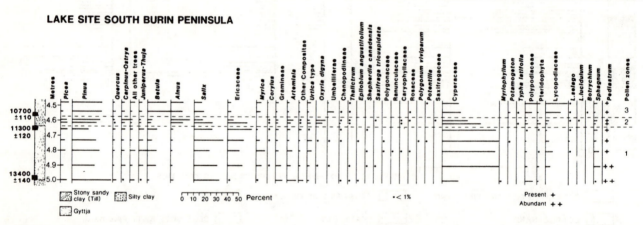

Figure 13. Percentage pollen diagram of basal lake sediments in Burin Peninsula, Newfoundland, from Anderson (1983).

clearly indicate marine inundation of the coastal areas of western Prince Edward Island, at least, and possibly northeastern New Brunswick.

Picea shows significant increases in the pollen records at several sites after about 12,000 B.P. indicating spruce had migrated into the central Maritime Provinces region by this time. Spruce woodland in association with shrub birch and herbs was apparently present at Canoran Lake, south-central Nova Scotia, by 11,700 ± 160 B.P. (GSC-1486) (Railton, 1972) and at Diligent River, northeastern Nova Scotia, by 11,555 ± 230 B.P. (DAL-301) (R. J. Mott, unpubl. GSC Palynological Report 79-15, in Wightman, 1980). Spruce became more widespread and by 11,300 B.P. an open spruce forest is believed to have existed in the southwestern New Brunswick and south-central Nova Scotia regions according to pollen evidence at Basswood Road and Canoran Lakes. Elsewhere, tundra may have been

present in the newly, deglaciated lowland and coastal parts of Nova Scotia, Cape Breton Island and New Brunswick. Prince Edward Island and presumably eastern New Brunswick probably still remained ice covered and/or eustatically depressed.

Dramatic and abrupt changes characterize the pollen stratigraphy in the Atlantic Provinces and Newfoundland over the period 11,300 to 10,000 B.P. These changes span clay layers within organic sediments at several lake sites and at localities where mineral sediment overlies organic beds. For example, shrub pollen of *Salix,* Ericaceae and *Myrica* were replaced by significant increases in herb pollen of *Oxyria digyna, Artemisia* and *Saxifraga* across a clay layer dating between 10,700 and 11,300 B.P. in Burin Peninsula, Newfoundland (Fig. 13). *Picea* pollen decreased from maximum values while *Betula,* Cyperaceae and Gramineae increased in a similar clay interval in basal gyttja at Basswood Road Lake

BASSWOOD ROAD LAKE, N.B.

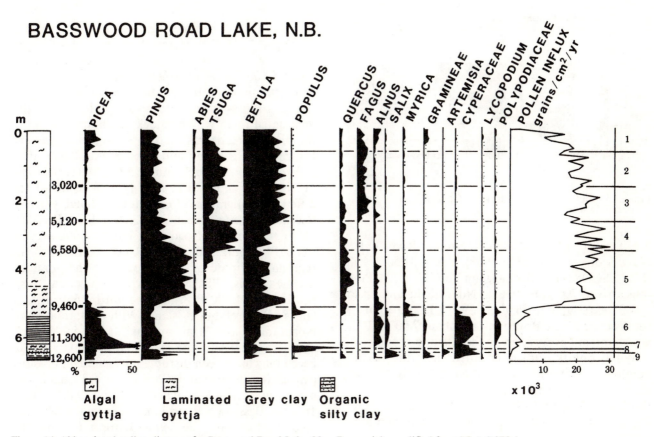

Figure 14. Abbreviated pollen diagram for Basswood Road Lake, New Brunswick, modified from Mott (1975a).

(Fig. 14). Gyttja below the clay was dated at 11,300 ± 180 B.P. (GSC-1645) and gyttja above the clay ranges between about 10,460 and 10,760 B.P. from estimates on the rate of deposition of the basal sediment. At Leak Lake, Nova Scotia, an inter-gyttja clay layer contains lower *Picea* and higher Cyperaceae and Polypodiaceae values than the adjacent gyttja; gyttja immediately above the clay was dated at 9,715 ± 200 B.P. (Dal-314) (R. J. Mott, unpubl. GSC Palynological Report 79-13, in Wightman, 1980). In a similar clay layer at Gillis Lake, southern Cape Breton Island (Livingstone and Livingstone, 1958), *Picea* and *Betula* are retarded but Cyperaceae reaches maximum values. A date of 10,340 ± 220 B.P. (Y-524) on the gyttja above the clay layer provides a date for the regional spruce pollen rise. The pollen stratigraphy was similarly altered at section sites throughout Nova Scotia where mineral sediment or a diamicton overlie organic beds which date to about 11,000 B.P. (Mott, 1983).

The overall biostratigraphic and lithostratigraphic changes reflect a reduced or altered vegetative cover brought about as a result of widespread climatic deterioration in the North Atlantic between about 10,000 and 11,000 B.P. (Ruddiman and McIntyre, 1981). Mott (1983) has tentatively correlated the climatic reversal with the well known Allerød-Younger Dryas climatic oscillation of northwest Europe and the British Isles.

The climatic oscillation is not evident in pollen diagrams from Prince Edward Island. Tundra vegetation developed soon after deglaciation and persisted unchanged until about 10,000 B.P. on the basis of peaks in non-arboreal birch, willow, *Artemisia,* sedge and grass pollen at the base of Portage Bog (Anderson, 1980) and MacLaughlin Pond (Fig. 15). Shortly after 10,000 B.P., however, spruce and birch dominate and total pollen influx increases two-fold corresponding with a change from tundra to spruce-woodland vegetation. The rise in spruce is dated at 9,670 ± 130 B.P. (GSC-2891) at MacLaughlin Pond which corresponds almost exactly with a date of 9,650 ± 150 B.P. (GSC-333) for the same horizon at Silver Lake, Nova Scotia (Fig. 16). A date of 9,460 ± 220 B.P. (GSC-1643) was obtained for the top of the spruce zone at Basswood Road Lake. Thus from about 9,500 to 9,700 B.P. spruce was widespread throughout the Atlantic Provinces; hence pollen zonation was synchronous everywhere in the region for the first time in the Holocene.

Spruce gave way to increases in pine at 9,460 ± 220 B.P. (GSC-1643) at Basswood Road Lake (Fig. 14),

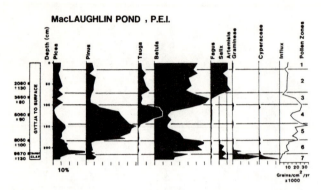

Figure 15. Abbreviated pollen diagram for MacLaughlin Pond, Prince Edward Island.

8,770 ± 150 B.P. (GSC-336) at Salmon River Lake, Nova Scotia (Livingstone, 1968), and at 8,050 ± 100 B.P. (GSC-2887) at MacLaughlin Pond (Fig. 15). Pine thus arrived 700 years later in Nova Scotia and up to 1,400 years later in Prince Edward Island than in southwestern New Brunswick; by about 7,500 B.P. it was dominant everywhere in the Atlantic Provinces.

Synchroneity continued until the time of the hemlock maximum, i.e. to about 5,000 B.P. Four dates ranging from 6,290 ± 140 B.P. (I-7078) (Hadden, 1975) to 7,140 ± 140 B.P. (GSC-772) (Fig. 16) for the first rise in hemlock indicate hemlock

arrived in the Maritime region as a whole between 6,000 and 7,000 B.P. Hemlock declined shortly after reaching a maximum at 5,000 B.P. An average age of about 4,000 B.P. is obtained from five dates on the hemlock minimum in the Atlantic region.

The middle- to late-Holocene pollen zones are dominated mainly by hemlock, birch and beech but the birch percentages tend to be higher in eastern Prince Edward Island and Cape Breton Island than elsewhere. Zonal correlation on a regional scale is less precise after 5,000 B.P. mainly as a result of the differential migration of certain taxa, notably beech. Beech increased noticeably at 5,120 ± 220 B.P. (GSC-1595) at Basswood Road Lake but this date may be too old by some 300 years when compared with the regional estimate of 4,800 years for the hemlock decline (Davis, 1981). Beech arrived in central Nova Scotia after 4,500 B.P. but not until 3,660 B.P. or later in Prince Edward Island. With the formation of Northumberland Strait approximately 5,000 years ago (Kranck, 1972), the late-Holocene migration of beech may have been disrupted which might possibly explain its late arrival on Prince Edward Island. The second rise in hemlock, dated 3,020 ± 150 B.P. (GSC-1693) at Basswood Road Lake, reflects the hemlock recovery following the hemlock minimum, but the

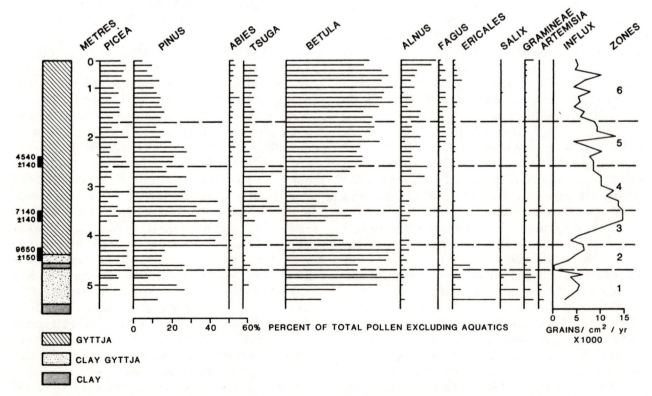

Figure 16. Abbreviated pollen diagram for Silver Lake, Nova Scotia, modified from Livingstone (1968). Aquatics = aquatic plants.

lower hemlock percentages in most sites suggest hemlock did not regain its previous potential in the mixed forest at the time.

The post-10,000 B.P. record in Newfoundland (Fig. 17) is characterized initially by a lower shrub zone dominated by *Salix* and Ericales, by an upper shrub zone dominated by *Betula, Myrica* and Ericales and an intervening herb zone of sedge and *Oxyria digyna* at Sugarloaf Pond (Macpherson, 1982). The sequence is stratigraphically but not temporally similar to that in Burin Peninsula discussed earlier. The upper shrub zone is bracketed by a lower date of 9,270 ± 150 B.P. (GSC-2601) and by an upper inferred age of 8,300 B.P. Arboreal birch, spruce and fir are the dominant taxa after 8,300 B.P. Fir increases upwards in pollen diagrams from Avalon Peninsula; in Goulds, Whitbourne and Clarenville Bogs (Terasmae, 1963) fir averages 20-30% maximum from about 3,600 B.P. to present.

Total pollen influx at Basswood Road Lake (Fig. 14) is low (6,500 grains/cm^2/year maximum) in the pre-pine zones and increases four-fold to 26,000 grains/cm^2/year maximum at the increase in pine at 9,460 B.P. Influx rates decline slightly after 3,000 B.P. A similar trend occurs at MacLaughlin Pond (Fig. 15) where total pollen influx increases gradually from about 9,700 B.P. and jumps two-fold at the increase in pine at 8,050 B.P. Maximum rates occur between about 8,000 and 4,000 B.P. followed by lower values from 3,660 B.P. to present. In the Sugarloaf Lake diagram (Fig. 17) influx estimates increase dramatically at approximately 9,000 B.P.

Late Wisconsin-Holocene Records, eastern Ontario and southern Quebec

The largest concentration of Late Wisconsin and Holocene pollen records in eastern Canada occurs in this region. Most of the sites are located between the Appalachian Highlands to the southeast and the Laurentide Highlands to the north and northwest of the St. Lawrence-Ottawa River valleys. Five sites were chosen as being representative of the pollen stratigraphy of this region, Lambs Pond (Fig. 18) in the extreme south, Ramsay Lake (Fig. 19) (Mott and Farley-Gill, 1981), Mont Shefford site (Fig. 20) (Richard, 1978) and Lac Colin (Fig. 21) (Mott, 1977) in the centre, and Lake Irene (Fig. 22) to the north (Ignatius, 1956).

As in the Atlantic Provinces, the earliest pollen zones are dominated by a herb assemblage consisting mainly of *Salix, Artemisia,* Cyperaceae and Gramineae and by an overlying shrub assemblage of *Salix, Betula, Alnus,* and *Juniperus* accompanied by declining herbs. High values of *Picea, Pinus, Quercus* and other thermophilous hardwoods in association with herb pollen are a familar feature of the basal-most zones; they are attributed to distant dispersal from sources to the south and deposition in a treeless environment. Total pollen influx is consistently low (300 grains/cm^2/yr maximum at Lambs Pond, 2,200 at Mont Shefford and 1,000 or less at Lac Colin) but it increases strongly in the overlying zones.

Dates on these zones are oldest to the south and east and youngest to the west and northwest. The top of the shrub zone is estimated at 11,700 B.P. in Lambs Pond and is dated at 11,100 ± 230 B.P. (I-8839) at Mont Shefford. At Lac Colin the herb-shrub boundary dates 11,100 ± 180 B.P. (GSC-2282). Dates on correlative herb zones at Ramsay Lake, St. Agathe (Savoie and Richard, 1979) and Sud du Lac du Noyer (Richard, 1977) to the west and northwest are 10,800 ± 180 B.P. (GSC-1963), 10,170 ±530 B.P. (I-10093) and 9,670 ± 190 B.P. (I-8497) respectively, and 10,710 ± 215 B.P. (GX-5328) from within the equivalent shrub zone at Lac à L'Ange (Labelle and Richard, 1981a). Dates of 12,640 ± 190 B.P. (GSC-312) and 14,800 ± 220 B.P. (GSC-1339) were obtained on the herb zones at Petit Lac Terrien and Unknown Pond (Maine) respectively, but these dates are considered anomalous by Mott (1977).

The shrub zones are succeeded first by a *Populus* zone and then by a *Picea* zone as at Ramsay Lake and Lac à L'Ange (Labelle and Richard, 1981a), or by a *Picea-Populus* zone often with high *Alnus crispa* as at Mont Shefford or with high *Juniperus* as in Lambs Pond, or by a *Picea* zone with high *Alnus crispa* as at Lac Colin and other sites in northeast St. Lawrence Lowlands (Mott, 1977). The dominant taxa of the preceding zones decline gradually but some (Gramineae, Cyperaceae and *Artemisia*) often remain relatively high. Pollen influx increases two-fold across the shrub-tree boundary at Mont Shefford, three-fold at Ramsay Lake and five-fold at Lambs Pond and Lac Colin.

Three dates are present on the *Picea-Populus* zone at Lambs Pond, 12,300 ± 230 B.P. (GSC-3088) for the rise in spruce (base of zone), 10,500 ± 110 B.P. (GSC-3273) for the spruce maximum and 10,200 ±160 B.P. (GSC-3259) at the decline in spruce (top of zone). Dates of 9,900 ± 100 B.P. (GSC-3127) and 10,500 ± 140 B.P. (GSC-3146) on wood and gyttja, respectively, were obtained for the spruce maximum in nearby Waterton Bog. The disparity between the

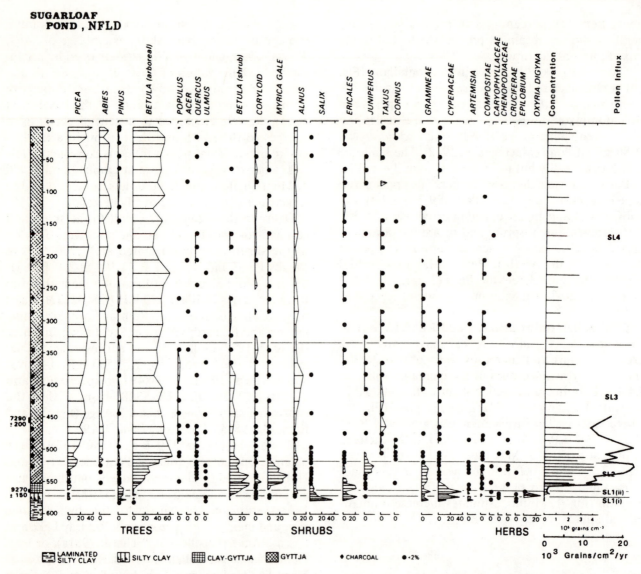

Figure 17. Abbreviated pollen diagram for Sugarloaf Pond, Avalon Peninsula, Newfoundland, modified from Macpherson (1982).

wood and gyttja dates suggests that the gyttja date is too old by some 600 years. By inference, the 10,500 B.P. date on the spruce maximum in Lambs Pond is also too old by about 600 years. Applying the same reasoning to the other basal dates in Lambs Pond reduces the 12,300 B.P. date to 11,700 years and the 10,200 B.P. date to 9,600 years.

The oldest reliable age for the base of the spruce zone in northeast St. Lawrence Lowlands is 11,200 ± 160 B.P. (GSC-1294) at Lac Dufresne but spruce peaked prior to this date at Boundary Pond (Mott, 1977). The upper boundary of the spruce zone is dated at 9,430 ± 140 B.P. (GSC-8) at Northfield Bog in eastern Ontario (Terasmae, 1965) and at 9,660 ± 140 B.P. (GSC-2345) at Lac Dufresne (Mott, 1977). Interpolation between radiocarbon dates gives an

estimate of 9,500 B.P. for the top of the spruce zone at Ramsay Lake (Mott and Farley-Gill, 1981).

Ages for the *Populus* zone become progressively younger to the north and northwest. Increases in *Populus* occur at 11,100 B.P. at Mont Shefford, 10,700 B.P. at Ramsay Lake, 10,000 B.P. and later at sites north of the Lowlands (9,855 B.P. at Lac à L'Ange; 9,700 B.P. at St. Agathe and Sud du Lac du Noyer) but not until about 8,300 B.P. at Lac Clo in the James Bay Lowland (Richard, 1980).

The *Picea* or *Picea-Populus* zones give way to a *Pinus* or *Pinus-Betula* zone at most sites, but a distinct *Abies* zone precedes the rise in pine in the eastern St. Lawrence Lowland sites (Mott, 1977; Richard, 1978). The base of the pine zone is placed at about 9,500 B.P. in the eastern Ontario-southeastern

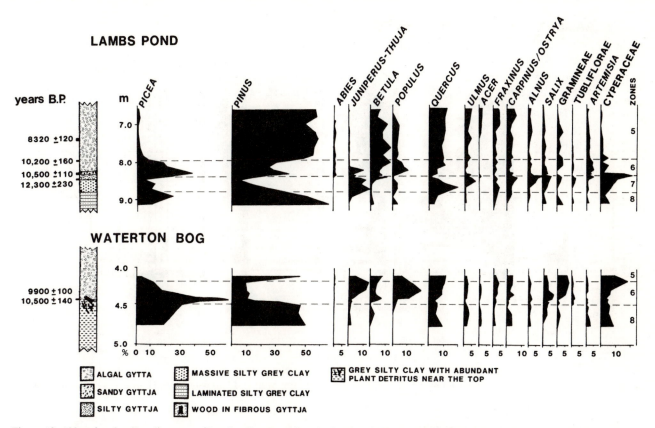

Figure 18. Abbreviated pollen diagrams of basal sediments at Lambs Pond and Waterton Bog, Ontario.

Quebec region; the regional rise in *Abies* is dated at 8,990 ± 100 B.P. (GSC-2325) at Lac Colin. *Pinus banksiana/resinosa* type *(P. divaricata)*dominate initially and are replaced by significant increases in *Pinus strobus* type at most sites in eastern Ontario and southeastern Quebec by 7,500 to 8,000 B.P. Maximum *Pinus strobus* percentages were reached by about 6,500 B.P. in the southeast but not until about 5,000 B.P. in the northwest (Terasmae and Anderson, 1970).

The transition to pine was accompanied by almost a two-fold increase in total pollen influx at Ramsay Lake (Fig. 19) and a six-fold increase at Lac Colin (Fig. 21). At Mont Shefford (Fig. 20) total pollen influx drops substantially from maximum values in the *Picea-Populus* zone to minimum rates in the *Abies* zone. At Ramsay Lake total pollen influx estimates reflect maximum influx rates of *Pinus strobus*.

By about 8,000 to 6,000 B.P. pollen records were changing dramatically in eastern Ontario and southern and south-central Quebec. Birch populations expanded to the north and south of the St. Lawrence River (Webb *et. al*, 1983), and from about 8,000 B.P. to present, pollen zones at sites from the Laurentide Highlands and eastern St. Lawrence Lowlands are ubiquitously dominated by birch. *Acer* and *Tsuga* grew in moderate numbers throughout eastern Ontario and southeastern Quebec as early as 7,000 B.P. (Webb *et. al*, 1983). *Tsuga* is documented by a bimodal profile. The first rise in *Tsuga* is dated at 7,620 ±170 B.P. (GSC-179) in southeastern Ontario (Terasmae, 1980), 6,420 ± 140 B.P. (GSC-2110) at Ramsay Lake and 6,360 ± 110 B.P. (GSC-2329) at Lac Colin. *Tsuga* declines at approximately 4,800 B.P. throughout its range (Davis, 1981) at the expense of the hardwoods and increases again but it never achieves the peak levels of the earlier hemlock maximum. The second rise in hemlock is dated at 3,210 ± 90 B.P. (GSC-2125) at Ramsay Lake and 3,360 ± 100 B.P. (GSC-2337) at Lac Colin. Beech increases regionally at many sites during the hemlock minimum.

The pollen stratigraphy of south-central Quebec and the James Bay Lowlands postdates the drainage of proglacial Lake Ojibway in this region. The earliest pollen assemblage at Lac Clo (Richard, 1980) consists of relatively low percentages of *Populus, Picea, Betula* and herbs and higher amounts of *Pinus divaricata* type *(P. banksiana/resinosa)* all of which are indicative of open-forest conditions. *Betula* and *Populus* dominate from about 8,300 B.P. to 7,200 B.P. and

RAMSAY LAKE SECTION ,QUEBEC

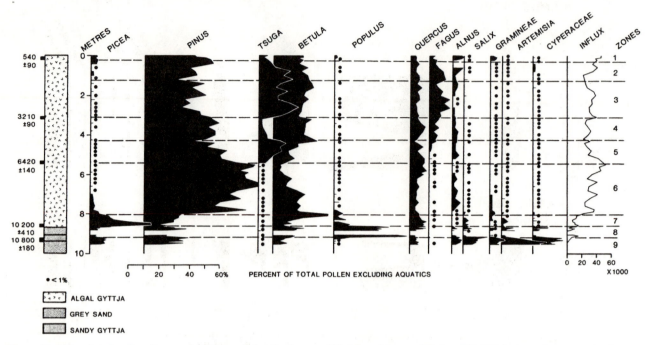

Figure 19. Abbreviated pollen diagram for Ramsay Lake, Quebec, modified from Mott and Farley-Gill (1981). Aquatics = aquatic plants.

MONT SHEFFORD, QUEBEC

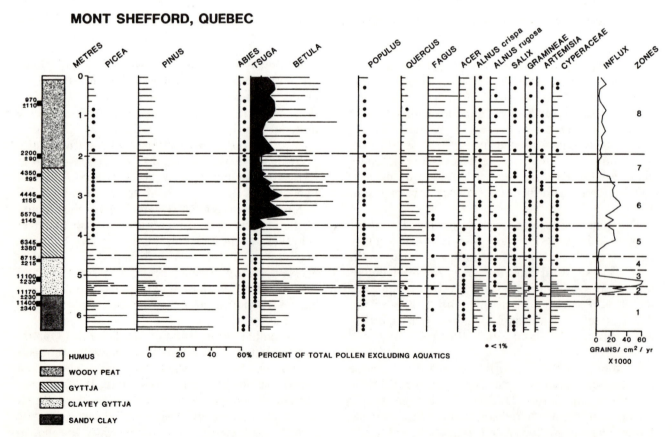

Figure 20. Abbreviated pollen diagram for Mont Shefford site, Quebec, modified from Richard (1978). Aquatics = aquatic plants.

LAC COLIN, QUEBEC

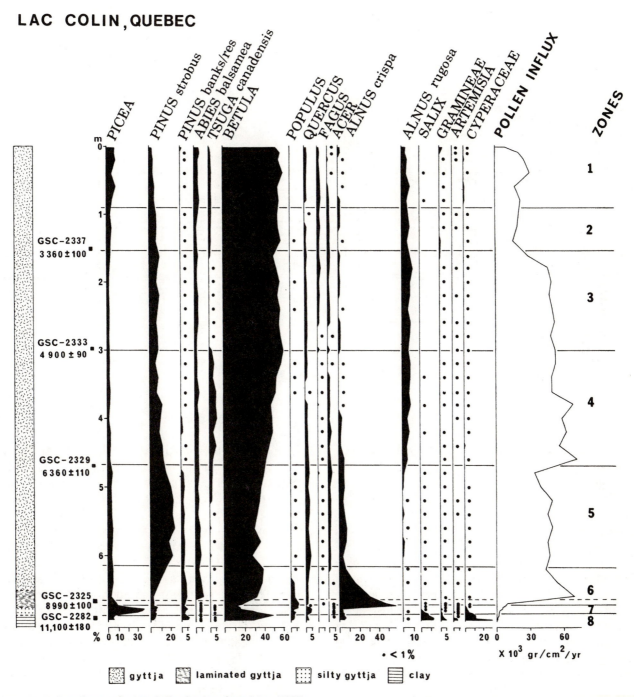

Figure 21. Pollen diagram for Lac Colin, Quebec, from Mott (1977).

Pinus strobus and *Betula* from about 7,200 B.P. to 3,200 B.P. A peak in juniper between 6,000 B.P. and 3,200 B.P. suggests the forest had remained open but black spruce and jack pine increase strongly at Lac Clo and Lake Irene (Fig. 22), (Ignatius, 1956) after 3,000 to 4,000 B.P. indicating closed forest conditions prevailed since this time.

The inferred vegetation of the eastern Ontario-south-central Quebec region consists of an initial herb-tundra phase, succeeded by shrub tundra which in turn gave way to an open spruce and poplar woodland. Pine (initially *Pinus banksiana/resinosa* and later *P. strobus*) replaced woodland in the more southerly parts while pine and birch succeeded open woodland in the northerly areas. A hemlock-maple-beech association dominated in the south from middle-Holocene time to the present; in the north birch dominated along with white pine and spruce.

LAKE IRENE SECTION, QUEBEC

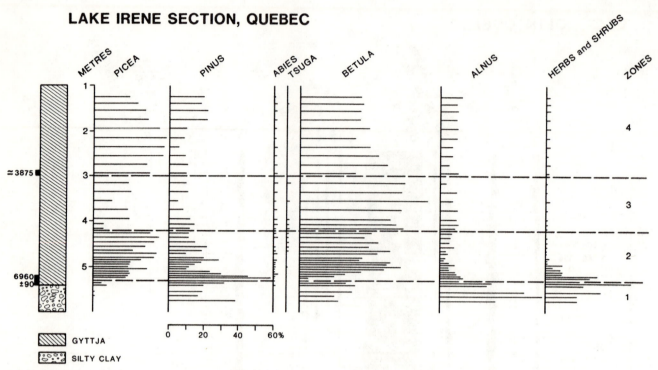

Figure 22. Pollen diagram for Lake Irene, Quebec, modified from Ignatius (1956).

Late Wisconsin-Holocene Records, Central Quebec, Nouveau-Québec and Labrador

The pollen stratigraphy of central Quebec–Nouveau–Québec–Labrador region is represented by pollen diagrams from Lake "LD" (Fig. 23) in the Sept-Îles area, east-central Quebec (Mott, 1976), the Bereziuk site (Fig. 24) in west-central Quebec (Richard, 1979), Lac Delorme-2 (Fig. 25) in central Nouveau-Québec (Richard *et. al*, 1982), Rivière aux Feuilles-1 (Fig. 26) in northern Nouveau-Québec (Richard, 1981b), Churchill Falls South (Fig. 27) in central Labrador (Morrison, 1970), Eagle Lake (Fig. 28) in southeast Labrador (Lamb, 1980), Ublik Pond (Fig. 29) in northeast Labrador (Short and Nichols, 1977) and Vieux-Port-Burwell (Fig. 30) at the northern tip of Nouveau-Québec east of Ungava Bay (Savoie et Gangloff, 1980). The earliest pollen zones at sites in Gaspé, coastal Labrador and Ungava region are dominated by herb-tundra-like assemblages consisting of *Betula, Salix, Artemisia,* Cyperaceae, Gramineae and other herbs. Pine is as high as 40% and 45% (Rivière aux Feuilles-1 site and Lac Turcotte, Gaspé) indicating the openness of the landscape. The earliest reliable dates or maximum estimates on these assemblages range from 10,550 ± 290 B.P. (SI-3139) and 10,260 ± 360 B.P. (SI-2739)

from the Labrador coast (Lamb, 1980; Short and Nichols, 1977), 10,360 ± 170 B.P. (DIC-2165) in Gaspé (Labelle and Richard, 1981b), 7,700 B.P. to 7,900 B.P. in central and northern interior Labrador (Stravers, 1981) to as late as 4,090 ± 250 B.P. (I-9067) in northeast Nouveau-Québec (McAndrews and Samson, 1977).

The herb-tundra-like assemblages give way to shrub-tundra assemblages characterized by high percentages of *Betula* and *Alnus.* The transition to shrub tundra occurs as early as 9,000 B.P. at Eagle Lake (Fig. 29), southeast Labrador, whereas in northeast Labrador it tends to be synchronous at 6,500 to 7,000 B.P. (Short and Nichols, 1977). It ranges from 7,400 to 8,000 B.P. in central and northern interior Labrador (Stravers, 1981; Grayson, 1956), to about 5,200 B.P. in Ungava (Richard, 1981a), to as late as 4,100 B.P. in northeastern Nouveau-Québec (McAndrews and Samson, 1977).

Total pollen influx is generally low in the lowermost herb-tundra-like zones (1,000 grains/cm²/yr maximum at Eagle Lake) and increases slightly in the shrub-tundra zones (between 1,000 and 2,000 grains/cm²/yr at Eagle Lake). Total influx averages about 5,000 grains/cm²/yr in the equivalent shrub-tundra zone at Rivière aux Feuilles-1 and increases to a maximum 20,000 grains/cm²/yr at Lac des Roches Moutonnées in northeast Nouveau-Québec (McAndrews and Samson, 1977).

"LD" LAKE, QUEBEC

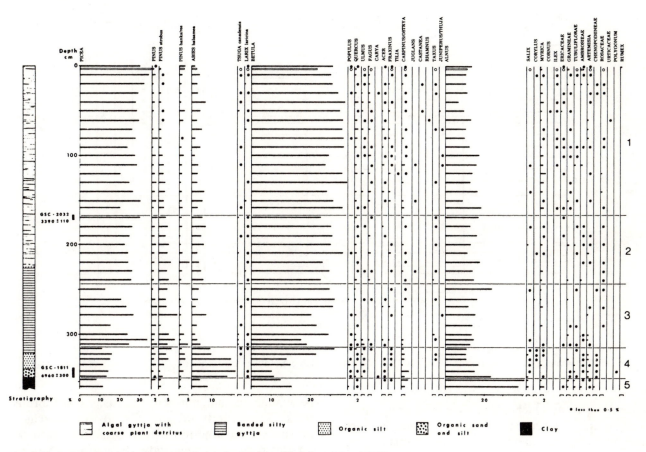

Figure 23. Pollen diagram for Lake "LD" site, Quebec, modified from Mott (1976).

Arboreal pollen zones distinguished by increasing percentages and influx of *Picea* succeed shrub tundra at most sites in Labrador and in west and northeast Nouveau-Québec. *Larix* and *Alnus crispa* dominate the earliest assemblages in the newly uncovered parts of central Nouveau-Québec (Richard *et. al,* 1982). An *Abies* zone precedes abrupt increases in birch and spruce at Lake "LD" (Fig. 24) and spruce at Eagle Lake (Fig. 29); a *Populus* zone precedes spruce at the Bereziuk site (Fig. 25). The rise in spruce occurs earlier southeast and southwest of Nouveau-Québec (6,400 B.P. at Lake "LD", 6,200 B.P. at Eagle Lake and 6,600 B.P. at Bereziuk) than in Nouveau-Québec and Labrador (5,200 B.P. at Lac Delorme-2, 5,400 B.P. at Churchill Falls South, between 4,000 and 4,500 B.P. at sites in northern Labrador, 3,700 B.P. at Lac des Roches Moutonnèes, northeast Nouveau-Québec). Birch dominates pollen records from the Gaspé from about 8,200 B.P. to present (Labelle and Richard, 1981b); birch and spruce from about 6,400 B.P. to present at Lake "LD"; spruce, birch and alder

from about 4,000 B.P. to present at most sites in Nouveau-Québec, and spruce from 4,000 to 4,500 B.P. until present in Labrador. Late-Holocene pollen sequences from northernmost Nouveau-Québec west and east of Ungava Bay (Fig. 30) are dominated exclusively by herbs. Spruce increases to minor percentage and influx peaks after 4,000 B.P. and shows a slight decline towards the present (Richard, 1981a; Savoie et Gangloff, 1980).

Pollen influx increases at most sites from low values in the basal zones to generally higher values in the arboreal zones. This increase is most dramatic at Eagle Lake where there is almost an eight-fold increase at the shrub-spruce transition. Rates were at a maximum at about 4,000 B.P. and declined after about 3,000 B.P. reflecting declines in influx of spruce, birch and alder.

Briefly, the inferred Holocene vegetation was dominated initially by herb tundra in the coastal parts of Labrador and Ungava region, by alder (mainly *Alnus crispa*) in east-central Quebec and

BEREZIUK, QUEBEC

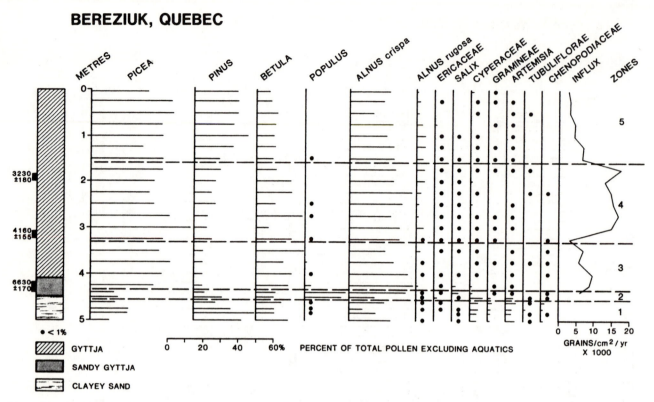

Figure 24. Abbreviated pollen diagram for the Bereziuk site, Nouveau-Québec, modified from Richard (1979). Aquatics = aquatic plants.

central Nouveau-Québec and by poplar or aspen in west-central Quebec. Shrub tundra (primarily alder and birch) replaced herb tundra and eventually gave way to spruce woodland. Spruce dominated everywhere except in the extreme north from about 4,000 B.P. to present.

DISCUSSION AND CONCLUSIONS

The pre-Late Wisconsin sites occur mainly in Nova Scotia, Magdalen Islands and St. Lawrence Lowlands, Quebec, and as isolated occurrences in the Ottawa Valley and Newfoundland. The distribution of Late Wisconsin and Holocene sites, on the other hand, is widespread. A cursory examination of Figure 6 shows areas where Late Wisconsin and Holocene pollen stratigraphic data is most abundant and areas where there are obvious deficiencies. The heavy concentration of sites in southern Quebec shows that this area is the most intensely studied in eastern Canada. Less concentrated areas are central Labrador and the Atlantic Provinces. The largest of the deficient areas is central Quebec between Gulf of St. Lawrence and James Bay. Other areas where data is lacking are south and west of Ungava Bay, central and western

Newfoundland and northern and eastern New Brunswick. The paucity of core sites in these areas is presumably a function of inaccessibility into the interior parts. On-going studies by various investigators in parts of interior Quebec, northeast Labrador, central Newfoundland and northeast New Brunswick will help rectify this situation.

Similarities exist amongst several of the pre-Late Wisconsin sites. Four sites (Portage-du-Cap, East Milford, Addington Forks, Leitches Creek) show hardwood pollen assemblages which are indicative of warm interglacial (Sangamon) conditions. Five sites (Hillsborough, Whycocomagh, River Inhabitants, Miller Creek, the upper record at East Milford) are distinguished by boreal assemblages signifying cooler conditions possibly associated with a late phase of the Sangamonian Interglacial. Two other sites (Bay St. Lawrence and Woody Cove) contain pollen assemblages that are ascribed to a tundra-boreal-tundra sequence. The Bay St. Lawrence sequence is interpreted as an interstade but it is not too dissimilar from the interglacial record at Woody Cove. Differences between the sites and between Bay St. Lawrence and other sites from inland Nova Scotia perhaps reflect contrasting environmental conditions in the coastal areas and between coastal and inland sites at the time. Pollen assemblages that characterize

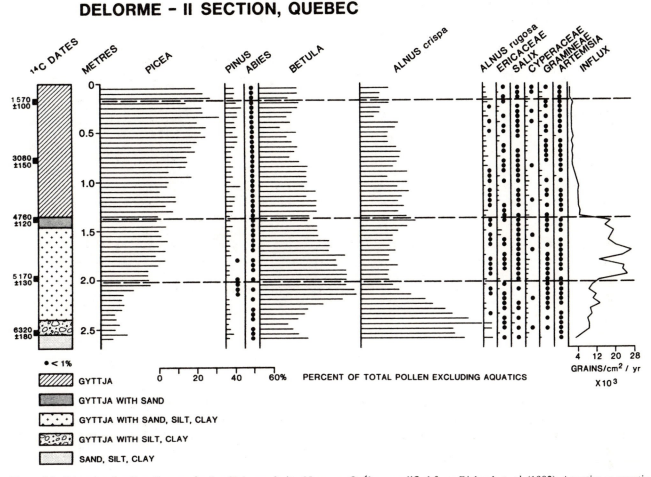

Figure 25. Abbreviated pollen diagram for Lac Delorme-2 site, Nouveau-Québec, modified from Richard *et. al,* (1982). Aquatics = aquatic plants.

the Early Wisconsin St. Pierre beds of the St. Lawrence Lowlands and the correlative deposits at Pointe-Fortune in the Ottawa Valley imply cooler-than-present (interstadial) conditions. Interestingly, these assemblages are not too unlike the Late Sangamonian record at East Milford. Further pollen studies and better dating techniques (thermoluminescence or Uranium-Thorium dating) may eventually resolve the contention of interstadials versus interglacials.

The lack of any Middle Wisconsin non-glacial sediments and hence pollen records of the same age in the Atlantic Provinces is apparent. The paucity of terrestrial organic sites dating to the Middle Wisconsin may be attributed to the presence of widespread glaciation at this time as is implied from the offshore records.

Late Wisconsin-Holocene pollen profiles were combined to form two pollen stratigraphic transects

extending from the Atlantic Provinces to Newfoundland (Fig. 31) and from eastern Ontario to northernmost Nouveau-Québec and Labrador (Fig. 32). Composite pollen stratigraphies were developed for some areas by combining representative pollen profiles of a particular region. For example, the older record at Canoran Lake (Railton, 1972) was combined with the younger record at Silver Lake (Fig. 16) to obtain a composite pollen stratigraphy for south-central Nova Scotia. Likewise, diagrams from Burin Peninsula (Fig. 13) and Sugarloaf Lake (Fig. 17) were joined for a representative pollen stratigraphy of eastern Newfoundland. Combined profiles from Lambs Pond (Fig. 18) and Atkins Lake (Terasmae, 1980) yielded a common stratigraphy for eastern Ontario. In a similar way, the Mont Shefford (Fig. 20) and Lac Colin (Fig. 21) sites were combined for southern Quebec, south of the St. Lawrence, and Lake Irene (Fig. 22) and Lac Clo (Richard, 1979) diagrams were combined for south-central Quebec.

RIVIERE AUX FEUILLES I ,QUEBEC
58°14'00" N, 72°04'00" W

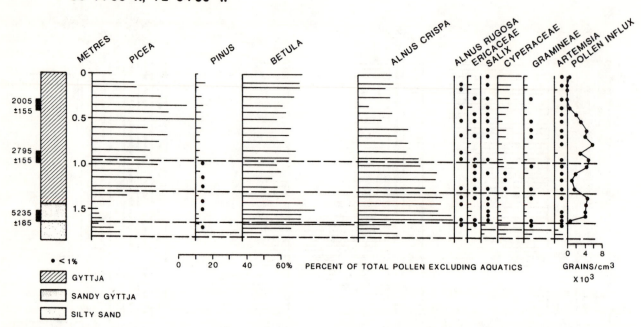

Figure 26. Abbreviated pollen diagram for Rivière aux Feuilles-1 site, Nouveau-Québec, modified from Richard (1981a). Aquatics = aquatic plants.

The Late Wisconsin-Holocene pollen stratigraphy, although varied from site to site, is clearly time-transgressive from the coastal areas towards the interior

at sites in coastal Labrador and the youngest in eastern Nouveau-Québec and interior Labrador. The chrono-palynostratigraphy corresponds generally

CHURCHILL FALLS SOUTH, LABRADOR

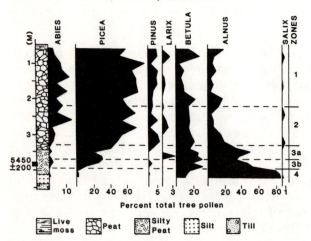

Figure 27. Percentage pollen diagram for Churchill Falls South site, Labrador, from Morrison (1970).

EAGLE LAKE, LABRADOR

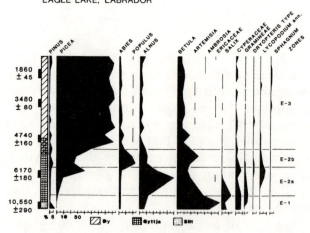

Figure 28. Percentage pollen diagram for Eagle Lake, southeast Labrador, modified from Lamb (1980).

parts and from south to north. The transect in Figure 31 shows that the oldest basal stratigraphy occurs along the outer coasts of Newfoundland and in southwestern New Brunswick and the youngest in northern New Brunswick and Prince Edward Island. In Figure 32 the oldest basal records are in eastern Ontario and

with the pattern of ice retreat in eastern Canada in that the oldest basal pollen records occur in areas that were deglaciated early while the youngest denote areas that were characterized by the last remnants of ice.

Eastern Canada can be divided into three separate regions in terms of Holocene pollen stratigraphy (Fig.

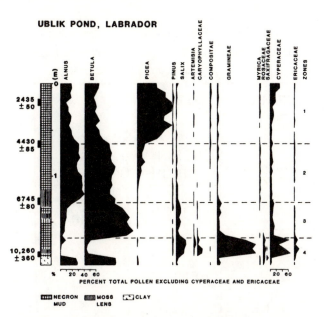

Figure 29. Percentage pollen diagram for Ublik Pond, northeast Labrador, modified from Short and Nichols (1977).

31 and 32). Region 1 encompasses the Atlantic Provinces, eastern Ontario and south-central Quebec from Gaspé Peninsula to James Bay. The stratigraphic sections of this region indicate a mixed forest was present throughout much of the Holocene. Region 2 encompasses Newfoundland, Labrador, central Quebec and Nouveau-Québec to Hudson Bay except northernmost Nouveau-Québec. The pollen zones of this region show the Holocene forest was always boreal in character. Region 3 is the northernmost tip of Nouveau-Québec east and west of Ungava Bay. This region is distinguished by herb assemblages indicating the persistence of tundra throughout the Holocene.

The pollen stratigraphy is transitional at the regional boundaries and reflects north-south movements of the forest or particular elements of the forest across the boundaries at certain times in the Holocene. For example, zones dominated by white pine characterize the pollen stratigraphy of south-central Quebec (Fig. 32) from about 7,000 to 3,200 B.P. These zones mark the northward migration of the white pine ecotone in response to Hypsithermal warming. The succeeding spruce zones represent a revertence from a mixed forest to a boreal-type forest after about 4,000 B.P. when the spruce ecotone advanced southward in response to post-Hypsithermal cooling.

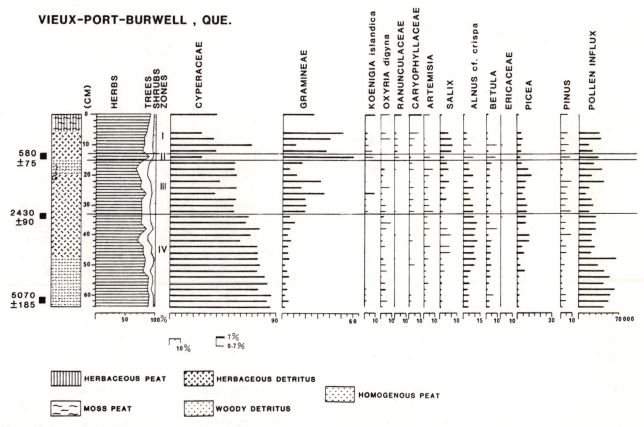

Figure 30. Percentage pollen diagram of a pulsa at Vieux-Port-Burwell, northernmost Nouveau-Québec, modified from Savoie et Gangloff (1980).

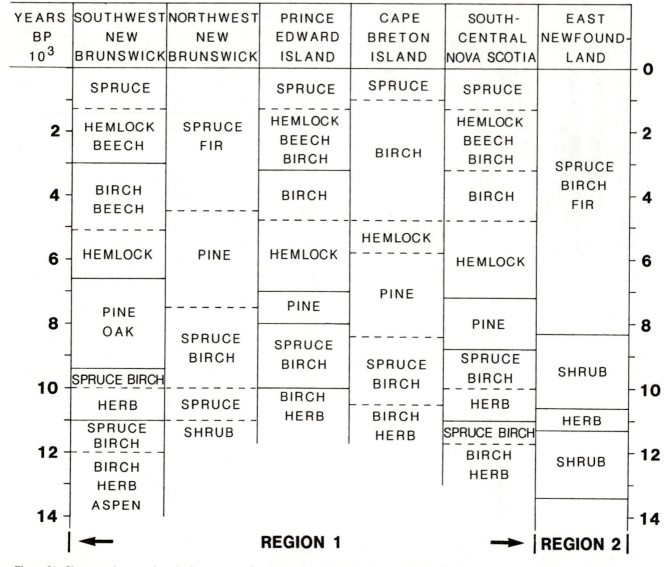

Figure 31. Chrono-palynostratigraphy in a transect from the Atlantic Provinces to Newfoundland.

Pollen influx trends from eastern Canada are useful criteria in recognizing the middle-Holocene Hypsithermal interval and in determining its upper and lower limits. Total pollen influx estimates increase several-fold between 9,000 and 10,000 B.P. at ten sites in the Atlantic region, southern Quebec and eastern Ontario. These increases represent the widespread proliferation of the vegetation as a result of regional climatic warming at this time. The thermal maximum may have been achieved everywhere in the south by about 6,000 to 8,000 B.P. but ice persisted in the Nouveau-Québec–western Labrador region until about 6,000 B.P. which delayed warming from reaching the north. Influx values at sites in central Quebec, Nouveau-Québec and Labrador increase gradually to maximum proportions by

about 4,000 B.P. reflecting the northward migration of forest, notably spruce, in response to the warmer conditions. The influx of spruce pollen (increasing trends at some sites, decreasing trends at others) reflects an overall cooling tendency commencing between 3,000 and 4,000 B.P.

ACKNOWLEDGMENTS

The author gratefully acknowledges the following for providing up-dated lists of site investigations and for unpublished pollen stratigraphic data: R. J. Mott for New Brunswick and Nova Scotia, J. B. Macpherson for Newfoundland, P. J. H. Richard for Quebec, D. R. Engstrom for southwest Labrador–east-central

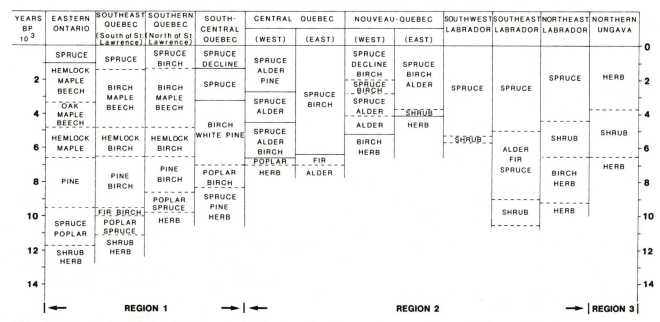

Figure 32. Chrono-palynostratigraphy in a transect from eastern Ontario to northernmost Nouveau-Québec and Labrador.

Quebec, S. K. Short for northeast Labrador, and T. Webb III for sites in eastern Canada as a whole. I am indebted to J. Dale for assistance in compiling the pollen data and for modifying and drafting pollen diagrams and to T. Gerrard for drafting pollen diagrams. I acknowledge with thanks the helpful suggestions and constructive criticism of the manuscript by R. J. Mott and W. Blake, Jr.

References Cited

ANDERSON, T. W.
1980   Holocene vegetation and climatic history of Prince Edward Island, Canada. *Canadian Journal of Earth Sciences*, 17: 1152-1165.
1983   Preliminary evidence for late Wisconsinan climatic fluctuations from pollen stratigraphy in Burin Peninsula, Newfoundland. *In: Current Research, Part B, Geological Survey of Canada, Paper 83-1B*: 185-188.

AUER, V.
1930   Peat bogs in southeastern Canada. *Geological Survey of Canada, Memoir*, 162: 1-32.

BANFIELD, C. C.
1981   The climatic environment of Newfoundland. *In:* Macpherson, A. G. and Macpherson, J. B. (eds.), *The Natural Environment of Newfoundland, Past and Present*. Department of Geography, Memorial University of Newfoundland, p. 83-153.

BARENDSEN, G. W., DEEVEY, E. S. and GRALENSKI, L. J.
1957   Yale Natural Radiocarbon Measurements III. *Science*, 126: 908-919.

BARTLEY, D. D. and MATTHEWS, B.
1969   A palaeobotanical investigation of postglacial deposits in the Sugluk area of northern Ungava (Quebec, Canada). *Review of Palaeobotany and Palynology*, 9: 45-61.

BOSTOCK, H. S.
1970   Physiographic subdivisions of Canada. *In:* Douglas, R. J. W. (ed.), *Geology and Economic Minerals of Canada, Economic Geology Report No. 1, Geological Survey of Canada*, p. 10-30.

BOWMAN, P. W.
1931   Study of a peat bog near the Matamek River, Quebec, by the method of pollen analysis. *Ecology* 12: 694-708.

BROOKES, I. A., MCANDREWS, J. H., and VON BITTER, P. H.
1982   Quaternary interglacial and associated deposits in south-west Newfoundland. *Canadian Journal of Earth Sciences,* 19: 410-423.

BRYSON, R. A.
1966   Air masses, streamlines, and the boreal forest. *Geographical Bulletin,* 8: 228-269.

BRYSON, R. A. and HARE, F. K.
1974   *Climates of North America.* Elsevier Scientific Publishing Company, New York, 420 p.

CALKIN, P. E.
1982   Glacial geology of the Erie Lowland and adjoining Allegheny Plateau, western New York. *Field Trips Guidebook for New York State Geological Association 54th Annual Meeting,* p. 121-139.

CANADA DEPARTMENT OF ENERGY, MINES AND RESOURCES.
1969   *National Atlas of Canada. Growing Season, Frost.* Map 51-52.
1973   *The National Atlas of Canada. Vegetation Regions.* Map 45-46.

COMTOIS, P.
1982   Histoire holocène du climat et de la végétation à Lanoraie (Québec). *Canadian Journal of Earth Sciences,* 19: 1938-1952.

DAVIS, M. B.
1981   Outbreaks of forest pathogens in Quaternary history. *Proceedings IV International Palynological Conference,* 3: 216-227.

de VERNAL, A., MATHIEU, C. and GANGLOFF, P.
1983   Analyse stratigraphique d'un lobe de gélifluxion des Torngats centrales, Labrador. *Géographie physique et Quaternaire,* XXXVII: 205-210.

de VERNAL, A., RICHARD, P. J. H. and OCCHIETTI, S.
1983   Palynologie et paléoenvironments du Wisconsinien de la région de la baie Saint-Laurent, Île du Cap Breton. *Géographie physique et Quaternaire,* XXXVII: 307-322.

DENTON, G. H. and HUGHES, T. J.
1981   *The Last Great Ice Sheets.* John Wiley and Sons, Toronto. 484 p.

FRANKEL, L. and CROWL, G. H.
1961   Drowned forests along the eastern coast of Prince Edward Island, Canada. *The Journal of Geology,* 69: 352-357.

GADD, N. R.
1971   Pleistocene geology of the Central St. Lawrence Lowlands. *Geological Survey of Canada, Memoir 359:* 1-153.

GADD, N. R., RICHARD, S. H. and GRANT, D. R.
1981   Pre-last-glacial organic remains in Ottawa Valley. *In: Current Research, Part C, Geological Survey of Canada,* Paper 81-1C; 65-66.

GRANT, D. R.
1969   Surficial deposits, geomorphic features, and Late Quaternary history of the terminus of the northern peninsula of Newfoundland and adjacent Quebec-Labrador. *Maritime Sediments,* 5: 123-125.
1977   Glacial style and ice limits, the Quaternary stratigraphic record, and changes of land and ocean level in the Atlantic Provinces, Canada. *Géographie physique et Quaternaire,* XXXI: 247-260.

GRANT, D. R. and KING, L. H.
1984   A stratigraphic framework for the Quaternary history of the Atlantic Provinces, Canada. *In:* Fulton, R. J. (ed.), *Quaternary Stratigraphy of Canada – A Canadian Contribution to IGCP Project 24, Geological Survey of Canada,* Paper 84-10, p. 173-191.

GRAYSON, J. F.
1956   The postglacial history of vegetation and climate in the Labrador-Quebec region as determined by palynology. Unpublished thesis, University of Michigan, Ann Arbor, Michigan. 252 p.

GREEN, D. G.
1976   Nova Scotian forest history – evidence from statistical analysis of pollen data. Unpublished thesis, Dalhousie University, Halifax, Nova Scotia. 155 p.
1981   Time series and postglacial forest ecology. *Quaternary Research,* 15: 265-277.

HADDEN, K. A.
1975   A pollen diagram from a postglacial peat bog in Hants County, Nova Scotia. *Canadian Journal of Botany,* 53: 39-47.

HARDY, L.
1982   La Moraine Frontal de Sakami, Québec subarctique. *Géographie physique et Quaternaire,* XXXVI: 51-61.

HARE, F. K. and THOMAS, M. K.
1974   *Climate Canada.* Wiley Publishers of Canada Limited, Toronto. 256 p.

HILLAIRE-MARCEL, C., GRANT, D. R. and VINCENT, J.-S.
1980   Comment on "Keewatin Ice Sheet – Re-evaluation of the traditional concept of the Laurentide Ice Sheet" and "Glacial erosion and ice sheet divides, northeastern Laurentide Ice Sheet, on the basis of the distribution of limestone erratics." *Geology,* 8: 466-468.

HILLAIRE-MARCEL, C. and OCCHIETTI, S.
1980   Chronology, paleogeography and paleoclimatic significance of the late and postglacial events in eastern Canada. *Zeitschrift für Geomorphologie,* 24: 373-392.

HILLS, L. V. and SANGSTER, E. V.
1980   A review of paleobotanical studies dealing with the last 20,000 years; Alaska, Canada and Greenland. *In:* Harington, C. R. (ed.), *Climatic Change in Canada, National Museums of Canada, Syllogeus Series,* 26: 73-246.

HOUDE, A.
1978   *Atlas climatologue de Québec.* Température-precipitation. Service de la Meteorologie, Ministère des Richesses Naturelles, Quebec.

IGNATIUS, H. G.
1956   Late-Wisconsin stratigraphy in north-central Quebec and Ontario, Canada. Unpublished thesis, Yale University, New Haven. 84 p.

JORDON, R. H., JR.
1975   Pollen diagrams from Hamilton Inlet, central Labrador, and their environmental implications for the Northern Maritime Archaic. *Arctic Anthropology,* 12: 92-116.

KARROW, P. F., ANDERSON, T. W., CLARKE, A. H., DELORME, L. D. and SREENIVASA, M. R.
1975   Stratigraphy, paleontology, and age of Lake Algonquin Sediments in southwestern Ontario, Canada. *Quaternary Research,* 5: 49-87.

KING, L. H.
1969   Submarine end moraines and associated deposits on the Scotian shelf. *Geological Society of America Bulletin,* 80: 83-96.

KRANCK, K.

1972 Geomorphological development and post-Pleistocene sea level changes, Northumberland Strait, Maritime Provinces. *Canadian Journal of Earth Sciences,* 9: 835-844.

LABELLE, C. et RICHARD, P. J. H.

1981a Végétation tardiglaciaire et postglaciaire au sud-est du Parc des Laurentides, Québec. *Géographie physique et Quaternaire,* XXXV: 345-359.

1981b Palynological data from the Mont Saint-Pierre-Mont Jacques-Cartier region, Gaspé. *In: Excursion and Conference in Gaspé, Quebec. Weathering Zones and the Problem of Glacial Limits.* The Canadian Association for Quaternary Studies and the Quebec Association for Quaternary Studies, p. 137-150.

LAMP, H. F.

1980 Late Quaternary vegetational history of southeastern Labrador. *Arctic and Alpine Research,* 12: 117-135.

LASALLE, P.

1966 Late Quaternary vegetation and glacial history in the St. Lawrence Lowlands, Canada. *Leidse Geologische Mededelingen,* 38: 91-128.

1984 Quaternary stratigraphy of Quebec: A review. *In:*Fulton, R. J. (ed.), *Quaternary Stratigraphy of Canada – A Canadian Contribution to IGCP Project 24, Geological Survey of Canada,* Paper 84-10, p. 155-171.

LASALLE, P. and ELSON, J. A.

1975 Emplacement of the St. Narcisse morine as a climatic event in eastern Canada. *Quaternary Research,* 5: 621-625.

LAURIOL, B., GRAY, J. T. et CYR. A

1979 Le Cadre chronologique et paléogeographique de l'évolution marine depuis la déglaciation dans la région d'Aupaluk, Nouveau-Québec. *Géographie physique et Quaternaire,* XXXIII: 189-203.

LIVINGSTONE, D. A.

1968 Some interstadial and postglacial pollen diagrams from eastern Canada. *Ecological Monographs,* 38: 87-125.

LIVINGSTONE, D. A. and ESTES, H.

1967 A carbon-dated pollen diagram from the Cape Breton plateau, Nova Scotia. *Canadian Journal of Botany,* 45: 339-359.

LIVINGSTONE, D. A. and LIVINGSTONE, B. G. R.

1958 Late-glacial and postglacial vegetation from Gillis Lake in Richmond County, Cape Breton Island, Nova Scotia. *American Journal of Science,* 256: 341-359.

LOUCKS, O. L.

1962 A forest classification for the Maritime Provinces. *Proceedings of the Nova Scotian Institute of Science,* 25, Part 2: 1-167.

MACNEILL, R. H.

1969 Some dates relating to the dating of the last major ice sheet in Nova Scotia. *Maritime Sediments,* 5: 3.

MACPHERSON, J. B.

1981 The development of the vegetation of Newfoundland and climatic change during the Holocene. *In:* Macpherson, A. G. and Macpherson, J. B. (eds.), *The Natural Environment of Newfoundland, Past and Present.* Department of Geography, Memorial University of Newfoundland, p. 189-217.

1982 Postglacial vegetational history of the eastern Avalon Peninsula, Newfoundland, and Holocene climatic change along the eastern Canadian seaboard. *Géographie physique et Quaternaire,* XXXVI: 175-196.

MCANDREWS, J. H. et SAMSON, G.

1977 Analyse pollinique et implications archéologiques et géomorphologiques, Lac de la Hutte Sauvage (Mushuaunipi), Nouveau-Québec. *Géographie physique et Quaternaire,* XXXI: 177-183.

MCDONALD, B. C.

1971 Late Quaternary stratigraphy and deglaciation in eastern Canada. *In:* Turekian, K. K. (ed.), *The Late Cenozoic Glacial Ages.* Yale University Press, New Haven, p. 331-353.

MCDONALD, B. C. and SHILTS, W. W.

1971 Quaternary stratigraphy and events in southeastern Quebec. *Geological Society of America Bulletin,* 82: 683-698.

MORRISON, A.

1970 Pollen diagrams from interior Labrador. *Canadian Journal of Botany,* 48: 1957-1975.

MOTT, R. J.

1971 Palynology of a buried organic deposit, River Inhabitants, Cape Breton Island, Nova Scotia. *In: Report of Activities, Part B, Geological Survey of Canada, Paper 71-1B:* 123-125.

1973 Buried Quaternary organic deposits from Cape Breton Island, Nova Scotia, Canada. Abstract, *Geoscience and Man,* VII: 122.

1975a Palynological studies of lake sediment profiles from southwestern New Brunswick. *Canadian Journal of Earth Sciences,* 12: 273-288.

1975b Postglacial history and environments in southwestern New Brunswick. *In:* Ogden, J. G., III and Harvey, M. J. (eds.), *Environmental Change in the Maritimes. Proceedings of the Nova Scotia Institute of Science 27,* Supplement 3: 67-82.

1975c Palynological studies of peat monoliths from L'Anse aux Meadows Norse site, Newfoundland. *In: Report of Activities, Geological Survey of Canada, Paper 75-1A:* 451-454.

1976 A Holocene pollen profile from the Sept-Îles area, Quebec. *Le Naturaliste Canadien,* 103: 457-467.

1977 Late-Pleistocene and Holocene palynology in southeastern Quebec. *Géographie physique et Quaternaire,* XXXI: 139-149.

1983 Late-glacial climatic change in Maritime Canada. *Program and Abstracts, Critical Periods in the Quaternary Climatic History of Northern North America,* National Museum of Natural Sciences, Climatic Change in Canada Project, Ottawa, p. 28.

MOTT, R. J., ANDERSON, T. W. and MATTHEWS, J. V.

1981 Late-glacial paleoenvironments of sites bordering the Champlain Sea based on pollen and macrofossil evidence. *In:* Mahaney, W. C. (ed.), *Quaternary Paleoclimate.* GeoAbstracts, England, p. 129-172.

1982 Pollen and macrofossil study of an interglacial deposit in Nova Scotia. *Géographic physique et Quaternaire* XXXVI: 197-208.

MOTT, R. J. and CAMFIELD, M.

1969 Palynological studies in the Ottawa area. *Geological Survey of Canada, Paper 69-38:* 1-16.

MOTT, R. J. and FARLEY-GILL, L.

1981 Two late Quaternary pollen profiles from Gatineau Park, Quebec. *Geological Survey of Canada, Paper 80-31:* 1-10.

MOTT, R. J. and PREST, V. K.
1967 Stratigraphy and palynology of buried organic deposits from Cape Breton Island, Nova Scotia. *Canadian Journal of Earth Sciences*, 4: 709-724.

OCCHIETTI, S.
1982 Synthèse lithostratigraphique et paleoenvironments du Quaternaire au Québec mèridional. Hypothése d'un centre d'englacement Wisconsinen au Nouveau-Québec. *Géographic physique et Quaternaire*, XXXVI: 15-49.

OGDEN, J. G., III
1960 Recurrence surfaces and pollen stratigraphy of a postglacial raised bog, Kings County, Nova Scotia. *American Journal of Science*, 258: 341-353.

OSVALD, H.
1970 Vegetation and stratigraphy of peatlands in North America. *Acta Universitatus Upsaliensis, Series 5-C*, 1: 1-96.

OUELLET, M. et POULIN, P.
1975 Quelques aspects peléoécologiques de la tourbière et du Lac Desauliniers et quelques spectres sporolliniques modernes du bassin de la Grande Rivière, Baie James. *L'Institut national de la recherche scientifique – Eau, Rapport Scientifique*, No. 58: 1-48.

1976 Etudes paléoécologiques des sédiments du Lac Waterloo, Québec. *L'Institut national de la recherche scientifique – Eau, Rapport Scientifique*, No. 64: 1-87.

PAYETTE, S.
1983 The forest tundra and present tree-lines of the northern Québec-Labrador Peninsula. *Nordicana*, 47: 3-23

POLUNIN, N.
1948 Botany of the Canadian Eastern Arctic – Part III. Vegetation and Ecology. *National Museum of Canada, Bulletin* 104: 1-304.

POTZGER, J. E.
1953 Nineteen bogs from southern Quebec. *Canadian Journal of Botany*, 31: 383-401.

POTZGER, J. E. and COURTEMANCHE, A.
1954a A radiocarbon date of peat from James Bay, Quebec. *Science*, 119: 908-909.

1954b Bog and lake studies on the Laurentian Shield in Mont Tremblant Park, Quebec. *Canadian Journal of Botany*, 32: 549-560.

1956a A series of bogs across Quebec from the St. Lawrence valley to James Bay. *Canadian Journal of Botany*, 34: 473-500.

1956b Pollen study in the Gatineau Valley, Quebec. *Butler University Botanical Studies*, 13: 12-23.

PREST, V. K.
1970 Quaternary geology of Canada. *In*: Douglas, R. J. W. (ed.), *Geology and Economic Minerals of Canada, Economic Geology Report No. 1 Geological Survey of Canada*, p. 676-764.

1977 General stratigraphic framework of the Quaternary in eastern Canada. *Géographie physique et Quaternaire*, XXXI: 7-14.

1983 Canada's heritage of glacial features. *Geological Survey of Canada, Miscellaneous Report*, 28: 1-119.

PREST, V. K., GRANT, D. R., BORNS, H. W., JR., BROOKES, I. A., MACNEILL, R. H., OGDEN, J. G., III, JONES, J. F., LIN, C. L., HENNIGAR, T. W. and PARSONS, M. L.
1972 Quaternary geology, geomorphology and hydrogeology of the Atlantic Provinces. *Guidebook Excursion A61-C61*. XXIV International Geological Congress. Montreal. 79 p.

PREST, V. K., TERASMAE, J., MATTHEWS, J. V. and LICHTI-FEDEROVICH, S.
1976 Late-Quaternary history of Magdalen Islands, Quebec. *Maritime Sediments*, 12: 39-59.

PUTNAM, D. F.
1940 The climate of the Maritime Provinces. *Canadian Geographical Journal*, 21: 134-147.

RADFORTH, N. W.
1945 Report on the spore and pollen constituents of a peat bed in the Shipshaw area, Quebec. *Proceedings and Transactions of the Royal Society of Canada, Third Series, Section V*, XXXIX: 131-142.

RAILTON, J. B.
1972 Vegetational and climatic history of southwestern Nova Scotia in relation to a South Mountain ice cap. Unpublished thesis, Dalhousie University, Halifax, Nova Scotia, 146 p.

RICHARD, P. J. H.
1971 Two pollen diagrams from the Quebec City area, Canada. *Pollen et Spores*, 13: 523-559.

1973a Histoire postglaciaire de la végétation dans la région de Saint-Raymond de Portneuf, telle que révélée par l'analyse pollinique d'une tourbière. *Le Naturaliste Canadien*, 100: 561-575.

1973b Histoire postglaciaire comparée de la végétation dans deux localités au nord du Parc des Laurentides, Québec. *Le Naturaliste Canadien*, 100: 577-590.

1973c Histoire postglaciaire comparée de la végétation dans deux localités au sud de la ville de Québec. *Le Naturaliste Canadien*, 100: 591-603.

1975a Contribution à l'histoire postglaciaire de la végétation dans la plaine du Saint-Laurent: Lotbinière et Princeville. *La Revue de Géographie de Montréal*, 29: 95-107.

1975b Contribution à l'histoire postglaciaire de la végétation dans les Cantons-de-l'est: ètude des sites de Weedon et Albion. *Cahiers de Géographie de Québec*, 19: 267-284.

1977 Histoire post-Wisconsinienne de la végétation du Québec méridional par l'analyse pollinique. Tome 1: 1-312; Tome 2: 1-141. *Gouvernement du Québec, Service de la Recherche, Direction Générale des Forêts, Ministère des Terres et Forêts*.

1978 Histoire tardiglaciaire et postglaciaire de la végétation au Mont Shefford, Québec. *Géographie physique et Quaternaire*, XXXII: 81-93.

1979 Contribution à l'histoire postglaciaire de la végétation au nord-est de la Jamésie, Nouveau-Québec. *Géographie physique et Quaternaire*, XXXIII: 93-112.

1980 Histoire postglaciaire de la végétation au sud du lac Abitibi, Ontario et Québec. *Géographie physique et Quaternaire*, XXXIV: 77-94.

1981a Paléophytogéographie postglaciaire en Ungava par l'analyse pollinque. *Collection Paléo-Québec*, 13: 1-153.

1981b Paleoclimatic significance of the late-Pleistocene and Holocene pollen record in south-central Québec. *In*: Mahaney, W. C. (ed.), *Quaternary Paleoclimate*, GeoAbstracts, England, p. 335-350.

RICHARD, P. J. H., LAROUCHE, A. et BOUCHARD, M.
1982 Age de la déglaciation finale et histoire postglaciaire de la vegetation dans la partie centrale du Nouveau-Québec. *Géographie physique et Quaternaire*, XXXVI: 63-90.

RICHARD, P. J. H. et POULIN, P.
1976   Un Diagramme pollinque au Mont des Èboulements, région de Charlevoix, Québec. *Canadian Journal of Earth Sciences,* 13: 145-156.

ROGERSON, R. J.
1982   The glaciation of Newfoundland and Labrador. *In:* Davenport, P. H. (ed.), *Prospecting in Areas of glaciated terrain, Geology Division, The Canadian Institute of Mining and Metallurgy,* p. 37-56.

ROWE, J. S.
1977   *Forest regions of Canada.* Department of Fisheries and the Environment, Canadian Forestry Service, Publication 1300, 172 p.

RUBIN, M. and SUESS, H. E.
1955   United States Geological Survey radiocarbon dates II. *Science,* 121:485.

RUDDIMAN, W. F. and MCINTYRE, A. F.
1981   The North Atlantic Ocean during the last deglaciation. *Palaeogeography, Palaeoclimatology, Palaeoecology,* 35: 145-214.

SAVOIE, L. et RICHARD, P. J. H.
1979   Paléophytogéographie de l'episode de Sainte-Narcisse dans la région de Sainte-Agathe, Québec. *Géographie physique et Quaternaire,* XXXIII: 175-188.

SAVOIE, L. et GANGLOFF, P.
1980   Analyse pollinique d'une palse au site archéologique de Vieux-Port-Burwell (Killiniq), Territoires de Nord-Ouest. *Géographic physiqe et Quaternaire,* XXXIV: 301-320.

SHORT, S. K.
1980   Pollen analyses, northeastern Labrador Coast. *In: Climatic Reconstructions of Late- and Post-Glacial Environments: Eastern Canadian Arctic. Final Report to the National Science Foundation. Institute of Arctic and Alpine Research,* University of Colorado, p. 126-134.

SHORT, S. K. and NICHOLS, H.
1977   Holocene pollen diagrams from subarctic Labrador-Ungava: Vegetational history and climatic change. *Arctic and Alpine Research,* 9: 265-290.

SKINNER, R. G.
1973   Quaternary stratigraphy of the Moose River Basin, Ontario. *Geological Survey of Canada, Bulletin* 225: 1-77.

STEA, R. and HEMSWORTH, D.
1979   Pleistocene stratigraphy of the Miller Creek Section, Hants County, Nova Scotia. *Nova Scotia Department of Mines and Energy, Paper,* 79-5: 1-16.

STRAVERS, L. K. S.
1981   Palynology and deglaciation history of the central Labrador-Ungava Peninsula. Unpublished thesis, University of Colorado, Boulder, Colorado, 171 p.

STUIVER, M., HEUSSER, C. J. and YANG, I. C.
1978   North American glacial history extended back to 75,000 years ago. *Science,* 200: 16-21.

STUIVER, M., DEEVEY, E. S., JR. and ROUSE, I.
1963   Yale Natural Radiocarbon Measurements VIII. *Radiocarbon,* 5: 312-341.

TERASMAE, J.
1958   Contributions to Canadian palynology. Part 1: The use of palynological studies in Pleistocene stratigraphy; Part 2: Non-glacial deposits in the St. Lawrence Lowlands, Quebec; Part 3: Non-glacial deposits along the Missinaibi River, Ontario. *Geological Survey of Canada, Bulletin* 46: 1-35.

1960   Contributions to Canadian palynology No. 2, Part I: A palynological study of post-glacial deposits in the St. Lawrence Lowlands. *Geological Survey of Canada, Bulletin,* 56: 1-22.

1963   Three C-14 dated pollen diagrams from Newfoundland. *Advancing Frontiers of Plant Sciences (New Delhi),* 6: 149-162.

1965   Surficial geology of the Cornwall and St. Lawrence Seaway Project areas, Ontario. *Geological Survey of Canada, Bulletin,* 121: 1-54.

1973   Notes on late Wisconsin and early Holocene history of vegetation in Canada. *Arctic and Alpine Research,* 5: 201-222.

1974   Deglaciation of Port Hood Island, Nova Scotia. *Canadian Journal of Earth Sciences,* 11: 1357-1365.

1980   Some problems of late Wisconsin history and geochronology in southeastern Ontario. *Canadian Journal of Earth Sciences,* 17: 361-381.

TERASMAE, J. and ANDERSON, T. W.
1970   Hypsithermal range extension of white pine (*Pinus strobus* L.) in Quebec, Canada. *Canadian Journal of Earth Sciences,* 7: 406-413.

TERASMAE, J. and LASALLE, P.
1968   Notes on late glacial palynology and geochronology at St. Hilaire, Quebec. *Canadian Journal of Earth Sciences,* 5: 249-257.

TERASMAE, J. and MOTT, R. J.
1971   Postglacial history and palynology of Sable Island, Nova Scotia. *Geoscience and Man,* 3: 17-28.

TERASMAE, J., WEBBER, P. J. and ANDREWS, J. T.
1966   A study of late-Quaternary plant-bearing beds in north-central Baffin Island, Canada. *Arctic,* 19: 296-318.

TUCKER, C. M. and MCCANN, B.
1980   Quaternary events on the Burin Peninsula, Newfoundland and the islands of St. Pierre and Miquelon, France. *Canadian Journal of Earth Sciences,* 17: 1462-1479.

VINCENT, J. S.
1973   A palynological study for the Little Clay Belt, northwestern Québec. *Le Naturaliste Canadien,* 100: 59-70.

WALKER, I. R. and PATERSON, C. G.
1983   Post-glacial chironomid succession in two small, humic lakes in New Brunswick-Nova Scotia (Canada) border area. *Freshwater Invertebrate Biology,* 2: 61-73.

WEBB, T., III, RICHARD, P. J. H. and MOTT, R. J.
1983   A mapped history of Holocene vegetation in southern Quebec, *In:* Harington, C. E. (ed.), *Climatic Change in Canada 3, National Museums of Canada, Syllogeus Series,* 49: 273-336.

WENDLAND, W. M. and BRYSON, R. A.
1981   Northern hemisphere airstream regions. *Monthly Weather Review,* 109: 255-270.

WENNER, C. G.
1947   Pollen diagrams from Labrador. *Geografiska Annaler,* 29: 137-364.

WERNSTEDT, F. L.
1972   *World climatic data.* Climatic Data Press, Lemont, Pennsylvania. 522 p.

WIGHTMAN, D. M.
1980   Late Pleistocene glaciofluvial and glaciomarine sediments on the north side of the Minas Basin, Nova Scotia. Unpublished thesis, Dalhousie University, Halifax, Nova Scotia, 426 p.

WILSON, C. V.

1971   The climate of Québec in two parts: Part 1, Climatic
        atlas. *Canadian Meteorological Service, Climatological
        Studies II,* Information Canada.

1973   The climate of Québec, climatic atlas. *Canadian Meteor-
        ological Service,* Environment Canada.

QUATERNARY POLLEN RECORDS FROM THE WESTERN INTERIOR AND THE ARCTIC OF CANADA

JAMES C. RITCHIE
Scarborough College
University of Toronto
Scarborough, Ontario
M1C 1A4, Canada

Abstract

This paper represents an overview of the palynological records from the Canadian Arctic Archipelago, the continental Northwest Territories, the Yukon Territory, the eastern segment of the Province of British Columbia, and all of the Provinces of Alberta, Saskatchewan, and Manitoba. Pollen data from 218 sites have thus far been published for this area and they are mentioned in this article as to their name, location, palynologist who worked at the site, time span covered by the examined sediments, and supporting paleoecological data, if present. This summary article discusses a wide variety of sites ranging from a few early and middle-Pleistocene sites to a large number of Holocene records; thus the treatment of each site has, by necessity, been condensed. A final section mentioning some of the needs for future research in this region also is included.

INTRODUCTION

The area dealt with in this summary includes the Canadian Arctic Archipelago, the continental Northwest Territories, the Yukon Territory, the eastern segment of the Province of British Columbia, and all of the Provinces of Alberta, Saskatchewan, and Manitoba. (Fig. 1). This immense region, roughly 6.3 million sq. km., includes the full range of vegetation, fauna and soils associated with regional climates that range from semi-arid continental through humid continental to polar continental and polar maritime.

It consists of four physiographic regions: the Cordillera, a north-south complex of several major mountain ranges with relief up to 3,000 m; the Interior Plains forming a wide arc in the western region, composed of thick glacial deposits with relief from > 2,000 m to 300 m; the Shield region, flanking Hudson's Bay, a mosaic of predominantly bedrock-controlled landscapes with relief from 100 to 300 m; and the Borderlands, consisting of the Arctic lowlands, made up primarily of the western and central arctic islands, with relief up to 600 m, and the

Inuitian region centered on Ellesmere and Baffin Islands, with extensive montane relief from 1,500 to 2,000 m, with maximum up to 2,800 m.

The climate displays steep gradients along a line running roughly from the southwest corner of the area to North Ellesmere Island, with the following values at the two extremities: from 2,500 to 25 degree days of growing season, defined as the number of days with the average daily temperature greater than 5.6°C; from 4.9°C to −15°C for the mean annual daily temperature; and a period of from 240 days to 50 days per year when lakes are ice-free. Average annual precipitation is low in the southwest corner (12 cm), higher in the central zone (30.5 cm near the northern boundary of Saskatchewan), and very low on Ellesmere Island in the far northeast (5 cm).

The major vegetation zones along this broad transect are grasslands in the southwest, a broad band of boreal forest, forest-tundra, low shrub tundra and high arctic herb-shrub tundra and polar desert, with regional variants in the Cordilleran foothills (Fig. 1).

The extent and chronology of Pre Wisconsin glaciations is poorly understood. All of the area, with the exception of much of the Yukon Territory, Banks Island and parts of the uplands between the mouths of the Mackenzie and Coppermine Rivers, was covered by Late Wisconsin Cordilleran or Laurentide ice until 13,500 yrs. B.P. By 10,000 the western limit of Laurentide ice had retreated to a line running approximately from western Ellesmere Island, through central Victoria Island, Great Bear Lake, Great Slave Lake, Lake Athabasca, to just North of Lake Winnipeg in Manitoba.

The density of sites with Quaternary pollen records probably is less than in any other area dealt with in this book, partly because much of the terrain is accessible only with great difficulty and considerable

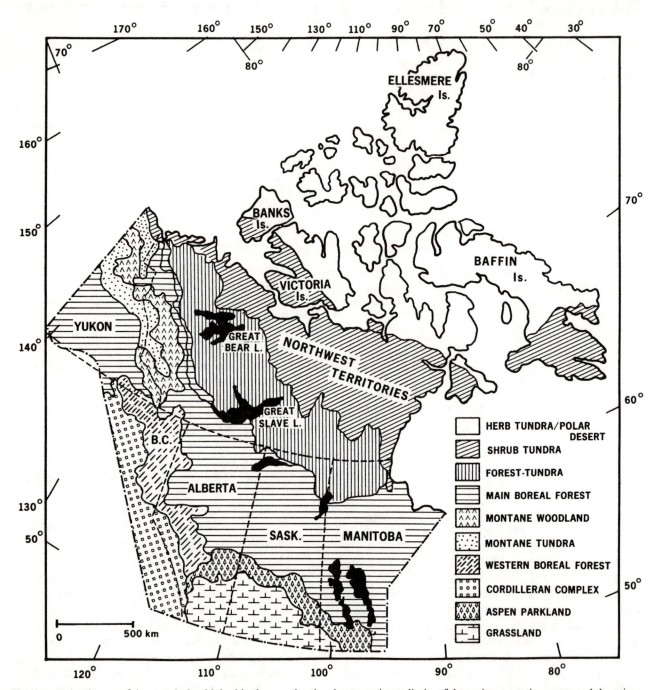

Figure 1. A sketch map of the area dealt with in this chapter, showing the approximate limits of the major vegetation zones and the primary geographical regions referred to in the text.

expense, and partly because the number of active centres and investigators pursuing research in this area is relatively small.

Pollen data from 218 sites have been published, their locations are shown on Figure 2, and they are listed in Table 1 with the site name, geographical location, author and year, time period covered by the record, data category, and any supporting paleoecological data. The data category classification is:

1– a continuous pollen record with several ^{14}C estimates, PAR and percent data,

2 – a continuous record with several ^{14}C estimates and only percent data,

3 – a continuous record with one or two ^{14}C estimates and percent data,

4 – a continous record with (no age estimate and) percent data only,

5 – a discontinuous or truncated or single sample record with or without ^{14}C ages, presented as percent data or raw pollen counts.

Sixteen sites have been selected for more detailed treatment, including a summary percentage pollen diagram, using two criteria – the site is representative of a broad biogeographic region within the area, and it provides the most complete record available. Some subjectivity has been involved in this selection.

Early and middle-Pleistocene sites will be considered first, followed by the much larger number of late-Pleistocene and Holocene records which are grouped into broad biogeographic zones. The final section offers a brief interpretive commentary, with some intimations of needs for future research. I include only those sites recorded in the literature before October 1, 1983. The overall treatment is relatively condensed because parts of the area have been reviewed in detail recently (Nichols, 1974, 1975; Ritchie, 1976, 1984) and surveys of the literature for all of Canada have been published by Hills and Sangster (1977) and Liu (1981a, 1981b), both in the context of evidence for climatic change.

The following abbreviations are used throughout this chapter: NAP = non-arboreal pollen types; PAR = pollen accumulation rate; BP = age before present.

Table 1.

A summary of sites reviewed in this chapter.

Site numbers appear on Fig. 2. Abbreviations: bryo. = bryophytes; chem. = chemical analysis; coll. = colluvial deposits; diat. = diatoms; fluv. = fluvial sediments; foram. = foraminifera; glac = glacial; H = Holocene; insec. = insects; lac = lacustrine; L-P = Late-Pleistocene; L-Q = Late-Quaternary; macr. = plant macrofossil; moll. = molluscs; ost. = ostracods; P = Pleistocene; sol. = paleosol

Site #	Author(s)	Site Name	Co-ordinates	Time Period	Data Class	Sediment Type	Supporting Evidence
1	Alley 1976	Okanagan Valley	49° 56′N 119° 23′W	L-P, H	2	bog	none
2	Strong 1977	Southern Alberta	50°-52° 44′N 110° 23′ - 113° 34′W	H	5	lac	none
3-27	Hansen 1955	Hart Highway B.C.	47° -56°N 120°-127°W	H	4	bog	none
28	MacDonald 1982	Kananaskis Valley	51° 04′N 115° 03′W	H	3	bog, lac	moll.
29	Mott & Jackson 1982	Chalmers Bog	50° 37′N 114° 25′W	H	1	bog, lac	none
30	Holloway *et al.* 1981	Lake Wabamun	53° 30′N 114° 30′W	L-P, H	1	lac	none
31	Schweger 1972	Robinson	49° 55′N 110° 05′W	L-P, H	5	sol	chem.
32-109	Hansen 1949-1953	N.A.	52°-65°N 113°-148°W	L-P, H	4	bog	none
135	White & Mathewes 1982	Fiddlers' Pond	56° 15′N 121° 20′W	H	1	lac	macr.
136	White 1983	Boone Lake	55°34′N 119° 24′W	L-P, H	1	lac	none
136a	White 1983	Spring Lake	55° 31′N 119° 35′W	L-P, H	1	lac	none
137	Harington *et al.* 1974	Babine Lake	55° 0′N 127° 14′W	P	5	bog	bones
138	Lichti-Federovich 1972	Alpen Siding	54° 27′N 113° 00′W	L-P, H	2	lac	none
139	Lichti-Federovich 1970	Lofty Lake	54° 44′N 112° 29′W	L-P, H	2	lac	none
140	Kupsch 1960	Herbert	50° 23′N 107° 13′W	L-P, H	3	lac	none
141	Mott 1973	Clearwater Lake	50° 52′N 107° 56′W	H	3	lac	none
142	Ritchie & deVries 1964	Hafichuk	50° 20′N 105° 48′W	L-P	2	lac	macr.
143	Ritchie 1976	Crestwynd	50° 10′N 105° 40′W	H	3	lac	none
144	Terasmae unpub. Ritchie 1976	Scrimbit	50° 05′N 105° 03′W	L-P, H	3	lac	none
145	Mott 1973	Lake-A	53° 14′N 105° 43′W	L-P, H	3	lac	none

146	Mott 1973	Lake-B	53° 48′N 106° 05′W	L-P, H	3	lac	none
147	Mott 1973	Cycloid Lake	55° 16′N 105° 16′W	L-P, H	3	lac	none
148	Ritchie 1976	Sewell Lake	49° 35′N 99° 15′W	L-P, H	3	lac	none
149	Ritchie & Lichti-Federovich 1968	Belmont Lake	49° 26′N 99° 26′W	L-P, H	3	lac	none
150	Ritchie & Lichti-Federovich 1968	Glenboro	49° 26′N 99° 17′W	L-P, H	2	lac	none
151	Teller & Last 1979	Lake Manitoba	50° 15′N 98° 25′W	L-P, H	2	lac	none
152	Ritchie 1964 & 1969	Riding Mountain	50° 43′N 99° 39′W	L-P, H	1	lac	macr.
155	Ritchie 1967 & 1976	Russell	50° 50′N 101° 05′W	L-P, H	3	lac	none
156	Nichols 1969	Porcupine Mountain	52° 31′N 101° 15′W	H	2	bog	none
157	Ritchie & Hadden 1975	Grand Rapids	53° 02′N 99° 43′W	H	2	lac	diat.
158	Ritchie 1976	Flin Flon	55° 45′N 102° 05′W	L-P, H	2	lac	none
159	Ritchie 1976	Thompson	56° 10′N 97° 50′W	H	3	lac	none
160	Nichols 1967a, b, c	Lynn Lake	56° 50′N 101° 03′W	H	2	bog	none
161	Ritchie 1976	Reindeer Lake	56° 40′N 102° 35′W	H	3	lac	none
162	Nichols 1969	Clearwater Bog	53° 59′N 101° 12′W	H	2	bog	none
163	Rampton 1971	Antifreeze Pond	62° 21′N 140° 50′W	P, H	2	lac	macr.
164	Terasmae & Hughes 1966	Chapman Lake	64° 55′N 138° 23′W	L-P, H	2	bog	none
164a	Terasmae & Hughes 1966	Gill Lake	65° 30′N 139° 45′W	H	4	lac	none
165	Ritchie 1982	Lateral Pond	65° 57′N 135° 31′W	L-P, H	1	lac	none
166	Ritchie 1982	Tyrell Lake	66° 03′N 135° 39′W	H	3	lac	none
167	Ritchie et al. 1982	Bluefish Cave	67° 09′N 140° 45′W	L-P, H	3	loess	bones
168	Ovenden 1982	Polybog	67° 48′N 139° 48′W	L-P, H	1	bog, lac	macr.
169	Cwynar 1982	Hanging Lake	68° 23′N 138° 23′W	P, H	1	lac	none
170	Matthews 1975	near Clarence Lagoon	69° 36′N 140° 36′W	H	5	bog	insec. macr.
171	Ritchie & Hare 1971	Tuktoyaktuk 5	69° 03′N 133° 27′W	L-P, H	2	lac	none
172	Ritchie 1972	Tuktoyaktuk 6	69° 06′N 133° 25′W	L-P, H	4	lac	macr.
173	Spear 1983	Sleet Lake	69° 17′N 133° 35′W	L-P, H	1	lac	macr.
174	Hyvärinen & Ritchie 1975	Eskimo Lakes	69° 24′N 131° 40′W	L-P, H	2	lac	none
175	Hyvärinen & Ritchie 1975	Hendrickson Is.	69° 32′N 133° 34′W	L-P, H	1	lac	none
176	Ritchie 1972	Richard Is.	69° 26′N 134° 30′W	L-P, H	3	lac	none
177	Ritchie 1972	Hooper Is.	69° 42′N 134° 52′W	L-P	3	lac	none
178	Mackay & Terasmae 1963	Eskimo Lakes	69° 15′N 132° 20′W	H	3	bog	none
179	Delorme et al. 1977	Parsons	68° 57′N 133° 45′W	H	5	pingo	ost. moll.
180	Delorme et al. 1977	Site 1	67° 16′N 135° 14′W	H	5	coll.	ost. moll.
181	Delorme et al. 1977	Caribou	66° 02′N 135° 06′W	L-P, H	5	coll.	ost. moll.

182	Delorme *et al.* 1977	Sans Sault Rapids	65° 44′N 128° 42′W	H	5	coll.	ost. moll.
183	Ritchie 1977	Maria Lake	68° 06′N 133° 28′W	L-P, H	1	lac	none
184	Mackay and Terasmae 1963	Twin Lake	68° 20′N 134° 00′W	H	4	bog, lac	none
185	Ritchie 1984	Twin Tamarack	68° 09′N 133° 29′W	L-P, H	1	lac	none
186	Ritchie 1984	Sweet Little Lake	67° 38′N 132° 00′W	L-P, H	1	lac	none
187	MacDonald 1983	Natla Bog	63° 00′N 129° 05′W	H	1	bog	macr.
188	Slater 1978	Eildun Lake	63° 08′N 122° 46′W	L-P, H	2	lac	none
189	Matthews 1980	John Klondike	60° 21′N 123° 39′W	H	2	bog, lac	macr. insec.
190	Nichols 1972, 1974	Colville	67° 06′N 125° 47′W	H	2	bog	none
191	Nichols 1975	Port Radium	66° 07′N 117° 50′W	H	1	bog	none
192	Nichols 1975	Coppermine	67° 50′N 115° 05′W	H	2	bog	none
193	Nichols 1975	Thompson Landing	62° 58′N 110° 35′W	H	1	bog	none
194	Ritchie 1979	Porter Lake	67° 10′N 131° 00′W	H	2	lac	none
195	Terasmae *et al.* 1966	Gordon Bay	66° 49′N 107° 06′W	H	3	peat, silt	macr.
196	Kay 1979	Nicol Lake	61° 35′N 103° 29′W	H	3	bog	none
197	Kay 1979	Long Lake	62° 38′N 101° 14′W	H	2	bog	none
198	Nichols 1967a, b, c 1972, 1974	Ennadai Lake	61° 10′N 100° 55′W	H	2	bog	none
200	Kay 1979	Slow River	63° 02′N 100° 45′W	H	3	bog	none
201	Kay 1979	Grant Lake	64° 43′N 100° 28′W	H	3	bog	none
202	Nichols 1970, 1972	Pelly Lake	66° 05′N 101° 04′W	H	2	bog	none
203	Nichols 1970, 1972, 1974, 1975	Drainage Lake	66° 08′N 101° 04′W	H	3	bog	none
204	Terasmae 1967	MacAlpine Lake	66° 35′N 103° 15′W	H	4	bog	none
205	Hegg 1963	Axel Heiberg Is.	79° 25′N 90° 30′W	H	5	bog	none
206	Hyvärinen and Blake 1981	Baird Inlet	78° 29′N 76° 46′W	L-P, H	1	lac	none
207	Jankovska and Bliss 1977	Truelove Lowland	75° 33′N 84° 40′W	H	4	bog	none
207a	Lichti-Federovich 1975 Mathewes 1984 McAndrews 1984	Devon Island Ice Cap	75° 45′N 83° 00′W	P, H	3	ice	none
209	Mode 1980	Patricia Bay	70° 30′N 68° 20′W	H	2	lac	none
210	Boulton *et al.* 1976	Maktak Fiord	67° 23′N 64° 58′W	H	1	bog	macr.
211	Nichols 1975	Windy Lake	66° 25′N 65° 30′W	H	1	bog	none
212	Short and Andrews 1980	Padloping Island	67° 06′N 62° 21′W	H	1	bog	none
213	Short and Andrews 1980	Broughton Is.	67° 33′N 64° 03′W	H	1	bog	none
214	Short and Andrews 1980	Idjuniving Is.	67° 55′N 64° 50′W	H	1	bog	none
215	Short and Andrews 1980	Quajon Fiord	67° 40′N 65° 00′W	H	1	bog	none
216	Short and Andrews 1980	Owl River	67° 00′N 64° 50′W	H	1	bog	none
217	Short and Andrews 1980	Pass Head	67° 00′N 64° 35′W	H	1	bog	none

219	Davis 1980	Iglutalik Lake	66° 00'N 64° 30'W	H	1	lac	none
220	Mott and Christiansen 1981	Martens Slough	52° 02'N 106° 29'W	H	2	lac	none
I	Klassen *et al.* 1967	Roaring River Man.	51° 51'N 101° 08'W	P	4	intertill seds.	moll. ost.
XI	Miller *et al.* 1977	Clyde Foreland	70° 30'N 68° 20'W	P	5	drifts, tills	macr. moll. chem.
XII	Terasmae *et al.* 1966	Barnes Ice Cap	70° 30'N 74° 30'W	P	5	fluv.	macr.
II	Westgate *et al.* 1972	Watino	55° 45'N 117° 44'W	P	4	fluv, lac	ost. moll. insec. macr.
III	Lichti-Federovich 1974	Porcupine River	67° 28'N 139° 54'W and 67° 31'N 140° 15'W	P	2	fluv.	none
IV	Lichti-Federovich 1973	Old Crow River	67° 45'- 68° 15'N 139° 45'- 140° 30'W	L-Q	2	fluv.	none
V	Hughes *et al.* 1981	Hungry Creek	65° 55'N 135° 20'W	P	5	glac lacust.	macr. insec. bryo. bones
VIII	Rampton 1982	Kay Pt.	69° 17'N 138° 23'W	P	5	fluv.	macr. ost. foram. insec.
VII	Rampton 1982	Stokes Pt.	69° 22'N 138° 48'W	P	5	fluv.	macr. ost. foram. insec.
VI	Rampton 1982	King Pt.	69° 05'N 137° 50'W	P	5	fluv.	macr. ost. foram. insec.
X	Mackay 1963	Eskimo Lakes	68° 46'N 133° 15'W	P	4	bog	macr. ost. foram. insec.
IX	Terasmae 1959	East Channel	approx. 69°N 134° 30'W	P	5	bog	none
XIII	Rampton 1971	Antifreeze Pond	62° 21'N 140° 51'W	P	2	lac	none

EARLY AND MIDDLE-PLEISTOCENE SITES

Manitoba and Alberta Sites

Roaring River, Manitoba, Site I. Klassen *et al.*, (1967) report on one of the most important middle-Pleistocene sites in Canada, discovered last century by the celebrated Canadian explorer and geologist, J. B. Tyrrell (1892) near the northern edge of the Duck Mountain upland, in Manitoba. The pollen record consists of a percentage diagram with 23 taxa from 20 samples spaced roughly equally throughout a 2.2 m unit of fossiliferous silt, below a till unit that caps the 45 m thick riverbank section. A simplified version of the original pollen diagram is shown here (Fig. 3). Four pollen zones were distinguished by Klassen *et*

al., (1967). The lowest (Zone 1) consists of a single sample dominated by *Picea.* Zones 2 and 3 are dominated by NAP, with high percentages of *Artemisia,* Gramineae and Cheno-Am. Zone 3 is distinguished by the high proportions of *Quercus* and Zone 4 is marked by a decline to low values of NAP and *Quercus* and an increase of *Picea* frequencies to roughly 60%. The uppermost zone is separated by a decline in *Picea* values and increases in *Betula* and NAP. The interpretation of the pollen data is significantly complemented by detailed ostracod and mollusc analyses. The authors propose a sequence of paleoenvironments corresponding to the pollen zone sequence, from a cool climate supporting a boreal forest (Zone 1), through a warm episode characterized by grasslands (Zone 2) and oak savannah (Zone

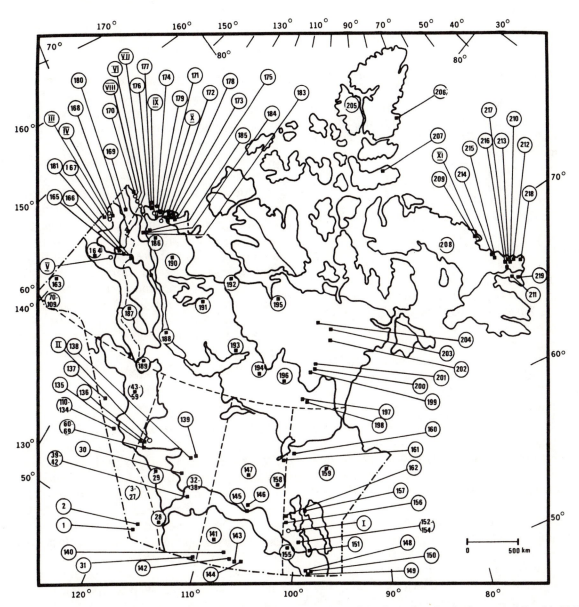

Figure 2. The same base map as in Fig. 1, showing the approximate locations of the sites referred to in the text and listed in Table 1.

3), and a later reversion to boreal or subarctic climates. "Carbonaceous debris" in the fossiliferous unit gave a ^{14}C age estimate of > 37,760 yrs. B.P.

This notable section has been resampled recently by Drs. A. C. Ashworth and D. P. Schwert of the University of North Dakota at Fargo, and their future analysis of the rich beetle remains in the sediments should provide further interesting paleoecological insights (personal communication from Dr. Ashworth).

The Watino Site, Alberta, Site II. A riverbank exposure on a tributary of the Peace River in west-central Alberta was investigated by a team of Quaternary scientists over a decade ago, reported in abstract form by Westgate *et al.*, (1972). The section consists

of roughly 30 m of variably fossiliferous alluvial sediments, composed of gravels, silts and laminated silts and clays that spanned the period between 27,400 and at least 43,500 radiocarbon years B.P., on the basis of five dated levels. The pollen record is preliminary, showing only five categories in 19 sampled levels, and it is clear that further investigation of this site is required. The general paleoenvironmental conclusions, based on analyses of mollusca, ostracodes, beetles and plant macrofossils (Westgate *et al.*, 1972), are that the fossiliferous units were laid down in a floodplain environment under climatic conditions very similar to those prevailing at the site today.

North Yukon Sites

Porcupine River Site III. The main contributions to knowledge of early and middle-Pleistocene palynology in northwest Canada are based on sites exposed along the Porcupine and Old Crow Rivers, near the village of Old Crow in the Yukon Territories. The most important pollen data were presented a decade ago by Lichti-Federovich (1974) and her Porcupine River diagram is shown here in greatly simplified format, with the addition of some recent chronostratigraphic information published by Briggs and Westgate (1978) and by Pearce *et al.,* (1982) (Fig. 3). Lichti-Federovich (1974) identified 73 pollen taxa in 70 levels of a ~65 m section. The pollen assemblages in the lower unit, apparently with an age greater than 700,000 yrs. B.P. are significantly different from late-Pleistocene and Holocene spectra from western North America in that they show an association of *Picea, Pinus* and *Betula* with *Corylus, Alnus* and *Salix.* In addition the lowest levels produced both pollen and cones of a spruce very similar to the extinct late-Tertiary taxon, *Picea banksii,* recorded first from Banks Island by Hills and Ogilvie (1970).

The upper part of the section, above a sedimentary disconformity, shows a tripartite zonation, with two pollen zones dominated by chiefly arctic herbs and shrub birch, separated by a zone dominated by *Picea,* all occurring within the period 80,000 to 30,000 yrs. B.P.

Old Crow River Site IV. Lichti-Federovich (1973) published pollen diagrams for six sites along the Old Crow River, where late-Pleistocene sediments are exposed by the downcutting of the meandering river. The silts and sands below the Glacial Lake Kutchin lacustrine unit, apparently coeval with the ~18,000 yr. B.P. maximum of Laurentide glaciation, yielded a tripartite sequence of pollen zones, with a *Picea-Betula* zone separating two zones with high NAP values. More detailed discussions of the Porcupine and Old Crow River sections can be found in the original papers, and in a recent collation (Ritchie, 1984).

Hungry Creek, Site V. Hughes *et al.,* (1981) present a tabular account of 34 pollen taxa recorded in six levels of an exposure of fluvial sediments with a single radiocarbon estimate of 36,900 yrs. B.P. The site is near the confluence of Hungry Creek and Wind River, which flows into the Peel River near the southern extremity of the Richardson Mountains. The pollen percentages are dominated by *Picea, Betula* and *Alnus* and resemble closely modern boreal forest spectra from the northwestern sector, from which

pine is absent. Detailed bryological analyses of the same section support the paleoecological reconstruction.

North Yukon Coastal Sites, VI to VIII. Rampton (1982) includes pollen records from three main sites on the North Yukon coast, at King Point, Stokes Point and Kay Point, all close to, and east of, Herschel Island. These sediments lie below a till unit described as the Buckland formation, ascribed to an early glacial maximum event of uncertain age. Organic units within these sections produced variably fossiliferous samples. Some of the pollen spectra are based on low pollen sums (< 100) with relatively few taxa (12). They fall roughly into two types: those dominated by boreal tree taxa, particularly *Picea* and *Betula,* associated with *Alnus;* and those with high percentages of Gramineae and Cyperaceae pollen, associated with shrub birch. The plant macrofossil, insect, and ostracod analysis support the general conclusion that both boreal forest and shrub-tundra episodes occurred on the North Yukon coastal plain at certain times in the middle-Pleistocene.

Mackenzie Delta Area

East Channel Site IX. Terasmae (1959) recorded pollen spectra dominated by *Picea, Betula* and *Alnus* with *Pinus* and *Larix* from a bed of peat exposed in the bank of the East Channel of the Mackenzie River between Inuvik and Tuktoyaktuk. He interprets these as interglacial boreal forest assemblages reflecting a climate warmer than modern conditions.

Eskimo Lake Site X. A similar bed of undated peat yielded pollen spectra with comparable frequencies of the same taxa as in the East Channel site, except that pine was absent, reported briefly in Mackay (1963) and in more detail, following further analysis, in Ritchie (1984).

Baffin Island

Clyde Foreland, Baffin Island, Site XI. Miller *et al.,*(1977) have collated the results of several important palynological analyses of sediments discovered during the long and continuing series of investigations of many aspects of the Quaternary history of Baffin Island, led by Dr. John T. Andrews. The important middle-Pleistocene beds are from the Isortoq River and Flitaway Lake sites reported by Terasmae *et al.,* (1966) and from the several Clyde

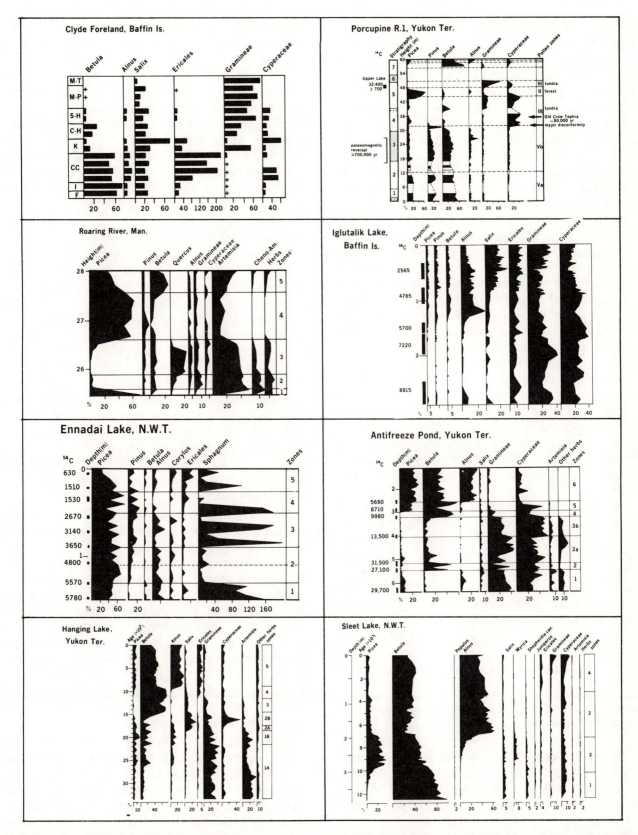

Figure 3. Summary percentage pollen diagrams for eight representative sites, referred to in the text. Clockwise from the Porcupine R. Site, they are: Site III (Lichti-Federovich 1974), Site 219 (Davis, P. T. 1980), Site XIII (Rampton 1971), Site 173 (Spear 1983), Site 169 (Cwynar 1982), Site 198 (Nichols 1967 b-d, 1972, 1974), Site I (Klassen *et al.*, 1967) and Site XI (Miller *et al.*, 1977). Note that the Site XI record from the Clyde Foreland is a composite of several sedimentary units, elaborated in the original publication.

Foreland sections described by Miller *et al.,* (1977) with later commentary on their possible chronology by Andrews, Miller *et al.,* (1981). The main pollen assemblages identified at these sites are reproduced here in histogram form (Fig. 3).

The Cape Christian unit, assigned tentatively to Isotope Stage 5e (Miller *et al.,* 1977), is characterized by high frequencies (~70%) of *Betula* pollen, associated with *Alnus, Salix* and Ericaceae. Macrofossil records of *Betula nana* and *B. glandulosa* support the conclusion that the birch was represented by the dwarf shrub taxa. Miller *et al.,* (1977) concur with Terasmae's (1966) interpretation of the same unit at the nearby Barnes Icecap sites (Isortoq and Flitaway) and suggest that "Baffin Island, at least to 71°N, had a low arctic shrub tundra cover during the last interglaciation." They also point out that the Cape Christian assemblages differ markedly from Holocene spectra from the same area – the former are dominated by shrub taxa (birch, heaths, willows), the latter by grass and sedge pollen. Quantitative analysis of the Cape Christian unit, including a comparison with modern pollen spectra, led Miller *et al.,* (1977) to propose a modern analogue in spectra from North Labrador or West Greenland, where shrub tundras occur dominated by birch and ericads and where the mean July temperature is ~3°C above modern values at Clyde Foreland. That conclusion is corroborated by analyses of paleosols and mollusca from the same sediments.

The Kuvinilk organic beds at the Clyde Foreland are tentatively correlated with a Clyde "glacial event" at about 115,000 yrs. B.P. The samples yielded very low PAR values and are dominated by *Salix* and Gramineae pollen. Miller *et al.,* (1977) stress the difficulties of interpreting this assemblage which is generally intermediate between the Cape Christian and the modern spectra.

LATE-PLEISTOCENE AND HOLOCENE

Grassland, Parkland and Southern Boreal and Montane Forest Zones

Okanagan Valley Site 1. A bog in the Okanagan Valley of British Columbia in the ponderosa pine-bunchgrass biogeoclimatic zone, yielded 2.75 m of peat and basal clay from which Alley (1976) reports 49 pollen taxa in 5 cm interval samples. Two radiocarbon age estimates and two identified ash layers provide a secure chronology for the percentage diagram. Alley (1976) recognises three pollen zones

and further subdivides the middle one into two subzones and the uppermost into five subzones. With the exception of *Pseudotsuga* frequencies, which increase gradually through time, the arboreal taxa show highly variable patterns and the author depends heavily on the taxa of local origin (Cyperaceae, *Typha*) and the NAP in establishing the pollen zonation. As he emphasises, it is clear that additional sites and PAR data are required to establish securely the pollen stratigraphy of this difficult, mountainous region, following which useful paleoclimatic reconstructions might be feasible.

Sites 3-27. Hansen (1955) reports the pollen stratigraphy of 25 bog sites located near highways in south-central, central and north-eastern British Columbia. The documentation of the pollen data is unsatisfactory – in particular the imprecision of site locations, the uncritical identification of volcanic ash in 14 sections with a single event, the uncritical use of pollen size in separating pine species, the absence from the analysis of deciduous tree, shrub and many NAP taxa, and the combining of grass, chenopod and composite pollen in one curve. A maximum of eight taxa are recorded from any one site and the sampling interval was 20 cm, even in bogs only 1.2 m in depth. The vague and discursive commentary on the Holocene vegetation and paleoclimate of the area bears little relation to the primitive supporting pollen diagrams, which are in fact almost totally devoid of useful information.

Chalmers Bog, Site 29. Mott and Jackson (1982) have published a detailed (48 taxa) percentage diagram from a 10.5 m section of peat and clay, with three radiocarbon dates and a tephra horizon. Total PAR is given for all 30 levels analysed. The site is in southwest Alberta, in the Foothills region near the transition between mixed conifer forest and grassland. The record is notable as the first reliable evidence of an "ice-free" corridor area, represented in this diagram as a tundra pollen assemblage radiocarbon-dated at 18,000 yrs. B.P. The early-Holocene has a rise of pine pollen, followed by a spruce increase at 8,220 yrs. B.P. with little significant change throughout the rest of the diagram.

Kananaskis Valley, Site 28. MacDonald (1982) presents percentage and concentration diagrams for 15 pollen taxa recorded in two adjacent sites in the Kananaskis valley of southwestern Alberta. Both sites were sampled at 10 cm intervals and a tephra layer plus three radiocarbon dates on wood in the sediment provide a tentative chronology. The two pollen stratigraphies are very similar, divided into a lower zone characterised by *Artemisia-Salix-Juniperus* between the time of deglaciation and roughly 10,000 yrs. B.P.; and an upper zone dominated by *Picea* and *Pinus* from ~9,400 yrs. B.P. to the present. Molluscan analysis supports the author's paleoecological interpretation of an early cool period followed at roughly 9,400 yrs. B.P. by a warming, with some evidence of conditions warmer than present during the mid-Holocene.

Lake Wabamun, Site 30. Holloway *et al.,* (1981) present percentage and PAR diagrams for a 15.5 m section of lake sediment sampled at 10 cm intervals, giving frequency curves for 18 taxa. The site is 48 km west of Edmonton, Alberta. Eleven radiocarbon dates provide the basis for establishing the age/depth relationships of the section. It appears that the rate of sedimentation has varied greatly throughout the depositional history of this large (roughly 6,000 ha) Lake – from 0.15 mm. yr. between 16,200 and 11,700 yrs. B.P. to 3.03 mm. yr. between 11,700 and 9,500 yrs. B.P. The pollen diagrams are not zoned but are divided into late-glacial, early post-glacial, Hypsithermal and post-hypsithermal periods on the basis of radiocarbon age. Pollen preservation in the Hypsithermal (9,000 to 4,800 yrs. B.P.) was poor and only one sample was usable. The late-glacial (16,180 to 11,750) is described as tundra on the basis of low influx values, although the pollen assemblage is dominated by *Picea* and *Pinus,* associated with *Artemisia, Alnus,* Cyperaceae, *Salix* and Gramineae. The early post-glacial (11,750 to 9,000 yrs. B.P.) is interpreted as a mixed boreal forest assemblage and the post-hypsithermal as a mixed boreal forest with declining conifer representation. The uncertainties of the chronology, no explanation of the method of calculating PAR, the undocumented assertion that Lake Wabamun is meromictic and therefore that the laminae between 1,180 and 729 cm are annual varves, and problems of pollen preservation at certain levels suggest that this site presents too many equivocal results to sustain confidently the authors' conclusions.

Site 2. this aggregate site consists of nine small lakes in southern Alberta from which Strong (1977) has recovered short cores (12 to 20 cm length). He presents tabular and histogram data for eight taxa to illustrate the modern and immediately pre-settlement spectra of this part of the grassland and parkland vegetation zones.

Site 31. Schweger (1972) published a pollen diagram for a site near Robinson, South Alberta, showing the percentages of 15 pollen taxa recorded in each of five samples of a buried soil, radiocarbon-dated at 10,230 yrs. B.P. The spectra are dominated

by NAP and Schweger (1972) suggests that they represent an open steppe environment. Schweger *et al.,* (1981) have compiled interesting Holocene pollen records, supported by detailed diatom analysis, from several lake sites in central Alberta, but none of these investigations has yet been published.

Lofty Lake, Site 139. This site is roughly 150 km north-northeast of Edmonton, Alberta, on a morainic ridge, near the southern boundary of the boreal forest. Lichti-Federovich (1970) presented detailed pollen analysis (59 taxa from 5 cm interval samples) from a section of sediment with five radiocarbon age estimates and a securely identified tephra layer. A summary pollen diagram based on the original detailed diagram is published here (Fig. 4) because this site remains the most thoroughly documented, published record from central Alberta. Five local pollen assemblage zones were recognised, as follows: Zone 1, dominated by *Populus,* associated with *Salix, Artemisia* and *Shepherdia canadensis.* Zone 2, delineated by the *Picea* rise at ~10,800 yrs. B.P. with *Populus, Shepherdia* and Cyperaceae. Zone 3 is dominated by *Betula* and *Populus* with significant *Corylus* representation. Zone 4 has high NAP percentages associated with *Betula* and *Alnus,* and the uppermost Zone 5 is dominated by *Picea, Betula* and *Alnus.* It is unfortunate that PAR was not determined for these samples to help clarify the interpretation. The main features of the diagram are that a boreal assemblage dominated by spruce was established at 10,800 yrs. B.P., that deciduous woodlands, locally grassland vegetation replaced it in the early and mid-Holocene, and that the modern mixed boreal forest was established at roughly 4,000 yrs. B.P.

Alpen Siding, Site 138. A pollen sequence very similar to Lofty Lake was recorded from this site, 50 km south of Lofty Lake, by the same author (Lichti-Federovich, 1972).

Sites 32 to 38, 39 to 42, 43 to 59, 60 to 69 and 70 to 109 are reported on here together, although not all are confined to this biogeographic zone. They represent a survey of bog pollen stratigraphy reported by Hansen (1949a, 1949b, 1950, 1952, 1953) from sites along highways in central Alberta, Northern British Columbia, the Yukon Territory and Alaska. All pollen diagrams show only two, three or occasionally four tree pollen taxa. Field and laboratory methods are described in cursory fashion. Quantitative data, including pollen sums are never shown. Although this activity, sustained over 15 years, is referred to politely as having "demonstrated the potential of palynological studies for the region", in fact it is difficult to comprehend how such ill-conceived research

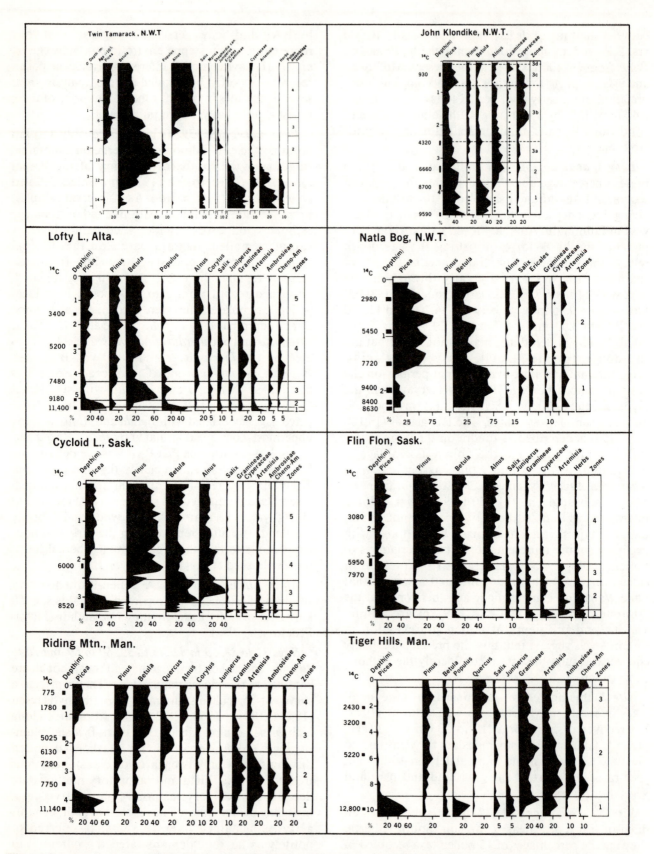

Figure 4. Summary percentage pollen diagrams from eight representative sites, referred to in the text. Clockwise from the John Klondike site they area, Site 189 (Matthews 1980), Site 187 (MacDonald 1983), Site 158 (Ritchie 1976), Site 150 (Ritchie and Lichti-Federovich 1968), Site 152 (Ritchie 1964, 1969), Site 147 (Mott 1973), Site 139 (Lichti-Federovich 1970) and Site 185 (Ritchie 1984).

could have been pursued unchanged over this long period, far less been funded and published in the leading journals, at a time when pollen analysis had advanced rapidly in both Europe and America.

Site 137. Harington *et al.,* (1974) record a single pollen spectrum of 28 taxa from organic silt associated with a mammoth skeleton near Babine Lake, central British Columbia. The sediment has a radiocarbon age of about 43,000 yrs. B.P. and yielded a pollen spectrum dominated by dwarf birch, grass and *Artemisia.*

Kearney and Luckman (1983) provide some preliminary intimations of the results of pollen analyses of Holocene sediments sampled in the Jasper National Park as part of a study of treeline fluctuations, in which they use the ratios of fir and pine as indicators of treeline changes. A detailed publication on the pollen data is anticipated.

Fiddler's Pond, B.C. Site 135. White and Mathewes (1982) provide a detailed analysis and interpretation of a core of Holocene pond sediment from the Alberta Plateau district. Both PAR and percentage diagrams are given, showing 39 taxa recorded in 45 sampled levels, with spruce size measurement histograms to separate *Picea glauca* and *P. mariana.* Four pollen zones are described and interpreted. The lowest (from 7,250 to 5,500 yrs. B.P.) is dominated by pine and is interpreted as a closed coniferous forest. Zone 2 is distinguished by an increase in sedge pollen, interpreted in terms of local changes in the pond vegetation; the surrounding vegetation shows the addition of alder to the conifer elements. Zone 3 is described as transitional to Zone 4 which begins at 1,200 yrs. B.P. and is distinguished primarily by a minor rise in alder frequencies.

Boone and Spring Lake, B.C., Sites 136-136a. White (1983) has completed a detailed investigation of the pollen stratigraphy of two small lake sites in the Upper Peace River region of northeast British Columbia and northwestern Alberta. Several radiocarbon dates, detailed pollen identification, percentage and PAR data and varied statistical applications are provided. Deglaciation occurred at 12,000 yrs. B.P. and a short-lived treeless episode of uncertain composition was followed by a *Populus-Salix-Artemisia*-Gramineae zone, replaced at 10,800 by a boreal forest assemblage dominated by *Picea* and *Pinus.* Tree birch became important in the landscape at 8,500, followed by a pine maximum at 7,400 yrs. B.P. At least one published account of this thesis is anticipated so detailed comment is not appropriate until after that event.

The following six sites in the grassland and aspen parkland zone of the Western Interior of Canada yielded rich, organic lake muds derived from former or in some cases persisting small lakes or ponds.

Herbert, Saskatchewan, Site 140. Kupsch (1960) published a pollen diagram based on analysis of a 1 m unit of buried lacustrine sediment with a basal date of 10,050 yrs. B.P. A percentage diagram of 21 taxa shows an abrupt transition at ~10,000 from a spruce-dominated assemblage to an NAP-dominated assemblage.

Hafichuk Ranch, Saskatchewan, Site 142. A similar buried organic lake sediment was excavated near Moose Jaw, Saskatchewan and pollen analysis was supplemented by detailed vascular plant macrofossil analysis (Ritchie and de Vries, 1964) and later by bryophyte analysis (de Vries and Bird, 1965). 36 pollen taxa were recorded in 34 levels of a 2.5 m section that included several units of laminated gyttja. Three radiocarbon age estimates provide a reasonable chronology and suggest that this unit was deposited between 11,650 and 10,000, probably in a supraglacial pond. The pollen spectra are dominated by *Picea,* with continuous low frequencies of *Salix* and *Shepherdia canadensis,* along with *Artemisia*(~20%), Gramineae and Chenopodiineae. The upper 80 cm of the unit show a decline of spruce percentages and increases of NAP followed by an increase of *Picea* in the uppermost samples to values similar to those of the lower levels. The suggestion of the authors that this apparently tripartite sequence might be correlative with the North Western European late-Pleistocene pollen stratigraphy has not been corroborated by any other evidence from west-central or eastern North America. The notion was abandoned in a later review (Ritchie, 1976).

Scrimbit Farm, Kayville, Saskatchewan, Site 144. A simplified version of the original unpublished pollen diagram prepared by Dr. J. Terasmae from this site was presented in a review paper (Ritchie, 1976) and it shows a pattern similar to the previous two sites. A 2 m unit of buried organic lake mud with two radiocarbon dates of 11,700 near the bottom and 10,400 at the top was analyzed at 20 cm intervals and yielded a lower unit dominated by *Picea* (~40%) with high *Artemisia* and an upper unit with decreased *Picea* and increases in all NAP taxa.

Clearwater Lake, Saskatchewan, Site 141. This small (50 ha) shallow (maximum depth 6 m) lake in the aspen parkland-grassland transition zone yielded 12.7 m of marly gyttja, analysed in detail by Mott (1973). 50 pollen taxa were recorded in 34 unevenly spaced samples. The basal sediment was radiocarbon dated at 7,580 yrs. B.P. and the 60-70 cm level, where European settlement indicators first appear in the record, yielded an age of 1,170 yrs. B.P. Mott (1973) concludes that both dates are too old due to old carbonate effects, and he further suggests that the absence of the Mazama ash from the section indicates an age younger than 6,400 yrs. B.P. The entire pollen record is dominated by NAP (> 60%) and only minor changes are apparent. The uppermost zone is interesting being characterized by elevated frequencies of Chenopodiineae (~30%) and *Selaginella densa* (~15%) which Mott suggests reflect cattle grazing effects.

Crestwynd, Saskatchewan, Site 143. A very similar record was recovered from a small depression in the Missouri Coteau. The detailed pollen diagram, with two radiocarbon dates and 43 pollen taxa is unpublished but a simplified version was reported by Ritchie (1976). The diagram has a basal date of 9,390 yrs. B.P. and demonstrates that spectra dominated by NAP were prevalent in southern Saskatchewan by at least that time.

Martens Slough, Saskatchewan, Site 220. A small shallow (1-2 m) saline pond in the northern fringe of the Grassland zone yielded 6.5 m of polleniferous sediment and the lower non-calcareous unit provided three samples for radiocarbon age estimation, of 8,350, 10,240 and 11,070 yrs. B.P. A detailed pollen diagram (59 taxa in 12 levels) provides the first record from the Canadian prairies that includes the late-Pleistocene and most of the Holocene (Mott and Christiansen, 1981). The authors recognise four pollen zones. The basal zone is dominated by *Shepherdia canadensis* associated with various NAP taxa, followed by a narrow (one sample) zone dominated by *Picea,* and two NAP-dominated zones separated by different proportions of the dominant *Artemisia* and Chenopodiineae pollen. The transition from the *Picea* zone to the NAP zone is radiocarbon dated at 10,240.

Lake A, Saskatchewan, Site 145. The following three sites lie on a south to north transect of the boreal forest and provide an excellent, detailed record of the Holocene history of the forest-parkland transition of the Western Interior. For example, Mott (1973) provides a detailed percentage pollen diagram for Site 145 with a single basal radiocarbon date (11,560 yrs. B.P.) from a small (25 ha) shallow (1 m) lake near the southern limit of the mixed conifer-deciduous forest sector of the boreal forest. Forty irregularly spaced levels were sampled in 6 m of marly gyttja and 58 pollen taxa were identified. Five pollen zones are recognised. The lowest is dominated

by *Picea* and Cyperaceae, followed by a *Betula*-NAP zone, an *Artemisia*-Gramineae zone, a *Betula-Artemisia* zone and finally a *Pinus* zone. The general reconstruction proposed by Mott (1973) is a late-Pleistocene spruce woodland replaced by grassland in the mid-Holocene with revertence to boreal forest of the modern type in the late-Holocene.

Lake B, Saskatchewan, Site 146. A similar small (25 ha) shallow (3 m) lake farther north, in the mixed forest zone of the boreal forest, yielded 5 m of gyttja and marl. A single basal radiocarbon determination provided an age of 10,260 yrs. B.P. near the base. Fifty seven pollen taxa were recorded in 50 sampled levels, and the resulting percentage diagram is divided into five assemblage zones – a basal *Picea* zone, a *Picea*-NAP zone, a *Pinus banksiana* - NAP zone, a *Pinus-Betula* zone and finally a *Pinus* zone. Mott (1973) suggests a sequence from a spruce woodland in the late-Pleistocene to an early-Holocene open parkland with scattered trees, and a late-Holocene establishment of the modern forests dominated by spruce, pine and birch.

Cycloid Lake, Saskatchewan, Site 147. A small (20 ha) shallow (2 m) lake in the Precambrian Shield, within the main Boreal Forest zone, yielded a 3.4 m core of silty gyttja. Thirty-nine levels were sampled and 49 pollen taxa recorded in a percentage diagram with five pollen assemblage zones, reproduced here in summary form (Fig. 4). The lowest two zones are dominated by *Picea* (15-60%) with variable proportions of *Artemisia, Shepherdia canadensis* and *Salix. Populus* is used to distinguish the lowermost zone. A marked rise of *Pinus* at 6,000 yrs. B.P. forms the lower boundary of the two uppermost zones while Zone 3 is distinguished by the dominance of *Betula* and *Alnus* percentages. The modern dominance of *Picea, Pinus, Betula* and *Alnus* pollen characterizes the uppermost zone. Mott (1973) suggests that a late-Pleistocene/early-Holocene spruce woodland was the earliest vegetation in the area, followed by spruce-birch forests and later (6,000 yrs. B.P.) spruce, birch and pine forests as at present.

Glenboro and Belmont Lakes, Manitoba, Sites 149 and 150. These are considered together as they were investigated at the same time by the same workers (Ritchie and Lichti-Federovich, 1968) and produced identical sequences. Two small, shallow lakes (Glenboro 10 ha and 3 m, Belmont 8 ha, 3 m) in the Tiger Hills moraine area of south-central Manitoba, within the aspen parkland vegetation zone, yielded respectively 10 m and 7.4 m of polleniferous sediment. The Glenboro diagram is reproduced here in summary form (Fig. 4). Five radiocarbon dates provide reasonable chronological control. Ninety-two levels gave 57 taxa in a percentage diagram, divided into five zones which I have simplified to four by combining the original two lowest zones into Zone 1. It is dominated by *Picea* associated with *Populus, Juniperus, Salix* and *Artemisia*. Zone 2 is dominated by the three main NAP categories, *Artemisia,* Chenopodiineae and Gramineae with scattered occurrences of prairie indicators. Zone 3 has a significant proportion of *Quercus* associated with the NAP, while the most recent zone shows increased percentages of Ambrosieae and Chenopodiineae. The Belmont site has only two radiocarbon dates, and 63 sampled levels of the 7.3 m section produced 56 taxa with the same zonation as Glenboro. The general vegetation reconstruction suggested by the authors is, a late-Pleistocene spruce woodland replaced at 10,000 yrs. B.P. by grasslands, followed by the establishment of oak savannah or gallery oak woodlands by about 3,000 yrs. B.P. with little change to the present except for the effects of European settlement and agriculture.

Sewell Lake, Manitoba, Site 148, lies on a large deltaic feature formed as an underflow fan by drainage into Glacial Lake Agassiz through the Assiniboine spillway (Fenton *et al.,* 1983). It is a small (3 ha), shallow (4 m) pond from which a 4.6 m core of organic sediment yielded 55 pollen taxa tabulated in 71 levels spaced at 5 or 10 cm intervals. A summary diagram (Ritchie, 1976) has five pollen zones – a basal *Picea-Betula* zone, followed by a *Picea-Juniperus* zone, an *Artemisia*-Gramineae-Chenopodiineae zone, a *Betula-Quercus*-NAP zone and finally a *Picea*-Gramineae-Ambrosieae zone. A single basal radiocarbon date (13,900 yrs. B.P.) is interpreted as being too old by roughly two millenia by some Pleistocene geologists (Fenton *et al.,* 1983) and accurate by others (Klassen, 1975, Christiansen, 1979). The area is of considerable phytogeographic interest because it supports a modern vegetation consisting of both boreal and prairie elements, forming unique physiognomic and floristic associations. The detailed pollen diagram remains unpublished, mainly because of the uncertainties noted above about the chronology of deglaciation; further coring of the site is planned.

Lake Manitoba, Site 151. Pollen analysis of long cores of sediment from Lake Manitoba has been done by Nambudiri (in Teller and Last, 1979), as part of a large, multi-disciplinary investigation of the Pleistocene history of the lake. As one might expect from sediment of such a large (4,700 km²) and relatively shallow lake, the pollen stratigraphic record lacks sharp zone boundaries. A summary pollen diagram

(Teller and Last, 1979) of one core has been divided into three zones – Zone I at the base is dominated by *Picea;* Zone II is primarily characterized by NAP; Zone III is dominated by *Betula, Quercus, Acer* and *Alnus*. The complete pollen record is expected in published form soon. A broad correlation is evident in the above tripartite stratigraphy with sites from the surrounding uplands (Ritchie, 1983).

Riding Mountain, Manitoba, Sites 152-154. Percentage and PAR data from three ponds in the Riding Mountain uplands have been reported (Ritchie, 1964, 1969). The ponds are small (2-3 ha) and shallow (1.5-2.5 m). One (E-Lake) has been studied in detail, with PAR data based on 8 radiocarbon age estimates. Fifty-two to 60 pollen taxa were recognised in each section, sampled at 2.5 or 5 cm intervals. A summary diagram is presented here (Fig. 4). A primary tripartite zonation is clear in all diagrams, although further subdivision was attempted in the original publications. At the base is a *Picea*-dominated zone, from −11,000 to 10,000 yrs. B.P. followed between 10,000 and 3,500 yrs. B.P. by an NAP zone with increasing proportions of *Quercus* and *Corylus* towards the top; the uppermost zone is dominated by *Picea-Pinus-Betula-Alnus.*The reconstructed vegetation is, an initial spruce woodland, differing markedly from modern boreal forest chiefly in the absence of pine, birch and alder; a replacement at 10,000 of the woodland by grassland and, later in the Holocene, oak savannah; and finally, at roughly 3,500 yrs. B.P., a reversion to boreal forest, dominated as at the present day by *Picea, Betula* and *Pinus*. Various numerical manipulations of these data (Ritchie and Yarranton, 1978a) have served to confirm the basic conclusions outlined above, and an application of a calibration technique to reconstruct paleoclimate has generated a growing-season temperature curve that shows partial agreement with independent interpretations (Ritchie, 1983).

Russell, Manitoba, Site 155. A small (4 ha), shallow (2 m) pond in hummocky disintegration moraine was cored and a 4.2 m section of sediment sampled at 2 and 10 cm intervals produced a three zone summary pollen diagram of the complete sequence (Ritchie, 1976, Fig. 1b) and a detailed analysis of the early-Holocene transition from forest to grassland (Ritchie, 1967). The complete diagram has not been published because only a single radiocarbon estimate is available and adequate cores for further dating no longer exist. The lowest pollen zone is dominated by *Picea* with *Juniperus* and *Artemisia;* Zone II (undated) is dominated by Chenopodiineae with high values of *Artemisia* and *Salix,* and the top zone has *Picea-*

Pinus-Betula and *Alnus* dominance although the site is actually in the Aspen Parkland zone, but only a few kilometers west of the Mixed Boreal Forest boundary. The significant feature of the site is the evidence in both the sediments (clay-silt) and pollen spectra of Zone II for an episode in the early-Holocene of lowered water levels, presumably in response to increased effective aridity. This site is being re-investiaged currently as part of a more general coring program in the prairie-aspen parkland region, based at the University of Toronto.

Central and Northern Boreal Forest including Tundra Transition Zone

Grand Rapids, Manitoba, Site 157. A small pond on The Pas moraine that terminates at the west shore of Lake Winnipeg, yielded 265 cm of lake mud, sampled at 5 cm intervals, with three radiocarbon dates (Ritchie and Hadden, 1975). The 64 pollen taxa were recorded in a three zone pollen diagram; the lowest zone, from 7,300 to 6,200 yrs. B.P. is dominated by *Picea, Juniperus* and NAP; Zone II, from 6,200 to 3,500 by *Pinus, Picea* and *Betula;* and Zone III is separated by increases in *Picea* and *Betula* and decreases in *Pinus* and NAP. A second core, with two radiocarbon dates, produced similar zones and chronology. The vegetation interpretation was tentative – an early open spruce woodland, invaded at about 6,200 yrs. B.P. by jackpine, with increases in spruce and birch at the expense of pine in the later-Holocene. The age of the basal sediments of this pond, which was covered by the final stages of Glacial Lake Agassiz, was suggested as the "minimum age for the recession of Glacial Lake Agassiz II" (Ritchie and Hadden, 1975). However, subsequent investigations (reported in Fenton *et al.,* 1983) show that the suggested minimum age (7,220 yrs. B.P.) is probably too young by roughly two millennia.

Flin Flon, Saskatchewan, Site 158. A 5.2 m core of organic mud was recovered by R. J. Mott from a small (9 ha) shallow (5 m) lake in the central boreal forest region and made available to the present author for analysis. A summary pollen diagram has been published (Ritchie, 1976, 1979, Ritchie and Yarranton, 1978b) and Figure 4 but the detailed diagram with 67 taxa recorded from 55 sampled levels remains unpublished. Four pollen zones were distinguished – the lowest, dominated by NAP (*Artemisia,* Gramineae and Cyperaceae), is followed by a *Picea*-NAP zone, a *Picea-Betula* zone, and finally

at ~7,000 yrs. B.P. the modern pollen spectra become established, dominated by *Picea, Pinus, Betula* and *Alnus.* The suggested interpretation (Ritchie, 1976), elaborated by numerical comparison with a large set of modern pollen data (Ritchie and Yarranton, 1978b), is that a brief period of tundra vegetation, possibly forming a narrow zone adjacent to the Laurentide ice, was followed by establishment of the early version of the spruce boreal forest recognised elsewhere in the Western Interior; this forest complex was modified in the later-Holocene by the migration of tree birch and later jackpine and alder, producing the modern boreal forest mosaic at roughly 6,500 yrs. B.P.

Porcupine Mountain, Manitoba, Site 156. Nichols (1969) reported on the analysis of a 220 cm section of bog and pond sediment from a site in the southern boreal forest region of western Manitoba. The percentage diagram shows 40 taxa and eight radiocarbon dates and the vegetational and climatic reconstruction are based on radiocarbon-dated intervals rather than pollen zones. Nichols (1969) suggests that spruce forest was replaced by grassland at about 6,700 yrs. B.P. and that there was a maximum prairie period at about 5,500 yrs. B.P. The evidence for such an interpretation is not entirely convincing and the diagram differs considerably from those from nearby sites in the southern boreal forest, particularly Site 145 (Mott, 1973) and Site 152 (Ritchie, 1964, and Fig. 4 in the present paper). As Nichols points out (1969) the first 3,000 years of the Holocene record are missing from this site.

Clearwater Bog, Manitoba, Site 162. Nichols (1969) presents a pollen diagram based on analysis of 5 cm interval samples of an 80 cm peat section from a treed bog near The Pas, Manitoba. Thirty-one pollen taxa are recorded and there are five radiocarbon dates from the site, the oldest of which is 1,280 ± 75. The diagram reveals no pollen stratigraphy of any significance, as the author notes.

Thompson, Manitoba, Site 159. A 1.7 m section of a peat bog in the central boreal forest region of Manitoba was analyzed at 10 cm intervals; 34 pollen taxa were recorded and a single radiocarbon determination made. A summary pollen diagram was published (Ritchie, 1976, Fig. 2b) but the complete record remains unpublished because of inadequate chronological control and an apparently uninformative sequence. The basal samples show a spruce-birch-juniper-sagebush assemblage, separated as a pollen zone, but the upper zone in a spruce-pine-birch-alder assemblage is of doubtful validity. The most notable feature of the record is the rapid increase in pine pollen frequencies at about 6,500 yrs. B.P., indicating immigration of this tree to the region.

Lynn Lake, Manitoba, Site 160. The results of pollen analysis of a 155 cm peat monolith overlying detritus mud and blue-grey glacial lake clays from a site in the central boreal forest of Manitoba were extensively reported by Nichols (1967a, 1967b, 1967c, 1975). Samples were analyzed at 5 cm intervals; 32 pollen taxa were recognised and the chronology is based on eight radiocarbon determinations. Although the author delineates three pollen zones in the diagram, he devotes little attention to questions of pollen stratigraphy – the main thrust of the paper is paleoclimate, and particularly the question of transatlantic correlations. The most interesting pollen stratigraphic feature is an apparent pine rise at the 62 cm level, but the chronology of this event is obscured by the reversal of one radiocarbon age determination.

Reindeer Lake, Saskatchewan, Site 161. A 2 m core of sediment was raised from the South Bay of this vast lake and 20 cm interval samples yielded 28 pollen taxa. The complete record remains unpublished but a summary diagram was included in an earlier review of the pollen stratigraphy of the Western Interior (Ritchie, 1976, Fig. 2b). The only notable feature of the record is the pine rise, which delimits the second of three pollen zones, dated at roughly 5,800 yrs. B.P. The lowest zone is dominated by spruce, birch, juniper, willow and sagebush, and an uppermost zone (III) is separated on grounds that were described as "rather tenuous" (Ritchie, 1976: 1805) and seem in retrospect better depicted as quite inadequate.

Ennadai Lake, Northwest Territories, Site 198. This is perhaps one of the best known records from Canada, having been published in full at least five times (Nichols 1967a, 1967b, 1967c, 1972, 1974, 1975) and referred to widely in the paleoecological literature. Little need be added here. A 150 cm monolith of peat, sampled at 5 cm intervals, supported by 12 radiocarbon dates, yielded 32 pollen taxa. A simplified version is presented here (Fig. 3). Although three pollen zones with sub-zones are delineated, no attempt was made to describe the zonation *per se,* probably because the author was interested primarily in the paleoclimatic significance of the record. The only pollen stratigraphic point that might be made to supplement the very full commentary that exists is that the pine pollen percentages reach >10% at about 4,800 yrs. B.P. and continue to rise to >25% by −2,500 yrs. B.P., then decline sharply. A second peat profile at Ennadai Lake was sampled in 1972 and later published (Nichols, 1975). It shows the same, basically complacent pollen stratigraphic record as the first, several curiously inverted radiocarbon ages, and a numerical pollen zonation scheme that bears very little relation to the zonation of the original Ennadai diagram (Nichols 1967c, Fig. 2), and is not referred to in the text.

Nicol Lake, Site 196, Long Lake, Site 197, Grant Lake, Site 201, and Slow River, Site 200. These four sites, reported by Kay (1979), are all in the low arctic tundra of the central Northwest Territories, and in each case a peat monolith was sampled for pollen. Only the Long Lake section is presented as a pollen diagram, showing both percentage and concentration values for nine taxa recorded at (apparently) 1 to 2 cm intervals. The Long Lake site has two radiocarbon dates but the other sites have only a single value. The pollen data are treated by principal component analysis and paleoclimates are reconstructed by a transfer function procedure using modern pollen and climate data. The percentage diagram, and to a lesser extent the concentration curves, show a change from a spruce dominated assemblage before about 3,700 yrs. B.P. to an assemblage similar to the modern (surface) spectrum. The calibration procedure shows this change and Kay (1979: 137-138) notes that "The clearly marked discontinuity in vegetation and climate at Long Lake indicates that the site was close to the mean summer frontal zone prior to about 3,700 yrs. B.P. . . . ."

Drainage Lake, Northwest Territories, Site 203. A 30 cm peat monolith from a site in the low arctic tundra yielded a pollen diagram with two radiocarbon dates and 16 pollen taxa recorded from seven levels (Nichols, 1970, 1972, 1974, 1975). In addition, curves showing the numbers of grains per dry weight of peat are plotted. No pollen zones are described and the pollen stratigraphy shows changes of only local significance, primarily caused by fluctuations in the amounts of sedge and ericad pollen.

Pelly Lake, Northwest Territories, Site 202. A similar peat monolith from a nearby site yielded a comparably complacent pollen diagram (Nichols, 1970, 1972), based on 14 sampled levels from the 35 cm sequence, with 22 pollen taxa and three radiocarbon dates. Curves of numbers of grains per unit dry weight of peat are shown and although no significant pollen stratigraphy is indicated, the author suggests treeline movements based on the record, particularly a maximum extension northward at 900 yrs. B.P., apparently correlated with paleoclimatic events in northwest Europe.

MacAlpine Lake, Northwest Territories, Site 204. Terasmae (1967) reported the pollen percentages of five samples from a 55 cm peat section collected at a site in the low arctic tundra zone. The pollen stratigraphy is uninformative and the original collector of the samples urges "great caution in the use of the diagram" as there is "a very good chance that peat of different ages is mixed" (Blake in Nichols, 1972: 339).

Gordon Bay, Northwest Territories, Site 195. A similar injunction to caution on the basis of questionable dates and the likelihood of reworked sediment (Blake in Nichols, 1972: 339) suggests that the pollen diagram from this site (Terasmae *et al.,* 1966), in itself rather uninformative, should be simply noted in passing as the only available data from this vast, poorly known region.

Coppermine River, Northwest Territories, Site 192. Two profiles were sampled from this tundra site near the mouth of the Coppermine River and percentage and absolute diagrams of each are available (Nichols, 1975). The profiles are 2,500 and 3,715 radiocarbon years old respectively and they show little significant pollen stratigraphic patterns, with the exception of a late-Holocene episode of local vegetation change reflected in the large variations in the abundances of birch, ericad and willow pollen. This area was deglaciated about 10,000 years ago and it is unfortunate that sediment from small lakes could not have been sampled during the fieldwork. Such sediment would have produced a long record and might have included data on the spruce rise, probably the single most interesting pollen stratigraphic question remaining unanswered from this remote locality.

Porter Lake, Northwest Territories, Site 194. A small (4 ha), shallow (5 m) lake near the northeast shore of Porter Lake was cored and yielded a 160 cm section of sediment. The full record, a percentage diagram with 27 taxa recorded at 5 or 10 cm intervals and three radiocarbon dates, has not been published but a summary version was included in a review paper (Ritchie, 1979). Three pollen zones are readily distinguished – the basal from ~7,000 to 6,700 yrs. B.P. dominated by *Picea, Betula, Salix* and NAP; the middle zone, from 6,700 to ~5,200 dominated by *Picea* and *Alnus;* and the uppermost from 5,200 to the present by *Pinus, Picea, Betula* and *Alnus.* The zone 1-2 boundary at about 6,700 yrs. B.P. marks the immigration of pine to the region. *Myrica* percentages peak between 6,700 and 5,800, and decline thereafter to their modern sporadic occurrence. The record also indicates that alder arrived in this area at about 6,700 yrs. B.P.

Thompson Landing, Northwest Territories, Site 193. A 160 cm peat profile was recovered from this site in

the transition zone between forest-tundra and open boreal woodlands along the north shore of the northeast arm of Great Slave Lake. Percentage and absolute diagrams show 27 taxa recorded in 65 levels with four radiocarbon dates, and, remarkably enough, 12 pollen zones. (Nichols, 1975). The diagrams lack any significant pollen stratigraphy and in particular it is surprising that the pine rise and alder rise events, recorded in nearby small lake records (Site 194, Ritchie, 1979, MacDonald personal communication, 1984) were not registered.

Port Radium, Northwest Territories, Site 191. Two peat profiles were sampled from this site in the open boreal woodland zone. Percentage and absolute pollen diagrams for the longer (120 cm) have 31 pollen taxa, three radiocarbon determinations and six pollen zones determined by a clustering technique (Nichols, 1975). The second profile is shorter. No reference is made in the text to the pollen zonation but the interested reader can find quite detailed descriptions of the author's paleoclimatic interpretation of the data. The oldest level is ~5,600 years, so that the lower half of the Holocene record from this site was not sampled.

Colville Lake, Northwest Territories, Site 190. This site lies in the northern boreal forest zone where continuous but open canopy spruce woodlands prevail. A 215 cm peat monolith was sampled at variable intervals and a percentage pollen diagram published (Nichols, 1972, 1974). Seven radiocarbon dates are provided, the oldest being 6,790 yrs. B.P. and 23 pollen taxa were registered. The author makes no attempt to zone the diagram, appropriately, as it shows very little pollen stratigraphy if one discounts the *Sphagnum* curves. Largely on the basis of these, Nichols (1974) does propose that the initially open spruce forest at the site closed up between 6,600 and 3,700 yrs. B.P. in response to a supposed climatic warming and then opened up again at about 3,700 as the climate cooled.

Eildun Lake, Northwest Territories, Site 188. A 268 cm core of sediment from a small lake in the boreal forest zone of the central Mackenzie River valley yielded an excellent pollen record with good radiocarbon control (Slater, 1978). The full record has yet to be published, but a clear pollen stratigraphy has been established beginning with a late-Pleistocene NAP zone dominated by *Artemisia,* Gramineae and Cyperaceae, "succeeded by a *Betula-Salix* shrub-tundra, followed by a *Betula-Populus* forest-tundra. A *Picea-Betula-Alnus* dominated forest became established by approximately 6,000 yrs. B.P. and has persisted since then with little change." (Slater, 1978). As the author notes, this sequence is identical to that established for the Mackenzie Delta region to the immediate north (Ritchie, 1977, 1984).

Antifreeze Pond, Yukon Territory, Site 163. Rampton (1971) reported the results of detailed pollen analysis of 5.2 m of organic pond sediments cored from a 140 m diameter, shallow (2 m) pond at 720 m elevation in the southwest Yukon. Samples at roughly 10 cm intervals yielded 43 pollen taxa. Seven radiocarbon age determinations provided a secure chronology, with one age out of sequence. A continuous record from ~31,000 yrs. B.P. provided the first long record with detailed analyses for this entire region, and a simplified version of the percentage diagram is shown here (Fig. 3) with the original zonation.

Pollen Zone 1 is of great interest because, in addition to the dominant NAP types (Cyperaceae, Gramineae, *Artemisia, Thalictrum*), *Picea* and *Alnus* show continuous curves each reaching 10%. Pollen Zone 2 is distinguished by a double peak in *Betula*. Zone 3 has a dominance of NAP taxa with reduced *Picea* and *Alnus* and a continuous curve of *Betula* between 5 and 20%. Rampton (1971) subdivides it into a lower subzone (3a) with reduced Cyperaceae and increased Gramineae and an upper (3b) with higher *Artemisia* and *Phlox* frequencies. The Holocene zones (4 to 6) are distinguished by what has become a basic pattern for diagrams from this northwest, boreal-subarctic region as a whole – a *Betula* zone (4) followed by *Picea-Betula* (Zone 5, beginning at 8,710 yrs. B.P.) and finally at 5,690 yrs. B.P., Zone 6 characterized by a sharp *Alnus* rise with continued high percentages of *Picea* and *Betula*.

This fine pollen record was supplemented by data on Holocene spruce logs found above the present treeline and by tree-ring studies of treeline spruces. The interpretation of the pollen diagram was facilitated by reference to a suite of modern samples from different vegetation zones in the area. Rampton (1971) proposes a sedge-moss or fell-field vegetation for Zone 1 and ascribes the *Picea* and *Alnus* frequencies to redeposition from older sediments. Zone 2 (27,000 to 31,000) is interpreted as a shrub tundra in a slightly less severe climatic episode. Zone 3, from 27,000 to 10,000 is reconstructed as a sedge-moss tundra landscape with cold summers. Zones 4 to 6 (10,000 to present) are interpreted as a transition from shrub tundra through spruce woodland to spruce forest, with a corresponding amelioration in climate, and it is suggested that treeline was higher than its modern position at least three times since 5,700 yrs. B.P.

Natla Bog, Northwest Territories, Site 187. MacDonald (1983) presents the results of pollen analysis of a 2.3 m peat lens exposed along the Natla River in the Selwyn Mountains. The site lies above but close to the limit of spruce forest, in the transitional zone between forest and shrub tundra. Six samples were radiocarbon dated and pollen accumulation rates were calculated for all 26 taxa recorded in the 10 cm interval samples. The chronology was strengthened by the identification of the White River Ash in the uppermost 20 cm of the section. The diagram is divided into zones using numerical techniques, and a set of 15 surface pollen samples was used to develop a principal component analysis comparison of the fossil and modern spectra. A simplified version of the original percentage diagram is presented here (Fig. 4) because this site fills a large gap in the record and is rigorously documented. The lower pollen zone (1) is dominated by *Betula*, probably *B. glandulosa,* associated with *Salix, Artemisia,* Gramineae and Ericales, and MacDonald (1983) suggests that a dwarf birch tundra was widespread at this time (8,630 to 7,700 yrs. B.P.) in the entire valley. The upper zone (2) is dominated by high percentages and PAR of *Picea* and it is likely that the modern vegetation type was established by about 7,000 yrs. B.P.

John Klondike Bog, Northwest Territories, Site 189. A percentage pollen diagram is reported by Matthews (1980) for a 4.7 m core of peat and gyttja recovered from a small bog near Fort Liard in the southwest District of Mackenzie. Coring difficulties prevented sampling the entire section but a simplified diagram is reproduced here (Fig. 4) because this site fills a gap in the published record for the Mackenzie River valley. MacDonald (personal communication, 1984) has recovered longer records from sites in the same general area so we can expect, as Matthews anticipates (1980: 5), that the John Klondike Bog record will be significantly expanded in the next few years.

The percentage pollen diagram (24 taxa with a sampling interval of 10 cm) is divided into three regional zones, and four subzones reflecting local changes. The lowest Zone (1) from 8,600 to ~6,000 is delimited by the *Alnus* rise, with lower *Betula* frequencies but still high *Picea* values. The boundary between Zone 2 and Zone 3 is marked by the rise in *Pinus* from negligible values to about 10%. In general, the data from this site are more informative about local environmental change than about regional events, and analysis of molluscs, insects and plant macrofossils complement that reconstruction effectively.

Chapman Lake, Yukon Territory, Site 164. This first effective investigation of late-Quaternary pollen stratigraphy of the unglaciated Yukon was published by Terasmae and Hughes (1966), from a site in the Ogilvie Mountains. A total of 4.2 m of frozen peat were cored at Chapman Lake, yielding 24 irregularly spaced samples in which 32 pollen types were enumerated. Three radiocarbon age estimates showed that the section encompassed the period from 13,870 to present. Three pollen zones were recognised, the lowest dominated by NAP, in particular Cyperaceae and Gramineae; Zone II delimited by an abrupt *Betula* increase; and Zone III a *Picea-Alnus* assemblage.

Gill Lake, Yukon Territory, Site 164a. The second site investigated by Terasmae and Hughes (1966) produced a percentage diagram very similar to that from Chapman Lake except that it lacks the lower, NAP zone. However, the basic pattern, subsequently repeated from many sites in the northwest, was recognized at these two sites, namely, a sequence of an NAP zone followed by abrupt increases in the frequencies of *Betula,* then *Picea* and finally *Alnus.*

Northern Yukon and Mackenzie Delta region, Sites 165 to 186. The pollen stratigraphy, vegetation reconstruction and paleoecology of this region have been described and discussed fully, including summary pollen diagrams, in a very recent review (Ritchie, 1984). Twenty sites have been recorded and they are listed in Table 1, where the reader can find references to the original publications. This large data set will not be described here to avoid major duplication of material. However, three pollen sequences representative of tundra, treeline and northern boreal forest sites are reproduced here in summary form (Fig. 4). The Hanging Lake site, based on Cwynar (1982) illustrates the pollen stratigraphy of a locality in the modern tundra of the North Yukon, showing a full-glacial herb zone followed by a transition in the early-Holocene to heath-dwarf birch tundras typical of the present day. The Sleet Lake Site based on Spear (1983) represents localities north of, but close to, the modern limit of tree occurrence, and demonstrates that a late-Pleistocene dwarf birch pollen zone was followed by an early-Holocene spruce zone, replaced in the mid-Holocene by the modern pollen assemblage type. The Twin Tamarack sequence (Site 185), based on Ritchie (1985), is typical of sites immediately south of the northern limit of the boreal woodland zone. It shows (Fig. 4) a late-Pleistocene herb tundra zone followed by a dwarf birch zone with three sub-zones, a spruce zone and finally the modern

type (spruce-birch-alder) established by 6,000 yrs. B.P.

Eastern Arctic Sites

Baffin Island Sites have been investigated by a multi-disciplinary effort led by Dr. John T. Andrews for over a decade, involving mainly workers from the Institute of Arctic and Alpine Research at the University of Colorado. A complete compilation in book form of all the Quaternary research on Baffin Island by this group will be published shortly, including a chapter on the Holocene pollen record by Short *et al.,* (1984); therefore the following account will be brief, to avoid excessive duplication.

In spite of this large field and laboratory effort over several years, the pollen record remains intractable, as Short *et al.,*(1984) emphasize. Holocene pollen diagrams have been compiled for two lake sites and 16 peat sections, supplemented by over 50 radiocarbon determinations. Interpretation of the pollen data has been attempted using 175 modern polster spectra and various standard transfer function approaches. Tauber trap studies were completed at one site, by Davis (1980b).

Iglutalik Lake, Site 219, was studied in detail by Davis (1980a, 1980b), also reported in Davis *et al.,* (1980), and although the percentage and concentration diagrams have yet to be published in full, this investigation is the most detailed and informative of the Holocene records from Baffin. A summary diagram is reproduced here (Fig. 3), derived, with Dr. Davis' permission, from his thesis and later amendments to the analysis. A 290 cm core of lake sediment was recovered from Iglutalik Lake and 34 pollen taxa were tabulated from samples at 2 cm intervals. Pollen percentage and concentration diagrams were prepared, and the chronology was established by five radiocarbon dates. The lake lies 1 km inland from Cumberland Sound and is about 1 km in length; water depth at the coring sites ranged between 3.5 and 6 m. The early-Holocene (9,200 to ~5,500) is characterised by Cyperaceae and Gramineae among the pollen of local origin, but Davis (1980a; 61) notes that "exotic tree pollen *(Alnus, Picea, Pinus)* was higher between 9,200 and 8,700 B.P. than anytime until 5,800 B.P., and suggests strong summer airflows from the Boreal Forest . . . .". Increased pollen concentration of the local taxa (Cyperaceae, Gramineae, *Betula*) between 5,800 and 4,700 B.P., associated with increased organic content and pollen concentration led Davis (1980a; 61) to conclude that "vegetation became well established. . . . ", and, with increases of exotic pollen at this time *(Alnus, Picea, Pinus)*, that there was a "locally late and short-lived hypsithermal". *Salix* increased markedly at ~4,000 years ago, and dominates the spectra with reduced proportions of sedge and grass pollen. Gramineae increases to over-all dominance from ~1,200 to the present day.

Patricia Bay Lake, Site 209, was investigated by Mode (1980) by pollen analysis of a 1.9 m core of lake sediment recovered from a small (4 ha) lake. The record was interrupted at ~6,500 yrs. B.P. by a brief period of marine inundation. Four radiocarbon dates show an inversion of the lower two (8,810 overlies 6,320). Thirty two pollen taxa were enumerated in 2.5 cm interval samples. Four pollen zones were described: the lowest (A), between 6,800 and 5,600 yrs. B.P., dominated by *Betula,* Filicales and *Lycopodium clavatum/annotinum,* is interpreted as a tundra with dwarf birch close to the site, reflecting maximal Holocene warming; a Caryophyllaceae zone (B) on the basis of high percentages of taxa ascribed to that family; an *Alnus* zone (C) from 4,500 to 3,100 yrs. B.P. with a maximum of alder percentages associated with high ericad frequencies. The appearance of exotic *Picea* in this zone, along with alder and birch, is interpreted as due to northward advances of the treeline at stations in mainland Canada. The *Salix* zone (D) from 3,100 to the present day, is characterized by maximum willow percentages (40%) interpreted as cooler climates favouring chionophilous species of *Salix*.

Holocene peat monoliths, Sites 210 to 217, were investigated by Miller *et al.,* (1977), at Clyde Foreland, Mode (1980) at Qivitu Cliffs near Clyde (Site 209), by Short and Andrews (1980) from six sites on the northern Cumberland Peninsula (Padloping Island, Site 212; Broughton Island, Site 213; Owl River, Site 216; Pass Head, Site 217; Quajon Fiord (Site 215) and Idjuniving Island (Site 214) and at Maktak Fiord (Site 210) by Boulton *et al.* (1976) and at Windy Lake (Site 211) by Nichols (1975) and later by Davis (1980). Short *et al.* (1984) summarise these findings as follows:

"The most important trend observed is the change from a mixed and relatively diverse shrub and heath assemblage that characterised the period between about 5,500 B.P. (Pass Head site) and 1,500 B.P. (Owl River, Windy lake, Maktak Fiord, and the lower part of Padloping Island sites) and the predominantly graminoid (grass and sedge) phase that characterized

the last 1,500 years (Quajon Fiord, Idjuniving Island, Broughton Island, and the upper part of the Padloping Island sites). The combined record also suggests that even the shrub assemblage from the period between 3,000 and 2,000 B.P. records a climatic deterioration from the local climatic optimum, registered at Pass Head, which is characterized by a much greater pollen productivity with higher *Salix* values."

In other words, the Holocene record from Baffin remains at a fairly primitive level of understanding. As Short *et al.* (1984) imply, the early preoccupation of the INSTAAR pollen group with peat samples rather than sediments from small lakes, influenced presumably by Nichols' (1967a) wariness of arctic lacustrine samples, has not produced a very informative record of vegetation history. By contrast, the following investigation, based on a single field season, illustrates the value of sampling small, deep lakes in arctic situations.

Baird Inlet, Ellesmere Island, Site 206 yielded an informative pollen record from a 56 cm core recovered from a small (~3 ha), relatively deep (14.5 m) closed basin lake. Three radiocarbon age determinations indicate that the sedimentation began at 9,000 yrs. B.P. Twenty pollen taxa were recognized in 2.5 cm interval samples. The full diagram has not yet been published, but an abstract by Hyvärinen and Blake (1981: 35) states that:

"Rates of sedimentation vary from 0.07 to 0.04 mm/yr. Pollen concentration in samples analysed at 2.5 cm intervals varies from 1,500-2,000 per cm^3 in the lower part of the core to 100-200 in the upper part: calculated pollen influx is between 14 and 0.8 grains per cm^2 per yr. The basic total of *ca.* 100 grains/sample includes exotic derived pollen (tree pollen and indeterminables) and excludes spores. The amount of exotic/derived pollen is low (2-10%) throughout most of the core, with some rise toward the top. The basal sandy-silt is devoid of local pollen but does contain some obviously derived pollen, mainly degraded betuloid and coniferous types. Four local pollen zones reflect an early pioneer phase (grass-sedge-*Oxyria/Rumex*) in the lowermost organic sediment, followed by a spread of *Salix* and then, some 7,000 to 6,560 years ago, a rise in Ericales. The topmost zone shows some increase in indicators of bare ground and fell-field vegetation *(Saxifraga,* Ranunculaceae, Caryophyllaceae, *Dryas),* hence deterioration of local conditions during the last 4,000 years."

Axel Heiberg Island, Site 205. Several peat profiles were sampled by Hegg (1963) during the fieldwork of the Jacobsen-McGill Arctic Research Expedition to

Axel Heiberg Island. Several radiocarbon age determinations indicated that the peats ranged in age from 4,200 years ago to the present. However the pollen results were not particularly informative, lacking detail and any evidence of clear stratigraphy.

Truelove Lowland, Devon Island, Site 207. Cores recovered from an ice-wedge polygon deposit were analysed with supporting radiocarbon dates, reported by Jankovska and Bliss (1977). The main profile had a basal age of 2,450 radiocarbon years. The pollen results were inconclusive, and should at best be regarded as quite preliminary.

Devon Island Ice Cap, Site 207a. Meltwater samples from a 299 m ice core were analysed, initially by Lichti-Federovich (1975) and later by McAndrews (1984). Forty pollen taxa were identified in 112 core segments that were consolidated by McAndrews (1984) into 21 time intervals to increase the pollen sum in each sample. A percentage diagram shows some rough stratigraphy over the 130,000 year period spanned by the core. McAndrews (1984) concludes that "The late-Holocene and interglacial assemblages were dominated by *Alnus* (alder), whereas the early-Holocene and Wisconsinan were dominated by *Betula* (birch) and *Artemisia* (sage). During the Holocene and probably the last interglaciation most of the pollen and spores were blown a minimum of 1,000 km from low arctic shrub tundra and adjacent subarctic *Picea* (spruce) forest; these areas were dominated by the arctic air mass during the summer pollinating season. During the Wisconsinan-early-Holocene, glacier ice and arctic air was more widespread and pollen sources were more distant; thus, at this time relatively little pollen was incorporated into the ice".

CONCLUDING COMMENTS

The area reviewed above is too vast and diverse, and the pollen record is too unevenly distributed and variable in quality, to justify any attempt here to synthesise the pollen stratigraphy or the reconstructions of vegetation and environment. A few comments are offered below on the quality of the record and on useful future directions.

A striking feature of the pollen data from the more than 200 sites is the immense variation in their quality and therefore scientific value. Partly this can be explained as a reflection of the process of refinement of the methods and approaches of pollen analysis over the past 40 years, in ways that are familiar. In other respects it indicates weaknesses in the design of

projects and, in some instances, of course, pollen data were collected as minor parts of a larger investigation. It appears in certain cases that useful, precise questions were not addressed before field coring trips were undertaken, but that even when field sampling was defective for whatever reason, the resulting pollen diagrams were published and occasionally republished. One explanation of this tendency might be that pollen analysis remains labour-intensive, particularly at the phase of microscopic work, and therefore the production of a pollen diagram seems to be a more significant and thus publishable event that it would be if measured critically in terms of its useful scientific content. Until automated pollen counting becomes an effective and cheap technique, if ever, the compulsion will probably remain among some investigators to publish pollen diagrams simply because they represent the expenditure of much time and resources, regardless of their useful content.

In the area reviewed in this chapter, effective future palynology should be directed along the following lines:

1. Completing the stratigraphic record. Several large gaps remain in both space and time. Older records, from Pleistocene sediments deposited before the latest glacial cycle, are lacking from most regions. The late-Pleistocene and Holocene record is very poorly known from the entire arctic, and particularly the vast mid-arctic swathe from Banks and Victoria Islands east to western Baffin Island. The majority of the several peat bog sites reported on above from the subarctic region west of Hudson Bay are truncated and otherwise of limited value and a well designed coring project(s) straddling treeline from the Mackenzie River Delta eastward to Keewatin would answer many outstanding questions about the survival and migration of spruce in relation to the shrinking of Laurentide ice; about the vegetation history of the modern forest-tundra zone; about the northern extension of pine; and about the late-Pleistocene and Holocene paleoenvironments of the region. The western segment of the boreal forest, overlapping roughly with the range of lodgepole pine, has notable gaps in the record, though G. M. MacDonald and L. C. Cwynar (personal communication) have several unpublished sites with significant new records. Palynological (not to say geological!) evidence for ice-free corridor vegetation remains very uncertain. The southwest plains region remains poorly understood, particulary in terms of the postglacial origins and migration of the main tree species (spruces, pines and birches). A continuous, detailed and well dated record of late-Pleistocene and Holocene pollen from the Canadian Prairies has yet to be obtained.

2. *Investigating long-term changes in vegetation.* Careful site selection (Jacobson & Bradshaw, 1981) is essential if useful pollen data are to be obtained at scales appropriate to the ecological questions being posed. Basin size, sediment type and the scale of local landform-vegetation features are critical factors that must be evaluated thoroughly before sites are selected. A recent essay by Prentice (1983) provides both stimulating and cautionary comment on the capacity of pollen analysis to address paleoecological problems, using various approaches (isopol mapping, simulation studies, calibration functions).

The allocation of resources to palynological investigations is a central issue requiring careful planning, because of the high costs of both the field and the laboratory phases of the research. If the idea to be tested or explored requires precise information about past vegetation composition, then detailed analysis of a single site is justified, involving the tabulation of pollen taxa in the most exact taxonomic detail possible, at many closely spaced sample levels, and with very high (>2,000) counts per level. (*e.g.,* Cwynar

1982). On the other hand, if the basic pollen stratigraphic record of a region is unknown, or if hypotheses about post-glacial species migration are being tested, then a number of evenly spaced, similar sites is required and smaller pollen counts at wider intervals with less attention to the identification of uncommon taxa would be appropriate. In short, the basic question being posed should be clarified and elaborated fully before the design of data-collection and analysis is decided. This may seem too obvious an exhortation to mention here, but a scrutiny of the variably useful pollen diagrams reviewed in this chapter will indicate that it is a procedure not universally practised.

ACKNOWLEDGEMENTS

The support of the National Sciences and Engineering Research Council of Canada is acknowledged (Grant A6320). I am indebted to Kathleen Hadden who assisted with the collation of data and arrangement of illustrations, and to Danielle Carbone who typed the manuscript and table.

References Cited

ALLEY, N.
   1976   The palynology and paleoclimatic significance of a dated core of Holocene peat, Okanagan Valley, southern British Columbia. *Canadian Journal of Earth Sciences,* 13: 1131-1144.

ANDREWS, J. T., MILLER, G. H., NELSON, A. R., MODE, W. N., and LOCKE, W. W.
   1981   Quaternary Near-Shore Environments on Eastern Baffin Island, N.W.T. *In:* Mahaney, W. C. (ed.), *Quaternary Paleoclimate.* Geological Abstracts, Norwich, U.K., p. 13-14.

BOULTON, G. S., DICKSON, J. H., NICHOLS, H., NICHOLS, M., and SHORT, S. K.
   1976   Late Holocene glacier fluctuations and vegetation changes at Maktak Fiord, Baffin Island, N.W.T., Canada. *Arctic and Alpine Research,* 8: 343-356.

BRIGGS, N. D., and WESTGATE, J. A.
   1978   Fission-track age of tephra marker beds in Beringia. *American Quaternary Association, 5th Biennial Meeting, Edmonton, Alberta, Abstracts,* 190 p.

CHRISTIANSEN, E. A.
   1979   The Wisconsinan deglaciaton of Southern Saskatchewan and adjacent areas. *Canadian Journal of Earth Sciences,* 10: 913-938.

CWYNAR, L. C.
   1982   A Late-Quaternary vegetation history from Hanging Lake, Northern Yukon. *Ecological Monographs,* 52: 1-24.

DAVIS, P. T.
   1980a   Holocene vegetation and climate record from Iglutalik Lake, Cumberland Sound, Baffin Island, N.W.T., Canada. *Abstracts and Program, Sixth Biennial Meeting, American Quaternary Association, Orono, Maine,* 61 p.

   1980b   Late Holocene glacial, vegetational, and climatic history of Pangnirtung and Kingnait Fiord area, Baffin Island, Canada. Ph.D. thesis, Department of Geological Sciences, University of Colorado, 366 p.

DAVIS, P. T., NICHOLS, H., and ANDREWS, J. T.
   1980   Holocene vegetation and climate record from Iglutalik Lake, Baffin Island. *Abstracts, Fifth International Palynological Conference, Cambridge,* 105 p.

DELORME, L. D., ZOLTAI, S. C., and KALAS, L. L.
   1977   Freshwater shelled invertebrate indicators of paleoclimate in northwestern Canada during late glacial times. *Canadian Journal of Earth Sciences,* 14: 2029-2046.

De VRIES, B., and BIRD, C. D.
   1965   Bryophyte subfossils of a late-glacial deposit from the Missouri Coteau, Saskatchewan. *Canadian Journal of Botany,* 43: 947-953.

FENTON, M. M., MORAN, S. R., TELLER, J. T. and CLAYTON, L.

1983   Quaternary Stratigraphy and History in the Southern Part of the Lake Agassiz Basin. *In:* Teller, J. T. and Clayton, L. (eds.), *Glacial Lake Agassiz,* Geological Association of Canada Special Paper 26: 49-74.

HANSEN, H. P.

1949a   Postglacial forests in west central Alberta, Canada. *Torrey Botanical Club,* 76: 278-289.

1949b   Postglacial forests in south-central Alberta, Canada. *American Journal of Botany,* 36: 54-65.

1950   Postglacial forests along the Alaska Highway, British Columbia. *Proceedings of the American Philosophical Society,* 94: 411-421.

1952   Postglacial forests in the Grande Prairie-Lesser Slave Lake region of Alberta, Canada. *Ecology,* 33: 31-41.

1953   Postglacial forests in the Yukon Territory and Alaska. *American Journal of Science,* 251: 505-542.

1955   Postglacial forests in South Central and Central British Columbia. *American Journal of Science,* 253: 640-658.

HARINGTON, C. R., TIPPER, H. W., and MOTT, R. J.

1974   Mammoth from Babine Lake, British Colulmbia. *Canadian Journal of Earth Sciences,* 11(2): 285-303.

HEGG, O.

1963   Palynological Studies of a Peat Deposit in front of the Thompson Glacier. *Axel Heiberg Island Research Report.* McGill University, Montreal, p. 217-219.

HILLS, L. V., and OGILVIE, R. T.

1970   *Picea banksii* n. sp. Beaufort Formation (Tertiary) northwestern Banks Island, Arctic Canada. *Canadian Journal of Botany,* 48: 457-464.

HILLS, L. V., and SANGSTER, E. V.

1977   A review of paleobotanical studies dealing with the last 20,000 years; Alaska, Canada and Greenland. *In:* C. R. Harington (ed.),*Climatic Change in Canada.* Syllogeus 26. National Museums of Canada. p. 73-224.

HOLLOWAY, R. G., BRYANT, V. M., and VALANT, S.

1981   A 16,000 year pollen record from Lake Wabamun, Alberta, Canada. *Palynology,* 5: 195-208.

HUGHES, O. L., HARINGTON, C. R., JANSSENS, J. A., MATTHEWS, J. V., Jr., MORLAN, R. E., RUTTER N. W., and SCHWEGER, C. E.

1981   Upper Pleistocene stratigraphy, paleoecology and archaeology of the Northern Yukon Interior, Eastern Beringia. I. Bonnet Plume Basin. *Arctic,* 34: 329-365.

HYVÄRINEN, H., and BLAKE, H.

1981   Lake sediments from Baird Inlet, East-Central Ellesmere Island, Arctic Canada: Radiocarbon and Pollen data. *In: Abstracts, 3rd International Symposium on Paleolimnology, Joensuu, Finland,* 35 p.

HYVÄRINEN, H., and RITCHIE, J. C.

1975   Pollen stratigraphy of Mackenzie pingo sediments N.W.T., Canada. *Arctic and Alpine Research,* 7: 261-272.

JACOBSON, G. L., and BRADSHAW, R. H. W.

1981   The Selection of Sites for Paleovegetational Studies. *Quaternary Research,* 16: 80-96.

JANKOVSKA, V., and BLISS, L. C.

1977   Palynological analysis of a peat from Truelove Lowland. *In:* Bliss, L. C. (ed.), *Truelove Lowland Devon Island, Canada: a High-Arctic Ecosystem.* University of Alberta Press, Edmonton, p. 139-142.

KAY, P. A.

1979   Multivariate Statistical Estimates of Holocene Vegetation and Climate Change, Forest-Tundra Transition Zone, N.W.T., Canada. *Quaternary Research,* 11: 125-140.

KEARNEY, M. S., and LUCKMAN, B. M.

1983   Holocene Timberline Fluctuations in Jasper National Park, Alberta. *Science,* 221: 261-263.

KLASSEN, R. W.

1975   Quaternary Geology and Geomorphology of Assiniboine and Qu'Appelle Valleys of Manitoba and Saskatchewan. *Geological Survey of Canada,* Bulletin 228, 61 p.

KLASSEN, R. W., DELORME, L. D., and MOTT, R. J.

1967   Geology and paleontology of Pleistocene deposits in southwestern Manitoba. *Canadian Journal of Earth Sciences,* 4: 433-477.

KUPSCH, W. O.

1960   Radiocarbon-dated organic sediment near Herbert, Saskatchewan. *American Journal of Science,* 258: 282-292.

LICHTI-FEDEROVICH, S.

1970   The pollen stratigraphy of a dated section of late-Pleistocene lake sediment from central Alberta. *Canadian Journal of Earth Sciences,* 7: 938-945.

1972   Pollen Stratigraphy of a Sediment Core From Alpen Siding, Alberta. *Geological Survey of Canada,* Report of Activities, Paper 72-1B: 113-115. Ottawa.

1973   Palynology of six sections of late Quaternary sediments from the Old Crow River, Yukon Territory. *Canadian Journal of Botany,* 51: 553-564.

1974   Palynology of two sections of late Quaternary sediments from the Porcupine River, Yukon Territory. *Geological Survey of Canada,* Paper 74-23. Ottawa.

1975   Pollen analysis of ice core samples from the Devon Island Ice Cap. *In:* Report of Activities, *Geological Survey of Canada,* Paper 75-1, p. 441-444.

LUI, KAM-BIU

1981a   Pollen Evidence of Late-Quaternary Climatic Changes in Canada: A review. Part I. Western Canada. *Ontario Geography,* 15: 83-101.

1981b   Pollen Evidence of Late-Quaternary Climatic Changes in Canada: A review. Part II. Eastern Arctic and Subarctic Canada. *Ontario Geography,* 17: 61-82.

MacDONALD, G. M.

1982   Late Quaternary paleoenvironments of the Morley Flats and Kananaskis Valley of southwestern Alberta. *Canadian Journal of Earth Sciences,*19: 23-35.

1983   Holocene Vegetation History of the Upper Natla River Area, Northwest Territories, Canada. *Arctic and Alpine Research,* 15: 169-180.

MACKAY, J. R.

1963   The Mackenzie Delta Area, N.W.T. Ottawa, *Geographical Branch, Memoir 8,* 202 p.

MACKAY, J. R., and TERASMAE, J.

1963   Pollen diagrams in the Mackenzie Delta area, N.W.T. *Arctic,* 16: 228-238.

McANDREWS, J. H.

1984   Pollen analysis of the 1973 Ice Core from Devon Island Glacier, Canada. Quaternary Research, *in press.*

MATTHEWS, J. V., JR.
1975 Incongruence of macrofossil and pollen evidence: a case from the late Pleistocene of the Northern Yukon coast. *Geological Survey of Canada,* Paper 75-1B: 139-145.
1980 Paleoecology of John Klondike Bog, Fisherman Lake Region, Southwest District of Mackenzie. *Geological Survey of Canada,* Paper 80-22: 12 p.

MILLER, G. H., ANDREWS, J. R., and SHORT, S. K.
1977 The last interglacial-glacial cycle, Clyde Foreland, Baffin Island, N.W.T.: stratigraphy, biostratigraphy and chronology. *Canadian Journal of Earth Sciences,* 14: 2824-2857.

MODE, W. N.
1980 Quaternary stratigraphy and palynology of the Clyde Foreland, Baffin Island, N.W.T., Canada. Ph.D. thesis, Department of Geological Sciences, University of Colorado. 219 p.

MOTT, R. J.
1973 Palynological Studies in Central Saskatchewan. Pollen Stratigraphy From Lake Sediment Sequences. *Geological Survey of Canada,* Paper 72-49. Ottawa.

MOTT, R. J., and CHRISTIANSEN, E. A.
1981 Palynological Study of Slough Sediments from Central Saskatchewan. Current Research B, *Geological Survey of Canada,* paper 81-1B: 133-136.

MOTT, R. J., and JACKSON, L. E., JR.
1982 An 18,000 year palynological record from the southern Alberta segment of the classical Wisconsinan "Ice-free Corridor". *Canadian Journal of Earth Sciences,* 19: 504-513.

NICHOLS, H.
1967a The suitability of certain categories of lake sediment for pollen analysis. *Pollen et Spores,* 9: 615-630.
1967b The post-glacial history of vegetation and climate at Ennadai Lake, Keewatin, and Lynn Lake, Manitoba, Canada. *Eiszeitalter und Gegenwart,* 18: 176-197.
1967c Pollen diagrams from Sub-Arctic Canada. *Science* 155: 1665-1668.
1969 The late Quaternary history of vegetation and climate at Porcupine Mountain and Clearwater Bog, Manitoba. *Arctic and Alpine Research,* 1: 155-167.
1970 Late Quaternary pollen diagrams from the Canadian arctic Barren Grounds at Pelly Lake, northern Keewatin, N.W.T. *Arctic and Alpine Research,* 2: 43-61.
1972 Summary of the palynological evidence for late-Quaternary vegetation and climatic change in the central and eastern Canadian Arctic. *In:* Vasari, Y., Hyvärinen, H., and Hicks, S. (eds.), *"Climatic Changes in Arctic Areas during the Last Ten-Thousand Years".* Acta Universitatis Ouluensis, Series A, Scientiae Rerum Naturalium No. 3, Geologica No. 1, p. 309-339.
1974 Arctic North America palaeoecology: The recent history of vegetation and climate deduced from pollen analysis. *In:* Ives, J. D. and Barry, R. G. (eds.), *"Arctic and Alpine Environments".* Methuen, London, p. 637-668.
1975 Palynological and Paleoclimatic Study of the Late Quaternary Displacement of the Boreal Forest in Keewatin and Mackenzie, N.W.T., Canada. *Occasional Paper 15.* Boulder: Institute of Arctic and Alpine Research, University of Colorado.

OVENDEN, L. E.
1982 Vegetation history of a polygonal peatland, northern Yukon. *Boreas,* 11: 209-224.

PEARCE, G. W., WESTGATE, J. A., and ROBERTSON, S.
1982 Magnetic reversal history of Pleistocene sediments at Old Crow, northwestern Yukon Territory. *Canadian Journal of Earth Sciences,* 19: 919-929.

PRENTICE, I. C. M.
1983 Postglacial climatic change: vegetation dynamics and the pollen record. Progress in Physical Geography, *in press.*

RAMPTON, V. N.
1971 Late Quaternary Vegetational and Climatic History of the Snag-Klutlan area, Southwestern Yukon Territory, Canada. *Geological Society of America Bulletin,* 82: 959-978.
1982 Quaternary geology of the Yukon Coastal Plain. Bulletin 317, *Geological Survey of Canada,* Ottawa 49 p.

RITCHIE, J. C.
1964 Contributions to the Holocene Paleoecology of West-Central Canada. I. The Riding Mountain Area. *Canadian Journal of Botany,* 42: 181-197.
1967 Holocene Vegetation of the Northwestern Precincts of the Glacial Lake Agassiz Basin. *In:* Mayer-Oakes, W. J. (ed.). University of Manitoba Press. *Review Paleobotany and Palynology,* 3: 255-266.
1969 Absolute pollen frequencies and Carbon-14 age of a section of Holocene lake sediment from the Riding Mountain Area of Manitoba. *Canadian Journal of Botany,* 47: 1345-1349.
1972 Pollen analysis of late-quaternary sediments from the arctic treeline of the Mackenzie River Delta region, Northwest Territories, Canada. *In:* Vasari, Y, Hyvärinen, H., and Hicks, S. (eds.), *"Climatic Changes in Arctic Areas during the Last Ten-Thousand Years."* Acta Universitatis Ouluensis, Series A, Scientiae Rerum Naturalium No. 3, Geologica No. 1, p. 253-271.
1976 The late-Quaternary vegetational history of the western interior of Canada. *Canadian Journal of Botany,* 54: 1793-1818.
1977 The modern and late Quaternary vegetation of the Campbell-Dolomite Uplands near Inuvik, N.W.T., Canada. *Ecological Monographs,* 47: 401-423.
1979 Towards a Late-Quaternary Palaeoecology of the Ice-Free Corridor. *Canadian Journal of Anthropology,* 1: 15-28.
1982 The modern and Late-Quaternary vegetation of the Doll Creek area, North Yukon, Canada. *New Phytologist,* 90: 563-603.
1983 The Paleoecology of the Central and Northern Parts of the Glacial Lake Agassiz Basin. *In:* Teller, J. T. and Clayton, L. (eds.), *Glacial Lake Agassiz.* Geological Association of Canada Special Paper 26, 1983, p. 156-170.
1984 Past and Present Vegetation of the Far Northwest of Canada. University of Toronto Press, Toronto, 251 p.
1985 Quaternary climatic and vegetational change in the Lower Mackenzie Basin, Northwest Canada. *Ecology,* 65, *in press.*

RITCHIE, J. C., CINQ-MARS, J., and CWYNAR, L. C.
1982 L'environnement tardiglaciaire du Yukon septentrional, Canada. *Geographie Physique et Quaternaire,* 36: 241-250.

RITCHIE, J. C., and De VRIES, B.
1964 Contributions to the Holocene paleoecology of west central Canada. II. A late-glacial deposit from the Missouri Coteau. *Canadian Journal of Botany,* 42: 677-692.

RITCHIE, J. C., and HADDEN, K. A.

1975   Pollen stratigraphy of Holocene sediments from the Grand Rapids area, Manitoba, Canada. *Review of Palaeobotany and Palynology,* 19: 193-202.

RITCHIE, J. C., and HARE, F. K.

1971   Late-Quaternary vegetation and climate near the Arctic treeline of northwestern North America. *Quaternary Research,* 1: 331-342.

RITCHIE, J. C, and LICHTI-FEDEROVICH, S.

1968   Holocene pollen assemblages from the Tiger Hills, Manitoba. *Canadian Journal of Earth Sciences,* 5: 873-880.

RITCHIE, J. C., and YARRANTON, G. A.

1978a   Patterns of change in the Late-Quaternary vegetation history of the Western Interior of Canada. *Canadian Journal of Botany,* 56: 2177-2182.

1978b   The Late-Quaternary history of the boreal forest of central Canada. *Journal of Ecology,* 66: 199-212.

SCHWEGER, C. E.

1972   Pollen diagram from Ah horizon of buried Orthic Humic Gleysol at Robinson. *International Geological Congress, Field guide C-2,* 58 p.

SCHWEGER, C. E., HABGOOD, T., and HICKMAN, M.

1981   Late Glacial-Holocene changes of Alberta – the Record from lake sediment studies. *In: The Impacts of Climatic Fluctuations on Alberta's Resources and Environment, Report WAES – 1-81:* 47-60.

SHORT, S. K., and ANDREWS, J. T.

1980   Palynology of six Middle and Late Holocene Peat Sections, Baffin Island. *Geographie Physique et Quaternaire,* 34: 61-75.

SHORT, S. K., MODE, W. N., and DAVIS, P. T.

1984   The Holocene Record from Baffin Island: Modern and Fossil Pollen Studies. Chapter Manuscript, *in press,* Institute of Arctic and Alpine Research, University of Colorado, Boulder, Colorado.

SLATER, D. S.

1978   Late Quaternary pollen diagram from the central Mackenzie corridor area. *Abstracts, American Quaternary Association. 5th biennial meeting, Edmonton.*

SPEAR, R. W.

1983   Paleoecological approaches to a study of treeline fluctuation in the Mackenzie Delta region, Northwest Territories: preliminary results. *In:* Morrisset, P. and Payette, S. (eds.), *Treeline Ecology, Proceedings of the Northern Quebec Tree-Line Conference, Nordicana,* 47: 61-72.

STRONG, W. L.

1977   Pre- and postsettlement palynology of southern Alberta. *Review of Palaeobotany and Palynology,* 23(5): 373-387.

TELLER, J. T., and LAST, W. M.

1979   Post-Glacial Sedimentation and History of Lake Manitoba. *Manitoba Department of Mines, Natural Resources and Environment, Report* 79-41, 185 p.

TERASMAE, J.

1959   Palaeobotanical study of buried peat from the Mackenzie River Delta Area, Northwest Territories. *Canadian Journal of Botany,* 37: 715-717.

1967   Recent pollen deposition in the northeastern district of Mackenzie (Northwest Territories, Canada). *Palaeogeography, Palaeoclimatology, Palaeoecology,* 3: 17-27.

TERASMAE, J., and HUGHES, O. L.

1966   Late-Wisconsinan chronology and history of vegetation in the Ogilvie Mountains, Yukon Territory, Canada. *The Palaeobotanist,* 15: 235-242.

TERASMAE, J., WEBBER, P. J., and ANDREWS, J. T.

1966   A study of late-Quaternary plant-bearing beds in north-central Baffin Island, Canada. *Arctic,* 19: 296-318.

TYRRELL, J. B.

1892   Report on northwestern Manitoba with portions of the adjacent districts of Assiniboia and Saskatchewan. *Geological Survey of Canada,* Annual Report 5, Part 1E, Ottawa.

WESTGATE, J. A., FRITZ, P., MATTHEWS, J. V., JR., KALAS, L., DELORME, L. D., GREEN, R., and AARIO, R.

1972   Geochronology and palaeoecology of Mid-Wisconsin sediments in west-central Alberta. *Abstracts, International Geological Congress, 24th session, Montreal.* 380 p.

WHITE, J. M., and MATHEWES, R. W.

1982   Holocene vegetation and climatic change in the Peace River district, Canada. *Canadian Journal of Earth Sciences,* 19: 555-570.

WHITE, J. M.

1983   Late Quaternary Geochronology and Palaeoecology of the Upper Peace River District, Canada. Unpublished Ph.D. Thesis, Simon Fraser University, Burnaby, B.C., Canada. 146 p.

QUATERNARY PALYNOLOGY AND VEGETATIONAL HISTORY OF ALASKA

THOMAS A. AGER
U.S. Geological Survey
National Center, M.S. 970
Reston, Virginia 22092

LINDA BRUBAKER
College of Forest Resources
University of Washington
Seattle, Washington 98195

Abstract

During several intervals of the Quaternary, eustatic lowering of sea level exposed large tracts of the shallow Bering and Chukchi Sea floors. This vast plain formed a broad land connection (Bering land bridge) between northeastern Asia and North America, permitting extensive biotic interchange between the continents. Half the area of Alaska remained unglaciated throughout the Quaternary, and therefore provided important refugia for plants and animals. Much of the palynological research in Alaska has been directed towards reconstructing the history of the Beringian environment, and the postglacial development of modern vegetation. Pollen data suggest that boreal forest and tundra vegetation had developed in Alaska by late-Pliocene time. A few sites from which pollen data of middle-Pleistocene age are available suggest that tundra vegetation existed in areas now covered by boreal forest. Late-Pleistocene records suggest that tundra and boreal forest environments co-existed in Alaska during the Sangamon Interglacial, but the severe climate of the Early Wisconsin glacial interval probably reduced or eliminated forests and replaced them with herbaceous tundra. The long Middle Wisconsin interstadial was characterized by oscillating climates and widespread tundra vegetation, with boreal forest or forest-tundra restricted to interior Alaska during the warmer intervals of the interstadial. The Late Wisconsin glacial interval was cold and arid. Trees and shrubs were rare and herbaceous tundra covered most of unglaciated Alaska. By 14,000 yrs. B.P. shrub tundra began to replace herbaceous tundra as climatic warming began. By 11,000 yrs. B.P. *Populus* spread widely in Alaska, followed by *Alnus* in early-Holocene time. Boreal *Picea* appeared in interior Alaska 9,500 yrs. B.P. and spread to Cook Inlet by 8,000 yrs. B.P. and to western Alaska by 5,500 yrs. B.P.

INTRODUCTION

Present and past Cenozoic ecosystems of North America have been greatly influenced by the roles Alaska has played as a refugium and as a route for intercontinental exchange of plants and animals. Three features of Alaskan geography and geology have contributed to these roles. First, Alaska is the extreme northwest extension of North America, presently separated from Siberia by the Bering and

Chukchi Seas, generally less than 100 m deep. At the Bering Strait the continents are separated by a mere 80 km. This narrow seaway has posed little obstacle to the exchange of many birds and plants, and even humans, for many thousands of years.

Second, the shallow floors of the Bering and Chukchi Seas have been repeatedly exposed as land during late-Cenozoic glacial intervals, because of eustatic lowering of sea level (Hopkins, 1967, 1973, 1979). Land connections also developed several times during the late-Cenozoic as a result of tectonic events. This direct land connection (Bering land bridge) between continents permitted extensive exchange of biotas during many extended periods of the Cenozoic (Hultén, 1937; Wolfe and Leopold, 1967; Hopkins, 1967, 1979; Hopkins *et al.*, 1982; Matthews, 1979).

Third, much of interior and northernmost Alaska remained free of glacial ice during repeated late-Cenozoic intervals of cold climate (Fig. 1).

Mountainous regions of Alaska were extensively glaciated during these times, but acted as barriers that severely limited the amount of precipitation that could penetrate the interior from the northern Pacific (Péwé, 1975; Hamilton and Thorson, 1983; Porter *et al.*, 1983). Consequently, unglaciated parts of Alaska and adjacent Yukon Territory served an important role as refugia for many plants and animals that later spread into terrain exposed as glaciers waned.

Thus Alaska's Cenozoic geological history and the past roles it has played as a biological refugium and exchange route make it an important and interesting target area for palynological and other paleoecological research (Hopkins, 1967; Hopkins *et al.*, 1982).

Palynological researchers working in Alaska have been interested in several aspects of Quaternary

---

Figure 1. Map of Alaska and adjacent areas, showing maximum extent of Pleistocene glaciers (dot pattern), and the approximate extent of emergent sea floor when sea level was eustatically lowered ca. 100 m (modified from Coulter *et al.*, 1965; Hopkins, 1967, 1973).

vegetational history, including: 1) boreal forest history (Hansen, 1953; Matthews, 1974b, Ager, 1975, 1983); 2) coastal forest history (Heusser, 1960, 1983b); 3) vegetational reconstruction of Bering land bridge and adjacent unglaciated regions during glacial intervals (Colinvaux, 1964b, 1967c, 1981; Schweger, 1982; Matthews, 1970, 1974a, 1974b; Ager, 1975, 1982c, 1983; Anderson, 1982; Brubaker *et al.*, 1983); 4) vegetational reconstruction as a contribution to the interpretation of archeological sites (*e.g.*, Schweger, 1976, 1981; Ager, 1975, 1983; Ager and Sims, 1981); 5) tracing the long-term history of tundra ecosystems (Matthews, 1974a); and 6) reconstructing post-glacial vegetational development in glaciated mountain regions (Livingstone, 1955; Ager, 1983; Ager and Rubin, 1984; Brubaker *et al.*, 1983).

The major objective of this review is to provide a current summary of Quaternary vegetational history of Alaska based primarily on palynological evidence. Some discussion of pollen evidence of possible early and middle-Pleistocene age is included, but the major temporal emphasis of the review is on the late-Pleistocene and Holocene; this reflects the age range of most deposits that have been studied rather than a deliberate choice on the part of the authors.

In order to organize this review we have divided the state into regions with boundaries that are in some cases rather arbitrary. We attempt at least to mention briefly most significant research that is

currently available in the form of theses, dissertations and publications. Since space is limited, however, we have selected key sites from each region for more extended discussion. Key sites were selected with several criteria in mind. One criterion was to present pollen data from a variety of deposits to show their relative merits. Another was to select records that are particularly informative and whenever possible have good age control. Several obvious "key sites" were excluded from discussion because they were presented in recently published reviews.

For a historical perspective on the development of palynological research in Alaska the reader is referred to a series of past review papers on the subject (Heusser, 1957, 1965, 1983a; Wolfe and Leopold, 1967; Colinvaux, 1967c; Ager, 1982c; 1983).

This paper excludes discussion of the palynological records from the southern coastal regions of Alaska, extending from the Aleutian Islands to southeastern Alaska. The vegetational history of those regions is discussed elsewhere (Heusser, 1960, 1965, this volume; Ager, 1983).

Palynological research on Quaternary-age deposits was slow to develop in Alaska because of the logistics problems and high cost of conducting field studies in remote parts of the State. By the late 1960's the late-Quaternary vegetational histories of only a few parts of the state were known in even general terms. Few well-dated, detailed studies had yet been published, as can be seen from early reviews of Alaska palynology (Heusser, 1957, 1965; Colinvaux, 1967c).

The pace of palynological research in Alaska has accelerated considerably since the late 1960's, particularly during the past decade. Although the number of sites studied has greatly increased, Alaska palynology remains today in the "pioneering stage" in the sense that several large areas of the State remain unstudied or little known. However, recent improvements in methodology, growing sophistication of interpretation, and the emergence of detailed regional studies involving multiple, well-dated sites suggest that for some parts of Alaska, the pioneer stage is ending. The beginnings of a "regional synthesis" stage can be seen in parts of southern interior and south-central Alaska and the Brooks Range. The development of detailed regional histories permits evaluation of past vegetation variation within regions and chronologies of vegetation changes on a regional scale. From this is emerging a clearer understanding of past plant invasions (*i.e.*, rates, directions), and relationships between vegetation and climate change.

REGIONAL SETTING

Geography

Alaska has an area of 1,520,000 km^2, more than twice the area of the State of Texas. Mountains and uplands higher than 600 m cover 33% of the area of Alaska; glaciers now cover about 5% of the state, but during the Pleistocene, glaciers may have covered up to 50% of Alaska during the most extensive advances (Péwé, 1975; Fig. 1). Modern snowline is as low as 400 m along parts of the southern coast, but rises inland to between 1350 and 1800 m in central Alaska (Péwé, 1975, p. 29).

Permafrost underlies most of Alaska, but is generally discontinuous south of the Brooks Range and central Seward Peninsula; permafrost is absent along the southern coast (Péwé, 1975, p. 46).

There are an estimated three million lakes larger than eight hectares in Alaska; many of these owe their origin to thermokarst processes in areas of permafrost, but glacial kettle ponds, outwash- and moraine-dammed lakes, cirque lakes, and oxbow lakes are common in many areas of the state. Volcanic crater lakes are found in several regions (*e.g.*, Seward Peninsula, Yukon Delta, St. Michael Island, Pribilof Islands) and are important sites for preserving long sediment and pollen records.

Climate

Alaska is divided into four climate zones (Fig. 2) that reflect the influences of the State's broad latitudinal range (51°16' N to 71°23'29" N latitude), extensive mountainous terrain, and long coastline.

The Maritime Climate Zone covers the southern coastal regions of the state from the western Aleutian Islands to extreme southeastern Alaska. The Maritime Zone climate is typified by cool, wet summers and mild, wet winters. At sea level mean annual temperatures are above freezing and permafrost is absent.

The Continental Climate Zone covers much of interior Alaska between the Brooks Range in northern Alaska and the Alaska Range in the southern interior part of the state. This zone has mean annual temperatures below freezing (-4° to -10° C) with severely cold winters but warm summers. Permafrost is generally continuous in the northern part of the

Figure 2. Map of present-day climate zones of Alaska (adapted from Péwé, 1975, p. 5).

zone, but is discontinuous in the southern part. Precipitation is low (*ca.* 25-35 cm in lowlands) due in large part to the barrier effect of high mountains parallel to the southern coast. Little moisture penetrates this barrier to reach the interior.

The Transitional Climate Zone is intermediate in character between the Maritime and Continental Zones. The Transitional Zone extends from southern Seward Peninsula, across southwestern Alaska to Cook Inlet and Copper River Basin. Mean annual temperatures within this zone range from -4° to 2° C. Permafrost conditions vary from continuous in the northwest part of the zone to sporadic or absent in southern portions of the zone. Precipitation is lower than in areas along the southern coast, but higher than in interior Alaska.

The Arctic Climate Zone stretches across northern Alaska to include the arctic coastal plain, the arctic foothills, the Brooks Range, and northern Seward Peninsula. Mean annual temperatures are low in this zone, in the general range of -7° to -12° C, and permafrost in continuous. Cool summers and very cold winters typify this zone. Precipitation is low (12-20 cm on the arctic coastal plain) (Viereck and Little, 1975; Hartman and Johnson, 1978).

Alaskan Flora

The present-day vascular flora of Alaska includes about 1440 native species and subspecies. In the slightly more than two centuries since the beginnings of exploration and settlement of Alaska more than 200 non-native species of vascular plants have been introduced to the State's flora (Hultén, 1968). The Alaskan flora is described in Hultén (1968), Welsh

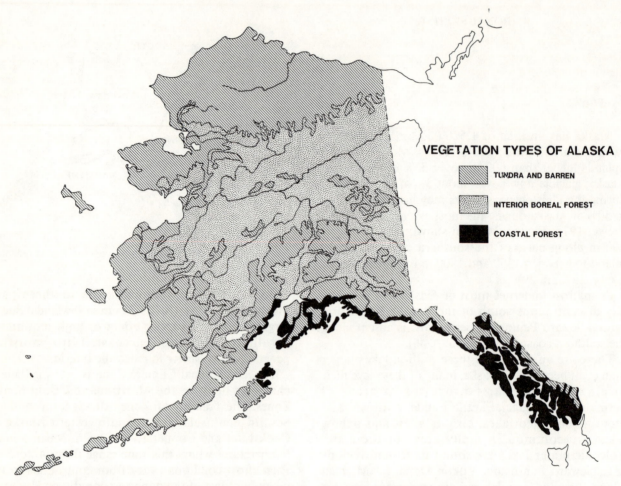

Figure 3. Map of present-day major vegetation types of Alaska (modified from Viereck and Little, 1972).

(1974) and Viereck and Little (1972). An atlas by Viereck and Little (1975) provides detailed distributional maps for Alaskan trees and shrubs.

Alaskan Vegetation

Only a fraction of 1% of the present vegetation of Alaska has been significantly disturbed by human activity. In many parts of North America, disturbance of vegetation has been drastic, making the task of finding realistic modern analogs of past vegetation a challenging task. Researchers in Alaska can generally begin with the assumption that modern vegetational communities in most areas can serve as valid analogs for at least Holocene age fossil assemblages.

Three broadly-defined vegetation types are recognized in Alaska (Fig. 3): 1) Sitka spruce-hemlock forests of the southern and southeastern coasts; 2) boreal spruce-birch forests of the interior; and 3) tundra, including lowland tundra of western and arctic Alaska and alpine tundra of mountainous

regions. Major Alaska vegetation types have been discussed by Viereck and Little (1972), Viereck and Dyrness (1980), Larsen (1980), and Bailey (1980).

Sitka spruce-hemlock forests. These forests are confined mostly to the southern coastal regions discussed by Heusser (1960; this volume). Some elements of this vegetation type extend into Cook Inlet and will therefore be discussed briefly here. Dominant trees in the south-central coastal region are Sitka spruce (*Picea sitchensis*) and mountain hemlock (*Tsuga mertensiana).* Western hemlock (*T. heterophylla*) is slowly invading this coastal region from the southeast, but is rarely encountered in the Cook Inlet region. Black cottonwood (*Populus trichocarpa*) and alder (*Alnus sinuata*) are important elements of the coastal vegetation.

Boreal spruce-birch forests. These forests cover much of interior Alaska and are composed primarily of white spruce (*Picea glauca*), black spruce (*P. mariana),* paper birch (*Betula papyrifera*), balsam poplar (*Populus balsamifera),*quaking aspen (*P. tremuloides),* and in some areas of the Tanana,

Kuskokwim, and Yukon River drainages, tamarack or larch *(Larix laricina).* Alders *(Alnus crispa, A. tenuifolia)* and willows *(Salix* spp.) are important shrubs in the boreal vegetation. Fluvial terraces of the interior lowlands are often occupied by well-developed forests of *Picea glauca* and *Populus balsamifera.* Lowlands with poorer drainage and cool north-facing upland slopes usually have a vegetation cover of open *Picea mariana* muskeg, sometimes along with *Larix laricina.*

Upland vegetation of the interior is a forest mosaic of mixed and pure stands of *Picea glauca, P. mariana, Betula papyrifera, Populus tremuloides, Alnus* spp., and *Salix* spp. This "mosaic" pattern is a result of frequent fires as well as microclimatic influences of slope aspect. Upland forests rarely extend higher than 1,000 m altitude in the interior.

Tundra. Tundra is a broadly defined term that encompasses a wide range of arctic and alpine treeless vegetation. Several categories of tundra vegetation occur in Alaska.

1) Wet lowland tundra is typified by wetland Cyperaceae, particularly *Carex* spp., non-tussock forming *Eriophorum* spp., Gramineae, and a rich assemblage of forbs *(e.g., Petasites* spp., *Pedicularis* spp., *Polemonium acutiflorum, Senecio congestus.)* Low woody shrubs include *Empetrum nigrum, Betula nana, Salix fuscescens,* and *S. pulchra.* This type of vegetation is extensive in western Alaska in the Yukon-Kuskokwim coastal lowland, and in northern Alaska on the arctic coastal plain. In the northernmost portions of the arctic coastal plain several of these taxa are eliminated from the tundra flora by severe climate *(e.g., Betula nana* and *Empetrum nigrum).*

2) Moist tundra vegetation is composed of Cyperaceae such as *Carex* spp. and tussock-forming *Eriophorum (e.g., E. vaginatum),* as well as Gramineae, Ericaceae *(e.g., Vaccinium* spp., *Ledum decumbens), Empetrum nigrum, Salix* spp., *Betula nana, B. glandulosa, Spiraea beauverdiana,* and numerous forbs. In some areas, thickets of *Alnus* spp. and *Salix* spp. occupy ravines and sheltered low areas within the tundra.

Moist tundra vegetation covers extensive areas of the northern slope of the Brooks Range, the foothills of the Alaska Range, central Seward Peninsula, and much of the Bristol Bay area and Alaska Peninsula.

3) Dry or alpine tundra forms a discontinuous, often sparse ground cover of primarily herbaceous plants adapted to severe climate and a very short growing season. Typical taxa include Gramineae, Cyperaceae, Saxifragaceae, Caryophyllaceae,

Oxytropis spp., *Artemisia* spp., *Primula* spp., *Campanula* spp., *Papaver* spp., and many other forbs, mosses, and lichens.

Low woody taxa include *Dryas* spp., *Empetrum nigrum, Salix reticulata, S. arctica, Vaccinium* spp., *Arctostaphylos* spp., and sometimes *Betula nana.* Dry tundra occurs generally in higher elevations in mountains and uplands, and is often associated with rocky, unstable soils.

Modern Pollen Flora and Pollen-rain Studies

Identification of pollen and spore types encountered in Alaskan Quaternary deposits is accomplished primarily by direct comparisons of specimens with reference slides of modern pollen and spore types made from herbarium specimens. Reference slides can be supplemented with pollen spore identification keys and photomicrographs published for other regions that share in common some of the taxa that are found in Alaska *(e.g.,* McAndrews *et al.,* 1973), and with a volume of color photomicrographs of 519 Alaskan pollen and spore types (Moriya, 1978).

Studies of modern pollen rain-vegetational relationships in Alaska are few, and until recently, have been very limited in terms of numbers of samples, geographic coverage of sampling points, and sophistication of interpretations, as discussed by Ager (1983, p. 131). Recent reports by Anderson (1982), Anderson *et al.* (1984), and Anderson and Brubaker (in press) provide results of the first regional-scale investigations that quantitatively relate surface pollen spectra from many lake sediment samples to large-scale vegetation and climate patterns in northern Alaska. Another recent study is based upon pollen and spore spectra from moss polsters collected along the Dalton Highway in north-central Alaska. These data provide an improved basis for understanding relationships between vegetation types and pollen rain in peat lands and bogs. The study should lead researchers to more sophisticated paleoenvironmental interpretations of palynological records from peat deposits (S. Short, unpublished data).

Aeropalynological studies have been undertaken recently by J. H. Anderson of University of Alaska. Initial results provide data on types, concentrations, and timing of release of airborne pollen and spores for the Fairbanks area (Anderson, 1984). Although the objectives of the investigation are related to understanding seasonality of airborne allergens, the data have useful applications to paleoenvironmental research in the region.

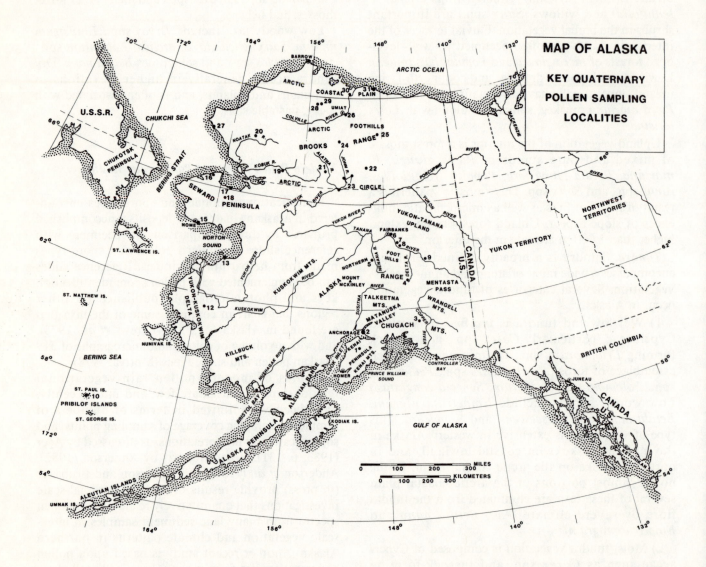

Figure 4. Map of Alaska showing major geographic place names and distribution of key sites where palynological data discussed in the text were collected. Numbered sites refer to the following localities and sources:

1) Hidden Lake, Kenai Peninsula (Ager, 1983; Ager and Sims, 1984); 2) Point Woronzof coastal bluffs, Anchorage, upper Cook Inlet (Ager, unpublished data); 3) "70 Mile Lake," northern Chugach Mountains (Ager, unpublished data); 4) Tangle Lakes, Gulkana Upland (Schweger, 1981; Ager and Sims, 1981); 5) Eightmile Lake, northern foothills of the Alaska Range (Ager, 1983); 6) Isabella Basin, Yukon-Tanana Upland (Matthews, 1974b); 7) Harding Lake, Tanana Valley (Ager, 1983); 8) Birch Lake, Tanana Valley (Ager, 1975, 1983); 9) Lake George, Tanana Valley (Ager, 1975); 10) Cagalog Lake, St. Paul Island, Pribilof Islands (Colinvaux, 1967c, 1981); 11) Kvichak Peninsula coastal bluffs Bristol Bay (Ager, 1982c); 12) Tungak Lake, Ingakslugwat Hills, Yukon Delta (Ager, 1982c); 13) Puyuk and Zagoskin Lakes, St. Michael Island (Ager, 1972c, 1983); 14) Flora Lake, St. Lawrence Island (Colinvaux, 1967a, 1967c); 15) Nome coastal plain deposits, Seward Peninsula (Hopkins et al., 1960); 16) Whitefish Lake, Seward Peninsula (Shackleton, 1982); 17) Cape Deceit coastal bluffs, Seward Peninsula (Matthews, 1974a; Giterman et al., 1982); 18) Imuruk Lake, Seward Peninsula (Colinvaux, 1964b; Colbaugh, 1968; Shackleton, 1982); 19) Epiguruk bluffs, Kobuk River (Schweger, 1976, 1982); 20) Kaiyak Lake, Noatak River (Anderson, 1982); 21) Ranger Lake, Alatna River area (Brubaker el al., 1983); 22) Rebel Lake, Upper Koyukuk River area (Edwards et al., in press); 23) Koyukuk River bluffs (Schweger, 1982); 24) Chandler Lake, Upper Chandler River (Livingstone, 1955); 25) Toolik Lake, arctic foothills (Bergstrom, 1984); 26) Umiat, Colville River (Livingstone, 1957); 27) Ogotoruk Creek, Cape Thompson (Heusser, 1963a); 28) Titaluk River bluffs, arctic foothills (Nelson, 1982); 29) Ikpikpuk River bluffs, arctic foothills (Nelson, 1982); 30) Ocean Point bluffs, Colville River (Nelson, 1979); 31) Prudhoe Bay area, arctic coastal plain (Walker et al., 1981).

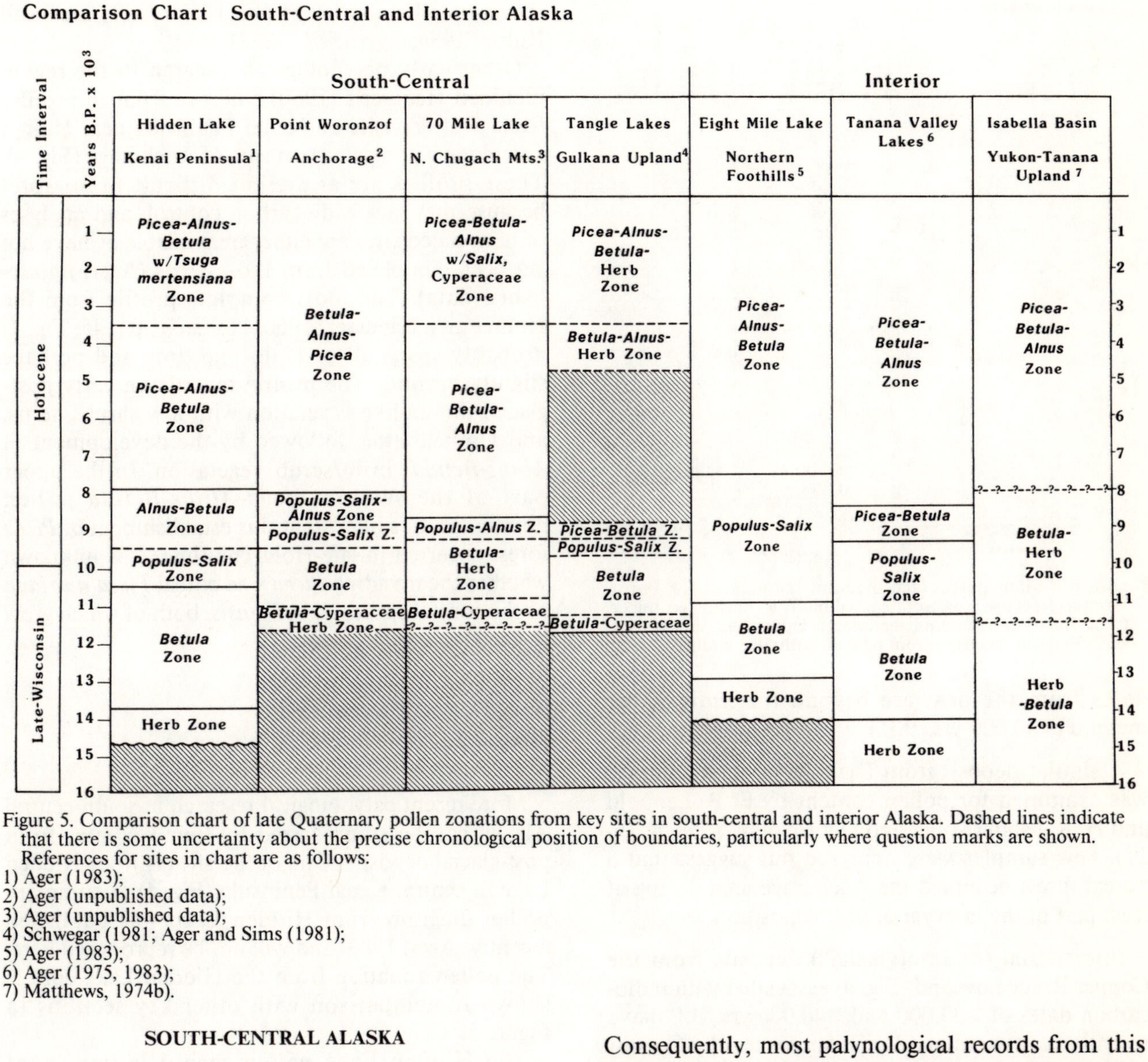

Comparison Chart   South-Central and Interior Alaska

Time Interval	Years B.P. x 10³	South-Central				Interior		
		Hidden Lake / Kenai Peninsula[1]	Point Woronzof / Anchorage[2]	70 Mile Lake / N. Chugach Mts.[3]	Tangle Lakes / Gulkana Upland[4]	Eight Mile Lake / Northern Foothills[5]	Tanana Valley Lakes[6]	Isabella Basin / Yukon-Tanana Upland[7]

Figure 5. Comparison chart of late Quaternary pollen zonations from key sites in south-central and interior Alaska. Dashed lines indicate that there is some uncertainty about the precise chronological position of boundaries, particularly where question marks are shown. References for sites in chart are as follows:
1) Ager (1983);
2) Ager (unpublished data);
3) Ager (unpublished data);
4) Schwegar (1981; Ager and Sims (1981);
5) Ager (1983);
6) Ager (1975, 1983);
7) Matthews, 1974b).

SOUTH-CENTRAL ALASKA

South-central Alaska extends from the crest of the Alaska Range south to the Gulf of Alaska (Fig. 4). This review excludes discussion of palynological sites along the southern coast except in the Cook Inlet area because the coastal region is summarized elsewhere (Heusser, 1960, this volume).

The mountain ranges of the region intercept most of the precipitation carried north from the Gulf of Alaska. The most extensive glaciers and ice fields in Alaska are found in this region today; during Pleistocene glacial intervals icefields covered nearly all of the region (Coulter *et al.,* 1965; Karlstrom, 1964; Hamilton and Thorson, 1983; Porter *et al.,* 1983).

Consequently, most palynological records from this region extend no farther back in time than early post-glacial, about 14,000 to 10,000 yrs B.P., depending on location. However, some tantalizing glimpses of palynological assemblages older than the post-glacial have been reported from the upper Cook Inlet region and Copper River lowland. A compressed peat unit overlain by glacial deposits of Wisconsin age (Naptowne Glaciation of Karlstrom, 1964) at Goose Bay, Knik Arm north of Anchorage was examined for pollen content (Benninghoff, 1957; Karlstrom, 1964, p.34-37; Reger and Updike, 1983, p.198; Ager, unpublished data). These investigations indicate that a boreal forest of *Picea, Alnus,* and *Betula* with Polypodiaceae existed in Upper Cook Inlet prior to the Naptowne Glaciation, when climate was probably quite similar to that of the present day. Radiocarbon

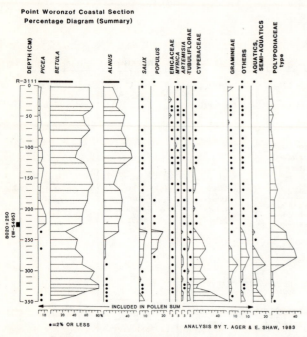

Figure 6. Pollen percentage diagram (summary) for Point Woronzof peat section from coastal bluffs at Anchorage, upper Cook Inlet (Ager, unpublished data). Uppermost sample is a surface pollen spectrum from near the sampling locality.

dates from the peat are beyond the limit of the method (>40,000 yrs. B.P.).

A similar deposit from Eagle River in Anchorage was examined for pollen content by E. B. Leopold and H. Ransom (Miller and Dobrovolny, 1959, p. 16-21). Few samples were analyzed but suggest that a boreal forest occupied the Anchorage area during at least part of the interval of peat deposition.

Interstadial (or interglacial?) deposits from the Copper River Lowland (Fig. 4) associated with radio-carbon dates of >33,000 and >40,000 yrs. B.P. have recently been examined for pollen content by Connor (1984). Four samples that yielded usable pollen assemblages indicate that a boreal forest of *Picea, Alnus,* and *Betula* with Ericaceae, Cyperaceae and herbs was established in the lowland prior to the last interval of glaciation.

Post-glacial deposits in the Copper River lowland and Matanuska Valley were first examined for pollen content by Hansen (1953). His published summary of this reconnaissance study suggests that most of his 20 undated shallow peat cores were late-Holocene records of *Picea*-dominated boreal forest vegetation. However, one site 24 km NE of Anchorage penetrated 6.9 m of peat; below a depth of 5.2 meters *Picea* and other "tree pollen" types were rare or absent. Present-day evidence of the regional pollen zonation indicates that his 6.9 m core probably

spanned all or most of the Holocene (Ager and Rubin, 1984; Ager *et al.,* 1984).

Other early palynological research in the region included Heusser's (1960) studies of pollen profiles from peat deposits on Kenai Peninsula near Homer and along Turnagain Arm east of Anchorage (Fig. 4). These profiles are at present difficult to interpret because they lack radiocarbon control, and analyses of new, dated sites are either unavailable or have not yet been completed from those areas (Ager, unpublished data). The most complete profile from the Homer area (Heusser, 1960, Homer 1, p. 145; Fig. 4) probably spans all of Holocene time and perhaps slightly beyond. The profile records an early post-glacial tundra-like vegetation with low shrubs, ferns, and Umbelliferae, followed by the development of *Alnus-Betula* shrub/scrub vegetation. In the upper part of the section *Picea-Alnus-Betula* pollen dominates. It is likely that this establishment of *Picea* forest occurred in late-Holocene time. It is unknown whether the invading *Picea* was boreal *Picea mariana* or the coastal spruce *P. sitchensis,* both of which grow in the Homer area today.

Hidden Lake

More recent palynological research in south-central Alaska can be summarized by four key sites from post-glacial deposits. One of the sites is from Hidden Lake in central Kenai Peninsula (Fig. 4). A summary pollen diagram from Hidden Lake was published recently (Ager, 1983) and will not be reproduced here. The pollen zonation from the Hidden Lake core is shown in comparison with other key sections in Figure 5.

The Hidden Lake pollen record is important because is represents the oldest continuous record available of vegetation development in south-central Alaska, following the onset of deglaciation *ca.* 14,000 yrs. B.P. The oldest pollen assemblages (Herb Zone) in the core are composed of primarily herbaceous taxa such as Gramineae, Cyperaceae, *Artemisia, Ambrosia,* Tubuliflorae, *Campanula* and Filicales. Pollen of shrub taxa is present, but in small amounts, except for *Salix* which is well-represented (up to 15% of the pollen sum). This early pioneer vegetation that developed on the newly exposed moraines and bedrock surfaces around Hidden Lake was an herba-ceous tundra with scattered low shrubs *(Salix, Betula,* Ericaceae).

A *Betula* Zone overlies the Herb Zone, and radio-carbon dates indicate that the zone was deposited

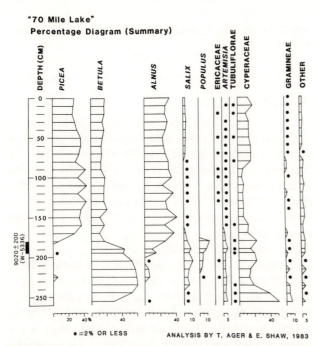

Figure 7. Pollen percentage diagram (summary) for "70 Mile Lake," northern Chugach Mountains (Ager, unpublished data).

between about 13,700 and 10,500 yrs. B.P. (Ager, 1983, p. 133). The pollen assemblages of this zone include *Betula, Salix,* Gramineae, Cyperaceae, small amounts of Ericaceae, and forbs. The *Betula* Zone is interpreted to have been derived from a dwarf birch shrub tundra.

A *Populus-Salix* Zone in the Hidden Lake core is dated at approximately 10,500 to 9,500 yrs. B.P.; the assemblage is similar to the *Betula* Zone type but with the important addition of *Populus* and abundant *Salix* pollen. The pollen assemblages from this zone were probably derived from a vegetation mosaic of shrub tundra and alpine tundra communities, thickets of *Salix* shrubs, and scattered stands of *Populus*. It is not yet known which of three species of *Populus* invaded at that time. All three now grow on Kenai Peninsula *(Populus balsamifera, P. tremuloides, P. trichocarpa)* (Viereck and Little, 1975).

An *Alnus* Zone occurs in the core, dated by extrapolation and regional correlations to represent the interval *ca.* 9,500-7,800 yrs. B.P. The pollen assemblages in the zone are dominated by *Alnus* pollen, and contain *Salix, Betula,* small amounts of Ericaceae, and herbaceous taxa. The vegetation represented by this assemblage is a mixture of shrub thickets and diminishing areas of shrub tundra.

The overlying zone is the *Picea-Alnus-Betula* Zone that spans the past *ca.* 7,800 yrs. B.P. *Salix,*

Ericaceae, and herbaceous taxa are minor components of the pollen assemblages. The pollen assemblage represents a boreal spruce forest that rapidly invaded central Kenai Peninsula from the north about 7,800 yrs. B.P. (Ager, unpublished data).

Small amounts of *Tsuga mertensiana* pollen occur in the upper 1 m of the core, and are most consistently represented in the upper *ca.* 60 cm. This probably reflects the late-Holocene invasion of coastal Sitka spruce-hemlock forest vegetation in the western Prince William Sound-Kenai Mountains area (Heusser, 1983b).

Point Woronzof

A previously unpublished pollen diagram (Fig. 6) from a peat deposit exposed in coastal bluffs near Point Woronzof in the Anchorage area (Fig. 4) is the first dated post-glacial pollen diagram available from upper Cook Inlet (Ager and Shaw, unpublished data). The radiocarbon chronology from the studied section is as yet incomplete, but a previously published date from the base of the peat at the site is 11,600 ± 300 yrs. B.P. (W-540) (Miller and Dobrovolny, 1959, p. 68). A recently obtained radiocarbon date from interval 222-227 cm in the peat section sampled for pollen analysis is 8,020 ± 250 yrs. B.P. (W-5495).

A simplified zonation for the Point Woronzof section is shown in Fig. 5. The basal part of the section (280-350 cm) is considered a single *Betula* zone, although pollen of Cyperaceae, aquatic taxa, and herbs are relatively abundant at the base of the section. In the upper 25 cm of the zone there are simultaneous increases in spores of Polypodiaceae and pollen of Gramineae, herb taxa, *Salix,* and aquatic-semiaquatic types. This may reflect a shift to a moister climate regime about 9,600 yrs. B.P.

As at Hidden Lake, a *Populus-Salix* zone is present, but the tentative chronology for the Point Woronzof section suggests the base of the zone is about 9,100 yrs. B.P. Other pollen data and radiocarbon dates from Upper Cook Inlet suggest *Populus* increased about 10,500 yrs. B.P. (Ager, unpublished data). An abrupt rise in *Alnus* percentages is recorded within the *Populus-Salix* Zone.

The upper pollen zone in the Woronzof section is a *Picea-Betula-Alnus* Zone. *Picea* percentages are rather low throughout the zone *(ca.* 5-15%), probably reflecting the relative predominance of deciduous taxa in the Anchorage area *(e.g., Betula papyrifera, Populus* spp., *Salix* spp., *Alnus*). The radiocarbon date from the base of the *Picea-Betula-Alnus* Zone indicates that

Picea invaded the Anchorage area about 8,000 yrs. B.P.

"70 Mile" Lake

The Copper River lowland (Fig. 4) was a little-known region of Alaska in terms of its vegetational history until recently. Several post-glacial samples examined for pollen content by Connor (1984) and pollen data from several lakes and exposures in the region (Ager and Rubin, 1984; Ager, unpublished data) are contributing to an emerging outline of vegetation changes. Preliminary results of palynological studies in the lower Copper River Valley (Sirkin *et al.*, 1971) help link the new records from the interior Copper River Lowland with the published coastal records (Heusser, 1960, 1983b; Sirkin and Tuthill, 1969).

A pollen diagram from a shallow lake in the northern Chugach Mountains provides an important glimpse of vegetation history slightly south of the Copper River Lowland (Ager, unpublished data). The diagram is from an unnamed lake at Mile 70, Richardson Highway, in an area of boreal spruce forest (Figs. 4 and 7). A sample from core interval 180-195 cm provides a radiocarbon date of 9,020 ± 200 yrs. B.P. (W-5336).

The age of the core base is estimated to be about 11,500-12,000 yrs. B.P. The pollen zonation from "70 Mile" Lake is remarkably similar to that seen from the Point Woronzof section, with a basal *Betula* Zone rich in Cyperaceae in the lower portion, and displaying an increase in the proportion of shrub taxa higher in the zone. The *Populus-Salix* Zone in "70 Mile" Lake has a pollen assemblage similar to that from the Woronzof section, but the estimated age of the base of the zone in the former is older, about 10,700 yrs. B.P. As at Point Woronzof, the top of the *Populus-Salix* zone shows an increase in *Alnus* percentages about 9,500 yrs. B.P.

The *Picea-Betula-Alnus* zone begins about 8,700 yrs. B.P. at "70 Mile" Lake (versus *ca.* 8,000 yrs. B.P. at Anchorage). The upper 70 cm of the zone displays a slight decrease in *Picea* percentages, and increases in *Salix* and Cyperaceae percentages. This portion of the core represents the past *ca.* 3,300 years of deposition. This subtle change in pollen assemblages may reflect possible depression of treeline in the Chugach Mountains during Neoglaciation. The coring site is located at an altitude of 560 m; altitudinal tree limit in the northern Chugach Mountains is about 700 m (Péwé, 1975, p. 23), therefore the site may be more

sensitive to tree line changes than sites from lower altitudes.

Tangle Lakes

Pollen data from two sites in the Tangle Lakes area in the Gulkana Upland (Fig. 4) form the basis of a fragmentary post-glacial vegetation record near the south flank of the Alaska Range (Schweger, 1981; Ager and Sims, 1981). Present-day vegetation in the Tangle Lakes is mesic shrub tundra and an open taiga of *Picea glauca, P. mariana, Betula papyrifera, Populus balsamifera,* and *Alnus.*

A summary of the combined pollen zonations from the Tangle Lakes sites is presented in Figure 5. The early post-glacial part of the record comes from pollen and plant macrofossil evidence collected from an exposure of lacustrine sediments along Rock Creek (Schweger, 1981). The base of the deposit yielded wood fragments that date to 11,800 ± 750 yrs. B.P. The top of the section contained a *Populus* log that dated to 9,100 ± 80 yrs. B.P., and a cone of *Picea glauca.*

The lower part (0.4-1.6 m depth) of the section from Rock Creek contains *Betula* Zone pollen assemblages with herbaceous taxa such as Cyperaceae, Gramineae, *Artemisia, Epilobium,* Ranunculaceae, Caryophyllaceae, and others.

The interval of 0.2-0.4 m records a *Populus-Salix* Zone with *Betula,* herbaceous taxa, and small amounts of *Alnus* pollen. The upper 0.2 m of the core is a *Picea-Betula* Zone with herbaceous taxa and *Sphagnum* spores.

No pollen data have yet been reported from the Tangle Lakes area for the interval between *ca.* 9,000 to 4,700 yrs. B.P. A lacustrine sediment core from Long Tangle Lake provides a pollen record from the area for the past *ca.* 4,700 years (Ager and Sims, 1981). The 5.5 m core is divided into two pollen zones, A and B. Zone B (5.5-3.4 m) is a *Betula-Alnus* zone with minor amounts (0-20%) of *Picea* pollen, and a diverse herb assemblage. The Zone B pollen data suggest that the *Picea* that invaded the area *ca.* 9,100 yrs. B.P. (Schweger, 1981) had become sparse at least locally by mid-Holocene time. Vegetation was probably predominantly shrub tundra with *Alnus* and *Salix* thickets.

Zone A (0-3.4 m) pollen assemblages are quite similar to those of Zone B, except that *Picea* percentages are significantly higher (20-40%). This suggests that a reinvasion of *Picea* occurred in the Gulkana Upland about 3,500 yrs. B.P., an event that seems puzzling in view of its apparent coincidence with Neoglacial cooling (Ager and Sims, 1981).

Pollen data from these and other sites in south-central Alaska provide a basis for the first attempt to synthesize regional vegetation history since deglaciation (Ager, 1983; Ager and Rubin, 1984; Ager *et al.,* 1984). Deglaciation occurred between 14,000 and 12,000 yrs. B.P. in most of the lowlands of southern Alaska. Ice retreat was soon followed by development of herb-shrub tundra and shrub tundra communities. The first tree to invade the region was *Populus,* between *ca.*10,500 and 8,000 yrs. B.P. *Alnus* appears to have invaded initially from the southern coast *ca.* 9,500 yrs. B.P., and to have spread rapidly northward and westward. Boreal species of *Picea (P. glauca,* probably followed by *P. mariana*) appear to have dispersed rapidly through South-Central Alaska from the Tanana Valley through the lowest mountain passes in the Alaska Range. *Picea* probably spread via the Delta River Valley into the Gulkana Upland by 9,100 yrs. B.P., and probably via Mentasta Pass (Fig. 4) into the Copper River lowland. *Picea* reached the northern Chugach Mountains by about 8,700 yrs. B.P., then spread west to upper Cook Inlet via Matanuska Valley by *ca.* 8,000 yrs. B.P., and south to central Kenai Peninsula by *ca.* 7,800 yrs. B.P.

Coastal Sitka spruce-mountain hemlock *(Picea sitchensis-Tsuga mertensiana)* forests in the Kenai Mountains developed in late-Holocene time as a result of gradual dispersal along the southern Alaskan coast from southeastern Alaska during post-glacial time (Ager, 1983; Heusser, 1960, 1983b, this volume).

INTERIOR ALASKA

For this review "interior" Alaska extends from the crest of the Alaska Range north to the Arctic Circle (Fig. 4). Most of this vast region was unglaciated throughout the Pleistocene (Fig. 1), except in the immediate vicinity of the Alaska Range and a few small upland sites farther inland (*e.g.,* parts of the Yukon-Tanana Upland) (Péwé, 1975). Interior Alaska formed part of an uninterrupted corridor of land extending from eastern Siberia across the Bering land bridge to central Alaska and into the large area of Yukon Territory that escaped glaciation (Hopkins, 1967; Péwé, 1975; Ritchie, 1984). Because of its history as a glacial refugium, and because a variety of deposits have been found to contain a rich record of past plant and animal life, interior Alaska has attracted a number of investigators who have

conducted palynological and/or other paleoecological studies in the region.

An early reconnaissance study of pollen preserved in shallow peat cores in the Tanana Valley (Hansen, 1953) marked the beginning of palynological work in the interior. The objectives of that study were limited to evaluating only major arboreal pollen types present in undated, but apparently late-Holocene, shallow peat cores from road-accessible muskeg and bogs.

A small number of pollen samples were collected by T. L. Péwé during the 1950's from perennially-frozen colluvial silt and peat deposits in the Yukon-Tanana Upland near Fairbanks. They were analyzed by E. S. Barghoorn and the results were summarized by Péwé (1975). Data from these samples provided some of the first botanical evidence that during Pleistocene intervals of harsher climate, dramatic lowerings of treeline occurred, effectively eliminating trees from many areas of interior Alaska.

A more detailed pioneering study of organic-rich frozen colluvium in the Yukon-Tanana Upland was reported by Matthews (1970). The oldest unit sampled was from Lost Chicken Mine in eastern Alaska near the Canadian border. The unit was initially interpreted to be of uncertain but pre-Wisconsin age (Matthews, 1970). Recent evidence (Matthews, 1980, p. 1091) suggests that the sample from Lost Chicken Mine may be as old as late-Pliocene, based on the age determination of an associated tephra. The pollen assemblage from Lost Chicken contains a "boreal" forest pollen assemblage of *Picea, Betula, Larix, Alnus,* Cyperaceae, and Gramineae, but also contains 15% *Pinus* pollen. *Pinus* does not grow in interior Alaska today; the nearest pines are *Pinus contorta* that grow in southeastern Alaska, British Columbia, and southeastern Yukon Territory (Viereck and Little, 1975; Hultén, 1968).

Other units of pre-Illinoian to Wisconsin age were sampled from several locations in the uplands near Fairbanks (Matthews, 1970). Most samples contain only small amounts of *Picea* pollen. The most common constituents include *Betula, Salix,* Cyperaceae, Gramineae, and herb taxa. *Alnus* percentages vary from zero to as much as 50% of the pollen sum. Matthews (1970) interprets these pollen spectra to indicate that treeline was considerably lower than at present in interior Alaska during deposition of several colluvial silt units in Pleistocene time. Some of the pollen samples from Eva Creek near Fairbanks are of Late Wisconsin age and are quite similar to Herb Zone pollen spectra seen in Wisconsin glacial

Tanana Valley Lakes

Pollen Percentage Diagram (Composite Summary)

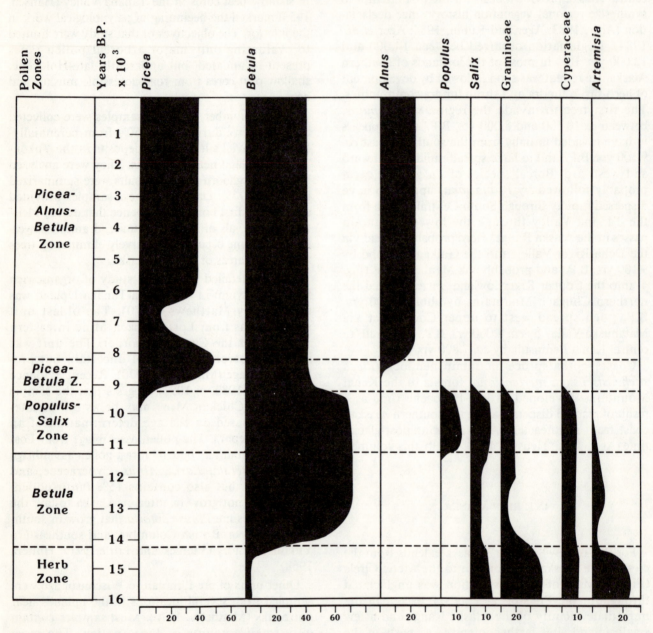

Figure 8. Pollen percentage diagram (composite summary) compiled from several pollen profiles from lacustrine sediment cores from the Tanana Valley (Ager, 1975, 1983).

deposits elsewhere in the state. Matthews interprets the herb zone assemblages to represent herbaceous tundra that developed under harsh glacial climate regimes.

The most detailed palynological studies in interior Alaska have been concerned with late-Quaternary deposits in the Tanana Valley and adjacent parts of

the Yukon-Tanana Uplands near Fairbanks, and in the northern foothills of the Alaska Range (Fig. 4). A variety of deposits have been examined from the area, including peat (Hansen, 1953; Ager, 1975), lacustrine sediments (Ager, 1975, 1982a, 1982b, 1983; Anderson, 1975), and frozen colluvial silts (Matthews, 1974b).

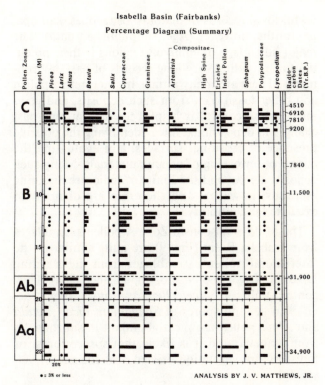

Figure 9. Pollen percentage diagram (summary) from a perennially-frozen valley fill in Isabella Basin, Yukon-Tanana Upland near Fairbanks. Redrafted from "Wisconsin environment of interior Alaska: Pollen and macrofossil analysis of a 27-meter core from the Isabella Basin (Fairbanks, Alaska), " by J. V. Matthews, Jr., *Canadian Journal of Earth Sciences*, 1974, 11: 828-841. Copyright © 1974, *Canadian Journal of Earth Sciences*. Reproduced by permission of the journal and the author.

Tanana Valley Lakes

Pollen records from lakes in the Tanana Valley provide a detailed continuous history of vegetation for the past *ca.* 16,000 years (Ager, 1975, 1983). At Harding Lake (Fig. 4) the pollen record from one of the cores is interpreted to extend into the Middle Wisconsin interstadial, but the core appears to have a thin or incomplete record of the Late Wisconsin glacial interval (Nakao *et al.,* 1980; Ager and Nakao, unpublished data). Summary pollen diagrams representing the Late Wisconsin and Holocene portion of the Harding Lake and Birch Lake records were recently published (Ager, 1983). Pollen zonations derived from several sites in the region are compared in Figure 5. A composite pollen diagram based on pollen assemblages from several lakes in the region provides an idealized example of regional palynological records derived from lacustrine sediments (Fig. 8).

The Middle Wisconsin interstadial record from the Harding Lake Sediment core is characterized by

pollen-spore assemblages composed primarily of *Picea, Betula, Alnus,* Ericaceae, Cyperaceae, and *Sphagnum* spores. A radiocarbon date at the top of the interstadial unit is 26,500±400 yrs. B.P. (W-4817). The interstadial pollen-spore assemblages are interpreted to have been derived from spruce muskeg-bog environments that rimmed Harding Lake at a time when the lake was probably only a shallow pond.

Full-glacial pollen spectra from Harding Lake and other lakes in the Tanana Valley are Herb Zone types composed of Gramineae, Cyperaceae, *Artemisia,* and tundra forbs such as Caryophyllaceae, Ranunculaceae, Cruciferae, Tubuliflorae and *Plantago.* Gramineae and *Artemisia* percentages are generally high in the Herb Zone samples from the Tanana Valley lakes, in comparison with most full-glacial samples from elsewhere in Beringia. Full-glacial vegetation of the Tanana Valley appears to have been composed of herbaceous tundra communities, with a shrub component of *Salix,* and very small amounts of *Betula* and Ericaceae.

Betula Zone assemblages replaced Herb Zone assemblages by about 14,000 yrs. B.P. in the Tanana Valley and slightly later in the Eightmile Lake area in the northern foothills (Figs. 4 and 5). *Betula* pollen overwhelmingly dominates the *Betula* Zone assemblages in both areas. Smaller amounts of pollen of Gramineae, Cyperaceae, *Salix,* and herbs are common constituents. The pollen data suggests that the *Betula* Zone vegetation was shrub tundra.

Populus-Salix Zone assemblages are well-preserved in only two cores from this region, but it is likely the vegetation type that produced these pollen was widespread in the region between *ca.* 11,5000-9,500 yrs. B.P. Vegetation was probably open *Populus* woodland and birch-shrub tundra.

Picea appears in the Tanana Valley about 9,500 yrs. B.P. The *Picea-Betula* Zone is probably the signature of open spruce-birch forest interspersed with shrub tundra and bog communities, and shrub thickets of *Salix.*

The *Picea-Betula-Alnus* Zone spans the past 8,400 yrs. B.P. in Tanana Valley, but at the higher elevation Eightmile Lake site near the Alaska Range, *Picea* and *Alnus* appear to have invaded simultaneously about 7,500 yrs. B.P. (Ager, 1982b, 1983). Lowland boreal forest vegetation similar to that of the present is suggested by this assemblage.

Cagaloq Lake, St. Paul Island
Percentage Diagram (Summary)

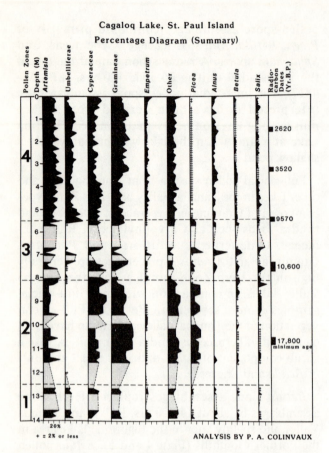

ANALYSIS BY P. A. COLINVAUX

Figure 10: Pollen percentage diagram (summary) for Cagaloq Lake, Pribilof Islands. Redrafted from "Historical ecology in Beringia: The south land bridge coast at St. Paul Island," by P. A. Colinvaux, *Quaternary Research,* 1981, 16: 18-36. Copyright © 1981, University of Washington. Reproduced by permission of the journal.

Isabella Basin

Comparison of pollen records from the Isabella Basin and Tanana Valley Lakes reveals some interesting differences (Figs. 8 and 9). Whereas pollen preservation is generally very good in the lake cores (usually less than 3% indeterminate pollen), preservation is often poor in the colluvial samples (20-30% indeterminate). The pollen record from the colluvial section does not contain clear evidence of several distinct pollen zones seen in the nearby lake records, perhaps as a result of depositional hiatuses and extensive reworking. However, the colluvial section has the advantage of preserving a rich variety of organic materials useful for paleoenvironmental studies (*e.g.,* wood, seeds, insects, fossil mammal bones) whereas such materials are relatively rare in cores from the region's lakes. The lake cores do sometimes preserve

useful fossil records of diatoms, ostracodes, and opal phytoliths, however (Ager, unpublished data). The major advantage of colluvial sections is their potential for providing very long records of past environments. There are many pitfalls in interpreting pollen and other fossil data from such deposits, and good radiocarbon control is helpful for determining past changes in depositional rates, identifying the possible presence of erosional surfaces, and evaluating the degree of reworking of organic materials. Many such deposits are beyond the limits of radiocarbon dating methods, however, and therefore are difficult to evaluate for continuity.

The Middle Wisconsin Zone A pollen assemblages from Isabella Basin (Fig. 9) are quite similar to the interstadial pollen assemblages from the Harding Lake core (Ager and Nakao, unpublished data). They contain pollen of *Picea, Betula, Alnus,* Cyperaceae, Gramineae, and spores of *Sphagnum. Picea, Betula* and *Alnus* pollen are most abundant in subzone Ab from Isabella, and it is likely that spruce forest or forest-tundra vegetation existed near the site 31,900 yrs. B.P.

Isabella Zone B spans the equivalent of the Herb and *Betula* Zones in the region's lakes. No *Populus-Salix* Zone or early-Holocene *Picea-Betula* Zone is in evidence. Lack of radiocarbon control between 10-18 m depth in the Isabella Basin record (Fig. 9) makes the presumed "full-glacial" portion of the record difficult to evaluate. At the U. S. Army CRREL Permafrost tunnel near Fairbanks, radiocarbon dates indicate that a major hiatus exists within Late Wisconsin age frozen colluvial silt deposits at that site. Sediments younger than about 30,000 yrs. B.P. and older than 14,000 yrs. B.P. are missing or have been extensively reworked into the latest Wisconsin deposits (Sellman, 1967). It is not clear if any major portion of the Late Wisconsin record is missing from Isabella Basin.

It was long assumed that lowland sites in interior Alaska served as a refugium for *Picea* and other boreal taxa during Pleistocene glaciations (*e.g.,* Hultén, 1937; Hopkins, 1967, p. 462). The near-absence of *Picea* pollen (Ager, 1975, 1983), and the lack of *Picea* wood or other spruce macrofossils in sediments of glacial age in Alaska (Hopkins *et al.,* 1981) called the assumption of a spruce refugium in interior Alaska into question. An alternative hypothesis suggested that spruce was eliminated in Alaska during the Late Wisconsin but re-invaded from refugia south of the Laurentide and Cordilleran ice sheets via an ice free corridor that developed in western Canada during deglaciation. According to that hypothesis, spruce would then have spread to Alaska from northwest Canada in earliest Holocene time (Hopkins *et al.,* 1981).

Preliminary reports of research in progress in the Canadian "ice-free" corridor (MacDonald, 1984a, 1984b) and in the Yukon Lowland north of the Yukon-Tanana Upland (Edwards and Brubaker, 1984; Edwards and McDowell, 1984) now make that hypothesis less likely. First appearances of *Picea* at these sites are somewhat later than would be expected if *Picea* advanced along the route through the corridor to northwest Canada, then northeast Alaska. It is again necessary to consider the real possibility that small populations of spruce persisted somewhere in unglaciated interior Alaska and/or Yukon Territory. Studies on the genetics, biochemistry, and phenotypic characteristics of modern *Picea glauca* populations of North America provide inconclusive evidence for persistence of spruce in Beringia during the last major glacial interval (Critchfield, 1984, p. 81-87).

SOUTHWESTERN ALASKA

Southwestern Alaska as defined here includes the region from Norton Sound south to the Bristol Bay area, and the Bering Sea islands that were once part of the Bering Land Bridge (*e.g.,* St. Lawrence Island, Pribilof Islands; Fig. 4). Pollen data from the Aleutians and Alaska Peninsula are discussed elsewhere (Ager, 1982c, 1983; Heusser, 1983c, this volume).

Palynological research conducted in the region prior to the mid-1960's was previously reviewed by Colinvaux (1967c) and will not be discussed here in detail. Key sites in southwestern Alaska are located in the Pribilof Islands (Colinvaux, 1981), the Kvichak Peninsula in the Bristol Bay area (Ager, 1982c), the Yukon Delta (Ager, 1982c), and St. Michael Island in southern Norton Sound (Ager, 1982c, 1983; Ager and Bradbury, 1982; Fig. 4).

Pribilof Islands

Cagaloq Lake is located within a volcanic crater on St. Paul Island in the Pribilof Islands (Fig. 4). A pollen record obtained from a 14 m sediment core from the lake has been described by Colinvaux (1967b, 1967c, 1981). The record is of particular importance because it is interpreted to penetrate to sediments of mid-Wisconsin age, and because it

provides evidence for the nature of the environment of the southern edge of the Bering land bridge during the Late Wisconsin glacial interval when sea level was eustatically lowered to between 90 to 120 m below modern level (Hopkins, 1982, p. 12-14).

A summary pollen diagram from Cagaloq Lake (Fig. 10) shows the pollen zonation defined by Colinvaux (1981). The basal part of the core (Zone 1) is undated, but is interpreted to be of Middle Wisconsin interstadial age, more than 25-30,000 yrs. B.P. The Zone 1 pollen assemblage is characterized by an herbaceous flora in which Umbelliferae and *Artemisia* are dominant components, along with Gramineae, Cyperaceae, and numerous forbs. The only shrubs that are interpreted to have grown locally are *Empetrum* and *Salix.* Small amounts of pollen of *Picea, Betula,*and *Alnus* are present but arrived from distant sources by wind transport. The Zone 1 pollen assemblage is interpreted to represent a maritime tundra vegetation quite similar to that of the present day on the Pribilof Islands. This implies that during at least part of the mid-Wisconsin interstadial, the Pribilofs were islands in the Bering Sea, apart from the Bering land bridge.

The Zone 2 pollen assemblages are interpreted to be of Late Wisconsin age, based in part on a radiocarbon date of 17,800 yrs. B.P., considered to be a minimum age due to modern groundwater contamination of other radiocarbon samples in the sandier parts of the core. The Zone 2 pollen assemblages are characterized by *Artemisia,* Gramineae, Cyperaceae, and a rather rich assemblage of forbs. *Picea* percentages are significant, but low pollen concentrations and pollen accumulation rate estimates suggest that the *Picea* pollen was derived from distant sources by wind. The pollen assemblages in Zone 2 are interpreted to represent an herbaceous tundra in an arid full-glacial climate when the Pribilofs were joined to the Bering land bridge.

Sand layers in the core are indications that the lake was intermittent, and that episodic eolian sand deposition occurred during glacial time. Other observations of glacial-age sand deposits on St. Paul Island suggest that a substantial snow cover accumulated in winter (Hopkins, 1982, p. 23).

Zone 3 pollen assemblages from Cagaloq Lake are characterized by significant amounts of *Picea, Betula,* and *Alnus* pollen along with *Artemisia,* Umbelliferae, Gramineae, numberous forbs, *Empetrum,* and *Lycopodium.* The assemblage is interpreted to represent a maritime herbaceous tundra that developed about 11,000 yrs. B.P. By that time the southern Bering land bridge had been undergoing inundation

Yukon Delta - Southern Norton Sound

Pollen Percentage Diagram (Composite Summary)

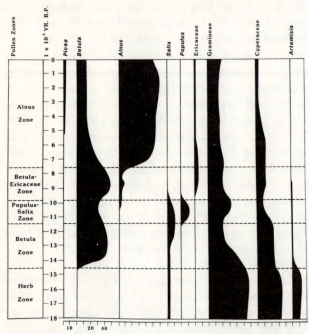

Figure 11. Pollen percentage diagram (composite summary) based upon pollen profiles from Tungak and Puyuk Lakes (Ager, 1982c) and Zagoskin Lake (Ager, 1983), Yukon Delta-southern Norton Sound area.

by rising sea level for several thousand years (Hopkins, 1982, p. 12); establishment of the modern climate regime of the Pribilofs apparently occurred about 11,000 yrs. B.P. The most curious feature of the Zone 3 pollen assemblage is the presence of significant percentages of *Picea, Alnus,* and *Betula.* Influx calculations for those taxa suggest pollen sources of those pollen types somewhere in the vicinity of what is now the Pribilof Islands during Zone 3 time (Colinvaux, 1981). Reconstruction of the approximate sea level position in Beringia during Zone 3 time indicates that the present sites of the Pribilofs were part of a single, larger island several hundred kilometers southwest of the retreating edge of the much-reduced land bridge (Hopkins, 1973). If *Picea* and other trees and shrubs were present locally the question remains where they survived during the glacial interval.

Zone 4 assemblages are characterized by pollen of *Artemisia,* Umbelliferae, Gramineae, Cyperaceae, numerous forbs, *Empetrum,* and *Salix* and spores of *Lycopodium.* These assemblages represent a maritime herb tundra that has undergone little or no change during the past 9,500 years. The occurrence of small amounts of pollen of *Picea, Alnus,* and *Betula* pollen is attributed to long distance wind transport

from distant mainland vegetation. Other palynological research in the Pribilof Islands by Parrish (1980) involves analysis of peat deposits of Holocene age. The data suggest minor vegetation changes occurred in middle and late-Holocene time that may be related to climate shifts (*e.g.,* Neoglaciation in the late-Holocene).

Yukon Delta-Norton Sound

The first pollen records from the Yukon Delta-southern Norton Sound area have been reported recently by Ager (1982c, 1983). Much of this area consists of deltaic wetlands where lakes are abundant but generally too young to yield long pollen records. Older lakes have been found in several volcanic fields of late Cenozoic age in the region, however. Lakes within volcanic craters in the Ingakslugwat Hills (Tungak Lake, Fig. 4) and on St. Michael Island (Puyuk Lake, Zagoskin Lake, Fig. 4) have yielded records that penetrate all or part of Late Wisconsin time. The present-day vegetation near all three lakes is moist tundra, with scattered thickets of *Salix* and *Alnus* in protected hollows and ravines.

Summary diagrams from these lakes have been published recently (Ager, 1982c, 1983) and therefore will not be reproduced here. A composite pollen diagram based upon features from the pollen diagrams for those three sites is presented in Figure 11.

Full-glacial pollen assemblages from the Yukon Delta-Norton Sound area (Fig. 11) are characterized by herbaceous taxa including Gramineae, Cyperaceae, smaller amounts of *Salix* and *Artemisia,* and assorted herb types that constitute about 10-20% of the pollen sum. Herb types include Caryophyllaceae, *Thalictrum* and other Ranunculaceae, Cruciferae, *Potentilla, Rumex/Oxyria, Polemonium,* and Compositae (Tubuliflorae, Liguliflorae). In some cores very small amounts of Ericaceae and *Betula* pollen and *Sphagnum* and Polypodiaceae spores are present, suggesting the persistence of some local mesic habitats in a landscape that was predominantly xeric. Herb Zone pollen assemblages are interpreted to represent herbaceous tundra communities adapted to dry cool full-glacial climates.

Betula Zone assemblages began to develop in this region between 14,000 and 15,000 yrs. B.P. These pollen assemblages are composed of *Betula* (*B. nana,* dwarf birch), Gramineae, Cyperaceae, *Salix,* small amounts of Ericaceae, and assorted tundra forbs. The assemblages represent shrub tundra communities.

A *Populus* Zone is now documented from three lakes in the region (Ager, 1982c, 1983, unpublished data). *Populus* and *Salix,* along with *Betula* and other *Betula* Zone elements, characterize this zone. The duration of this zone appears to have been brief, *ca.* 11,000-10,000 yrs. B.P. This zone is interpreted to represent a state-wide rapid dispersal event in which *Populus* appeared at many sites during that time interval, including some beyond the present limits of *Populus* distribution in Alaska. Vegetation cover in the region was probably a mosaic of shrub tundra communities within which groves of *Populus* (probably *P. balsamifera*) and thickets of *Salix* were scattered.

The overlying *Betula*-Ericaceae Zone is very similar to the *Betula* Zone except that the amounts of Ericaceae pollen and Polypodiaceae spores are somewhat greater in the former Zone in most sites examined thus far. The zone spans about 10,000 to 7,500 yrs. B.P. and the pollen assemblages appear to represent shrub tundra communities. *Populus* apparently disappeared from the coring sites during this interval; presumably this reflects a range retreat of *Populus* during early-Holocene time. *Populus* does not grow on or adjacent to St. Michael Island, and is rare in the Ingakslugwat Hills area.

The *Alnus* Zone spans the past *ca.* 7,500 yrs. B.P. in this region, and pollen assemblages from this zone are characterized by large percentages of *Alnus* pollen, along with *Betula* shrub tundra taxa similar to those seen in the *Betula* and *Betula*-Ericaceae zones. Small amounts of *Picea* pollen first appear in this zone in sediments of mid-Holocene age, *ca.* 5,500 yrs. B.P. The abrupt increase of *Alnus* pollen at the base of this zone reflects either a rapid invasion of *Alnus* from an adjacent region or the sudden expansion of a small, climatically-suppressed population of *Alnus crispa* that had survived the glacial interval within the region. *Picea* appears to have reached the eastern borders of this region, probably from interior Alaska, in mid-Holocene time (Ager, 1982c, 1983).

Bristol Bay

Coastal exposures on the Kvichak Peninsula in the Bristol Bay area (Fig. 4) provide a key pollen record dating to beyond the limits of the radiocarbon method. Samples were collected by D. M. Hopkins in 1972. Pollen samples were later analyzed for palynomorphs by R. E. Giterman. Some of these data were summarized by Ager (1982c, p. 88-90). A pollen diagram from one of the sections (72-17) was

presented in that review and will not be reproduced here. A summary of the pollen zonation from several sections is presented in Figure 12.

The large and irregular sampling interval in these sections results in a pollen record that provides only a broad outline of vegetation change rather than a detailed history. Pollen data indicate that the basal pollen samples associated with infinite radiocarbon dates (>33,000 yrs. B.P.) are probably attributable to the mid-Wisconsin interstadial. Pollen assemblages are composed of pollen of Cyperaceae, *Artemisia,* Gramineae, *Salix,* and small amounts of *Betula.* The assemblages suggest a somewhat mesic herb-shrub tundra vegetation; climate was probably cooler and drier than during the Holocene.

Full-glacial pollen samples, bracketed by radiocarbon dates of 12,760 ±300 yrs. B.P. and >33,000 yrs. B.P. are characterized by Gramineae, Cyperaceae, *Artemisia, Salix,* and other herb taxa. The assemblage is interpreted to represent a discontinuous vegetation cover of herbaceous tundra. Environmental conditions were probably cold and arid.

Samples of latest Wisconsin to early-Holocene age are poorly represented in these profiles. *Betula,* Gramineae, Cyperaceae and *Salix* are present in two samples from Section 72-42 that probably date to this interval (Hopkins and Giterman, unpublished data). These samples are probably equivalent to the *Betula* Zone seen in the Yukon Delta-Norton Sound area to the north (Figs. 11 and 12).

Holocene samples 7,600 yrs. B.P. and younger in age contain *Alnus, Betula,* Ericaceae, Gramineae, and Cyperaceae. *Picea* percentages range from zero to several percent. This assemblage represents shrub tundra with *Alnus,* and resembles the *Alnus* Zone observed elsewhere in western Alaskan Holocene-age deposits. (Ager, 1982c, 1983, Fig. 12).

Erratic *Picea* percentages suggest that spruce pollen probably was carried from distant sources by wind transport. Holocene-age pollen profiles from the nearby Naknek River area suggest that *Picea* invaded in late-Holocene time (Heusser, 1963b).

St. Lawrence Island

A lacustrine core was obtained from Flora Lake on St. Lawrence Island in the Bering Sea (Fig. 4) for pollen analysis by Colinvaux (1967a, 1967c). Basal samples from the core were interpreted to represent interglacial vegetation of presumed Sangamon age. These samples contain pollen of Gramineae, *Betula, Alnus, Picea,* and Ericaceae, and spores of

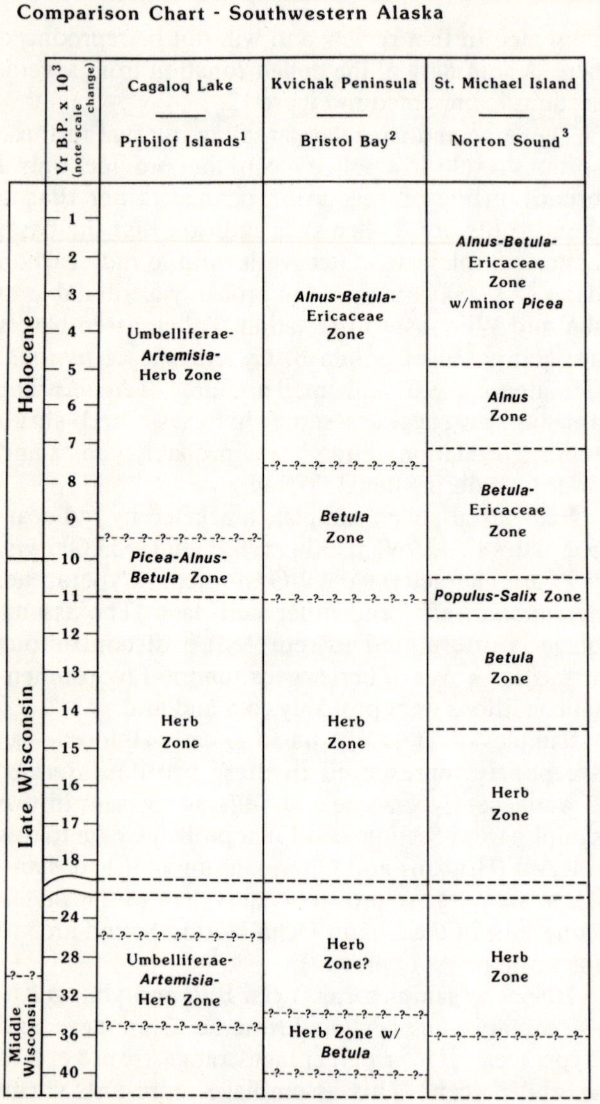

Comparison Chart - Southwestern Alaska

Yr B.P. x 10³ (note scale change)	Cagaloq Lake — Pribilof Islands[1]	Kvichak Peninsula — Bristol Bay[2]	St. Michael Island — Norton Sound[3]
Holocene 1			*Alnus-Betula-*Ericaceae Zone w/minor *Picea*
2			
3	Umbelliferae-*Artemisia-*Herb Zone	*Alnus-Betula-*Ericaceae Zone	
4			
5			
6			*Alnus* Zone
7			
8			*Betula-*Ericaceae Zone
9	*Picea-Alnus-**Betula* Zone	*Betula* Zone	
10			
11			*Populus-Salix* Zone
12			*Betula* Zone
13			
Late Wisconsin 14	Herb Zone	Herb Zone	
15			
16			Herb Zone
17			
18			
24			
28	Umbelliferae-*Artemisia-*Herb Zone	Herb Zone?	Herb Zone
Middle Wisconsin 32			
36		Herb Zone w/ *Betula*	
40			

Figure 12. Comparison chart of late Quaternary pollen zonations from key sites in southwestern Alaska. Dashed lines indicate that there is some uncertainty about the precise chronological position of some zone boundaries, particularly where question marks are shown. References for sites are as follows:
1) Colinvaux (1981);
2) Ager (1982c);
3) Ager (1983).
Note scale change within time scale.

Polypodiaceae. The late-Quaternary zonation from the core is inadequately dated, but the sequence parallels that seen in the Yukon Delta-Norton Sound area, with a full-glacial herb zone, overlain by a *Betula* Zone, and an upper *Alnus* Zone spanning the past 6,000 years. *Picea* pollen appears in mid-Holocene time. *Alnus* and *Picea* do not now grow on St. Lawrence Island, and the presence of those pollen types in the Flora Lake record is a result of long-distance wind transport from the mainland.

SEWARD PENINSULA

Nome

The first palynological investigation of Quaternary deposits on Seward Peninsula was based on samples collected from coastal plain sediments at Nome (Hopkins *et al.,* 1960; Fig. 4). These data have been discussed in previous reviews (Colinvaux, 1967c; Ager, 1982c, 1983) and therefore will be mentioned here only briefly. Three peat samples of Sangamon age were analyzed from the Nome deposits, and the pollen assemblages suggest that *Betula*-shrub tundra similar to that of the present day prevailed in the area during at least part of the interglacial, but that at the apparent peak of interglacial warm temperature, *Alnus* and probably *Picea* grew at or near Nome. A fragmentary late-Quaternary pollen record from Nome suggests that herbaceous tundra (Herb Zone) vegetation existed locally until at least 13,000 yrs. B.P., and that shrub tundra with *Betula* was established by about 10,000 yrs. B.P. The Holocene record is inadequately dated, but it suggests that shrub tundra vegetation with dwarf *Betula* and Ericaceae existed locally throughout the last 10,000 years, and *Alnus* pollen began to reach Nome by middle-Holocene time. *Alnus* pollen is considerably less abundant in the Nome deposits than in comparable Holocene-age samples on the south side of Norton Sound at St. Michael Island (Ager, 1982c; 1983; Fig. 11). Therefore, the base of the *Alnus* Zone at Nome is less clearly defined in the pollen diagram.

Whitefish Lake

Pollen from a sediment core from Whitefish Lake near the northwestern coast of Seward Peninsula (Fig.4) was analyzed by Shackleton (1982). Although there are problems with the radiocarbon dating of the core, it appears likely that the 5.3 m core was deposited during about the past 10,500 years. The pollen assemblages in the core are remarkably similar throughout, and are composed of *Alnus, Betula,* Cyperaceae, Gramineae, and smaller amounts of Ericaceae, *Salix,* and herbs. The lake occupies a maar; therefore, it is unlikely that the uniformity of the pollen profile is the result of sediment mixing such as occurs in thermokarst lakes. Pollen data from the Whitefish Lake core suggest that shrub tundra

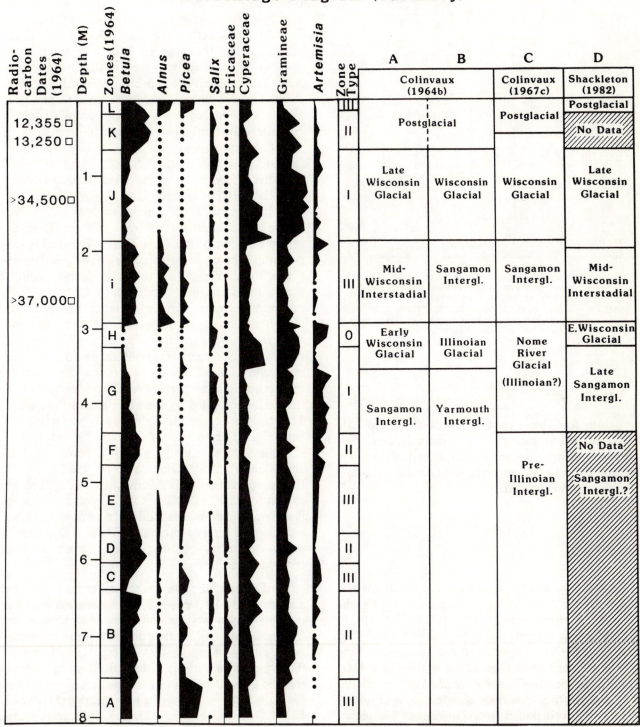

Figure 13. Pollen percentage diagram (summary) for Imuruk Lake, Seward Peninsula. Columns A-D show summaries of alternative interpretations of the zonation's chronology. Redrafted from "The Environment of the Bering Land Bridge." by P. A. Colinvaux, *Ecological Monographs*, 1964, 34: 297-329. Copyright © 1964 by *Ecological Monographs*. Reprinted by permission of the journal.

vegetation with *Alnus* has existed in the area through-out the time represented by the core. If the core does indeed span the past 10,500 years, it suggests either that northern Seward Peninsula may have been a

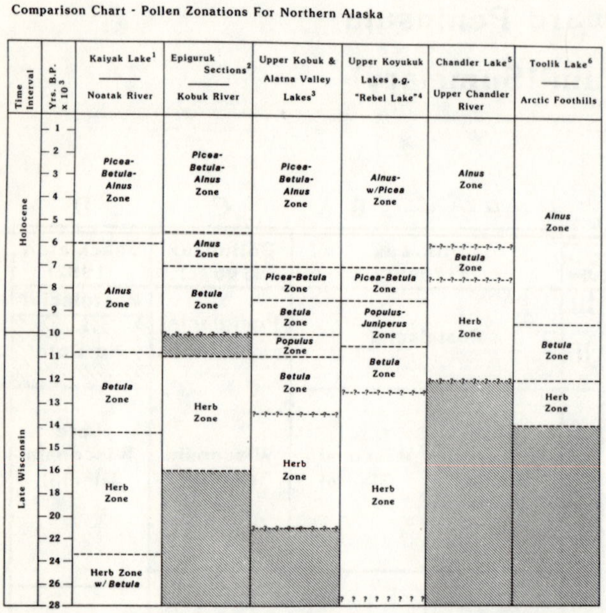

Comparison Chart - Pollen Zonations For Northern Alaska

Time Interval	Yrs. B.P. × 10³	Kaiyak Lake[1] / Noatak River	Epiguruk Sections[2] / Kobuk River	Upper Kobuk & Alatna Valley Lakes[3]	Upper Koyukuk Lakes e.g. "Rebel Lake"[4]	Chandler Lake[5] / Upper Chandler River	Toolik Lake[6] / Arctic Foothills
Holocene	1–5	Picea-Betula-Alnus Zone	Picea-Betula-Alnus Zone	Picea-Betula-Alnus Zone	Alnus-w/Picea Zone	Alnus Zone	Alnus Zone
	6		Alnus Zone			Betula Zone	
	7	Alnus Zone	Betula Zone	Picea-Betula Zone	Picea-Betula Zone		
	8–9			Betula Zone	Populus-Juniperus Zone	Herb Zone	
	10			Populus Zone			Betula Zone
	11	Betula Zone	Herb Zone	Betula Zone	Betula Zone		
Late Wisconsin	12–14						Herb Zone
	15–18	Herb Zone	Herb Zone	Herb Zone	Herb Zone		
	20–23						
	24–28	Herb Zone w/ Betula					

Figure 14. Comparison chart of late Quaternary pollen zonations from key sites in northern Alaska. Dashed lines indicate that there is some uncertainty about the precise chronological position of zone boundaries. Some zonations have been modified from original published versions to emphasize significant details. References for sites are as follows:
1) Anderson (1982);
2) Schweger (1976, 1982);
3) Brubaker *et al.* (1983);
4) Edwards *et al.* (in press);
5) Livingstone (1955, 1957);
6) Bergstrom (1984);

refugium for *Alnus,* possibly throughout the Late Wisconsin glacial interval, or alternatively, that *Alnus* invaded northern Seward Peninsula earlier than it is known to have appeared in substantial amounts elsewhere in western Alaska.

Cape Deceit[1]

Coastal exposures of Quaternary and perhaps late Pliocene age deposits have been studied at Cape Deceit, northern Seward Peninsula (Fig. 4), for pollen, plant macrofossils, and insect fossils by Matthews (1974a). Important reinterpretations of the stratigraphy and chronology of the sections have been published recently (Giterman *et al.,* 1982, p. 65-70); therefore the following discussion attempts to review the Cape Deceit pollen record in light of Matthews' most recent interpretations. The lowermost unit exposed in the coastal bluffs is the Cape Deceit Formation, which was originally interpreted to be of early Pleistocene age (>700,000 yrs. B.P.). More recent evidence suggests that the formation may actually be of late-Pliocene age (*ca.* 1.8-2.5 million

yrs. B.P.) (Giterman *et al.,* 1982). Pollen and some plant macrofossil evidence suggest that the deposit preserves a record of treeless tundra-like vegetation over most of the interval represented by the deposit. Several units within the Cape Deceit Formation contain pollen that suggests that at least *Larix* and possibly some *Picea* trees grew in the area during at least part of the time represented by the deposit. Recent interpretations suggest that the Cape Deceit Formation may represent a single complex interglacial interval of unknown duration (Giterman *et al.,* 1982). Assemblages of the Cape Deceit Formation include pollen of *Betula, Salix,* Ericaceae, small amounts of *Alnus, Larix,* and *Picea* in most samples, as well as Cyperaceae, Gramineae, *Artemisia,* Tubuliflorae, and other herbs such as Caryophyllaceae and *Thalictrum.* In view of the revised age interpretations of the Cape Deceit Formation, the pollen and macrofossil evidence are of great significance for understanding the early development of tundra or tundra-like vegetation in Alaska. If these interpretations are correct, vegetation that produced pollen assemblages quite similar to those from modern tundra communities appeared in western Alaska at least as early as late-Pliocene time. Furthermore, the data indicate that *Larix* may have formed the treeline in some parts of Alaska during part of the interval represented by the Cape Deceit Formation. The middle depositional unit at Cape Deceit, the Inmachuk Formation (Matthews, 1974a), is now given less importance because it is poorly exposed and of uncertain age (Giterman *et al.,* 1982, p. 65-70). Pollen spectra from the Inmachuk Formation are similar to those of the Cape Deceit Formation, except that *Betula* percentages tend to be somewhat higher, and *Larix* is rarely represented. These pollen spectra were probably produced by shrub tundra communities.

The Deering Formation is the uppermost unit, and according to Matthews (1974a), represents much of the time since a pre-Illinoian interglacial. More recent interpretations suggest that the base of the formation is no older than the last interglacial, however (Giterman *et al.,* 1982, p. 65-70). Trees apparently reached Cape Deceit during the last interglacial, when *Picea* and *Betula papyrifera* grew nearby (Matthews, 1974a; Giterman *et al.,* 1982). One of the units within the Deering Formation, Peat 5, was previously interpreted to be of Sangamon age, but is now interpreted to fall within the Middle Wisconsin interstadial, but older than 39,000 yrs. B.P. The Deering Formation pollen and macrofossil records suggest that tundra vegetation communities have predominated at Cape

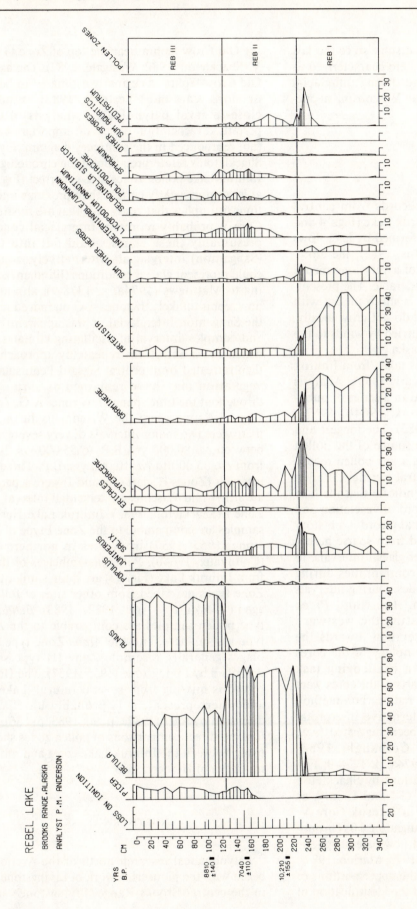

Figure 15. Pollen percentage diagram, Rebel Lake, Upper Koyukuk River Basin, Brooks Range. Reprinted from "Late Wisconsin and Holocene vegetation history of the Upper Koyukuk Region, Brooks Range, Alaska," By M. E. Edwards, P.M. Anderson, H.L. Garfinkel, and L.E. Brubaker, in press. Copyright© 1985 by *Canadian Journal of Botany.* Reproduced by permission of the journal.

Deceit over most of its known history since the last interglacial. These communities are characterized as shrub tundra, herb tundra, and during full-glacial intervals such as during the Late Wisconsin, steppe-tundra.

Imuruk Lake

The only available pollen records from central Seward Peninsula are from Imuruk Lake (Figs. 4 and 13). The lake occupies a basin formed by lava flows of mid-Pleistocene age (Hopkins, 1963); therefore, the potential for accumulation of a very long record of sediments and pollen is considerable. The present-day vegetation of the lake area is shrub tundra with sedge tussocks. *Alnus* and *Picea* do not grow locally, but pollen from those taxa are carried by winds to the lake area from other areas of Alaska.

A series of sediment cores was taken from Imuruk Lake by P. A. Colinvaux in the early 1960's. Analysis of pollen from the cores resulted in a series of publications and reports (Colinvaux, 1964b, 1967c; Colbaugh, 1968; Shackleton, 1982). The longest and oldest core from the lake is the source of the pollen diagram (Fig. 13) that serves as a key pollen record for Seward Peninsula and central Beringia. Pollen profiles and radiocarbon dates indicate that the lake does indeed preserve a long record of vegetation and climate change for the region. That record is a history of tundra vegetation that shifted from sparse herbaceous tundra during cooler, drier glacial intervals, to various types of shrub tundra communities during somewhat milder climatic episodes. During intervals of greatest apparent warmth, *Alnus* and *Picea* percentages increased, suggesting the westward expansion of boreal forest vegetation towards the lake. Interpretation of the chronology of the cores from Imuruk Lake has been a challenging task because all but the uppermost parts of the cores were too old to date by means of the radiocarbon method (Colinvaux, 1964b). Several alternative interpretations for the chronology have been suggested (*e.g.,* Colinvaux, 1964b, 1967c; Colbaugh, 1968; Shackleton, 1982; Matthews, 1974a, p. 1278-9; Fig. 13). The most recent interpretations by Shackleton (1982) are applicable to zones G-J and L only, as those were the zones present in Imuruk Core V, which was the object of his investigation. His chronological revisions are based upon new paleomagnetic data (Noltimier and Colinvaux, 1976; Marino, 1977), additional radiocarbon dates in the upper portions of the cores (Shackleton, 1982), and the identification of

the Old Crow tephra near the top of Zone G in Core V (Shackleton, 1982; Westgate, 1982). The age of the Old Crow tephra is currently estimated to be 80,000 or more years old (Westgate, 1982). In addition, close-interval palynological analysis of Core V permitted recognition of several important subzones (not reproduced in the summary diagram, Fig. 13). If Shackleton's reinterpretation of the chronology of the Imuruk Lake pollen zones G-L is correct (Fig. 13), or at least more nearly correct than previous interpretations, it implies that zones A-F that are missing from Core V probably represent interglacial conditions; presumably these zones would fall into the last (Sangamon) interglacial. Alternatively, zones A-G could represent glacial conditions (Illinoian?), according to Matthews (1974a, p. 1378-9), although this now seems unlikely. If zones A-G do indeed represent the Sangamon Interglacial, it was apparently a long and complex interval of oscillating climates during which *Picea* and *Alnus* repeatedly approached and then retreated from central Seward Peninsula. Local vegetation may have remained as shrub tundra throughout the time spanned by zones A-G. Zones H-J (or H-K) represent the Wisconsin, during which there were two major intervals of very severe climate, between *ca.* 80,000 yrs. B.P. to 55,000 yrs. B.P. and from *ca.* 35,000 to *ca.* 10,000 yrs. B.P. These glacial intervals (Zones H, part of i, and J) were separated by a warmer mid-Wisconsin interstadial interval, part of Zone i (Shackleton, 1982). Imuruk Lake Herb Zone samples are comparable to the Zone I type of Livingstone (1955, 1957) from sites in northern Alaska (Colinvaux, 1964b). Pollen assemblages of this type from Imuruk Lake (Fig. 13) are quite similar to Herb Zone types described from other sites of full-glacial age in Alaska (*e.g.,* Ager, 1982c, 1983). *Betula* Zone-type pollen samples are comparable to the Zone II type assemblages, and the *Alnus* Zone-type assemblages generally resemble Zone III type samples described by Livingstone (1955, 1957). The Holocene record is missing from several Imuruk Lake cores, and where present, it is probably unreliable, as discussed by Ager (1982c, p. 90; 1983, p. 136). Therefore, no direct comparisons of pollen zones should be made between the Imuruk Lake cores and other sites in Alaska of Holocene age.

NORTHERN ALASKA

Palynological research north of the Arctic Circle began with the pioneering work of Livingstone (1955) in the central Brooks Range. Livingstone's longest, but undated, pollen record from Chandler Lake (Figs. 4 and 14) displayed three major zones, dominated by pollen of herbaceous taxa (Zone I) in the oldest part of the core, then *Betula* (Zone II), and finally *Alnus,* usually with at least some *Picea* (Zone III) in the upper part. Livinstone interpreted these zones to reflect a south-to-north shift of modern vegetation zones (herb tundra, shrub birch tundra, and boreal forest) in response to ameliorating climates during late-glacial and Holocene times. The extreme simplicity of this pollen stratigraphy suggested that pollen analysis was a rather imprecise tool for studying the detailed vegetational history of the arctic. It also suggested that the history of vegetation in this region of Alaska was relatively uneventful when compared to that of more temperate regions.

The general characteristics and sequence of these pollen zones have now been identified in lake and bog sediments throughout much of northern and western Alaska and northwest Canada, but regional chronologies differ. In addition, recent studies have added new detail to the three-zone sequence and, in turn, new complexity to the interpretation of past vegetation in this region. The increased resolution of the pollen record is due primarily to refinements in palynological and radiocarbon-dating techniques since the time of Livingstone's initial work. Three factors stand out as most important: 1) radiocarbon dating of lake sediments with low organic content, 2) increased taxonomic resolution of pollen identifications, and 3) refinements in comparisons between fossil and modern pollen spectra.

Radiocarbon dates have revealed the ages of important pollen-stratigraphic boundaries and sometimes have permitted the calculation of pollen accumulation rates. Radiocarbon dates from northern Alaska now clearly demonstrate that vegetation changes have been time-transgressive, usually from both east to west and south to north. Pollen accumulation rates have provided important insights about the nature of vegetation cover, particularly during the waning phases of the last glacial interval.

The taxonomic resolution of pollen identification in northern Alaska has improved considerably in recent years. The increased resolution is due in part to simple improvements in techniques (*e.g.,* Cwynar *et al.,* 1979) for separating pollen from the highly inorganic sediments that characterize lakes in this region. The efficient removal of clay- and silt-sized particles has aided in the detection of problem pollen types, such as *Juniperus* and *Populus,* which are commonly recorded in recent pollen diagrams but were absent from early records from northern Alaska.

Taxonomic resolution of *Picea* and *Betula* pollen identifications has improved through measurements of pollen-grain dimensions. Separation of species within these genera is especially important to the interpretation of the pollen records because so few arboreal species are present in the flora of northern Alaska. Pollen measurements have generally shown that species within these genera have had very different vegetation histories.

Modern pollen spectra are more abundant than at the time of Livingstone's work. Pollen spectra from moss polsters, which are most useful for interpreting pollen records from peat sections, are available from the Dalton Highway (S. Short, unpublished data) and Arctic slope (Nelson, 1979, 1982). Pollen samples from recently deposited lake sediments, which are preferable for interpreting lake pollen records, are now available from a grid of sites covering much of northern Alaska (Anderson and Brubaker, unpublished data; Anderson, 1982).

In summary, recent studies employing new techniques show that the quality of pollen records at high latitudes is comparable to that of those from lower latitudes and that the vegetational history of northern Alaska is more complex than suggested by early diagrams.

Brooks Range

Late-glacial and Holocene pollen records are available from several lakes in the Kobuk, Alatna and Koyukuk River drainages in the central Brooks Range. The region was last covered by ice during the Walker Lake glaciation (Hamilton, 1982), which consisted of two advances between approximately 24,000 and 13,000 years ago. The longest records come from high elevation cirque-basin lakes surrounded by dwarf shrub *Betula* tundra. The basal portions of these cores typically penetrate more than a meter of sediment dominated by herb taxa. Unfortunately, age control in the herb zone is poor because the sediments are very inorganic and tend to yield unreliable radiocarbon dates. Lakes at lower elevations are often in kettle depressions and have shorter records, possibly because slow melting of buried ice blocks delayed lake formation at such sites.

Rebel Lake. The vegetation history of this region is illustrated with a previously unpublished pollen diagram form Rebel Lake (Edwards *et al.,* in press) located in the upper Koyukuk River drainage (Figs. 4, 14, and 15). Radiocarbon dates for this lake are not reliable because they showed reversals and large

Epiguruk-III, Kobuk River, Alaska

Percentage Diagram (Summary)

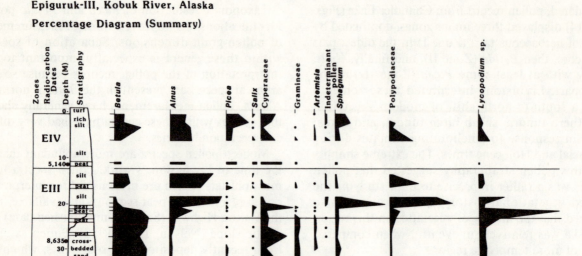

ANALYSIS BY C.E. SCHWEGER

Figure 16. Pollen percentage diagram (summary) from Epiguruk III, a bluff exposure of alluvial deposits, Kobuk River. Redrafted from *Late Quaternary Paleoecology of the Onion Portage Region, Northwestern Alaska,* by C. E. Schweger, 1976, Ph.D. Dissertation, University of Alberta, Edmonton. Copyright © 1976 by C. E. Schweger. Reproduced by permission of the author.

discrepancies with dates from two nearby lakes. Approximate dates are assigned to pollen zone boundaries in the Rebel Lake pollen diagram based on dates for corresponding zone boundaries at nearby sites.

This pollen sequence is comparable to Livingstone's (1955) pollen diagram from Chandler Lake and several recently reported records from the central Brooks Range area (Bergstrom, 1984; Brubaker *et al.,* 1983; Edwards *et al.,* in press). Zone I at Rebel Lake is dominated by pollen of Cyperaceae, Gramineae, *Artemisia, Salix* and several herb taxa, and extends from an unknown basal age to about 12,000-13,000 yrs. B.P. *Artemisia* pollen percentages are greatest near the base of the zone while Cyperaceae and *Salix* percentages are greatest near the top. Gramineae shows no strong trend within the zone. The vegetation during this period is difficult to interpret because modern lake samples from herb tundra zones of northern Alaska show much lower percentages of *Artemisia* pollen, but contain higher percentages of shrub and tree pollen, which is blown in from the shrub tundra and boreal forest (Anderson and Brubaker, unpublished data).

Betula pollen, in the size range of shrub *Betula,* rises to and maintains values of nearly 70% in Zone II sediments. Small peaks in *Juniperus* and *Populus* pollen occur in the middle of this zone. Such peaks are sometimes prominent at low-and mid-elevation lakes, and at some lakes they are accompanied by increases in *Salix* pollen percentages (Edwards *et al.,* in press). *Picea* pollen, identified as predominantly *Picea glauca,* increased about 8,000-8,500 yrs. B.P. The vegetation represented by this zone was probably a shrub *Betula* tundra, possibly with a higher shrub cover than is found today in shrub-tundra communities. The increase in *Juniperus* and *Populus* pollen undoubtedly represents an increase in these taxa in the landscape, but the climatic interpretation of such changes is unclear. Citing range extensions of several species, including *Picea* in northwestern Canada, Ritchie *et al.* (1983) propose that summers during this period (*ca.* 10,000-9,000 yrs. B.P.) may have been warmer than at present. *Picea glauca* probably arrived in the central Brooks Range at low and midelevations about 8,000 years ago. However, no evidence has yet been found to suggest that *Picea* was more widespread than today at any time during the Holocene in or near the Brooks Range.

Alnus percentages increased to more than 40% at the beginning of Zone III, approximately 7,000 yrs. B.P. The increase in *Alnus* pollen depressed percentages of *Picea* and *Betula,* but the pollen accumulation rates of the latter taxa remain constant across the Zone II-III boundary, implying little or no changes in the size of those populations. *Alnus* shrubs became established in the area at this time, therefore, but apparently did not compete directly with *Picea* and shrub *Betula.*Measurements of *Betula* and *Picea* pollen within the *Alnus* zone indicate that *Picea mariana* (black spruce) and *Betula papyrifera* (a tree birch) arrived in the central Brooks Range during the middle Holocene.

Kaiyak Lake. Anderson (1982, in press) has recently described two long (26,000 and *ca.* 37,000 years) pollen sequences from lakes in the Noatak and Kobuk River valleys, in the western Brooks Range. This region was glaciated during the Itkillik I Glaciation but large areas were ice free during the Walker Lake Glaciation (Hamilton, 1982).

Kaiyak Lake (Anderson, 1982, in press; Figs. 4 and 14), located beyond treeline in the Noatak Valley, has a pollen record that spans at least 37,000 years. It is the longest continuous pollen record yet available from northern Alaska. The herb zone at Kaiyak Lake is dominated by Cyperaceae, Gramineae, and *Salix,* but *Artemisia* pollen percentages never reach values comparable to those in lakes in the central Brooks Range (*ca.* 40%). The basal pollen zone assemblages at Kaiyak Lake may represent Gramineae and Cyperaceae meadow at low elevations in mountain valleys, and thus contrast with the sparse vegetation cover inferred for upland and mountain sites in the central Brooks Range and northwest Canada. The low, but consistent, values of *Betula* pollen percentages in this zone may indicate the local persistence of *Betula* shrubs in moist protected sites in the valleys of the western Brooks Range. The Kaiyak Lake pollen zonation is summarized in Figure 14.

The timing of percentage increases of key pollen taxa differ between the western and central Brooks Range. The change from herb- to *Betula*-dominated pollen assemblages takes place around 14,000 yrs. B.P. in the west, about 1,000-2,000 years earlier than in the central Brooks Range. The rise in *Alnus* pollen percentages is also approximately 1,000-2,000 years earlier in the west, but the increase in *Picea*in the west lags behind its increase in the central Brooks Range by about 4,000 years. Overall, the data available from the Brooks Range, central Alaska, and

northwest Canada support the idea that *Picea* migrated westward across northern Alaska, from Canada or central Alaska. The western limit of *Picea* apparently remained fairly stationary from about 8,000 yrs. B.P. to sometime in mid-Holocene time, and then spread rapidly, reaching its current location in the western Brooks Range approximately 5,000 yrs. B.P.

Kobuk and Koyukuk Rivers. Pollen assemblages from exposures of sediments of primarily alluvial origin were examined from six localities along the Kobuk and Koyukuk Rivers south of the Brooks Range (Schweger, 1976, 1982; Fig. 4). These sections range in age from Early Wisconsin to Holocene. Pollen sample processing techniques for these sediments were modified from those developed for alluvial sediments in the arid western U.S. (Schweger, 1976).

Schweger's investigations of Alaskan alluvial sections reveal that such deposits may preserve important palynological records, and that the pollen assemblages reflect the strong influence of floodplain vegetation. Some of the units investigated contained little or no pollen; where present preservation varied from quite good (a few percent indeterminates) to poor (30-40% indeterminates). Stratigraphy of alluvial sections is often complex, requiring careful site interpretation and subsampling. Depositional hiatuses are common, and sedimentation rates highly variable. Reworking could be a significant problem in some units. In spite of these potential problems, the alluvial sections permit the piecing together of vegetational histories that span very long intervals, and allow reconstruction of past vegetation from areas where suitable lakes are unavailable or are too young for more traditional approaches to obtaining pollen records.

An example of a pollen record from an alluvial deposit from the Kobuk River is presented in Figure 16. This site is Section III from Epiguruk near the Onion Portage archeological site (Schweger, 1976; Fig. 4). The section represents an alluvial infilling of a floodplain slough during the Holocene. The radiocarbon date of 8,635 ± 210 yrs. B.P. from the middle of the section is apparently spurious, the result of fluvial reworking of older wood. Pollen preservation in Epiguruk Section III is considerably better than in some of the older alluvial sections described from the Koyukuk River (Schweger, 1982). Pollen Zone E II is an early-Holocene *Betula* Zone.

Zone E III is an *Alnus* Zone that began about 7,000 yrs. B.P. in this region, according to Schweger's data

(1976). Zone E IV is a *Picea* Zone representing the past *ca.* 5,500 yrs. B.P.

Early, Middle, and Late Wisconsin age pollen assemblages from the Koyukuk River alluvial sections suggest that sedge-herb tundra persisted for a very long interval (>40,000 to *ca.* 10,000 to 12,000 yrs. B.P.) in areas immediately to the south of the Brooks Range (Schweger, 1982).

A composite late-Quaternary pollen zonation for the alluvial sections from the Kobuk and Koyukuk Rivers is compared with other northern Alaskan zonations in Figure 14.

Arctic Slope.

The arctic slope of northern Alaska encompasses the arctic foothills north of the Brooks Range and the arctic coastal plain (Fig. 4). Most of the region escaped glaciation during the Pleistocene, and therefore the potential for obtaining long pollen records is considerable. Palynological investigations have been relatively few, however, due to the remoteness of the region and resulting complex and costly logistic arrangements required to conduct field work. Several other factors contribute to the difficulty of obtaining reliable pollen records from the arctic slope region, as discussed below.

The region is underlain by continuous permafrost (Péwé, 1975), and the lakes in the region are nearly all of thermokarst origin, which can result in extensive mixing of pollen-bearing sediments of different ages. Anderson (1982) conducted a reconnaissance study of Holocene palynological records from such lakes in northern Alaska. Her study suggests that some thermokarst lakes may preserve a usable record, but that pollen preservation is often poor and the "signal to noise" ratio from reworked pollen and spores is relatively unfavorable. The objective of reconstructing detailed vegetational histories from such deposits would be difficult to achieve.

Extensive fluvial, eolian, and, on the coastal plain, marine erosional and depositional processes have resulted in massive reworking of Cretaceous, Tertiary, and Quaternary palynomorphs in many deposits. To further complicate interpretations, the tundra vegetation of the arctic slope, particularly in the coastal plain and northernmost foothills, produces little pollen. Low local pollen production combined with deposition of long-distance wind-transported pollen of *Picea, Pinus, Alnus,* and *Betula,* add to the difficulties of interpreting the regional pollen record.

Early attempts to outline the late-Quaternary history of arctic slope vegetation by means of pollen analysis of peat and silt deposits were published by Livingstone (1957) for the Umiat area (Fig. 4), Heusser (1963a) for Ogotoruk Creek, near Cape Thompson (Fig. 4), and Colinvaux (1964a, 1967c) for the Barrow area and a few other sites on the arctic slope.

Umiat. A frozen valley-fill from Umiat in the arctic foothills (Fig. 4) provided the first radiocarbon-dated pollen profile from northern Alaska (Livingstone, 1957). The pollen assemblage zones from the Umiat section were interpreted to parallel closely those from the undated Chandler Lake core in the north-central Brooks Range (Livingstone, 1955; Fig. 4). The chronology from the Umiat Site was accepted for many years and was applied, at least tentatively, to undated sites as remote as Ogotoruk Creek (Heusser, 1963a; Fig. 4). Ager (1983, p. 135-136) has questioned the validity of applying the Umiat chronology to the Chandler Lake site and other northern Alaskan sites. The Umiat pollen record is highly truncated, spanning only the interval of *ca.* 8,200-5,500 yrs. B.P. Correlations of Zone I (Herb Zone) pollen assemblages between Umiat and Chandler Lake are questionable (Livingstone, 1957; Ager, 1983). The Umiat record suggests tentatively that the transition from Zone I to Zone II (*Betula* Zone) occurred about 7,500 yrs. B.P. and the Zone II-Zone III (*Alnus* Zone) transitition about 6,000 yrs. B.P. (Livingstone, 1957). More recent research in the Arctic foothills at Ikpikpuk River (Nelson, 1982; Fig. 4) suggests that *Betula* shrubs were a significant component of the regional tundra vegetation at least as early as 9,700 yrs. B.P. At Toolik Lake on the northern edge of the Brooks Range (Bergstrom, 1984; Figs. 4 and 14), pollen data and radiocarbon dates indicate that the basal Herb Zone-*Betula* Zone boundary dates to about 12,000 yrs. B.P., and the *Betula* Zone-*Alnus* Zone boundary is dated at 9,500 yrs. B.P. The Umiat chronology therefore should no longer be applied uncritically to other undated (or inadequately dated) sites in northern Alaska.

Ocean Point. Recent studies by R. E. Nelson (1979) involved palynological investigations of the Gubik Formation exposed along the Colville River near Ocean Point (Fig. 4). A lower marine unit from the Gubik Formation, now thought to be of late-Pliocene age, (Repenning, 1983) contains pollen of *Picea, Betula,* and Ericaceae; *Pinus* and *Abies* pollen are present in small amounts. Nelson interprets the pollen assemblage to represent a boreal spruce-birch forest with a rich ericaceous ground cover that grew in the region during a warmer than present interglacial interval.

An upper fluvial unit in the Gubik Formation is interpreted to be of mid-Pleistocene age. Pollen of Gramineae and Cyperaceae predominate throughout the unit. The unit was divided into three pollen subzones that suggest that two intervals with herbaceous tundra occured during glacial episodes, separated by an interstadial or interglacial interval during which shrub-herb tundra developed.

Beaufort Sea. Palynomorph assemblages from sediments of the Flaxman Formation were analyzed by Nelson (1979) from samples obtained from offshore bore holes in the Beaufort Sea. These sediments are tentatively dated as Sangamon Interglacial in age, based on associated marine fossils. Pollen assemblages from this formation display relatively high percentages of *Picea* (2-8%), *Alnus* (15-20%), and *Betula* (20-30%), along with significant amounts of Cyperaceae and Gramineae. Nelson suggests that *Picea* was present on the Arctic Slope during deposition of the Flaxman Formation, but that the trees did not reach the Beaufort Sea shore. Regional vegetation was primarily tussocky shrub tundra.

Titaluk River. Nelson (1982) studied late Quaternary fluvial sediments along the Titaluk River at the arctic coastal plain-foothills boundary (Fig. 4).

The major portion of this section is comprised of highly organic silts and lenses of peat and is thought to represent ancient flood plain environments of the Titaluk River. Dating of this deposit is insecure because radiocarbon dates from wood, coarse peat and fine peat in adjacent samples yield disparate ages due in large part to fluvial recycling of organics, especially wood. The most reasonable interpretation of available radiocarbon dates is that the section encompasses the period from about 42,000 to 25,000 yrs. B.P. A capping layer of peat dates from the late-Holocene (2,500 yrs. B.P.).

Pollen percentages in basal fluvial sediments (Pollen Zone TA) are dominated by Cyperaceae (50-60%) and Gramineae (10-20%), but small quantities of *Betula, Alnus* and *Picea* are consistently present. Nelson interprets Zone TA as the end of a mid-Wisconsin warm period, during which vegetation was rather similar to today's; dwarf *Betula* was present, *Alnus* was rare, and the nearest *Picea* was south of the Brooks Range. Upland vegetation is interpreted to have been less continuous than today's, however, and the climate was apparently somewhat drier.

The upper fluvial sediments (Pollen Zone TB) contain rather similar pollen assemblages but show less *Alnus* and *Betula* and no *Picea.*This deposit is believed to represent the transition between mid-Wisconsin interstadial and Late Wisconsin glacial climates. Pollen and other evidence suggest a long-term gradual decline in temperatures and moisture. The Zone TB record ends about 25,000 yrs. B.P.

Pollen Zone TC is derived from a thick thaw lake deposit representing the past 2,500 years. Pollen assemblages indicate that vegetation similar to that of the present day has existed near the site over that time interval. Gramineae, Cyperaceae, *Betula, Alnus,* and *Picea* are important constituents of Zone TC pollen assemblages (Nelson, 1982).

Prudhoe Bay. Late-Holocene pollen profiles from the arctic coastal plain at Prudhoe Bay (Fig. 4) were published recently by Walker *et al.* (1981). These samples provide a record of the past *ca.* 3,800 years. Cyperaceae pollen dominates the pollen assemblages from herbaceous tundra. Small amounts of pollen of *Alnus, Betula, Picea,* and *Pinus* are present in most samples and reflect the influence of wind-transported pollen from distant sources. The presence of *Pinus* pollen in these samples suggests sources in northwestern Canada.

SUMMARY AND DISCUSSION

Very little is yet known about the vegetational history of Alaska during early and middle-Pleistocene time. Several sites where palynological investigations were conducted were originally interpreted to be of early-Pleistocene age, but are now thought to be of probable late-Pliocene age. Late-Pliocene vegetation at Lost Chicken in eastern interior Alaska was boreal forest of *Picea, Betula,* and *Alnus,* but with the addition of *Pinus* and *Larix* which do not grow there today. At Cape Deceit on Seward Peninsula, the Cape Deceit Formation, now believed to be of late-Pliocene age contains a pollen record that suggests tundra-like vegetation near present-day sea level in an area where tussocky shrub tundra vegetation grows today. *Larix* and *Picea* may have reached the Cape Deceit area occasionally during the interval represented by the deposit. At Ocean Point on the arctic coastal plain of northern Alaska, a marine unit of the Gubik Formation is now interpreted to be of late-Pliocene age, and contains pollen of *Picea, Betula,* and Ericaceae along with small amounts of *Pinus* and *Larix.* This suggests that boreal forest vegetation grew in northern Alaska at that time, whereas today the vegetation is wet tundra.

Although these sites tell us nothing about the details of early-Pleistocene vegetation history, they establish that both "boreal forest" and tundra or tundra-like vegetation had developed in Alaska by late-Pliocene time. The similarities in pollen assemblages of late-Pliocene and late-Quaternary age may be more apparent than real, since little is yet known about the species-level composition of vegetational communities of the late-Pliocene. The available data suggest, however, that it is likely that early-Pleistocene vegetation types in Alaska consisted of boreal forest and tundra communities.

Few middle-Pleistocene sites have yet been studied. At Cape Deceit, Seward Peninsula, the insecurely dated Inmachuk Formation (middle-Pleistocene?) contains a pollen record that suggests local shrub tundra vegetation. In the arctic coastal plain at Ocean Point, a unit within the Gubik Formation is of probable middle-Pleistocene age. The pollen data from that unit suggest that the vegetation shifted between shrub tundra and herbaceous tundra during the interval represented by the deposit. A few samples from interior Alaska thought to be of middle-Pleistocene age suggest that treeline was significantly lower than it is today in the region, and that tundra and shrub communities covered much of the landscape.

Most palynological investigations in Alaska have concentrated on the relatively accessible late-Pleistocene and Holocene record, and consequently our understanding of late-Quaternary vegetation history is relatively good in comparison with older records. Pollen data of probable Sangamon interglacial age are available from several areas of the State. The available data suggest that the Sangamon interglacial was a complex climatic event during which boreal forests of interior Alaska expanded into and retreated from tundra zones paralleling the western and northern coasts. The Goose Bay peat north of Anchorage may be of Sangamon age, and suggests that boreal forest vegetation occupied upper Cook Inlet during the interglacial.

Early Wisconsin records are rare, but suggest that the climatic episode was very cold and arid, and sparse herbaceous tundra vegetation developed in probably most of unglaciated Alaska. The Middle Wisconsin was a long interval of oscillating climates that ended about 26,000 yrs. B.P. Tundra vegetation was widespread during the Middle Wisconsin, but *Picea* forests and forest-tundra existed in the valleys of interior Alaska during at least the warmest intervals of the interstadial.

The Late Wisconsin glacial interval was a cold, arid climatic interval during which most trees and shrubs were nearly eliminated from Alaska. Some genera such as *Picea* may possibly have been eliminated entirely. Pollen records from all areas of Alaska that have been investigated thus far suggest that the full-glacial vegetation of the refugia in Alaska consisted of herbaceous tundra communities. Interpretation of the pollen and plant macrofossil data from Late Wisconsin-age sites has played a central role in recent debates on the nature of the ecosystem in Beringia during that time. These arguments have been discussed extensively in several recent publications (*e.g.*, Hopkins *et al.*, 1982; Cwynar and Ritchie, 1980), so will not be repeated here in detail. One aspect of the controversy is whether or not the herbaceous pollen assemblages of Late Wisconsin age represent vegetation types significantly different than present-day tundra communities (a no-analog situation), or whether certain fell-field or polar desert communities serve as valid modern analogs (*e.g.*, Cwynar and Ritchie, 1980; Matthews, 1982).

By 14,000 years ago climate change towards warmer, moister conditions began in Alaska, and shrub tundra vegetation with dwarf *Betula* spread rapidly into many parts of the State where only herb tundra or glacial ice had existed previously. By *ca.* 11,000 yrs. B.P. *Populus* stands developed in many parts of the State, even in areas beyond the present tree limit in western and northern Alaska. This nearly simultaneous spread of *Populus* suggests that a climatic threshold had been crossed that permitted small populations of *Populus* (probably *Populus balsamifera*) that had presumably survived in scattered sites in Beringia during the Late Wisconsin to quickly expand into previously unsuitable habitats. By about 9,500 yrs. B.P. *Alnus* began to invade southern Alaska from the coast. A population of *Alnus crispa* probably survived the Late Wisconsin in western Alaska and began expanding its range in latest Wisconsin and early-Holocene time. *Picea* first appeared in Holocene age pollen records in eastern interior Alaska about 9,500 yrs. B.P., then spread southward, reaching Cook Inlet by *ca.* 8,000 yrs. B.P., and westward, nearing the present tree limit by *ca.* 5,500 yrs. B.P. In general, vegetation types quite similar in composition to those of today reached their approximate modern distributions in many regions of the State by mid-Holocene time.

Much has been learned about the Quaternary vegetational history of Alaska through palynological investigations, sometimes supplemented by studies of plant macrofossil assemblages. Much palynological research remains to be done in the State, but the results reported thus far demonstrate that there is a wide range of deposits from which pollen records can be extracted. These represent a great variety of environments of deposition, including lacustrine, bog, alluvial, colluvial, eolian, and marine environments. Each type of deposit presents interpretational challenges for palynologists because of differences in pollen abundance, quality of preservation, and the relative influence of reworking and local vs. regional vegetation. The many potential sources of pollen

records, coupled with the fact that much of Alaska escaped glaciation during the Pleistocene suggest that there is a great opportunity to piece together a nearly complete record of Alaskan vegetational history for the Quaternary. The importance of such a record is great, because of the profound influence Alaska has had on North American biology as a result of its past roles as ice-age refugium and dispersal corridor connecting North American and northeastern Asian ecosystems.

References Cited

AGER, T. A.

1975 *Late Quaternary Environmental History of the Tanana Valley,* Alaska. Ohio State University Institute of Polar Studies, Report 54, Columbus, 117 p.

1982a Pollen studies of Quaternary-age sediments in the Tanana Valley. *In* Coonrad, W. L. (ed.), United States Geological Survey in Alaska: Accomplishments during 1980. *U.S. Geological Survey Circular* 844: 66-67.

1982b Quaternary history of vegetation in the North Alaska Range. *In* Coonrad, W. L. (ed.), United States Geological Survey in Alaska: Accomplishments during 1980. *U.S. Geological Survey Circular* 844: 109-111.

1982c Vegetational history of western Alaska during the Wisconsin glacial interval and the Holocene. *In* Hopkins, D. M., Matthews, J. V., Jr., Schweger, C. E., and Young, S. B., (eds.), *Paleoecology of Beringia.* Academic Press, New York, 75-93.

1983 Holocene vegetational history of Alaska. *In* Wright, H. E., Jr., (ed.), *Late-Quaternary Environments of the United States.* Vol. 2, *The Holocene.* University of Minnesota Press, Minneapolis, 128-140.

AGER, T. A., and BRADBURY, J. P.

1982 Quaternary history of vegetation and climate of the Yukon Delta-Norton Sound area. *In* Coonrad, W. L. (ed.), The United States Geological Survey in Alaska: Accomplishments during 1980. *U.S. Geological Survey Circular* 844: 78-80.

AGER, T. A., and SIMS, J. D.

1981 Holocene pollen and sediment record from the Tangle Lakes area, central Alaska. *Palynology,* 5: 85-98.

1984 Postglacial pollen and tephra records from lakes in the Cook Inlet region, southern Alaska. *In* Coonrad, W. L., and Elliott, R. L., (eds.), The United States Geological Survey in Alaska: Accomplishments during 1981. *U.S. Geological Survey Circular* 868: 103-105.

AGER, T. A., and RUBIN, M.

1984 Vegetation changes in South-Central Alaska since deglaciation. *American Quaternary Association, Eighth Biennial Conference, Boulder, Program and Abstracts,* 1

AGER, T. A., RUBIN, M., and RIEHLE, J. H.

1984 History of vegetation in the Cook Inlet region, South-Central Alaska since deglaciation. *American Association of Stratigraphic Palynologists, Seventeenth Annual Meeting, Arlington, Virginia, Program and Abstracts,* 1.

ANDERSON, J. H.

1975 A palynological study of late Holocene vegetation and climate in the Healy Lake area, Alaska. *Arctic,* 28: 29-62.

1984 A survey of allergenic airborne pollen and spores in the Fairbanks area, Alaska: Annals of Allergy, v. 52, p. 26-31.

ANDERSON, P. M.

1982 Reconstructing the past: The synthesis of archaeological and palynological data, northern Alaska and northwestern Canada. Ph D. Thesis, Brown University, 388 p.
      Late-Quaternary Vegetation Change in the Kotzebue Sound Area, Northwestern Alaska. *Quaternary Research.* In press.

ANDERSON, P. M., BARTLEIN, P. J. and BRUBAKER, L. B.

1984 Spatial patterns of modern pollen and temperature in Northern Alaska: Their use in a preliminary reconstruction of seasonal temperature variations during the Holocene. *American Quaternary Association, Eighth Biennial Meeting, Boulder, Program and Abstracts,* 2.

BAILEY, R. G.

1980 Description of the Ecoregions of the United States. U.S. Department of Agriculture Miscellaneous Publication 1391, 77 p.

BERGSTROM, M. F.

1984 Late Wisconsin and Holocene History of a Deep Arctic Lake, North-Central Brooks Range, Alaska. M.S. Thesis, The Ohio State University, Columbus, 112 p.

BENNINGHOFF, W. S.

1957 Recent contributions to Quaternary vegetation history in Alaska. *Proceedings, Fifth Alaska Science Conference,* Anchorage, 1954, 28.

BRUBAKER, L. B., GARFINKEL, H. L., and EDWARDS, M. E.

1983 Late-Wisconsin and Holocene vegetation change in the Walker Lake/Alatna Valley Region of the Brooks Range, Alaska, U.S.A. *Quaternary Research,* 20: 194-214.

COLBAUGH, P. R.

1968   The Environment of the Imuruk Lake Area, Seward Peninsula during Wisconsin time. M.S. thesis, Ohio State University, Columbus, 118 p.

COLINVAUX, P. A.

1964a   Origin of ice ages: pollen evidence from Arctic Alaska. *Science*, 145: 707-708.

1964b   The environment of the Bering Land Bridge. *Ecological Monographs*, 34: 297-329.

1967a   A long pollen record from St. Lawrence Island, Bering Sea, Alaska. *Palaeogeography, Palaeoclimatology, and Palaeoecology*, 3: 29-48.

1967b   Bering Land Bridge: Evidence of spruce in Late Wisconsin Times. *Science*, 156: 380-383.

1967c   Quaternary vegetational history of Arctic Alaska. *In* Hopkins, D. M. (ed.), *The Bering Land Bridge*. Stanford University Press, Stanford, p. 207-231.

1981   Historical ecology in Beringia: The south land bridge coast at St. Paul Island. *Quaternary Research*, 16: 18-36.

CONNOR, C. L.

1984   Late Quaternary Glaciolacustrine and Vegetational History of the Copper River Basin, South-Central Alaska. Ph.D. Dissertation, University of Montana, 115 p.

COULTER, H. W., HOPKINS, D. M., KARLSTROM, T. N. V., PÉWÉ, T.L., WAHRHAFTIG, C., and WILLIAMS, J. R.

1965   Map Showing extent of glaciations in Alaska. *U.S. Geological Survey Miscellaneous Geological Investigations Map* I-415.

CRITCHFIELD, W. B.

1984   Impact of the Pleistocene on the genetic structure of North American conifers. *In* Lanner, R. M. (ed.), *Proceedings of the Eighth North American Forest Biology Workshop*. Utah State University, Logan, 70-118.

CWYNAR, L. C., and RITCHIE, J. C.

1980   Arctic steppe-tundra: A Yukon perspective. *Science*, 208: 1375-77.

CWYNAR, L. C., BURDEN, E., and MCANDREWS, J. H.

1979   An inexpensive sieving method for concentrating pollen and spores from fine-grained sediments. *Canadian Journal of Earth Sciences*, 16 (5): 1115-1120.

EDWARDS, M. E., and BRUBAKER, L. B.

1984   A 23,000 year pollen record from northern interior Alaska. *American Quaternary Association, Eighth Biennial Meeting, Boulder, Program and Abstracts*, 35.

EDWARDS, M. E., and MCDOWELL, P. F.

1984   Quaternary environmental history in the southern Yukon Lowland, N.E. interior Alaska. *Sixth International Palynological Conference, Calgary, Abstracts*, 40.

EDWARDS, M. E., ANDERSON, P. M., GARFINKEL, H. L., and BRUBAKER, L. B.

        Late Wisconsin and Holocene vegetation history of the Upper Koyukuk region, Brooks Range, Alaska. *Canadian Journal of Botany*. In press.

GITERMAN, R. E., SHER, A. V., and MATTHEWS, J. V., JR.

1982   Comparison of the development of tundra-steppe environments in west and east Beringia: Pollen and macrofossil evidence from key sections. *In* Hopkins, D. M., Matthews, J. V., Jr., Schweger, C. E., and Young, S. B. (eds.), *Paleoecology of Beringia*. Academic Press, New York, 43-73.

HAMILTON, T. D.

1982   A late Pleistocene glacial chronology for the southern Brooks Range—Stratigraphic record and regional significance. *Geological Society of America Bulletin*, 93: 700-716

HAMILTON, T. D., and THORSON, R. M.

1983   The Cordilleran ice sheet in Alaska. *In* Porter, S. C., (ed.), *Late-Quaternary Environments of the United States*. Vol. 1, *The Late Pleistocene*. University of Minnesota Press, Minneapolis, 38-52.

HANSEN, H. P.

1953   Postglacial forests in the Yukon Territory and Alaska. *American Journal of Science*, 251: 505-542.

HARTMAN, C. W., and JOHNSON, P. R.

1978   *Environmental Atlas of Alaska*. 2nd Edition. University of Alaska Institute of Water Resources, Fairbanks, 95 p.

HEUSSER, C. J.

1957   Pleistocene and postglacial vegetation of Alaska and Yukon Territory. *In* Hansen, H. P. (ed.), *Arctic Biology*. Oregon State College, Corvallis, 131-151.

1960   Late-Pleistocene Environments of North Pacific North America. American Geographical Society Special Publication 35, 308 p.

1963a   Pollen diagrams from Ogotoruk Creek, Cape Thompson, Alaska. *Grana Palynologica*, 4: 149-159.

1963b   Postglacial palynology and archaeology in the Naknek River drainage area, Alaska. *American Antiquity*, 29: 74-81.

1965   A Pleistocene phytogeographical sketch of the Pacific Northwest and Alaska. *In* Wright, H. E., Jr., and Frey, D. G. (eds.), *The Quaternary of the United States*. Princeton University Press, Princeton, 469-483.

1983a   Vegetational history of the northwestern United States including Alaska. *In* Porter, S. C. (ed.), *Late Quaternary Environments of the United States*. Vol. 1, *The Late Pleistocene*. University of Minnesota Press, Minneapolis, 239-258.

1983b   Holocene vegetation history of the Prince William Sound region, South-Central Alaska. *Quaternary Research*, 19: 337-355.

1983c   Pollen diagrams from the Shumagin Islands and adjacent Alaska Peninsula, southwestern Alaska. *Boreas*, 12: 279-295.

HOPKINS, D. M.

1963   Geology of the Imuruk Lake area, Seward Peninsula, Alaska. *U.S. Geological Survey Bulletin* 1141-C, 101 p.

1967   *The Bering Land Bridge*. Stanford University Press, Stanford, 495 p.

1973   Sea level history in Beringia during the last 250,000 years. *Quaternary Research*, 3: 520-40.

1979   Landscape and climate of Beringia during late Pleistocene and Holocene time. *In* Laughlin, W. S., and Harper, A. B., (eds.), *The First Americans: Origins, Affinities and Adaptations*. Gustav Fischer, New York, 15-41.

1982   Aspects of the paleogeography of Beringia during the late Pleistocene. *In* Hopkins, D. M., Matthews, J. V., Schweger, C. E., and Young, S. B., (eds.). *Paleoecology of Beringia*. Academic Press, New York, 3-28.

HOPKINS, D. M., MACNEIL, F. S., and LEOPOLD, E. B.

1960   The Coastal Plain at Nome, Alaska: A late Cenozoic type section for the Bering Strait region. *Report of the 21st International Geological Congress*, part 4, Copenhagen, 44-57.

HOPKINS, D. M., MATTHEWS, J. V., JR., SCHWEGER, C. E., and YOUNG, S. B. (eds.)

1982 *Paleoecology of Beringia.* Academic Press, New York, 489 p.

HOPKINS, D. M., SMITH, P. A., and MATTHEWS, J. V., JR.

1981 Dated wood from Alaska and the Yukon: Implications for forest refugia in Beringia. *Quaternary Research,* 15: 217-249.

HULTÉN, E.

1937 *Outline of the History of Arctic and Boreal Biota during the Quaternary Period.* Borlags Aktiebolaget Thule, Stockholm, 165 p.

1968 *Flora of Alaska and Neighboring Territories.* Stanford University Press, Stanford, 1008 p.

KARLSTROM, T. N. V.

1964 Quaternary Geology of the Kenai Lowland and Glacial History of the Cook Inlet Region, Alaska. *U.S. Geological Survey Professional Paper* 443, 69 p.

LARSEN, J. A.

1980 *The Boreal Ecosystem.* Academic Press, New York, 500 p.

LIVINGSTONE, D. A.

1955 Some pollen profiles from Arctic Alaska. *Ecology,* 36: 587-600.

1957 Pollen analysis of a valley fill near Umiat, Alaska. *American Journal of Science,* 255-254-260.

MACDONALD, G. M.

1984a Post-glacial plant migration and vegetation dynamics in the western Canadian boreal forest. *Sixth International Palynological Conference, Calgary. Abstracts,* 96.

1984b Postglacial vegetation changes in the western Canadian 'ice free' corridor region. *American Quaternary Association, Eighth Biennial Meeting, Boulder, Program and Abstracts,* 74.

MARINO, R. J.

1977 Paleomagnetism of two lake sediment cores from Seward Peninsula, Alaska, M.S. Thesis, Ohio State University, Columbus, 183 p.

MATTHEWS, J. V., JR.

1970 Quaternary environmental history of interior Alaska: Pollen samples from organic colluvium and peats. *Arctic and Alpine Research,* 2: 241-251.

1974a Quaternary environments at Cape Deceit (Seward Peninsula, Alaska): Evolution of a tundra ecosystem. *Geological Society of America Bulletin,* 85: 1353-1384.

1974b Wisconsin environment of interior Alaska: Pollen and macrofossil analysis of a 27-meter core from the Isabella Basin (Fairbanks, Alaska). *Canadian Journal of Earth Sciences,* 11: 828-841.

1979 Tertiary and Quaternary environments: Historical background for an analysis of the Canadian insect fauna. *In* Danks, H. V. (ed.). *Canada and its Insect Fauna.* Memoirs of the Entomological Society of Canada, 108: 31-86.

1980 Tertiary land bridges and their climate: backdrop for development of the present Canadian insect fauna. *The Canadian Entomologist,* 112: 1089-1103.

MCANDREWS, J. H., BERTI, A. A., and NORRIS, G.

1973 *Key to the Quaternary Pollen and Spores of the Great Lakes Region.* Royal Ontario Museum, Life Sciences Miscellaneous Publication, 61 p.

MILLER, R. D., and DOBROVOLNY, ERNEST

1959 Surficial geology of Anchorage and vicinity, Alaska. *U.S. Geological Survey Bulletin* 1093, 128 p.

MORIYA, K.

1978 *Flora and Palynomorphs of Alaska:* Kodansha Publishing Co., Tokyo, 367 p. (in Japanese).

NAKAO, KINSHIRO, LAPERRIERE, J., and AGER, T. A.

1980 Climate changes in interior Alaska. *In* Nakao, K. (ed.), *Climate Changes in Interior Alaska,* Hokkaido University Department of Geophysics, Sapporo, 16-23.

NELSON, R. E.

1979 Quaternary Environments of the Arctic Slope of Alaska. M.S. thesis, University of Washington, Seattle, 141 p.

1982 Late Quaternary Environments of the Western Arctic Slope, Alaska. Ph.D. thesis, University of Washington, 90 p.

NOLTIMIER, H., and COLINVAUX, P. A.

1976 Geomagnetic excursion from Imuruk Lake, Alaska. *Nature,* 259: 197-200.

PARRISH, L.

1980 A Record of Holocene Climate Changes from St. George Island, Pribilofs, Alaska. *Ohio State University, Institute of Polar Studies Report* 75, 45 p.

PÉWÉ, T. L.

1975 Quaternary Geology of Alaska. *U.S. Geological Survey Professional Paper* 835, 145 p.

PORTER, S. C., PIERCE, K. L., and HAMILTON, T. D.

1983 Late Wisconsin mountain glaciation in the western United States. *In* Porter, S. C. (ed.), *Late-Quaternary Environments of the United States.* Vol. 1, *The Late Pleistocene.* University of Minnesota Press, Minneapolis, 71-111.

REGER, R. D., and UPDIKE, R. G.

1983 Upper Cook Inlet Region and the Matanuska Valley. *In* Péwé, T. L., and Reger, R. D. (eds.), *Guidebook to Permafrost and Quaternary Geology along the Richardson and Glenn Highways between Fairbanks and Anchorage, Alaska.* Fourth International Conference on Permafrost, July 18-22, 1983, Fairbanks, Alaska. Guidebook 1: 185-263.

REPENNING, C. A.

1983 New evidence for the age of the Gubik Formation Alaska North Slope. *Quaternary Research,* 19: 356-372.

RITCHIE, J. C.

1984 *Past and Present Vegetation of the Far Northwest of Canada.* University of Toronto Press, Toronto, 251 p.

RITCHIE, J. C., CWYNAR, L. C., and SPEAR, R. W.

1983 Evidence from north-west Canada for an early Holocene Milankovitch thermal maximum. *Nature,* 305 (5930): 126-128.

SCHWEGER, C. E.

1976 Late Quaternary paleoecology of the Onion Portage region, northwestern Alaska. Ph.D. dissertation, University of Alberta, Edmonton, 183 p.

1981 Chronology of Late Glacial events from the Tangle Lakes, Alaska Range, Alaska. *Arctic Anthropology,* 18: 97-101.

1982 Late Pleistocene vegetation of eastern Beringia: Pollen analysis of dated alluvium. *In* Hopkins, D. M., Matthews, J. V., Jr., Schweger, C. E., and Young, S. B. (eds.), *Paleoecology of Beringia.* Academic Press, New York, 95-112.

SELLMAN, P. V.

1967 *Geology of the USA CRREL Permafrost Tunnel, Fairbanks, Alaska.* U.S. Army Cold Regions Research & Engineering Laboratory, Hanover, N.H. Technical Report 199, 24 p.

SHACKLETON, J.

1982   Environmental Histories from Whitefish and Imuruk Lakes, Seward Peninsula, Alaska. *Ohio State University Institute of Polar Studies, Report* 76, 50 p.

SIRKIN, L. A., and TUTHILL, S. J.

1969   Late Pleistocene palynology and stratigraphy of Controller Bay region, Gulf of Alaska. *Etudes sur le Quaternaire dans le Monde,* Eighth Congress of the International Association for Quaternary Research, Paris, 197-208.

SIRKIN, L. A., TUTHILL, S. J., and CLAYTON, L. S.

1971   Late Pleistocene history of the lower Copper River Valley, Alaska. *Geological Society of America Abstracts with Programs,* 3 (7): 708.

VIERECK, L. A., and DYRNESS, C. T.

1980   *A Preliminary Classification System for Vegetation of Alaska.* U.S. Department of Agriculture Forest Service, Pacific Northwest Forest and Range Experiment Station. General Technical Report PN W-106, 38 p.

VIERECK, L. A., and LITTLE, E. L., JR.

1972   *Alaska Trees and Shrubs.* U.S. Department of Agriculture Forest Service. Handbook 410, 265 p.

1975   *Atlas of United States Trees.* Vol. 2, *Alaskan Trees and Common Shrubs.* U.S. Department of Agriculture Forest Service, Miscellaneous Publication 1293, 127 p.

WALKER, D. A., SHORT, S. K., ANDREWS, J. T., and WEBBER, P. J.

1981   Late Holocene pollen and present-day vegetation, Prudhoe Bay and Atigun River, Alaskan North Slope. *Arctic and Alpine Research,* 13 (2): 153-172.

WELSH, S. L.

1974   *Anderson's Flora of Alaska and Adjacent Parts of Canada.* Brigham Young University Press, Provo. 724 p.

WESTGATE, J. A.

1982   Discovery of a large-magnitude, late Pleistocene volcanic eruption in Alaska. *Science;* 218 4574 789-790.

WOLFE, J. A., and LEOPOLD, E. B.

1967   Neogene and early Quaternary vegetation of Northwestern North America and Northeastern Asia. *In* Hopkins, D. M. (ed.), *The Bering Land Bridge.* Stanford University Press, Stanford, 193-206.

QUATERNARY PALYNOLOGY OF MARINE SEDIMENTS IN THE NORTHEAST PACIFIC, NORTHWEST ATLANTIC, AND GULF OF MEXICO

LINDA E. HEUSSER
Lamont-Doherty Geological Observatory of
Columbia University
Palisades, New York 10964

Abstract

This review of Quaternary marine palynology of North America, a 30-year old discipline, summarizes published and unpublished data from the Gulf of Mexico, Gulf of California, northwest North Atlantic, and northeast North Pacific Oceans. Results from the last ten years confirm and expand earlier observations regarding marine pollen sedimentation and stratigraphy. Empirical studies of pollen distribution in surface sediments on the continental margin include ~600 samples. Pollen analyses from forty cores yield correlative marine and terrestrial environmental data from the past 30,000 years. Two cores cover the last 90,000 years and nine cores extend past 900,000 years. Several studies use new techniques, principal component analysis and regression equations, to synthesize and calibrate marine pollen data, or are multidisciplinary approaches which integrate marine palynology, geochemistry, micropaleontology, and sedimentology.

INTRODUCTION

In the 1950's-1960's, reports of abundant pollen in modern marine sediments and the recovery of numerous deep-sea cores (Fig. 1) stimulated North American Quaternary marine pollen research (Cross *et al.,* 1966; Traverse and Ginsburg, 1966; Stanley, 1965). Stratigraphic applications developed (Elsik, 1969; Groot and Groot, 1966); however, interest subsequently declined because of major differences in interpretations of marine pollen sedimentation (particulary pollen-vegetation relationships), difficulties in extracting pollen from marine sediments, and advantages of other marine microfossils (foraminifera, radiolaria, diatoms) as stratigraphic and climatic indicators.

Improved laboratory techniques for concentrating pollen, and development of global, detailed Quaternary marine stratigraphy (Shackleton and Opdyke, 1973) re-interested palynologists and other Quaternary scientists in obtaining lengthy, uninterrupted records of continental vegetation and climates from deep-sea cores (Heusser and Shackleton, 1979). Several continuous pollen records from marine cores extend through the Quaternary (Florer-Heusser, 1975), and new drilling techniques (hydraulic piston coring and long coring facilities) will provide more.

POLLEN DISTRIBUTION IN RECENT SEDIMENTS

As on land, pollen in marine sediments reflects vegetation, and physical, chemical and biological processes involved from initial pollen dispersal to final deposition. Fluvio-marine sedimentation is of primary importance, although aeolian transport is important along the coast (Brush and DeFries, 1981; Heusser, 1978b; Stanley, 1969).

Recent Pollen Distribution on the Shelf

Pollen is rapidly deposited in shallow coastal muds, ~50,000 grains/cc in eastern Canadian estuaries and in the Mississippi delta (Darrell and Hart, 1970; Groot, 1966; Heusser *et al.,* 1975a; Mudie, 1982). Coarse-grained sediments of high-energy environments contain variable amounts of pollen, usually less than the slope and upper rise (Heusser and Balsam, 1977; Stanley, 1969). Where water turbulence is minimal, as in deep basins and on the Great Bahama Bank, fine-grained sediment contains moderate to very high concentrations of pollen (Heusser, 1978a; Mudie, 1982; Pack, 1980; Traverse and Ginsburg, 1966).

Along with sedimentary factors (distance from source, deposition rate, particle size, and presence of carbonates and organic matter), Engelhardt (1963)

Figure 1. Piston-coring in the North Atlantic Ocean.

Recent Pollen Distribution on the Outer Continental Margin

Northwest Atlantic Ocean. Concentration ranges from 23,000/gm on the slope to <10 grains on abyssal plains (Heusser, 1983b; Mudie, 1982). Isopolls parallel the coastline (Fig. 2), are highest near rivers with large suspended loads, and low near northern tundra and under major current systems. In surface samples from a transect off southeastern United States (Fig. 3), pollen increases from 1,000/cc on the upper slope to >2,300/cc at a depth of 2,900m. At 4,400m, pollen decreases abruptly to <225/cc, a ten-fold difference related to the Western Boundary Undercurrent (Heusser and Balsam, in press).

Under the Western Boundary Undercurrent, ~25% of the pollen in surficial sediments is recycled (L. Heusser, unpub. data). At ~36°N (Fig. 3), reworked pollen and spores, including distinctive Permo-Carboniferous types, are minimal (<6%) in slope sediments. At 4,400 m, under the high-velocity core of the Western Boundary Undercurrent, redeposited pollen and spores increase to 44%, a change associated with long-distance transport (Heusser and Balsam, in press; Needham *et al.,* 1969).

On the slope and rise, distribution of diagnostic pollen types of eastern North American plant formations corresponds with onshore distribution of vegetation (Heusser, 1983b; Mudie, 1982). Synthesis of marine pollen data using factor analysis yields pollen assemblages (Fig. 4) that are broad equivalents of major vegetation groups. Pollen from oak *(Quercus)* and other temperate deciduous trees is most abundant near deciduous forests (the *Pinus-Quercus* factor). Hemlock *(Tsuga canadensis)* pollen is essentially limited to sediments off mixed hemlock-oak forests, and boreal forests are characterized at sea by significant amounts of spruce pollen (the *Picea-Sphagnum* factor). Pine *(Pinus)* pollen is concentrated near southern pine forests and is over-represented in deep-sea sediments with low pollen influx, i.e., off tundra or thousands of kilometers from shore (Heusser and Balsam, 1977; Heusser, 1983b). Hydraulic efficiency of the bisaccate grain accounts for pine pollen far from trees, as in Baffin Bay, mid Atlantic Ridge, and Great Bahama Bank sediments (Hopkins, 1950; Traverse and Ginsburg, 1966).

Northeast Pacific Ocean. Processes of pollen distribution and sedimentation on the Pacific slope and rise are like those of the western North Atlantic (Heusser and Balsam, 1977) (Fig. 1). Fine-grained, high organic hemipelagic sediments contain large amounts of well-preserved pollen. Maximum values

stresses the role of preservation in determining marine pollen concentration and composition. In the Gulf of Mexico, fragmented, poorly-preserved pollen correlates positively with abundant fungal spores and high bacteria counts in bay muds. Pollen, fungal spore and dinoflagellate distribution are related to fluvio-marine sedimentation in the Gulf of California (Cross *et al.,* 1966).

Close to shore in the Atlantic and Gulf of Mexico, marine pollen represents regional and local vegetation (Groot, 1966; Mudie, 1982; Thompson, 1972), modified by selective effects of fluvio-marine sedimentation (Darrell and Hart, 1970). Gulf estuaries and bays contain more pollen from local brackish water plants than shelf sediments, in which pollen represents regional southeastern vegetation (Engelhardt, 1963). Surface deposits near heavily-logged California forests have large quantities of pollen from successional vegetation [5000/cc of alder *(Alnus)*]; offshore, pollen from regional forests increases (Gardner, *et al.,* 1984; L. Heusser, unpub. data).

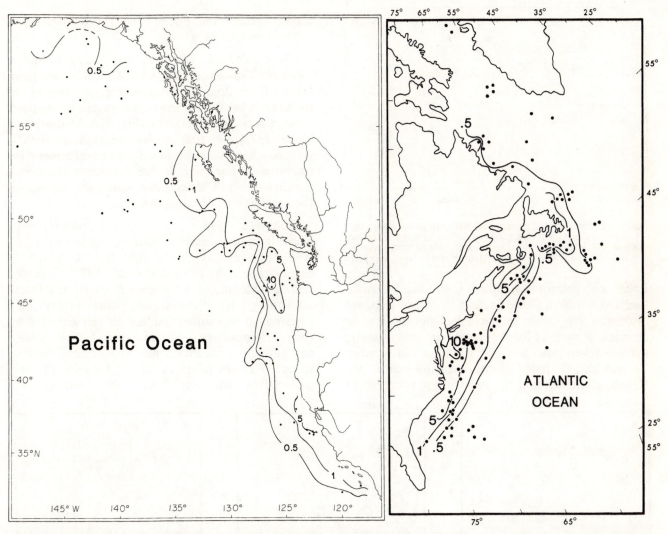

Figure 2. Left. Map of the northeast Pacific Ocean showing 1) pollen concentration (grains/cm³ x 10³) on the continental margin (exclusive of the shelf), and 2) location of coretops used in this data set, which include coretops cited by Heusser and Balsam (1977) and unpublished data. Right. Map of the northwest Atlantic Ocean showing 1) pollen concentration (grains/gm x 10³) on the slope, rise, and abyssal plains, and 2) location of coretops used in this data set (Heusser, 1983b; Mudie, 1982).

(>10,000/cc) coincide with the Columbia River plume; minimum values (1-100/cc) adjoin Alaskan herbaceous vegetation (Colinvaux, 1964; Heusser and Balsam, 1977; Sancetta, *et al.,* 1984).

Geographic distribution of individual marine pollen taxa, marine pollen assemblages (factors) and coastal northwest North American vegetation formations show good agreement (Fig. 5) (C. Heusser, this volume). Oak and herbs (Compositae and Chenopodiaceae) characterize marine sediments near arid southern California; oak and redwood *(Sequoia sempervirens)* are prominant off central California coastal redwoods and oak woodlands. North Pacific coast vegetation is clearly identified in marine deposits by pollen from diagnostic plants. Western hemlock *(Tsuga heterophylla),* spruce *(Picea),* Douglas fir *(Pseudotsuga),* and fir *(Abies)* pollen

characterize Pacific Coastal Forest; and alder, birch *(Betula),* sedge (Cyperaceae), grass (Gramineae), and *Sphagnum* identify Alaskan tundra (Heusser and Balsam, 1977; Gardner, *et al.,* 1984).

STRATIGRAPHIC PALYNOLOGY

Northeast Pacific Ocean

Regional and local marine pollen records are lengthy (>900,000 years) and detailed (Fig. 6; Table 1). Chronologic resolution between average sampling intervals of the last 150,000 years ranges from 1,500 to ~10 years (L. Heusser, unpub. data; 1983; Heusser and Florer, 1973; Heusser and Shackleton, 1979).

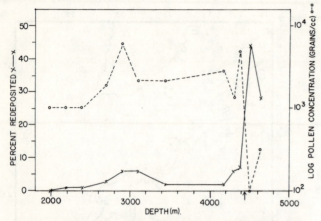

Figure 3. Distance plot of percent of redeposited pollen grains and pollen concentration (log scale) in surface samples from the northwest Atlantic Ocean. The transect extends from approximately 74° to 66° west longitude (Heusser and Balsam, in press).

Pollen assemblages near British Columbia, Washington, and northern Oregon reflect the broad response of coastal vegetation to glacial/nonglacial climatic changes, as well as local vegetational and climatic events–development of oak woodland in western Oregon and British Columbia during warm, dry Holocene climate, and rise of herbs following recent

logging and farming (Fig. 6; Table 1) (Florer-Heusser, 1975; Heusser, 1983a).

Cores TT63-13, Y6910-2, and Y7211-1. Pollen zones of ^{14}C-dated cores TT63-13 (Pacific Ocean) and HV-67 (Hoh Bog, Washington, Fig. 6, Table 1) are the same: a basal grass-sedge zone succeeded by pine, alder, and western hemlock (Fig. 7, 8; Heusser and Florer, 1973). Key stratigraphic events in this 16,000-yrs. sequence of equilibrium forest development are isochronous. For example, the distinctive alder peak terminates at a ^{14}C-dated ash horizon, that lies within the timeframe of Mazama ash.

From ~18,000 to ~110,000 yrs. B.P., pollen assemblages in marine and continental cores show correlative changes, as confirmed by ^{14}C and [18]O chronstratigraphy (Heusser, *et al.,* 1975b; Heusser and Shackleton, 1979). Pollen from climax forest alternates with successional plants (alder) and nonarboreal vegetation. Glacials are characterized by herbs (Compositae, Gramineae, and Cyperaceae), and subalpine mountain hemlock *(Tsuga mertensiana).* Western hemlock, alder, Douglas fir, and pollen from thermophilous vegetation (coastal

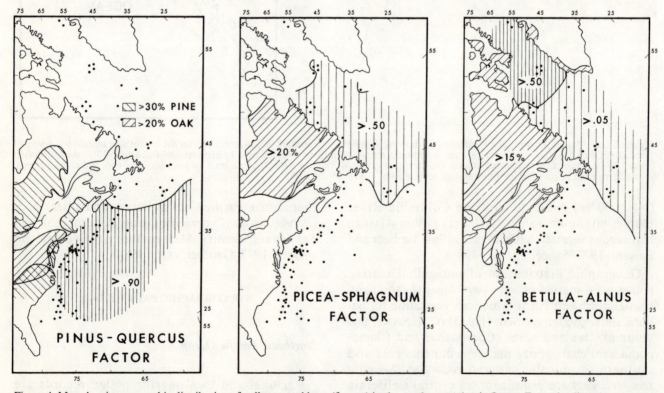

Figure 4. Map showing geographic distribution of pollen assemblages (factors) in the northwest Atlantic Ocean. Factor loadings (component scores) show the importance of each assemblage. The pine oak *(Pinus-Quercus)* factor accounts for 67% of the variance in the data; the spruce-*Sphagnum, (Picea-Sphagnum)* factor accounts for 19%, and the birch-alder *(Betula-Alnus)* factor accounts for 6%. Variables include: pine, spruce, hemlock, fir, oak, willow *(Salix),* birch, alder, grass, sedge, composites, ericads (Ericaceae), polypods (Polypodiaceae), lycopods (Lycopodiaceae), and *Sphagnum.* Spatial distributions of relative abundance of principal pollen types in eastern North America are shown for comparison (Heusser, 1983b).

Table 1.
Core locations and depths, and age of Quaternary samples discussed in text.
(Numbers 2-10 refer to core locations shown in Figure 6.)
Unpublished data of L. E. Heusser is indicated by a *.

Map Number	Core Identification	Latitude (N)	Longitude (W)	Water Depth (m)	Age (kyr)	Reference
2	Hoh Bog	47°48′	124°12′		0-15.6	Heusser and Florer, 1973
3	Quarry Short Core	48°34′	123°31′	200	0-00.2	Heusser, 1983a
	LaCiencia Cores	48°34′	123°30′	238	0-12.0	Heusser, 1983a
4	V20-75	48°12′	126°10′	1798	0-18.0	*
5	TT63-13	47°09′	125°17′	1502	0-15.0	Heusser and Florer, 1973
6	Y6705-7	46°04′	126°38′	2730	0-30.0	Florer-Heusser, 1975
7	BB326-36	46°29′	124°42′	658	0-08.0	Florer, 1973
8	Y6609-7	43°12′	126°44′	2911	0-32.0	Florer-Heusser, 1975
9	DSDP18-175	44°50′	125°15′	1999	Quaternary	Florer-Heusser, 1975; Musich, 1973
10	Y72-11-1	43°15′	126°22′	2913	4-130.0	Heusser and Shackleton, 1979
	V-13	70°45′	150°45′	<20	Recent	Nelson, in press
	V-17	70°35′	150°45′	<20	Recent	Nelson, in press
	PB-2	70°29′	148°20′		Quaternary	Nelson, in press
	V-12	70°20′	148°18′	<20	Recent	Nelson, in press
	V-1	70°16′	148°20′	<20	Recent	Nelson, in press
	TT-020 039 109	67°51′	169°10′	52	Quaternary	Colinvaux, written commun. 1974
	TT-020 026 057	67°30′	165°54′	41	Quaternary	Colinvaux, written commun. 1974
	Kotzebue Sound	67°30′	167°52′	45	Quaternary	Colinvaux, 1964
	TT-042 108 240	63°01′	173°02′	67	Quaternary	Colinvaux, written commun. 1974
	TT-042 020 048	60°55′	176°08′	52	Quaternary	Colinvaux, written commun. 1974
	RC14-121	54°51′	170°41′	2530	0-75.0	*
	TT53-21	46°13′	125°23′	1928	0-15.	*
	DSDP18-176	45°56′	124°37′	193	Quaternary	Florer-Heusser, 1975; Musich, 1973
	DSDP18-174	44°54′	126°21′	2815	Quaternary	Musich, 1973
	Y77 10A 26	44°50′	125°04′	1822	2-04.0	*
	Y6910-2	41°16′	127°01′	2615	0-110.0	Heusser et al., 1975
	DSDP18-173	39°58′	125°27′	2927	Quaternary	Musich, 1973
	CL-73-4	39°	122°30′		0-120.0	Adam, 1979
	V1 80 P3	38°26′	123°47′	1600	0-18.0	Gardner, et al, 1983
	Y71 10 117	34°14′	120°04′	570	0-12.0	Heusser, 1978a
	DSDP63-467	33°51′	120°46′	2127	Quaternary	Ballog and Malloy, 1981; Heusser, 1981
	DSDP64-480	27°54′	111°39′	655	Quaternary	Byrne, 1982; Heusser, 1982a
	DSDP64-479	27°51′	111°38′	757	Quaternary	Sirkin, 1982
	DSDP64-474	22°58′	108°59′	3033	Quaternary	Sirkin, 1982
	DSDP66-493	16°23′	98°56′	655	Quaternary	Fournier, 1981
	V18-338	15°08′	95°43′	5253	0-8.0	Habib et al, 1970
	V18-339	15°10′	95°37′	5369	0-8.0	Habib et al, 1970
	EN32-PC16	26°57′	91°21′	2280	0-2.9	*
	A185-35	24°34′	92°37′	3626	10.9	Stanley, 1966
	V3-126	23°45′	92°27′	3484	Quaternary	Stanley, 1966
	Gulf of Mexico				Quaternary	Elsik, 1969
	C8-1	34°38′	75°39′	165	0-5.0	Stanley, 1966
	V24-1	36°03′	72°23′	3942	0-10.8	Balsam and Heusser, 1976
	V26-176	36°30′	73°30′	3012	0-10.8	Balsam and Heusser, 1976
	M28	37°	76°		10-15	Harrison et al, 1965
	A164-1	38°44′	71°23′	2770	Holocene	Stanley, 1966
	A164-61	39°32′	68°47′	2725	Quaternary	*
	V4-1	38°53′	70°55′	2867	Quaternary	Balsam et al, 1977
	Hudson River	40°48′	74°00′	15	0-10.0	Weiss, 1974
	Texas Tower No. 3	40°58′	69°23′	20	Quaternary	Groot and Groot, 1964
	Texas Tower No. 12	40°58′	69°23′	30	Quaternary	Livingstone, 1964
	KN 10-1	42°25′	70°35′	81	0-18.9	Pack, 1980
	V17-178	43°23′	54°52′	4006	0-20.0	*
	GC79-11-01	44°42′	63°39′	60	0-8.0	Miller et al, 1982
	Linwood	45°50′	61°35′	20	0-10.0	Schafer and Mudie, 1980
	Pomquet	45°50′	61°45′	20	0-2.0	Schafer and Mudie, 1980
	Corehole 2	45°10′	52°50′	~420	Pleistocene	Williams and Brideaux, 1975
	Corehole 23	45°40′	53°30′	~340	Pleistocene	Williams and Brideaux, 1975
	77 034 13	49°30′	48°48′	2000	Quaternary	*
	BP-103	49°50′	51°14′	290	Quaternary	Piper et al, 1978
	Core 11,12	54°38′	56°13′	495	0-20.0	Vilks and Mudie, 1978

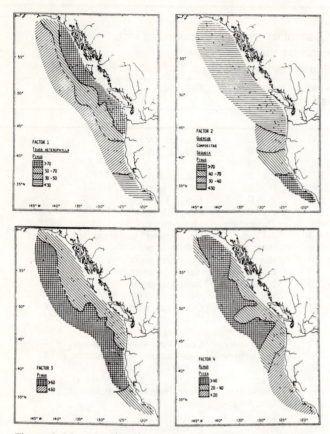

Figure 5. Map showing the geographic distribution of pollen assemblages (factors) in the northeast Pacific Ocean. Factor loadings (component scores) indicate the importance of each assemblage. Factor 1, the western hemlock-pine factor *(Tsuga heterophylla-Pinus)* accounts for 39% of the variance; factor 2, the oak-composite-redwood-pine factor *(Quercus-Compositae-Sequoia-Pinus)* accounts for 14%; factor 3, the pine factor accounts for 32%; and factor 4, the alder-pine factor *(Alnus-Pinus)* accounts for 13%. Variables included in the Q-mode factor analysis are pine, spruce, fir, western hemlock, Douglas fir, alder, grass, sedge, oak, coastal redwood, composites, and chenopods (Heusser and Balsam, 1977).

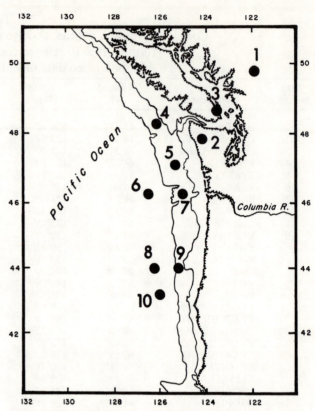

Figure 6. Map showing core locations discussed in the text from the northeast Pacific Ocean, Washington, and British Columbia: 2. Hoh Bog, 3. Quarry Short Core and LaCiencia cores, Saanich Inlet, Vancouver, British Columbia. 4. V20-75, 5. TT63-13, 6. Y6705-7, 7. BB326-36, 8. Y6609-7, 9. DSDP18-175, and 10. Y7211-1. Numbers on the map correspond to core numbers in Table 1. (Florer-Heusser, 1973; Heusser, 1983a; Heusser and Florer, 1973; Heusser and Shackleton, 1979; Musich, 1973).

redwood and oak) are not well-represented. Fluctuations in temperate conifer (western hemlock factor) and nonarboreal (grass factor) pollen mirror global climatic oscillations in isotope stage 3 and 4 (~early and mid-glacial) (Figs. 8, 9). Progressively increasing amounts of western hemlock, oak, Douglas fir, and redwood characterize basal sediments. Pollen assemblages of the last interglacial (substage 5e) are comparable with Holocene assemblages. The distinctive alder peak (Figs. 8 and 9) precedes the western hemlock rise and leads the ^{18}O minimum of substage 5e in exactly the same manner as the alder peak precedes the rise of western hemlock and ^{18}O minimum in stage 1.

Relating marine and continental interglacial pollen assemblages is complicated by Pacific Northwest Quaternary stratigraphy (particularly units deposited >40,000-50,000 yrs. B.P.) which is not as firmly

established as in Europe or central North America (S. Coleman and K. Pierce, written comm., 1984). On land, stratigraphic relations are usually obscured and numerically datable materials are rare; however, pollen spectra from western Washington interglacial sedimentary units [the Whidbey Formation (Heusser and Heusser, 1981), and a unit tentatively assigned to the last interglacial from Aberdeen (L. Heusser, unpub. data)] appear correlative with pollen zone 9 (substage 5e) in core Y7211-1.

Regression equations relating modern climatic variables and pollen data from the northwest coast to fossil pollen data from cores TT63-13, V20-75, and Y7211-1, produce temperature estimates in which general trends parallel those onshore (L. Heusser, unpub. data; Heusser, *et al.,* 1980; Mathewes and Heusser, 1982). Estimated temperatures from the last interglacial are the same as Holocene temperature maxima, and moderately cool glacial conditions precede minimal temperatures ~18,000-23,000 yrs. B.P.

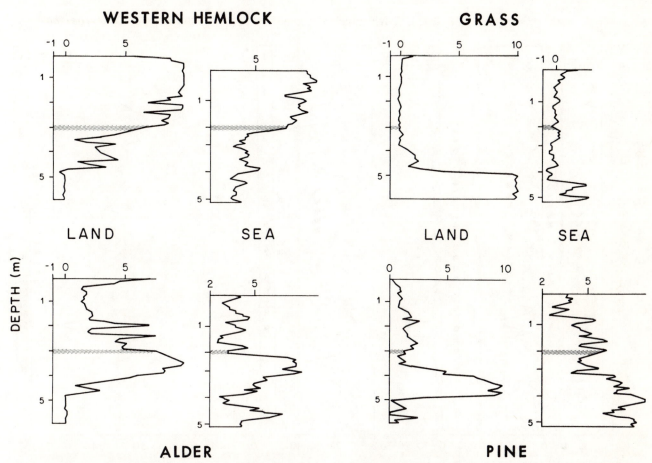

WESTERN HEMLOCK

GRASS

DEPTH (m)

LAND   SEA   LAND   SEA

ALDER

PINE

Figure 7. Depth plots of factor loadings (component scores) of late-glacial and postglacial pollen assemblages of the Pacific Northwest coast of North America in core TT63-13 from the continental slope of the northeast Pacific and a core from Hoh Bog, Washington. In each of the four factors (identified by the dominant component in each factor), data from the land core are on the left and data from the marine core are on the right. Chronology is based on 17 ^{14}C dates. The dotted line at 3m depth shows the position of an ash horizon (dated about 6,600-6,700 yrs. B.P., within the time of Mazama Ash deposition) in both cores (Heusser and Florer, 1973).

Joint pollen and radiolaria analyses from core Y7211-1 yield correlative continental and oceanographic paleoclimatic data (Pisias, *et al.,* 1981). Comparison of four pollen and seven radiolaria groups (representing modern vegetation formations and North Pacific water masses, respectively), shows temperate coastal forest highly correlated with subtropical and transition zone faunas and absence of subpolar fauna. A grass (tundra) pollen assemblage correlates with a subpolar radiolarian assemblage. Time series analysis indicates that changes in these groups are not all synchronous; changes in temperate flora follow those in transition fauna, while abundance peaks in tundra pollen and subpolar radiolarian assemblages are synchronous. Comparison with oxygen isotope data from the same sediment samples as the pollen and radiolarian data, indicates variations in these marine and land records can be associated with global climate changes at ~41,000-year periods, the period of the orbital tilt cycle.

DSDP Sites 18-174, 18-175, and 18-176. Pollen spectra from these sites (Fig. 6, Table 1) are similar to pollen spectra in other northeast Pacific cores. In the last 900,000 years, temperate coastal forest assemblages alternate with nonarboreal and subalpine assemblages, punctuated by distinctive alder peaks. Relating continental pollen zones and geologic-climatic units is hindered by lack of detailed chronostratigraphy in the DSDP cores and uncertainties regarding Pacific Northwest Quaternary stratigraphy. Marine and continental paleoclimates are broadly correlated. Pollen representing cool onshore conditions is associated with foraminifera indicating low sea surface temperatures (SST), and pollen from temperate continental environments is associated with higher SST (Florer-Heusser, 1975; Musich, 1973).

DSDP Sites 18-173 and 63-467. Quaternary pollen data in DSDP cores off California (Table 1) are less detailed than in DSDP cores to the north. Changes in the southernmost extension of lowland Pacific

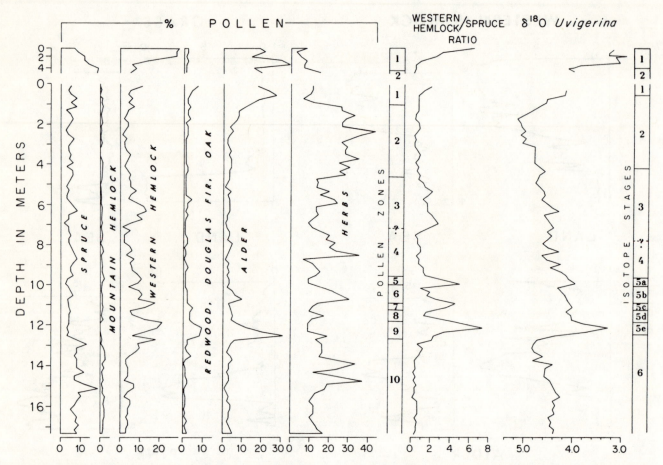

Figure 8. Depth plot of pollen and oxygen isotope data from marine cores Y7211-1 (bottom) and TT63-13 (top) from the northeast Pacific Ocean. (Core locations are shown in Figure 6 and described in Table 1). Pollen percentages of selected pollen types, pollen zones, and the western hemlock/spruce pollen ratio are on the left; oxygen isotope ^{18}O ratios and oxygen isotope stages are on the right. Pollen and oxygen isotope data from core TT63-13 were averaged to approximately the same time interval as represented by data from core Y7211-1 (Heusser and Shackleton, 1979).

Coastal Forest are monitored in DSDP18-173 (Musich, 1973), and pollen from DSDP63-467 shows subtle differences in southern California coastal sage scrub, chaparral, oak-grassland, and oak woodland (Ballog and Malloy, 1981; Heusser, 1981).

Core V1-80-P3. This 18,000-year record contains pollen from interior oak woodlands and coastal redwood, mixed evergreen, and montane conifer-dominated forests (Heusser, 1983c). Depth plots of selected faunal and floral data are shown in Fig. 10 (Table 1) (Gardner *et al,* 1983; 1984; Heusser *et al,* 1981). A subpolar marine fauna and a montane pollen assemblage (dominated by pine and small amounts of other conifers) suggest low ocean and land temperatures from 18,000 to 14,000 yrs. B.P. An increase in transition and upwelling foraminifera, coccoliths, and diatoms indicates subsequent northward movement of warmer water and upwelling. On shore, coastal redwoods (presently restricted to foggy, warm-temperate areas near upwelling areas) and oak woodlands replace pine-dominated conifers at the

same time. The abrupt rise in *Sequoia* pollen corresponds with increased SST (indicated by subtropical/temperate diatoms).

Pollen assemblages of cores V1-80-P3, M-1 (from a small marsh near cypress-pine and coastal redwood forests), and CL73-4 (from a large lake surrounded by pine-oak woodlands) are correlative (Figs. 10, 11) (Adam, 1979; Heusser, 1982b). Montane forest pollen (pine and other conifers) dominates from 23,000 to 14,000 yrs. B.P. Oak then increases in importance, rising rapidly at Clear Lake (CL73-4) and more gradually on the coast (M-1). Rapid expansion of the *Sequoia-Quercus* assemblage is ^{14}C-dated at 8,000 yrs. B.P. in both land and marine cores.

Core Y71-10-117P. Varved sediments of Santa Barbara Basin in the southern California Borderland contain undisturbed records of the past 12,000 years (Heusser, 1978a) (Table 1). Upland conifers are replaced by lowland communities, with optimal oak and composite (Asteraceae) development ~5,700 yrs. B.P. In the last few thousand years, chaparral and

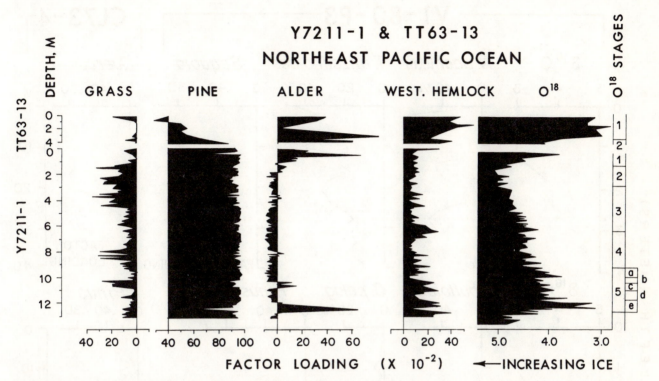

Figure 9. Depth plot of pollen factor loadings and oxygen isotope stratigraphy ^{18}0 of marine cores Y7211-1 (bottom) and TT63-13 (top) from the northeast Pacific Ocean. (Core locations are shown in Figure 6 and described in Table 1). The four pollen factors on the left (grass, pine, alder, and western hemlock) are the same factors as in Figure 7; however, in this figure, pollen and isotope data points from core TT63-13 are averaged to the same approximate time interval as data points from core Y7211-1 for ease of comparing data from the two cores (Heusser and Shackleton, 1979).

coastal sage scrub dominate. Radiolaria show major changes in marine environments and atmospheric circulation of southern California during the past 8,000 years (Pisias, 1978, 1979). From ~5,400 yrs. B.P. to 2,000 yrs. B.P., cool surface water coincides with warm, dry onshore climate. Before 5,400 yrs. B.P., warm SSTs correspond with mesic conditions in coastal California. These relationships are consistent with present-day climatic patterns: warm sea surface temperatures are associated with high precipitation, and cool surface waters with dry conditions on land.

DSDP Site 66-493, and cores V18-338 and V18-339. On the Mexican continental margin, Quaternary pollen spectra from Site 493 are distinguished by large numbers of fern spores, oak and alder (Fournier, 1981) (Table 1). Diminished Holocene precipitation and stream activity are inferred from decreased pollen diversity. No specific correlations are made between Quaternary pollen and marine microfossils, but the entire Miocene-Quaternary record suggests that climatic and tectonic events affect land floras instantaneously and are delayed in marine biota.

Using pollen diversity and oak, alder, and fir abundance, Habib *et al,* (1970) identify four Holocene climatic events in cores V18-338 and V18-339 from the Middle America Trench (Table 1).

Climate of the basal zone (dominated by pine) varies from cool dry to cool moist. Warm, dry climate of the oak, alder, *Hedyosmum* zone (6,000 to 3,500 yrs. B.P.) precedes the driest interval (pine oak zone). The last 2,000 years (pine, oak, alder zone) are interpreted as moist. Pollen zones are correlated with pollen zones and climatic events elsewhere in the world; no correlations are made with local marine environments.

Gulf of California

Pollen in samples from DSDP sites 474, 479, and 480 derives from vegetation of northwest Mexico: low-elevation subtropical desert and mesquite grassland, higher elevation pine-oak forests, and boreal forests of pine, fir, and alder on moist mountain heights. South of 40°N, tropical forests and mangrove swamps are increasingly important (Table 1) (Byrne, 1982; Heusser, 1982a; Sirkin, 1982).

In Guaymas Basin (Site 480), Pleistocene pollen spectra show minor changes in upland and lowland forests, savannahs, and coastal wetlands and saltpans (Sirkin, 1982). Site 479 hydraulic piston core samples show variations in concentration and percentages of pollen from woodland and shrub-herb communities

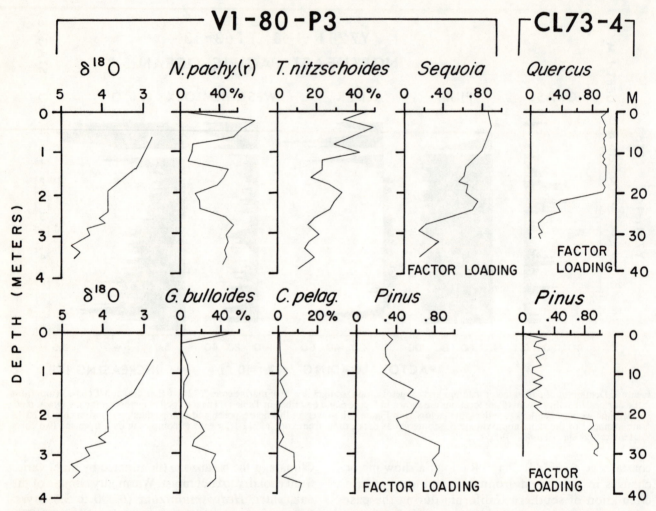

Figure 10. Depth plots of oxygen isotope stratigraphy and pollen, forminifera *(Neogloboguadrina pachyderma* right and *Globigerina bulloides),* diatom *(Thalassionema nitzschoides)* and coccolith *(Coccolithus pelagicus)* data from marine core V1-80-P3 taken in the northeast Pacific Ocean off the mouth of the Russian River, and pollen data from core CL73-4, Clear Lake, California (Adam, 1979; Gardner *et al.,* 1983a, 1984; Heusser *et al.,* 1981). Marine microfossil data are shown in percentages; pollen data are factor loadings of pollen factors (assemblages) dominated by pine *(Pinus* factor) and by coastal redwood and oak *(Sequoia* and *Quercus).* (See Table 1 for core locations.)

(Heusser, 1982a). Large fluctuations in chenopod-amaranth (Chenopodiaceae/Amaranthaceae) and composite assemblages occur in the Holocene. Spruce, an abundance of *Juniperus,* and variations in oak and pine pollen characterize glacial sediments (Byrne, 1982; Heusser, 1982a). Apparently, desert, subtropical woodland, and cool, temperate woody species grew near the central Gulf of California during the last glacial maximum. Pollen assemblages in older sediments show fluctuations comparable to those of the past 24,000 years.

Gulf of Mexico

In one of the few Quaternary publications from this region, Elsik (1969) interprets palynomorph fluctua-

tions in two late Neogene cores from the northern shelf (Table 1) in terms of Pleistocene climatic and sedimentologic changes. Increased conifers show cooling, and increased alder shows warming. In the southern gulf (Table 1), reworked pollen forms a stratigraphic marker identifying glacial intervals (Stanley, 1966).

Core EN32-PC6. High-resolution pollen data from the continental slope of the northern Gulf (Orca Basin) document sedimentary and climatic changes during deglaciation (Table 1, Fig. 12) (L. Heusser, unpub. data). Vegetation drained by the Mississippi

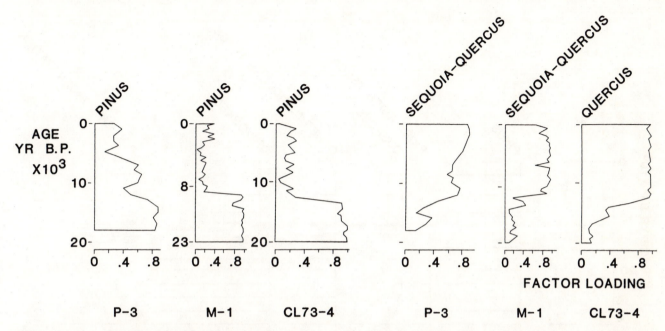

Figure 11. Comparison of pine *(Pinus)* and redwood-oak *(Sequoia-Quercus)* pollen assemblages (factors) in marine core P-3 (V1 80 P3), core M-1 from McCrary's Marsh, California, and core CL73-4 from Clear Lake, California (Adam, 1979; Gardner *et al.,* 1983; 1984; Heusser, 1982b). Pine-dominated factors are on the left; redwood-oak on the right. Age in thousands of years B.P. is on the Y axis; factor loadings are on x axes. Core locations are given in Table 1.

and other rivers flowing into the northern Gulf is the main pollen source. Local coastal wind-blown pollen is probably less important (Heusser, 1978b; Stanley, 1969). The Mississippi drains conifer forests, grasslands, and deciduous forests, and smaller southeastern rivers flow through eastern deciduous forests and pine savannas (Bryant and Holloway, and Delcourt and Delcourt, this volume). Glacial vegetation may differ from modern plant associations (Delcourt and Delcourt, 1983). On the Gulf coast, warm-temperate evergreen forests were relatively stable. To the north, boundaries between temperate and conifer forests, prairie, and tundra shifted in response to changing ice sheets and climate. In central Texas, northern, western, and eastern vegetational elements merged, forming open woodland deciduous forest with conifers (Bryant, 1977).

Orca Basin glacial sediments are characterized by conifer pollen, reworked pollen and spores, and low concentration of contemporaneous pollen (Fig. 12). Pine and inaperturate conifers form >50% of glacial pollen spectra, spruce contributes 10%, and fir and hemlock occur sporadically. Deciduous pollen [oak, birch, alder and elm *((Ulmus)]* is low, fern spores are numerous, and chenopod pollen is rare. These pollen spectra apparently represent southeastern pine and floodplain forests, as well as boreal plant associations.

The deglacial pollen assemblage contemporaneous with lightening of ^{18}O is distinctive (Leventer, *et al*, 1982). Pine decreases to less than a third of full-glacial values, other conifer pollen is low, and oak, ash *(Fraxinus),* sweetgum *(Liquidambar),* elm, hickory *(Carya),* tupleo *(Nyssa),* and tuilp tree *(Liriodendron)* increase in diversity and abundance. Herbs, primarily Compositae, increase and fern spores decrease. In the ^{18}O spike, concentration of contemporaneous grains triples and reworked pollen decreases by half (Fig. 12). During deglaciation, pollen derives from vegetation adjusting to processes associated with climatic changes. Higher temperature and precipitation are reflected in the increased deciduous tree pollen, including taxa from warm temperate forests. Contraction of boreal forests is indicated by lower input of pollen from cool, moist environments (spruce, sedges, and fern spores).

Late-Holocene pollen (pine, *Taxodium*-type, oak and other deciduous tree types) apparently reflects development of modern vegetation of southeastern United States. Grass and chenopod are more important, and spruce, fir, and hemlock are virtually absent. Pollen concentration gradually reaches modern values.

An overall decrease in reworked pollen indicates changes in Gulf pollen sedimentation. In surficial sediment from Orca Basin, recycled pollen is five times higher than in rivers entering the Gulf (Engelhardt, 1963). This suggests that some recycled pollen is transported by the Mississippi, by other rivers not studied by Engelhardt, and/or by redeposition of

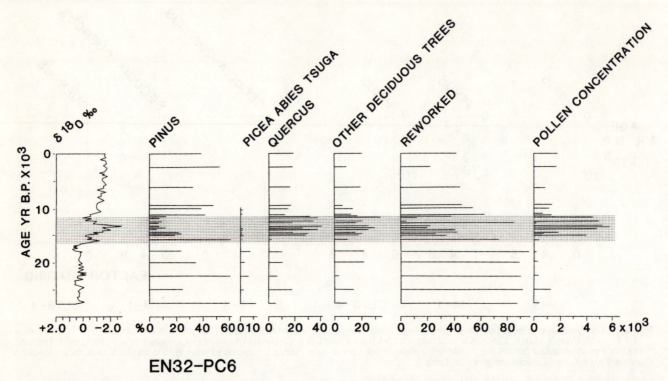

EN32-PC6

Figure 12. Depth plot of oxygen isotope stratigraphy and selected pollen data from marine core EN32-PC6, taken from Orca Basin in the Gulf of Mexico (Table 1). Percentages of conifer and deciduous pollen types illustrated in this pollen diagram are based on the total number of contemporaneous pollen. Reworked pollen percentages are calculated outside the basic pollen sum. Pollen concentration (grains/cc) refers to contemporaneous pollen. Age in thousands of years is shown on the Y axis. The stippled bar identifies the meltwater spike, fluctuations in the ^{18}O curve which reflect meltwater discharge from the southern margin of the Laurentide ice sheet during the last deglaciation (Leventer et al., 1982).

sediments upslope in Orca Basin or elsewhere in the Gulf.

Recycled pollen in glacial samples probably derives from glacial erosion and/or reworking of older sediment in the Mississippi drainage or Gulf of Mexico. The relatively small concentration of redeposited pollen in the ^{18}O spike suggests rapid deposition of contemporaneous terrigenous sediment under conditions of increased rainfall plus increased flow of the Mississippi River, as well as decreased erosion of older continental and marine sediments. High concentrations of contemporaneous pollen suggest an actual increase in non-recycled pollen due to increased pollen production of expanded temperate vegetation and/or more efficient fluvial pollen transport.

Chukchi, Beaufort, and Bering Seas

Two marine surface samples dominated by grass, sedge, and birch (TT20 039 PC 109, Table 1) resemble one Arctic Alaskan surface sample and differ from another willow (Salix)-dominated sample (Nelson, in press). In the Beaufort Sea, >50% of pollen in core

tops is sedge; grass, alder and birch account for 5-15%, spruce for <10%, and reworked pollen from 30-60%. These pollen spectra reflect local vegetation and fluvio-marine transport. Older samples show climatic change. High alder, birch, and Artemisia, suggest Arctic Coastal Plain vegetation growing in a climate warmer than present, probably a former interglacial.

Colinvaux (1964) published the first pollen diagram from an Alaskan marine core (Table 1). Six samples dominated by birch, sedge, and grass assigned to the last glacial underlie a postglacial sample with high percentages of alder. From pollen assemblages and submarine topography, Colinvaux concludes that rivers carried pollen from grass and tussock tundra in which birch, and later alder, was important. Pollen in a nearby core shows gradual spruce development <6,000 yrs. B.P. (Table 1) (P. Colinvaux, written commun., 1974).

Other Bering Sea cores show similar results (P. Colinvaux, written commun., 1974; Table 1). Birch, sedge, and grass dominate older samples, and alder is prominent in younger samples. Spruce varies between 3% to 22% (Table 1). Fluctuations in pollen, spores, and reworked pollen are interpreted as sedimentary changes. Sediments with high spruce and no hemlock probably originate in the Kuskokwim drainage; allocthonous pollen is reworked from shelf sediments at times of low sea level.

Core RC14-121. Integrated pollen, diatom, oxygen isotope, and clay mineral analyses from a 14m ^{14}C-dated core document four major climatic and oceanographic events during the past 75,000 years (Sancetta *et al.,* 1984) (Table 1 and Fig. 13). In the oldest interval when extensive sea ice covered cold, low-salinity waters, mesic, herbaceous tundra-like vegetation (represented by the moss-sedge factor) covered exposed shelf. Reworked pollen in Unit 4, derived from eroded marine or continental sediments (including loess), reflects increased aeolian and/or fluviomarine transport.

Mild interstadial climate and higher sea level are recorded in basal samples of Unit 3. Herbaceous tundra grew on shore, with shrub tundra or deciduous scrub-forest on high ground. Marine waters show high seasonal variability. In the following stadial, characterized by cold surface waters and greater winter sea ice, increased moss spores suggest expansion of wet tundra vegetation onshore. Two unusual events in the middle part of Unit 3, a marked decrease in ^{18}0 and increase in clay minerals from the Aleutian Islands, may relate to continental runoff and seismic activity, respectively.

During the late glacial, maximum sea ice occurred from ~18,500 to 16,000 yrs. B.P., when scattered tundra-like vegetation overlying a high permafrost table covered exposed continental shelf. Strong winds and alpine glaciation eroded and transported large amounts of reworked pollen to the marine environment. Between 12,800 and 11,900 yrs. B.P., isotopic ratios reflect a major meltwater event.

Surface water temperature and salinity, and pollen of birch and alder increase during the Holocene. Wet and mesic coastal tundra is minimal, and shrub-tundra or scrub forest expand on coastal uplands. Regional vegetation inferred from marine pollen data correlates with local Alaskan vegetational reconstructions (Ager, 1982).

Northwest Atlantic Ocean

Most Quaternary pollen records from the northwest Atlantic are short (<30,000 yrs. B.P.). Pleistocene pollen spectra are identified in cores off northeastern Newfoundland, Hudson Canyon, and southeastern United States, and pre-Wisconsin pollen

assemblages occur in marine sediments on the southern Atlantic Coastal Plain (Tables 1 and 2) (Balsam, *et al.,* 1977; Cronin *et al.,* 1981; Needham *et al.,* 1969; Piper, *et al.,* 1978; Stanley, 1966).

Pollen and ostracods from outcrops correlate continental and marine environments of five interglacials (Cronin *et al.,* 1981). These Pleistocene assemblages, characterized by oak and pine with less hickory, sweet gum, and black gum, resemble Holocene pollen assemblages from the Atlantic Ocean, Virginia, North Carolina, and South Carolina, (Balsam and Heusser, 1976; Stanley, 1966).

Cores V26-176 and V24-1. East of North Carolina on the continental margin, pollen and foraminifera yield correlative marine (SSTs) and continental (vegetation) paleoclimatic evidence (Table 1) (Balsam and Heusser, 1976). SST increased ~12,000 yrs. B.P., ~2,000 years before temperatures rose on land (temperate deciduous forests replacing boreal forest; Table 2). Estimated SSTs and vegetational changes appear more concordant in younger sediments; cooling of coastal waters ~4,000 yrs. B.P. precedes cooling on land by ~500 years (shown by increased pine and decreased oak).

Cores M28 ,A164-1, A164-61, and V4-1. Pollen stratigraphy of ^{14}C-dated peats cored in Chesapeake Bay (Harrison *et al.,* 1965) (Tables 1 and 2) correlates with pollen stratigraphy in cores V26-176 and V24-1. Late-glacial conifer pollen (pine, spruce, and fir) is replaced by oak ~10,000 yrs. B.P. Chronology in a deeper-water core differs slightly. Stanley (1966) places the base of a pine-oak pollen assemblage at ~5,000 yrs. B.P. (Table 2). Apparent differences in ages of Holocene marine pollen zones off southern United States may be artifacts of ^{14}C dating, or may reflect local vegetational differences.

On the slope and rise, correlations of marine and continental pollen zones becomes more difficult as chronostratigraphic control and sampling density decrease (Table 1). Holocene oak assemblages in cores from the continental marine generally correlate with the C-zone, uppermost pine assemblages with the B-zone, and youngest spruce assemblages with the A-zone (Balsam *et al.,* 1977; Stanley, 1966).

Hudson River Cores. A paleoecologic study of estuarine sediments uses late-Pleistocene marine pollen stratigraphy to correlate marine faunal (foraminifera) zones in six cores (Tables 1 and 2) (Weiss, 1984). Hudson River pollen assemblage zones are isochronous with standard northeastern United States pollen zones; *i.e.,* the pine zone succeeds the spruce-fir zone at 10,000 yrs. B.P., and

Correlation table of marine cores from the Northwest Atlantic.
Radiocarbon dated levels are indicated by a *.
Core locations are given in Table 1.

Age (kyr)	C8-1 Stanley, 1966	V26-176 / V24-1 Balsam & Heusser, 1976	M-28 Harrison et al., 1965	Hudson River 1-5 Weiss, 1974	KN10-1 Pack, 1980	V17 178 Heusser unpub.	GC79-111-1 Miller et al., 1982	St. George's Bay Schafer & Mudie, 1980	Labrador 11-12 Vilks & Mudie, 1978
0	*			C-3			C3-C	C3-C	birch
				oak	hemlock		C3-B	C-3	alder
1				pine	oak		birch		
		pine		hemlock	birch		fir	hemlock	spruce
2	pine	oak							
	oak						C3-a	oak	
3		*		C-2			birch	beech	
				oak	oak		fir		spruce
4				pine	pine	spruce			
				hickory			C2		
5	*						hemlock	*	
				C-1	hemlock				
6		oak		oak	oak		C1	hemlock	
		hemlock		pine			mixed forest	oak	
7		pine		hemlock					
				*					
				B-2	pine	pine	woodland		birch
8	*			pine-oak	spruce-oak	hemlock oak	marsh		
9				B-1					
			oak	pine	pine				
10			*						
		pine	pine	A-4		spruce			sedge
11				A-3		*Sphagnum*			shrub
		hemlock	birch	pine	pine				tundra
12		*							
		spruce		spruce-oak					
		fir		A1-2	spruce				
13			pine	pine	* fir				*
			spruce	spruce	birch				
14				fir					
		pine			pine				
		spruce			spruce				
15		fir	*		fir				
16									

RC 14-121

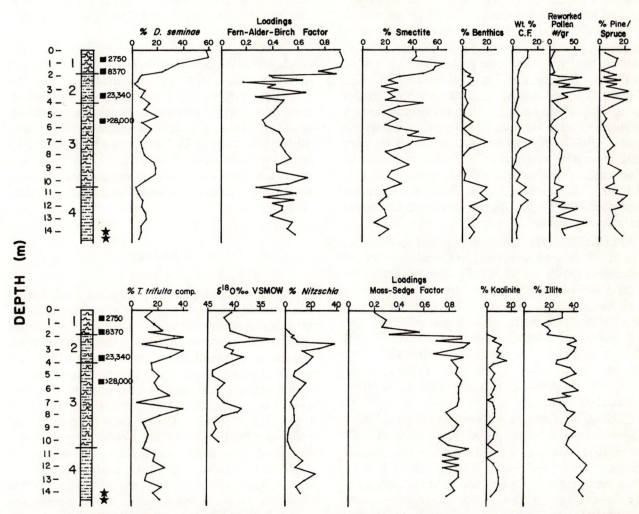

Figure 13. Depth plots of significant components of core RC 14-121 from the Bering Sea include percentages of diatoms *(Denticulopsis seminae, Thalassiosira trifulta, Nitzschia* spp., and benthics), factor loadings of pollen components (factors), pollen concentration (#/gm), pine/spruce pollen ratio, percentages of clay minerals (smectite, kaolinite, and illite), weight percent coarse fraction, and oxygen ^{18}O ratios from diatoms. Chronologic control is provided by radiocarbon dates, shown on the left, and by correlation with diatom stratigraphy of the north Pacific Ocean (Sancetta *et al.,* 1984). Core location is given in Table 1.

the base of the oak-pine-hemlock zone is dated 8,000 yrs. B.P.

Cores TT-3, TT-11, and TT-12. Samples from Nantucket Shoals with high percentages of oak and other deciduous pollen are placed in Holocene pollen zone C-3, or in an earlier interglacial (Livingstone, 1964) (Table 1). Groot and Groot (1964) (Table 1) interpret hemlock, fir, spruce, and birch pollen associated with a ^{14}C date of 11,000 yrs. B.P. as evidence of cool-temperate, post glacial climate.

Core KN10-1. Stellwagen Basin in the Gulf of Maine has a high-resolution 15,000-year pollen record, fifteen meters of silty clay analyzed at ~100-year intervals (Pack, 1980) (Table 1). A tripartite pollen sequence consists of spruce-hardwood, followed by pine, and hemlock-mixed hardwood

(Table 2). Regional forest development inferred from marine pollen correlates with late-glacial and Holocene development of New England forests (Pack, 1980). Recent pollen spectra from the Gulf show New England deforestation during early settlement by a rise in ragweed *(Ambrosia)* and other herbs. Climatic changes on land correspond with changes in the marine environment. Foraminifera from KN10-1 indicate warmer water ~9,000 yrs. B.P., at the same time the prominence of pine pollen suggests relatively warm, dry conditions in New England.

Core V17-178. Basal pollen spectra in a deep-sea core from the Laurentian Fan show spruce park-tundra in the Canadian Maritimes ~14,000 yrs. B.P.

(Tables 1 and 2) (L. Heusser, unpub. data). Expansion of open pine woodland with birch precedes a spruce rise ~7,000 yrs. B.P. Similar trends appear in pollen diagrams from Nova Scotia (Railton, 1975). Marine floral and faunal changes are synchronous. At 14,000 yrs. B.P., benthic and planktonic foraminifera show specific and isotopic terminations, SST changes from cool to warm, and red muds dominated by reworked pollen (including Permo-Carboniferous types) and dinoflagellates are replaced by gray sediments with little recycled pollen and modern dinoflagellate concentrations (W. L. Balsam and S. S. Streeter, oral commun., 1979). This event records a major sedimentologic (opening of the St. Lawrence) and climatic change.

Core GC79-11-1. Holocene pollen, foraminifera, and dinoflagellate assemblages in Canadian estaurine sediments (Miller *et al.,* 1982) reflects changes in coastal waters and vegetation. Mixed forest replaces park-woodland ~7,000 yrs. B.P. Hemlock forests develop in which birch and fir increase ~2,000 yrs. B.P. Pollen and dinoflagellates show effects of recent disturbance (land clearing, basin dredging, and filling). Marine pollen stratigraphy (Table 2) correlates with ^{14}C-dated pollen stratigraphy of Nova Scotian lake sediments.

Linwood and Pomquet Cores. Short cores show European weed pollen assemblage zone (C3-C) associated with marine sedimentary and faunal changes (Schafer and Mudie, 1980) (Table 2). In A.D. 1760, an increase in fine-grained sediment accompanies a change in the calcareous/arenaceous foraminifera ratio, and a rise in *Ambrosia, Rumex acetosella,* grass, alder, and *Pteridium.* Correlations of older Holocene pollen assemblages with Nova Scotian pollen stratigraphy vary with fine-grained muds providing better regional biostratigraphy and chronology than coarse-silts or sands.

Core BP-102, 11, and 12. Closed-crown boreal forest development ~4,000 yrs. B.P. is inferred from pollen spectra in shelf cores near Newfoundland (Piper *et al.,* 1978). Open boreal woodland interspersed with tundra (4,000-6,000 yrs. B.P.), was preceded by shrub-herb tundra. Similar trends occur in pollen data from a deep-water core 77034-13 (Heusser, unpub. data), and from Labrador Shelf cores (Tables 1 and 2).

Pollen spectra from cores 11 and 12 show boreal forest of the past 6,500 years (spruce and birch-alder-spruce) preceded by birch-shrub tundra and an older sedge-shrub tundra, which may have begun ~21,000 yrs. B.P. (Table 2). Foraminifera indicate open water (at least during part of the summer) throughout the

time of sediment deposition. Vilks and Mudie (1978) interpret these data as evidence for early deglaciation at 55°N.

Coreholes 2 and 23. Pleistocene sediment from two coreholes on the Grand Banks is characterized by low diversity pollen assemblages dominated by bisaccate pinaceous pollen with a large derived (reworked) component, which includes Paleozoic spores, pollen, and acritarchs; Cretaceous spores, pollen, and dinoflagellates; and Tertiary pollen and dinoflagellates (Williams and Brideaux, 1975). The most important factor controlling composition of these pollen assemblages is sedimentological rather than climatic.

SUMMARY

Pollen assemblages in Pleistocene and Recent hemipelagic sediments on North American continental margins reflect changes in both continental and marine environments. Marine pollen spectra reflect vegetation and climates of regions from which they are derived, and sedimentary processes of regions in which they are deposited. Spatial patterns of pollen assemblages in contemporary marine sediments in the northeast Atlantic Ocean, Gulf of Mexico, and northeast Pacific Ocean are similar to geographic distributions of pollen assemblages and vegetation onshore. Downcore, marine and continental Quaternary pollen stratigraphic units are comparable. Pollen in North American marine cores directly correlates regional and global continental and marine stratigraphic and paleoecologic data. Where North American pollen depositional sites are limited or lacking, as in glaciated or arid regions, pollen in offshore sediments provides vegetation/climatic records which otherwise would not be available. Other applications of North American Quaternary marine palynology include determining relative sea level, identifying current transport, and establishing marine sediment provenance.

ACKNOWLEDGMENTS

My thanks to those who contributed to this paper, particularly W. Balsam, D. Engelhardt, C. Heusser, J. Kennett, and C. Stock. Core samples were provided by Bedford Institute of Oceanography; Deep Sea Drilling Project; Lamont-Doherty Geological Observatory of; School of Oceanography, Oregon State University; School of Oceanography, University of

Rhode Island; Department of Oceanography, University of Washington; and Woods Hole Oceanographic Institution. NSF grant DEB 81-09916 provided support for research in California.

References Cited

ADAM, D. P.
  1979   Raw pollen counts from Core 4, Clear Lake, Lake County, California. *United States Geological Survey Open File Report:* 79-663.

AGER, T. A.
  1982   Vegetational history of western Alaska during the Wisconsin glacial interval and the Holocene. *In:* Hopkins, D. M., Matthews, J. V., Jr., Schweger, C. E., and Young, S. B. (eds.). *Paleoecology of Beringia.* Academic Press, New York, p. 75-93.

BALLOG, R. A., and MALLOY, R. E.
  1981   Neogene palynology from the southern California continental borderland, site 467, Deep Sea Drilling Project Leg 63. *In:* Yeats, R. S., Haq, B. U., *et al., Initial Reports of the Deep Sea Drilling Project,* Volume LXIII, Washington (U.S. Government Printing Office), p. 565-576.

BALSAM, W. L., and HEUSSER, L. E.
  1976   Direct correlation of sea surface paleotemperatures, deep circulation, and terrestrial paleoclimates: foraminiferal and palynological evidence from two cores off Chesapeake Bay. *Marine Geology,* 21: 21-147.

BALSAM, W. L., HEUSSER, L. E., PANDEL, R. G., ROBINSON, W. J., and MULCAHY, S. A.
  1977   Estimating paleo-environment from pollen in marine cores: an example from the western North Atlantic. *The Geological Society of America Abstracts with Programs,* 11: 383.

BRUSH, G. S., and DEFRIES, R. S.
  1981   Spatial distributions of pollen in surface sediments of the Potomac estuary. *Limnology and Oceanography,* 26: 295-309.

BRYANT, V. M., JR.
  1977   A 16,000 year pollen record of vegetational change in central Texas. *Palynology,* 1: 143-156.

BYRNE, R.
  1982   Preliminary pollen analysis of Deep Sea Drilling Project Leg 64 Hole 480 (Cores 1-11). *In:* Curray, J. R., Moore, D. G., *et al.,*(eds.), *Initial Reports of the Deep Sea Drilling Project,* Volume LXIV, Part 2, Washington (U.S. Government Printing Office), p. 1225-1238.

COLINVAUX, P. A.
  1964   The environment of the Bering Land Bridge. *Ecological Monographs,* 34: 297-329.

CRONIN, T. M., SZABO, B. J., AGER, T. A., HAZEL, J. E., and OWENS, J. P.
  1981   Quaternary climates and sea levels of the U.S. Atlantic Coastal Plain. *Science,* 211: 233-240.

CROSS, A. T., THOMPSON, G. G., and ZAITZEFF, J. B.
  1966   Source and distribution of palynomorphs in bottom sediments, southern part of the Gulf of California. *Marine Geology,* 4: 467-524.

DARRELL, J. A., and HART, G. F.
  1970   Environmental determinations using absolute miospore frequency, Mississippi River delta. *Geological Society of America Bulletin,* 81: 2513-2518.

DELCOURT, P. A., and DELCOURT, H. R.
  1983   Late-Quaternary vegetational dynamics and community stability reconsidered. *Quaternary Research,* 12: 265-271.

ELSIK, W. C.
  1969   Late Neogene palynomorph diagrams, northern Gulf of Mexico.*Transactions – Gulf Coast Association of Geological Societies,*Volume XIX: 509-528.

ENGELHARDT, D. W.
  1963   Palynology of Recent sediments in southeastern United States. *Pan American Petroleum Corporation Report.*

FLORER-HEUSSER, L. E.
  1975   Late-Cenozoic marine palynology of northeast Pacific Ocean cores. *In:* Suggate, R. P., and Cresswell, M. M. (eds.), *Quaternary Studies.* The Royal Society of New Zealand, Wellington, p. 133-138.

FOURNIER, G. R.
  1981   Palynostratigraphic analysis of cores from Site 493, Deep Sea Drilling Project Leg 66. *In:* Watkins, J. S., Moore, J. C., *et al,*(eds.), *Initial Reports of the Deep Sea Drilling Project,* Volume LXVI, Washington (U.S. Government Printing Office), p. 661-670.

GARDNER, J. V., BARRON, J. A., DEAN, W. E., HEUSSER, L. E., KLISE, D. H., POORE, R. Z., QUINTERO, P. J., and STONE, S. M.
  1983   Quantitative microfossil, sedimentologic, and geochemical data on cores V1-80-P3, V1-80-G1, and V1-80-P8 from the continental slope off northern California. *United States Geological Survey Open-File Report,* 83-83. 51p.

GARDNER, J. V., BARRON, J. A., DEAN, W. E., HEUSSER, L. E., POORE, R. Z., QUINTERO, P. J., STONE, S. M., and WILSON, C.
  1984   Quantitative microfossil, sedimentologic, and geochemical data on core L13-81-G138 and surface samples from the continental shelf and slope off Northern California. *United States Geological Survey Open-File Report,* 84-369. 118p.

GROOT, C. R., and GROOT, J. J.
  1964   The pollen flora of Quaternary sediments beneath Nantucket Shoals. *American Journal of Science,* 262: 488-493.

GROOT, J. J.
1966   Some observations of pollen grains in suspension in the estuary of the Delaware River. *Marine Geology, 20:* 409-416.

GROOT, J. J., and GROOT, C. R.
1966   Marine palynology: possibilities, limitations, problems. *Marine Geology, 4:* 387-396.

HABIB, D., THURBER, D., ROSS, D., and DONAHUE, J.
1970   Holocene palynology of the Middle America Trench near Tehuantepec, Mexico. *The Geological Society of America,* Memoir 126: 233-261.

HARRISON, W., MALLOY, R. J., RUSNAK, G. A., and TERASMAE, J.
1965   Possible Late Pleistocene uplift Chesapeake Bay entrance. *The Journal of Geology, 73:* 210-229.

HEUSSER, C. J., and FLORER, L. E.
1973   Correlation of marine and Quaternary pollen records from the northeast Pacific and western Washington. *Quaternary Research, 3:* 661-670.

HEUSSER, C. J., and HEUSSER, L. E.
1981   Palynology and paleotemperature analysis of the Whidbey Formation, Puget Lowland, Washington. *Canadian Journal of Earth Sciences, 18:* 136-149.

HEUSSER, C. J., HEUSSER, L. E. and STREETER, S. S.
1980   Quaternary temperatures and precipitation for the north-west coast of North America. *Nature, 286:* 702-704.

HEUSSER, L. E.
1978a   Pollen in Santa Barbara Basin, California: a 12,000-yr record. *Bulletin of the Geological Society of America, 89:* 673-678.

1978b   Spores and pollen in the marine realm. *In:* Haq, B., and Boersma, A. (eds.), *Introduction to Marine Micropaleontology.* Elsevier, New York. p. 327-339.

1981   Pollen analysis of selected samples from Deep Sea Drilling Project Leg 63. *In:* Yeats, R. S., Haq, B. U., *et al,* (eds.), *Initial Reports of the Deep Sea Drilling Project,* Volume LXIII, Washington (U.S. Government Printing Office), p. 559-563.

1982a   Pollen analysis of laminated and homogeneous sediment from the Guaymas Basin, Gulf of California. *In:* Curray, J. R., Moore, D. G., *et. al,* (eds.), *Initial Reports of the Deep Sea Drilling Project,* Volume LXIV, Washington (U.S. Government Printing Office), p. 1217-1223.

1982b   Quaternary palynology of northwest California and southwest Oregon. *American Quaternary Association, Program and Abstracts.* University of Washington, p. 104.

1983a   Palynology and paleoecology of postglacial sediments in an anoxic basin, Saanich Inlet, British Columbia. *Canadian Journal of Earth Sciences, 20:* 873-885.

1983b   Pollen distribution in the bottom sediments of the western North Atlantic Ocean. *Marine Micropaleontology, 8:* 77-88.

1983c   Contemporary pollen distribution in coastal California and Oregon. *Palynology, 7:* 19-42.

HEUSSER, L. E., and BALSAM, W. B.
1977   Pollen distribution in the northeast Pacific Ocean. *Quaternary Research, 7:* 45-62.

HEUSSER, L. E., and BALSAM, W. B.
1985   Sedimentation of pollen on the North American continental margin: effects of the Western Boundary Undercurrent. In press.

HEUSSER, L. E., and SHACKLETON, N. J.
1979   Direct marine-continental correlation: 150,000-year oxygen isotope-pollen record from the North Pacific. *Science, 204:* 837-839.

HEUSSER, L. E., BARRON, J. A., POORE, R. Z., GARDNER, J. V., STONE, S. M.
1981   Correlation of continental and marine paleoclimatic records of central California and the adjacent slope of the Pacific Ocean. *The Geological Society of America Abstracts with Programs,* 13: 473.

HEUSSER, L. E., HEUSSER, C. J., and WEISS, D.
1975a   Man's influence on the development of the estuarine marsh, Flax Pond, Long Island, New York. *Bulletin Torrey Botanical Club,* 102: 61-66.

HEUSSER, L. E., SHACKLETON, N. J., MOORE, T. C., and BALSAM, W. L.
1975b   Land and marine records in the Pacific Northwest during the last glacial interval. *The Geological Society of America Abstracts with Programs,* 7: 1113-1114.

HOPKINS, J. S.
1950   Differential flotation and deposition of coniferous and deciduous tree pollen. *Ecology,* 31: 633-641.

LEVENTER, A., WILLIAMS, D. F., and KENNETT, J. P.
1982   Dynamics of the Laurentide ice sheet during the last deglaciation: evidence from the Gulf of Mexico. *Earth and Planetary Science Letters,* 59: 11-17.

LIVINGSTONE, D. A.
1964   The pollen flora of submarine sediments from Nantucket Shoals. *American Journal of Science,* 262: 479-487.

MATHEWES, R. W., and HEUSSER, L. E.
1982   A 12,000 year palynological record of temperature and precipitation trends in southwestern British Columbia. *Canadian Journal of Botany,* 59: 707-710.

MILLER, A. A. L., MUDIE, P. J., and SCOTT, D. B.
1982   Holocene history of Bedford Basin, Nova Scotia: foraminifera, dinoflagellate, and pollen records. *Canadian Journal of Earth Sciences,* 19: 2342-2367.

MUDIE, P. J.
1982   Pollen distribution in recent marine sediments, eastern Canada. *Canadian Journal of Earth Sciences,* 19: 729-747.

MUSICH, L. F.
1973   Pollen occurence in eastern North Pacific sediments. Deep Sea Drilling Project Leg 18. *In:* Kulm, L. D. *et al.,* (eds.), *Initial Reports of the Deep Sea Drilling Project,* Volume 18. Washington (U.S. Government Printing Office), p. 799-815.

NEEDHAM, H. D., HABIB, D., and HEEZEN, B. C.
1969   Upper Carboniferous palynomorphs as a tracer of red sediment dispersal patterns in the northwest Atlantic. *Journal of Geology,* 77: 113-120.

NELSON, R. E.
1985   Appendix IV. A) Final report on pollen from Borehole PB-2. B) Modern pollen rain on the Chukchi and Beaufort Sea Coasts, Alaska. *In:* Hopkins, D. M., and Hartz, R. W. (eds.), *Offshore permafrost studies, Beaufort Sea, (In: Environmental Assessment of the Alaska Continental Shelf, Annual Reports of Principal Investigators for the year ending March, 1978).* In press.

PACK, S. K.
1980   A late Quaternary marine pollen record from Stellwagen Basin in the Gulf of Maine. M.S. thesis, New York University, 36 p.

PIPER, D. J., MUDIE, P. J., AKSU, A. E., and HILL, P. R.

1978   Late Quaternary sedimentation, 50°N, north-east Newfoundland shelf. *Geographie physique et Quaternaire,* XXXII: 321-332.

PISIAS, N. G.

1978   Paleoceanography of the Santa Barbara Basin during the last 8,000 years. *Quaternary Research,* 10: 366-384.

1979   Model for paleoceanographic reconstructions of the California Current during the last 8,000 years. *Quaternary Research,* 11: 373-386.

PISIAS, N. G., HEUSSER, L. E., and SHACKLETON, N. J.

1981   Direct comparison of marine and continental climate records. *The Geological Society of America Abstracts with Programs,* 13: 529.

RAILTON, J. B.

1975   Post-glacial history of Nova Scotia. *Proceedings of the Nova Scotia Institute of Science,* 27, Supplement 3: 37-41.

SANCETTA, C., HEUSSER, L., LABEYRIE, L., NAIDU, A., and ROBINSON, S.

1984   Wisconsin-Holocene paleoenvironment of the Bering Sea: evidence from diatoms, pollen, oxygen isotopes and clay minerals. *Marine Geology,* 62: 55-68.

SCHAFER, C. T., and MUDIE, P. J.

1980   Spatial variability of foraminifera and pollen in two nearshore sediment sites, St. Georges Bay, Nova Scotia. *Canadian Journal of Earth Sciences,* 17: 312-324.

SHACKLETON, N. J., and OPDYKE, N.

1976   Oxygen isotope and palaeomagnetic stratigraphy of equatorial Pacific core V28-238: oxygen isotope temperatures and ice volumes on a 10^5 year and 10^6 year scale. *Quaternary Research,* 3: 39-55.

SIRKIN, L.

1982   Preliminary palynology of Pleistocene sediments from Deep Sea Drilling Project Sites 474 and 479. *In:* Curray, J. R., Moore, D. G., *et. al,* (eds.) *Initial Reports of the Deep Sea Drilling Project,* Volume LXIV, Washington (U.S. Government Printing Office), p. 1211-1216.

STANLEY, E. A.

1965   Abundance of pollen and spores in marine sediments off the eastern coast of the United States. *Southeastern Geology,* 7: 25-33.

1966   The application of palynology to oceanology with reference to the northwestern Atlantic. *Deep-sea Research,* 13: 921-939.

1969   Marine Palynology. *Oceanography and Marine Biology: Annual Review,* 7: 277-292.

THOMPSON, D. E.

1972   Paleoecology of the Pamlico Formation, Saint Mary's County, Maryland. Ph.D. thesis, Rutgers University. 179 p.

TRAVERSE, A., and GINSBURG, R. N.

1966   Palynology of the surface sediments of the Great Bahama Bank, as related to water movement and sedimentation. *Marine Geology,* 4: 417-460.

VILKS, G., and MUDIE, P. J.

1978   Early deglaciation of the Labrador Shelf. *Science,* 202: 1181-1183.

WEISS, D.

1974   Late Pleistocene stratigraphy and paleoecology of the lower Hudson River Estuary. *Geological Society of America Bulletin,* 85: 1561-1570.

WILLIAMS, G. L., and BRIDEAUX, W. W.

1975   Palynologic analyses of Upper Mesozoic and Cenozoic Rocks of the Grand Banks, Atlantic Continental Margin. *Geological Survey of Canada,* 236: 1-163.

A BIBLIOGRAPHY OF QUATERNARY PALYNOLOGY IN ARIZONA, COLORADO, NEW MEXICO, AND UTAH

STEPHEN A. HALL
Department of Geography
University of Texas at Austin
Austin, Texas 78712

The following is a bibliography of 548 published and unpublished papers (including theses, dissertations, cultural resource management reports) that describe or discuss the Southwestern pollen record. One-half of the studies are unpublished or limited distribution reports and are identified by institution or agency where they are on file or may be inspected. Reports originating from the Castetter Laboratory for Ethnobotanical Studies, Department of Biology, University of New Mexico, Albuquerque, are listed by their CLES Technical Series number.

Many pollen analysts have contributed to the compilation of the bibliography. Special acknowledgements are extended to V. L. Bohrer, V. M. Bryant, Jr., J. R. Bushman, A. C. Cully, O. K. Davis, G. J. Dean, P. L. Fall, S. K. Fish, J. W. Gish, R. H. Hevly, R. G. Holloway, B. F. Jacobs, W. C. Johnson, D. B. Madsen, V. Markgraf, P. J. Mehringer, Jr., M. K. O'Rourke, K. L. Petersen, J. Schoenwetter, L. J. Scott, and S. K. Short.

AASEN, D. K.
1984 Pollen, macrofossil, and charcoal analyses of Basketmaker coprolites from Turkey Pen Ruin, Cedar Mesa, Utah. M.A. thesis, Washington State University, Pullman, 73 p.

ADAM, D. P.
1970 Some palynological applications of multivariate statistics. Ph.D. dissertation, University of Arizona, Tucson, 132 p.

ADAM, D. P., FERGUSSON, C. W., and LaMARCH, V. C., JR.
1967 Enclosed bark as a pollen trap. *Science,* 157: 1067-1068.

ADAM, D. P., and MEHRINGER, P. J., JR.
1975 Modern pollen surface samples—an analysis of subsamples. *Journal of Research U.S. Geological Survey,* 3: 733-736.

ADAMS, K. R.
1980 Pollen, parched seeds, and prehistory: a pilot investigation of prehistoric plant remains from Salmon Ruin, a Chacoan pueblo in northwestern New Mexico. *Eastern New Mexico University Contributions in Anthropology* 9, 94 p.

AGENBROAD, L. D., MEADE, J. I., MARTIN, P. S., DAVIS, O. K., and BOLEN, C. W.
1984 A late Pleistocene record of the habitat and diet of *Mammuthus,* from Bechan Cave, Utah. Geological Society of America, Cordilleran Section, 80th ann. mtg., Anchorage, Alaska, *Abstracts with Programs 1984,* 265.

ANDERSON, R. Y.
1955 Pollen analysis, a research tool for the study of cave deposits. *American Antiquity,* 21: 84-85.

ANDREWS, J. T., CARRARA, P. E., KING, F. B., and STUCKENRATH, R.
1975 Holocene environmental changes in the alpine zone, northern San Juan Mountains, Colorado: evidence from bog stratigraphy and palynology. *Quaternary Research,* 5: 173-197.

ARMS, B. C.
1960 A silica-depressant method for concentrating fossil pollen and spores. *Micropaleontology,* 6: 327-328.

BAUCHHUBER, F. W.
1970 The effects of differential preservation in a pollen analytical study. Geological Society of America, ann. mtg., Milwaukee, *Abstracts and Program,* 2(7): 486.

1971 Paleolimnology of Lake Estancia and the Quaternary history of the Estancia Valley, central New Mexico. Ph.D. dissertation, University of New Mexico, Albuquerque, 238 p.

BAER, J. L., ROBERTS, H. H., and BUSHMAN, J. R.
1968 Recent sediments and associated microflora from Sevier Playa, Millard, County, Utah. *Geological Society of America Special Paper* 115, 404.

BAKER, R. G.
1983 Holocene vegetational history of the western United States. *In:* Wright, H. E., Jr. (ed.), *Late Quaternary Environments of the United States, Vol. 2, The Holocene* (Wright, H. E., Jr., ed.). University of Minnesota Press, Minneapolis, 109-127.

1984 Quaternary history of the tundra biome, central North America. Sixth International Palynological Conference (IPC), Calgary, Alberta, Canada, *Abstracts,* 4.

BATCHELDER, G. L.
   1974   Pollen evidence for post-pluvial ecotone shifts of pinyon
          pine *(Pinus edulis* Englem.) near Montezuma Well on the
          southern Colorado Plateau, Arizona. American Quater-
          nary Association (AMQUA), 3rd biennial mtg., Madison,
          Wisconsin, *Abstracts,* 101.

BATCHELDER, G. L., and CARLSON, A. M.
   1976   Late-glacial and post-glacial pollen and its accumulation
          density, Walker Lake, Coconino Co., Ariz. American
          Quaternary Association (AMQUA), 4th biennial mtg.,
          Tempe, Arizona, *Abstracts,* 123-124.

BATCHELDER, G. L., and MERRILL, R. K.
   1976   Late Quaternary environmental interpretations from
          palynological data, White Mountains, Arizona. American
          Quaternary Association (AMQUA), 4th biennial mtg.,
          Tempe, Arizona, *Abstracts,* 125.

BENNETT, P. S.
   1967   Pollen and soil analysis at I:14:30. Report to Museum of
          Northern Arizona, Flagstaff.

BENT, A. M.
   1960   Pollen analysis at Deadman Lake, Chuska Mountains,
          New Mexico. M.S. thesis, University of Minnesota,
          Minneapolis, 22 p.

BENT, A. M., and WRIGHT, H. E., JR.
   1963   Pollen analyses of surface materials and lake sediments
          from the Chuska Mountains, New Mexico. *Geological
          Society of American Bulletin,* 74: 491-500.

BERLIN, G. L., AMBLER, J. R., HEVLY, R. H., and SCHABER,
G. G.
   1977   Identification of a Sinagua agricultural field by aerial
          thermography, soil chemistry, pollen/plant analysis, and
          archaeology. *American Antiquity,* 42: 588-600.

BERRY, R. W., McCORMICK, C. W., and ADAM, D. P.
   1982   Pollen data from a 5-meter upper Pleistocene lacustrine
          section from Walker Lake, Coconino County, Arizona.
          U.S. Geological Survey Open-File Report 82-0383, 108 p.

BETANCOURT, J. L., and DAVIS, O. K.
   1984   Packrat middens from Canyon de Chelly, northeastern
          Arizona: paleoecological and archaeological implications.
          Quaternary Research, 21: 56-64.

BOHRER, V. L.
   1968   Paleoecology of an archaeological site near Snowflake,
          Arizona. Ph.D. dissertation, University of Arizona,
          Tucson.

   1970   Ethnobotanical aspects of Snaketown, a Hohokam village
          in southern Arizona. *American Antiquity,* 35: 413-430.

   1972   Paleoecology of the Hay Hollow site, Arizona. Field
          Museum of Natural History, *Fieldiana, Anthropology,*
          63(1): 1-30.

   1975   Pollen and flotation samples from pits, hearths, and
          ovens, Arizona P:14:1. Report, Department of Anthro-
          pology, University of Arizona, Tucson.

   1979   Pollen studies at Arroyo Hondo Site (LA 12) near Santa
          Fe, New Mexico. Report to School of American
          Research, Santa Fe.

   1980   Plant evidence bearing on medicinal ceremonial usage.
          In: Irwin-Williams, C., and Shelley, P. H. (eds.), *Investi-
          gations of the Salmon Site: the structure of Chacoan
          society in the northern Southwest.* Final report to funding
          agencies. Vol. 3, 163-535. Manuscript, Eastern New
          Mexico University, Portales.

   1981   Methods of recognizing cultural activity from pollen in
          archaeological sites. *The Kiva,* 46: 135-142.

   1982   Plant remains from rooms at Grasshopper Pueblo. *In:*
          Longacre, W. A., Holbrook, S. J., and Graves, M. W.
          (eds.), Multidisciplinary research at Grasshopper Pueblo,
          Arizona. *Anthropological Papers of the University of
          Arizona,* 40: 97-105.

   1982   One step in the search for pueblo fields on Mesita del
          Buey. *In:* Steen, C. R., *Pajarito Plateau Archaeological
          Surveys and Excavations, II.* Los Alamos National
          Laboratory, Los Alamos, New Mexico, 57-59.

   1984   Prehistoric plant remains from the Upper Bajadas and
          the Mesa Canyon complex of the Central Arizona
          Ecotone Project. *In:* Spoerl, P., and Gumerman, G. (eds.),
          Archaeological investigations near New River, cultural
          development and conflict in central Arizona. *Southern
          Illinois University, Center for Archaeological Investiga-
          tions, Occasional Papers.*

BOHRER, V. L., and ADAMS, K. R.
   1977   Ethnobotanical techniques and approaches at Salmon
          Ruin, New Mexico. *Eastern New Mexico University
          Contributions in Anthropology,* 8 (San Juan Valley
          Archaeological Project, Technical Series 2), 214 p.

BRADFIELD, M.
   1973   Rock-cut cisterns and pollen "rain" in the vicinity of Old
          Oraibi, Arizona. *Plateau,* 46: 68-71.

BRYANT, V. M., JR.
   1983   Preliminary analysis of 15 samples from the Placitas
          Arroyo sites. *In:* Morenon, E. P., and Hays, T. R.,
          *Archaeological Investigations in the Placitas Arroyo, New
          Mexico.* Draft report, Institute of Applied Sciences,
          North Texas State University, Denton, D-1 to D-3.
          (Dona Ana County)

BRYANT, V. M., JR., and MORRIS, D. P.
   1985   Uses of ceramic vessels and grinding implements: the
          pollen evidence. *In:* Morris, D. P., and Bancroft, J. M.
          (eds.), Archeological Investigations at Antelope House.
          National Park Service Publications in Archaeology Series,
          Washington, D.C. In press.

BRYANT, V. M., JR., and WEIR, G. H.
   1985   Pollen analysis of floor sediment samples from Antelope
          House: a guide to room use and function. *In:* Morris, D.
          P., and Bancroft, J. M. (eds.), Archeological Investiga-
          tions at Antelope House. *National Park Service Publica-
          tions in Archaeology Series,* Washington, D.C. In press.

BUCHMANN, S. L., O'ROURKE, M. K., and SHIPMAN, C. W.
   1983   Aerobiology of pollen abundance and dispersal in a
          native stand of jojoba *(Simmondsia chinensis)* in
          Arizona. *In:* Elias-Cesnik, A. (ed.), *Jojoba and its uses
          through 1982.* Office of Arid Lands Publications, 79-92.

BUCK, P., and LEVETIN, E.
   1984   Aeroallergens in a subalpine environment. Sixth Interna-
          tional Palynological Conference (IPC), Calgary, Alberta,
          Canada, *Abstracts,* 16.

BUGE, D. E.
   1971   Pollen analysis of the Folsom type locality: preliminary
          report. Palynology Laboratory, Department of Anthro-
          pology, Arizona State University, Tempe, 4 p., 1 fig.

BUGE, D. E., and SCHOENWETTER, J.
   1977   Pollen studies at Chimney Rock Mesa. *In:* Eddy, F. W.,
          Archaeological investigations at Chimney Rock Mesa:
          1970-1972. *Memoirs of the Colorado Archaeological
          Society* 1, 77-80.

BUSHMAN, J. R.

1980 The rate of sedimentation in Utah Lake and the use of pollen as an indicator of time in the sediments. *Brigham Young University Geology Studies,* 27 (Part 3): 35-43.

CALLEN, O. C., and MARTIN, P. S.

1969 Plant remains in some coprolites from Utah. *American Antiquity,* 34: 329-331.

CARLSON, A. M., BATCHELDER, G. L., and UPDIKE, R. G.

1976 A recent pollen diagram from Walker Lake, Coconino Co., Ariz. American Quaternary Association (AMQUA), 4th biennial mtg., Tempe, Arizona, *Abstracts,* 129-130.

CARRARA, P. E., MODE, W. N., RUBIN, M., and ROBINSON, S. W.

1984 Deglaciation and postglacial timberline in the San Juan Mountains, Colorado. *Quaternary Research,* 21: 42-55.

CLARY, K. H.

1980 Pollen analysis of LA 20896, a rockshelter in Socorro Co., New Mexico. Ms. on file (CLES 18), Office of Contract Archeology, University of New Mexico, Albuquerque, 11 p.

1980 Pollen analysis of outlying structures of Nuestra Senora de Dolores Pueblo (LA 677) in Bernalillo, NM. Ms. on file (CLES 22), Office of Contract Archeology, University of New Mexico, Albuquerque, 13 p.

1980 Pollen analysis at two Puebloan sites on the La Jolla Game Refuge, New Mexico. Ms. on file (CLES 44), Office of Contract Archeology, University of New Mexico, Albuquerque, 9 p.

1981 Pollen analysis at an historic site, Forked Stick Hogan, LA 20239, on the Navajo Indian Irrigation Project, Block IX, New Mexico. Ms. on file (CLES 45), Laboratory of Anthropology, Museum of New Mexico, Santa Fe, 11 p.

1981 Summary of botanical remains (pollen and flotation) from the Little Water archeological sites, the Chuska Valley, New Mexico (LA 16020, 16026, 16029, 20975). Ms. on file (CLES 46), Laboratory of Anthropology, Museum of New Mexico, Santa Fe, 59 p.

1981 Pollen analysis of coprolites from GBN 1, a Gallina Phase site, north central New Mexico. Ms. on file (CLES 47), University of New Mexico, Albuquerque, 16 p.

1981 Pollen analysis of coprolites from Chaco Canyon, New Mexico. Ms. on file (CLES 52), Chaco Center, National Park Service, Albuquerque, 37 p.

1981 Pollen analysis of an infant burial: HUTR-10, a White Mound phase pithouse village, Ganado, Arizona. Ms. on file (CLES 56), Branch of Indian Cultural Resources, National Park Service, Southwest Regional Office, Santa Fe, 4 p.

1982 Pollen analysis of 11 samples from PM 203 and PM 205. *In:* Allen, C. G., and Nelson, B. A. (eds.), *Anasazi and Navajo land use in the McKinley mine area near Gallup, New Mexico. Vol. 1. Archaeology.* Part 2. University of New Mexico, Office of Contract Archaeology, 787-803.

1982 Pollen analysis of a 20th century trading post (LA 20237), San Juan Co., New Mexico. Ms. on file (CLES 73), Laboratory of Anthropology, Museum of New Mexico, Santa Fe, 8 p.

1983 An investigation of pollen remains from two small sites along State Road 117, west-central New Mexico. Ms. on file (CLES 82), Laboratory of Anthropology, Museum of New Mexico, Santa Fe, 6 p.

1983 An analysis of prehistoric coprolite remains from Chaco Canyon, New Mexico: inferences for Anasazi diet and subsistence. Ms. on file (CLES 85), Chaco Center, National Park Service, University of New Mexico, Albuquerque.

1983 Pollen analysis of Archaic and Pueblo period sites on Block XI, Navajo Indian Irrigation Project, the San Juan Basin, northwestern New Mexico. Ms. on file (CLES 100), Navajo Nation Cultural Resources Management Program, Window Rock, Arizona, 8 p.

1983 An investigation of pollen remains from two pot wash samples from burials at the Chamisal Site (LA 22765), Albuquerque, NM. Ms. on file (CLES 101), Chamisal Project, Albuquerque, 5 p.

1983 An analysis of pollen from a Basketmaker II and a 20th century Navajo archaeological site (AZ-I-25-7 and AZ-I-25-10) from the Red Rock Valley, Arizona, the N-13 Project. Ms. on file (CLES 103), Navajo Nation Cultural Resources Management Program, Window Rock, Arizona, 15 p.

CLARY, K. H., and CULLY, A. C.

1979 Pollen analysis of Block III Mitigation Project, Navajo Indian Irrigation Project, San Juan County, New Mexico: Final report. Ms. on file (CLES 1), Navajo Nation Cultural Resources Management Program, Window Rock, Arizona, 86 p.

CLISBY, K. H., FOREMAN, F., and SEARS, P. B.

1957 Pleistocene climatic changes in New Mexico, U.S.A. *Veroff. Geobotanisches Inst. Rubel in Zurich,* 34: 21-26.

CLISBY, K. H., and SEARS, P. B.

1956 San Augustin Plains—Pleistocene climatic changes. *Science,* 124: 537-539.

COOLEY, M. E., and HEVLY, R. H.

1964 Geology and depositional environment of Laguna Salada, Arizona. *In:* Martin, P. S., *et al.,* Chapters in the Prehistory of eastern Arizona, II. Chicago Natural History Museum, *Fieldiana: Anthropology,* 55: 188-200.

CULLY, A. C.

1977 Relation of pollen analysis to archaeological excavations, Chaco Canyon, New Mexico. M.S. thesis, University of New Mexico, Albuquerque.

1977 Paleoclimatic variability in the North-Middle Rio Grande, New Mexico. *In:* Biella, J. V., and Chapman, R. C. (eds.), *Archeological investigations in Cochiti Reservoir, New Mexico, Vol. 1, A survey of regional variability.* Office of Contract Archeology, University of New Mexico, Albuquerque, 97-101.

1978 Pollen analysis. *In:* Windes, T. C., Stone circles of Chaco Canyon, north-western New Mexico. *Reports of the Chaco Center* 5, National Park Service, Division of Chaco Research, Albuquerque, Appendix B, 108-109.

1979 Some aspects of pollen analysis in relation to archeology. *The Kiva,* 44: 95-100.

1980 Pollen analysis at Pierre's Site, New Mexico. Report prepared for Chaco Center, National Park Service, University of New Mexico, 3 p.

1980 Two pollen wash samples from Starkweather Area A. Report prepared for Cultural Resources Management Division, New Mexico State University, Las Cruces, 11 p.

1981 Pollen analysis at AZ P:24:1, Window Rock, Arizona. Report prepared for Navajo Nation Cultural Resources Management Program, Window Rock, Arizona, 9 p.

1981   Pollen analysis at 29SJ629: a comparison of two sites in Marcia's Rincon, Chaco Canyon, New Mexico. Report prepared for Division of Chaco Research, National Park Service, Albuquerque, 64 p.

1982   Pollen samples from the Two Rivers Survey, Hondo Valley, Lincoln Co., New Mexico. Ms. on file (CLES 59), New World Research, Tucson, Arizona, 3 p.

1982   Pollen analysis from sites on Blocks VIII-XI, Navajo Indian Irrigation Project, San Juan Co., New Mexico. Ms. on file (CLES 61), Navajo Nation Cultural Resources Management Program, Window Rock, Arizona, 17 p.

1982   Pollen analysis from the Pictured Cliffs sites, near Farmington, New Mexico. Ms. on file (CLES 64), Laboratory of Anthropology, Museum of New Mexico, Santa Fe, 9 p.

1982   Prehistoric subsistence in the Bis sa'ani community: evidence from pollen analysis. *In:* Breternitz, C. D., Doyel, D. E., and Marshall, M. P. (eds.), Bis sa'ani: a late Bonito Phase community on Escavada Wash, northwest New Mexico. Window Rock, Arizona, *Navajo Nation Papers in Anthropology* 14, 1181-1208.

1982   Pollen analysis at Keystone Dam Site 32, El Paso, Texas. Ms. on file (CLES 77), Prewitt and Associates, Austin, Texas, 10 p.

1982   Pollen analysis at Gallina Sites near Llaves and Cebolla, Rio Arriba Co., New Mexico. Ms. on file (CLES 78), Ghost Ranch Museum, Abiquiu, New Mexico, 12 p.

1983   The distribution of corn pollen in small sites and large structures, Chaco Canyon, New Mexico. Ms. on file (CLES 84), Chaco Center, National Park Service, University of New Mexico, Albuquerque, 12 p.

1983   Prehistoric subsistence at Pueblo Alto, Chaco Canyon, New Mexico: evidence from pollen analysis. Ms. on file (CLES 89), Chaco Center, National Park Service, University of New Mexico, Albuquerque, 58 p.

1983   Pollen analysis from the West Mesa sites, Dona Ana Co., New Mexico. Ms. on file (CLES 92), Cultural Resources Management Division, Department of Sociology and Anthropology, New Mexico State University, Las Cruces, 35 p.

1983   Forty-four pollen samples from Archaic, Basketmaker, and Navajo sites along the Shell CO$_2$ pipeline route, northwestern New Mexico. Ms. on file (CLES 94), Office of Contract Archeology, Albuquerque, 19 p.

1983   The analysis of seven pollen samples from Rocky Mountain National Park, Colorado. Ms. on file (CLES 106), National Park Service, Midwest Archeological Center, Lincoln, Nebraska, 18 p.

1983   Pollen evidence for prehistoric use of wild plant resources at the RATSCAT Advanced Measurement Site, White Sands Missile Range, New Mexico. *In:* Eidenbach, P. L. (ed.), *The prehistory of Rhodes Canyon, N.M. Survey and mitigation.* Human Systems Research, Inc., Tularosa, New Mexico, 55-61.

1984   Analysis of a pollen sample from site 42UN1103, a rock shelter in Dinosaur National Monument, Utah. Ms. on file (CLES 108), National Park Service, Midwestern Archeological Center, Lincoln, Nebraska, 13 p.

1984   Pollen evidence of past subsistence and environment at Chaco Canyon, New Mexico. Ms. on file (CLES 113), Chaco Center, National Park Service, Albuquerque, 151 p.

1984   Twelve pollen samples from Glen Canyon National Recreation Area, Utah-Arizona: metate washes and sediment samples from sites 42KA1991, 2000 and 2001. Ms. on file (CLES 114), National Park Service, Midwestern Archeological Center, Lincoln, Nebraska, 9 p.

CULLY, A. C., and CLARY, K. H.
1980   Pollen analysis at Keystone Dam, Site 33. *In:* O'Laughlin, T. C., The Keystone Dam Site and Other Archaic and Formative Sites in Northwest El Paso, Texas. *El Paso Centennial Museum, University of Texas at El Paso, Publications in Anthropology* 8, 250-269.

1981   Pollen analysis at nine sites, Navajo Indian Irrigation Project Blocks VI and VII. Ms. on file (CLES 36), Navajo Nation Cultural Resources Management Program, Window Rock, Arizona, 48 p.

1984   Pollen analysis. *In:* Hogan, P. C., and Winter, J. C. (eds.), *Economy and Interaction along the Lower Chaco River.* Office of Contract Archeology, University of New Mexico, Albuqueruqe, 18-1 to 18-24.

DaCOSTA, V.
1976   Pollen analysis at Santa Rosa Wash. *In:* Raab, L., The Structure of Prehistoric Community Organization at Santa Rosa Wash, Southern Arizona. Ph.D. dissertation, Arizona State University, Tempe.

1977   Analysis of pollen samples from Copper Basin. *In:* Jeter, M. D., Archaeology in Copper Basin, Yavapai County, Arizona: model building for the prehistory of the Prescott region. *Arizona State University Anthropological Research Papers* 11.

DALLEY, G. F.
1972   Palynology of the Evans Mound deposits. *In:* Berry, M. S., *The Evans site.* A Special Report, Department of Anthropology, University of Utah, Salt Lake City, 187-194.

1976   Palynology of the Swallow Shelter deposits. *In:* Dalley, G. F., Swallow Shelter and associated sites. *University of Utah Anthropological Papers* 96.

DAVIS, O. K.
1984   Late Holocene vegetation change at low elevation in central Arizona: human and climatic causes. Sixth International Palynological Conference (IPC), Calgary, Alberta, Canada, *Abstracts,* 30.

DAVIS, O. K., AGENBROAD, L., MARTIN, P. S., and MEAD, J. I.
1984   The Pleistocene dung blanket of Bechan Cave, Utah. *In:* Genoways, H. H., and Dawson, M. R. (eds.), *Contributions in Vertebrate Paleontology: A Volume in Memorial to John E. Guilday.* Special Publication of Carnegie Museum of Natural History 8, 267-282.

DICKEY, A. M.
1971   Palynology in Hay Hollow Valley. M.S. thesis, Northern Arizona University, Flagstaff, 55 p.

DIXON, H. N.
1962   Vegetation, pollen rain, and pollen preservation, Sangre de Cristo Mountains, New Mexico. M.S. thesis, University of New Mexico, Albuquerque, 69 p.

EARL, R. A.
1983   Paleohydrology and paleoclimatology of the Skunk Creek Basin during Holocene time. Ph.D. dissertation, Arizona State University, Tempe.

EULER, R. C., GUMERMAN, G. J., KARLSTROM, T. N. V., DEAN, J. S., and HEVLY, R. H.
1979   The Colorado Plateaus; cultural dynamics and paleoenvironment. *Science,* 205: 1089-1101.

FALL, P. L.
1981 Modern pollen spectra and their application to alluvial pollen sedimentology. M.S. thesis, University of Arizona, Tucson, 63 p. (Canyon de Chelly, Arizona)

1984 Late Quaternary vegetation and climate of the central Colorado Rocky Mountains. Sixth International Palynological Conference (IPC), Calgary, Alberta, Canada, *Abstracts,* 44.

1984 Development of modern alpine tundra and subalpine forest during the late Quaternary and their climatic implications. American Quaternary Association (AMQUA), 8th biennial meeting, Boulder, Colorado, *Program and Abstracts,* 41.

FALL, P. L., KELSO, G., and MARKGRAF, V.
1981 Paleoenvironmental reconstruction at Canyon del Muerto, Arizona, based on principal component analysis. *Journal of Archaeological Science,* 8: 297-307.

FISH, S. K.
1977 Mule Shoe Bend pollen analysis. *In:* James, C. D., The Mule Shoe Bend Site. *Museum of Northern Arizona Research Paper* 9, 39-44.

1978 The pollen record of prehistoric sites and suspected fields at Bandelier National Monument. Report to Bandelier Archaeological Project, National Park Service, Southwest Region, Santa Fe.

1978 Pollen analysis. *In:* Moffitt, K., Archaeological investigations along the Navajo-McCullough transmission line, southern Utah and northern Arizona. *Museum of Northern Arizona Research Paper* 10, 175-180.

1980 Agricultural features and their pollen record at NA 11,504. *In:* Fiero, D. C., Munson, R. W., McClain, M. T., Wilson, S. M., and Zier, A. H., The Navajo Project. *Museum of Northern Arizona Research Paper* 11, 261-270.

1981 Palynological results from Las Colinas. *In:* Schreiber, K. J., McCarthy, C. H., and Byrd, B. Report of the testing of Interstate 10 corridor prehistoric and historic archaeological remains between Interstate 17 and 30th Drive. *Arizona State Museum Archaeological Series* 156, 245-251.

1981 Pollen analysis. *In:* Westfall, D. A., Prehistory of the St. Johns area, east-central Arizona: the TEP St. Johns Project. *Arizona State Museum Archaeological Series* 153, 321-328.

1982 Pollen analysis at AZ BB:13:146. *In:* Brew, S. A., Archaeological Test Excavations in Southern Arizona. *Arizona State Museum Cultural Resources Management Division Archaeological Series* 152, 140-142.

1982 Palynological investigations at six sites in the Gallo Wash mine lease. *In:* Simmons, A. H., Prehistoric adoptive strategies in the Chaco Canyon region, northeastern New Mexico. Volume I. Introduction, environmental studies, and analytical approaches. *Navajo Nation Papers in Anthropology* 9, 129-142.

1983 Pollen analysis. *In:* Doelle, W. H., Archaeological and historical investigations at Nolic Papago Indian Reservation, Arizona. *Institute for American Research Anthropological Papers* 2, 145-151.

1984 Archeological disturbance floras of the lower Sonoran Desert and their implications. Sixth International Palynological Conference (IPC), Calgary, Alberta, Canada, *Abstracts,* 47.

FISH, S. K., BARBER, R., and MIKSICEK, C.
1982 Environment and subsistence studies. *In:* Teague, L. S., and Crown, P. L. (eds.), Hohokam archaeology along the Salt-Gila Aqueduct, Central Arizona Project. Volume I. Research design. *Arizona State Museum Archaeological Series* 150, 129-140.

FISH, S. K., and MIKSICEK, C.
1982 Sampling guidelines for pollen and flotation. *In:* Teague, L. S., and Crown, P. L. (eds.), Hohokam archaeology along the the Salt-Gila Aqueduct, Central Arizona Project. Volume I. Research design. *Arizona State Museum Archaeological Series* 150, 169-173.

FOREMAN, F., and CLISBY, K. H.
1961 Crane Lake pollen and sediment correlation. *In:* Wendorf, F., Paleoecology of the Llano Estacado. Museum of New Mexico Press, Santa Fe. *Fort Burgwin Research Center Publication* 1, 92-93. (pollen analysis by U. Hafsten)

FOREMAN, F., CLISBY, K. H., and SEARS, P. B.
1959 Plio-Pleistocene sediments and climates of the San Augustin Plains, New Mexico. New Mexico Geological Society, 10th Field Conference, *Guidebook of West-Central New Mexico,* 117-120.

FORESTER, R. M., and MARKGRAF, V.
1984 Late Pleistocene and Holocene seasonal climatic records from lacustrine ostracode assemblages and regional (pollen) vegetational patterns in southwestern U.S.A. American Quaternary Association (AMQUA), 8th biennial meeting, Boulder, Colorado, *Program and Abstracts,* 43-45.

FREDLUND, G.
1984 Palynological analysis of sediments from Sheep Camp and Ashislepah shelters. *In:* Simmons, A. H. (ed.), Archaic Prehistory and Paleoenvironments in the San Juan Basin, New Mexico: The Chaco Shelters Project. *University of Kansas, Museum of Anthropology, Project Report Series* 53, 186-209.

FREDLUND, G. G., and JOHNSON, W. C.
1984 Palynological evidence for late Quaternary paleoenvironmental change in the San Juan Basin. American Quaternary Association (AMQUA), 8th biennial meeting, Boulder, Colorado, *Program and Abstracts,* 46.

FREEMAN, C. E.
1968 A pollen study of some post-Wisconsin alluvial deposits in Dona Ana County, New Mexico. Ph.D. dissertation, New Mexico State University, Las Cruces.

1972 Pollen study of some Holocene alluvial deposits in Dona Ana County, southern New Mexico. *Texas Journal of Science,* 24: 203-220.

FRY, G. and HALL, H. J.
1975 The human coprolites from Antelope House: preliminary analysis. *The Kiva,* 41: 87-96.

GASSER, R. E., and SCOTT, L. J.
1982 Pollen analysis in selected rooms at Walpi and along modern transects near First Mesa. *In: Walpi Archaeological Project—Phase II, Vol. 7, Archaeobotanical Remains.* Museum of Northern Arizona, Flagstaff.

GISH, J. W.
1975 Pollen—AA:1290. *In:* Kinkade, G. M., and Fritz, G. L., *The Tucson Sewage Project: Studies at Two Archaeological Sites in the Tucson Basin.* Arizona State Museum, University of Arizona, Tucson, 52-53. (Pima County)

1975 Preliminary report on pollen analysis from Hecla I, II, and III. *In:* Goodyear, A. C., Hecla II and III: An interpretative study of archaeological remains from the Lake Shore Project, Papago Reservation, south central Arizona. *Arizona State University Anthropological Research Paper* 9, 254-270. (Pima County)

1977 A pollen study of the El Rito Site, New Mexico. Report to the USDA Forest Service, Southwest Region, Albuquerque, 6 p., 1 fig. (Rio Arriba County)

1977 Palynological investigations at Bog Hole, Arizona. Report to the USDA Forest Service, Southwest Region, Albuquerque, 10 p., 1 fig. (Santa Cruz County)

1977 An archaeological pollen analysis of the Jicarilla Site, New Mexico. Report to the USDA Forest Service, Southwest Region, Albuquerque, 8 p., 1 fig. (Rio Arriba County)

1977 Archaeological pollen samples from the Santa Rosa Valley, southern Arizona. Report to the Office of Cultural Resource Management, Department of Anthropology, Arizona State University, 5 p., 1 fig. (Pinal County)

1977 Palynological investigations of the Carefree Land Exchange Project. Report to the Office of Cultural Resource Management, Department of Anthropology, Arizona State University, 16 p., 2 figs., 1 table. (Maricopa County)

1978 McGaffey #1, an intensive palynological study. Report to the USDA Forest Service, Southwest Region, Albuquerque. 11 p., 1 fig., 2 tables.

1978 Six pollen samples from a Mimbres Site, New Mexico. Report to the USDA Forest Service, Southwest Region, Albuquerque, 6 p., 1 table. (Catron County)

1978 Four pollen samples from ENM 10390, northeastern Arizona. Report to Eastern New Mexico University Agency for Conservation Archaeology, Portales, 4 p., 1 table. (Apache County)

1978 Pollen results from the Los Hornos Site, central Arizona. Report to the Office of Cultural Resource Management, Department of Anthropology, Arizona State University, 19 p., 2 figs. (Maricopa County)

1978 Palynological investigations at three archaeological sites in the Cibola National Forest, New Mexico. Report to Northern Arizona University, Flagstaff, and the USDA Forest Service, Southwest Region, Albuquerque, 49., 4 figs., 4 tables. (Valencia County)

1978 Eight pollen samples from MAV-4 and MAV-37, Arizona. Report to the Arizona State Museum, Tucson, 10 p., 1 table. (Yuma County)

1978 Pollen analysis of seven samples from AZ BB:14:73. Report to Archaeological Research Services, Tempe, 6 p., 1 table. (Pima County)

1978 Archaeological pollen samples from the Upper Pecos River, New Mexico. Report to the Center for Anthropological Studies, Albuquerque, 28 p., 2 figs., 4 tables.

1978 Palynology of the Reno-Park Creek Project. *In:* Jeter, M. D., The Reno-Park Creek Project: Archaeological investigations in Tonto Basin, Arizona. *Arizona State Museum Contribution to Highway Salvage Archaeology in Arizona* 49, 115-121. (Gila County)

1979 Pollen results from Nawthis Village, central Utah. Report to the USDA Forest Service Archaeology Laboratory, Ogden, 11 p., 1 table. (Sevier County)

1979 Palynological research at Pueblo Grande Ruin. *The Kiva*, 44: 159-172.

1979 Pollen results from Los Aumentos, Phoenix, Arizona. Report to the Museum of Northern Arizona, 9 p., 1 fig. (Maricopa County)

1979 An archaeological pollen analysis of five sites along the San Carlos River, Arizona. Report to Archaeological Research Services, Tempe, 25 p., 2 tables. (Gila County)

1979 Cave Buttes alternative dam mitigation, Arizona. A palynological perspective. *In:* Henderson, T. K., and Rodgers, J. B., Archaeological Investigations in the Cave Creek Area, Maricopa County, south-central Arizona. *Arizona State University Anthropological Research Papers* 17, 158-179. (Maricopa County)

1980 Pollen results from AZ T:12:3, central Arizona. Report to Archaeological Research Services, Tempe, 7 p., 2 tables. (Maricopa County)

1980 Pollen results from three sites in Rio Arriba County, New Mexico. Report to the School of American Research, Santa Fe, 14 p., 1 table.

1980 The Three Rivers pollen analysis. Report to the Bureau of Land Management, Las Cruces, 16 p., 3 tables. (Otero County)

1981 Pollen results from the Desert Gold sites, Arizona. *In:* Rice, G. E., and Dobbins, E., Prehistoric Community Patterns in the Western Desert of Arizona. *Arizona State University Anthropological Field Studies* 2, 81-85. (Maricopa County)

1981 Pollen results from four Fremont sites in Sevier County, central Utah. Report to the USDA Forest Service, Ogden, 19 p., 2 tables.

1981 Preliminary pollen results from four sites in central Arizona. Report to the Museum of Northern Arizona, Flagstaff, 10 p., 1 table. (Maricopa County)

1981 Pollen results from the Jones Ranch Road Project, northwestern New Mexico. Report to the Zuni Archaeology Program, Pueblo of Zuni, 32 p., 2 figs., 2 tables. (McKinley County)

1981 Floral resources—pollen studies. *In:* Antieau, J. M., The Palo Verde Archaeological Investigations, Hohokam Settlement at the Confluence: Excavations along the Palo Verde Pipeline. *Museum of Northern Arizona Research Paper* 20, 335-345. (Maricopa County)

1981 Pollen studies in the Verde Valley, Arizona. *In:* Stebbins, S., Weaver, D. E., Jr., and Dosh, S. G., Archaeological Investigations at the Confluence of the Verde River and West Clear Creek. *Museum of Northern Arizona Research Paper* 24, 130-141. (Yavapai County)

1982 Pollen results from The Crawford Site. *In:* Whitten, P., Excavations at the Crawford Site, A Basketmaker-Pueblo I Site Near Crownpoint, New Mexico. *San Juan County Archaeological Research Center and Library, Contributions to Anthropology Series* 307, 161-166. (McKinley County)

1982 Pollen results from two small sites in San Juan County, northwestern New Mexico. *In:* Whitten, P., Excavations at Four Pueblo II Sites on the San Juan Coal Lease, northwestern New Mexico. *San Juan County Archaeological Research Center and Library, Contributions to Anthropology Series* 515, 110-116.

1982 Results from the Prewitt Project, northwestern New Mexico. *In:* Beal, J. D., Archaeological Investigations in the Eastern Red Mesa Valley: The Plains/Escalante Generating Station. *School of American Research Report* 005, 280-305. (McKinley County)

1982 Pollen results from NM:12:V2:98 and NM:12:V2:108, and a summary of the Jones Ranch Road Project pollen study, northwestern New Mexico. Report to the Pueblo of Zuni Archaeology Program, Zuni, 39 p., 2 figs., 4 tables. (McKinley County)

1982 Palynological results of the Coronado Project. *In:* Gasser, R. E., The Coronado Project Archaeological Investigations, the specialists' volume: biocultural analyses. *Museum of Northern Arizona Research Paper 23,* 96-224. (Apache County)

1983 Pollen results from the Gamerco Project, New Mexico. *In:* Scheick, C., The Gamerco Project: Flexibility as an Adaptive Response. *School of American Research Report* 71, 672-699. (McKinley County)

1983 Pollen results from AZ T:12:42 (ASM), Phoenix, Arizona. Report to Soil Systems, Inc., Phoenix, 12 p., 1 fig., 1 table. (Maricopa County)

1983 Preliminary pollen results from seven pre-historic and historic sites in the San Juan Basin of northwestern, New Mexico. Report to the School of American Research, Santa Fe, 14 p., 2 tables. (Valencia County)

1983 Palynology of the Calmark Site, central Arizona. Report to Soil Systems, Inc., Phoenix, 31 p., 4 figs., 5 tables. (Maricopa County)

1983 Palynology of Site HR 100 near La Plata, northwestern New Mexico. *In:* Scheick, C., HR 100: Excavations in the Dead Zone. *School of American Research Report* 102, 71-78. (San Juan County)

1983 Pollen results from Project 096, San Juan Basin, northwestern New Mexico. Report to the School of American Research, Santa Fe, 46 p., 7 figs., 8 tables. (San Juan County)

1983 Pollen results from LA 30949, a Mimbres site on the Mescalero Apache Indian Reservation, south-central New Mexico. Report to the Agency for Conservation Archaeology, Eastern New Mexico University, Portales, 8 p., 1 table. (Otero County)

1983 Pollen results from NA 15,909, the Adobe Dam Site. *In:* Bruder, J. S., Archaeological Investigations in the Adobe Dam Project Area. *Museum of Northern Arizona Research Paper* 27, 253-267.

1983 Pollen analysis. *In:* Anyon, R., Collins, S. M., and Bennett, K. H. (eds.), Archaeological Investigations between Manuelito Canyon and Whitewater Arroyo, Northwest New Mexico. Vol. 2. *Zuni Archaeology Program Report* 185, 649-698.

1984 Pollen results from the New River Project, central Arizona, and a discussion of pollen sampling strategies for agricultural systems. Report to Soil Systems, Inc., Phoenix. 145 p., 6 figs., 8 tables.

1984 Pollen results from the Wilson Project sites, near Flagstaff, Arizona. Report to Northland Research, Flagstaff, 16 p., 1 fig., 1 table.

1984 Pollen results from Project 117, McKinley County, northwestern New Mexico. Report to the School of American Research, Santa Fe, 23 p., 3 tables.

GRAY, J.

1961 Early Pleistocene paleoclimatic record from Sonoran Desert. *Science,* 133: 38-39.

GREEN, F. E.

1961 Discussion of the pollen and stratigraphic data. *In:* Wendorf, F., Paleoecology of the Llano Estacado. Museum of New Mexico Press, Santa Fe. *Fort Burgwin Research Center Publication* 1, 94-97. (pollen analysis by U. Hafsten)

GREENHOUSE, R., GASSER, R. E., and GISH, J. W.

1981 Cholla bud roasting pits: an ethnoarchaeological example. *The Kiva,* 46: 227-242.

HAFSTEN, U.

1959 Bleaching + HF + acetolysis—a hazardous preparation process. *Pollen et Spores,* 1: 77-79.

1961 Pleistocene development of vegetation and climate in the Southern High Plains as evidenced by pollen analysis. *In:* Wendorf, F., Paleoecology of the Llano Estacado. Museum of New Mexico Press, Santa Fe, *Fort Burgwin Research Center Publication* 1, 59-91.

1964 A standard pollen diagram for the Southern High Plains, USA, covering the period back to the Early Wisconsin Glaciation. International Union for Quaternary Research (INQUA), Report of VI Congress, Warsaw, Poland, Vol. 2, Paleobotanical Section, 407-420.

HAGER, M. W.

1975 Late Pliocene and Pleistocene history of the Donnelly Ranch vertebrate site, southeastern Colorado. University of Wyoming, *Contributions to Geology, Special Paper* 2, 62 p. (pollen analysis by J. Beiswenger)

HALL, S. A.

1973 Pollen inventory of two horizons of the Ruby-1 Site, near Sapello, New Mexico. Unpublished report, University of Michigan, Ann Arbor, 2 p. (San Miguel County)

1975 Stratigraphy and palynology of Quaternary alluvium at Chaco Canyon, New Mexico. Ph.D. dissertation, University of Michigan, Ann Arbor, 66 p.

1977 Late Quaternary sedimentation and paleoecologic history of Chaco Canyon, New Mexico. *Geological Society of America Bulletin,* 88: 1593-1618.

1981 Deteriorated pollen grains and the interpretation of Quaternary pollen diagrams. *Review of Palaeobotany and Palynology,* 32: 193-206.

1981 Holocene vegetation at Chaco Canyon: pollen evidence from alluvium and pack rat middens. Society for American Archaeology, 46th Ann. Mtg., San Diego, *Program and Abstracts,* 61.

1982 Reconstruction of local and regional Holocene vegetation in the arid Southwestern United States based on combined pollen analytical results from *Neotoma* middens and alluvium: International Union for Quaternary Research (INQUA), XI Congress, Moscow, U.S.S.R., *Abstracts,* 1: 130.

1983 Palynology of the Middle Pleistocene Alamosa local fauna, southern Colorado. Manuscript, North Texas State University, Denton, and Adam State College, Alamosa.

1983 Holocene stratigraphy and paleoecology of Chaco Canyon. *In:* Wells, S. G., Love, D. W., and Gardner, T. W. (eds.), Chaco Canyon Country. American Geomorphological Field Group, 1983 Field Trip Guidebook, 219-226.

1984 Pollen analysis of Rocky Arroyo Site, a pilot study. Report to Museum of New Mexico, Santa Fe. (Chaves County)

1984   Pollen analysis of the Garnsey Bison Kill Site, southeastern New Mexico. *In:* Parry, W. J. and Speth, J. D., The Garnsey Spring Campsite: late prehistoric occupation in southeastern New Mexico. *Museum of Anthropology, University of Michigan, Technical Reports* 15, 85-108.

1984   Pollen influx and distribution in the Southern Rockies and SW Plains (U.S.A.). Sixth International Palynological Conference (IPC), Calgary, Alberta, Canada. International Association for Aerobiology Symposium, *Abstracts.*

1984   A late Holocene record of short-term vegetation change in the semiarid grassland of SE New Mexico (U.S.A.). Sixth International Palynological Conference (IPC), Calgary, Alberta, Canada, *Abstracts,* 59.

1984   A major drought in the Southwest Plains during the 15th century A.D.: evidence from pollen analysis. American Quaternary Association (AMQUA), 8th biennial meeting, Boulder, Colorado, *Program and Abstracts,* 55.

HANSEN, B. S., and CUSHING, E. J.

1973   Identification of pine pollen of late Quaternary age from the Chuska Mountains, New Mexico. *Geological Society of America Bulletin,* 84: 1181-1200.

HESTER, J. J.

1972   Blackwater Locality No. 1, A stratified early man site in eastern New Mexico. *Publication of the Fort Burgwin Research Center* 8, 239 p. (pollen analysis by P. B. Sears)

HEUSSER, C. J.

1977   A survey of Pleistocene pollen types of North America. *In:* Elsik, W. C. (ed.), Cenozoic Palynology. *American Association of Stratigraphic Palynologists Contribution Series* 5A, 111-129.

HEVLY, R. H.

1962   Pollen analysis of Laguna Salida. *In:* New Mexico Geological Society, 13th Field Conference, *Guidebook of the Mogollon Rim region, east-central Arizona,* 115-117.

1964   Pollen analysis of the Quaternary archaeological and lacustrine sediments from the Colorado Plateau. Ph.D. dissertation, University of Arizona, Tucson, 124 p.

1964   Paleoecology of Laguna Salada. *In:* Martin, P. S., *et al.,* Chapters in the prehistory of eastern Arizona, II. Chicago Natural History Museum, *Fieldiana: Anthropology,* 55: 171-187.

1968   Studies of the modern pollen rain in northern Arizona. *Journal of the Arizona Academy of Science,* 5: 116-127.

1968   Sand Dune Cave pollen studies. *In:* Lindsay, A. J., Jr., Ambler, J. R., Stein, M. A., and Hobler, P. M., Survey and excavations north and east of Navajo Mountain, Utah, 1959-1962. *Museum of Northern Arizona Bulletin* 45 (Glen Canyon Series no. 8), 393-397.

1970   Botanical studies of sealed storage jar cached near Grand Falls, Arizona. *Plateau,* 42: 150-156.

1974   Recent paleoenvironments and geological history at Montezuma Well. *Journal of the Arizona Academy of Science,* 9: 66-75.

1979   Paleocology of Holocene and Pleistocene lacustrine sediments from Zion National Park, Utah. *In:* Linn, R. M. (ed.), Proceedings of the First Conference on Scientific Research in the National Parks, Volume 1. *U.S. National Park Service Transactions and Proceedings Series* 5, 151-158.

1981   Pollen production, transport, and preservation: potentials and limitations in archaeological palynology. *Journal of Ethnobiology,* 1: 39-54.

1983   High altitude biotic resources, paleoenvironments, and demograhic patterns: southern Colorado Plateaus, A.D. 500-1400. *In:* Winter, J. C. (ed.), High Altitude Adaptations in the Southwest. U.S. Dept. of Agriculture, Forest Service, Southwest Region, *Cultural Resources Management Report* 2.

1984   Quaternary environments: Walker Lake, Coconino County, Arizona. Sixth International Palynological Conference (IPC), Calgary, Alberta, Canada, *Abstracts,* 63.

HEVLY, R. H., HEUETT, M. L., and OLSEN, S. J.

1978   Paleoecological reconstruction from an upland Patayan rock shelter, Arizona. *Journal of the Arizona-Nevada Academy of Science,* 13: 67-78.

HEVLY, R. H., HULBERT, C., and JEFFERS, C.

1980   Palynology. *In:* Kirkpatrick, D. T. (ed.), Prehistory and History of the Ojo Amarilla. New Mexico State University, Las Cruces, *Cultural Resource Mangement Report* 276.

HEVLY, R. H., KARLSTROM, T. N. V.

1974   Southwest paleoclimatic and continental correlations. *In:* Karlstrom, T. N. V., Swann, G. A., and Eastwood, R. L. (eds.), *Geology of northern Arizona with notes on archaeology and paleoclimate.* Geological Society of America, Rocky Mountain Section mtg., Flagstaff, Arizona, 257-295.

1975   Southwest paleoclimate and continental correlations. *In:* Horie, S. (ed.), *Paleolimnology of Lake Biwa and the Japanese Pleistocene.* Vol. 3, 455-493. (reprint of Hevly and Karlstrom, 1974)

HEVLY, R. H., KELLY, R. E., ANDERSON, G. A., and OLSEN, S. J.

1979   Comparative effects of climatic change, cultural impact, and volcanism in the paleoecology of Flagstaff, Arizona, A. D. 900-1300. *In:* Sheets, P. D., and Grayson, D. K. (eds.), *Volcanic Activity and Human Ecology.* Academic Press, N.Y., 487-523.

HEVLY, R. H., and MARTIN, P. S.

1961   Geochronology of Pluvial Lake Cochise, southern Arizona: I. pollen analysis of shore deposits. *Journal of the Arizona Academy of Science,* 2: 24-31.

HEVLY, R. H., MEHRINGER, P. J., JR., and YOCUM, H. G.

1965   Modern pollen rain in the Sonoran Desert. *Journal of the Arizona Academy of Science,* 3: 123-135.

HEVLY, R. H., and RENNER, L. E.

1979   Atmospheric pollen and spores in Flagstaff, Arizona. Unpublished manuscript.

HILL, J. N., and HEVLY, R. H.

1968   Pollen at Broken K Pueblo: some new interpretations. *American Antiquity,* 33: 200-210.

HOLBROOK, S. J., and MACKEY, J. C.

1976   Prehistoric environmental change in northern New Mexico: evidence from a Gallina Phase archaeological site. *The Kiva,* 41: 309-317.

HOROWITZ, A., GERALD, R. E., and CHAIFFETZ, M. S.

1981   Preliminary paleoenvironmental implications of pollen analyzed from Archaic, Formative, and Historic Sites near El Paso, Texas. *Texas Journal of Science,* 33: 61-72.

IBERALL, E. R.

1972   Paleoecological studies from fecal pellets: Stanton's Cave, Grand Canyon, Arizona. M.S. thesis, University of Arizona, Tucson.

JACOBS, B. F.

1983 Past vegetation and climate of the Mogollon Rim area, Arizona. Ph.D. dissertation, University of Arizona, Tucson, 166 p.

1984 Identification of Southwestern pine species applied to a pollen record from Hay Lake, Arizona. Sixth International Palynological Conference (IPC), Calgary, Alberta, Canada, *Abstracts,* 71.

JELINEK, A. J.

1966 Correlation of archaeological and palynological data. *Science,* 152: 1507-1509. (pollen analysis by P. S. Martin)

1967 A prehistoric sequence in the Middle Pecos Valley, New Mexico. *Museum of Anthropology, University of Michigan, Anthropological Papers* 31. (pollen analysis by P. S. Martin)

JONES, J. G.

1984 Kiet Siel coprolite analysis. Report to Navajo National Monument, National Park Service, Arizona.

KAPP, R. O.

1965 Illinoian and Sangamon vegetation in southwestern Kansas and adjacent Oklahoma. University of Michigan, *Contributions from the Museum of Paleontology,* 19 (14): 167-255.

1969 *How to Know Pollen and Spores.* Wm. C. Brown Co. Publishers, Dubuque, Iowa, 249 p.

KAUFFMAN, E. G. and McCULLOCK, D. S.

1965 Biota of a late glacial Rocky Mountain pond. *Geological Society of America Bulletin,* 76: 1203-1232.

KELSO, G. K.

1970 Hogup Cave, Utah: comparative pollen analysis of human coprolites and cave fill. *In:* Aikens, C. M., Hogup Cave. *University of Utah Anthropological Papers* 93, 251-262.

1976 Absolute pollen frequencies applied to the interpretation of human activities in northern Arizona. Ph.D. dissertation, University of Arizona, Tucson, 170 p.

1982 Two pollen profiles from Grasshopper Pueblo. *In:* Longacre, W. A., Holbrook, S. J., and Graves, M. W. (eds.), Multidisciplinary research at Grasshopper Pueblo, Arizona. *Anthropological Papers of the University of Arizona* 40, 106-109.

KING, F. B.

1977 An evaluation of the pollen contents of coprolites as environmental indicators. *Journal of the Arizona Academy of Science,* 12: 47-52.

1984 A Holocene pollen record from Thomas Lake bog, Mt. Sporis, Colorado. American Quaternary Association (AMQUA), 8th biennial meeting, Boulder, Colorado, *Program and Abstract,* 66.

KING, J. E.

1964 Modern pollen rain and fossil profiles, Sandia Mountains, New Mexico. M.S. thesis, University of New Mexico, Albuquerque, 50 p.

1967 Modern pollen rain and fossil pollen in soils in the Sandia Mountains, New Mexico. *Papers of the Michigan Academy of Science, Arts, and Letters,* 52: 31-41.

KING, J. E., and SIGLEO, W. R.

1973 Modern pollen in the Grand Canyon, Arizona. *Geoscience and Man,* 7: 73-81.

KING, J. E., and VAN DEVENDER, T. R.

1977 Pollen analysis of fossil packrat middens from the Sonoran Desert. *Quaternary Research,* 8: 191-204.

KREMP, G. O. W.

1978 Pliocene palynological literature: five hundred implemented references. *Paleo Data Banks* 9, 50 p.

KURTZ, E. B., JR., and TURNER, R. M.

1957 An oil-flotation method for the recovery of pollen from inorganic sediments. *Micropaleontology,* 3: 67-68.

LAUDERMILK, J. D., and MUNZ, P. A.

1938 Plants in the dung of *Nothrotherium* from Rampart and Muav Caves, Arizona. *In: Studies on Cenozoic Vertebrates of Western North America.* Carnegie Institute of Washington Publication 487, 273-281.

LEBOWITZ, M. D., O'ROURKE, M. K., DODGE, R., HOLBERG, C. J., CORMAN, G., HO SHAW, R., PINNAS, J. L., BARBEE, R. A., and SHELLER, M. R.

1982 The adverse health effects of biological aerosols, other aerosols, and indoor microclimate on asthmatics and nonasthmatics. *Environment International,* 8: 375-380.

LEGG, T. E.

1977 Palynology of middle Pinedale sediments in Devlins Park, Boulder County, Colorado: M.S. thesis, University of Iowa, Iowa City.

LEGG, T. E., and BAKER, R. G.

1980 Palynology of Pinedale sediments, Devlins Park, Boulder County, Colorado. *Arctic and Alpine Research,* 12: 319-333.

LEOPOLD, E. B.

1969 Late Cenozoic palynology. *In:* Tschudy, R. H., and Scott, R. A. (eds.), *Aspects of Palynology. An Introduction to Plant Microfossils in Time.* Wiley-Interscience, N. Y., 377-438.

LEOPOLD, E. B., and WHEAT, J. B.

1972 Palynology of the Olsen-Chubbuck Site. *In:* Wheat, J. B., The Olsen-Chubbuck Site, A Paleo-Indian Bison Kill. *American Antiquity,* 37 *(Society for American Archaeology Memoir,* 26), 178-180.

LEOPOLD, L. B., LEOPOLD, E. B., and WENDORF, F.

1963 Some climatic indicators in the period A. D. 1200-1400 in New Mexico. *In:* UNESCO, *Changes of Climate.* Arid Zone Research 20, 265-270.

LEWIS, W. H., VINAY, P., and ZENGER, V. E.

1983 *Airborne and Allergenic Pollen of North America.* Johns Hopkins University Press, Baltimore, 254 p.

LIMON, A. E.

1983 Determination of chronology and room function at AZ J:6:1 (ASM) using pollen analysis. M.A. thesis, Arizona State University, Tempe, 68 p.

LINDSAY, A. J., JR.

1958 Fossil pollen and its bearing on the archaeology of the Lehner mammoth site. M.S. thesis, University of Arizona, Tucson, 81 p.

LINDSAY, L. W.

1974 Preliminary palynological studies on Clear Mesa. *In:* Wilson, C. J. (ed.), *Highway U-95 archaeology: Comb Wash to Grand Flat, Vol. 2.* A Special Report, Department of Anthropology, University of Utah, Salt Lake city, 155-176.

1975 Palynological analysis and paleoecology of Innocents Ridge. *In:* Schroedl, A. R. and Hogan, P. F., Innocents Ridge and the San Rafael Fremont. Utah State Historical Society, Salt Lake City, *Antiquities Section Selected Papers,* 1(2): 61-64.

1976   Site paleoecology, palynology, and macrofossil analysis. *In:* Lindsay, L. W., and Lund, C. K., Pint-size Shelter. Antiquities Section, Division of State History, State of Utah, *Antiquities Section Selected Papers* 10, 67-74.

1980   Palynology of Cowboy Cave cultural deposits. *In:* Jennings, J. D., Schroedl, A. R., and Holmer, R. N., Cowboy Cave. *University of Utah Anthropological Papers* 104.

1981   Pollen analysis of Sudden Shelter Site deposits. *In:* Jennings, J. D., Schroedl, A. R., and Holmer, R. N., Sudden Shelter. *University of Utah Anthropological Papers* 103.

1981   Big Westwater Ruin. Bureau of Land Management, Salt Lake City, Utah, *Cultural Resource Series* 9. (San Juan County)

1982   Pollen analysis of soil samples from cultural features at site 42Sa9937, San Juan County, Utah. Environmental Studies Group, Inc., Salt Lake City.

1983   Pollen analysis of La Cabrana Site deposits and cultural features. Environmental Studies Group, Inc., Salt Lake City.

1983   Pollen analysis of soil samples from site 42Sa6565 deposits and cultural features. Environmental Studies Group, Inc., Salt Lake City.

1983   Pollen analysis of Black Rock Cave deposits and cultural features. *In:* Madsen, D. B., Black Rock Cave Revisited. Bureau of Land Management, *Cultural Resource Series* 15.

1983   Pollen analysis of Antelope Cave cultural deposits and coprolite samples. *In:* Janetski, J., and Hall, M. J., *An Archaeological and Geological Assessment of Antelope Cave in Mohave County, northwestern Arizona.* A Special Report, Cultural Resource Management Services, Department of Anthropology, Brigham Young University, Provo.

LINDSAY, L. W., and DYKMAN, J. L.
1978   Westwater Ruin, San Juan County, Utah. A Special Report. Antiquities Section, Division of State History, Salt Lake City.

LINDSAY, L. W., and LUND, C. K.
1976   Pint-size Shelter. *Antiquities Section Selected Papers* 3(10), Salt Lake City.

LINDSAY, L. W., and MADSEN, D. B.
1984   Pollen analysis of Recapture Wash archeological sites. Environmental Studies Group, Inc., Salt Lake City.

LIPE, W. D., BREED, W. J., WEST, J., and BATCHELDER, G.
1975   Lake Pagahrit, southeastern Utah, a preliminary research report. *In:* Fassett, J. E. (ed.), *Canyonlands County.* Four Corners Geological Society Guidebook, 8th Field Conference, 103-110.

LYTLE-WEBB, J.
1974   Pollen analysis. *In:* Reynolds, W. E., Archaeological investigations at Arizona U:9:45 (ASM): a limited activity site. Report to Arizona Archaeological Center, National Park Service.

1978   The environment of Miami Wash, Gila County, Arizona. Ph.D. dissertation, University of Arizona, Tucson.

1978   Pollen analysis in Southwestern archaeology. *In:* Grebinger, P. (ed.), *Discovering Past Behavior, Experiments in the Archaeology of the Southwest.* Gordon and Breach Press, 13-28.

1981   Pollen analysis of irrigation canals. *The Kiva,* 47: 83-90.

MACKEY, J. C., and HOLBROOK. S. J.
1978   Environmental reconstruction and the abandonment of the Largo-Gallina area, New Mexico. *Journal of Field Archaeology,* 5: 29-49.

MADOLE, R. F., BACHHUBER, F. W., and LARSON, E. E.
1973   Geomorphology, palynology, and paleomagnetic record of glacial Lake Devlin, Front Range. Geological Society of America, Rocky Mountain Section, 26th Annual Meeting, *Guidebook,* Trip 1, 25 p.

MADOLE, R. F., BAKER, R. G., ROSEBAUM, J. G., and LARSON, E. E.
1984   Geology of a late Quaternary lake and Pinedale glacial history, Front Range, Colorado. American Quaternary Association (AMQUA), 8th biennial meeting, Boulder, Colorado. *Guidebook for Field Trip* 3, 28 p.

MADSEN, D. B.
1971   O'Malley Shelter. M.A. thesis, University of Utah, Salt Lake City.

1972   Paleoecological investigations in Meadow Valley Wash, Nevada. *In:* Fowler, D. D., (ed.), Great Basin Cultural Ecology: A Symposium. *Desert Research Institute Publications in the Social Science,* 8, 57-65.

1973   The pollen analysis of O'Malley Shelter. *In:* Fowler, D. D., Madsen, D. B., and Hattori, E. M., Prehistory of Southeastern Nevada. *Desert Research Institute Publications in the Social Sciences,* 6, 137-142.

1976   Pluvial—post-pluvial vegetation changes in the southeastern Great Basin. *In:* Elston, R. (ed.), Holocene environmental change in the Great Basin. University of Nevada, *Nevada Archaeological Survey Research Paper* 6, 104-119.

MADSEN, D. B., and CURREY, D. R.
1979   Late Quaternary glacial and vegetation changes, Little Cottonwood Canyon area, Wasatch Mountains, Utah. *Quaternary Research,* 12: 254-270.

MADSEN, D. B., and KAY, P. A.
1982   Late Quaternary pollen analysis in the Bonneville Basin. American Quaternary Association (AMQUA), 7th biennial conference, Seattle, Washington, *Program and Abstracts,* 128.

MAHER, L. J., JR.
1961   Pollen analysis and postglacial vegetation history in the Animas Valley region, southern San Juan Mountains, Colorado. Ph.D. dissertation, University of Minnesota, Minneapolis, 85 p.

1963   Pollen analyses of surface materials from the southern San Juan Mountains, Colorado. *Geological Society of America Bulletin,* 74: 1485-1504.

1972   Absolute pollen diagram of Redrock Lake, Boulder County, Colorado. *Quaternary Research,* 2: 531-553.

1973   Pollen evidence suggests that climatic changes in the Colorado Rockies during last 5,000 years were out of phase with those in the northeastern United States. International Union for Quaternary Research (INQUA), IX Congress, Christchurch, New Zealand, *Abstracts,* 227-228.

MARKGRAF, V., BRADBURY, J. P., FORESTER, R. M., McCOY, W., SINGH, G., and STERNBERG, R.
1983   Paleoenvironmental reassessment of the 1.6-million-year-old record from San Agustin Basin, New Mexico. New Mexico Geological Society Guidebook, 34th Field Conference, Socorro Region, 291-297.

MARKGRAF, V., BRADBURY, J. P., FORESTER, R. M., SINGH, G., and STERNBERG, R. S.

1984   San Agustin Plains, New Mexico: age and paleoenvironmental potential reassessed. *Quaternary Research*, 22: 336-343.

MARKGRAF, V., BRADBURY, J. P., SINGH, G., and STERNBERG, R.

1982   Reassessment of the 1.6 m yr record from San Augustin Plains, New Mexico. American Quaternary Association (AMQUA), 7th biennial Conference, Seattle, Washington, *Program and Abstracts*, 130.

MARKGRAF, V., and SCOTT, L.

1981   Lower timberline in central Colorado during the past 15,000 years. *Geology*, 9: 231-234.

MARTIN, P. S.

1962   On attempting to explain the post-13th century pine rise. *Pollen et spores* 4: 364.

1963   Early man in Arizona, the pollen evidence. *American Antiquity*, 29: 67-73.

1963   *The Last 10,000 years, A Fossil Pollen Record of the American Southwest.* University of Arizona Press, Tucson, 87 p.

1963   Geochronology of Pluvial Lake Cochise, southern Arizona. II. pollen analysis of a 42-meter core. *Ecology*, 44: 436-444.

1964   Pollen analysis in the Glen Canyon area. *In:* Sharrock, F. W., *et al.* (eds.), 1962 Excavations, Glen Canyon Area. *University of Utah Anthropological Papers* 73 (Glen Canyon Series 25), 176-195.

1964   Pollen analysis and the full glacial landscape. *In:* Hester, J. J., and Schoenwetter, J., The Reconstruction of Past Environments. *Fort Burgwin Research Center Publication* 3: 66-75.

1967   Pollen analysis of prehistoric middens near Ft. Sumner, New Mexico. *In:* Jelinek, A. J., A Prehistoric Sequence in the Middle Pecos Valley, New Mexico. *Museum of Anthropology, University of Michigan, Anthropological Papers* 31, 130-134.

1969   Pollen analysis and the scanning electron microscope. *In:* Jahari, C. O. (ed.), *Scanning Electron Microscopy/1969; Proceeding of the 2nd Annual Scanning Electron Microscope Symposium*, IIT Research Institute, Chicago, 89-102.

1970   Vegetation of the Southwest between 14,000 and 9,000 years ago. American Quaternary Association (AMQUA), 1st mtg., Bozeman, *Abstracts*, 87.

1980   Pollen stratigraphy of Long House. *In:* Cattanach, G. S., Jr., Long House, Mesa Verde National Park, Colorado (Wetherill Mesa Studies). *National Park Service, Publications in Archaeology* 7-H, 401-405.

1983   Pollen profile from the east bank of Cienega Creek. *In:* Eddy, F. W., and Cooley, M. E., Cultural and environmental history of Cienega Creek, southeastern Arizona. *Anthropological Papers of the University of Arizona* 43, 42-44.

MARTIN, P. S., and BYERS, W.

1965   Pollen and archaeology at Wetherill Mesa. *In:* Osborne, D., Contributions of the Wetherill Mesa Archaeological Project. *American Antiquity*, 31 *(Society for American Archaeology Memoir* 19), 122-135.

MARTIN, P. S., and DREW, C. M.

1969   Scanning electron photomicrographs of Southwestern pollen grains. *Journal of the Arizona Academy of Science*, 5: 147-176.

1970   Additional scanning electron photomicrographs of Southwestern pollen grains. *Journal of the Arizona Academy of Science*, 6: 140-161.

MARTIN, P. S., and GRAY, J.

1962   Pollen analysis and the Cenozoic. *Science*, 137: 103-111.

MARTIN, P. S., and MEHRINGER, P. J., JR.

1965   Pleistocene pollen analysis and biogeography of the Southwest. *In:* Wright, H. E., Jr., and Frey, D. G. (eds.), *The Quaternary of the United States.* Yale University Press, New Haven, 433-451.

MARTIN, P. S., and MOSIMANN, J. E.

1965   Geochronology of Pluvial Lake Cochise, southern Arizona, III. pollen statistics and Pleistocene metastability. *American Journal of Science*, 263: 313-358.

MARTIN, P. S., SABELS, B. E., and SHUTLER, D., JR.

1961   Rampart Cave coprolite and ecology of the Shasta ground sloth. *American Journal of Science*, 259: 102-127.

MARTIN, P. S., and SCHOENWETTER, J.

1960   Arizona's oldest cornfield. *Science*, 132: 33-34.

n.d.   Pollen stratigraphy of a great kiva from Chaco Canyon. Geoscience Department, University of Arizona, Tucson.

MARTIN, P. S., SCHOENWETTER, J., and ARMS, B. C.

1961   Palynology and prehistory: the last 10,000 years. Geochronology Laboratories, University of Arizona, Tucson, 119 p. (multilithed)

MARTIN, P. S., and SHARROCK, F. W.

1964   Pollen analysis of prehistoric human feces: a new approach to ethnobotany. *American Antiquity*, 30: 168-180.

McANDREWS, J. H., and KING, J. E.

1976   Pollen of the North American Quaternary: the top twenty. *Geoscience and Man*, 15: 41-49.

MEHRINGER, P. J., JR.

1965   Late Pleistocene vegetation in the Mohave Desert of southern Nevada. *Journal of the Arizona Academy of Science*, 3: 172-188.

1967   Pollen analysis of the Tule Springs Site area, Nevada. *In:* Wormington, H. M., and Ellis, D. (eds.), Pleistocene studies in southern Nevada. *Nevada State Museum Anthropological Papers* 13, 129-200.

1967   The environment of extinction of the late-Pleistocene megafauna in the arid southwestern United States. *In:* Martin, P. S., and Wright, H. E., Jr. (eds.), *Pleistocene Extinctions, The Search for a Cause.* Yale University Press, New Haven, 247-266.

1967   Pollen analysis and the alluvial chronology. *The Kiva*, 32: 96-101.

1977   Great Basin late Quaternary environments and chronology. *In:* Fowler, D. D. (ed.), Models and Great Basin Prehistory: A Symposium. *Desert Research Institute Publications in the Social Sciences* 12, 113-167.

MEHRINGER, P. J., JR., ADAM, D. P., and MARTIN, P. S.

1971   Pollen analysis at Lehner Ranch arroyo. *In:* American Association of Stratigraphic Palynologists, Field Trip Guide, *Lehner Early Man-Mammoth Site*, 10-26.

MEHRINGER, P. J., JR., and HAYNES, C. V., JR.

1965   The pollen evidence for the environment of early man and extinct mammals at the Lehner mammoth site, southeastern Arizona. *American Antiquity*, 31: 17-23.

MEHRINGER, P. J., JR., MARTIN, P. S., and HAYNES, C. V., JR.

1967   Murray Springs, a mid-postglacial pollen record from southern Arizona. *American Journal of Science,* 265: 786-797.

MERRILL, R. K.

1974   The late Cenozoic geology of the White Mountains. Ph.D. dissertation, Arizona State University, Tempe.

MILLINGTON, A. C.

1976   Late Quaternary paleo-environmental history of the Mary Jane Creek Valley, Grand County, Colorado: M.A. thesis, University of Colorado, Boulder, 194 p.

MOBLEY, C. M.

1978   Archaeological research and management at Los Esteros Reservoir, New Mexico. Archaeology Research Program, Southern Methodist University, Dallas, Texas. (pollen analysis by S. A. Hall and J. Zauderer)

MURRY, R. E., JR.

1983   Pollen analysis of Anasazi sites at Black Mesa, Arizona. M.A. thesis, Texas A & M University, College Station, 177 p.

NELSON, A. R., MILLINGTON, A. C., ANDREWS, J. T., and NICHOLS, H.

1979   Radiocarbon-dated upper Pleistocene glacial sequence, Fraser Valley, Colorado Front Range. *Geology,* 7: 410-414.

NICHOLS, H.

1982   Review of late Quaternary history of vegetation and climate in the mountains of Colorado. *In:* Halfpenny, J. C. (ed.), Ecological studies in the Colorado alpine, A festschrift for John W. Marr. University of Colorado, *Institute of Arctic and Alpine Research, Occasional Paper* 37, 27-33.

NICHOLS, H., SHORT, S., ELIAS, S., and HARBOR, J.

1984   Holocene sedimentation, palynology, and paleoecology of alpine and subalpine lakes and peat bogs, Colorado Front Range. American Quaternary Association (AMQUA), 8th biennial meeting, Boulder, Colorado, *Guide to Field Trip* 11, 21 p.

O'LAUGHLIN, T. C.

1980   The Keystone Dam Site and Other Archaic and Formative Sites in Northwest El Paso, Texas. El Paso Centennial Museum, University of Texas at El Paso, *Publications in Anthropology* 8, 283 p. (pollen analysis by A. C. Cully and K. H. Clary)

OLDFIELD, F.

1975   Pollen-analytical results, Part II. *In:* Wendorf, F., and Hester, J. J. (eds.), Late Pleistocene Environments of the Southern High Plains. *Publication of the Fort Burgwin Research Center* 9, 121-147.

OLDFIELD, F., and SCHOENWETTER, J.

1964   Late Quaternary environments and early man on the Southern High Plains. *Antiquity,* 38: 226-229.

1970   Pollen analysis of late Pleistocene deposits in West Texas and eastern New Mexico. American Quaternary Association (AMQUA), 1st mtg., Bozeman, *Abstracts,* 102.

1975   Discussion of the pollen-analytical evidence. *In:* Wendorf, F., and Hester, J. J. (eds.), Late Pleistocene Environments of the Southern High Plains. *Publication of the Fort Burgwin Research Center* 9, 149-177.

O'ROURKE, M. K.

1980   Pollen samples from site AZ AA:16:57. *In:* Nickerson, T., Archaeological investigations at AZ AA:16:57. *Arizona State Museum Cultural Resource Management Section Archaeological Series* 139.

1980   Pollen disposal and its relationship to respiratory illness. *In:* Federal Environmental Agency (ed.), *Proceedings: First International Conference on Aerobiology.* E. Schmidt, Berlin, 81-88.

1981   Pollen samples from AZ Z:1:8. *In:* Teauge, L. S., Test Excavations at Painted Rock Reservoir: Sites AZ Z:1:7, Z:1:8, and AZ S:16:36. *Arizona State Museum Cultural Resources Management Division Archaeological Series* 143.

1982   Pollen content and character of adobe brick from Fort Lowell. *In:* Huntington, F. (ed.), Archaeological data recovery at AZ BB:9:72 (ASM), the Band Quarter's kitchen and corral wall at Fort Lowell, and AZ BB:9:54 (ASM), a Rincon Phase habitation site, Craycroft Road, Tucson, Arizona. *Arizona State Museum Archaeological Series* 163: 75-85.

1983   Pollen from adobe brick. *Journal of Ethnobiology,* 3: 39-48.

1984   Airborne pollen from an arid region: differences between native and urban environments. Sixth International Palynological Conference (IPC), Calgary, Alberta, Canada, *Abstracts,* 120.

O'ROURKE, M. K., and LEBOWITZ, M. D.

1984   A comparison of regional atmospheric pollen with pollen collected at and near homes. *Grana* 23: 55-64.

O'ROURKE, M. K., MEADE, J. I., and MARTIN, P. S.

1984   Late Pleistocene and Holocene pollen records from three caves in the Grand Canyon of Arizona, USA. Sixth International Palynological Conference (IPC), Calgary, Alberta, Canada, *Abstracts,* 121.

PENNAK, R. W.

1963   Ecological and radiocarbon correlations in some Colorado mountain lake and bog deposits. *Ecology,* 44: 1-15.

PETERSEN, K. L.

1975   Exploratory palynology of a subalpine meadow, La Plata Mountains, southwestern Colorado. M.A. thesis, Washington State University, Pullman, 39 p.

1981   10,000 years of climatic change reconstructed from fossil pollen, La Plata Mountains, southwestern Colorado: Ph.D. dissertation, Washington State University, Pullman, 211 p.

1982   Review of "Lower timberline in central Colorado during the past 15,000 years." *Southwestern Lore,* 48: 22-24.

1984   Palynology of three sites in Montezuma County, southwest Colorado, U.S.A. Sixth International Palynological Conference (IPC), Calgary, Alberta, Canada, *Abstracts,* 127.

1984   Man and environment in the Dolores River valley, SW Colorado: some pollen evidence. American Quaternary Association (AMQUA), 8th biennial meeting, Boulder, Colorado, *Program and Abstracts,* 102.

PETERSEN, K. L., and MEHRINGER, P. J., JR.

1972 Paleobotany: pollen. *In:* Irwin-Williams, C. (ed.), The structure of Chacoan society in the northern Southwest, Investigations at the Salmon Site—1972. *Eastern New Mexico University Contributions in Anthropology* 4(3): 125-130.

1976 Postglacial timberline fluctuations, La Plata Mountains, southwestern Colorado. *Arctic and Alpine Research,* 8: 275-288.

PIPPIN, L. C.

1979 The prehistory and paleoecology of Guadalupe Ruin, Sandoval County, New Mexico. Ph.D. dissertation, Washington State University, Pullman, 435 p.

POTTER, L. D.

1967 Differential pollen accumulation in water-tank sediments and adjacent soils. *Ecology,* 48: 1041-1043.

POTTER, L. D., and ROWLEY, J.

1960 Pollen rain and vegetation, San Augustin Plains, New Mexico. *Botanical Gazette,* 122: 1-25.

POTTER, L. D., SCHOENWETTER, J., and OLDFIELD, F.

1975 Pollen-analytical procedures and methods of preservation. *In:* Wendorf, F., and Hester, J. J. (eds.), Late Pleistocene Environments of the Southern High Plains. *Publication of the Fort Burgwin Research Center* 9, 97-102.

PRICE, C. R.

1971 Preliminary paleopalynological analysis of Alamosa Formation sediments. *In:* James, H. L. (ed.), *Guidebook of the San Luis Basin, Colorado.* New Mexico Geological Society, 22nd Field Conference, 219-220.

RANKIN. A. G.

1978 Palynological investigations: Seneca Lake. *In:* Stafford, C. R., Archaeological investigations at Seneca Lake, San Carlos Indian Reservation, Arizona. *Arizona State University Anthropological Research Paper* 14.

1980 Pollen analytical studies in Corduroy Creek. *In:* Stafford, C. R., and Rice, G. E. (eds.), Studies in the prehistory of the Forestdale region, Arizona. Arizona State University Department of Anthropology, Tempe, *Anthropological Field Studies* 1.

1982 Analysis of pollen from AZ BB:9:54. *In:* Huntington, F., Archaeological data recovery at AZ BB:9:72 (ASM), the Band Quarters Kitchen and corral wall at Fort Lowell, and AZ BB:9:54 (ASM), a Rincon Phase habitation site, Craycroft Road, Tucson, Arizona. *Arizona State Museum Cultural Resource Management Division Archaeological Series* 163, 139-144.

ROGERS, K. L.

1984 A paleontological analysis of the Alamosa Formation (south-central Colorado: Pleistocene: Irvingtonian). *In:* New Mexico Geological Society Guidebook, 35th Field Conference. *Rio Grande Rift: Northern New Mexico,* 151-155. (pollen analysis by S. A. Hall)

ROHN, A. H.

1971 Mug House, Mesa Verde National Park, Colorado (Wetherill Mesa Excavations). *National Park Service Archeological Research Series* 7-D, 280 p. (pollen analysis by P. S. Martin and W. Byers.

ROSENBERG, B. H.

1977 An archaeological pollen study in Big House Canyon, New Mexico. M.A. thesis, Arizona State University, Tempe, 111 p.

ROSENBERG, B., and GISH, J. W.

1975 Preliminary pollen analysis of sediments from Gallinas Springs, Cibola National Forest, New Mexico. Report to Palynology Laboratory, Department of Anthropology, Arizona State University, Tempe.

RUPPE, T.

1978 A comparative pollen study of two Hohokam sites in the Salt River Valley. Manuscript, Palynology Laboratory, Department of Anthropology, Arizona State University, Tempe.

SCHOENWETTER, J.

1960 Pollen analysis of sediments from Matty Wash: M.S. thesis, University of Arizona, Tucson, 72 p.

1961 Pollen stratigraphy of the Wetherill Mesa region. Report, Geochronology Laboratories, University of Arizona, Tucson. (Long House)

1962 The pollen analysis of eighteen archaeological sites in Arizona and New Mexico. *In:* Martin, P. S., *et al.,* Chapters in the prehistory of eastern Arizona, I. Chicago Natural History Museum, *Fieldiana: Anthropology,* 53: 168-209.

1964 The palynological research. *In:* Schoenwetter, J., and Eddy, F. W., Alluvial and Palynological Reconstruction of Environments, Navajo Reservoir District. *Museum of New Mexico Papers in Anthropology* 13: 63-107.

1964 Phenology of allergen pollen of Santa Fe (1964). New Mexico Department of Public Health, Santa Fe, *Communicable Disease Summary* 11, 39.

1964 Pollen analysis of Cochiti Project materials. Report to Laboratory of Anthropology, Museum of New Mexico, Santa Fe.

1964 Site 222 + 50: palynological analysis. Report to National Park Service, Santa Fe, New Mexico.

1964 Palynological analysis of the Tohatchi-Crown Point project. Report to Laboratory of Anthropology, Museum of New Mexico, Santa Fe.

1964 Pollen analysis of LA 9152. Report to Laboratory of Anthropology, Museum of New Mexico, Santa Fe.

1965 Pollen studies at the Sapawe Site: preliminary report. Report to Department of Anthropology, University of New Mexico, Albuquerque. (New Mexico)

1965 Pollen studies at Picuris Pueblo: preliminary report. Report to Department of Anthropology, Adam State College, Alamosa, Colorado. (New Mexico)

1965 Preliminary palynological investigations on the Archaic horizon. Report to Department of Anthropology, Eastern New Mexico University, Portales. (New Mexico)

1965 Pollen analysis at Arizona 1:15:18. Report to Department of Anthropology, Northern Arizona University, Flagstaff.

1965 Pollen analysis of sediments from northeastern Colorado: preliminary report. Report to Department of Anthropology, University of Colorado, Boulder.

1965 Pollen statistics from the Gallisteo Basin. Report to Laboratory of Anthropology, Museum of New Mexico, Santa Fe. (Santa Fe County, New Mexico)

1965 Pollen studies at Reeves Ruin and the Davis Ranch Site: preliminary report. Report to R. E. Gerald, Department of Anthropology, University of Texas at El Paso. (southeastern Arizona)

1965 Utah W:5:50 palynological analysis. *In:* Schroeder, A. H., Salvage excavations at Natural Bridges National Monument. *University of Utah Anthropological Papers* 75, 102-104.

1966   A re-evaluation of the Navajo Reservoir pollen chronology. *El Palacio,* 73: 19-26.

1967   Pollen survey of the Chuska Valley. *In:* Harris, A. H., Schoenwetter, J., and Warren, A. H., An Archaeological Survey of the Chuska Valley and the Chaco Plateau, New Mexico. *Museum of New Mexico Research Records* 4 (Part I, Natural Science Studies), 72-103.

1967   Report on palynological investigations at Hopi Buttes. Report to U.S. Geological Survey, Flagstaff, Arizona. (northeastern Arizona)

1967   Preliminary palynological investigations on West Mesa. Report to T. Rinehart, Department of Anthropology, University of New Mexico, Albuquerque.

1967   Palynological analysis of sediment from the Largo-Blanco sites. Report to Laboratory of Anthropology, Museum of New Mexico, Santa Fe. (New Mexico)

1967   Pollen studies in Taos County. Report to Museum of Anthropology, University of New Mexico, Albuquerque. (New Mexico)

1968   Archaeological pollen studies at U:9:100. Palynology Laboratory, Department of Anthropology, Arizona State University, Tempe. (Maricopa County, Arizona)

1968   Palynological studies. *In:* Wilson, J. P., The Sinagua and Their Neighbors. Ph.D. dissertation, Harvard University, Cambridge.

1969   Pollen analysis in the Walnut Creek Basin, Arizona. Palynology Laboratory, Department of Anthropology, Arizona State University, Tempe.

1969   Pollen studies at AZ N:4:6. Palynology Laboratory, Department of Anthropology, Arizona State University, Tempe. (Arizona)

1970   Archaeological pollen studies of the Colorado Plateau. *American Antiquity,* 35: 35-48.

1973   Pollen studies at Wide Reed Ruin. Report to National Park Service, Western Archaeological Center, Tucson, Arizona. (Arizona)

1974   Palynological records of Joe's Valley Alcove: a multicomponent site in southwest Utah. Report to the U.S.D.A. Forest Service, Ogden, Utah.

1975   Pollen-analytical results, Part I. *In:* Wendorf, F., and Hester, J. J. (eds.), Late Pleistocene Environments of the Southern High Plains. *Publication of the Fort Burgwin Research Center* 9, 103-120.

1976   A test of the Colorado Plateau pollen chronology. *Journal of the Arizona Academy of Science,* 11: 89-96.

1976   Pollen report: Cibola area sites. Data to Institute of Archaeology, UCLA, California. (New Mexico)

1976   Pollen records of AZ:U:1:30 and :31 (ASU): an assessment. Palynology Laboratory, Department of Anthropology, Arizona State University, Tempe.

1977   A pollen study of the Cave Buttes locality. *In:* Rodgers, J. B., Archaeological investigation of the Granite Reef aquaduct, Cave Creek Archaeological District, Arizona. Arizona State University, Tempe, *Anthropological Research Papers* 12.

1977   The Seneca Lake pollen study: preliminary report. Report to Office of Cultural Resources Management, Arizona State University, Tempe.

1978   Report on the palynology of two Hohokam sites. Report to Archaeological Research Services, Tempe, Arizona. (Arizona)

1979   Initial assessments of the palynological record: Gila Butte-Santan region. *In:* Rice, G., Wilcox, D., Rafferty, K., and Schoenwetter, J., An archaeological test of sites in the Gila Butte-Santan region, south-central Arizona. Arizona State University, Tempe. *Anthropological Research Papers* 18, *Technical Paper* 3.

1980   Palynological test of AZ:U:16:6 (ASU). Report to Department of Anthropology, Arizona State University, Tempe. (Arizona)

1980   The Los Hornos pollen study. Report to Department of Anthropology, Arizona State University, Tempe. (Arizona)

1980   Archaeological pollen study of AR-4. *In:* Schaafsma, C. F., *The Cerrito Site (AR-4): a Piedra Lumbre Phase settlement at Abiquiu Reservoir.* School of American Research, Santa Fe. (New Mexico)

1980   Palynological studies. *In:* Thompson, M. (ed.), Implications of archaeological collections, tests, and excavations in the Carlsbad area. *New Mexico State University Cultural Resources Management Division Report* 433. (New Mexico)

1980   A pollen study of the Abiquiu Reservoir. *In:* Beal, J. D., 1979. *Sample and site specific testing program at Abiquiu Reservoir.* School of American Research Contract Archaeology Program, Santa Fe. Report to Corps of Engineers, Albuquerque District.

1982   Pollen records of the Walhalla Glades survey. Report to Grand Canyon National Park. (Arizona)

1983   Pollen statistics from Pratt Cave. El Paso Archaeological Society, *The Artifact,* 21: 155-159. (New Mexico)

SCHOENWETTER, J., and BALGEMAN, W. H.

1960   Pollen analysis of Dark Canyon Cave, New Mexico. Contribution 39, Program in Geochronology, University of Arizona, Tucson.

SCHOENWETTER, J., and DaCOSTA, V.

1976   Apache-Sitgreaves palynology. Report to U.S.D.A. Forest Service. (Arizona)

1976   Pollen studies in the Marble Canyon Area, Arizona. Report to Grand Canyon National Park.

SCHOENWETTER, J., and DITTERT, A. E., JR.

1968   An ecological interpretation of Anasazi settlement patterns. *In:* Meggers, B. J. (ed.), *Anthropological Archaeology in the Americas.* Anthropological Society of Washington, 41-66.

SCHOENWETTER, J., and DOERSCHLAG, L. A.

1970   Surficial pollen records for central Arizona: I. *Arizona State University Anthropological Research Paper* 3, 12 p.

1971   Surficial pollen records from central Arizona, I: Sonoran Desert scrub. *Journal of the Arizona Academy of Science,* 6: 216-221.

SCHOENWETTER, J., and MARTIN, P. S.

n.d.   Pollen analysis of alluvium from Binne-Ettini Canyon. Report, University of Arizona, Tucson. (New Mexico).

SCHOENWETTER, J., and RANKIN, A. G.

1976   Archaeological pollen study of two dune sites in New Mexico. Report to Laboratory of Anthropology, Museum of New Mexico, Santa Fe.

1977   Palynological investigations of the inundation studies program: 1976-77. Report to National Park Service, Santa Fe. (New Mexico)

SCHOENWETTER, J., and STEWART, E.

1978   Palynological chronology and antiquity estimation of Chavez Pass Ruin. Report to Department of Anthropology, Arizona State University, Tempe. (Arizona)

SCOFIELD, J. A. R.

1973   Pollen analysis of the late Wisconsin sediments from the Willcox Basin, Arizona. M.S. thesis, University of Arizona, Tucson.

SCOTT, D. D., and SCOTT, L. J.

1984   Analyses of the historic artifacts and evidence from the pollen, fibers, and hair from 42UN1225. *In:* Fike, R. E., and Phillips, H. B., II, Nineteenth century Ute burial from northeast Utah. Bureau of Land Management, Salt Lake City, *Cultural Resource Series* 16.

SCOTT, L. J.

1972   Folsom Cave palynological analysis. Ms. on file with University of Colorado Museum, Boulder.

1972   Pollen analysis of Mummy Lake. Ms. on file with Department of Anthropology, University of Colorado, Boulder.

1974   Palynological analysis of Site 5MTUMR2344. Ms. on file with Department of Anthropology, University of Colorado, Boulder.

1975   Palynological analysis of sites 5MTUMR2343 and 5MTUMR2346. Ms. on file with the Department of Anthropology, University of Colorado, Boulder.

1976   Analysis of pollen from the Skull Creek Basin—a feasibility study. Ms. on file with the Craig District Office, Bureau of Land Management.

1976   Hoy House—a palynological study, *In:* Nickens, P. R., *The Johnson-Lion Canyon Project.* Report of Investigation III, Mesa Verde Research Center, University of Colorado, Boulder, 8-49.

1977   Pollen analysis of South Pueblo at Pecos National Monument, New Mexico. Ms. on file with National Park Service, Southwest Region, Santa Fe.

1977   Paleoclimate and plant utilization as reflected in the pollen analysis of four sites in the Chuska Valley, New Mexico. Ms. on file with the Laboratory of Anthropology, Museum of New Mexico, Santa Fe.

1977   A study of ethnobotanic pollen from 5MTUMR2785, Mancos Canyon, Colorado. *In:* Emslie, S. D., *Excavation at Site 5MTUMR2785, Mancos Canyon, Ute Mountain Homelands, Colorado.* Department of Anthropology, University of Colorado, Boulder.

1978   Pollen analysis at sites LA 14695, 14702, 14703, 14704, and 14705, west of Chaco Canyon National Monument, New Mexico. Laboratory of Anthropology, Museum of New Mexico, Santa Fe, 42 p.

1978   Pollen analysis at site LA 15867, Union County, New Mexico. Laboratory of Anthropology, Museum of New Mexico, Santa Fe, 16 p.

1978   Palynological investigations at sites LA 13659 and LA 12117, Bandelier National Monument, New Mexico. Southwest Region, National Park Service, Santa Fe, 26 p.

1978   Pollen analysis of three sites in the La Mesa Fire Study Area. Southwest Region, National Park Service, Santa Fe, 19 p. (Bandelier National Monument, New Mexico)

1978   Analysis of pollen from Sites 5MF480 and 5MF607, Moffat County, Colorado. Ms. on file with Laboratory of Public Archaeology, Colorado State University, Fort Collins.

1978   An analysis of pollen from 5MTUMR2837, Mancos Canyon, Colorado. Ms. on file with Department of Anthropology, University of Colorado, Boulder.

1978   Pollen analysis of Midden 2 at Inscription House, Arizona. *In:* Breternitz, D. A., *Inscription House, 1977, Part I,* Department of Anthropology, University of Colorado, Boulder.

1978   Palynological investigations at 5ME217: a rock shelter in western Colorado. *In:* Lutz, B. J., The test excavations of 5ME217: a rockshelter in Mesa County, Colorado. Ms. on file with Bureau of Land Management, Grand Junction District.

1978   Palynological investigations of four sites at Curecanti National Recreation Area, Colorado. *In:* Euler, T. B., and Stiger, M. A., 1978 test excavations at five archeological sites in Curecanti National Recreation Area, Intermountain Colorado. Ms. on file with National Park Service, Midwest Archeological Center.

1979   Dietary inferences from Hoy House coprolites: a palynological interpretation: *The Kiva,* 44: 257-281.

1979   Jurgens Site palynological analysis. *In:* Wheat, J. B., The Jurgens Site. *Plains Anthropologist (Memoir 15),* 24(2): 149-151.

1979   Pollen analysis of Dominguez Ruin. *In:* Reed, A. D., The Dominguez Ruin: A McElmo Phase Pueblo in southwestern Colorado. Part 1. *Cultural Resources Series,* No. 7. Bureau of Land Management, Denver.

1979   Pollen analysis in Tijeras Canyon: sites LA 14261, LA 14857, LA 14258, and LA 10794. *In:* Oakes, Y. R., Excavation at Deadman's Curve, Tijeras Canyon, New Mexico: New Mexico State Highway Department Projects I-040-3(55)171 and I-040-3(36)169. Museum of New Mexico, Santa Fe, *Laboratory of Anthropology Note* 137.

1979   Palynological investigations of fifteen vessels from MV820, Mesa Verde National Park. Ms. on file with Mesa Verde National Park, Colorado.

1979   An initial study of the environment and subsistence in the Dolores Archaeological Project Study Area: the pollen record. *Dolores Archaeological Program Technical Series,* Vol. 6, Ch. 8.

1980   Pollen analysis and environmental model for lower Tijeras Canyon. *In:* Wiseman, R. N., The Carnue Project: Excavation of a Late Coalition Period Pueblo in Tijeras Canyon, New Mexico. Museum of New Mexico, Santa Fe, *Laboratory of Anthropology Note* 166, 28-32.

1980   Palynological investigations at MV1936. Ms. on file with Mesa Verde National Park, Colorado.

1980   Pollen analysis at 5WL453: a Woodland site in northeastern Colorado. Ms. on file with Department of Anthropology, University of Northern Colorado, Greeley.

1980   Palynological investigations at DeBeque Rockshelter, (5ME82) in western Colorado. *In:* Reed, A. D., and Nickens, P. R., *Archaeological investigations at the DeBeque Rockshelter: a stratified Archaic site in west-central Colorado.* Prepared for Bureau of Land Management, Grand Junction District Office.

1980   Preliminary paleo-environmental interpretations based on archeological pollen samples from Alkali Creek. *In:* Baker, S. G., Baseline cultural resource surveys and evaluations in primary impact areas of the Mt. Emmons Project: 1978 and 1979 field seasons. Ms. on file with Bureau of Land Management, Montrose District Office.

1980   Pollen analysis at 42EM959 and 42EM960. Ms. on file with Archaeological-Environmental Research Corporation, Bountiful, Utah.

1980   Pollen analysis from sites 5FN189 and 5FN349 for the Cyprus Mines Hansen Project, Fremont County, Colorado. Ms. on file with Gordon and Branzush, Inc., Boulder.

1980   Palynological analysis of sites LA 16297, LA 18436, and LA 2315 in Lincoln County, New Mexico. Ms. on file with Laboratory of Anthropology, Museum of New Mexico, Santa Fe.

1981   Palynological investigation at Curecanti National Recreation Area, Colorado. *Proceedings of the Second Conference on Scientific Research in the National Parks.*

1981   Pollen analysis of two sites in the Canyon Pintado Historic District, Rio Blanco, Colorado. *In:* Creasman, S. D., Archaeological investigations in the Canyon Pintado Historic District, Rio Blanco County, Colorado. Phase I —Inventory and Test Excavations. *Reports of the Laboratory of Public Archaeology No. 34.* Laboratory of Public Archaeology, Ft. Collins.

1981   Pollen analysis at the Coral Lodge Site, 5LP264, Colorado. *Southwestern Lore,* 47(4): 23-27.

1981   Pollen analysis of Glenwood Canyon, Colorado. Ms. on file with Colorado Department of Highways, Archaeology Lab, Boulder.

1981   Pollen analysis of 42KA1969. *In:* Nickens, P. R., and Kvamme, K. L., Archaeological investigations at the Kanab Site, Kane County, Utah. Ms. on file with Bureau of Land Management, Cedar City District Office.

1981   Pollen analysis of five sites along the Yampa River in Moffat County, Colorado. Ms. on file with Laboratory of Public Archaeology, Fort Collins.

1981   Pollen analysis of groundstone from sites 5MF435 and 5MF436 Moffat County, Colorado. Ms. on file with Laboratory of Public Archaeology, Fort Collins.

1981   Palynological record of treeline movement at 5PA153, Park County, Colorado. Ms. on file with Laboratory of Public Archeology, Colorado State University, Ft. Collins.

1981   Pollen analysis at two high altitude sites (5LK372 and 5ST114) near Climax, Colorado. Ms. on file with Laboratory of Public Archaeology, Colorado State University, Ft. Collins.

1981   The pollen record at eight Anasazi sites in Dolores, Montezuma, and La Plata counties, Colorado. *In:* Testing and excavation report, MAPCO's Rocky Mountain Liquid Hydrocarbons Pipeline, southwest Colorado. Ms. on file with Woodward-Clyde Consultants, San Francisco.

1981   Pollen and macrofossil analysis at 5ML45 in Mineral County, Colorado. *In:* Alan D. Reed, Nickens and Associates, Archaeological investigations of two Archaic campsites located along the Continental Divide, Mineral County, Colorado. Ms. on file with Division of Wildlife, Denver.

1981   Pollen analysis of eight sites in Blocks IV and V, Navajo Indian Irrigation Project, New Mexico. *In:* Simmons, A. H. (ed.), Archaeological investigations into the prehistory of northwestern New Mexico: data recovery on Blocks IV and V of the Navajo Indian Irrigation Project. Ms. on file with Professional Analysts, Eugene, Oregon, and National Park Service, Santa Fe, New Mexico.

1982   Pollen and fiber analysis of the McEndree Ranch Site, 5BA30, southeastern Colorado. *Southwestern Lore,* 48: 18-24.

1982   The pollen analyses. *In:* Wiseman, R. N., The Tsaya Project, archeological excavation near Lake Valley, San Juan County, New Mexico. Museum of New Mexico, Santa Fe, *Laboratory of Anthropology Note* 308.

1982   Pollen analysis in the Pinon-Forest Lake Region. *In:* Linford, L. D. (ed.), Kayenta Anasazi archaeology on central Black Mesa, northeastern Arizona: The Pinon Project. *Navajo Nation Papers in Anthropology* No. 10. Navajo Nation Cultural Resource Management Program, Window Rock, Arizona.

1982   Pollen analysis at 5PT86. Ms. on file with Cultural Resource Consultants, Denver.

1982   Pollen analysis of an Archaic campsite (AR 03-10-08-442) in Los Alamos County, New Mexico. Ms. on file with the National Park Service, Southwest Region, Santa Fe.

1982   Pollen analysis of a hearth from 42SA10685, Southeastern Utah. Ms. on file with La Plata Archaeological Consultants, Dolores, Colorado.

1982   Pollen analysis at 5LP630, La Plata County, Colorado. Ms. on file with Department of Anthropology, Fort Lewis College, Durango.

1982   Paleoenvironment and subsistence at four sites on Battlement Mesa, west-central Colorado: the pollen, microscopic fiber, and seed evidence. Ms. on file with Grand River Institute, Grand Junction.

1982   Pollen analysis of 5RB1872, a late Archaic site in northwestern Colorado. Ms. on file with Grand River Consultants, Inc., Grand Junction. (Rio Blanco County)

1982   Pollen and macrofloral analysis of four hearths (5RB2372) near Douglas Creek in western Colorado. Ms. on file with Grand River Consultants, Grand Junction. (Rio Blanco County)

1983   Paleoenvironmental interpretations in the Douglas Creek Drainage, northwestern Colorado. *In:* Arthur, C., Final Report on the Archaeological Monitoring of Northwest Pipeline Corporation's Trunk "D" Pipeline in the Canyon Pintado Historic District. *Cultural Resource Management Report No. 9.* Archaeological Services, Western Wyoming College, Rock Spring.

1983   Pollen report for site 5MT2191. *In:* Breternitz, D. A., *Dolores Archaeological Program: Field Investigations and Analysis—1978.* Bureau of Reclamation, Denver.

1983   Pollen report for site 5MT2198. *In:* Breternitz, D. A., *Dolores Archaeological Program: Field Investigations and Analysis—1978.* Bureau of Reclamation, Denver.

1983   Pollen and macrofloral analysis at Trough Hollow, Sevier County, Utah. *In:* Class II Cultural Resource Inventory and Test Excavation Program of the Trough Hollow-Emery Coal Lease Tracts within the Ivie Creek-Emery Area, Emery and Sevier Counties, Utah. Ms. on file with Bureau of Land Management, Richfield.

1983   Pollen analysis at Cedar Siding Shelter (42EM1533), Emery County, Utah. Ms. on file with Grand River Institute, Grand Junction.

1983   Pollen and macrofloral analyses at Abiquiu Reservoir. *In:* Reed, A. D., and Tucker, G. C., Jr., Archaeological investigations at four sites in the Abiquiu Multiple Resource Area, New Mexico. Ms. on file with Corps of Engineers, Albuquerque District.

1983   Pollen and macrofloral analysis at 42SA14187P: an Anasazi burial in southeastern Utah. *In:* Davis, W. E., 42SA14187: A Pueblo II Anasazi Burial in Westwater Canyon, San Juan County, Utah. Ms. on file with Abajo Archaeology and the City of Blanding, Utah.

1983   Pollen and macrofloral analysis at NMAS 5476: a limited activity site in Eddy County, New Mexico. Ms. on file with New Mexico Archaeological Services, Inc., Carlsbad.

1983 Pollen analysis from selected sites in the eastern Red Mesa Valley, New Mexico. Ms. on file with School of American Research, Santa Fe.

1983 Pollen analysis at sites FA-3-6, FA-1-6, FA-3-3, and FA-2-8 in northwest New Mexico. Ms. on file with United States Forest Service, Albuquerque.

1983 Pollen analysis of FA-2-13. Ms. on file with United States Forest Service, Albuquerque.

1983 A model for the interpretation of pit structure activity areas at Anasazi sites (Basketmaker III-Pueblo I) through pollen analysis. M.A. thesis, University of Colorado, Boulder.

1984 Pollen analysis at selected sites in the Atrisco Grant along the Rio Puerco, New Mexico. Ms. on file with United States Forest Service, Albuquerque.

1984 Pollen analysis at two sites in the Placitas Parcel, New Mexico. Ms. on file with United States Forest Service, Albuquerque.

1984 Pollen analysis at a slab-lined storage cist, 5RB2636, Rio Blanco County, Colorado. Report to Nickens and Associates, Montrose, Colorado.

1984 Report of the pollen, macrofloral, and fiber analyses at Pinon Canyon, 1983 field season. Ms. on file with Archaeological Research Institute, University of Denver, Denver, Colorado.

1984 Pollen analysis at site 5DL775, Dolores County, Colorado. Ms. on file with U.S. Forest Service, Durango, Colorado.

1984 Pollen analysis at Fort Utah (42UT150), a prehistoric component. Ms. on file with Dept. of Anthropology, Brigham Young University, Provo, Utah.

1984 Pollen appendix for 5MT4683, Singing Shelter. Ms. on file with the Dolores Archaeological Project, Dolores, Colorado.

1984 Pollen analysis at 5RB2448 and 5RB2449, Douglas Creek drainage, northwestern Colorado. Ms. on file with Grand River Consultants, Grand Junction, Colorado.

1984 Pollen and macrofloral interpretations from the Vermilion Cliffs area, southwestern Utah. Ms. on file with Abajo Archaeology, Bluff, Utah.

1984 Pollen and macrofloral interpretations from a slab-lined cist in southeastern Utah. Ms. on file with Abajo Archaeology, Bluff, Utah.

1984 Pollen and macrofloral interpretations from the Sevier Desrt, western Utah. Ms. on file with Gilbert/Commonwealth, Englewood, Colorado.

1984 Synthetic report of the pollen analysis of sites 5MT4475, 5MT4477, 5MT4479, 5MT5106, 5MT5107, and 5MT5108 in the McPhee Community cluster, Dolores Archaeological Project, Colorado. Ms. on file with Dolores Archaeological Project, Dolores, Colorado.

1984 Pollen analysis at selected sites near Farmington, New Mexico. Ms. on file with U.S. Forest Service, Albuquerque, New Mexico.

1984 Pollen analysis of four sites in the Michael's land exchange, northwest New Mexico. Ms. on file with U.S. Forest Service, Albuquerque, New Mexico.

1984 Pollen analysis at the Indian Mountain Site, a tipi ring in Boulder County, Colorado. Ms. on file with Plano Archaeological Consultants, Longmont, Colorado.

1984 Pollen analysis of middens on the Pajarito Plateau, New Mexico. Ms. on file with Dept. of Anthropology, University of California, Los Angeles.

SCOTT, L. J., and ROOD, R. J.
1984 Subsistence data from the pollen, macrofloral, and pollen records at the Texas Creek Overlook Site. Ms. on file with Western Wyoming College, Rock Springs, Wyoming, and Bureau of Land Management, Craig District, Colorado.

SCOTT, L. J., and SEWARD, D. T.
1981 Pollen and macrofossil analysis of five hearths in Moffat County, Colorado. Ms. on file with Powers Elevation Company, Denver.

SCOTT, L. J., and STIGER, M. A.
1981 Pollen analysis of cultural features and coprolites from Glen Canyon National Recreation Area. *In:* Schroedel, A. R., Archeological research in Glen Canyon, 1977. Ms. on file with National Park Service, Midwest Archeological Center, Lincoln.

SCOTT, L. J., and WHEELER, L. A.
1979 Pollen analysis at Jerry Creek. Ms. on file with Grand River Institute, Grand Junction.

SEARS, P. B.
1937 Pollen analysis as an aid in dating cultural deposits in the United States. *In:* MacCurdy, G. G. (ed.), *Early Man.* J. B. Lippincott Co., London, 61-66.

1950 Pollen analysis in Old and New Mexico. *Bulletin of the Geological Society of America,* 61: 1171.

1953 Climate and civilization. *In:* Shapley, H. (ed.), *Climatic Change: Evidence, Causes, and Effects.* Harvard University Press, Cambridge, 34-50.

1961 Palynology and the climatic record of the Southwest. *Annals of the New York Academy of Sciences,* 95: 632-641.

SEARS, P. B., and CLISBY, K. H.
1952 Two long climatic records. *Science,* 116: 176-178.

SHORT, S. K.
1978 Analyses of pollen from the Tamarron Site. *Southwestern Lore,* 44: 64-73.

1980 Pollen analysis, 5LP110 and 5LP111, Durango, Colorado. *In:* Gooding, J. D. (ed.), The Durango South Project. Archaeological Salvage of two late Basketmaker III sites in the Durango District. *Anthropological Papers of the University of Arizona* 34, 149-156.

1981 Palynological analysis of Vail Pass Bog. *In:* Gooding, J. D., The archaeology of Vail Pass Camp. A multicomponent base camp below treelimit in the Southern Rockies. Colorado Department of Highways. *Highway Salvage Report* 35, 147-153.

SHORT, S. K., and ELIAS, S. A.
1984 Holocene vegetational history of the Colorado Front Range. Sixth International Palynological Conference (IPC), Calgary, Alberta, Canada, *Abstracts,* 153.

SMITH, L. D.
1974 Archaeological and paleoenvironmental investigations in Cave Buttes area north of Phoenix, Arizona. M.A. thesis, Arizona State University, Tempe.

SOLOMON, A. M., BLASING, T. J., and SOLOMON, J. A.
1982 Interpretation of floodplain pollen in alluvial sediments from an arid region. *Quaternary Research,* 18: 52-71.

SOLOMON, A. M., and HAYES, H. D.
1972 Desert pollen production, I: qualitative influence of moisture. *Journal of the Arizona Academy of Science,* 7: 52-74.

1980 Impacts of urban development upon allergenic pollen in a desert city. *Journal of Arid Environments,* 3: 169-178.

SOLOMON, A. M., KING, J. E., MARTIN, P. S., and THOMAS, J.

1973   Further scanning electron photomicrographs of Southwestern pollen grains. *Journal of the Arizona Academy of Science,* 8: 135-157.

SOLOMON, A. M., and WEBB, J. L.

1974   Human disturbance in arid lands: pollen evidence of prehistoric land use. *Bulletin of the Ecological Society of America,* 55: 28.

SPAULDING, W. G.

1974   A preliminary statement on the pollen analysis from the Escalante Ruin group. *In:* Doyel, D. E. (ed.), Excavations in the Escalante Ruin group, southern Arizona. *Arizona State Museum Archaeological Series* 37, 262-268.

SPAULDING, W. G., LEOPOLD, E. B., and VAN DEVENDER, T. R.

1983   Late Wisconsin paleoecology of the American Southwest. *In:* Wright, H. E., Jr. (ed.), *Late-Quaternary Environments of the United States. Vol. 1, The Late Pleistocene* (Porter, S. C., ed.). University of Minnesota Press, Minneapolis, 259-293.

SPAULDING, W. G., and PETERSEN, K. L.

1980   Late Pleistocene and early Holocene paleoecology of Cowboy Cave. *In:* Jennings, J. D. (ed.), Cowboy Cave. *University of Utah Anthropological Papers* 104, 163-177.

STEARNS, T. B.

1981   Palynological evidence of the prehistoric effective environment. *In:* Baker, C. and Winter, J. C. (eds.), High Altitude Adaptations along Redondo Creek, The Boca Geothermal Anthropological Project. University of New Mexico, Office of Contract Archaeology, 25-39.

STIGER, M. A.

1977   Anasazi diet: the coprolite evidence. M.A. thesis, University of Colorado, Boulder.

STRUEVER, M. B.

1977   Relation of pollen and flotation analyses to archaeological excavations, Chaco Canyon, New Mexico: M.A. thesis, University of New Mexico, Albuquerque, 161 p. (pollen analysis by A. C. Cully)

THOMPSON, R. S.

1984   Palynology and packrat middens in the western United States. Sixth International Palynological Conference (IPC), Calgary, Alberta, Canada, *Abstracts,* 165.

THOMPSON, R. S., and KANTZ, R. R.

1985   Pollen analysis of Gatecliff Shelter. *In:* Thomas, D. H., (ed.), The Archeology of Gatecliff Shelter and Moniter Valley: *American Museum of Natural History Anthropological Papers.* In press.

THOMPSON, R. S., VAN DEVENDER, T. R., MARTIN, P. S., FOPPE, T., and LONG, A.

1980   Shasta ground sloth *(Nothrotheriops shastense* Hoffstetter) at Shelter Cave, New Mexico: environment, diet, and extinction. *Quaternary Research,* 14: 360-376.

TOLL, M. S., and CULLY, A. C.

1984   Archaic subsistence in the Four Corners area: evidence for an hypothesized seasonal round. *In:* Hogan, P., and Winter, J. C. (eds.), *Economy and interaction along the lower Chaco River.* Office of Contract Archeology, University of New Mexico, Albuquerque.

TRENCH, N. R.

1978   The geomorpology and paleoenvironmental history of the Lake City landslide complex, southwest Colorado. M. A. thesis, University of Colorado, Boulder, 149 p.

VAN DEVENDER, T. R., and KING, J. E.

1971   Late Pleistocene vegetational records in western Arizona. *Journal of the Arizona Academy of Science,* 6: 240-244.

VAN DEVENDER, T. R., SPAULDING, W. G., and PHILLIPS, A. M., III

1979   Late Pleistocene plant communities in the Guadalupe Mountains, Culberson County, Texas. *In:* Genoways, H. H., and Baker, R. J. (eds.), Biological Investigations in the Guadalupe Mountains National Park, Texas. National Park Service, *Proceedings and Transactions Series* 4, 13-30.

VAN DEVENDER, T. R., and TOOLIN, L. J.

1983   Late Quaternary vegetation of the San Andres Mountains, Sierra County, New Mexico. *In:* Eidenbach, P. L. (ed.), *The prehistory of Rhodes Canyon, N.M. Survey and mitigation.* Human Systems Research, Inc., Tularosa, New Mexico, 33-54. (pollen analysis by O. K. Davis)

WEIR, G. H.

1976   Palynology, flora and vegetation of Hovenweep National Monument: implications for aboriginal plant use on Cajon Mesa, Colorado and Utah. Ph.D. dissertation, Texas A & M University, College Station, 215 p.

WENDORF, F.

1961   An interpretation of late Pleistocene environments of the Llano Estacado. *In:* Wendorf, F., Paleoecology of the Llano Estacado. Museum of New Mexico, Santa Fe. *Fort Burgwin Research Center Publication 1,* 115-133. (pollen analysis by U. Hafsten)

1975   Summary and conclusions. *In:* Wendorf, F., and Hester, J. J. (eds.), Late Pleistocene Environments of the Southern High Plains. *Publication of the Fort Burgwin Research Center* 9, 257-278. (pollen analysis by F. Oldfield and J. Schoenwetter)

WERNER, W. I., REED, W., and STORMFELS, E. L.

1947   Hay-fever plants of Albuquerque, New Mexico: a preliminary report. *Annals of Allergy,* 3: 47-54, 57.

WEST, G. J.

1978   Recent palynology of the Cedar Mesa area, Utah. Ph.D. dissertation, University of California, Davis, 175 p.

WETTERSTROM, W. E.

1976   The effects of nutrition on population size at Pueblo Arroyo Hondo, New Mexico. Ph.D. dissertation, University of Michigan, Ann Arbor. (pollen analysis by V. L. Bohrer)

WHITEHEAD, D. R.

1959   Fossil pollen and spores from the LaDaisKa Site, Colorado. *In:* Irwin, H. J., and Irwin, C. C., Excavations at the LaDaisKa Site in the Denver, Colorado, area. *Denver Museum of Natural History Proceedings* 8, 114-118.

WHITESIDE, M. C.

1964   Paleoecological studies of Potato Lake and its environs: M.S. thesis, Arizona State University, Tempe, 59 p.

1965   On the occurrence of *Pediastrum* in lake sediment. *Journal of the Arizona Academy of Science,* 3: 144-146.

1965   Paleoecological studies of Potato Lake and its environs. *Ecology,* 46: 807-816.

WIGAND, P. E., and MEHRINGER, P. J., JR.

1985   Pollen and seed analyses. *In:* Thomas, D. H. (ed.), The Archaeology of Hidden Cave. *American Museum of Natural History, Anthropological Papers.* In press.

WILLIAMS-DEAN, G. J.

1975 Pollen analysis of prehistoric human coprolites from Antelope House Ruin, Canyon de Chelly National Monument, Arizona. M.A. thesis, University of Texas at Austin.

1985 Pollen analysis of prehistoric human coprolites from Antelope House Ruin. *In:* Morris, D. P., and Bancroft, J. M. (eds.), Archeological Investigations at Antelope House. *National Park Service Publications in Archeology Series,* Washingtion, D.C. In press.

WILLIAMS-DEAN, G., and BRYANT, V. M., JR.

1975 Pollen analysis of human coprolities from Antelope House. *The Kiva,* 41: 97-111.

WILMSEN, E. N.

1974 *Lindenmeier: A Pleistocene Hunting Society.* Harper & Row, Publishers, N. Y., 126 p. (pollen analysis by V. L. Bohrer)

WILMSEN, E. N., and ROBERTS, F. H. H., JR.

1978 Lindenmeier, 1934-1974; concluding report on investigations. Smithsonian Institution Press, *Smithsonian Contributions to Anthropology* 24. (pollen analysis by V. L. Bohrer and S. K. Fish)

WOOSLEY, A. I.

1977 Farm field location through palynology. *In:* Winter, J. C. *et al.* (eds.), Hovenweep 1976. *San Jose State University, Archaeological Report* 3, 133-150.

WRIGHT, H. E., JR.

1971 Late Quaternary vegetational history of North America. *In:* Turekian, K. K. (ed.), *The Late Cenozoic Glacial Ages.* Yale University Press, New Haven, 425-464.

WRIGHT, H. E., JR., BENT, A. M., HANSEN, B. S., and MAHER, L. J., JR.

1973 Present and past vegetation of the Chuska Mountains, northwestern New Mexico: *Geological Society of America Bulletin,* 84: 1155-1180.

WYCKOFF, D. G.

1977 Secondary forest succession following abandonment of Mesa Verde. *The Kiva,* 42: 215-231.

THE AUTHORS

David P. Adam, a Geologist with the U.S. Geological Survey in Menlo Park, California, received his B.A. from Harvard and his M.S. and Ph.D. from The University of Arizona. He is presently directing a multidisciplinary study of a 334-meter sediment core from Tule Lake, California, that spans the past 2.5-million years. Other research interests include chrysophyte cysts, the theory of ice ages, computer graphics, and data applications in paleontology.

Thomas A. Ager, received his M.S. degree in geology from The University of Alaska in 1972, and his Ph.D. in geology from The Ohio State University in 1975. He is currently employed by the U.S. Geological Survey at Reston, Virginia. The geographic areas where most of his research efforts have been concentrated include Alaska and the coastal plain of the southeastern U.S.

Thane W. Anderson, is a Research Scientist at the Geological Survey of Canada, Ottawa, Ontario, Canada. He received his B.S. from Dalhousie University, Halifax, Nova Scotia, Canada, and his M.S. and Ph.D. from The University of Waterloo, Waterloo, Ontario. His research interests are in Quaternary pollen and plant macrofossils. He has carried out paleoenvironmental studies in the Great Lakes, St. Lawrence-Ottawa Lowland region of Ontario and Quebec and in Atlantic Canada.

Richard G. Baker, is a Professor of Geology at The University of Iowa. He received his Ph.D. at The University of Colorado. His research interests are Quaternary palynology, plant macrofossils, and paleoecology in the Midwest and Rocky Mountain areas.

R. Ben Brown, received his Ph.D. in Anthropology from The University of Arizona in 1984. His research interests include the archaeology and paleoecology of the northern frontier of Mesoamerica, and tropical palynology.

Linda B. Brubaker, is an Associate Professor of Forestry at The University of Washington. She received her Ph.D. from The University of Michigan in 1973. Her current research includes palynological studies of late-Quaternary vegetation of northern Alaska and dendrochronological investigation of recent forest history in the Pacific Northwest.

Vaughn M. Bryant, Jr., is a Professor of Anthropology and Biology and Head of the Department of Anthropology at Texas A&M University. He is the former Managing Editor and the current President of A.A.S.P. His degrees include a B.A., M.A., and Ph.D. from The University of Texas at Austin. His research interests include paleoenvironmental reconstruction, the role of palynology in archaeology, and the reconstruction of prehistoric diets.

Hazel R. Delcourt, received her Ph.D. from The University of Minnesota in 1978. At present, she holds an appointment as Research Assistant Professor in the Department of Botany and the Graduate Program in Ecology at The University of Tennessee, Knoxville. Her research focus is Quaternary palynology and paleoecology of the southeastern United States and the application of paleoecological techniques to contemporary questions in plant ecology.

Paul A. Delcourt, received his Ph.D. from The University of Minnesota in 1978, and currently holds a joint appointment as Assistant Professor with the Department of Geological Sciences and the Graduate Program in Ecology, at The University of Tennessee, Knoxville. His research interest involves the calibration of modern pollen with vegetation in order to quantitatively map changes in vegetation through the late-Quaternary.

Knut Faegri, is a Professor (emeritus) in various departments at The University, Bergen, Norway. He received his doctorate degree from The University of Oslo in 1934, and his h.c. from Uppsala. His Major scientific interests include pollen analysis and pollination ecology. He has written two major textbooks in these fields (both are in their 3rd edition), has written a number of popular books, and over 400 listed (and nobody knows how many unlisted) research papers and articles in all sorts of fields. He is still the editor of a popular scientific journal (and has been for 31 yrs.), he also is a foreign, invited, or honorary member of several international scientific soicieities (including A.A.S.P.), and is a member of several academies. He has served as a past board member (general secretary, president) of various international scientific organizations, including the International Union of Biological Sciences.

Denise C. Gaudeau, is an Assistant Professor of Geology at Southampton College, New York. She is currently completing her Ph.D. from Yale University, (Department of Biology), and has worked as a Research Assistant in the Department of Geological Sciences at Brown University. Her palynological research interests center on the Holocene vegetational history of the northeastern United States.

Stephen A. Hall, is currently an Associate Professor of Geography at The University of Texas at Austin. He received a B.S. from The University of Oklahoma, an M.S. from The University of Iowa, and a Ph.D. in 1975 from The University of Michigan. He has worked with Cretaceous and Tertiary material from Antarctica and is presently investigating Quaternary environments of the Southwest, Rocky Mountains, and Plains, using the fields of pollen analysis, geomorphology, stratigraphy, malacology, aerobiology, and archaeological geology.

Calvin J. Heusser, is Professor of Biology at New York University. He received his Ph.D. from Oregon State College (1952) where he studied under Henry P. Hansen. Following a postdoctoral Theresa Seessel Fellowship at Yale University, he was a Research Associate at the American Geographical Society of New York before joining New York University on a full-time basis in 1967. His research interests include late-Cenozoic vegetation, glaciation, and environments of north Pacific North America and Southern South America. He is currently (1985) a Visiting Fellow at Clare Hall, University of Cambridge, England.

Linda E. Heusser, is a Senior Research Associate (part-time) at Lamont-Doherty Geological Observatory of Columbia University, Palisades, New York. She received her B.A. from Wellesley College, M.A. (geology) from Columbia University, and Ph.D. (geology) from New York University (1971). The primary focus of her research is marine palynology, ranging from the present distribution of pollen in the northwest Atlantic Ocean, northeast and northwest Pacific Ocean, Sea of Japan, and Bering Sea, to the Tertiary and Quaternary distribution of palynomorphs in marine sediments off New Zealand, the United States, Canada, and Mexico. Current research includes Quaternary sediments in the Orca Basin, Gulf of Mexico; stream transport of pollen in California; and comparison of pollen and lignin in recent fluvial and marine sediments from Washington.

Richard G. Holloway, is a Research Scientist with the Anthropology Department at Texas A&M University. His research interests include paleoenvironmental reconstruction, paleolimnology,

and pollen preservation. He received his M.S. and Ph.D. (1981) degrees in botany from Texas A&M University. His doctoral research centered on modern and fossil pollen degradation.

Peter J. Mehringer, Jr., is a Professor of Anthropology at Washington State University where he also holds an appointment in the Deparment of Geology. He received his Ph.D. from The University of Arizona in 1968. His research interests include Quaternary chronology, biogeography, paleoecology and the role of palynology in archaeology. He has conducted studies throughout the Western U.S., in Egypt, and the Sudan.

James C. Ritchie, received his B.S. in botany from The University of Aberdeen in 1941 and his Ph.D. from The University of Sheffield in 1953. In 1962, he was awarded the D.S. degree by The University of Aberdeen. Currently, since 1975, he is a Professor of Botany at The University of Toronto, Scarborough College. His research interests include the history of the vegetation and environment of north-west Canada (boreal, arctic) with particular reference to the dynamics of plant populations in response to post-glacial climatic change, and past vegetation and environmental change in the Circum-Mediterranean zone with particular reference to (a) the evidence for changes in the African monsoon position during the Holocene, and (b), the nature of pre-Holocene Mediterranean vegetation. This research continues to be supported by the Natural Sciences and Engineering Research Council of Canada.

Kirk A. Waln, is a graduate student in the Geology Department at The University of Iowa where he is finishing his MS degree. His thesis is on the palynology of the Oligocene-age Lincoln Creek Formation, Western Washington.

Thompson Webb III, received his B.A. in botany from Swarthmore College and his Ph.D. in meteorology from The University of Wisconsin-Madison in 1971. He is currently a Professor of Geological Sciences at Brown University. His research interests include: Quaternary palynology, paleoecology, and paleoclimatology. Since 1972 he has published many articles in journals such as *Quaternary Research, Ecology,* and *Review of Paleobotany and Palynology.*